INTRODUCTION TO ECONOMIC ZOOLOGY

OTHER BOOKS BY THE SAME AUTHORS

- **Introduction to General Zoology**
  [Volumes I and II]

INTRODUCTION TO
ECONOMIC ZOOLOGY

Supriti Sarkar MSc, PhD

Associate Professor, Dept. of Zoology, City College, Kolkata

Gautam Kundu MSc, PhD

Associate Professor, Dept. of Zoology, City College, Kolkata,
Formerly Sr. Lecturer in Zoology, St. Joseph's College, Dargeeling

Korak Kanti Chaki MSc, PhD, FZS (Cal.)

Associate Professor and Head, Dept. of Zoology, City College, Kolkata
Formerly Lecturer in Zoology, Bangabashi College (Day), Kolkata
Senior Lecturer and Head in Zoology, Uluberia College, Howrah

New Central Book Agency (P) Ltd

Co-Published by

Bijay Publications

E-mail : bijaypublications@gmail.com

Exclusively Distributed by

Bijay Pustak Bhandar (P) Ltd

90/6A M. G. Road, Bhawani Dutta Lane, Kolkata-700007

✆ : (033) 4605-7856, 8777898576, 8274916104

E-mail : bijaypustakbhandar@gmail.com

NCBA

REGD OFFICE
8/1 Chintamoni Das Lane, Kolkata 700 009, India
email: ncbakolkata@gmail.com

OVERSEAS
NCBA (UK) Ltd, 149 Park Avenue North, Northampton, NN3 2HY, UK
email: ncbauk@yahoo.com

CORPORATE OFFICE
212 Shahpur Jat, New Delhi 110 049
email: delhincba@gmail.com

BIJAY PUBLICATIONS
90/6A M. G. Road, Bhawani Dutta Lane, Kolkata-7
E-mail : bijaypublications@gmail.com

BIJAY PUSTAK BHANDAR (P) LTD
90/6A M. G. Road, Bhawani Dutta Lane, Kolkata-7
E-mail : bijaypustakbhandar@gmail.com

INTRODUCTION TO ECONOMIC ZOOLOGY • Supriti Sarkar, Gautam Kundu and Korak Kanti Chaki

First Published: June 2014
Reprinted: September 2016, **2025**

PUBLISHER • TYPESETTER • PRINTER
New Central Book Agency (P) Ltd
8/1 Chintamoni Das Lane, Kolkata 700 009

CO-PUBLISHED BY
Bijay Publications
90/6A M G Road, Bhawani Dutta Lane, Kolkata 700007

TECHNICAL EDITOR
Dr Mita Sen, MBBS
Dipan Roy

PROJECT TEAM
Nochiel Radhakrishnan, Ashit Ghosh and Sumanto Mohanto

ISBN: 978 81 7381 899 8

Price ₹ 850.00

Dedicated

to

our beloved students

Contents

Contents

Contents

Contents

Contents

Contents

Contents

Contents

Contents

Contents

Preface

Our long endeavour to write a book on 'Introduction to Economic Zoology' has finally taken shape. A number of publications on this subject are in circulation, most of which have focussed on one aspect of culture or the other. We have, therefore, taken up the onus to pen a book on the subject that would cover all probable culture practices which have gained importance in the Indian context in the past decade. This will certainly widen up the value of this book—for which we have left no stone unturned.

The Government of India has been extravagant in diverting funds in promoting this branch of science and encouraging the youth to take up self-employment through building small scale industries in these sectors. The FAO (Food and Agriculture Organisation) too, in their world conferences, have stressed and highlighted the importance of training of personnel, technology development through research, and wide spreading of information (from the laboratory to the field). This would lead to rapid and orderly development of the various sectors of economic zoology. Today, keeping in view the national interest, all the Indian Universities have not only included 'Economic Zoology' in their syllabi of Zoology and Agriculture but also stressed upon field training and submission of field report.

The past two decades have witnessed profuse growth of economic zoology from stature to stature. Today it has blended into a multidiscipline with subjects like ecology, genetics, cytology, hydrography, molecular biology, microbiology, biotechnology, gastronomy, statistics, economics, commerce, and so on, adding much weight, importance, and variety. It is not too far when this subject would be promoted to a full-fledged discipline with an applied flavour—leading to self-employment.

This book covers a wide array of topics. It is ornamented with 13 chapters—ranging from popular and well-known farmings such as sericulture, apiculture, prawn culture, dairy, aquaculture, lac culture, duck and poultry farming, pearl culture, etc., to lesser known cultures such as rabbit farming, quail and emu farming, vermiculture, goat and sheep farming. This book also contains vivid description of a number of pests that have been gaining importance as they have successfully established themselves as dire competitors of human beings for consumption of natural resources. Their eradication without the use of pesticides have become a great concern of this decade.

The three of us have pooled all our variegated resources and long experience and poured them unstintingly into this book. We have taken help from various quarters and have consulted journals and experts in the respective fields. We have meticulously incorporated the various technological advancements in the various fields and, as a result, the tome has become voluminous but more valuable.

This book will not only suffice the purpose of Undergraduate and Postgraduate students but can also be used by those interested—for one reason or another—in Economic Zoology. Small-scale industrialists would also be profusely benefited from the multifarious topics dealt with in this book.

We are indebted and thankful to our various well-wishers who have been pecking and encouraging us to write this book by extending their helping hands. It has taken us nearly half a decade to formulate and gather our multifarious resources and bring them out in print. The onus lies on Sri Amitabha Sen, Director of New Central Book Agency (P) Ltd, who had given us full liberty and free time for the completion of this book. We also thankfully acknowledge the relentless help extended to us by the entire production unit of New Central Book Agency (P) Ltd. They have—individually and collectively—done splendid work to help us fulfil our dream project.

Bengali New Year's Day
15 April 2014
Kolkata

Supriti Sarkar
Gautam Kundu
Korak Kanti Chaki

Chapter One : Pest

Any organism detrimental to man or his property in causing damage of significant economic importance, is termed as **pest**. Usually insect and disease producing organisms are recognised as pest by common people. Other than insects, the pest may be arachnids, nematodes, mollusca and disease producing pathogens including fungi, bacteria, viruses, mycoplasma,weeds, angiospermic parasites, rodents,birds, mammals and other animals. However, pest conveys the idea or concept that they are controllable by suitable methods. The pest control method is described in later chapter.

DAMAGES CAUSED BY PESTS

There are three main types of damage done by pests which are related to their mode of feeding. These are — (i) damages by pests with biting and chewing mouthparts, (ii) damages by pests with piercing and sucking mouth parts, (iii) damages by pests that are vectors of pathogens.

(i) **Damages by pests with biting and chewing mouth parts** : They feed by biting pieces of plant material and chewing. It may be adult or larvae of insects. The type of damage done by them may be generalized as follows :

(a) Loss of photosynthetic tissues; in extreme cases defoliation may result.

(b) Destruction of buds and shoots.

(c) Destruction of flowers, fruits and seeds.

(d) Boring and tunnelling of stems; interruption of sap flow and physical weakening of the stem; stem breakage may result.

(e) Eating or boring of roots and tubers in the soil.

(f) Destruction of seedlings and young plants.

(g) Formation of galls* on all or any parts of the plant body.

(ii) **Damages by pests with piercing and sucking mouth parts** : These pests have part or all of the mouthparts modified into a piercing proboscis or stylet. Sap is sucked either from the phloem or xylem or from general tissues of foliage, roots or fruits. Within this category there are two basically different types of damage.

(a) **Pests without toxic saliva** : These insects and mites remove sap, causing tissue wilt, leaf curl, stunting and in extreme condition causes death of the host plant.

(b) **Pests with toxic saliva** : In this case toxic saliva causes a disproportionate amount of damage in relation to insect number. The toxins cause death of the cells and if injected into a young shoot, the entire shoot, distal to the feeding site, may die.

(iii) **Damages by pests that are vectors of pathogens** : These pest are extremely harmful, as a very small number of infective individuals may be responsible for severe outbreak of disease. Disease control by destruction of vectors is very difficult. These pests are of two types :

(a) **Indirect vectors** : This category includes those insects that make feeding punctures, which later becomes infected by aerial spores; this occurs in fruits, tunnelled by fruit fly larvae (Tephritidae) and fruit worms (Lepidoptera).

* Abnormal growth on plants or trees, caused by the presence of insect larvae, mites or fungi.

(b) **Direct vectors** : These insects and nematodes are sometimes called **'biological vectors'**. They are responsible for active transmission, as they are often intermediate hosts. The pathogens are plant microbes, which can only be transmitted through vectors.

The parts of plant damaged

Pest damage can be categorized according to the parts of the plant body attacked. These are generally of six categories :

1. **Sown seeds and seedlings** : The most common pests of this type are bean seed fly maggot, mice and other small rodents which dig up freshly sown seeds. Many birds feed on the seeds sown in the field. Dead heart disease caused by *Scirpophaga incertulus.*
2. **Fruits and seeds** : Cereal panicles* sometimes have only a few grains developed, like sorghum infested with sorghum midge larvae. Some fruits and seeds, including nuts, are bored by weevil larvae, e.g., cotton boll weevil, hazel nut, mango weevil, etc.
3. **Flower and buds** : Blister beetles chew the petals of many plants. Anther eater pollen beetles are also evident.
4. **Leaves** : The leaf margin – notched, irregularly eaten, regularly cut away, leaf–holed, windowed, skeletonized, galled, etc. are the regular activities of different larvae of insects and adult molluscs.
5. **Stems** : Stem borers are common examples in paddy, jute and vegetable plants.
6. **Roots and tubers** : Roots of economic importance, tubers, etc. are severely attacked by various pests.

SOME INSECT PESTS OCCURING IN INDIA

FIELD PEST OF PADDY

I. ***Scirpophaga (syn. Tryporyza) incertulus (Walker)***

Scirpophaga incertulus (Fig. 1.1) is commonly known as **yellow stem borer** of paddy. They thrive only on paddy plant, so they are called **monophagous pest**.

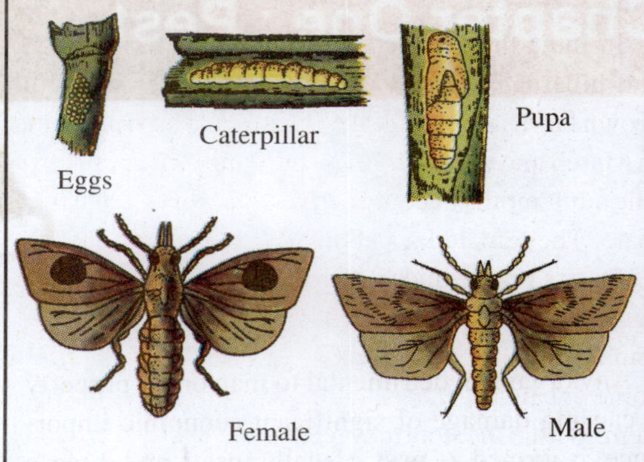

Fig. 1.1. *Scirpophaga (Tryporyza) incertulus*

DISTRIBUTION

In India, this is a major pest in Andhra Pradesh, Orissa, Tamil Nadu and West Bengal. This is widely distributed throughout the oriental region.

MORPHOLOGICAL FEATURES

(a) **Egg** : The freshly laid eggs are oval, flattened and pearly white in colour, which become darker and turn brown at the time of hatching.

(b) **Larva/Caterpillar** :The tiny (1.5 mm) newly hatched caterpillar is white in colour and black headed. They bear powerful mouth parts for boring the stem. When fully grown they measure about 20 mm and are dirty white or greenish yellow in colour, having brown coloured head and pronotum.

(c) **Pupa** : The full grown larva stops eating within the stem and covers itself with fibres secreted from their own saliva. These fibres are deep grey in colour and 11 mm to 12 mm in length.

(d) **Adult** : The adults have a wing expanse of 25 to 45 mm and the wings are yellowish white with orange-yellow coloured apical margin. The female moth is bigger than the male and has a centrally situated black spot on each of the fore wings. The female have a prominent tuft of brownish yellow silken hair at the tip of their abdomen.

* *Panicle* is a loose branching cluster of flowers

Life history

In India this pest is active from April to October and hibernates from November to March as a full grown larva in rice stubble. Pupation starts sometime in March and the emergence of moths begin in April. The adult moths become active after dusk, when they mate. The female lays about 120 to 150 eggs in 2 to 5 clusters on the underside of the paddy leaves. The eggs are covered with yellowish brown hair of the female tuft. They hatch in 6 to 7days. The caterpillars come out and start boring the stem and enters into the stem. Within the stem they feed on the plant tissues.The larval period lasts for 20 to 30 days passing through six stages of different instars. After larval stage they pupate inside the attacked plant stem. Before going to pupation it constructs an emergence hole on the stem for the future adult. This emergence hole is always located above the water level. Within 9 to 12 days the pupa emerges out as an adult moth. The life cycle is completed within 31 to 46 days.

Mode of damage

The caterpillar alone is destructive. They start boring into the stem of the paddy plant downwards up to the root causing destruction of the pith. This makes the stem hollow and causing **dead heart** of the central shoots, which easily comes off when pulled. The plants attacked in early stages dry up altogether but when infection occurs during flowering stage, the ear heads dry up and grains are not formed. This is known as the **white ear heads**.

Control measure

(a) **Cultural process**

As this pest is monophagous, alternative cultivation of paddy and other crops will prevent the stem borer to continue its generation.

(b) **Physical process**

(1) The removal and **destruction of stubble** at the time of the first ploughing after harvesting the crop is necessary. It decreases the chances of carry-over to the next crop.

(2) Light attracts this pest — therefore, **light trap** can be used to kill the adults.

(c) **Chemical process**

(1) The **spraying** of 0.08% Methyl parathion, 0.05% Phosphamidon, 0.6% Diazion granules in the paddy field after 20 to 30 days of interval can control the pest.

(2) **Fumigation** by methylamide or calcium cyanide gives good result.

II. *Nilaparvata lugens*

Nilaparvata lugens (Fig. 1.2) is commonly known as **brown plant hopper**.

Distribution

These are highly harmful in the paddy field of Korea, Japan, China and several countries of South-east Asia. In India this pest infest high yielding variety of paddy plant. It is common in Punjab, Uttar Pradesh, Madhya Pradesh, West Bengal, Orissa and Andhra Pradesh.

Morphological features

(a) **Egg** : Cylindrical greyish eggs possess two black spots. Eggs can be seen in clusters on the leaf. Incubation period of the egg is 4 to 11 days.

(b) **Nymph** : Typical miniature insect like nymph bears all the characters of an adult, except reproductive organs. These nymphs are deep brown in colour. Nymph possesses five instar stages; each stage is slightly larger than its previous stage.

(c) **Adult** : Blackish grey insect is 3.4 to 4.5 mm in length. Eyes are reddish in colour. The adults possess grey wings with black vertical lines. Sexual

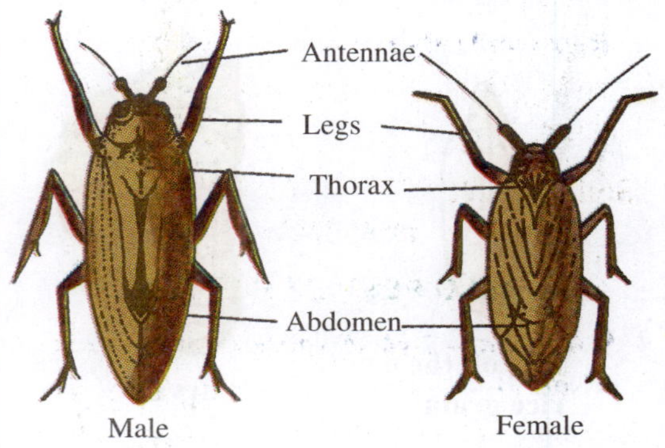

Fig. 1.2. *Nilaparvata lugens* (Brown plant hopper)

dimorphism is present. Adult males are larger in size than females.

Life history

Generally adults stay in group near the base of the paddy plant. Adult *Nilaparvata* becomes active during the month of August and September in India. Females lay numerous eggs in cluster on the leaf of the paddy plant. There are as many as 124 to 158 clusters at a time. A female insect generally lays about 680 eggs in its lifetime. The incubation period of the egg is 4 to 11 days, depending upon the atmospheric temperature and humidity. After incubation period nymphs come out and start eating the paddy leaf. The nymph completes its five instar stages by 18 to 35 days. After completion of every instar it moults and during the whole nymph stage it moults five times. The nymph period varies upon environmental temperature — during hot period the moulting occurs rapidly. After nymph stage (fifth instar) the adult comes out and become sexually mature within 5 to 10 days. The adult female immediately lay eggs after copulation under favourable conditions.

Mode of damage

Adult and nymph both are harmful to paddy plant. They suck cell sap from the leaf and gradually the leaf becomes colourless and yellow. The paddy plant attacked by brown plant hopper becomes weak and its height becomes small. In favourable condition they can cause minimum 10% to maximum 70% destruction of the crop. Moreover, this pest is the agent of transmitting **grassy stunt virus** of the paddy plant.

Control measure

(a) **Nursery treatment**

Spraying of 0.03% Endosulphan 35E or 0.03% Phosphamidon 100EC or 0.03% Phenthion 20EC at the rate of 500 litres per hectare after 15 days of rowing gives good result.

(b) **Treatment after transplantation**

Before transplantation, 3% Carbofuran or 10% Phoretat at the rate of 10 to 15 kg/hectare should be mixed in the soil. Then, 7 days after transplantation 0.05% Phosphamidon at the rate of 700 litre/hectare should be sprayed. Again after 45 to 65 days same spray at same rate will completely remove brown plant hoppers from the field.

(c) **Light trap**

This physical method can remove adult brown plant hopper to some extent.

(d) **Cultivation of resistant variety**

Ptb-33, Ptb-21, ARC-6650, HR-1523 varieties of paddy are resistant to brown plant hopper.

STORED GRAIN PEST

I. *Sitophilus oryzae*

Sitophilus oryzae (Fig. 1.3) is commonly known as **rice weevil**.

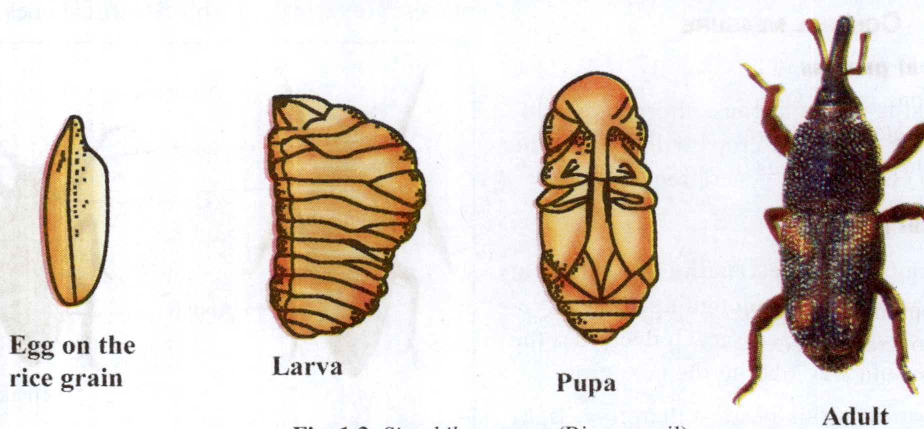

Fig. 1.3. *Sitophilus oryzae* (Rice weevil)

Distribution

It is found worldwide. In India it is found in all the states. It is probably the most destructive pest of rice in store houses. These pests are strong fliers, and it has been reported to fly from the godown to the field in vicinity, where it begins to infest the grain in field.

Morphological Features

(a) **Egg** : White coloured oval egg is laid in the pore made by the female in the rice grain. The egg is soft and plastic enough to fit into the cavity. A single female lays abaout 400 eggs in a breeding season. The eggs measure 0.6 to 0.7 mm in diameter.

(b) **Larva** : The full-grown larva is about 5 mm in length. It is plump, fleshy and legless creature. Larva is with white body and yellow brown head.

(c) **Pupa** : The pupa at first is dirty white, but later on becomes dark brown.

(d) **Adult** : The adult is a small reddish brown beetle, about 3 mm in length. It possesses a cylindrical body and a long, slender curved rostrum. Its elytra or upper pair of wings are hard and bear four light reddish or yellowish spots. The thorax is marked with round depressions. The metathoracic wings are very well developed.

Life history

Breeding period of this pest is from April to October. The adult hibernates in winter inside the cracks and crevices of the store house. During breeding period, the female makes a depression on the grain with the help of their mandibles and lay eggs in it. One egg is laid in one grain. After egg laying the hole is sealed with a gelatinous secretion. A single female may lay as many as 400 eggs in a season.

The eggs hatch in 6 to 7 days and the young larvae bore directly into the grain. They feed voraciously on the starch material of the rice grain but the shell of the grain is kept intact. At the end of the larval stage they become covered by a covering, called **puparium** inside the rice grain.

The pupa stage starts with the cessation of eating. The pupa stage persists for 6 to 14 days to become an adult.

On emergence, the adult weevil cuts its way out of the grain and lives for about 4 to 5 months. The weevil completes 3 to 4 generations in a year.

Mode of damage

Both adult and larva cause damage to rice grains. In store house the weevil causes heavy damage to rice. They make the grains hollow and unsuitable for human consumption. Not only rice they also damage other grains like wheat, maize and sorghum grains particularly in monsoon. It has also been reported to feed on oats, barley, cotton seed, linseed and coca. The weevils spoil more than what they eat.

Control measure

The weakest point of the life history of this pest is that it is unable to breed in the grain if the grain moisture content is less than 9%. Therefore, keeping the grain in dry condition is the best way of controlling the pest.

If keeping the grain in dry condition is not practicable, then fumigation is the other alternative proposition as soon as the infestation reaches an intolerable level. The common fumigant used is methyl bromide, ethylene bromide, ethylene dichloride, carbon tetrachloride, etc.

II. *Tribolium castaneum*

Tribolium castaneum is commonly called the **red-rust flour beetle** (Fig. 1.4).

Distribution

This is a common pest worldwide of wheat-flour. The adults are active creatures capable of short flights but are mostly found concealed in flour.

Morphological Features

(a) **Egg**: The eggs are white, transparent and cylindrical in shape. The surface of freshly laid eggs is sticky and thus flour or dust particles easily adhere to them making a white cover.

(b) **Larva** : The newly hatched larva is a yellowish white worm-like creature and measures about 1 mm in length. As it matures, it turns reddish yellow, becomes hairy and measures over 6 mm in length. Its head appendages and last abdominal segments are darker in colour.

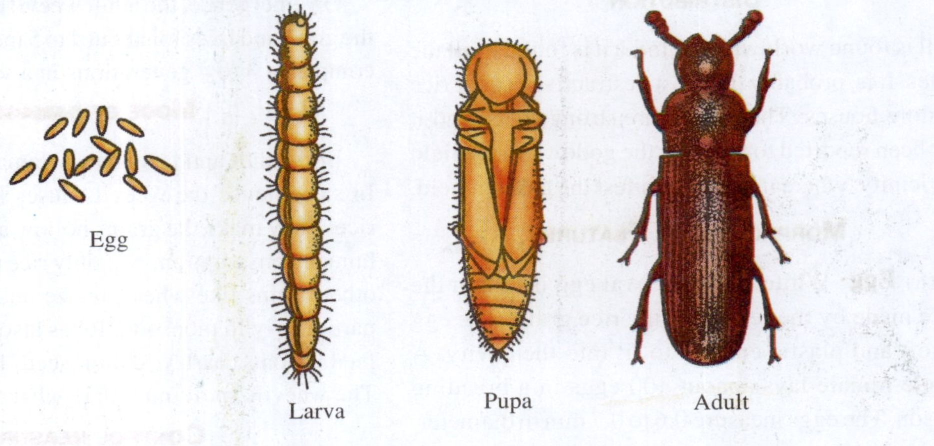

Fig. 1.4. Different stages of life-cycle of *Tribolium castaneum*

The larvae are negatively phototactic and are always found hidden in the flour. The infested flour becomes greyish and mouldy, so the pests are commonly known as **red-rust flour beetle**.

(c) **Pupa** : The pupa has a yellowish and hairy worm-like structure.

(d) **Adult** : The adult is a small reddish brown beetle, measuring about 3.5 mm in length and 1.2 mm in width. Its antennae are bent and bear a distinct club, formed by the three enlarged terminal joints. The last antennal segment is transversely rounded.

Life history

The beetle passes winter season as adult and breeds during April to October. During breeding season, the adults copulate one or two days and then females lay eggs in the flour. A female can lay 300 to 900 eggs in a season. The incubation period lasts for 4 to 10 days.

After hatching, worm-like larvae come out. They eat the flour voraciously and undergo 6 to 7 moultings. The larvae become full grown in 22 to 25 days in favourable temperature of 30°C. At the end of larval stage they stop feeding and make a puparium in the flour with the help of salivary secretion.

Pupation takes place in the flour and the stage lasts for 5 to 9 days.

The development period from egg to adult is 26 to 30 days in summer, but is longer under unfavourable conditions of temperature and food.

Mode of damage

Both the larva and adult cause damage to the flour, flour products and also to the grains damaged by other pests. The insect is incapable of feeding on healthy and sound grain. The greatest damage is made during the hot and humid monsoon season. In severe infestation, the flour turns greyish and mouldy and has a pungent, repealing odour making it unsuitable for human consumption.

Control measure

The pest is unable to breed in the flour if the moisture content of the flour is less than 9%. Therefore, keeping the flour in dry condition is the best way of controlling this pest.

The other way of controlling the pest is fumigation before storing the flour in the godown. The common fumigant used is methyl bromide, ethylene bromide, ethylene dichloride, carbon tetrachloride, etc.

WHEAT PEST

Tanymecus indicus

Tanymecus indicus, commonly known as **'gujhia weevil'** (Fig. 1.5) is a very common wheat pest.

Distribution

In India, the most serious pests of the wheat crop

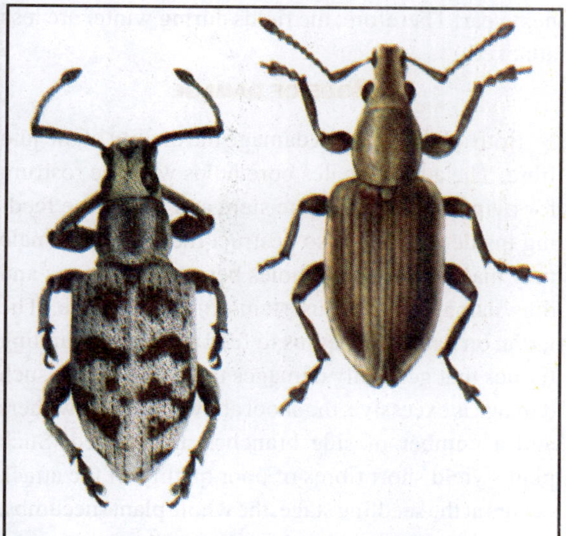

Fig. 1.5. Adult *Tanymecus indicus*

are the termites. Among other pests which have caused concern to cultivators in some areas during cultivation, is *Tanymecus indicus.* It causes great concern in Uttar Pradesh. In recent years it has also been reported from Punjab, Madhya Pradesh, Assam, Bengal, Tamil Nadu and Maharashtra.

Morphological Features

(a) **Egg** : The eggs are whitish, oval and compressed in appearance.

(b) **Larva** : It is whitish in colour with blackish head. The body is smooth and flabby.

(c) **Pupa** : Pupa is initially greyish in colour. Later it becomes darker as development proceeds.

(d) **Adult** : It is a small blackish weevil, 4 to 8 mm in length and 2 to 3 mm in wing span. Their fore-wings are oblong and hind-wings are more or less triangular in shape. Mouth parts are light brown and are concealed beneath the head.

Life History

The adult remains active from June to December. They pass rest of the year as pupa in the soil. The adult becomes sexually mature during October. They mate frequently and lay 65 to 75 eggs in 5 to 10 instalments. The weevil lays eggs in the soil under the clods or in crevices in the ground. The eggs hatch in 6 to 7 weeks. The larvae enter the soil where they feed on soil humus. The larva attains full growth at about 10 to 18 days. After that they pupate in earthen chambers constructed in the soil at the depth of 15 to 60 cm. Pupa stages are spent in the soil for 7 to 9 weeks. The adults emerge next year in June. Only the adult stage causes damage to the aerial portions of the plant. Sometimes adults are also found lurking below clods of earth. The pest has only one generation in a year.

Mode of Damage

It is a polyphagous pest attacking wheat, barley, gram, pea, poppy, maize, rice, etc. Its severe epidemic nature has often resulted in re-sowing of various crops.

The insects damage the roots of the wheat plants both during larval and adult stages. They cut the seedlings at or below the soil surface. The damage is particularly grave during October and November when the Rabi crops are germinating.

Control Measure

Soil treatment : Treatment of soil with persistent insecticides should be made. At the time of sowing, application of Aldrin, Heptachlor or Chlorodane dust at the rate of 20 kg per hectare are useful to protect the crop.

Dusting : The crop should be dusted with 5% BHC at the rate of 15 to 20 kg per hectare, when adult weevil can be seen.

JUTE PEST

I. *Apion corchori*

Apion corchori (Fig. 1.6) is commonly called as **jute Apion** or **stem borer weevil** (snouted beetle). They hamper the quality and size of the jute fibre.

Distribution

It is recorded from all the jute growing regions of India and Bangladesh. In India, the States of West Bengal, Assam and Bihar are large scale producer of jute.

Morphological Features

(a) **Egg** : The oval egg measures about 0.43 mm in length and 0.33 mm in breadth.

Fig. 1.6. Adult *Apion corchori*

(b) **Larva** : A full grown larva is about 2.85 mm in length and 0.98 mm in breadth.

(c) **Pupa** : It measures about 2.07 mm in length and 1.08 mm in breadth.

(d) **Adult** : The small weevil is 1.8 mm in length and 0.8 mm in breadth with a very conspicuous snout. The adult stem borer weevil is brown or dull black in colour and possesses small whitish setae on its body. Its life span is usually 120 to 200 days.

LIFE HISTORY

During April to June the weevils make holes with the help of its powerful rostrum in the apical part of the stem. The base of the petiole is generally the favoured site of making a hole. The female weevil lays eggs in the hole. A single female may lay as many as 675 eggs during its lifetime. The egg laying stage during optimum condition is usually three days. The larvae emerge after 3 to 5 days and start feeding on the surrounding tissues. The larvae become full grown at about 8 to 11 days. Then they make a rough chamber in the stem and pupate. The pupal stage is completed in about 4 to 6 days. The adult weevil emerges from the pupal chamber through an exit, which is either made by the larva before pupation or by the adult weevil itself. The life cycle is completed in 15 to 24 days. A number of overlapping generations are completed during the jute season. The adult stage during winter, seeks shelter in bushes, shrubs and hedges. It starts laying eggs on the new crop in the next year. Therefore, the fields during winter are less attacked.

MODE OF DAMAGE

Both adult and larva damage the quality of the jute fibre. The adult females bore holes with the rostrum for oviposition, in the jute stem and the larvae feeding inside the stem also destruct the fibre. A female may make a number of holes before laying eggs and thus damage numerous stems in her lifetime. The larva, on hatching, begins to feed on the surrounding tissues that generally damages the fibre. Where such damage is excessive, the shoot above this point withers and a number of side branches are formed. Such plants yield short fibres of poor quality. If the attack occurs at the seedling stage, the whole plant succumbs.

The quality of the jute fibre deteriorates very much due to a number of reaction changes brought about at the site of injury. A mucilaginous substance oozes out of the site of damage and it firmly cements the adjacent tissues with the insect's excreta into a hard structure, which is very resistant to the retting process. Further, a kind of corky tissue develops around the wound, which actually represents the plant's reaction to localize the injury. During this process the fibres get cemented with a good deal of surrounding tissues into a kind of knot.

During late period of infestation the basal portion of the plant becomes injured. The whole of the bark is found riddled by the grubs, resulting in a hard compact mass, with the bark sticking to the fibre. All these deformities resist retting so much that even after prolonged processing, when the fibres are washed, they are found to be either actually knotty or at least specky.

CONTROL MEASURES

Cultural measures : The damage in the apical portion is comparatively less in late-sown crops, mainly because from July onwards the weevil prefers to oviposit in the basal region.

Nitrogenous manures increase infestation but potash and phosphate fertilizers decrease the same.

Mechanical measures : The pest may be controlled by the removal and destruction of infested plants at the time of thinning the crop and by collecting and destroying the stubble after harvest.

Chemical control: In case of severe infestation a number of powerful persistent contact insecticides are now available, like 0.03% Methyl Parathion for grubs or 0.02 % Endrin for adults. But strong contact insecticides have a long residual effect.

II. *Anomis sabulifera*

Anomis sabulifera **(Fig. 1.7)** is commonly known as **semilooper** of jute and are considered as a major pest of jute.

Distribution

This pest has been recorded from all the jute growing parts of India, Bangladesh, Sri Lanka, Burma and Africa. Among the states of India, West Bengal, Assam and Bihar are large scale producer of jute and this pest is a regular visitor in these states.

Morphological Features

(a) **Egg**: The eggs are small, transparent and have the appearance of a water droplet.

(b) **Larva**: The larva of semilooper is green in colour and easily hides over the green leaves of the jute plant. Its head is slightly yellowish in colour. The elongated larva moves with loop like formation, hence it is called semilooper. The loop is formed because its five pairs of sucker feet are not well-developed, and when it moves it humps its back into an arch like structure. There are small warts over its dorsal side and the lateral sides possess yellow stripes. Each segment bears short hair on small white ringed black papillae. The full grown stage attains a length of 4 cm.

(c) **Pupa**: The pupa is inactive, soft bodied and brownish soil like colour. It is found naked in the soil as no pupal chamber is formed. The larva shrinks to about 3 cm in length at the end of pupal stage.

(c) **Adult**: The adult has a typical moth like structure. The wings are yellowish in colour, with black spots. The females are larger than males. These moths are not visible during day time but come out after the sun sets.

Life history

The moth passes winter in soil, in the pupal stage – this phenomenon is called **diapause**. They become adult during May and June, when the crop begins to grow in the field. Females lay eggs singly on the underside of the young leaves. There may be several eggs in one leaf. A female can lay 150 eggs in a month. The egg stage lasts for two days, after which a small green caterpillar hatches out with three pairs of tiny sucker feet. Subsequently fourth pair of legs develops. The fifth pair of legs is not well developed and in this region a hump is formed during movement. On emerging from the egg, the larva starts feeding on the apical soft leaves and buds. The larva attains full length at about 17 days after five moults. Before the last moult the larva generally goes down to the soil, where it usually pupates. The pupation may take place on the soil or on the plant. In summer, the pupae emerge in about a week, but those, which diapause, spend the entire winter in that stage. The life cycle is completed in about one month and several generations are completed in a year.

Mode of damage

The attack is severe on half grown plants of one

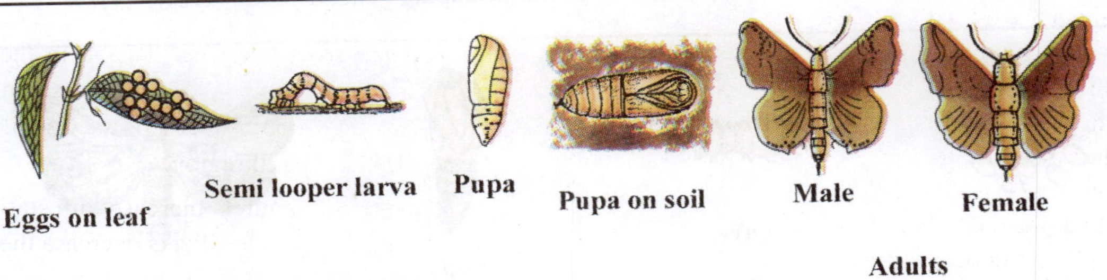

Fig. 1.7. Life cycle stages of *Anomis sabulifera*

metre high. The infestations come about in three successive waves of which the middle one is the most damaging. During this wave about 90% of the leaves are eaten up. The apical buds of the plant are most vulnerable and the intensity of infestation tends to vary inversely with the age of the foliage. Because of the green colour of the larva it is very difficult to identify the larvae on the leaves. But the holes and the marginal cuts of the leaves give a characteristic appearance to the foliage and presence of these symptoms is sure indication of the infestation. Due to heavy infestation the plant growth is adversely affected, resulting in a considerable reduction in the yield of the fibre.

CONTROL MEASURES

Mechanical measure : Thorough ploughing after harvesting gives good result of destruction of pupa.

The caterpillars can be dislodged into kerosenized water by drawing a rope across the young jute plants.

Chemical measure : Spraying of Endosulphan (0.075% in 1000 ml water) or 0.075 % Phosalone in 1000 ml water gives good result. During heavy infestation, spraying of 0.075 % Endosulphan in 500 ml water is prescribed.

VEGETABLE PEST

I. *Leucinodes orbonalis*

Leucinode orbonalis **(Fig. 1.8)** is a pest well known as **caterpillar pest** of **brinjal**.

DISTRIBUTION

This pest is widely distributed in Malaysia, Burma, Sri Lanka, India, Pakistan, Germany and East Africa.

MORPHOLOGICAL FEATURES

(a) **Egg** : Elongated in shape and scaly in nature.

(b) **Larva** : These are creamy white when young and light pinkish violet in colour when full grown. They measure about 18 to 23 mm in length.

(c) **Pupa** :The pupa stage is passed hidden in the fallen leaves in an inactive state. A silky cocoon is formed among the dry leaves.

(d) **Adult** : This is a white moth. The dorsum of the thorax and abdomen has pale brown or black spots. The wings are white with a pinkish or bluish tinge and are ringed with small hair along the apical and anal margins. There are black, pale and light brown markings on forewings. The adult moth measures about 20–22 mm across the spread wings.

LIFE HISTORY

The moth appears during March and April. The female lays elongated eggs singly or in batches of 2 to 4, generally scattered on the leaf surface. Each female can lay 150 eggs in its lifetime. The eggs hatch in about 3 to 6 days. The tiny larvae immediately enter the brinjal tissue by making a small hole on the surface. The larvae matures through five stages. The larval period varies within 9 to 28 days. The full grown larva leaves the fruit to pupate in some comparatively dry area. The cocoon is tough and silken, mingled among the fallen leaves. The pupal period is generally 6 to 17 days. The life cycle is completed in 20 to 43 days during active season. The adult survives for only few days.

The pest continues its successive overlapping generations, depending on the environmental conditions, without any obligatory diapause. There may be five overlapping generations in a year.

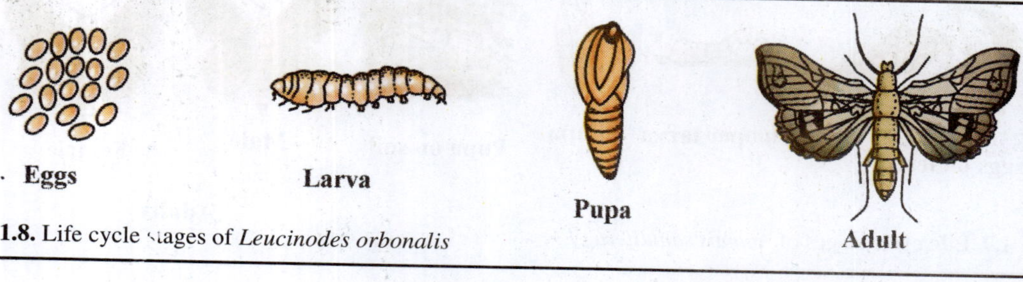

Fig. 1.8. Life cycle stages of *Leucinodes orbonalis*

Mode of damage

The damage is done by caterpillars. The tiny larvae enter into the fruit by making a tiny hole, which is not easily noticeable. It harbours within it and destroys the whole internal tissues. A healthy looking brinjal may contain several larvae. When these larvae enter the leaf petiole or young shoot, the infested portion wilts and droops.

Control measure

Precautionary measure : Special precaution is necessary to avoid the infestation of the brinjal pest. Uprooting and burning of the old brinjal plants, harbouring the borers, must be made before planting the new plants. These infected old plants would otherwise carry the infection from crop to crop and season to season.

Chemical measure : Insecticidal application can be expected only to check further infestation and not to kill most of the borers, which have already entered the plant tissues. The chemical control should be resorted to only when the initial precaution has been overlooked or when infestation by sap-sucking bugs or the *Epilachna* beetle is also involved.

The best course is to keep watch in the early stage of the crop when attack on the young shoot can be easily spotted and should be nipped in the bud.

Natural measure : There are some parasitoids associated with the larvae of this pest, they are —*Pristomerus testaceus, Cremastus flavoor bitalis, Bracon* sp., *Shirakias choenobii* and *Iphiaulax* sp. etc.

II. *Trichoplusia ni*

Trichoplusia ni (Fig. 1.9) is a moth pest, commonly called as **cabbage semilooper**. It is a minor pest of cabbage.

Distribution

This pest is recorded throughout India. It becomes major pest in northern states of India. Actually this is a polyphagous moth. It can also attack cauliflower, sunflower, and groundnut plants.

Morphological Features

(a) **Egg** : Eggs are minute and brownish in colour. They are laid either singly or in cluster, on the inner surface of the leaves of the cabbage.

(b) **Larva** : The larva is light green in colour and usually hides itself under the green leaves of the cabbage. The head is slightly blackish in colour. The elongated larva moves with loop formation, hence it is called semilooper. The loop is formed because its five pairs of feet are not equally developed, so when it moves, it humps its back into an arch like structure. Each segment of the larva bears groups of short hair on the dorsal side. The full grown stage attains a length of 4 cm.

(c) **Pupa** : The inactive larva, pupates either within the leaves of the cabbage or in soil. No pupal chamber is formed. The larva becomes stumpy and attains a length of 3 cm.

(d) **Adult** : The adult has a typical moth-like structure. It is 40 mm in spread wing. The wings are whitish in colour and ringed with hair.

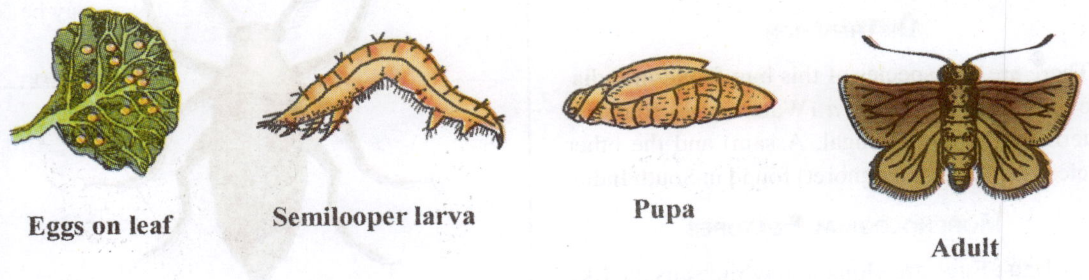

Fig. 1.9. Life cycle stages of *Trichoplusia ni*

Life history

This moth remains active during the month of October to February. The female lays eggs on the leaves of the cabbage. After 3 to 6 days the egg hatches into larva. Immediately after hatching the larva bores into the leaf leaving a mark of minute pore. They pass five instar stages to attain the full grown size. It usually takes 12 to 14 days to complete larval stage. Then it pupates either in the folds of the cabbage leaf or in the soil. The pupa stage lasts for 6 to 7 days. The adult remains dormant during summer and cannot be traced.

Mode of damage

The semilooper larva eats the leaves of the cabbage making holes on the leaves. Thus they destroy the cabbage totally. Several caterpillars may attack a cabbage at a time. These caterpillar-attacked cabbages have no market value.

Control measures

(a) **Mechanical measure** : During initial stage of cultivation the semilooper larvae can be picked and destroyed mechanically. For this purpose cultivator must have experienced eyes.

(b) **Chemical measure** : 50 EC Malathion or 1% Endosulphan may be sprayed over the field in every 10 days. However, well before the harvest, the chemical spray must be stopped.

TEA PEST

Helopetis theivora

Helopetis theivora (Fig. 1.10), commonly known as **tea mosquito bug**, causes much damage to the tea plantation.

Distribution

There are two species of this bug found in India. One species is *H. theivora* (Waterhouse) found in eastern India (West Bengal, Assam) and the other species is *H. antonii* (Signoret) found in South India.

Morphological Features

(a) **Egg** : Eggs are elongated white sausage like structure. Each egg has two fibre-like structures projecting out from the shell.

(b) **Nymph** : Nymphs are wingless. Appendages are very long.

(c) **Adult** : The adult bug is 6 to 8 mm in length, with powerful wings and a strong rostrum. The cephalothoracic region of the male bug is black and that of female bug is orange in colour. They fly during night and spread over the field. They hide under leaves during daytime and prefer hot and humid atmosphere.

Life history

They are bivoltine bugs, lay eggs twice in a year — once in March, April and another in October, November. After copulation female bug lays eggs on the buds of the tea plant. They also lay on the axis or broken end of the branch of the tea plant. At a time nearly 500 eggs are laid by a female in batches. During hot period hatching takes 5 to 15 days time while in winter it may take maximum 27 days. The newly born wingless nymph moults five times to become an adult. Therefore, the nymph passes through five instar stages during its development. The duration of nymph stage varies from two weeks in summer to eight weeks in winter.

Mode of damage

Both nymph and adult devours the bud, green leaves and new branches of the tea plant. They suck the cell sap by digging the plant parts with the help of powerful rostrum. Their salivary secretion destroys

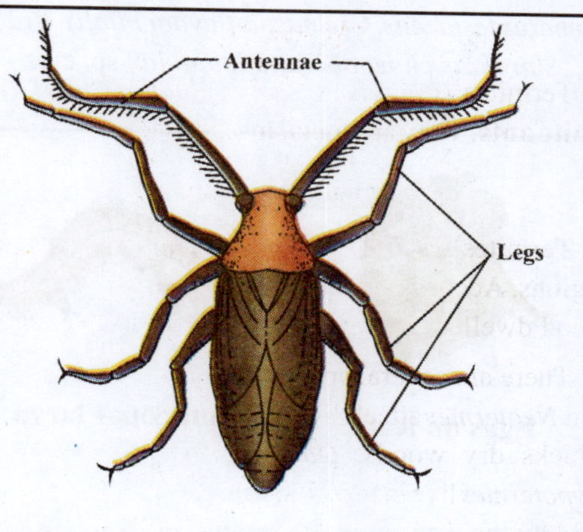

Fig. 1.10. *Helopetis theivora* (tea mosquito bug)

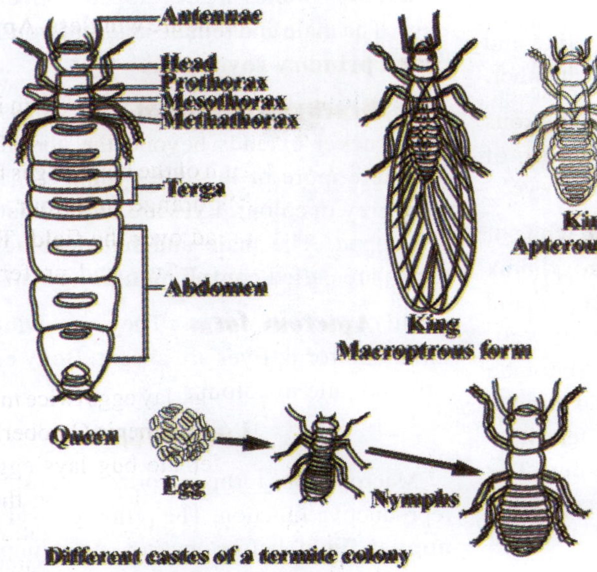

Fig. 1.11. Different castes of a termite colony

the soft plant tissues. The leaves become black, the branches become dry and ultimately shredded off from the main branch.

CONTROL MEASURES

Physical process : Everyday during dawn and dusk capturing of nymphs and bugs by hand net and killing them by fire is the best practice.

Chemical process : 0.1% Malathion spray at the rate of 500 litres per hectare can give good result during breeding season.

TERMITE

Termites (Fig. 1.11) are generally referred to as **white ants**. They are social insects and live in colony.

DISTRIBUTION

Termites are distributed in temperate and tropical regions. According to habitat they are of two types — wood dwellers and ground dwellers.

There are several species of termites. *Kalotermes* and *Neotermes* attack live woods, while *Cryptotermes* attacks dry woods. *Odontotermes, Macrotermes, Hypotermes* live in **termitarium** (Fig. 1.11A) constructed by the worker on the ground.

Fig. 1.11A. A large termitarium in a forest area of Sambalpur, Orissa, attains a height of about 7 feet.

Morphological Features

Egg : The eggs are elongated capsule like and grey in colour. Queen lays egg in special chamber.

Nymph : These are miniature forms of apterous reproductive form. They possess one pair of antenna and small ill-developed compound eye.

Adult : An adult termite is soft bodied, white or grey coloured insect. They possess head, thorax and abdomen. The head bears a pair of compound eyes and one pair of antenna. Mouth parts are well-developed for cutting and chewing. Abdomen consists of ten segments. As in other social insects, there are three castes in the life cycle of termite – so their colony is called polymorphic colony. The three castes are — **worker, soldier** and **reproductive** castes. The characteristics of the different castes are given below :

(a) **Worker** : These are small wingless forms having strong and sharp mandible, suitable for boring woods. These are infertile forms, although developed from fertilized eggs. They protect and serve the colony to run smoothly. They take care of queen. Generally they prefer dark, cold places.

(b) **Soldier** : Their head is large and hard. Cephalic appendages are developed for protecting the colony. Soldiers are of two types on the basis of the structure of mandible—**mandibulate,** with developed chisel-shaped mandible, and **nasute,** with degenerated mandible and developed rostrum. They are developed from unfertilised eggs. They generally reside on the outer sides of the termite mound and fight with the external enemies. Nasute generally spray toxin on enemies through their rostrum.

(c) **Reproductive caste** : **Drones** or **king** or males and **queens** or females are the members of the reproductive caste. They only perform reproductive function for the colony. They are of three types on the basis of their reproductive capability.

(i) **Macropterous form** : They are primary reproductive castes. Their wings are large and extend beyond the abdomen. Body is covered by deep grey chitin. They have highly developed compound eyes. Males are developed from unfertilised eggs. The male and female macropterous forms are called **primary royal pair.**

(ii) **Brachypterous form** : They possess wings but it never extends beyond the abdomen. They resemble more or less the nymph stage. Body is light grey in colour. Eyes are present but not well-developed. The male and female brachypterous forms are called **complementary royal pair.**

(iii) **Apterous form** : They are wingless and minute in form. Eyes are absent. Body covering is soft and pale in colour.

Life History

Macropterous forms are primarily responsible for reproductive function. The primary royal pair starts **nuptial flight** before mating. After nuptial flight when they return to the colony they lose their wings. Then they mate. They are generally monogamous. Queen starts laying eggs after mating. Usually queen lays 1000 eggs per day. Both fertilised and unfertilised eggs are laid by the queen. Male are developed from the unfertilised eggs. Brachypterous and apterous forms do not go for nuptial flight. They mate only when macropterous forms are not sufficiently successful in reproduction. The number of eggs they lay is very few in comparison to that laid by macropterous form.

The larva comes out of egg after two or three days. Nymphs are developed from larva after first moult. After second moult the nymph becomes worker and after fourth moult soldiers are formed from the destined nymph.

Mode of Damage

They attack any type of plant and plant product used by human being. A flagellate protozoon, *Trychonympha,* remains as symbiont in their intestine. This protozoa can digest cellulose which the termite ingests in their food.

Control Measures

(a) **Physical process** : After harvesting the stubbles of the field should be burnt. Deep ploughing can destroy the termitarium.

Fig. 1.12. Adult *Bandicota bengalensis*

(b) **Chemical process** : Before sowing 5% Endrin spray on the field will keep termite away.

MAMMALIAN PEST

Rats are considered to be the most important mammalian pest, both in field and store house. There are several types of rats found in India. About 110 species of rat are reported in India of which 70 species belong to the Family *Muridae* and Order *Rodentia*. According to their habit and habitat they are divided into three types — (a) Field rats, (b) House rats, and (c) Rats who cause damage both in house and field as well. Some of the important species of rats are mentioned below.

1. *Rattus rattus* (house rat)
2. *Rattus norvegicus* (Norway rat)
3. *Rattus meltada* (Soft fur field rat)
4. *Mus musculus* (Albino rat)
5. *Mus platythrix* (Brown spiny mouse)
6. *Mus booduga* (Common Indian field mouse)
7. *Bandicota indica* (Large Bandicoot rat)
8. *Bandicota bengalensis* (Indian mole rat)
9. *Tatera indica* (Indian gerbil)
10. *Nesokia indica* (Short tailed mole rat)
11. *Golundaelli oti* (Indian bush rat)
12. *Millardia meltada* (Grass rat)

All the above mentioned rats are to some extent considered as pest in different regions of India. Here we restrict our discussion to only about *Bandicota bengalensis.*

BANDICOTA BENGALENSIS (BANDICOOT RAT)

This vertebrate pest is commonly known as **mole rat/bandicoot rat/lesser bandicoot (Fig. 1.12)**.

Distribution

This rat is recorded from India, Sri Lanka, Myanmar, Indonesia, Bangladesh, Vietnam, Formosa, etc.

Morphological Features

The dorsal surface of the rat is covered with thick coarse fur of greyish brown colour. The ventral side of the body is covered with greyish white fur. The tail is sometimes shorter than and sometimes as long as the body.

Habit

This species commonly inhabits the fields of crop, vegetables and fruits. It is a fierce animal and grunts in anger. They are very good swimmers. The bandicoot rats are well adapted to adjust themselves everywhere because they are provided with highly developed sense of smell, taste, hearing and touch. Therefore, they keep themselves away from any type of danger. They always move along a definite pathway. They identify their routes with the help of dark greasy

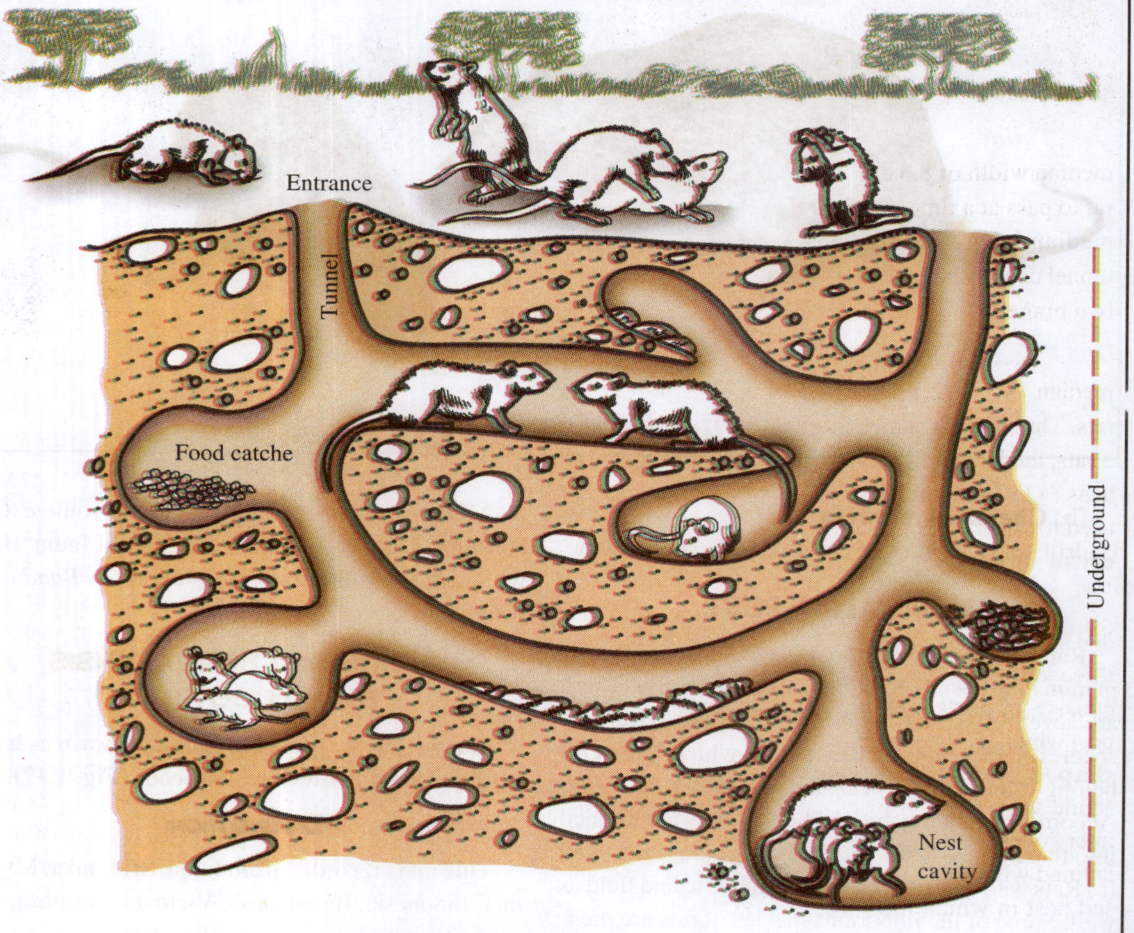

Fig. 1.13. Burrow system of rat

marks left in their way. They are nocturnal and highly engaged in collecting food. They stock the food grains inside the burrows for future when food would not be available. The home range of the males is greater than the females and varies from 50 to 100 feet from their burrow.

After harvesting of crops, in unfavourable climatic conditions, the rat starts migration from one field to another or one store to another.

Habitat

They inhabit in a well-formed burrow in the field. The rats vigorously survey the field, position of crop and humidity of the soil, before making the burrow.

Rat burrows

Rats dig and live in underground burrows (Fig. 1.13). Burrows can be simply a chamber connected to the outside by a short tunnel, or can be a large complex of interconnecting tunnels, passages, and cavities. Burrows are not unchanging entities. Burrows grow, change, and fall into disrepair as their inhabitants dig new tunnels, fill in old ones, or stop maintaining certain sections. In fact, the structure of a burrow system is related to the social relationships of the rats, who dig and maintain it. If the social structure of the group breaks down, the rats are no longer able to maintain a complex burrow system and it falls into disrepair.

Parts of a burrow

◆ **Entrance** : Usually placed in a sheltered and/or sloped location. May be sealed with cut grass or dirt.

◆ **Tunnel** : Straight tunnel segments have a median width of 8.3 cm. This width allows only one rat to pass at a time. Straight tunnel segments have a median length of 29.8 cm. After that distance, a tunnel usually bends, ends in a nest cavity, splits into two tunnels, or dead-ends.

◆ **Chambers** : Small chamber of varying size, median 18.5 × 22.1 cm, which could accommodate 7 rats. The smallest chambers could accommodate only 3 rats, the largest 11 rats. Chambers can have several uses : (1) **Nest cavity** : contains bedding material, used for sleeping and rearing young litters. (2) **Food-cache** : used for food storage

◆ **Nest** : Built in a nest cavity, it consists of shredded material and serves as a sleeping place. There are three kinds of nest : (1) **Pad** : Simplest kind of nest consists of just a few flat objects (leaves, bits of paper) which elevate one rat above the floor. (2) **Cup-shaped nest** : Larger nest made in a cup shape. Made with finer texture grass or shredded paper interwoven a bit to form the walls of the cup. The cup is lined with flat objects. (3) **Hooded nest** : Organised nest in which the walls grow so high that they form a ceiling, and the nest becomes a hollow sphere with just one opening. These nests are sometimes built by mothers for litters, but tend to be rather rare.

Growth of a burrow system : Burrows are frequently started by a pregnant rat just before giving birth. Rats like to start burrows against vertical surfaces (e.g., a man-made wall), under flat surfaces (e.g., under a board placed on the ground), under overhead cover (e.g., shrub, overhanging concrete, raised man-made floors), on slopes, and near sources of food and water. Dirt excavated from the initial tunnel is usually piled around the tunnel entrance in a burrow mound.

The initial tunnel usually ends in a nest cavity. After a few days, a second entrance is added by digging from below. This second hole has no excavated dirt around it and is called a bolt hole, and may serve as escape exits in case the burrow is invaded. Later expansions of the burrow system follow this same pattern of tunnel/cavity/bolt hole. The second cavity often becomes a food-cache, and is added by the mother rat shortly before the young are weaned. This allows the young to start on solid food in the safety of their home burrow. Once established, these simple burrows serve as centres for the elaboration of extensive burrow systems.

Modification of the burrow : Rats prefer to keep a certain distance between themselves and the ceiling of the nest cavity. As the rat brings in nesting material, the rat finds itself closer to the ceiling than it likes, so it removes dirt from the roof and carries it out. Over time, the bedding gets soiled and the rat brings more bedding in, and the process is repeated. Eventually, the nest cavity approaches the surface and its thin ceiling collapses.

Rats seal entrances temporarily with cut vegetation. These seals are easy to remove or squeeze through. Lactating mothers, especially those who have just given birth, may seal burrows in this way.

Rats also permanently seal entrances with loose dirt and lumps of dirt pushed up from below. They are packed into place with the forefeet. They may also carry dirt out another entrance and pack it into the neighboring one which is being sealed. A rat may seal a burrow when it feels a draft and when it is of low social rank. The seal of a low ranking rat may prevent higher ranking rats from coming in and harassing it. The burrows of low-ranking rats also tend to leak more, so such seals may help counteract such leaks.

Trail systems : Outside the burrow, rats tend to confine their movements to the same routes every day. Gradually, trails form on the surface.

The main aim of burrow formation is to make protection, favourable micro climate, storage of food and breeding facilities. The burrow of the rat also consists of different fauna and flora; such as earthworm, crickets, earwigs, beetles, ants, pseudo-scorpion, millipede, mites, etc.

Life history

The rats start living in separate burrows after attaining maturity. The breeding period is in two seasons; one is March to May and another is August to October. At the age of 3 to 6 months they start breeding and gestation period lasts for 21 to 25 days. The female comes in heat after every 4 to 5 days for about 2 to 3 days and breeds commonly twice in a year coinciding with the maturity of the crops. The female produces 8 to 10 young ones in a litre. Young one are hairless, reddish and with closed eyes. After two weeks from birth, their eyes open and thery generally begin to feed on grains after 3 weeks. The sex ratio generally found is 50:50. The life span of an adult female is longer than a male.

Mode of damage

The major damage is the consumption and destruction of crops and crop plants. It is found that a rat consumes about 10% of their body weight daily. Moreover, they destroy the crop by taking to their store chamber. It has been recorded that maximum storage was 5 kg in a chamber.

The burrows made by rats weaken the foundation of buildings, causes seepage in channels, damage railway tracks in yards, etc. They can cause fire by damaging electric wires.

There are about thirty two diseases of man and other animals that are known to be transmitted by rats. They are reported to carry 18 different kinds of lice, fleas, ticks and mites. Different types of viruses, bacteria, fungi, protozoa, nematodes, arthropods carried by the rats can cause several diseases like bubonic plague, typhus fever, infectious jaundice, rat bite fever, food poisoning, etc.

CONTROL MEASURE

A. Physical measures

The types of physical measures practised to control the rats are as follows :

(1) **Ultrasonic sound** : Ultrasonic sound emitter is introduced in the mouth of the burrow made by the rat. The charged ultrasound destroys the ear drums of the rats, which are ultimately killed. However, the wide range use of ultrasonic waves in the field is under investigation, especially with reference to its effect on the farmer and high cost of this method for field use.

2. **Electrocuting** :Sometimes iron fencing around the field are electrified, so that rats when come out of their burrows in search of food and move across the fencing gets electrocuted. This method proved useful when there are no rats inside the fenced area. This has the inherent danger of electrocution of men and animals. Such a device should not be suggested for highly populated countries.

3. **Traps** : These are the traditional methods to catch the rats by various mechanical devices and kill them. The formations of traps are devised differently in different localities. But this method takes much time and man power, and is very difficult to apply in large field area. However, in small area, like house and small godown, this method may be applied. Traps may be divided into two categories : ***live traps*** and ***kill traps***. In live traps, rats are caught alive and then killed by immersion in water, whereas in kill traps, such as break-back guillotine, rats are immediately killed as soon as they come in contact with them. The basic principle of the trap is to allure the rats to move into the contrivances from which they will not be able to escape or in which they will be killed. Some of the commonly used traps are described below :

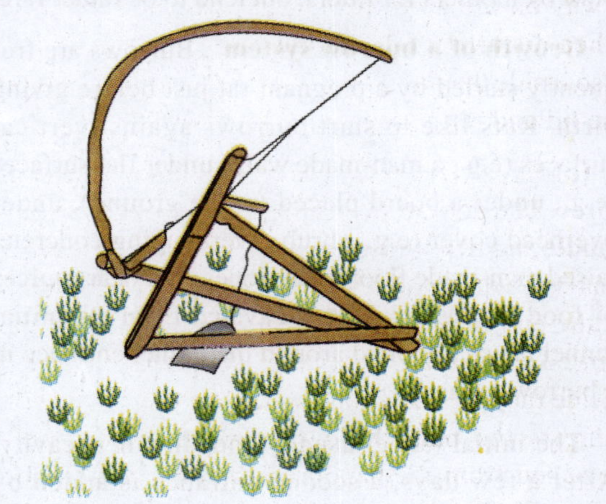

Fig. 1.14. Bamboo bow and arrow trap

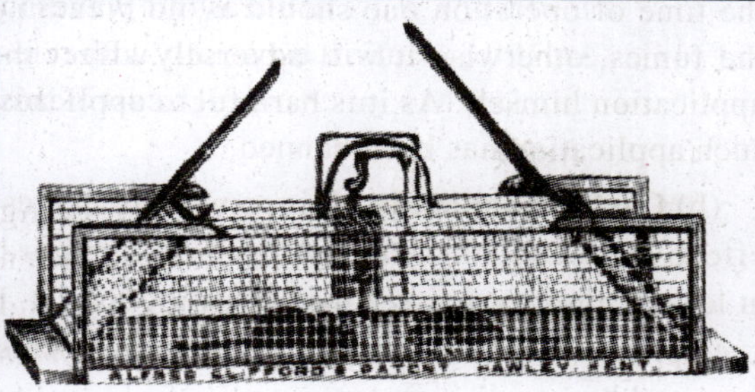

Fig. 1.15. Cage trap

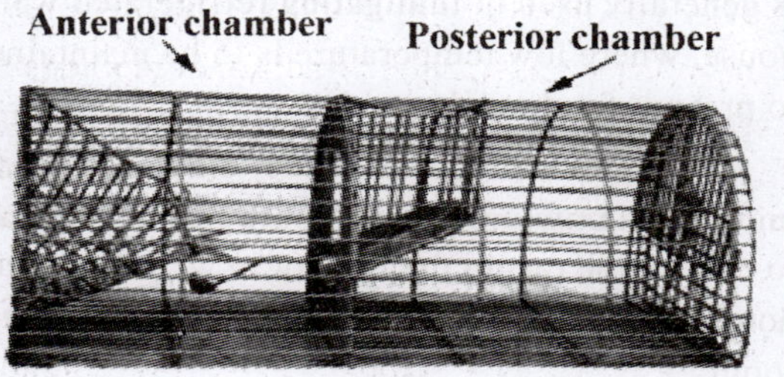

Fig. 1.16. Wonder trap (Monarch type)

(a) **Bamboo bow and arrow trap (Fig. 1.14)** : In this trap a triangular bamboo stick is fixed on the bunds, tracks and near the burrow of the rats. When the rats get in to take food, the bamboo sticks snap automatically, the rat is trapped and ultimately gets killed. It is very common in south India.

(b) **Pot trap** : It is a round earthen pot containing a lid and wood pegs. The lid is adjusted in such a way on the ground, that the pot falls on the lid when disturbed. The bait is placed on the lid and also inside the pot. The bait when disturbed falls on the lid and rat is trapped.

(c) **Cage trap (Fig. 1.15)** : It is made of wood/ wires. The door of the trap is provided with a long handle and two springs. It is kept open by engaging a curved wire on the roof of the box which is fixed at the end of the handle. On the other end of the curved wire the bait is kept hanging suspended inside the box . As soon as the rat entering the box tampers the bait, the curved wire or hook gets detached from the handle of the door and the lid falls off. In one setting only one rat can be caught.

(d) **Wonder trap (Fig. 1.16)** : Wonder trap is a small wire made cage divided into two chambers. A smaller chamber with tubular narrow opening, through which the rat enters and a larger one connected with the smaller one through another tubular passage, whose floor is moveable. This tubular passage allows the rat to pass only when the base of the passage is drawn downwards under the weight of the rat and the rat rolled down into the bigger chamber containing baits. The opening is covered by one way gate. Inside the cage baits are given. The rat enters inside the trap for the bait but cannot come out. In this trap more than one rat can be caught in one setting.

(e) **Back-breaking trap (Fig. 1.17 & 1.18)** : It consists of an iron plate having toothed edges

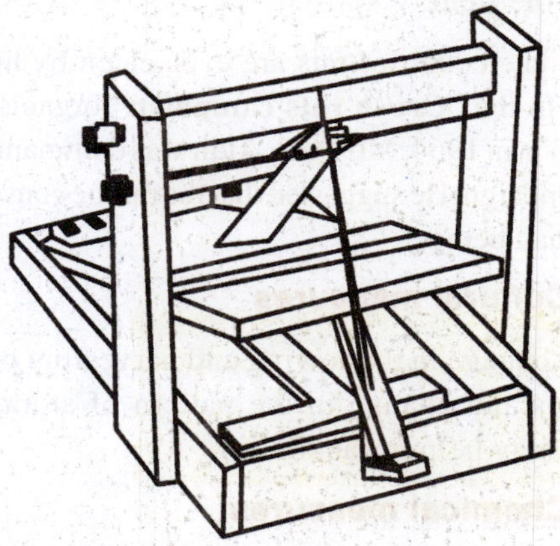

Fig. 1.17. Wooden back-breaking trap

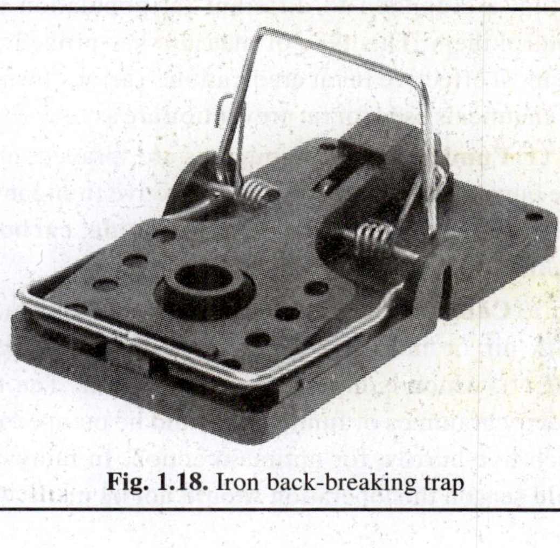

Fig. 1.18. Iron back-breaking trap

pointing upwards at its distal end. A U-shaped striker of an iron wire is attached with a spring on which the bait is fixed. As soon as the rat disturbs the bait for feeding, the trigger is released, crushing the rat between the edges of the basal plate and retreating U-shaped striker. Only one rat can be killed at one time. Sometime wooden traps are also indigenously used.

B. Biological measures

The natural enemies of rat are cat, dog, snake, mongoose and a number of birds. They catch the rat actively and eat them up. Many offices and housekeepers keep cats for protection from rats. Snakes are traditionally regarded as the most effective predator of field rats and also of house rats. Among birds, kite, owl and hawk are active predators of the rats in the field.

In some countries rats are even eaten by human being. In the **"Crush rats campaign"** organised in Indonesia in 1964, Mr. D. N. Aidit was commander of that campaign. He suggested that the rats be consumed by human beings.

C. Cultural measures

The change in the sowing and harvesting period of crops and change in the pattern of storage of grains may help in rat control.

D. Chemical measures

Several chemicals of various nature, origin and action are applied to different rat population to control them. This type of measure has proved to be most effective in rat eradication. Various forms of chemicals used for rat prevention are as follows :

(1) Fumigant : The fumigants are sprayed into the burrow of the rat. The most effective fumigants for rat eradication are : **calcium cyanide, carbon monoxide and carbon dioxide**.

(a) **Calcium cyanide** : This fumigant on contact with air or moisture, releases hydrocyanin gas (HCN), which is highly poisonous to rat. Ten to twenty grammes of fumigant should be pumped in each live burrow for optimum action. In rainy or cold season this operation should not be made. At the time of operation one should avoid breathing the fumes, otherwise it will adversely affect the application himself. As it is harmful to applicator such application has been banned.

(b) **Carbon monoxide** : It is comparatively less effective in action than calcium cyanide. It requires at least 5 minutes acting time for killing the rat. It is employed in rat burrow with the help of exhaust powered by gasoline motor.

(c) **Carbon dioxide** : In the form of dry ice carbon dioxide is applied into the burrow of the rat. It is generally used in fumigating refrigerated warehouse, where low temperature is to be maintained to prevent food spoilage.

(d) **Others** : Methyl bromide, ethylene di-bromide and aluminium phosphide are used as fumigant to control the rats in the field or in well-ventilated houses, but these should never be used in the bedroom.

(2) **Attractant and repellent** : Saffron, kewra, lemon, etc. are attractants of the rats. Poison baits are mixed with these attractants to kill the rats.

Among the repellent, lauro-nitritite, actidione, cycloheximide, triphenyltin and malathion are considered good. But with these repellent rats are kept away for sometime only. So repellents are totally temporary measures of rat control.

(3) **Chemosterillants** : Male and female rats are sterilised by feeding furadantin and colchicine respectively. Application of 0.25 gm furadantin and 0.001 gm of colchicine with the attractant give good result for sterilising the rat population. Cadmium chloride and glyzophorol (containing butandial-bismethanesulphonic acid estar) is recently introduced as rat steriliser.

(4) **Rodenticides** : The Rodenticides are grouped into acute and chronic poisons. The common acute poisons are arsenic trioxide, barium carbonate, zinc phosphide, norbromide, thallium sulphate and sodium fluroacetate. The common chronic poisons are derivatives of caumarin and are anticoagulants, e.g., Warfarin, Dethmore, Ratafin and Racumin.

SOME FREQUENTLY ASKED QUESTIONS (FAQ) ON RATS

Q.1. What are the differences between rat and mouse?

Ans. See Fig. 1.19. Adult mice are much smaller than adult rats. Adult mice weigh about 30 gm. The body of adult mice is 3 to 4 inches long with 3 to 4 inches tail. Adult rats are far heavier and longer than mice. The females weigh about 350 to 450 gm and the male 450 to 650 gm. They have 9 to 11 inches long body and 7 to 9 inches long tail. Due to larger body size rat faeces are also larger in size than mice.

Q.2. What are the differences between *Rattus rattus* and *Rattus norvegicus* ?

Ans. The differences are given along with the figure 1.20 below.

Fig. 1.19. Pictorial comparison between rat and mouse

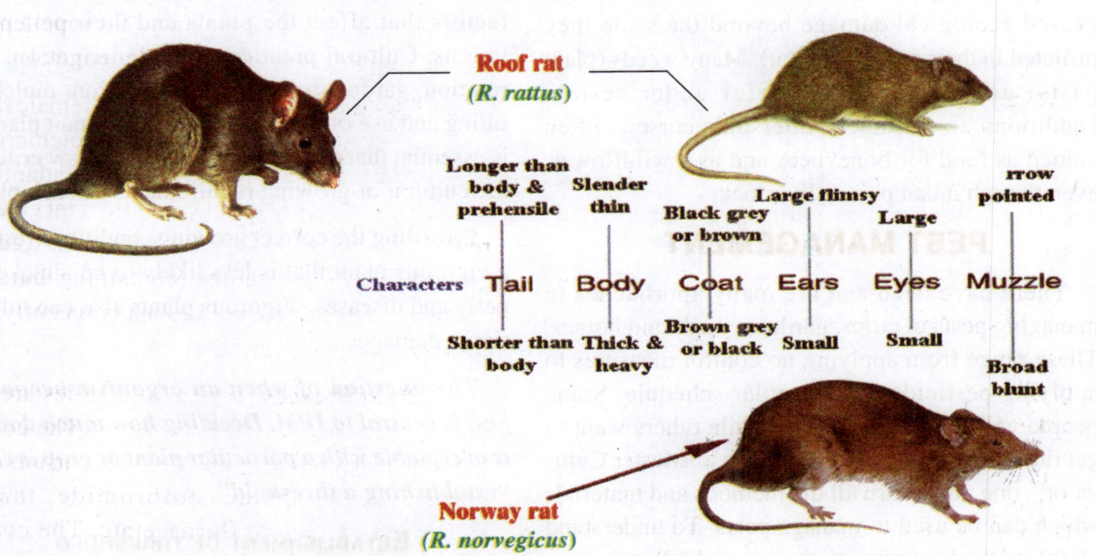

Fig. 1.20. Characteristic differences between *Rattus rattus* (Roof rat) and *Rattus norvegicus* (Norway rat)

Pest Management

A **pest** is an organism which has characteristics that are regarded by humans as injurious or unwanted. This is most often because it causes damage to agriculture through feeding on crops or parasitising livestock, such as codling moth on apples, or boll weevil on cotton. An animal can also be a pest when it causes damage to a wild ecosystem or carries germs within human habitats. Examples of these include those organisms which vector human disease, such as rats and fleas which carry the plague disease, or mosquitoes which vector malaria.

The term pest may be used to refer specifically to harmful animals, but is also often taken to mean all harmful organisms including fungi and viruses. Pesticides are chemicals that are used to control or protect other organisms from pests.

It is possible for an animal to be a pest in one setting but beneficial or domesticated in another (for example, European rabbits introduced to Australia caused ecological damage beyond the scale they inflicted in their natural habitat). Many weeds (plant pests) are also seen as useful under certain conditions, for instance Patterson's curse is often valued as food for honeybees and as a wildflower, even though it can poison livestock.

PEST MANAGEMENT

There have been and are many approaches to managing pests in crops, gardens, yards and homes. These range from applying no control measures to applying pesticides on a regular schedule. Some people will not use pesticides while others want to get rid of the pest by any means. As a "Master Cultivator", one must learn all the methods and materials which can be used to manage pests. To understand all the methods of managing pests and select the best method for a particular situation, people presently are inclining towards Integrated Pest Management (IPM).

PHILOSOPHY OF INTEGRATED PEST MANAGEMENT (IPM)

Integrated Pest Management or IPM provides a systematic approach to pest control. It begins with using the best cultivation practices. Following the correct selection and culture of plants, reduces the potential for pests and diseases.

When problems arise, the essential step in an IPM program is determining when a pest is causing enough damage to warrant control. It also helps the cultivator to determine which of the many possible control measures is most effective and appropriate in a particular situation. IPM can be applied to all floral and faunal pest groups.

Correct cultural practices

Cultural practices should be designed such that the cultivator has ways and means of implementing correct practices as per changing environmental factors that affect the plants and their pest populations. Cultural practices include irrigation, crop rotation, garden sanitation, soil aeration, mulching, tilling and use of disease- and pest-resistant plants. It is essential that the cultivators should know critically the cultural or growing requirements of each plant.

Providing the correct growing conditions result in a vigorous plant that is less likely to be attacked by pests and diseases. Vigorous plants also can tolerate tissue damages.

The question of when an organism becomes a pest is central to IPM. Deciding how much damage is acceptable with a particular plant or crop is called "establishing a threshold".

Establishment of threshold

Several factors influence how each cultivator will arrive at a pest threshold. These factors include the following :

- Amount of damage that can be tolerated.
- Pest population size required to cause significant damage.
- The plants' stage of development.
- The vigour of the threatened plant(s).
- The life cycle and habits of the pest.

The amount of damage that can be tolerated varies with the crop and personal preferences. A few holes on a collard leaf do not require control. However, if most of the leaf has been eaten, the crop is destroyed.

Less than 12 white grubs in a square foot of turf grass can be tolerated as long as the lawn is growing vigorously. However, if the density of grubs rises above 12 per square foot, damage will be significant.

Colorado potato beetles devastate a potato crop early in the season. However, in the latter part of the season, they feed on the foliage and practically no damage is done to the crop.

Peach tree borers are extremely difficult to control once they bore into the trunk of the tree. Knowing that the adults lay eggs in mid-July to mid-August and that the young larvae are susceptible during this period, the cultivators are advised to effectively apply control measures.

Establishing thresholds in the field involves observing the plant and its total environment.

There is also the question : Does the cure process cause more damage than the pest? There is a risk involved in using the more toxic pesticides. This risk must be weighed against the potential benefit.

Inspection and Monitoring

One must monitor the field to determine when action is necessary. Thorough inspection of the plant includes – the underside part of leaves, buds, new growth and even the roots in extreme cases. Regular inspection allows the cultivator to identify a problem before major damage occurs to the plant. Controls will be more effective if the problem is caught early. Infrequent inspection results in extensive damage. Once extensive damage occurs, it is often too late to do anything.

Inspection of the plant's entire environment for clues to the problem is a good practice for a good cultivator. Too much or too little moisture, wind, humidity and light may cause stress that results in susceptibility to damage from pests and diseases. Observing and keeping records of weather conditions provide clues to growth patterns and problems.

Certain insect pests can be monitored by using traps. Coloured, sticky traps, pheromone traps, pitfall traps and light traps are used with some pests. Whiteflies and aphids are attracted to bright yellow, sticky cards and are trapped on the sticky surface.

Identifying the Cause of the Problem

The problem may be caused by insects or related pests, diseases or cultural practices. Often, the problem is caused by a combination of these factors. Identification of the pest or disease and the conditions that allow organism to become a problem is of prime importance. Reference materials are guide to common problems for a particular plant. They list the insects and diseases that attack each plant. The ORTHO PROBLEM SOLVER is an excellent resource for identification. A questioning attitude and close inspection are required to make an accurate diagnosis. A magnifying or hand lens is helpful. Close examination of the specimen for feeding or disease patterns is essential. If insects are involved, collection of the pest is essential. Identification of insects and diseases is available through **local Extension offices** and the **Plant and Pest Identification Organizations**.

Application of Control Measures

To apply a control measure, one must understand the pest organism, its life cycle and the type of damage it causes. All control options should be considered. The most effective and appropriate option or combination of options should be selected. Control measures should be taken as fast as possible and at the right time. Insects are more vulnerable at certain stages of their life cycle. Eggs may be controlled with oil sprays. Larval stages are particularly vulnerable. It is critical to control insects before they reproduce and increase their population. When an insect is protected by the pupal case, control is nearly impossible.

Re-application of control measure may be necessary for many pests to kill the young as they emerge from eggs. It is important to know, how long the pesticide remains effective on the plant. This is also important in food crops. Cultivator must observe the prescribed time lapse between the last application and harvest.

Diseases are more difficult to control. Most of the fungicides prevent infection or at best stops an infection from spreading. Netting is not effective if there are unprotected areas. Pesticides must be applied to both the top and underside of leaves. Surfactants or spreader/stickers help the pesticide adhere to and spread over the foliage surface. Application rates are critical. The proper dilution is required to kill the pest. If it is too concentrated, plant damage may occur. Application of control measures should be as instructed by the authentic organization.

Control options

Control options can be arranged by their mode of action and their impact on the environment. Generally, these are grouped from least to highest impact : (a) **cultural and mechanical controls,** (b) **biological controls** and (c) **chemical controls (soaps and oils, botanical pesticides and synthetic pesticides)**.

(a) Cultural and mechanical controls

Keeping plants healthy and growing vigorously should be the first line of defense against pests. The next level of control includes *hand removal, use of screens, barriers, traps, freezing and crushing*. These are excellent options for the home cultivator. Mechanical methods generally have little or no negative effects on the environment.

(a) **Hand removal** : This method is suitable for larger insects. Shake pests into a can that contains a small amount of water and detergent. Scrape off and crush egg masses.

(b) **Screens and barriers** : Any material that is fine enough to keep pests out can be used as a barrier. Screens can keep out large insects, birds and rabbits. Floating row covers well anchored to the ground can prevent attack by nearly all insects. These will also prevent pollination by insects, resulting in lack of fruit. Cardboard and metal collars will prevent cutworms from reaching young transplants. Sticky bands placed on tree trunks to trap cankerworm and elm leaf beetles. Diatomaceous earth is advertised as a barrier to keep slugs from damaging plants. The sharp particles, or diatoms, are supposed to cut the slugs' underbody. However, recent reports indicate that it is not effective. The small particle size quickly becomes incorporated into the soil, especially after a rain or irrigation. Copper strips also are offered as a control. They supposedly react with the slugs' slime to shock them.

(c) **Trapping** : Traps usually serve as a monitoring system, warning of the presence or increase in undesirable pest numbers. Traps also can be useful in timing control measures. They are sometimes used to control numbers. Yellow, sticky traps attract whiteflies, aphids, thrips, leafhoppers and other small flying insects. Traps that use pheromones or attractive scents to tantalize adult insects are best used as a way to check presence and numbers. Pheromones are used in Japanese beetles traps.

Pitfall traps are cups or jars placed into the ground. Weevils, cutworms and other pests fall into the traps. Light traps capture insects using a black light (ultraviolet). These are used by advanced cultivators to monitor night-flying pests. "Bug zappers" are not considered useful because they kill beneficials as well as pests.

(d) **Syringing** : A hard stream of water from a hose, washes aphids, spider, mites and other small insects from plant foliage. This must be done frequently since it does not kill insects or eggs. It does not prevent some insects from crawling back onto plants.

(b) Biological controls

Nearly every species of plant-feeding insect has another insect that is its predator or parasite. Biological methods of control utilize natural enemies of pest insects to keep their populations under control. Biological controls involve several options : **predators and parasites, microbes,** etc.

(1) **Predators and parasites**

Predatory and parasitoidal insects and mites are often termed "beneficials". Beneficials include predators such as lady beetles, lacewings, syrphid flies, wasps and praying mantises as well as parasitoids, including certain wasps, flies and mites. There are some parasitoids associated with the larvae of *Leucinodes orbonalis* (Brinjal pest), they are — *Pristomerus testaceus, Cremastus flavoorbitalis, Bracon* sp., *Shirakia schoenobii* and *Iphiaulax* sp.

Cultivator should learn to recognize beneficial insects. The use of NON SELECTIVE insecticides kills these helpful insects. Rather than trying to introduce beneficial insects into the landscape, avoiding the killing of those that are already there, is a good prescription. A small pest population must continue or the "beneficial" will die from lack of food. Purchasing "beneficial" is only recommended for use in greenhouses.

Parasitic insects are less well known than predators, but equally effective. They lay their eggs in a pest species. When the eggs hatch, the larvae feed on the pest insect, killing it. Most of the parasitic insects are tiny wasps.

Many birds and bats eat insect pests. It is advisable to provide water and shelter so as to encourage them to visit your garden.

(2) **Microbes**

Bacteria, fungi, viruses, protozoans and parasitic nematodes are microorganisms that attack insects. These microbial "insecticides" are generally effective against very specific pests. As they pose little risk to humans and the environment, microbials are likely to be used more in the future with better preventive measures.

Many cultivators are familiar with a popularly known, microbial-based insecticide called as **"Bt"** or ***Bacillus thuringiensis***, bacteria which produces a toxin that kills specific caterpillars. The larval pest usually dies within 4 to 7 days. It is a commercially available product (Dipel ™, Thuricide ™) used to control many kinds of moth and butterfly larvae. There are many STRAINS of "Bt", each type controlling specific pests.

Parasitic nematodes are very effective against certain pests that live in the soil. However, the nematodes require moist conditions to survive.

Few fungi, viruses and protozoa are commercially available, however, these living organisms are difficult to raise, store and apply.

(c) Chemical controls

Controlling pest by chemical substances is called chemical control and the chemical is called 'pesticide'. Pesticides are any substances used to kill a pest. "*Cide*" means to kill; therefore, an *insecticide* kills insects and an *herbicide* kills plants. We also have miticides, rodenticides, molluscides, fungicides, etc. Pesticides include repellents, attractants, growth regulators, defoliants, anti-desiccants and antibiotics.

Properties of a pesticide

1. An ideal insecticide should control a particular pest and not kill other insects or animals.
2. It should DEGRADE or leave the environment quickly and safely and not be accumulated in the environment.
3. It should have low TOXICITY for humans and other mammals.

The way a pesticide affects the pest is important in choosing the correct pesticide. On the basis of working principle, pesticides are categorized in following ways :

(A) **CONTACT** pesticides kill by touching the pest. These are used for soft-bodied insects like aphids; e.g., Phosphamidon, Endosulphan, Phenthion 20EC, methyl parathion, Endrin, Phosalone, etc.

(B) **STOMACH** poisons must be eaten. They are used to control insects that chew foliage; e.g., Aldrin, Heptachlor, Chlorodane, BHC, etc.

(C) **SYSTEMIC** pesticides are used in the field during cultivation along with the manures so that they can be taken into the sap of the plant. The pest then ingests the sap and get poisoned; e.g., Carbofuran, Phoret, etc.

(D) **FUMIGANTS** are gases that are sprayed over the field, which are inhaled or absorbed by the pest; e.g., Nethylamide, Calcium cyanide, Methyl bromide, Ethylene bromide, Ethylene dichloride, Carbon tetrachloride, etc.

(E) **PROTECTANTS** form a barrier on the plant. Fungicides protect the plant from germinating fungal spores.

EXAMPLE OF SOME SPECIAL PESTICIDES

Insecticidal soaps

Insecticidal soaps are made from the salts of fatty acids. Fatty acids are components of the fats and oils found in plants and animals. These soaps should not be confused with ordinary cleaning soaps. Insecticidal soaps are CONTACT INSECTICIDES and are, therefore, effective against soft-bodied pests: aphids, thrips, crawler stage scales, whiteflies, leafhoppers and mites. Insecticidal soaps may cause burning on some plants. This is likely if the foliage is hairy. Test of insecticidal soap susceptibility for a plant is always recommended. Burning of plants will usually occur within 24 hours.

Horticultural oils

Horticultural oils are highly refined mineral oils that are used at very low rates. They appear to work by suffocating the pest and damaging the membranes of the pest. Dormant oils are less refined than horticultural oils. They are applied only when the plant is dormant and the temperature is above 4°C.

Both horticultural soaps and oils remain in our environment for a short time. Until recently, there has been no evidence of insects developing resistance to soaps and oils.

Botanical pesticides

Botanical pesticides are made or derived from plants. Botanical insecticides include *pyrethrins, rotenone, sabadilla, ryania, nicotine, citrus oil extracts* and *neem.*

Advantages of botanical pesticides : These are popular pesticides because of the following characteristics :

- They rapidly degrade in the environment and are less of a threat to other organisms. Mostly applied to food crops that are closer to harvest.
- They act rapidly to stop feeding by insects, although they may not kill the pest for hours or days.
- They generally have low-to-moderate toxicity for mammals.

Disadvantages of botanical pesticides : They must be applied more frequently. They may be more costly, may be difficult to obtain, and although generally less toxic, but they are toxic. Nicotine, in particular, is extremely toxic to mammals. Botanical pesticides have not been thoroughly researched for effectiveness and toxicity.

Synthetic pesticides

These are synthetically produced from byproducts of different chemical industries.

Advantages of synthetic pesticides : These synthetic compounds are less expensive, more effective, faster acting and may be longer lasting than the botanical pesticides. They include many of the common pesticides *: 2,4-D lindane, Malathion, Carbaryl, Diazinon, Kelthane* and *Captan.*

Disadvantages of synthetic pesticides : Synthetic pesticides may be longer lasting, they are generally nonselective and are capable of harming other organisms.

Use of synthetic chemicals should be the last resort in controlling pests. In some cases, synthetics are the only effective control. Both synthetic and "natural" pesticides are capable of harming humans, wildlife and the environment, if mishandled. Always follow instructions on the label when using any pesticide.

Pesticide safety
(Measures for safety use of Pesticides)

(1) **Following label instruction**

When using pesticides, safety is critical. Before application of a pesticide, reading and following the directions of pesticide label exactly is crucial. The label is a legal document. Using pesticides in ways other than listed on the label may result in legal action. On the label special instructions are given for the following measures :

1. Plants that may be treated
2. Application rates and timing
3. Application methods
4. Toxicity level: warning, caution, danger
5. Precautions to take when applying
6. Danger to other animals or nearby vegetation
7. Environmental concerns
8. Timing of application relative to harvest
9. Instructions to physicians when poisoning occurs
10. Disposal and cleaning of clothing used during application
11. Disposal and storage of all pesticide chemicals

(2) **Wearing protective clothing**

Wearing of protective clothing when applying pesticides is must. This includes long-sleeve shirt, long pants, shoes, socks, rubber or neoprene gloves, protective eye gear and a respirator or dust mask when necessary. Clothing exposed to pesticides should be washed in detergent and hot water.

(3) **Avoiding spilling and drift**

Spilled pesticides can be absorbed with kitty litter. If one is exposed to pesticides during mixing or application, he/she must wash it immediately and follow label instructions. Pesticides contacting the eyes, nose or ears should be rinsed with water. To avoid exposure to pesticides, application must be done when there is no wind to avoid PESTICIDE DRIFT.

(4) **Use of proper applicator**

It is advised to find the correct applicator. The label on applicator will provide specific instructions. These applicators include compressed-air sprayers, trombone sprayers, hose-end sprayers and hand-dusters. A cultivator should have separate sprayer equipment for herbicides and insecticides. Herbicide residues in a sprayer used to apply insecticides or fungicides may damage plants.

(5) **Taking correct amount of pesticide**

Only the required amount of a pesticide should be prepared to complete the job. Mixing exposes the applicator to the most concentrated form of the pesticide. Carefully measure concentrates with measuring spoons or cups reserved and labelled for that use only. No chemicals should be mixed together unless the label says they are specifically compatible.

If more pesticide is mixed than that which is needed, it should be applied to plants that are on the label. Never pour a pesticide down the drain or storm sewer. Rinse the sprayer with freshwater, applying the rinse water to plants that are listed on the label. A second and third rinse is recommended.

(6) Recording application

Keep records of pesticide applications for reference, otherwise problems may arise. This helps one to remember the date of application for the days before harvesting of edible crops is to be undertaken.

(7) **Storing pesticides**

Pesticides should be stored in their original containers in a locked cabinet. The storage location should not be exposed to heat or cold. Liquid pesticides should never freeze.

At this time there is no appropriate method for storing or disposing unused pesticides. Therefore, purchase of only required amount is suggested and will be used in only one growing season. Empty containers should be wrapped in several layers of newspaper and dumped in the trash.

PESTICIDE TOXICITY

Pesticide toxicity is measured most often by LD_{50}* and LC_{50}*. LD_{50} is the dose of a toxic substance required to kill 50% of a test population of animals. LD_{50} is the concentration of a toxic substance required to kill 50% of a test population of animals. LD_{50} is the abbreviation for "**median lethal dose**" and LC_{50} is the abbreviation of "**median lethal concentration**". The lower the LD_{50} of a pesticide, the more toxic is the substance. A high LD_{50} means that more of a substance is required to make a toxic

*LD = Lethal dose; *LC = Lethal concentration

dose. Rotenone has an LD_{50} of 132. Malathion's LD_{50} is 1,375, while table salt has an LD_{50} of 3,750. From this it can be said that rotenone, an organic pesticide, is much more toxic than malathion or salt. LD_{50} only indicates the acute toxicity or immediate effect of a substance. Also of concern are the long-term effects, called **chronic toxicity**. This is of much greater concern to applicators handling chemicals on a daily basis.

In toxicity test, it is also possible to determine the highest doses or concentrations which cause no toxicity. These are — **no observed effect dose (NOED)** and **no observed effect concentration (NOEC).** These values can only be determined in situations where a higher dose or concentration has produced an effect like the same toxicity test of LC_{50} or LD_{50}. If a test is carried out where the end point is an adverse response other than death then an **effective dose (ED_{50})** or **effective concentration (EC_{50})** is determined. Here, the concentration or dose producing the effect in 50% of the population is determined.

Pesticide Poisoning

Pesticide poisoning occurs from direct contact with the pesticide. This includes absorption through the skin, mouth, nose and eyes. Pesticides will soak into clothing and cause prolonged skin contact if the clothing, including leather or canvas shoes and gloves, are not removed immediately.

Symptoms of pesticide poisoning include headache, dizziness, dilated eyes, sweating, blurred vision, nausea, vomiting, cramps, diarrhoea, fatigue and weakness. If any of these symptoms appear, immediately contact physician or the poison control centre.

INTEGRATED PEST MANAGEMENT (IPM) IN PRACTICE

Definition

Integrated Pest Management (IPM) is a process consisting of the balanced use of cultural, biological, and chemical procedures that are environmentally compatible, economically feasible, and socially acceptable to reduce pest populations to tolerable levels.

Integrated means to combine. It is a combination of many strategies that are used to avoid or solve a pest problem. These strategies come from different disciplines, such as disease information from plant pathologists, weed information from agronomists, and insect information from entomologists. **Pests** are unwanted organisms that are a nuisance to man or domestic animals, and can cause injury to humans, animals, plants, structures, and possessions. **Management** is the process of making decisions in a systematic way to keep pests from reaching intolerable levels. Small populations of pests can often be tolerated; total eradication is often not necessary.

Why Practice IPM?

It is a pertinent question to ask, why should one consider IPM when chemical pesticides so often succeed at controlling pests ? There are some reasons for having a broader approach to pest management than just the use of chemicals. These are given below :

(a) **To keep a balanced ecosystem**

Every ecosystem is made up of living things and their non-living environment. The living and non-living constituents of the ecosystem maintain a highly tuned balance in nature. The actions of one creature in the ecosystem usually affect other organisms of same or different species. The introduction of chemicals into the ecosystem can change this balance, destroying certain species and allowing other species (sometimes pests themselves) to dominate. Beneficial insects such as the ladybird beetle and lacewing larvae, both of which consume pests, can be killed by pesticides, leaving few natural mechanisms of pest control.

(b) **Pesticides can be ineffective if not used in IPM method**

Chemical pesticides are not always effective. Pests can become resistant to pesticides. In fact, some 600 cases of pests, developing pesticide resistance varieties have been documented to date, including common lamb's-quarter, houseflies, the Colorado potato beetle, the Indian meal moth, Norway rats, and the greenhouse whitefly. Furthermore, pests may survive in some situations where the chemical does not reach it and is washed off. It may be applied at an improper rate, or is applied at an improper life stage of the pest.

(c) **IPM is not difficult**

If one have figured out the problem (the pest), determined the extent of the damage and decided on the action to take, he has already done most of the important steps in IPM. These steps are the ones used in IPM. Therefore, it is very easy to practise.

(d) **IPM saves money**

IPM can save money through avoiding crop loss (due to pests), and avoiding unnecessary pesticide expenses. For example, onion growers who followed IPM recommendations in 1987 saved more than ₹ 1000.00 per acre in insecticide costs. Golf course superintendents who replace fungicides with organic fertilizers or composts can save up to ₹ 60000.00 every time. Applicators are able to save on sprays because the calendar is not the basis for spraying; the need is the basis.

(e) **To promote a healthy environment**

Nowadays a severe problem to mankind is arising with the persistent presence of different chemicals in the environment along with their effect on living creatures. Reports of several cases of contaminated groundwater appear each year. Disposal of containers and unused pesticides pose challenges for applicators. Even though long-term documentation on the effects of all pesticides is still unavailable, it is generally agreed that fewer pesticides means less risk to surface water and groundwater, and less hazard to wildlife and humans.

(f) **To maintain a good public image**

Recent public outcry about the presence of pesticide residues on produce and the use of growth regulators has heightened pesticide applicator awareness of the level of public concern about chemicals. Consumers are pressurizing food stores, which in turn are pressurizing producers for products that has been grown with as few pesticides as possible. Growing food under IPM can help allay public concerns. Structural pest control professionals can suggest improvements in housekeeping or structural modifications as substitutes for chemical control.

THE BASIC STEPS OF IPM

IPM is based on taking preventive measures, monitoring the crop, assessing the pest damage, and choosing appropriate actions. Many different tactics are used in IPM, including cultural practices, biological control agents, chemical pesticides, pest-resistant varieties, and physical barriers. All of the components of an IPM approach can be grouped into four major steps.

1. The first step is taking preventative measures to prevent pest build-up.
2. The second is monitoring.
3. The third step is assessing the pest situation.
4. The fourth is determining the best action to take.

These steps are thoroughly discussed below :

1. FIRST STEP OF IPM : PREVENTATIVE MEASURES

Many IPM practices are used before a pest problem develops to prevent or stall the build-up of pests.

Cultural controls are those that disrupt the environment of the pest. Ploughing, crop rotation, removal of infected plant material, sanitation of greenhouse equipment and effective manure management, are all cultural practices that are employed to deprive pests of a comfortable habitat. The management of urban and industrial pests has improved when sanitation programs have been improved, pest harborages eliminated, garbage pickup frequency increased.

Structural modifications such as by preventing support timbers from soil contact, damages from several different wood destroying pests can be avoided. Wood absorbs moisture and is more susceptible to attack by carpenter ants and termites when in direct contact with the soil.

Construction site sanitation — Removing of tree stumps and lumber scraps from construction sites, which are prime food sources for subterranean termites, can prevent problems in the future.

Biological controls by using natural enemies (biological control agents) to keep pests in check can be put into place before pest problems increase. Examples of biological control agents are beneficial mites that feed on other mite pests in orchards, the milky spore disease that kills harmful soil grubs and *Encarsia formosa*, a wasp that parasitizes the greenhouse whitefly. Many biological control agents are commercially available.

Physical barriers such as netting over small fruits and screening in greenhouses can prevent crop loss. Physical barriers are important in termite, housefly, and rodent control.

Use of pheromones (natural insect scents) has become widely used in pest management. Sometimes a manufactured "copy" of the pheromone that a female insect emits to attract mates can be used to confuse males and prevent mating. This technique is used in curbing damage from the grape berry moth.

Pest-resistant varieties of cultivable plants are those, which are less susceptible than other varieties to certain insects and diseases. Use of resistant varieties often means that growers do not need to apply as many pesticides as is with susceptible varieties. Potato growers control the golden nematode by planting resistant cultivars. Apple growers can save up to eight fungicide applications a year by growing Liberty and Freedom cultivars, which resist diseases. Farmers growing alfalfa and wheat keep several pests at bay by planting resistant varieties.

Once a pest manager has taken precautions to prevent pest infestations, it is important to watch regularly for the appearance of insects, weeds, diseases, and other pests.

2. Second step of IPM : Monitoring (Scouting)

Monitoring pests involves :

(a) Regular checking of the infested area.

(b) Early detection of pests.

(c) Proper identification of pests.

(d) Identification of the effects of biological control agents.

Regular checking of a warehouse, bakery, restaurant, field, greenhouse, golf course, or other areas and **early detection** of pests can function together like an early warning system for pests, helping to avoid or prevent a pest problem.

Proper identification of pests is an extremely important prerequisite for handling problems effectively. For example, the brown banded cockroach and German cockroach can be easily confused with each other. Identification is important because certain management practices may control only one species and not the other. Correct identification enables to manage the real source of the problem and avoid merely treating the symptoms (or controlling non-pests). Some pests cause similar evidence. Unless the pest is identified, the control programme may have the wrong pest as its target. Identification enables to cure the pest problem and avoid injury to non-target organisms, particularly if :

(a) Use a pesticide that is specific to the pest.

(b) Control the pest effectively during the most susceptible stage of its life cycle.

(c) Consider the use of a non-chemical control.

Identifying the effects of biological control means knowing which creatures are helpful and determining if pests are affected by the beneficial organisms. Sometimes pests are kept in check naturally, and at other times the pest populations increase sharply.

3. Third step of IPM : Assessment

Assessment is the process of determining the potential for pest populations to reach an **economic threshold** or an **intolerable level**. There are important differences between the assessment of crop and urban pests.

Forecasting can help to determine weather conditions will be favourable for the development of diseases and insect pests. For example, by "plugging in" values (such as the number of rainy days and the

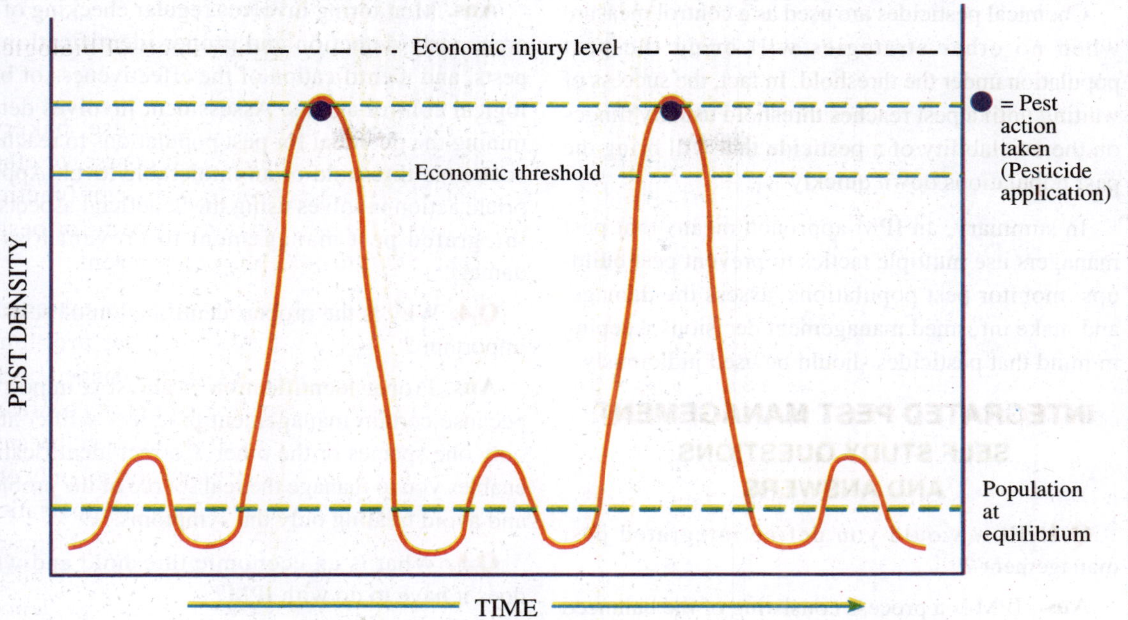

Fig. 1.21. Relationship between economic threshold, pest population density, and pesticide application point is shown

temperatures for those days), growers can predict outbreaks and spray only when conditions are favourable for diseases. Growers who have kept good records of pests in previous years can use these records to help determine if problems such as weeds, insects, and diseases will re-occur. They might be able, for example, to apply the most effective herbicides at the proper time for early control of a problem.

Thresholds, or more specifically economic thresholds, are levels that mark the highest point a pest population can reach without risk of economic loss. Populations above these thresholds can reach the economic injury level, where they cause enough damage for the grower to lose money. At the economic injury level, the cost of control is equal to the loss of yield or quality that would result otherwise (see Fig. 1.21).

Thresholds for many crops and pests have been scientifically determined. The advantage of thresholds is that if a pest has not reached threshold, there is no risk of economic loss. Therefore, there is no need to apply any control measure. Once the pest density (number of pests per unit area) has reached threshold, action should be taken. The costs of control must be less than or at least equal to the estimated losses that the pests would cause if left uncontrolled.

Zero thresholds : Urban pest thresholds are often related to aesthetics rather than economic considerations. Where health concerns or individual sensitivities exist, the tolerable level of the pest may be zero. A zero threshold forces action, even if only one pest has been detected. Zero thresholds exist in hospitals, food production, warehousing, and retail facilities.

4. Fourth step of IPM : Action (Control Measures)

Once a pest has reached the economic threshold, or intolerable level, action should be taken. In some situations, cultural controls can destroy pests. One example is early harvesting to avoid pest problems, which prevents crop loss and can sometimes be more economical than a pesticide application.

Chemical pesticides are used as a control measure when no other strategies will bring the pest population under the threshold. In fact, the success of waiting until a pest reaches threshold usually hinges on the availability of a pesticide that will bring the pest populations down quickly.

In summary, an IPM approach means that pest managers use multiple tactics to prevent pest build-ups, monitor pest populations, assess the damage, and make informed management decisions, keeping in mind that pesticides should be used judiciously.

INTEGRATED PEST MANAGEMENT
SELF STUDY QUESTIONS AND ANSWERS

Q.1. How would you define integrated pest management ?

Ans. IPM is a process consisting of the balanced use of cultural, biological, and chemical procedures that are environmentally compatible and economically feasible to reduce pest populations to tolerable levels.

Q.2. What are some of the benefits of integrated pest management?

Ans. IPM is beneficial in that it is easy, saves money, promotes a healthy environment and maintains a good public image.

Q.3. IPM is based on monitoring, assessment, and appropriate action. What does each of these steps involve?

Ans. Monitoring involves regular checking of an area, early detection and proper identification of pests, and identification of the effectiveness of biological control agents. Assessment involves determining the potential for pest populations to reach an economic threshold or an intolerable level. Appropriate action involves using the beneficial aspects of integrated pest management to prevent loss or damage.

Q.4. Why is the proper identification of pests so important ?

Ans. Proper identification of a pest is important because certain management practices will control only one species or the other. Correct identification enables you to manage the real source of the problem and avoid treating only the symptoms.

Q.5. What is an economic threshold and what does it have to do with IPM?

Ans. Economic thresholds are levels that mark the highest point a pest population can reach without risk of economic loss. It is important because it is a major factor in preventing loss of yield or quality.

Q.6. Name three different types of controls used in IPM and give an example of each.

Ans. Cultural controls disrupt the environment of a pest; an example is removal of infected material. Biological controls use natural enemies to a pest, for example mites that feed on mite pests. Physical barriers are the use of artificial material, for example, netting over fruits.

Chapter Two : Apiculture (Bee-Keeping)

INTRODUCTION

Apiculture (from Latin *apis*, a bee) or bee-keeping is the practice of scientific maintenance of honeybee colonies, commonly in hives, by humans and collection of bee products and marketing them professionally. A **bee-keeper,** also called **apiarist,** may keep bees in order to collect honey and beeswax, or for the purpose of pollinating crops, or to produce bees for sale to other beekeepers. A location where bees are kept is called an **apiary**.

Till now more than 20,000 species of wild bees have been described by taxonomists. Many of these species are solitary (e.g., mason bees), and many others rear their young in burrows and small colonies, (e.g., bumble bees). Apiculture is concerned with the practical management of the ***social species*** of honey bees, which live in large colonies of up to 100,000 individuals.

Terminology : **Apiology** (from Latin *apis*,"bee", and Greek *logia*) is the scientific study of honey bees, a subdiscipline of **melittology**, which is itself a branch of entomology. Honeybees are often chosen as a study group to answer questions on the evolution of social systems. **Melittology** is the study of all bees, which comprise more than 17,000 species other than honeybees. **Apicology** is honeybee ecology. **Apidology** is a variant spelling of *apiology* used outside of the Western Hemisphere, primarily in Europe; it is sometimes used interchangeably with *melittology*.

History of Wild honey harvesting

Collecting honey from wild bee colonies is one of the most ancient human activities and is still practised by aboriginal societies in parts of Africa, Asia, Australia and South America. Before the discovery of artificial culture method in India, people usually went to the forest for collecting honey. Till now it is reported that several people are lost during collection of honey from the dense forest of Sunderbans and other forests, perhaps devoured by Royal Bengal Tigers or crocodiles. Gathering honey from wild bee colonies is usually done by subduing the bees with smoke and breaking open the tree or rocks where the colony is located, often resulting in the physical destruction of the nest location.

History of Domestication of wild bees

In India report of domestication of honey bees were found in prehistoric epics. At some point humans began to domesticate wild bees in artificial hives made from hollow logs, wooden boxes, pottery vessels and woven straw baskets or "skeps". Honeybees were kept in Egypt from antiquity. On the walls of the Sun temple of Nyuserre Ini, before 2422 B.C, workers are depicted blowing smoke into hives as they are removing honeycombs.

In ancient Greece, aspects of the lives of bees and bee-keeping were discussed at length by Aristotle. It is reported from prehistoric Greece (Crete and Mycenae) that there existed a system of high-status apiculture. It is found from the early writings that the hives, smoking pots, honey extractors and other bee-keeping paraphernalia in Knossos were regularly used. Bee-keeping was considered a highly valued industry controlled by **bee-keeping overseers.** Bee-keeping was also documented by the Roman writers Virgil, Gaius Julius Hyginus, Varro, and Columella.

Archaeological finds relating to bee-keeping have been discovered at Rehov, a Bronze and Iron Age archaeological site in the Jordan Valley, Israel. Thirty intact hives, made of straw and unbaked clay, were discovered by archaeologist, Amihai Mazar, of the Hebrew University of Jerusalem in the ruins of the city, dating about 900 B.C. The hives were found in orderly rows of three feet high, in a manner that could have accommodated around 100 hives. This apiary could hold more than 1 million bees. It had a potential annual yield of 500 kilograms of honey and 70 kilograms of beeswax (according to Mazar). It is also in evidence that an advanced honey industry existed in ancient Israel 3,000 years ago. Ezra Marcus, an expert from the University of Haifa, said the finding was a glimpse of ancient bee-keeping seen in texts and ancient art from the Near East.

History of study of behaviour and biology of honey bees

It is revealed from the record that European natural philosophers undertook the scientific study of bee colonies since 18th century to understand the complex and *hidden world* of bee biology. Prominent among these scientific pioneers were **Swammerdam**, **René Antoine Ferchault de Réaumur**, **Charles Bonnet**, and the blind Swiss scientist **Francois Huber**. Swammerdam and Réaumur were the first who intended to understand the internal biology of honeybees. Réaumur at first constructed a glass-walled *observation hive* to observe activities of bees within hives. He observed queens laying eggs in open cells, but still had no idea of how a queen was fertilized. As nobody had ever witnessed the mating of a queen and drone and many theories held that queens were **self-fertile**. Some others believed that a **vapour** or **miasma** emanating from the drones fertilized queens without direct physical contact.

The father of modern bee-science : Francois Huber (Fig. 2.1)

Réaumur's design of 'Observation hive' influenced **Francois Huber** to build improved glass-walled observation hives and sectional hives which could be opened, like the leaves of a book, to inspect individual wax combs. This greatly improved the direct observation of activity within a hive. Although he became blind before he was twenty, Huber employed a secretary, Francois Burnens, to make daily observations, conduct careful experiments and to keep accurate notes over a period of more than twenty years. Among so many other observations Huber confirmed the following facts :

- That a hive *consists of **one queen*** who is the mother of all the female workers and male drones in the colony.
- That mating with drones' takes place outside of hives.
- That queen is inseminated by a number of successive matings with male drones, high in the air at a great distance from their hive.

He and Burnens dissected bees and first described the ovaries and spermatheca of queens as well as the penis of male drones. **Huber** is universally regarded as **the father of modern bee-science** and his "*Nouvelles Observations sur Les Abeilles*" (*New*

Fig. 2.1. Francois Huber (*Father of Bee-science*)

Observations on the Natural History of Bees) revealed all the basic scientific truths for the biology and ecology of honeybees.

History and Development of Apiculture and Apiary in India

In Indian subcontinent uses of bees and honey are common from the pre-historic ages. In our country, Veda, Ramayana, Koran has mentioned different uses of honey. Former Kings and Sultans used the symbol of bee as a mark of glory. Some of the earliest evidence of gathering honey from wild colonies is from rock paintings, dating to around 13,000 BC from different countries. Since early days honeybees are not cultured for honey, instead honey is collected from wild natural hives.

It is recorded that in 1882, artificial culture of honeybees were introduced in undivided Bengal following European methodologies. In 1883-84 the process was initiated in Punjab. In 1894, India Government first circulate details of information regarding the bee culture as promotional measure. **Bee-keepers Association** was established in 1907 in Punjab with its Head Office at Simla. In 1939, **All India Bee-keeping Association** was established and very soon it spread its branches to most of the states and districts of India. Now it merges with **ICAR** (Indian Council of Agricultural Research) and has expanded its activities. In 1945, **Central Bee-keeping Research Station** was established. It expands its research centres to Coimbatore (Tamil Nadu), Ruptala (Andhra Pradesh), Sundar Nagar (Himachal Pradesh), etc.

In 1953 **Khadi and Village Industries Commission** and in 1956 **Bee-keeping Directorate** were established by the Central Government. In 1962 **Central Bee Research Training Institute** was developed. After that its branches were established in Kodaikanal, Mahabaleswar, Kangra, Kashmir and other places. In 1980 ICAR, DST, INSA, Khadi and Village development Directorate jointly organised **International Conference on Apiculture in Tropical Climate** for discussion on the development on the apiculture around the world.

Systematic position of honeybees

(According to E. E. Ruppert and R. D. Barnes, 1994)

Phylum – Arthropoda
Subphylum —Uniramia
Class—Insecta/Hexapoda
Subclass —Ectognatha
Genus —*Apis*

Different Species of Domesticated Honeybee

In Europe and America the species universally managed by bee-keepers is the Western honeybee (*Apis mellifera*). This species has several subspecies or regional varieties, such as the Italian bee (*Apis mellifera ligustica*), European dark bee (*Apis mellifera mellifera*), and the Carniolan honeybee (*Apis mellifera carnica*). In the tropics, other species of social bee are managed for honey production, including *Apis cerana*.

Non- apis bees : In North America, efforts have focused on the development of non-Apis species with significant success for the **alkali bee**, *Nomia melanderi*, various **mason bees**, *Osmia* spp., and the **alfalfa leafcutter bee**, *Megachile rotundata*. Among these, alkali bees and mason bees are successfully cultured, but for the alfalfa leafcutter bee, the detailed studies necessary to successfully integrate a native bee into a sustainable agricultural system has remained unattended.

Honeybee Colony Composition

Its colony contains one **queen bee** (fertile female); a few thousand **drone bees** or fertile males; and a large population of sterile female or **worker bees**. Anatomical terminologies for different external organs of a bee is given in Figure 2.2.

Queen bee : Usually every beehive contains one queen (Fig. 2.3). The size of the queen is comparatively larger than the other types. Its wings are smaller than its body. The abdominal end gradually tappers. One queen usually flights once in its life time, either for copulation or for swarming. Its only function is to lay eggs. Some people consider it as a machine for egg laying in a hive. It usually lays about

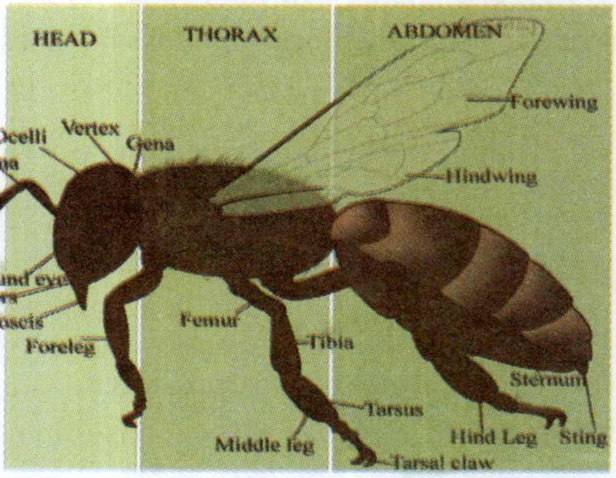

Fig. 2.2. Anatomy of a honeybee

Fig. 2.4. Queen, worker drone : Different castes of *Apis dorsata* — (A) worker (B) drone (C) queen

1500–3000 eggs per day, which includes both fertilised and unfertilised eggs. **Unfertilised eggs give rise to drones, and workers are developed from fertilised eggs**. A virgin queen develops from a fertilized egg. **The queens are prepared from the fertilized eggs** according to the requirement of the colony by feeding special types of food called **Royal Jelly**, a protein-rich secretion from glands on the heads of young workers. If the queen larva is not fed heavily with royal jelly, then it developes into a regular worker bee. All honey bee larvae are fed with royal jelly for the first few days after hatching but only queen larvae are fed on it exclu-sively throughout. As a result of the difference in diet, the queen will develop into a sexually mature female, unlike the worker bees.

Queens are raised in specially constructed **queen cells**. The fully constructed queen cells have a **peanut-like shape and texture**. Queen cells start out as **queen cups**. Queen cups are larger than the cells of normal brood comb and are oriented vertically instead of horizontally. Worker bees will build up the queen cup after the queen has laid an egg in a queen cup. In general, the **old queen starts laying eggs**

Fig. 2.3. *Apis mellifera* castes : Worker, queen and drone

into queen cups when conditions are right for swarming or **supersedure**. Swarm cells hang from the bottom of a frame while **supersedure queens or emergency queens** are generally raised in cells built out from the face of a frame.

As the young queen larva pupates with her head down, the workers cap the queen cell with beeswax. When ready to emerge, the virgin queen will chew a circular cut around the cap of her cell. Often the cap swings open when most of the cut is made, so as to appear like a hinged lid.

During swarming season, the old queen will leave with the prime swarm before the first virgin queen emerges from a queen cell.

A time frame of queen bee's developmental stages is given in Table 2.1.

Virgin queen bee

A virgin queen is a queen bee that has not mated with a drone. Virgins are intermediate in size between workers and mated queen or laying queens, and are much more active than the latter. Virgin queens have little queen pheromone and often do not appear to be recognised as queens by the workers. A virgin queen in her first few hours after emergence can be placed into the entrance of any **queen-less hive** or **nuc** and acceptance is usually very good. Whereas if a mated queen is placed into the entrance of any queen-less hive, it is repulsively recognised as a stranger and runs a high risk of being killed by the older workers.

Table 2.1 : METAMORPHOSIS OF THE QUEEN BEE

Egg	hatches on Day 3
Larva (several moultings)	Day 3 to Day 8 ½
Queen cell capped	~ Day 7 ½
Pupa	~ Day 8 until emergence
Emergence	~ Day 15 ½ – Day 17
Nuptial Flight(s)	~ Day 20 – 24
Egg Laying	~ Day 23 and onwards

When a young virgin queen emerges from a queen cell, she generally seeks out virgin queen rivals and attempts to kill them. Virgin queens will quickly find and kill (by stinging) any other emerging virgin queen (or be dispatched themselves) as well as any unemerged queens. Queen cells that are cut on the side indicate that a virgin queen was likely killed by a rival virgin queen. When a colony remains in swarm mode after the prime swarm has left, the workers may prevent virgins from fighting and one or several virgins may go with after-swarms. Other virgins may stay behind with the remnant of the hive. As many as 21 virgin queens have been counted in a single large swarm. When the after-swarm settles into a new home, the virgins will then resume normal behaviour and fight to the death until only one remains. If the prime swarm has a virgin queen and the old queen, the old queen will usually be allowed to live. The old queen continues laying. Within a couple of weeks she will die a natural death and the former virgin, now mated, will take her place. Unlike the worker bees, the queen's stinger is not barbed. The queen can sting repeatedly without dying.

Piping

Piping describes a noise made by virgin and mated queen bees during certain times of the virgin queens' development. Fully developed virgin queens communicate through vibratory signals : **quacking** from virgin queens in their queen cells and **tooting** from queens free in the colony, collectively known as **piping**. A virgin queen may frequently pipe before she emerges from her cell and for a brief time afterwards. Mated queens may briefly pipe after being released in a hive.

Piping is most common when there is more than one queen in a hive. It is postulated that piping is a form of **battle cry.** This is the announcement of competing queens and the workers for their willingness to fight. It may also be a signal to the worker bees which queen is the most worthwhile to support.

The piping sound is a G or A frequency of pitch. The adult queen pipes for a two-second pulse followed by a series of quarter-second toots. The queens of Africanized bees produce more vigorous and frequent bouts of piping.

Fig. 2.5. Queen bee in cup before emergence

Fig. 2.5A. Chambers — A frame showing different chambers for different members of the colony. Green curved arrow showing queen's chamber, wavy arrow (blue) for drone and straightblack arrow for worker bees.

Fig. 2.5B. Drone of *Apis dorsata*

Fig. 2.6. Queen cup on a bee hive

Drone bee : Drone or male bees (Fig. 2.4B) are the idlest of the three members of the colony. Their only work for the colony is to fertilize the female. These are somewhat larger in size and stouter than the workers, with powerful wings. They have greatly enlarged eyes, which cover most of the surface of the head. Abdomen of a male is blunt and flat. They do not possess any proboscis for pollen collection. Their tongue is developed for gathering nectar. They have no wax-producing gland. After 10 days of emergence from the cocoon they can fertilize the female. They can survive for 12 to 16 weeks. Not all but only some of the many male members can fertilize a female/ queen. However, the males, which succeed to copulate, die after copulation. Rest members remain for next chance in the hive and maintain the population balance of the hive. In a hive, they appear most plentifully in the early summer at swarming time, after which the workers try to drive them out of the hive.

Worker bee : Worker bees (Fig. 2.4A) are imperfectly developed females and are most numerous in a colony. They undertake all the household works of a hive. These are small bees but have very well-developed and powerful wings. A worker bee possesses wax glands on the ventral side of the abdomen and a pollen basket on her hind-legs. They have well-developed mandibles for work in the hive. A full-fledged sting is present in the last abdominal segment for protection as well as attack. However, these stings are for single use. If it is used, the bee will die immediately. Normally, the worker's ovaries are small and non-functional, and the spermatheca is rudimentary. But it is evident that if required a worker bee can lay eggs for the sake of the hive.

LIFE HISTORY/ CYCLE OF THE HONEYBEE

The honeybee is a eusocial insect. Its colony contains one **queen bee** or fertile female; seasonally up to a few thousand **drone bees** or fertile males; and a large seasonally variable population of sterile female **worker bees**. Details vary among the different species of honeybees, but common features in honeybees life history include :

(1) Virgin queens go out on **Nuptial flights** or **mating flights** away from their home colony and mate with multiple drones before returning either to the existing colony or fly outside to start a new colony. The drones die in the act of mating.

(2) Eggs are laid singly by the queen after mating flight, in a cell of the wax honeycomb, produced and shaped by the worker bees.

(3) Using her spermatheca, the queen actually can choose to fertilize the egg she is laying, usually depending on what cell she is laying.

(4) Drones develop from unfertilized eggs and are **haploid**, while females (queens and worker bees) develop from fertilized eggs and are **diploid**.

(5) Larvae are initially fed with **royal jelly** produced by worker bees, later switching to honey and bee bread or bee pollen. The exception is a larva fed solely on **royal jelly** will develop into a queen bee.

(6) The larva undergoes several moultings before spinning a cocoon within the cell and pupating.

(7) Young worker bees clean the hive and feed the larvae.

(8) When their **royal jelly** producing glands begin to atrophy, they begin building comb cells. They progress to other colony tasks as they become older, such as receiving nectar and pollen from foragers and guarding the hive.

(9) Later still, a worker takes her first orientation flights and finally leaves the hive and typically spends the remainder of her life as a forager.

(10) Worker bees cooperate to find food and use a pattern of "dancing" (known as the ***bee dance*** or ***waggle dance***) to communicate information regarding resources with each other. This dance varies from species to species, but all living species of *Apis* exhibit some form of this behavior. If the resources are very close to the hive, they may also exhibit a less specific dance, commonly known as the **Round Dance.**

(11) Honeybees also perform **tremble dances** which recruit receiver bees to collect nectar from returning foragers.

Swarming

New colonies are established by groups of bees known as **swarms** which consist of a mated queen and a large contingent of worker bees. This group moves en masse to a nest site that has been scouted by worker bees beforehand, this movement is called **swarming**. Once they reach the new site, they immediately construct a new wax comb and begin to raise new worker brood. Usually, stingless bees start new nests with large numbers of worker bees, but the nest is constructed before a queen is escorted to the site and this initial worker force is not a true swarm.

BEE LANGUAGE/ FORAGING

Some scout bees set out in the morning for foraging. On locating good sources of nectar (i.e., flowers) they return to their hive and perform characteristic movements (bee dances) at/near the comb (Fig. 2.7). In bee dance the middle course of the dance communicates to the other bees the angle from the hive with reference to the sun. Taking a hint from this angle they have to fly to reach the food source. After getting the information more and more worker bees are deployed in food gathering. The workers visit flower to flower, collect nectar and pollen and return to their own nest. During return

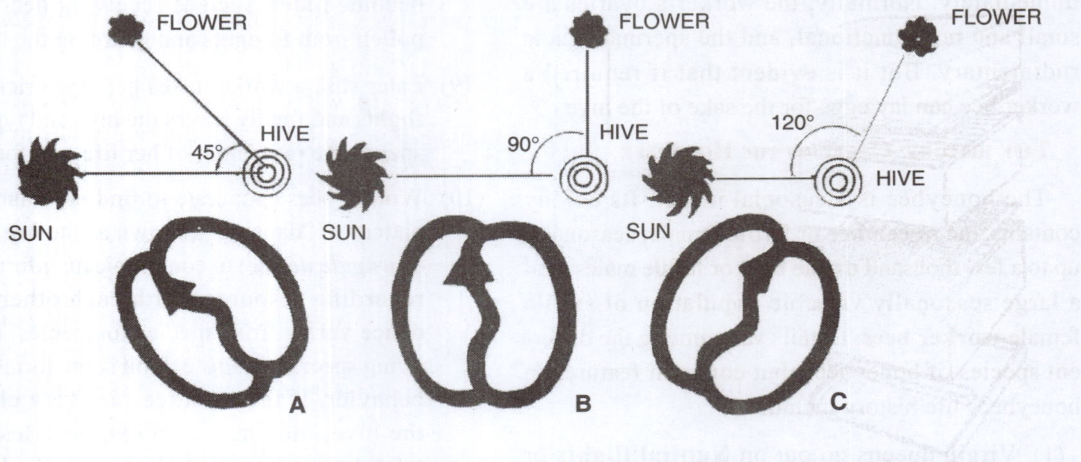

Fig. 2.7. The direction of bee dance near the hive communicate others about the source of food

journey they take clue from the position of the sun as well as by certain amount of memory. Finally, from the smell of their own hive they ascertain the landing place.

CULTURE PRACTICE

Genetic manipulation of bees

All the sub-species of *Apis mellifera* are capable of inter-breeding and hybridizing. Many bee breeding companies strive for selective breed and hybridised varieties to produce desirable qualities: disease and parasite resistance, good honey production, swarming behaviour reduction, prolific breeding, and mild disposition. Some of these hybrids are marketed under specific brand names, such as the Buckfast Bee or Midnite Bee. The advantages of the initial F1 hybrids produced by these crosses include: hybrid vigour, increased honey productivity, and greater disease resistance. The disadvantage is that in subsequent generations these advantages may fade away and hybrids tend to be very defensive and aggressive.

Indigenous methods of bee-keeping

Many villagers make wall or fixed types of hives in rectangular spaces in the walls with a small hole or movable types of hives in wooden boxes or earthen pitchers. The traditional bee-keepers catch clustered swarms from trees, bushes, etc. and transfer them to the above mentioned spaces. After sometime when the honey is ready, the bees are driven away from the comb usually by smoking the hive. Then the comb is cut away and the honey is squeezed out through a piece of large meshed cloth.

Modern hives – Langstroth Box

In modern bee-keeping, the combs are built on wooden frames that are moveable. This facilitates inspection and management of bee colonies. Three types of moveable frame hive are in common use: the Newton type along with its standardized version ISI Type A, the Jeolikote Villager and its counterpart ISI Type B, and the Langstroth type (Fig. 2.8). The beehive is made up of a series of square or oblong boxes without tops or bottoms, set one above the

Fig. 2.8. Modern Hive — Langstroth Box

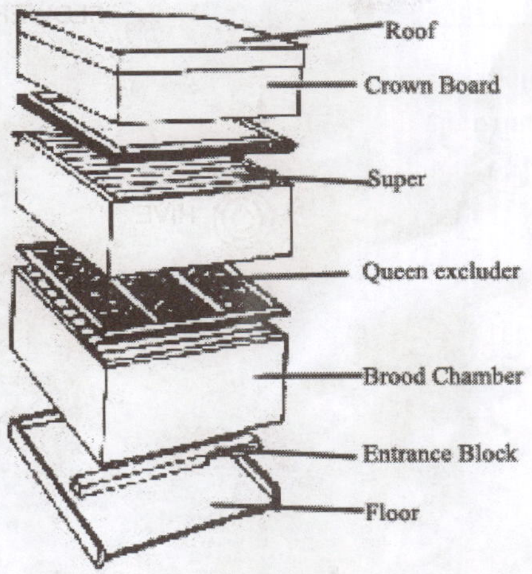

Fig. 2.9. Inner anatomy of the modern hive

Fig. 2.10. Skep with queen (Arrow)

other. This hive has the floor at the bottom with a **crown board** at the top and a **roof** over all. Inside these boxes, wooden frames are vertically hung, parallel to each other. The wooden frames are filled with sheets of **wax foundation** on which the combs are built by the bees. The only entrance to the hive is below the large bottom box **(brood chamber)**. The queen is usually confined to the brood chamber. The boxes termed **supers** are used for storage of honey. The queen is prevented from going to the **supers** by the **queen-excluder** that allows only the workers to move.

Catching a swarm and queen: Swarms, i.e., an old queen accompanied by huge population of workers, are collected from where they are settled. Special kind of container is needed to collect the queen, called **skep**. The container is usually a straw or a wire-net basket with a lid (Fig. 2.10).

Hiving a Swarm: It is the process in which the collected swarm is transferred to the hive to build up the colony and produce honey. It is operated in two ways :

(i) **Traditional method**

The hive is set up with brood chamber filled with its full number of frames. Each frame has a full sheet of foundation and there is a crown board with roof at the top. A sloping board with white sheet is set against the entrance of the hive. Bees in the skep (basket) are knocked out of it onto the slope. The instinct of the bees to move upwards onto the dark, drives them onto the hive through the entrance.

(ii) **Quick method**

In this method the crown board of the hive is taken off. Frames are also taken off and the entrance is closed. The skep is intimately united with the hive and the bees are poured into the brood chamber from the top. The frames containing the wax foundation are placed in the hive.

The crown board is put back in its position and the entrance is opened. It must be seen that the queen enters the hive. Sugar syrup is then fed to the swarm, as this feeding will help the bees to settle down to work in their new home.

Bee Pasturage: The plants that yield nectar and pollen are collectively termed "bee pasturage". The fruit trees, ornamental plants and forest trees comprise important bee pasturage. Nectar is the sweet secretion of the flowers. It is raw material for honey. Pollen provides the raw material necessary for the major food of the brood.

Extraction of honey : Successful commercial extraction of honey from the hive requires some instruments. These are described below.

Smoker : Smoker (Fig. 2.11) is a stainless steel flask with heat guard and a tube at the top for ejecting smoke. At the side of the flask there is a blower for blowing the smoke from inside to the hive. Various smoking agents are used in different parts of the country. During the time of honey extraction smoke is produced and sprayed on the hive to remove the workers from the hive. Then frames are taken out and put into the honey extractor.

Honey extractor : Honey extractor (Fig. 2.12) is a hand-operated roller drum with arrangement for holding frames in it. After keeping frames, the roller is rolled manually in a reasonable speed to extract honey, i.e., centrifugal force is applied on frames. Honey is collected in the drum until all the chambers become empty. At the end the honey is collected from a channel situated at the bottom of the chamber.

Products from a beehive

A. Honey

Honey is a food material for the bees and their larvae. Large quantities of honey are stored in the hive to meet the demands in scarcity. Chemically, honey is a viscous water solution of sugar. Its approximate composition in percentage is given in Table 2.2.

Table 2.2 : COMPOSITION OF HONEY

Nutrient	Percentage
Water	13-20
Fructose	40-50
Glucose	2-3
Sucrose	1-2
Maltose	7-8
Acids	0.5
Proteins	0.25
Minerals	Traces
Enzymes & Vitamins	2-3 (B1, B2, C)

Fig. 2.11. Smoker and its use

Fig. 2.12. Honey extractor

Composition of honey and its different flavours depend on the kinds of flowers from which the nectar is collected. Nectar is sucked from flowers and mixed with saliva. It is swallowed into a special region of the gut, called **honey stomach**. Nectar is a disaccharide (sucrose) it is hydrolyzed by the salivary amylase to produce monosaccharides (fructose and glucose).

Inside the hive the workers regurgitate the processed nectar. The honey thus produced is still very dilute. After placing this honey onto the storage cells of the hive the bees "fan" with their wings to evaporate the excess water and bring the honey to its required concentration.

Uses of Honey

Some uses of honey are as follows :

- **Food** : Honey is a nutritious food, rich in energy and vitamins.
- **Medicines** : It is used as a carrier in ayurvedic and unani-medicines. It acts as a laxative and prevents cold, cough and fever.
- **Religious** : It is used in religious ceremonies as a vital component, particularly in several Hindu rituals.
- **Drinks** : It is used in the making of alcoholic drinks.
- **Cosmetics** : Some skin and facial beauty lotions and soaps are made with honey.
- **Scientific** : In scientific research, some bacterial culture medium is prepared with honey making.
- **Pest control** : It is utilized for making poison baits for certain insect pests.

Purity Standards

There is no ready method to test the purity of honey by the customers. Homogenous granulation is a probable sign of its purity. Otherwise there are laboratory methods for testing (test for monosaccharides). In case of pure honey no sedimentation will appear if kept for several months.

B. Beeswax

Beeswax is secreted by the wax glands located on the underside of the last four abdominal segments (4th to 7th) of the worker bee. This wax is used in constructing bee combs in which the colony of the bees develops.

Uses of Beeswax

Some uses are as follows :

- Making of candles (the modern candles are made of paraffin wax, a petroleum product).
- Making pharmaceutical preparations.
- Preparation of varnishes and paints.
- Water proofing and waxing of threads.
- Formation of comb foundation (wax foundation in apiaries).
- It is used in pathological laboratories and scientific laboratories for preparation of tissues during micro-anatomical survey of soft tissues.

C. Honey dew

Honeybee collects enough nectar along with plant sap. They ingest the sap and nectar and processed these in the gut. Plant sap is mostly water and sugar with just a fraction of protein. After processing in the digestive system, the remains come out as black drops, which is called **honeydew**. Technically, the ingested substance cannot be called honeydew unless it has been run through bee's digestive tract. The honeydew that is expelled, lands on leaves, branches, needles, or even on the ground under the plant. Honeydew is also found from aphids, bugs and other insects. This processed sap is coveted by other creatures, including ants and bees. Although honeybees prefer floral nectar, during times of dearth – especially in the late summer — they will often collect the honeydew, transport it in their honey crop, and process it just like nectar. The liquid in honeydew evaporates quickly, so honeybees are more likely to collect it in the mornings or evenings. The bees treat the substance like nectar, so it is often mixed together with the nectar from flowers. As such, it is not really noticeable in the finished honey.

Honey made almost exclusively from honeydew is known as honeydew honey, forest honey, bug honey, flea honey, or tree honey. Sometimes it is named after its primary component, such as pine honey, fir honey, oak honey, etc. It is generally dark, strongly flavoured, less acidic, and less sweet than floral honey.

D. Bee Bread

Plant pollen collected by the field bees. As the bee is gathering nectar from flowers, the pollen from the stamen of the flower sticks on the hairs of the bee's body. Pollen is actually little dust-size particles that is the male gamete of the flower. When the bee arrives back at the hive, the house bees assist in "unloading" the pollen from the body hairs of the field bee. These gametes are collected in the pollen sacs on the legs of the bee. The house bees mix bee saliva, the plant pollen, and nectar. The bees secrete enzymes into the pollen. This mixture is known as **bee bread.** The composition of bee bread is given in Table 2.3.

Bee bread is bee pollen, and is different than plant pollen. Plant pollen has been known to increase reactions to allergies and hay fever, but some believe that bee pollen reduces those symptoms.

The picture below (in table 2.3) shows plant pollen collected by the bees and stored on sacs on their legs.

Table 2.3 : COMPOSITION OF BEE BREAD (in %)

Carbohydrate	55
Proteins	35
Minerals and vitamins	3
Fatty acids	2
Other ingredients	5

Bee pollen from local sources are especially helpful in reducing allergies and hay fever. There is one main thing to watch out for. If a person is allergic to bee stings, bee pollen may be anaphylactic (cause severe allergic reactions).

E. Propolis

Honeybees repair their damaged comb by a waxy substance, called propolis. It has some medicinal importance for human – medicines for tonsillitis, bronchitis and dental infections are made by this substance.

Fig. 2.13. Queen bee developing larva in royal jelly

F. Royal Jelly : Royal jelly (Fig. 2.13) is a milky substance and is produced specially for prospective queen larva. It is used as medicine for the treatment of human influenza, high blood pressure and respiratory infections. It is used as a component in some skin care and natural beauty products. It is believed to have anti-aging properties and used regularly in popular alternative folklore medicine.

Composition of Royal jelly : It is given in Table 2.4. Several water soluble vitamins are present, but none of the fat soluble vitamins A, D, E, K are present.

Table 2.4 : COMPOSITION OF ROYAL JELLY

Components	Amount
Water	67%
Crude protein	12.5%
Sugar (monosaccharides)	11%
Fatty acids	5%
Minerals enzymes Antibacterial and antibiotic components Pantothenic acid (Vitamin B_5) Pyridoxine (Vitamin B_6) Vitamin C Acetyle choline	All are present in traces, amounting 4.5% 

Royal-actin : The single protein **royal-actin,** present in royal jelly is responsible for developing a queen bee. This protein also induces similar phenotypical effect in *Drosophila melanogaster*.

G. Bee venom or Apitoxin : The poison of honeybee is used in the treatment of rheumatism, arthritis, etc. of man.

PESTS, PARASITES AND DISEASES OF BEE

(A) Microbial diseases

Bacterial diseases

1. **American foulbrood disease (AFB)** : Bee-

Fig. 2.14. American foulbrood disease (AFB)

keepers generally regard American foul brood (AFB) as possibly the most destructive microbial disease (Fig. 2.14) affecting bee brood. The disease did not originate in, nor is it confined to America. It is widely distributed wherever colonies of *Apis mellifera* are kept. In tropical Asia, where sunlight is abundant and temperatures are relatively high throughout the year, the disease causes severe damage to bee colonies. The disease is contagious and the pathogenic bacterium can remain dormant for more than 50 years.

Cause : American foulbrood disease is caused by a spore-forming bacterium, Paenibacillus larvae, which only affect bee brood; adult bees are safe from infection. The disease spread quickly within the colony and can spread to other colonies. Natural transfer mainly takes place within a radius of 1 km around the apiary.

Symptoms : At the initial stage of AFB infection, isolated capped cells from which brood has not emerged can be seen on the comb. The caps of these dead brood cells are usually darker than the caps of healthy cells, sunken and often punctured. On the other hand, the caps of healthy brood cells are slightly protruding and fully closed. As the disease spreads within the colony, a scattered, irregular pattern of sealed and unsealed brood cells (Fig. 2.14) can be easily distinguished from the normal, compact pattern of healthy brood cells observed in healthy colonies.

Control : A 'search and destroy' strategy may be adopted in an attempt to minimize damage to apiaries. The procedure involves regular hive inspections by qualified apiary inspectors. The entire honeybee population that is infected by American foulbrood is killed and hive materials belonging to the colony, are disinfected or destroyed by burning. The bees are usually killed by poisonous gas such as the burning of sulphur powder. The plastic hives should be cleaned and brushed with 3 to 5 per cent sodium hydroxide.

Chemotherapeutic methods of controlling AFB involve the administration of antibiotics or sodium sulfathiazole, in various formulations and is fed mixed with powdered sugar or sugar syrup. Antibiotics and sulfonamides prevent multiplication of the agent, though it will not kill the spores. Therefore, multiplication may begin again shortly after treatment. So the treatment must be repeated in much shorter intervals.

2. **European foulbrood disease (EFB)** : The range of distribution of European foulbrood disease is not confined to Europe alone and the disease is found in all continents where *Apis mellifera* colonies are kept. Reports from India indicate that *A. cerana* colonies are also subjected to EFB infection. The damage inflicted on honeybee colonies by the disease is variable. EFB is generally considered less virulent than AFB; although greater losses in commercial colonies have been recorded in some areas resulting from EFB.

Cause : The pathogenic bacterium of EFB is *Mellissococcus pluton*. It is lanceolate in shape and occurs singly, in chains of varying lengths, or in clusters. The bacterium is Gram-positive and does not form spores.

Symptoms : Honey bee larvae killed by EFB are younger than those killed by AFB. Generally, the diseased larvae die when they are four to five days old, or in the coiled stage. The colour of the larva changes as it decays from shiny white to pale yellow and then to brown. A sour odour can be detected from the decayed larvae. Another symptom that is characteristic of EFB is that most of the affected larvae die before their cells are capped. The sick larvae appear somewhat displaced in the cells.

Control : The choice of an EFB control method depends on the strength of the infection, i.e., how many brood cells and combs are infested. If the infection is weak, the change of hygiene behaviour of the bees by placing the hive to a good foraging site can control the disease. A better result is achieved if the individual combs are sprayed with a diluted honey solution. If the infestation is stronger, removal of number of pathogens in the colony will give good result. Since the bees' hygiene behaviour is genetically determined, replacement of the queen is possible. Requeening can strengthen the colony by giving it a better egg-laying queen, thus increasing its resistance to the disease and interrupting the ongoing brood cycle giving the house bees enough time to remove infected larvae from the hive. Sometimes chemotherapeutic measures such as application of antibiotics are called for, however, their application, always risks the danger of residues.

FUNGAL DISEASE

Chalkbrood disease : In Asia, Japan, temperate America and Europe the disease has been reported to cause serious problems to bee-keepers.

Cause : Chalkbrood is a disease caused by the fungus *Ascosphaera apis*. It affects honeybee brood. This fungus only forms spores during sexual reproduction. Infection by spores of the fungus is usually observed in larvae that is three to four days old. The spores are absorbed either via food or by the body surface.

Symptoms : Initially the dead larvae swell to the size of the cell and are covered with the whitish mycelia of the fungus. Subsequently, the dead larvae mummify, harden, shrink and appear chalk-like. The colour of the dead larvae varies with the stage of growth of the mycelia: first white, then grey and finally, when the fruiting bodies are formed, black (Fig. 2.15).

Control : As with other brood diseases, the bees remove the infested brood with their hygiene behaviour, which is especially effective for white mummies. Though as soon as the fruit bodies of *A. apis* have developed, cleaning honeybees spread the spores within the colony by this behaviour.

Fig. 2.15. White and black mummies

During the white mummy stage, the fungus continues to develop at the hive bottom. If the mummies are not removed quickly, the spores may enter the brood cells carried there by circulating air. At early stages of chalkbrood infection, adding young adult workers into the hive and hatching brood, combined with sugar-syrup feeding, proves to be helpful.

Till now there is no known successful chemical control against chalkbrood. In most cases, commercialized substances only show a positive effect because they are sprayed, or fed with sugar water as described above.

VIRAL DISEASES

At least 18 virus types and strains have been recorded as pathogenic to adult bees and bee brood. Most of these are RNA viruses. The damage caused to colonies by viral infection varies considerably according to a number of factors, which include the type and strain of virus involved, the strength of the colony, weather conditions, season and food availability. Among the viruses, Acute Paralyses Bee Virus (APBV) and Deformed Wing Virus (DWV) are becoming increasingly destructive. As not much is known about the life cycle and pathogenecity of most virus diseases, there are only a few ways to control them. Therefore, keeping this in mind, only the most widespread sacbrood is described.

Fig. 2.16. Sacbrood virus infected larva

Sacbrood disease : The virus is native to the Asian continents and is known for a long period. It was first discovered in Thailand in 1981. Since then it has been reported to be present in association with *A. cerana* in India, Pakistan, Nepal, and all other countries of Asia.

Cause : This disease is caused by *Morator aetotulas* and is perhaps the most common viral disease of honey bees. In Asia, at least two major types have been recorded. Sacbrood disease affects the common honeybee *Apis mellifera* and the Asian hive bee *A. cerana*. It is presumed that nurse bees are the vectors of the disease. Larvae are infected via brood-food gland secretions of worker bees.

Symptoms : Diseased larvae fail to pupate after four days. They remain stretched out on their backs within their cells. The anterior section of the larva, consisting of its head and thorax, is the first part of its body to change colour. It changes from white to pale yellow and finally to dark brown and black (Fig. 2.16). The skin of the larva becomes tough and internal organs become watery. The infected larva thus has the appearance of a small, watery sac.

Control : Field inspection to determine whether the pathogenic virus has infected a colony can be easily carried out following symptoms. No chemotherapeutic agent is effective in preventing or controlling sacbrood disease. The recovery mainly depends upon the hygiene behaviour of the bees, which may be stimulated with other brood diseases. The disease usually occurs when the colony is under stress, i.e., shortage of food, food-storage space, unfavourable climatic conditions (such as damp during the rainy or cold season), unhygienic hive interior, poor queen health, infestation with other diseases, etc. In severe cases, requeening the colony, removing infected brood combs and taking other management measures to restore colony strength, such as providing food and adding worker population generally gives good result.

Protozoan disease

Nosema disease (Nosemosis)

Nosema disease is generally regarded as one of the most destructive diseases of adult bees, affecting workers, queens and drones alike. Seriously affected worker bees are unable to fly and may crawl about at the hive entrance or stand trembling on top of the frames. The bees appear to age physiologically. Their life-span gets much shortened and their hypopharyngeal glands deteriorate – the result is a rapid dwindling of colony strength. Other important effects are abnormally high rates of winter losses and queen super-sedures.

Cause : The disease is caused by the protozoan, *Nosema apis*, whose spores infest the bees and are absorbed with the food. It germinates in the mid-gut. After penetration into the gut wall the cells multiply forming new spores that infect new gut cells. The nutrition of the bees is impaired, particularly protein metabolism.

Symptoms : In severe cases of infection, the abdomen of an infected worker becomes swollen and shiny in appearance. On dissection, the individual circular constrictions in the alimentary canal of uninfected bees are clearly visible, while the constrictions cannot be seen clearly in diseased bees. The most reliable method of detecting Nosema disease involves laboratory procedures using a microscope for diagnosis. The bees are killed, and their abdomens are removed and ground in water (2 to 3 ml per sample). A drop of the suspension of pulverized bee abdomens is then viewed under a microscope. If the disease is present, reasonably large individual bacilliform spores with bright, fluorescent edges will be observed. The microscopic inspection of queen's faeces makes it possible to verify the presence or absence of the disease.

Control : Nosema can best be controlled by keeping colonies as strong as possible and removing possible causes of stress. Colonies and apiaries should receive adequate ventilation and protection from cold and humidity. The bees should have the possibility of foraging regularly in order to defecate. This prevents spreading of the spores within the colony. Bee-keepers should also ensure that their colonies and queens come from disease-free stock. Hive equipment that is suspected of being contaminated by *Nosema apis* spores should be thoroughly decontaminated, preferably by heat and fumigation.

The only effective chemotherapeutic method currently available for treating *Nosema* is to feed the colony with fumagillin (25 mg active ingredient per litre of sugar syrup), preferably at a time when the colony is likely to encounter stress conditions, such as during a long winter or rainy season. The active ingredient of fumagillin is an antibiotic. It is of utmost importance that no medication be administered to colonies when there is a chance of contaminating the honey crop.

(B) Parasites

Parasitic bee mites

Bee-keepers in tropical countries generally agree that parasitic mites are among the most serious enemies of honeybees which they have to cope with. The success or failure of bee-keeping operations depends largely on mite control. Several major factors exacerbate bee-mite problems on the Asian continents. First, all known major species of parasitic honeybee mites are currently present in Asia, most being native to the continent. Second, the complete eradication of the mites from an apiary is impossible, because the feral nests of native bees infested by the parasites serve as reservoirs of mite re-infestation of domesticated honeybee colonies. Moreover, some mite species are able to survive, or even thrive on more than a single species of host bee. Several species of pollen-feeding mites are occasionally found in hives or attached to foragers. These phoretic mites are mostly innocuous to bee-keeping. Table 2.5 contains a list of parasitic and phoretic mites repr tedly found in association with honey bees in Asia.

Varroa mite (*Varroasis*)

This mite (*Varroa destructor*) is a native parasite of *A. cerana* throughout Asia. The effect of varroa infestation is to weaken the honey bee colonies and thus decrease honey production. This parasite is found throughout the world, except for Australia and New Zealand South Island. Varroasis is a brood disease. Occasional removal of *A. cerana* male brood combs and keeping the hive in healthy condition are the way of prevention of varroasis.

Cause : *Varroa destructor* is quite a large mite species, and can be seen with the naked eye. The shape of the adult female is distinctive. When observed from above the width of the body is clearly seen to be greater than the length, i.e., about 1.6 × 1.1 mm. The mite is reddish brown in colour and shiny. The body is dorso-ventrally flattened covered with short

Table 2.5 : Bee mites and their hosts

Mite	Mode of living	Hosts	Habitat
Varroa destructor	Parasite	*A. cerana**	Brood cell, adult bee
		A. mellifera	Brood cell, adult bee
Euvarroa sinhai	Parasite	*A. florae**	Brood cell, adult bee
Tropilaelaps sp.	Parasite	*A. dorsata**	Brood cell, adult bee
		A. mellifera	Brood cell, adult bee
Acarapis woodi	Parasite	*A. mellifera* *	Trachea of the adult bee
		A. cerana	Trachea of the adult bee
Neocypholaelaps sp.	Phoretic	*Apis sp.* *	Adult bee, pollen- storage cell

* Native host of the mite

Fig. 2.17. Bee larvae with varroa mites

hair (setae). Adult females of *V. destructor* are found inside brood cells (Fig. 2.17) or walking rapidly on comb surfaces. Adult males, and the immature stages of both sexes (egg, protonymph and deuteronymph), are not commonly seen outside the brood cells. All immature stages of the parasite live inside the brood cells. The immature mites are bright white and the adult females are brown.

Symptoms : Varroa causes injuries to honeybees by direct feeding. The adult female pierces the bees' inter-segmental membrane with their pointed chelicera and sucks the bees' haemolymph. Varroasis is a brood disease. If more than one parasitic female mite infests the brood cell the brood decays or deformations occur including shortened abdomen or deformed wings. The bee's behaviour may be disturbed in orientation or gathering food. The semen production of drones may get considerably reduced. Varroasis is a multi-factorial disease. Normally, the exoskeleton protects the bees from many virus infections. But the mite penetrates this natural barrier, transferring viruses or stimulating the multiplication of viruses with its saliva. In turn, viruses seem to speed the development of varroasis enhancing the parasite's virulence. Without treatment the colonies normally die after two to three years, management errors may also cause the collapse of colonies. Colonies destroyed by the varroa mite are often left with only a handful of bees and the queen. The other bees having died during foraging or having drifted to neighbouring colonies, where the mite population can increase before destroying these colonies also. The presence of adult bees with deformed wings, crawling on comb surfaces or near the hive entrance, usually indicate a late stage of heavy mite infestation. The bees are intoxicated and the mites crawl on the wall (Fig. 2.18).

Control : The control of *V. destructor* is one of the most difficult tasks facing apiculturists and bee-keepers throughout the world. The mite is a highly successful parasite, whose life history is well synchronized with that of its host. Two principal approaches to its control are currently available : (i) chemical control and (ii) hive manipulation techniques, sometimes referred to as **biological control**.

(i) **Chemical control** : Chemical control is by far the most popular method of varroa control among Asian bee-keepers and elsewhere. Although it creates the risk of honey contamination, the accumulation of residues within the hive and toxic effects to the bees, bee-keepers claim that chemotherapeutic treatment is the quickest and most reliable method of mite suppression. Among the commonly used mite-control agents are organic acids, ethereal oils, synthetic pyrethroids and amitraz. The application of chemical

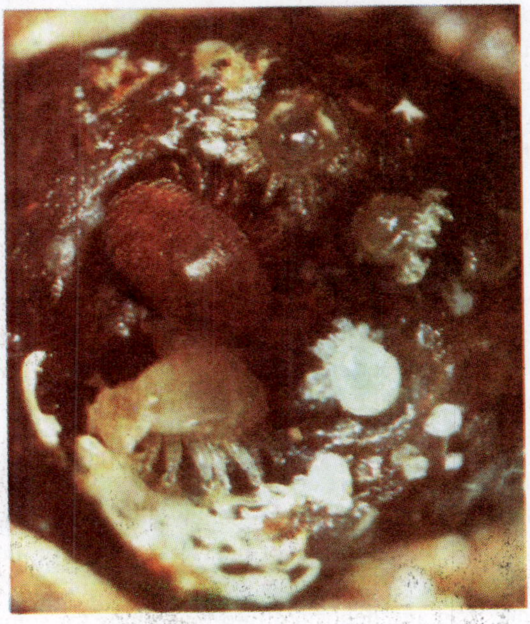

Fig. 2.18. Varroa mite showing nymphal stages and male

substances can only be started after honey harvest, This is the only way to avoid residues. Effectiveness of a chemical substance must be verified in the country side. The organic acids : formic acid, the ethereal oil thymol and the synthetic pyrethroids and amitraz may still be chosen to treat colonies with brood.

(1) **Formic acid** : Formic acid can kill some of the mites in the sealed brood cells. It is recommended that the formic acid be allowed to evaporate in colonies with sealed brood for at least two to three weeks. In this way, mites emerging from the brood will also be killed. The formic acid should be introduced into the colony only in the late afternoon to avoid damage to bees and brood. The application can be repeated three to four times at intervals of at least seven days.

(2) **Oxalic acid** : Contrary to formic acid oxalic acid does not act via evaporation but through contact with the bees. Thirty five grams of crystal oxalic acid (dihydrate) is thinned in one litre of sugar water (1:1). When handling crystal acid, special precautions must be taken because of health risks. Protective spectacles and acid-proof gloves must be used together with an adequate mouth protector. Depending on the size of the colony 20 to 30 ml of the suspension per chamber are dropped into the bee-ways. A repetition of the treatment can lead to damage of the bees.

(3) **Lactic acid** : Lactic acid is clearly better tolerated by bees and does not cause problems in warmer climatic zones. The disadvantage is that every single comb must be extracted to spray the bees with the acid. The dosage applied per comb side is 8 ml of 15 per cent acid. This treatment can be repeated several times at intervals of seven days.

(4) **Etheric oils** : The only etheric oil that is sufficiently effective against varroa mites is thymol. Thymol can be applied as a commercially available ready-made preparation or in crystal form. For this purpose, 0.5 mg thymol per bee-way are put into a gauze bag and deposited into the combs for some weeks. In this way mite emerging from the brood will be eradicated.

(5) **Synthetic chemicals**

Pyrethroids : Synthetic pyrethroids are contained in Apistan and Bayvarol, which have been developed for application on bees. The plastic strips are fixed in bee-ways in the brood nest. When the bees come into contact with them they transport the substance to the other bees, thus killing the parasitic mites on the bees. As the strips remain for several weeks in the colony, mites emerging from the brood are affected. Synthetic pyrethroids are highly effective, although there is the disadvantage that mites may rapidly develop resistance to them. Therefore, their effectiveness should be controlled regularly.

Amitraz : Taktic and Mitac are trade names of products containing amitraz at different concentrations. The recommended dosage for use on honey bee colonies is 1 ml of 12.5 per cent amitraz to 10 litres of water, sprayed lightly on bees, the comb surface of brood frames and hive walls. The amount of the solution to be sprayed at each application depends on the size of the colony, but is usually within the range of 80 to 250 ml. Amitraz can also be used as a hive fumigant. Fumigation should take place in the evening, when the foragers have returned to the hive.

(ii) **Control by hive manipulation** : The varroa mite depends on bee brood to complete its development cycle. Since the mite prefers drone brood to worker brood, empty frames are given to the colonies, which will rear drone brood in them. When the cells are sealed, the frames containing the mites get trapped inside the cells, can be removed and destroyed. The mites can also be trapped in worker-brood frames by using vertical queen-excluders in single-storey hives. The queen is confined between two excluders and allowed to lay eggs in one frame only. Female mites in the colony will be attracted to this brood frame which, when the cells are sealed, is removed from the colony so that the brood cells infested by the parasites can be destroyed.

Tropilaelaps mite

Modern bee-keeping with *Apis mellifera* in tropical and sub-tropical Asia frequently encounters problems caused by infestation with *Tropilaelaps* sp. The mite is a native parasite of the giant honeybee *A. dorsata*, widely distributed throughout tropical Asia, and whenever *A. mellifera* is kept within the range of distribution of *A. dorsata*, mite infestation of the colonies occurs.

Cause : *Tropilaelaps* mites are small in size. The adult female mite is light reddish-brown in colour, with an oval-shaped body about 0.96 mm in length and 0.55 mm in width. The mite's entire body is covered with short setae. A red streak runs longitudinally on the ventral surface of the adult female.

When the mites are present in a honey bee colony in large numbers, they can be observed walking rapidly on the surface of the comb. They are rarely found on adult bees. In all its immature stages, the mite lives within the brood cells of the bees, feeding on the brood's haemolymph. Fertilized adult females enter the cells before they are capped to lay their eggs. The stages of development of the mite are as follows : **egg, six-legged larva, protonymph, deutonymph, adult**. Adult males of *Tropilaelaps* do not feed, their chelicerae having been modified to transfer sperm. The life cycle of the mite is well synchronized with that of the host bee.

Symptoms : The abdomen of bees, surviving mite attacks, is reduced in size and they have a shorter life-span than healthy bees. In heavily infested colonies, bees with deformed wings can be observed crawling about the vicinity of the hive entrance and on the comb surfaces. Inspection of hives severely infested by *Tropilaelaps* reveals an irregular pattern of sealed and unsealed brood as found with all brood diseases. If mites are present, adult females will be seen walking rapidly out of the cells. To obtain a reasonably accurate estimate of the level of infestation, 100 to 200 cells should be opened and the brood removed with forceps for close inspection.

Control : Preventing infestation by the *Tropilaelaps* mite is nearly impossible. As is applicable to other bee diseases, robbery or a too large bee density should be avoided. In recent years, it is reported that the adult female of the mite can survive without bee brood as food, for only up to seven days.

Chemical control : Formic acid can be used successfully in its treatment. However, special attention must be paid, in tropical areas, regarding its dosage to avoid damage of the bees. The dosage per comb should not exceed 2 ml in a one-storey Langstroth-hive. Formic acid is placed onto a cloth deposited in the rear section of the hive. Applications of amitraz is very effective, either as a liquid sprayer on the surface of the brood comb and hive walls, or as a hive fumigant, in the same dosage. The treatment requires three to four applications at four-day intervals. All chemical treatments must be suspended at least eight weeks before the honey flow season arrives and amitraz must not be used in spray form, in the presence of large numbers of honeybee eggs and newly-hatched larvae.

Colony manipulation techniques

Many bee-keepers prefer not to use chemicals to control *Tropilaelaps*, but to manipulate the brood rearing cycle of their infested colonies in such a way that the mites are deprived of sealed and unsealed brood (their food), for at least three days. During this period, a large proportion of the mite population will starve to death. There are several means of creating this broodless situation in infested colonies. In smaller apiaries, the bee-keeper can simply remove the brood-comb frames (both sealed and unsealed) from the infested colonies and put them in new hives. Before the new larvae hatch, the hives manipulated in this way will be short of brood for two to three days, which is time enough to starve most of the mites. The best time of year to carry out these colony-manipulation techniques is during a heavy pollen-flow season, enabling the colonies to rear brood after the period of brood deprivation. Some bee-keepers prefer to combine chemical treatment with the brood-deprivation technique. In this approach, all sealed brood is removed from the mite-infested colonies, which are then fumigated. The adult female mites, having no capped brood cells in which to hide, are mostly killed by the fumigant, so that only one chemical treatment is required instead of three or four.

Tracheal mite (Acarapidosis)

This mite, *Acarapis woodi*, infests the tracheal system of adult bees (queens, workers and drones), all are equally susceptible to its attack. Although it was first reported in the *Apis mellifera* colonies in Europe way back in 1921, its reports from India and Pakistan indicate that the tracheal mite also cause severe losses of *A. cerana* colonies.

Cause : *A. woodi* is a very small mite (0.1 mm) species that lives and breeds within the thoracic tracheae of adult bees. The mite penetrates through the stigma (spiracles) into the first trachea pair of the thorax of a 10-day old honeybee. There it lays eggs at intervals of few days. After the deutonymph stage, male offsprings emerge after around 12 days and females after 13 to 16 days.

Symptoms : Unfortunately, there are no typically reliable symptoms of infestation in honeybees. Indeed, it has been demonstrated that bees severely infested with the mite can forage normally. Nevertheless, some differences exist with regard to the over-wintering capability of infested and healthy colonies. Infestation shortens the lifespan of the individual bees, so that severe infestation of colonies causes them to lose strength and thus increases a colony's susceptibility to winter losses. The most reliable diagnostic method is laboratory dissection. Samples of 20 or more bees found crawling near the hive and unable to fly are killed, their heads and legs removed and their thoraxes dissected for microscopic examination. If present, the mites are usually found at the end of the first pair of trachea in the thorax.

Control : Chemotherapeutic measures are widely adopted for mite control. Best results could be achieved with evaporating substances such as formic acid and ethereal oils.

Formic acid : Formic acid produces good results by applying it onto a cloth (20 ml of 65 per cent formic acid) four times at intervals of seven days. Treatment should be conducted during the period of low humidity and the temperature should not exceed 30°C.

Menthol or thymol : Menthol has a toxic effect on *A. woodi* in bee colonies. Crystalline menthol (50 g) or thymol (15 g) is placed in a gauze bag on the top of the bars to be kept there for one to two months. External temperatures should be around 21°C; otherwise the menthol vapours will not reach the mites in the trachea.

Parasitic beetles : There are several different beetles living in honeybee colonies. Most are harmless and feed on pollen or honey.

Small hive beetle (SHB) : This beetle is prevalent in Africa and America and there is a risk that the beetle may spread to Asia.

Cause : The beetle *Aethina tumida* (order : *Coleoptera*, family : *Nitidulidae*) is called the small hive beetle. This beetle lives and multiplies within and outside bee colonies. The beetle produces large cluster of eggs within a bee colony, in fissures and recesses out of reach of the bees. The larvae of this beetle preferred to live in pollen chamber of honeycombs. The adult larvae leave the hive to pupate in the earth in front of the apiary. The period of development from egg to adult beetle is at least four to five weeks.

Symptoms : The beetles and their larvae can infest bee brood and honeycombs in and outside the apiary. There, they form eating canals and destroy the cell caps and the honey starts to ferment. The adult beetle is dark brown to black, around 5 mm long and 3 mm wide. Any minor infestation is difficult to recognize because the beetles immediately hide in the dark.

Control : The best way to protect against an infestation of the small hive beetle is to keep strong colonies and to remove those that are weak from an apiary. The removed honey combs should be centrifugally extracted, one to two days after harvesting the honey. Currently, a successful control is made possible using a preparation named **Checkmate**. This product can kill more than 90 per cent of adult beetles and larvae. The beetle can live and multiply outside the beehive, where it seems to prefer rotting fruits (e.g., apples and bananas) as nesting sites. This is why reinvasion is always possible. The beetle is extremely quick moving and can fly, which contributes to its rapid spread among bee colonies and apiaries.

Ants : Ants are among the most common predators of honey bees in tropical and subtropical Asia. They are highly social insects and attack the hives *en masse*, taking virtually everything in them: dead or alive adult bees, the brood and honey. In addition to this destruction, they can also be a nuisance to bee-keepers and may sometimes cause pain from their bites. Apiaries of *Apis mellifera* when attacked by ant become aggressive and difficult to manage; weak colonies will sometimes abscond. Many ant genera and species are reported to cause problems to both traditional bee-keepings with *A. cerana* and to modern bee-keeping with *A. mellifera*. Among the most frequently recorded species of ants are the **weaver ant** (*Oecophylla smaragdina*), the **black ant** (*Monomorium indicum*), *Monomorium destructor*, *Oligomyrmex* sp, *Dorylus* sp, the **fire ants** (*Solenopsis sp*) and *Formica* sp.

Control : Bee-keepers have found that the most effective method of controlling weaver ants is to search systematically for the ants' nests in the vicinity of the apiaries and, when found, to destroy them by burning. General precautionary measure is to reduce ant nesting sites that include eliminating brush and rotten wood from the apiary and cutting the grass. A good general defense against ants in tropical apiaries is to place the hives on stands supported by posts, 30-50 cm high and to coat the posts with used engine oil or grease. Frequent inspection and renewed application of grease are both necessary. A more reliable method is to place the hive-stand posts in tin or plastic containers filled with either water or oil. Regular clean up is required to avoid the formation of patches of vegetation on earth that can be crossed by ants. Liquids need to be replenished frequently.

Wasps and Hornets

Nearly all countries in Asia report wasps and hornets as common enemies of their honeybees. Among the most frequently reported enemies are social wasps of the genus *Vespa*, which are widely distributed throughout the world. Colonies of both *A. cerana* and *A. mellifera* are frequently attacked by *Vespa*. Hornet invasion of *A. cerana* colonies generally causes the bees to abscond. Similar behaviour is also reported in weak colonies of *A. mellifera*. Predation by *Vespa* sp. on commercial apiaries is generally a seasonal problem. Apiaries situated near the foothills and tropical forests suffer more acutely than those on the plains. During attack an initial **hunting phase** is observed, during which a few hornets capture and kill slow-flying bees one at a time, usually near the entrance of a weak bee colony's hive. Later, a **slaughtering phase** sets in. Some 20 to 30 hornets attack a weak colony, using their strong jaws to maul the bees and dropping the dead and dying bees to the ground. Finally, when this phase has lasted long enough for the colony under attack to have lost most of its defender workers, the hornets invade the hive itself, the honey and brood nest and the wasps carry away any surplus brood to their nest.

Control : Due to the large body size the foraging range of *Vespa* sp can be comparatively of large area around their nests, which have populations of many thousands of individuals. Bee-keepers in Japan sometimes adopt methods such as bait-trapping, trapping at the hive entrance and using protective screens. It has been seen that the real damage inflicted by hornet attacks on honeybee colonies, occur during the slaughter and occupation phases. Killing hornets in the early stage of predation has the effect of disrupting the hunting phase and preventing the predation process from reaching the more destructive phases. Mass destruction of the colonies is thus prevented or, at the least, minimized.

Wax moths and other Lepidoptera

The greater wax moth (*Galleria mellonella*)

The greater wax moth is often reported to cause damage (**Galleriasis**) both to honeybee colonies and to bee products in tropical and sub-tropical Asia. Empty combs, rendered wax, comb foundation and bee collected pollen, if not properly stored and left unattended, almost always suffer considerable damage from wax moth infestation. The wax moth is a major pest of *A. cerana*, often causing colonies of bees to abscond. In wax moth attacks on colonies, the adult female enters the hive at night, through the entrance or cracks in the walls, deposits her eggs directly onto the combs or in narrow crevices that permit oviposition and offers protection against removal by worker bees. About 50 to 150 eggs are laid in each batch. They are glued together and adhere firmly to the surface on which they are laid. The newly-hatched *Galleria* larvae feed on honey and pollen. They then burrow into pollen storage cells or the outer edge of cell walls, later extending their tunnels to the midrib of the comb as they grow. At this stage the developing larvae are quite safe from the worker bees. As they advance into the combs, they leave behind them a mass of webs and debris. The complete destruction of unattended combs usually ensues within 10-15 days. In addition to stored pollen and comb wax, larvae of the greater wax moth will also attack bee brood when short of food. The development time of *Galleria* larvae depends on two factors – food availability and temperature. In tropical climates the larvae require only 18-20 days before spinning cocoons and becoming pupae. When weak colonies are infested, the symptom of **galleriasis** is frequently observed. The emerging adult worker and drone bees are unable to leave their cells because their bodies have been tied up by silken threads spun by the *Galleria* larvae.

Control : There are no easy or inexpensive chemotherapeutic measures for controlling the wax moth in living honeybee colonies once infestation has set in. The only possibility is treatment with *Bacillus thuringiensis*, in a watery suspension, sprayed onto the combs. The effect on the wax-moths larvae persists for several weeks. Preventive measures include ensuring that the colonies, whether of *A. cerana* or *A. mellifera*, are — (1) strong and have adequate food stores; (2) adapting the hive space to the strength of the colony; (3) reducing the hive entrance, sealing cracks and crevices in hive walls; (4) protecting the colonies against pesticide poisoning; (5) controlling pests and diseases that might otherwise weaken them; and (6) removing any wax and debris accumulated on the bottom boards of the hives. Several measures can be taken to prevent or control wax-moth infestation in stored combs and hive products.

Fumigation is the usual treatment, but new combs should be treated less frequently. Among the most commonly used fumigants are naphthalene, ethylene di-bromide and methyl bromide. All, including para-di-chlorbenzene, are very poisonous to bees and humans and, in addition, lead to residues in honey. The application of sulphur, however, is inoffensive. Sulphur powder is wrapped in newspaper and burned in a metal container. Liquid sulphur from sprayers can also be used. The development of wax moths can be interrupted for several months if the combs are heated at 48°C for three hours. All treatments should be repeated at intervals depending on the level of infestation.

The lesser wax moth (*Achroia grisella*)

The adult lesser wax moth, *Achroia grisella* are silver-grey in colour, with a distinct yellow head. The insect is quite small, with a slender body. Normal body lengths of adult female and male are about 13 and 10 mm, respectively. The life-span of the adult female is about seven days, during which she can lay 250 to 300 eggs. Infestation by the lesser wax moth usually occurs in weak honeybee colonies. The larvae prefer to feed on dark comb, with pollen or brood cells. They are often found on the bottom board among the wax debris. As larvae prefer to form small canals between the bottoms of the brood cells, the brood is lifted. The bees continue constructing cells heading upward leading to the typical scratched comb surface.

Control : The methods employed in controlling *Galleria mellonella* are equally effective for the control of *Achroia grisella*.

Other Lepidoptera

Other moth species are frequently recorded in association with bees and bee products. The Indian meal moth, *Plodia interpunctella,* is reported to feed on bee-collected pollen. Moths found dead on the bottom boards of beehives include death's head or hawk moths (*Acherontia atropos*), which invade the hives to feed on honey. Bee-keepers generally consider them to be minor pests.

Vertebrate pests

Amphibians

Bee-keeping in tropical climates frequently suffers from damage caused by amphibians – toads including *Bufo melanostictus* and *Kaloula pulchra* and frogs including *Rana limnocharis* and *Rana tigrina*. The detection of this problem generally requires close observation. Bee-keepers are normally unable to observe intense predation by amphibians on honey-bees in the daytime, when they are at work in their apiaries, because the heaviest attacks occur at night. Often the problem goes unrecognized until a substantial fall in colony populations is perceived. One sign indicating that toads and frogs are preying heavily on the colonies is the presence of the predators' dark brown droppings, scattered in front of the hive entrance. If the dry faecal deposits are spread apart (e.g., with a twig), the remains of bee parts can be seen. Continuous predation by toads and frogs, if not prevented, results in a loss of colony strength. While colonies with moderate or larger worker populations can withstand the predation and subsequently recover their full strength. However, weaker colonies are at considerable risk. Toads and frogs have similar attacking patterns. On arriving at the colony, the amphibians wait in the vicinity of the hive entrance, preying on passing bees. Colonies close to the ground provide easy access to the predators, for which guard bees at the hive entrance are easy prey.

Control : Although in some circumstances predation on honeybee colonies by amphibians cannot be overlooked, most bee-keepers perceive the problem as minor. Placing the hives on stands 40 to 60 cm high is usually a sufficient protective measure. Where large numbers of the predators tend to congregate near an apiary, it may be necessary to fence it with fine-mesh chicken wire. Other methods such as trapping, baiting or poisoning have not been advocated.

Reptiles : Geckos, skinks and other lizards are among the most commonly found reptiles in tropical Asian jungles, woods, grasslands and urban areas. Among the reptile species that are regularly recorded as present in commercial apiaries, are the tokay (*Gecko gecko*), which can be as much as 35 cm long, *Calotes* sp, *Acanthosaura* sp and the skink *Sphenomorphus* sp. Arboreal reptiles such as many geckos and skinks, can attack bees either near the hive entrance or on the limbs of flowering trees visited by forager bees. Smaller lizards, such as the gecko, *Hemidactylus frenatus*, often hide in the empty space between the outer and inner covers of the hive. In tropical areas, this type of predator causes sudden loss of queen from a weak colony.

Control : The bee-keeper can do little to prevent the loss of foragers to the highly mobile arboreal reptiles, usually well-hidden in the trees. Bee-keepers can destroy as many of them as possible when they are encountered, though this method is not recommended nor is it efficient. Hives placed on stands that are about 40-60 cm high are reasonably safe from reptiles attacking from the ground; coating the legs of the stands with used engine oil or grease may deter the reptiles from climbing up to the hive entrance. A well-kept bee yard that is frequently mowed, without dense bushes, shrubs and tall grass, that provide safe hiding places to the predators, has less chance of suffering losses from reptiles than an untended one. No reliable chemical control of reptiles is available for use in apiaries.

Birds : Birds prey upon many insect species and honey bees are no exception. Once airborne, the bees are virtually defenseless against birds, several species of which can tolerate their venomous stinging defense. The heavy traffic of bees flying in and out of the hives of commercial apiaries provides an exceptional opportunity for insectivorous birds, large numbers of which may be attracted by this situation. Birds that have been listed as attacking honey bees in Asia include bee-eaters (*Merops apiaster*, *Merops orientalis*), swifts (*Cypselus* sp., *Apus* sp.), drongos (*Dicurus* sp.) shrikes (*Lanius* sp.), woodpeckers (*Picus* sp.) and honey guides (Indicatoridae). The level of damage caused by apivorous birds varies. An attack by a single bird or by a few together rarely constitutes a serious problem, but when a large flock descends upon a few colonies or an apiary, a substantial decline in the worker population, in some or all the hives may be observed. Whereas the degree of damage to commercial apiaries caused by predatory birds depends largely on the number of the predators and the intensity of the attack, the mere presence of a few predators in apiaries engaged in queen rearing can inflict serious losses.

Control : While bee-keepers regard insectivorous birds as pests, other branches of agriculture generally do not consider them as problematic. In fact, birds that prey on insects are mostly considered to be beneficial to farming, in that they help in the control of insect pests. For this reason, before any attempt is made to solve the apiary's bird problems by mass killing of the predators, whether by chemical or physical means or by gunshot, the implications of this action on the environment must be seriously taken into account. Where heavy predation by birds on apiary bees tends to occur at fixed periods (e.g., during the migration season of swifts), the most practical means of solving the problem is usually to avoid the birds through careful site selection and by temporary relocation of the apiaries, at least until the migration period is over.

Mammals

Many groups of mammal may be considered as enemies of the honeybee. In general, they prey on colonies for honey and/or brood; some attacks are purely accidental or the result of animal curiosity. Such cases usually occur when apiaries are placed in or near forests and are not properly protected. In Asia, as well as in most other parts of the world, bee-keepers face the problem of colony destruction by bears. It has been said that once a bear has tasted honey and brood, it is almost impossible to keep it away from apiaries. Protecting colonies from bear attack is usually difficult, particularly when the animals are large and strong. Electrified barbed-wire fences are often used where bears represent a common problem; shooting and trapping them are other possibilities but very temporary control measures, which may go very much against efforts by others to manage and conserve sufficient numbers of large mammals in mostly declining populations. Moving hives closer to human habitation may be much more effective. In several tropical countries of Asia, monkeys and other primates have been mentioned as pests of honeybees, opening hives and consuming honey and brood. As a result, frames are destroyed and colonies may abscond. Discouraging such behaviour by wiring lids to boxes and boxes to each other may be a solution. Other options may include suspension of the colonies, as in Africa, particularly for small-scale bee-keepers. It is important to note that among the primate pests of honeybees, people are probably the most destructive. Honey crops may be stolen, or brood and combs consumed on the spot. Occasionally, entire hives are made off with. In areas where intensive modern agriculture is practised, the loss of bees through human misuse of pesticides is greater than loss from all other causes taken together.

Honeybee introduction and quarantine

Introduction of foreign bee species and races is unnecessary as native bees are generally better adapted to the climate, food and the pathogens present. It has often been observed that honeybee colonies that have been introduced have had disastrous effects in spreading bee diseases and parasites. However, in Asia it achieved a high degree of success in developing commercial bee-keeping by introducing colonies of European races of *Apis mellifera*. This provoked bee-keepers to introduce foreign bees in some countries and regions. But it requires proper legal regulation and enforcement to prevent the spread of disease; though it may not be sufficient. Providing quality stock and educating bee-keepers about the risks of buying foreign bees is a necessary and long term solution. In any of these cases the follow-ing basic guidelines should be respected :

- It is important to study and evaluate the potentials and limitations of the importing bees and their foraging ecosystems, including the availability of forage.

- If introduction of new bees are found necessary, the entire process – along with all requisite precautions, including quarantine and interim colony inspection should be carried out by, or closely supervised by well-trained personnel.

- Periodic evaluations of the current state of the local honeybee industry should be made to determine whether introductions continue to be necessary.

- In countries or regions in which bee-keeping is reasonably well-developed and there is no good technical reason to introduce new bee strains or races, then it should be halted. It is preferable that bee-keepers endeavor to improve their queen-rearing techniques and adjust their colony management schemes in order to obtain better yields should be encouraged.

All honeybee species native to Asia have their own diseases and enemies and it is unavoidable that introduced colonies will come into contact with bees from feral nests. Therefore, inter-species transmission of certain diseases and parasites will take place reciprocally. It is thus advisable that larger scale movements, or any scale of introduction, should either be prevented completely or reduced to a minimum.

SPACE FOR NOTES

Chapter Three : Sericulture

INTRODUCTION

Sericulture has originated from two French words, **"Seris"** meaning *silk* and **"Culture"** which means *rearing*. **Sericulture is the cultivation of silk through rearing of commercial silkworms. It is an agro-based industry that involves cultivation of food plants for silkworms, rearing of silk worms for production of cocoons, reeling and spinning of cocoons for production of yarn, etc. for value added benefits such as processing and weaving**. Sericulture also includes many other practical aspects such as increasing productivity of cultivationl as well as labour, improvement of cocoon production, silk yarn, fabric and generating profitable income, especially of rural people.

History of Sericulture

Sericulture or silk production from the moth, *Bombyx mori* (L), has a long and colourful history. According to well-established Chinese legend, **Empress Hsi Ling Shi,** wife of **Emperor Huang Ti (also called the Yellow Emperor),** was the first person to accidentally discover not only the means of raising silkworms, but also the manner of reeling the silk, and of employing it to make garments. She was later honoured with the name Seine-Than, or **"The Goddess of Silkworms"**. The secret of making silk remained in China for about 2,000 years. No one is sure when the secret first seeped out of China, but it is known to have reached Japan by way of Korea by the fourth century A.D. and said to have been brought there by four Chinese girls. Imperial law decreed death by torture to those who disclosed it. It is also said that silk was brought to India by a Chinese princess who hid eggs and mulberry seeds in the lining of her head dress. Thereafter, sericulture travelled Westward and had spread into Japan, Persia and gradually to Europe. Demand for this exotic fabric eventually created the lucrative trade route, now known as the ***silk road*** – some 4,000 miles long stretching from Eastern China to the Mediterranean Sea. In India, efforts were made to patronise silk during the British rule. Then it suffered a series of setbacks during the period of World Wars. After 1945, with the assistance of Govt. of India, this industry once again entered a period of prosperity with the implementation of various modern techniques.

Importance of sericulture

(1) **High Employment potential** : Only sericulture can generate vast employment*. No other industry generates this kind of employment. Hence, it is used as a tool for rural economic reconstruction.

(2) **Provides vibrancy to rural economy** : It is estimated that about 57% of the gross value of silk fabrics flow back to cocoon growers, who in general are rural people with share of income to different groups as follows : 56.8% to cocoon growers; 6.8% to the reelers, 9.1% to the twisters; 10.7% to the weavers; 16.6% to the trade. Thus, large amount of income from silk industry goes back to villages from the cities.

(3) **Low gestation, high returns** : Investment of only Rs. 12,000 to Rs. 15,000 (excluding cost of land and rearing space) for mulberry cultivation and silkworm rearing in one acre of irrigated land can

* At an estimated rate of 10-11 persons/day/kg production of raw silk.

generate net income levels up to Rs. 30,000/acre/annum. Mulberry takes only six months to grow for starting silkworm rearing and once planted it can support five crops in one year under tropical condition. Depending on the management, rearing at least for 15 years can be continued in the same land.

(4) **Women-friendly occupation** : Different activities of sericulture starting from mulberry garden management, leaf harvesting and silkworm rearing even reeling and weaving can effectively be supported by women workers. In India, it has been found that over 60% of scriculture employees constitutes the women folk.

(5) **Ideal programme for the weaker section of the society** : Low land holders even can start sericulture. Silkworm rearing in 3/4 acres of mulberry garden can supprot a family of three without hiring labour. Moreover, vast areas of tasar food plantation are available in different forests of India; if judiciously exploited for rearing tasar silkworm, these can offer a great supplementary employment to tribal Indians.

(6) **Eco-friendly properties** : (i) **Soil preservation** : Being perennial plant with green foliage, mulberry contributes to soil preservation and provides vast green covers. (ii) **Manures** : Wastes from silk worm rearing can be recycled for mulberry garden prepartion as manures. (iii) **Fuel source** : Dried mulberry twigs and branches can be used as fuels by the farmers, thus in turn can reduce the pressure on nearby vegetation or forest. (iv) **Pollution control** : Being an agro-based industry, it requires minimum use of fuel machinery, thus reduces air pollution. (v) **Integrated farming** : With other plantations like flower, vegetables, mulberry can be cultivated as intercrop, thus same land can be used for various productions. (vi) **Uses of land** : Besides cultivation lands, water shade areas, hill slopes, vacant lands, etc. can also be used for the deep-rooted mulberry plants.

(7) **Scopes for professional training** : Sericulture departments of various states arrange for training on different aspects of sericulture to the farmers with monthly stipend and field training.

At national level, the Central Silk Board (CSB) organises three kinds of programmes. The first is open to all, the second to in-service personnel and the third to the farmers. Short term courses, Intensive Training Programmes are also organised by CSB from time to time. The International Centre for Training and Research in Tropical Sericulture (ICTRETS), Mysore conducts three courses on tropical sericulture at international level. In the vocational scheme, sanctioned by UGC, students can take sericulture as their subject of choice. Thus, sericulture is providing a wide scope for choosing profession to all sorts of people in the society.

(8) **Facilities for Sericulturists** : Besides providing field and educational trainings, Regional Extension Centres and Technical Service Centres of both State and Central Sericulture Boards supply advanced-staged larvae to the rearers; rearing appliances and other raw materials like mulberry samplings are supplied at nominal cost to the rearers. Incentive bonus is offered to bivoltine rearers. Crop insurance schemes have been offered specially to bivoltine farmers to protect against any failure. In remote rural areas, farmers are supplied with pamphlets or other essential details of methods of rearing, use of bed disinfectants, insecticides, etc.

SERICULTURE

Sericulture generally comprises of three main branches : **(1) Silkworm rearing** — it has two branches : **Grainage technology** (egg production) and **Silkworm rearing** (larval stage management). **(2) Silk reeling and weaving** — it is the industrial aspects from cocoon onwards. **(3) Moriculture** (Mulberry cultivation) — it includes host plant cultivation, mainly Agronomic aspects. In India, following methods and technologies are being adopted towards the progression of sericulture.

MANAGEMENT OF EGG OF SILKMOTH

The most important as well as precarious stage of sericulture is the production of healthy and disease-free eggs and it plays a vital role on which the success of the industry depends. Silkworm eggs are also called **seeds** and for sericulture purpose the eggs are classified into two groups, viz., Reproductive seed and Industrial/Commercial seed. **Reproductive seed** : The seeds from which parents of different commercial seeds are produced. These are usually pure line of both local and high yielding races and rarely F1 hybrid. **Industrial/ Commercial seeds** : These are mostly F1 hybrids, produced by crossing between local hardy pureline female race and pure-line high yielding bivoltine (generally) males.

Egg/Seed production centre

(1) Breeding Stations : For maintaining racial purity, quality and diseasefreeness etc, reproductive seeds are produced as well as multiplicated in number in a series of breeding centre called Breeding Stations.

(2) Breeding centre : The number of pure breed parents required for a grainage are too large and that cannot be obtained by a single multiplication step from the stock. So generally, a 3 or 4-stage multiplication is carried out to maintain the racial purity as well as to produce quality eggs by technically qualified personnel in a series of breeding centers in the Breeding stations like P4, P3, P2 and P1. In India, usually 3 Tier system of seed production at P3 , P2, and P1 level is adopted as follows : **P4/P3 (Breeding centres)** — Centres where great great grand parents or the great grandparents of commercial grainage seeds are multiplied.They rear only pure breeds and no hybridisation is done here to maintain the purity of races. So each laying of each race is reared separately in different trays. These are the major germplasm bank and seeds produced here are called **Basic or Nucleus seed (Breeder's Stock Egg)**. **P2 (Breeding Centres)** – They receive seed cocoons from P4/P3 centres, so they rear the grandparents of commercial seeds and produce the **Foundation Stock Seeds** that also serve as germplasm bank. **P1 (Breeding Centres)** – Here the parents of commercial layings are reared. They receive the seed cocoons from P2 centres. The first hybridisation of double crosses is done here and the hybrids produced are called **Foundation hybrids**.

(3) Grainage : The establishments where the disease-free commercial seeds are produced in mass to be supplied to rearers are called **Grainages**. These are mostly owned by Government and may also be licensed private owned.

In grainages parental seed cocoons received from P1 centres are reared as well as preserved under optimum conditions. The sexes of either larvae or pupae are recognised and accordingly male and females are reared separately. Following their emergence, healthy adult male and female moths are allowed to mate and oviposit by different methods.

Egg/seed production methods : In grainages commercial eggs are produced by any of the following methods : (a) **Cellular bag method** : Each pairing moth is kept into a small bag of perforated parchment/paraffin paper or cloth where the female lays eggs. Mother moths are examined after oviposition for pebrine. (b) **Cellular card method** : Here, a card or craft paper is divided into 20 cellules or squares in 4 rows, each with 5 compartments. A moth funnel plate with 20 similar holes is placed on that paper. In each section of the hole, one coupling pair of moth is kept and covered by moth funnel. After egg laying, mother moths are removed and placed in an exact chamber of mother moth box with similar compartments. Then the moths are examined for pebrine. Diseased moth (if detected) and its eggs

Fig. 3.1. Laying of eggs on flatcard by the female mulberry moths.

are discarded. This method can produce eggs in large scale. (c) **Flat card method** : In this system, about 40 pairs of coupling moths are allowed to lay eggs on the rough surface of a flat card (Fig. 3.1). If pebrine is detected in any mother moth, all eggs are discarded. (d) **Loose eggs method** : Same as flat card method except that eggs are laid on smooth surface of paper. Healthy eggs are then loosened from the paper by soaking it in water. Unfertilised eggs which float on water are discarded. Healthy eggs are washed in water, dried and packed. Eggs prepared by any method are treated with 2% formalin solution for disinfection. Mother moths after egg laying are also checked for the transovarial infection of pebrine spores. If infection is detected, eggs of that layer are rejected immediately.

Storage of eggs : Silk moths lay two types of eggs (i) **Diapausing** or **hibernating** eggs and (ii) **non-diapausing** or **non-hibernating** eggs (see page 77 and Table 3.7 for details). The eggs are laid either in spring or in autumn.

Storage of Spring eggs : Spring eggs are stored at aestivation temperature (23-25°C) from June to August, then at room temperature till September or early October; then gradually lowered to 5°C in December for 50-60 days. This treatment awakens the embryo but arrests the development. The eggs are kept still at lower temperature (2.5°C) for another 40-50 days where these can also be stored for any length of time. Otherwise, in March, eggs are shifted at a temeprature of 10-15°C for 4-5 days before incubation.

Storage of Autumn eggs : Eggs laid in autumn (August) are kept at 23-25°C for 20 days; then the temperature is gradually lowered (to 2.5°C) as for spring eggs till natural temperature is reached.

Artificial breakage of diapause: In order to utilise available mulberry leaves, sometimes hibernating eggs are treated with diluted acid to break their diapause and to get more crops. Eggs are either kept in hot HCl acid at 25°C for 20-24 hours after oviposition or in dilute acid heated to 46°C for 5 minutes. Sometimes eggs are dipped in HCl at 15°C for about 50 minutes. After acid treatment, eggs should be washed in water.

Incubation: Incubation of eggs is done to achieve uniform hatching on a desired day by keeping the eggs at an optimum temperature of 25°C and 75-80% RH (Relative Humidity). Prior to incubation, eggs should be disinfected by treating with 2% formalin. Three methods of incubations are generally followed : (i) **Constant temperature incubation** where eggs are incubated at a constant temperature of 23-25°C till hatching. Non-hibernating and acid-treated eggs are incubated by this method. (ii) **Raised temperature incubation** where hibernating eggs are first kept at a temperature of 15°C, which is gradually increased by 0.5°C /day in the first week and then by 1°C in the second week till the temperature is reached to 27°C where the eggs are kept till hatching. (iii) **Embryonic method of incubation** where the eggs are kept during diapause at 15°C and then at 22–23°C till the revolution and then at 27°C till hatching. The incubation period varies with the races, voltinism and incubation temperature.

Transportation of eggs: The transportation of eggs plays an important role in the development of embryo and successful crop harvest. The disease-free eggs are collected in wet handbags with proper aeration in early morning or in late evening. Transportation may be done during aestivation period, hibernation or post-hibernation period or pre-incubation period.

MANAGEMENT OF SILKWORM / LARVAE

Being delicate and domesticated creatures, *B. mori* larvae cannot tolerate any environmental fluctuations. So in sericulture practice these are reared with very special cares. In India, the most common method of silkworm rearing is **Shelf Rearing Method,** where worms are reared (feeding, cocooning, etc.) on trays that are arranged one above the other in tiers held by rearing stand, so that a large number of worms can be reared at a time in a small

Fig. 3.2. Shelf rearing of mulberry silk worms. Rearing trays are arranged in tiers. Note leg of tray stand is placed on ant-well.

Fig. 3.3. Floor rearing of silkworms

Fig. 3.4. Shoot rearing of silkworms

place (Fig. 3.2)*. Other rearing methods include **floor rearing** that is carried out on fixed rearing seats arranged in 2/3 tiers kept on the floor of rearing house (Fig. 3.3); **Shoot rearing** where silkworms are also reared in fixed platform kept indoor or outdoor and fed with mulberry branches or shoots (Fig. 3.4).

Table 3.1 : Required temperature, humidity and space for different stages of silkworm

Stage of silkworms	Temperature (°C)	Humidity (%)	Spacing (for 100 DFLs*) in sq. ft
1st Instar	26 – 28	85 – 90	4 – 14
2nd Instar	26 – 28	85 – 90	15 – 45
3rd Instar	25 – 26	80 – 85	46 – 90
4th Instar	24 – 25	70 – 75	91 – 100
5th Instar	23 – 24	70	181 – 360

* DFL : Disease-free laying

Rearing Appliances

Being a rural-based agro-industry, appliances used in sericulture are mostly made of cheap and locally available materials and thus, vary from place to place. Moreover, researchers of CSB and other private rearers have developed many devices suitable for silkworm rearing. Here, appliances, required for shelf rearing, will be mentioned.

Shelf rearing appliances

(i) **Tray stand** : This bamboo-made or wooden stand is used to support the rearing trays (Fig. 3.5). It has a height of 2.5 m and has 10-12 cross-bars at a distance of 0-1.5 m to accommodate 10-12 tiers of rearing tray.

(ii) **Rearing trays** : These are circular bamboo trays (138 cm dia × 6.5 cm depth) (Fig. 3.5) or light wooden rectangular trays (0.7–0.9 m × 0.9–1.2 m), each of which can accommodate a large number of young larvae.The trays are often coated with cow-dung paste mixed with formalin, though this practice often invites many disease-causing agents.

(iii) **Ant wells** : Enamel / Aluminum bowl or stone / concrete blocks (Fig. 3.2) are used as ant wells which are kept filled with water except the central elevation on which legs of tray stands are placed. It prevents ants from crawling up the stand and harming the worms.

* The required temperature, humidity and space for different stages of silkworms by this method is mentioned in Table 3.1.

Fig. 3.5. Circular rearing trays are kept in a tray stand

(iv) **Paraffin paper** : To avoid evaporation of moisture and keep humidity on the rearing tray, pad of paraffin-coated papers are kept on rearing bed of young silkworm. Polythene sheets, dried banana leaves may also be used instead of paraffin-paper.

Fig. 3.6. Picking of worm by chop stick

(v) **Foam rubber strips**: To maintain the required humidity, sometimes thick foam pads soaked in water are placed all around the bed. Thick folding of newsprint paper or blotting paper soaked in water can also be used in place of foam strips.

(vi) **Chop sticks** : Forcep-like device, made of two thin bamboo sticks, joined at one end by thread is used to pick worms (Fig. 3.6).

(vii) **Feather** : Soft, white feathers are used for brushing newly hatched larvae from the egg card to rearing tray and to spread the young worms during spacing.

Devices used for feeding the worms

(i) **Leaf basket** : Bamboo baskets are used for keeping and transportation of collected mulberry leaves from the field to rearing place. Wet gunny bags, multilayered bags with gunny cloth are also used for this purpose.

(ii) **Leaf chamber** : For storing the harvested leaves, container made of strips of wooden reapers are used. Wet gunny cloth is placed wrapping all sides of this container to keep the leaves fresh.

(iii) **Chopping board** : Soft wood-made board is used for cutting the leaves for feeding the worms.

(iv) **Knife** : Sickle-shaped knife with sharp edge is used for chopping the leaves properly.

(v) **Mats** : Clean mats are placed under the chopping board prior to chopping to collect the cut leaves.

Devices to help the silkworm for spinning

The most important device that helps or supports the silkworms (larvae) for comfortable spinning their cocoon is called **cocoonage or mountage**. It determines both the quality and quantity of the cocoons. Different types of cocoonages are used in different parts of India. In general, these are made of wood, bamboo, cardbaord, plastic, grass, dry leaves, twigs, etc. However, ideal mountage should fulfil the following requirements :

(i) Convenient space with suitable dimension for spinning proper-sized cocoon.

(ii) Should not allow formation of double or malformed or flimsy cocoons.

(iii) Should have provisions for cleaning the excreta of spinning larvae.

(iv) Should be suitable for easy harvesting of cocoons.

(v) Should be cheap, durable and easy to handle.

In India following types of mountages are used :

(1) **Chandrika** : This round- or rectangular-shaped mountage is made of bamboo mat supported by split bamboo reapers on all sides (Fig. 3.7). On the mat, a bamboo tape of 4-5 cm width is wound in a spiral manner. About 1000 worms can be mounted on this mountage. These are easy to handle, repairable and cocoons spinned on Chandrika are of good quality. However, bamboo being a protected plant, nowadays Chandrika is rare and costly. It is prone to damage by rodents and requires a lot of area when not in use. Besides, occurrence of stained cocoons is more in chandrika. Chandrika is popular in South India and West Bengal.

(2) **Screen-type mountage** : It is made of bamboo or wooden or plastic reapers on which, instead of spiral bamboo tape, longitudinal strips with triangular peaks are placed. The screen can be folded and stored. This mountage can be kept clean and well-ventilated and hence, cocoons spun on this mountage are of good quality. It is more durable than chandrika. But occurrences of double cocoons are frequent in it.

(3) **Plastic mountage** : Like Chandrika, but is made from plastic instead of bamboo and hence, more durable, easy to clean, not prone to rodent

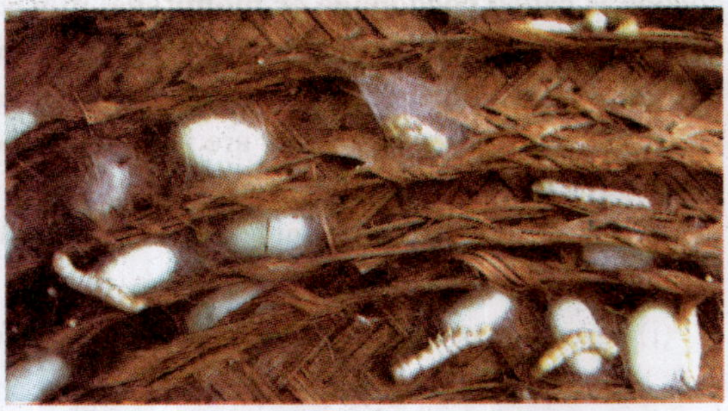

Fig. 3.7. Chandrika, made of bamboo mat (partly shown)

attack, and produce lesser number of double cocoons. Once invested, further maintenance, care or expenditure are not incurred. But these mountages are costly than Chandrika. The cocoons produced on these mountages are more flimsy and not of uniform size and hence not frequently used by farmers.

(4) **Japanese low cost mountage** : In this modified Japanese mountage, a wooden frame of 4 longitudinal rods is attached by means of cross-spokes at two ends to a central axis. Each rod has a number of pegs placed at equal distances. These pegs are connected by long threads of twisted rice straw in a regular pattern like that of charpoy. The size of frame and the number of pegs can be modified according to the requirement of the rearer. This mountage is cheap, more durable and make less chances of disease spread.

(5) **Bamboo Strip Mountage** : Made of bamboo stirps that are either nailed on wooden reaper or placed in grooves of wooden reapers. Several such frames are placed one above the other with the lower one keeping on four uniform bricks or wooden blocks. This mountages are cheap, durable, easy to handle, and harvest the cocoons.

(6) **Bottle brush mountage** : This recently introduced **mountage** consists of a thick coconut or jute fibre rope into which 6-9" sticks (midrib of coconut leaves) are inserted very closely. These are used by the worms as support. The worms spin their cocoons in the space between the sticks. This mountage is very cheap; can be made easily and occupies little space compared to Chandrika.

(7) **Plastic bottle brush** : This modified bottle brush was innovated by researchers of Central Sericulture Research and Training Institute (CSR and TI), Mysore with assistance from JICA. Here, each plastic rope has 8 branches of 1.5 cm length with 2 sub-branches of 9 cm length. The branches are equally distributed at a distane of 1 cm at the base and 4 cm at the ends of the branch in a circular fashion (Fig. 3.8). The circle formed by branches and sub-branches is 24 cm in diameter.Their surfaces are granulated (rough) in order to provide grip to the worms during spinning to minimise floss loss. Fifty

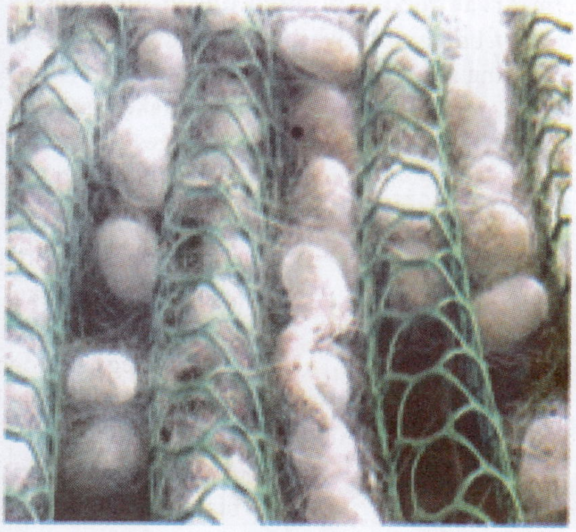

Fig. 3.8. Plastic bottle brush

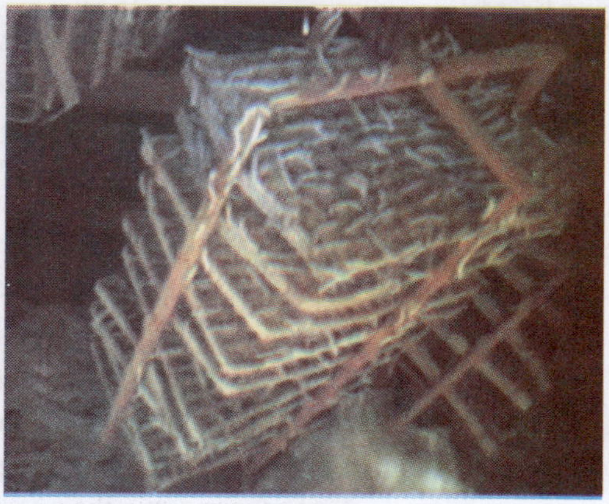

Fig. 3.9. Japanese rotary mountage

such individual pieces are jointed together with an iron rod and at the end of that rod, a steel stopper is fixed tightly, so as to keep the whole mountage in a correct position. This mountage is durable and can be reused after disinfection. It allows easy harvesting and also self mounting in shoot rearing method as well.

Besides the above mentioned mountages, many other mountages popular in some other countries are now being introduced in Indian sericulture. For example, folded straw mountage, straw rope mountage, rotary mountage (Fig. 3.9) from Japan, and folded bamboo strip mountage from China, etc.

Devices required for maintaining optimum environmental conditions and disinfection

Rearing set up needs thermometer to check temperature; hygrometer to measure humidity; charcoal stove / electric oven to warm up rearing room; door mat to soak the formalin sprayed; wash basin, etc. In addition to these, for disinfection, some items like sprayer, formalin solution, bleaching powder, etc. should be kept in rearing place.

REARING PROCESS OF SILKWORM

The rearing process of silkworm involves the following steps in general. However, details of these operational steps may vary from place to place to adjust the prevailing conditions and local tradition as well.

(1) **Disinfection** : It is the most important operation that to be carried out prior to the commencement of rearing. Disinfection of everything including rearing places is carried out by physical, chemical or radiation methods.

(i) **Physical methods** : These are cheap, convenient and easy to operate, e.g. (a) ***Sunlight*** : Drying of rearing appliances in sunlight can cause disinfection. However, sun drying cannot be carried out during winter and rainy seasons, and some appliances are likely to be damaged by exposure to sunlight. (b) ***Steam*** : Disinfection by steaming may be used for rearing room and some appliances (not made of bamboo or wood). However, initial cost for installing the steaming apparatus like boiler and pipeline is high. (c) ***Hot air*** : It is also a good sterilising method but cannot be used in routine sericulture because of its production cost.

(ii) **Chemical method** : The most commonly used disinfection method in sericulture is chemical method. Chemicals generally used are non-toxic to man and animals, have broad spectrum activity, stable and readily mixable with water and fair in cost. Most frequently used chemicals include chlorine as chloramine, iodine as iodophores, phenol as cresol and hexachlorophene, formaldehyde as formalin (2%), bleaching powder, etc. These are used as spray or fumigant. Precautions should be taken during and after the applications of such chemicals.

(2) **Brushing** : Brusing is the separation of newly hatched larvae from their egg shells and transferring them to rearing trays from the egg cards. The newly hatched larvae are black, bristly and called **ants**. Brushing is usually starts at 10 am when peak hatching occurs. Brushing can be done by various ways :

(a) **Brushing from loose eggs** : Fine meshed net or thin muslin cloth can be placed over the newly hatched larvae. Then freshly chopped mulberry leaves are sprinkled over that net / cloth. The larvae start to crawl up through the holes onto the leaves. After sometimes, the larvae along with leaves are gently tapped on the rearing bed.

(b) **Brushing from egg cards** : From egg cards, newly hatched larvae can be transferred by the following : (i) **Feather** : Here the egg card is held vertically above freshly prepared rearing bed and then by gentle strokes of a feather, the larvae are pulled out from the card on the rearing bed. However, this method is little bit crude and may cause some injury to the larvae. (ii) **Husk** : Here powdered husk is sprinkled over newly hatched larvae on the egg card. Then freshly cut mulberry leaves are sprinkled over the centre of husk. The larvae crawl up the husk to reach the leaves. After sometimes, the larvae are brushed from husk by means of a feather on the rearing bed.

However, in all cases of brushing, care should be taken not to touch the newly hatched larvae with hands.

(3) **Feeding the larvae** : Both the quality and size of the cocoons depend mainly on the quality of mulberry leaves fed by larvae during rearing. After a little practice, the amount of leaves that to be given per feeding to fulfil the appetite of the worms, is adjusted. The amount of food given also depends on races and voltinism of the moths. However, maximum amount should be given during the active feeding stage of instar and no food should be given during moulting. In Indian sericulture, nowadays four feedings per day is allowed. In case of shoot and floor rearings, three feedings per day are sufficient. In all larvae, several feeding stages can be noticed during their development, viz., *active feeding stage* when larvae feed maximum during the instar, *sparsely feeding stage* when they eat less, usually at premoulting stage when larvae empty their gut; *non-feeding stage* when the larvae do not take any food usually during moulting. Of the total ingestion during entire larval development, nearly 85% of food is taken during IVth and Vth instar stages. Leaf requirement during different instar stages are given in Table 3.2.

Table 3.2 : Leaf requirement (in kg) for 50 DFLs or 2000 eggs of different races of larvae

Larval instar	Uni-/bi-voltine race	Multivoltine race
1st Instar	1-2	1-2
2nd Instar	5-6	2-3
3rd Instar	20-55	15-20
4th Instar	80-90	35-50
5th Instar	450-475	300-325

During feeding, generally a gap of 2 hours is given before and after each moulting. Young worms are always fed with tender leaves while late stages are given mature mulberry leaves (Fig. 3.44). To enable the larvae to feed easily, young worms have to be given chopped leaves but for mature worms, full leaves or young branches or shoots may be given.

(4) **Spacing** : The silkworms grow very rapidly from age to age and increase many times their weight and size from the previous instar. The total increase in weight from hatching to the end of Vth instar is about 7,000 to 10,000 times. Crowded situation in rearing trays results in incrased humidity, heat, fermentation of litter, all of which will in turn cause under development of larvae, wastage of feeding leaf and unhygienic condition. To provide more and adequate space for the growing worm, the rearing space has to be extended at each stage and this is called **spacing.** Spacing is usually done along with bed cleaning and is given once a day. The number of trays and space required for each instar are given below in Table 3.3.

Table 3.3 : Number of trays and area of space required by different instars of silkworm during rearing

	No. of trays required for 100 DFLs (diameter of tray 3.5')		Space required (sq. ft) (Bi/multivoltine)	
Stage of silkworm	Begin-ning	End of stage	Begin-ning	End of stage
1st instar	2	2	4	15
2nd instar	2	5-6	15	45
3rd instar	5-6	10-12	46	90
4th instar	10-12	20	91	180
5th instar	20	40	181	360

(5) **Bed Cleaning** : The rearing tray of silk worms accumulates some unconsumed leaves after each feeding, exuviae after moulting, excreta, dead or diseased larvae, etc. All these if not cleaned, combine to form a thick and damp litter which promotes the growth of different micro-organisms, generation of heat and injurious gases and depletion of oxygen. Hence, it is very necessary to remove the litter periodically and the process of its removal is called bed cleaning.

Bed cleaning can be done by using paddy husk, straw and bed cleaning net (Fig. 3.10a). During 1st instar, bed cleaning should be done once during per moulting, during 2nd instar twice, once after moult and before next moult. During 3rd instar thrice, i.e. after moult, before next moult and once in the middle. During 4th and 5th instars once in a day in case of shelf rearing. However, in case of floor or shoot rearing, bed cleaning should be done once in each instar.

(6) **Caring during moulting** : In commercial races of silkworm, moulting occurs four times, lasting for 15-30 hours. During this time, the worm does not take any food, wriggles out of the old skin and comes out with a new, soft skin. Care taken during moulting includes stopping and resuming feeding at appropriate time to ensure uniform growth, keeping the bed dry and disinfected either by dusitng Resham Keed Oushad (RKO), formulated by CSR and TI, Mysore or by spraying Labex, formulated by Berhampur. Besides disinfecting action, RKO can reduce grasserie in different seasons and can increase growth rate of larvae leading to improved cocoon quality. Labex has anti muscardine effect and can inhibit early moulters from resuming feeding leading to uniform growth.

(7) **Mounting** : Mounting is the process of transferring the ripe worms to the mountages. On the mountage, the ripe worms exude silk, spin the cocoon around itself and transformed into the pupa inside it. The pupa after metamorphosing into adult moth comes out by piercing open the cocoon. The aim of sericulture is to rear the silkworm providing them optimum conditions and mountages so that they can spin good cocoon with high and best silk content. Mounting is done by following methods :

(i) **Hand picking** : Ripe worms are collected in a tray one by one by hand and then transferred to the mountages. Though some worms may be injured while picking and handling, but by this method, only ripe worms can be picked and distributed more uniformly in the mountages (Fig. 3.10b).

Fig. 3.10(a). Bed cleaning by net

Fig. 3.10(b). Mounting of ripe worm on Chandrika by hand-picking

(ii) **Simultaneous mounting** : In this method, a number of mature larvae is collected simultaneously and transferred to the mountage. Here, mature, immature and overmature worms are mounted together; hence, cocoon formed by them may not be uniform.

(iii) **Net method** : In the rearing tray, when worms are ripen, straw rope nets / rush nets or cleaning nets are spread over the rearing beds and left for some time. Ripe worms crawl alone on the nets while unripe worms continue feeding. The nets with ripe worms are then shaken on the mountages to transfer them without touching by hand.

(iv) **Branch method** : Here small branches of mulberry are spread over the rearing bed. Ripe worms crawling over them are then shaken off on the mountages. Besides branch, dried weeds (Russia) or cut straw (Japan) can also be used for transferring the ripe worms to mountages.

Number of ripe worms per mountage is very important. In general, one ripe worm requires an area that is the square of its body length for spinning its cocoon. Too wide spacing may cause wastage of silk for spinning the priliminary web. Again, too close spacing may result in formation of double cocoon (which are not reelable), staining of cocoons with excreta of the worms and also formation of damp cocoons. The optimum denisity for Chandrika is 50 worms per 0.1 m^2.

Precautions to be taken during mounting

(i) Only ripe worms should be mounted. Unripe worms spoil other cocoons with their excreta while overripe worms hastily spin cocoons which are malformed, flattened, sticky and inferior.

(ii) An optimum temperature (24°C) should be maintained in spinning place. Too low temperature causes delayed formation of cocoons, and affects colour, lustre and texture of the silk. Too high temperature results in the formation of deformed cocoons with thick filament.

(iii) The ideal humidity for spinning is 60-70%. Ventilation is needed to dry the wet silk into firm cocoon and to evaporate the water or excreta released by the worms during spinning.

(iv) The mountages should be disinfected before and after use.

The spinning worms should not be disturbed which otherwise would result suspension of spinning and breaking of thread.

MANAGEMENT OF COCOONS

(1) **Harvesting of cocoons** : Silkworms complete spinning in 2-4 days depending upon the race and climatic conditions provided. Harvesting or collection of mature cocoons must be done after the pupal cuticle has hardened and before the adult emerges. The recommended time of harvesting is fifth day of spinning for tropical races and seventh or eighth day for temperate races. Cocoons are harvested generally by hand. In advanced temperate country like Japan, simple devices are used for harvesting the cocoons from rotary mountages.

(2) **Marketing of cocoons** : Being an alive and also perishable commodity, cocoons should be marketed as early as possible after harvesting and before the moths emerge from them. Defective cocoons have no market value as most of them are not reelable. These include double cocoon, immature or premature and fragile cocoons, unsized or malformed cocoons, mute (with dead pupa sticking to inner shell) or black-stained cocoons, calcified cocoons, spotted and fungi-attacked cocoons and cocoons with thin middle part or thin ends.

(3) **Transport of cocoons** : During transportation cocoons are kept in perforated containers with enough aeration. The containers are so designed to prevent jolting and crushing of cocoons. The transport should be done quickly and for short distance.

(4) **Cocoon market** : To keep the price of cocoon uniform and moderate, all state Governments have set up centralised markting federations. They fix the floor price of cocoons and also formulate guidelines for cocoon transactions.

(5) **Commercial characters of cocoons** : The price structure of cocoons depends on the following physical properties :

(i) **Colour** : Commercially bright-coloured cocoons are preferred than dull colour.

(ii) **Shape** : Spherical, oval and moderately constricted cocoons are preferred for their easy reelability.

(iii) **Size** : Large-sized cocoons can be reeled easily than the smaller one.

(iv) **Compactness** : A tight-built cocoon has good reelability.

(v) **Grains or wrinkles** : Cocoons with grains or wrinkles on surface have high demand because of easy reelability.

(vi) **Weight** : Cocoons have optimum weight in the middle pupal period. So if sold in this stage can get maximum returns.

(vii) **Shell ratio** : Shell ratio is expressed as under:

$$\text{Shell ratio} = \frac{\text{Weight of cocoon shell}}{\text{Weight of the cocoon}} \times 100$$

For commercial purpose, the ratio should be in the range of 18-20% for bivoltine race.

(viii) **Non-breakable filament length (NBFL)** : Cocoon whose filament can be unwound without any break has best commercial value.

(ix) **Filament denier** : Denier is the unit used to express the thickness of silk filament. It is calculated by the formula given below :

$$\text{Denier} = \frac{\text{Weight of the filament (g)}}{\text{Length of the filament (m)}} \times 9000$$

Commercially approved denier varies from 1.7 to 2.8 g.

(x) **Renditta** : Number of kilograms of cocoons required for 1 kg of reeled silk is called renditta. Multivoltine has higher renditta than bivoltine races.

(xi) **Reelability** : The ratio of unbroken filament to the whole filament length in percentage is the reelability. Multivoltine has poor reelability than bivoltine.

SILK : ITS NATURE, PROPERTIES AND MANAGEMENT

Silk : *Silk is the natural protein fibre secreted by silk worm in making their cocoons and collected to make thread and fabric.* According to entomologists, there are about 500 different species of wild moths spinning silk cocoons, but only few have commercial value. Here the silk produced by the larvae of mulberry and non-mulberry moths will be discussed. However, the silk secreted by the silk worms to form the cocoon is little different from the silk thread used to weave. Because the latter lose some amount of sericin during processing. One thread of silk consists of 4-8 silk filaments and from one cocoon 1,000 yards of silk filament can be obtained.

Chemical Nature of Silk : Silk principally consists of two proteins, **Fibroin** and **Sericin,** in a ratio of 80:20 in case of mulberry silk or pure silk (see Tables 3.4 and 3.5). **Fibroin** ($C_{30} H_{46} N_{10} O_{12}$) is a yellow coloured fibrous protein, forming the core filament of silk thread (Tables 3.4 and 3.5). In the silk gland it is present as a helical, soluble protein. When this liquid is pressed or extruded, its helical structure is transformed into straight, crystalline chain structure. During this transformation various amino acids are also reoriented, imparting stability to the fibre. Fibroin is composed of 2 polypeptide chains, the heavy or H-chain (Mol. wt. 350 kD) and the light or L-chain (Mol. wt. 25 kD) which are connected with each other by disulfide bonds. **Sericin,** ($C_{30}H_{40}N_{10}D_{16}$), on the other hand, is a gelatinous protein remains as a covering around the fibroin fibre. Being water soluble, it gets dessovled to some extent while boiling the cocoons for silk extraction. Four types of serecins are generally present in silk like Sericin-I, Sericin–II, Sericin–III and Sericin–IV.

Table 3.4 : Percentage of proteins present in different silks

	Mulberry silk	Tasar silk	Muga silk	Eri silk
Fibroin	70-80%	80-90%	80-90%	80-90%
Sericin	20-30%	8-10%	8-10%	4-5%
Others	2-3%	3-5%	3-5%	3-5%

Besides fibroin and sericin, silk also cotains waxy matter, carbohydrates, minerals and some colouring substances. The golden silk contains xanthophyll, yellow silk contains caroteen and vialoxanthin, green silk has bombycin and bombycetin pigments. The colouring pigments of silk is due to the colouring pigments which were present in their food leaves.

Table 3.5 : Differences between fibroin and sericin

Fibroin	Sericin
1. Elastic protein forming the core of silk filament	1. Gelatinous protein forms the outer coat of silk filament
2. Secreted from the posterior part of silk gland	2. Secreted from the middle part of silk gland
3. Not soluble in water	3. Soluble in water
4. Pigments are absent	4. May contain pigments

Properties of silk

The ideal silk fibre should possess the following characteristics.

(A) PHYSICAL PROPERTIES

(i) **Length** : Being the longest natural fibre, the length of an unbroken filament may be up to 1000 metres. The diameter is around 0.013 mm to 0.08 mm.

(ii) **Density** : Raw silk has a density of 1.33 gm / cc and that of boiled off silk is around 1.25 gm / cc. It is lighter than cotton.

(iii) **Tensile strength** : Silk is one of the strongest natural textile fibre. The tenacity of raw silk is 4.5 – 4.8 gm / denier and that of degummed silk drops to 3.3 gm/ denier. Silk can be stretched from 15 to 20% before it breaks.

(iv) **Elasticity** : Silk elasticity is measured as the tension required stretching it to a definite length. It requires as high as 0.90 gm tension / denier to stretch 1%, in comparison to only 0.32 gm tension / denier required for wool.

(v) **Moisture take-up** : The moisture regain is 11% in case of raw silk and only 9% following degumming. Its hygroscopic capacity is about midway between cotton and wool.

(vi) **Electric properties** : Silk is a poor conductor of electricity, though can accumulate static charge by friction. The charge can be dissipated by high humidity.

(vii) **Scroop** : Scroop is the cracking sound emitted when silk is squeezed and pressed. It is not a inherent property but can be imposed by dipping it in dilute acetic or tartaric acid following by drying without rinsing.

(B) CHEMICAL PROPERTIES

(i) **Heating** : Silk can be heated to 140°C without any decomposition. It gets disintegrated at 170°C.

(ii) **Action with sunlight** : The strength of silk is reduced if exposed to UV rays of sunlight.

(iii) **Action with water** : Silk can readily be wetted by water; due to swelling in water it results an increase in weight by 30-55% and diameter by 16-18%. Boiling in water hydrolyses the fibroin more rapidly in alkaline pH.

(iv) **Action with acids** : The lustre of silk increases if treated with milk acid solution. Dilute nitric acid imparts a yellow colour which cannot be washed out easily. Strong acids, on the other hand, denature its protein structure.

(v) **Action with alkali** : Strong alkali dissolves the silk fibre, while mild alkaline solution may diminish its lustre.

(vi) **Action with metallic salts** : During mordanting, silk can absorb some salts like stannic chloride, resulting in weight increase. But salts like zinc chloride will dissolve silk. Sodium salt has no such effect but may cause tendering, if storage for long time.

PROCESSING OF SILK

Production of Raw Silk : Production of raw silk from the cocoon is done through following processes.

(1) **Drying / Stifling** : Drying or stifling is the killing of pupa inside the cocoon in order to make them suitable for storage. The cocoons are stifled either by direct method like sun drying or heating or indirect methods like steam or hot air drying. After drying the cocoons can be preserved or stored without any damage to the shell.

(2) **Storage** : The cocoons with dead pupae inside, can either be used for reeling or can be stored up to

6 months before reeling. During storing, cocoons should be protected from insect or fungal attack.

(3) **Sorting** : Before using cocoons are sorted on the basis of some physical characteristics like (i) colour, (ii) shape, (iii) weight, (iv) shell weight, (v) filament diameter, etc. Good cocoons are sorted from bad ones like stained cocoons, irregular cocoons, flossy cocoons and thin end cocoons.

(4) **Cooking** : The sorted cocoons are then further softened or cooked by hot soft water in order to swell and soften the sericin present in the cocoon without completely dissolving it. Cooking can be done by — (a) open pan (single basin) or (b) three pan (Fig. 3.11). In either case, boiling should not be done for long time and the pH of water is between 6.8–7.4.

Fig. 3.11. Three pan basin for cooking the cocoons

Fig. 3.12. Deflossing or brushing of cocoons by brush

Fig. 3.13. End picking from multiple cocoons before reeling

(5) **Deflossing/Brushing** : The unreelable, tangled mass of silk, found on the outside of cocoons is called floss. Deflossing is the removal of the floss to find out the actual reeling ends of the cocoon that can undergo continuous reeling. In India, deflossing is mostly done by peeling off the floss layer with hand. However, it can be done with simple device like brush or broom made of bamboo root sticks (Fig. 3.12) to save time and labour.

(6) **Reeling** : After boiling (cooking) and brushing, reeling started. **Reeling is the extraction of continuous silk thread of desired thickness from the cocoon without any break.** Before reeling, light brushing of the cocoon surface is done to find out the tip of the silk filament. Then the ends are all separated out and gathered together, this is called end picking (Fig. 3.13). Generally, filaments from multiple cocoons are combined together and transferred onto the reeling or spinning machine, which then automatically unravels the cocoons and simultaneously create a single strand of silk from the cocoons.

Throughout the reeling process, the cocoons remain in boiling water. When one cocoon gets exhausted, the second one is immediately substituted to reel a continuous and uniform yarn. So the important steps in reeling are brushing, end picking, casting, brinding and traversing. The individual filament must be interwined or twisted and cemented firmly so that the reeled filament does not break during subsequent processing of fabrics. This twisting and cementing is done with the help of a special device, called Croissure, which remains attached with the reeling machine.

Croissure helps in twisting and agglutinating the bave as well as it squeezes out excess water from the reeled thread to make it dry. Reeling can be done by 3 systems :

(A) **Charkha-Reeling** : In this manually operated system, inferior quality or defective cocoons are reeled. Here cooking and reeling are done in one basin. Following cooking at high temperature, the soften cocoons float in basin. The reeling ends are then taken out from 4-5 cocoons and twisted around themselves with the help of charkha reel (Fig. 3.14) to make uniform thread. Charkha reeling machine is traditionally home-made and used mainly in villages for small scale silk production. It uses chambon type of simple croissure where threads from two reeling ends are interwinded to form few spirals. So the silk reeled by charkha is of poor quality.

(B) **Cottage Basin System** : In this system, cooking and reeling are done separately. Boiling water basin is used for cooking while reeling is done in a hot water basin kept near the cooking basin. Cottage basin reeling machine uses travellette croissure. It consists of 3 pulleys to enable the thread to intertwine around itself and also better removal of water and gum spots from the thread (Fig. 3.15 a,b & c).

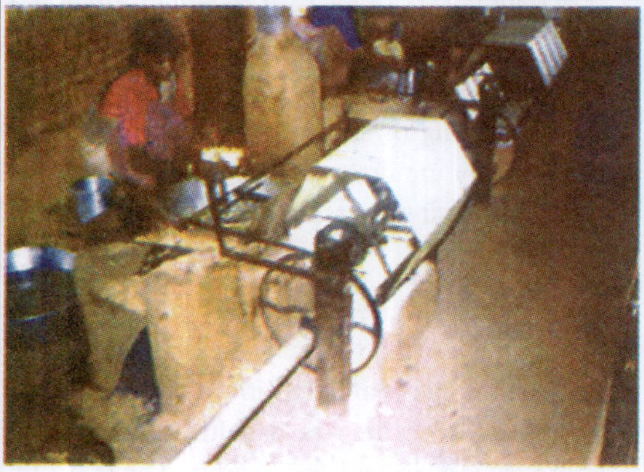

Fig. 3.14. Charkha reeling

Fig. 3.15(a). Reeling by college basin system. Note : cocoons are kept in hot water basin while reeling.

Cottage basin is also attached with traverse or distributor which is a modified hot rod. After being squeezed by the croissure, the thread is further dried as it passes on the traverse before reaching the reel. So cottage basin yields silk of good quality with uniform thickness. In India, as much as 806 tonnes silk are produced by 4000 cottage basin systems.

(C) **Filature System** : In this power-driven multi-end reeling device, cooking and reeling are done separately. It is a modern device where continuous reeling is done from cocoons of superior quality. The silk produced by filature is also of superior quality because of increased cleanliness and uniformity in thickness of the fibre. However, only 8% of the total silk production in Indis is contributed by filature.

(7) **Chemical processing of raw silk** : The raw silk thread, thus produced is rigid, stiff and not soft in touch and also lacks in lustre due to presence of

Fig. 3.15(b). Travellette croissure of cottege basin. Note the pulleys

Fig. 3.15(c). The distributor rod makes the thread more dry.

sericin and other impurities on the surface. So it needs following treatments :

(A) **Degumming of silk** : Degumming is the removal or elimination of sericin and other impurities from raw silk. It is carried out to impart the lustrous and softness on silk. Degumming is generally carried out by following treatments :

(a) **Extraction with water** : In this process, silk thread is treated in hot water (100-130°C) for 30 minutes to 4 hours, which can cause degumming of sericin. It is a safer process with minimum degradation to the silk fibre.

(b) **Treatment with alkali/acid** : Degumming with alkali or acid depends on pH, temperature, time of treatment, strength and nature of the reagents used. Alkalis used for degumming include sodium carbonate and bicarbonate, caustic soda, trisodium phosphate etc. Alkali treatment should be done for 30 minutes to 2 hours at a $pH > 8.5$. In a similar way acid treatment at $pH < 3.0$ can eliminate sericin.

(c) **Digestion with enzymes** : Enzymatic digestion by proteolytic enzymes like trypsin (pH 8.0), papain (pH 5.2), chymotrypsin, and pepsin at a temperature of 40-50°C were proved to be useful for degumming.

(B) **Weighting of silk** : Removal of sericin from the silk fibre results in weight loss of 22-25%. Thus the silk becomes further expensive. Therefore, such weight loss can be compensated by treating the silk with suitable compounds that adds weight or fix colour to the silk. Heavy metals like iron, tin, lead zinc, etc. and organic compounds like some tanning substances are used for weighting of silk. Weighting can be done by following steps. First the degummed silk is immersed in an acid solution of tin-tetrachloride. When the threads absorb the salt to the point of saturation, excess solution is then discarded by centrifugation. The material is then washed thoroughly with cold water to remove the acid. A treatment with hot solution of disodium phosphate then follows. It allows addition of PO_4 to the tin radical bound in the silk. This treatment is referred to as **pass**. If this pass is allowed for longer duration, the silk will absorb more tin-tetrachloride. Afterwards, the material is again treated with hot solution of Na silicate to increase the weight of silk still further and to fix the tin-salts uniformly. However, it should be done carefully as the temperature and concentration of silicate bath control the ultimate weighting of the silk in exact amount.

(8) **Re-reeling** : Sometimes silk after being reeled on device like charkha or cottage basin, may suffer from some defects like short length, broken threads, entanglements, hard gum spots, etc. These problems

may create wastages of thread during subsequent throwing and twisting. To overcome these defects silk is first reeled on small reels and later is transferred to standard sized hanks (Fig. 3.16). This process is called re-reeling.

(9) **Lacing and skeining** : Re-reeled silk hunk is then tied with a coloured silk thread by the two ends. The hunk is then divided into five parts and is laced with a cotton thread. This lacing prevents entangling of the silk in later operation.

The cleaned and laced hunks are then put on a skeining machine to be skeined. During this process, the hunk is twisted and folded upon itself to form a number of tight spirals or skeins (Fig. 3.17). The skeining prevents ruffling and further entanglement of filaments during packing.

Fig. 3.16. Rereeling by hank

Fig. 3.17. Spiral skeins of silk thread

The skeins are then made into books. In a book, eight skeins are placed in horizontal row and five in vertical row. Each book is then tied with separate cotton bands at three different places and wrapped with tissue paper. Books are kept in bales of twenty kg weight. Thus, silk threads are kept in store house as bales. Store house should be free from humidity and air.

BIOLOGY OF MULBERRY SILK MOTH

The mulberry silk moth, *Bombyx mori* belongs to the family *Bombycidae*. China is the native place of this moth, but now it is totally domesticated and

Fig. 3.18. Life cycle of *Bombyx mori*

(i) Egg; (ii) Newly hatched larvae; (iii) 2nd instar; (iv) 3rd instar; (v) 4th instar; (vi) 5th instar on a mulberry leaf; (vii) Cocoon; (viii) Pupa within cocoon; (ix) Adult moth (dorsal view); (x) Adult moth (ventral view); (xi) Silk thread

successfully reared in India, Japan, Korea, Italy, France and Russia. In India 92% of country's silk production comes from mulberry moth. The worm or larva of this moth solely feeds on leaves of mulberry plant, so named as **mulberry worm**. Its wild ancestor *B.mandarina* still occurs in some isolated areas of Himalaya's foothill.

Life cycle : *Bombyx mori* being a holometabolous insect, undergoes complete metamorphosis with four stages in its life cycle, viz., egg, larva, pupa and adult (Fig. 3.18). Sexual dimorphism is present in adult, pupal and larval stages of life cycle (Table 3.6, Fig. 3.19a, b, c). The duration of each stage varies according to inherited characters of the race and also with the available environmental conditions, quality of food provided, etc. Here, the life cycle of most prevailed race of *Bombyx mori* in India will be discussed with the following headings.

Adult moth: The adult *Bombyx* moths do not take any food and incapable to fly. Their life-span is only 3-6 days. The whole body along with wings remains covered with epidermal scales and the body is divisible into 3 parts, viz., head, thorax and abdomen.

Head: The small head bears mouth at ventral side and two prominent bipectinate and serrate antennae on two sides. Below the base of antennae, two

Table 3.6 : Differences in morphological characters between female and male mulberry moths during adult, pupal and larval stages

Female adult moth :	**Male adult moth :**
1. Body size larger (wing-span : 40-50 mm, length : 25-26 mm) than the male [Fig. 3.19a(i)].	1. Body size smaller (wing-span : 40-45 mm, length : 22-24 mm) than the female [Fig. 3.19a(ii)].
2. Head with antennae shorter than that of male.	2. Head with antennae longer than that of female.
3. Broad and long abdomen with seven visible segments.	3. Short and narrow abdomen with eight visible segments.
4. The caudal end bears median knob like hairy projection that secretes pheromone.	4. The caudal end bears a pair of hook-like harpens, which help in copulation.
5. Females are less active.	5. Males are more active.

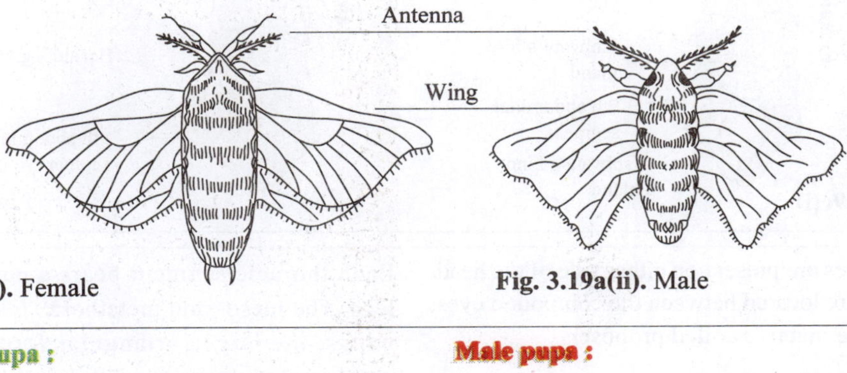

Fig. 3.19a(i). Female

Fig. 3.19a(ii). Male

Female Pupa :	**Male pupa :**
1. The body is broader than male pupa [Fig. 3.19b(i)].	1. The body is narrower than the female pupa [Fig. 3.19b(ii)].

Table 3.6 : Differences in morphological characters between female and male mulberry moths during adult, pupal and larval stages (contd. from previous page)

Female Pupa :	Male pupa :
2. On the ventral side of 8th abdominal segment, a vertical line is present.	2. No vertical line is present on 8th abdominal segment.
3. No spot is present on 9th abdominal segment [Fig. 3.19b(i)].	3. A small spot is present on 9th abdominal segment [Fig. 3.19b(ii)].

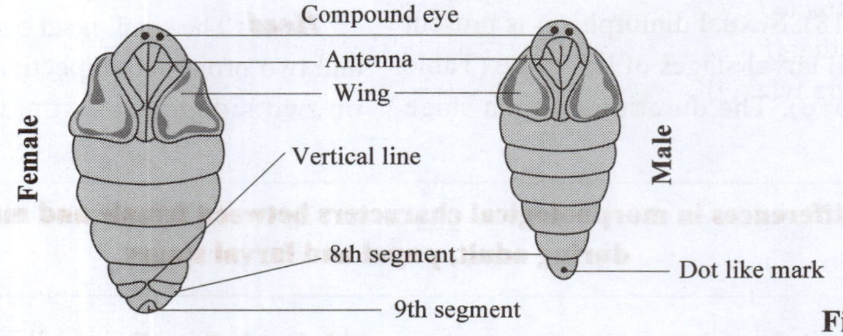

Fig. 3.19b(i) Fig. 3.19b(ii)

Female larva	Male larva
1. In the matured larva, one pair of Ishiwata's fore-gland and a pair of Ishiwata's hind glands are present at the ventral side of 8th and 9th abdominal segments respectively [Fig. 3.19c(i)].	1. The male larva possesses a single small median white gland — Herold's gland at ventral side, near the junction of 8th and 9th abdominal segments [Fig. 3.19c(ii)].
2. The female cocoon is heavier than the male cocoon.	2. The male coccon is lighter than the female cocoon.

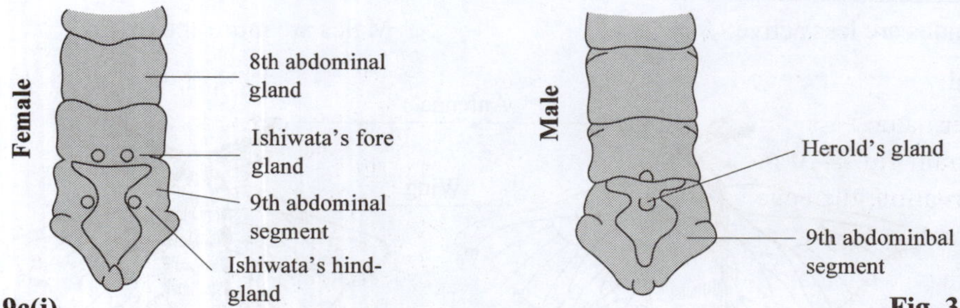

Fig. 3.19c(i) Fig. 3.19c(ii)

compound eyes are present on either side of the head. Mouthparts are located between the compound eyes, which include inactive coiled proboscis.

Thorax

Thorax has 3 segments — prothorax, mesothorax and metathorax of which mesothorax is the largest. Each thoracic segment bears a pair of 5-jointed legs. The meso- and metathorax bear two pairs of wings, the larger, triangular forewings remain attached to mesothorax while smaller hindwings are attached to the metathorax. Thorax also bears spiracles.

Abdomen : The male moth has eight abdominal segments with a pair of hooks 'harps' at the caudal end which help in copulation. The female has larger and fatter abdomen than that of the male which has seven segments. A knob like projection covered with sensory hairs at the caudal end of female is present. It can secrete pheromone.

Copulation : Following emergence from the cocoon, the females rest on the cocoon, spreading out her feeble wings and immediately begin to scent by exuding pheromone from her abdominal projection. The males being attracted by the pheromone find the female and start mating (Fig. 3.20). Thus they remain over for 12-24 hours when the females get inseminated by the male.

Fig. 3.20. Copulation between adult moths following emergence

Egg laying : After mating for 12-24 hours, the pair get separated and the female starts laying eggs while the male moths die. Most females lay 150-300 eggs over the course of 1-2 days, some can lay as many as 1000 also. According to the race, *B. mori* can lay two types of eggs–diapausing or hibernating and non-diapausing or non-hibernating eggs (for detail see Table 3.7).

Egg : The clustered eggs always remain covered with gelatinous secretion of the female moth. The colour of eggs depends on moth-race. In diapausing eggs, the colour changes from light yellow to deep brown when they enter diapause. In non-diapausing races, the colour of eggs does not change till the development is complete.

The eggs are ovoid, spherical or ellipsoid in shape and are flat on one side (dimple) through what eggs remain attached to the substratum. The centrolecithal eggs remain covered by chorion having two layers, outer exochorion and inner endochorion. Each egg has an opening, called micropyle at the anterior end.

Hatching : The hatching time of silk moth eggs varies with voltinism and environmental conditions. Diapausing eggs under natural conditions remain

Table 3.7 : Diapausing and non-diapausing eggs

Diapausing eggs/ Hibernating eggs	Non-diapausing eggs/Non-hibernating eggs
1. Usually temperate races lay diapausing eggs.	1. Usually tropical races lay non-diapausing eggs.
2. Under natural conditions, these eggs remain dormant for 4-10 months. Following the hibernation, the eggs can be hatched in the next spring.	2. Under natural conditions, these eggs can be hatched in 9-12 days after oviposition.
3. Artificial treatments with temperature or acid can break the dormancy if needed.	3. No such treatment is required.
4. Eggs are pigmented, actually the colour of eggs changes to deep brown when they enter into diapause.	4. Eggs are not pigmented and the colour of eggs remains same till development is complete and attains blue egg stage one day prior to hatching.
5. This egg can be preserved up to 6-10 months under different hibernation schedules.	5. This egg can not be preserved more than 20 days at 25°C.

dormant for months till next spring. If diapause is broken by acid treatment, the eggs have to be incubated at a temperature of 23-25°C for 11 to 14 days for completion of embryonic development and the larvae to hatch out. Non-diapausing eggs normally take 9-12 days to hatch into larvae.

Larvae : The newly hatched larvae are about 3 mm long, black in colour and is covered by bristles situated on tubercles in each body segment. Young larvae soon after hatching start feeding on mulberry leaves and grow. This period is called **first instar**. After about 3 days, larvae stop feeding and undergo ecdysis (shading off old cuticle). This period of ecdysis is called **moulting**. After moulting, the larvae start feeding again and grow rapidly in size. They repeat the moulting process usually for 4 times and thus the larvae or caterpillars get changed into second, third, fourth and fith instar stages during the whole larval period (Fig. 3.18). It takes about 21-25 days after hatching in case of multivoltine moths or 24-48 days in case of uni- or bivoltine moths. The fifth larval stage is the longest stage when the larvae show maximum consumption of mulberry leaves and high growth. Of the total ingestion during the entire larval period, nearly 85% is taken during 4th and 5th instars and a larva can take mulberry leaves weighing 30,000 times of its body weight to reach 5th instar stage from 1st instar.

The full grown 5th instar larvae are 7.5-10 cm in length. Its body is divisible into 3 parts, small *head* with special spinneret through which silk is exuded; *thorax* with 3 segments, each with a pair of clawed true legs which are not used for walking but for holding the leaves while feeding; and *abdomen* with 11 segments. The 9th, 10th and 11th abdominal segments are fused together to form the apparent 9th segment, anal plate and paired caudal prolegs. Each 3rd, 4th, 5th and 6th segment bears a pair of fleshy, unjointed muscular protuberances, known as pseudolegs or prolegs, each bearing curved hook on the outer edge of leg tip disc. The eighth abdominal segment has a projection on dorsal side, called *caudal horn* (Fig. 3.21a & b). The most characteristic feature of silkworm is their silk gland which is well-differentiated in 4th and 5th instar stages (discussed elsewhere). There are nine pairs of spiracles or stigmas placed laterally on either side of the larval body. Besides this, larva has a number of projections or nodules all over the body and markings on the skin which is race-specific.

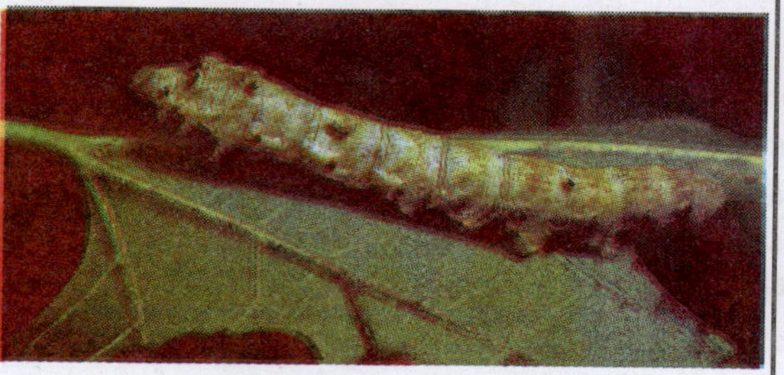

Fig. 3.21(a). A photograph of a ripe worm (*B. mori*)

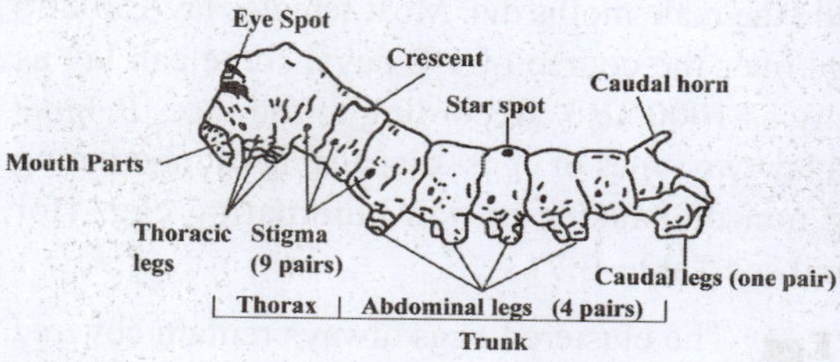

Fig. 3.21(b). Diagram of a 5th instar larva

The larvae have sexual dimorphism. In female larvae, the sexual marking appears as a pair of milky white spots known as *Ishiwatas' gland*, one each in the right and left sides of 8th and 9th abdominal segments. In male larvae, the reproductive gland appears as a small milky white protuberance at the centre of the vental side between 8th and 9th abdominal segments (Fig. 3.19c). This gland is known as *Herold's gland*.

Ripe larva : The full grown final 5th instar that has stopped feeding and ready to spin the cocoon is called *ripe worm* or *ripe larva*. This stage is attained after 6-7 days of feeding in 5th instar stage for bi- and multivoltine races in tropics and 7-9 days for uni- and bivoltine for subtropical races.

It is translucent as the gut discards all of its green content. The body of ripe worm shrinks in length and a visible constriction appears at 4th and 5th segments. The worm stops feeding and usually moves towards the edge of rearing tray.

Cocooning/spinning : The ripe worm after being shifted to mountage passes last excreta which are red in colour due to presence of tryptophan metabolites. After emptying its gut, it first secretes a tiny droplet of silk – the anchorage spot, which hardens and sticks onto the mountage. The larva then anchors itself to that spot and spins a loose hammock forming the framework of the cocoon. This loose, tangled mass of filament is called *blaze* or *floss* which is not reelable (Fig. 3.22a). After getting a good foothold in the hammock, the larva lays down the cocoon (Fig. 3.22b). During spinning, the larva moves its head in a typical pattern; so the filament is spun either in the shape of ~ or α. The former is found in the outer layer, whereas the latter type is common in middle and inner layers of cocoon. In this way, after forming a compact cocoon, the larva shrinks and covers itself with a thin layer of silk.Then it detaches itself from the shell, moults and transforms into pupa. The last layer of thin silk is called *pelade layer* of cocoon which is unreelable.

Fig. 3.22(a). Larva within floss

Fig. 3.22(b). Different stages of spinning by larva on Chandrika

During spinning, the secretions from two silk glands come out through the spinneret independently as fibres and called *brins*. The secreted sericin of both glands cements these two brins in a single thread, called *bave*. *Dragging* – a physical process is also responsible for the coagulation of liquid silk to the fibre. The process of spinning the cocoon takes about one to two days in case of multivoltine and two to three days in uni/bivoltine races.

Pupa : Once the cocoon formation is over, the larvae moult inside and transform into pupae (Fig. 3.18viii). This is known as pupation. Papal stage is also called as resting or inactive stage in the life cycle of silkmoth. Actually it is a transitional stage during which some changes take place in the body of mature larvae (ripe worms). Before transforming into proper pupa, the mature larvae pass through a pre-pupal stage during which dissolution **(histolysis)** of many larval organs like silkgland, moulting gland, abdominal appendages take place. It is followed by the formation of adult organs **(histogenesis)** during pupal stage when a pair of large compound eyes, a pair of large antennae, fore and hind wings, legs, etc. become prominent. The cuticle of newly formed pupa is soft and pale yellow in colour which gradually becomes darker.

In the female pupa, a fine longitudinal line joining the 8th and 9th abdominal segments is present on the ventral side. In male pupae such line is absent; instead a small opening is present in the 9th segment (Fig. 3.19b, Table 3.6). Usually female pupae are larger and heavier than the male pupae. The pupal stage continues about 10-12 days during which development of full grown moth is completed. The adult moth then secretes mild protease from their salivary gland to dissolve the cocoon membrane in the form of a small opening through which they emerge out.

In this way, *Bombyx mori* completes its life cycle through the above-mentioned stages and concomitant activities.

VOLTINISM

Voltinism can be defined as the number of broods or generations produced in a year in the life cycle of silk moth under natural conditions. Mulberry and other silk moths show different races based on voltinism (Table 3.8A). Characteristics of three different races of mulberry moth are shown in Table 3.8B.

(1) **Univoltine** : These silk moths hatch once a year, so have one generation per year. These moths lay only hibernating eggs. In these eggs, the embryo enters diapause after partial development and hatches out during the following spring. So they produce only one brood per year. These moths are found mainly in temperate region (Table 3.8B).

Table 3.8 A : Voltinisms shown by mulberry and other silk months

Common name	Scientific name	Voltinism (races)
Mulberry silk moth	*Bombyx mori*	Uni-, Bi-, and Multivoltine
Tasar silk moth (Tropical)	*Antheraea mylitta*	Uni-, Bi-, and Multivoltine
Tasar silk moth (Temperate)	*A. proyeli*	Bivoltine
Muga silk moth	*A. assamensis*	Multivoltine
Eri silk moth (Domesticated)	*Philosamia ricini*	Multivoltine
Eri silk moth (Wild)	*P. cynthia*	Uni- and Bivoltine

Table 3.8B : Comparison amongst univoltine, bivoltine and multivoltine races of mulberry moth

Univoltine	Bivoltine	Multivoltine
1. Produce only one generation per year.	1. Produce two generations per year.	1. Produce more than two generations per year
2. Moths lay diapausing eggs which remain dormant till the next spring.	2. Moths can lay both diapausing and non-diapausing eggs.	2. The moths lay only non-diapausing eggs.
3. Larvae of univoltine moth are very sensitive to temperature and other environmental conditions. So rearing of worms need strict hygienic environment.	3. Larvae are robust and can withstand environmental fluctuations. So rearing of larvae does not always need perfect conditions.	3. Larvae are resistant to high temperature and high humidity. So rearing can be done in tropical environment.
4. Larval period is long.	4. Larval period is long.	4. Larval duration is short.
5. Cocoon size is large and weight is good (2-2.4 gm).	5. Cocoon size is large and weight is good (2.25-2.39 gm).	5. Cocoon size is small and weight is low (0.9 gm).
6. Filament length is good up to 1700 m/cocoon. Threads are of good quality, lustrous and strong.	6. Filament length is good, i.e, up to 1600 m/cocoon; threads are strong and lustrous.	6. Filament length is poor, i.e, 350-400 m/cocoon; threads are of poor quality and less lustrous.
7. Silk yield is 20%.	7. Silk yield is 19%.	7. Silk yield is 6%.
8. Disease resistance power is weak.	8. Disease resistance power is good; resistant to pebrine.	8. Disease resistance power is good.
9. All European and some Chinese and Japanese races are univoltine.	9. Chinese and Japanese races are bivoltine.	9. Indian races are multivoltine.
10. Commercially very superior.	10. Commercially superior.	10. Commercially not superior.

(2) **Bivoltine** : These moths hatch twice a year, thus having two generations per year. The egg-laying behaviours of bivoltine moths are intermediate. They can lay non-hibernating eggs during first generation and hibernating eggs in the next generation from which the embryo hatches out in the following spring; thus they produce two generations in a year. These moths occur in temperate countries (Table 3.8B).

(3) **Multivoltine** : These moths can hatch several times in a year, so may have mutliple generations per year. They lay only non-hibernating eggs where the embryo develops without any diapause and hatches out in about 10 days.Thus, they can produce multiple generations in a year. These moths are found in tropical countries (Table 3.8B).

PREFERENCE OF BIVOLTINE OVER MULTIVOLTINE RACES

From various aspects, rearing of bivoltine moths proved to be more fruitful over the multivoltine in Indian sericulture. Yields from multivoltine races average 41-50 kg of cocoon /100 disease-free layings (DFL). But it goes up to 60-80 kg for bivoltine cocoon. For bivoltine reared by JICA (Japanese International Co-operative Agency), only 6-7 kg of cocoons are required to produce 1 kg of raw silk, compared to 8.5–9 kg of cocoons for 1 kg of raw silk by multivoltine cocoons. Therefore, rearing of bivoltine races are now being practised in India by adapting some new methods as follows :

(1) **Breaking diapause** : Bivoltines basically belong to temperate country like Japan. Their hibernating eggs pass through a period of diapause especially during winter for 6-8 months when availability of mulberry leave is a problem. The eggs, therefore, do not hatch and the youngs live on reserve energy drawn from glycogen in the yolk. Rearing bivoltine in tropical and sub-tropical countries (e.g, India), where mulberry is available round the year, requires a breaking of their natural diapause and inducing hatching through acid treatment of the eggs. Thus, bivoltine can be reared over a long period and even up to 5 hatchings in a year.

(2) **Adopting JICA programme** : To tropicalising bivoltine sericulture in India, Dr. Kawakami of JICA has introduced several modifications to traditional silkworm rearing. Multivoltines are adapted here for generations which are more and more hardy, whereas bivoltines are more susceptible to nuclear polyhedral virus (NPV) and hosts of other bacterial, protozoal and fungal pathogens. Rearing bivoltine in India thus needs some modifications like :

(i) **Shoot rearing** : Unlike traditional practice of feeding with mulberry leaves, the late stage larvae in JICA approach are subjected to 'shoot rearing', i.e., leaves are not separated from the mulberry shoot. This not only save labour and time, but also inhibits secondary contamination from pathogens by keeping the leaves fresh for a longer time.

(ii) **JICA–modified approach** : Here, the farmers directly rear the 3rd stage-larvae instead of eggs. The first two out of five instars are highly susceptible to diseases and demand high level of hygiene which the farmers cannot easily ensure. Therefore, instead of supplying eggs to farmer, 3rd stage larvae may be supplied to them. The young larvae are maintained in special Chawki centres with controlled temperature and humidity and then handed over to farmers.

CONTROL OF VOLTINISM

Voltinism of silk moths are controlled by genetic, environmental and hormonal factors :

(i) **Environmental factors** —Temperature and light are most important for voltinism. Incubating the eggs from bivoltine race at temperature of 15-18°C with short photoperiod (12 hours) can make the moths to appear and depose non-hibernating eggs (2 generations/year). If eggs are incubated at high temperature of 25-28°C with long photoperiod of 12-14 hours, moths appear and depose hibernating eggs (one generation / year).

(ii) **Genetic factors** : Voltinism is attributed to a number of 3 multiple sex alleles (Hs, Hs^2, hs) modified by a number of autosomal genes (H1, h1; H2, h2; H3, h3) (A-Matei, 2005).

(iii) **Hormonal factors** : Diapause hormone (DH) secreted from suboesophageal ganglia is known to control diapause or otherwise voltinism in silk moths.

Examples of Some Indian Commercial Races of Mulberry Silk Moth

In modern sericulture practice in India, no indigenous races of *Bombyx mori* are being cultivated, because of their poor productivity and low quality of silks produced by them. Exotic races are high yielding but not suitable to Indian climate. Hence, extensive hybridisation followed by selection has evolved numerable hybrids of both bivoltine and multivoltine races with hardy nature as well as high yielding characteristics. Some of such popular commercial races are as follows :

(a) Kalimpong A (KA) : Bivoltine race with oval cocoons.

(b) C. Nichi (NB4D2) : Bivoltine or multivoltine races, producing groundnut, shaped cocoons, with a constriction in the middle.

(c) Nandi : A hybrid between KA x NB4D2. The constriction in the cocoon is very shallow.

(d) Pure Mysore (PM) : Multivoltine. One side of the cocoons is broad and the other is pointed (i.e., egg-shaped)

(e) Tamil Nadu white (TNW) : Multivoltine producing white cocoons with pointed ends.

SILK MOTHS OTHER THAN *Bombyx mori*

Besides *Bombyx mori* that produces pure silk, other silk moths are collectively known as non-mulberry silk moths because of their different host plants. Their worms / larvae are reared for obtaining other kinds of silk. Except *Bombyx mori* that belongs to the family *Bombycidae*, non-mulberry silkmoths belong to the family *Saturnidae* under the order *Lepidoptera*. India, has the distinction of being the only country in the world, producing all the five commercially exploited silk varieties, viz., mulberry or pure silk, tropical *tasar*, temperate/oak *tasar*, *eri* silk and *muga* – golden silk. The non-mulberry silks (*tasar*, *muga* and *eri*) are now being popularised as *vanya* silk. Brief discription about these moths are given below :

(i) Arjun Tree

(ii) Asan Tree

Fig. 3.23. Host plants of tasar worm — (i) Arjun Tree and (ii) Asan Tree

Tasar silkmoth : Tasar silk is generated by the worm of the silk moth *Antheraea mylitta.* It accounts for 2.0% of total non-mulberry silk production in India.

Host Plants : The larvae of *Antheraea mylitta* are polyphagus and can take leaves of several host plants like Arjun [Fig. 3.23(i)], Asan [Fig. 3.23(ii)], Ber, Sal, Jam, Sidha, etc.

Distribution : This silk moth is available in China, Srilanka in addition to India. In India, it is available in the states of Jharkhand, Chattisgarh, Orissa, A. P. and West Bengal.

Life History : The *tasar* silkworms are grown only in the wild. So they are cultivated in places where their host plants are available.

Fig. 3.24(a). Adult tasar moth (male

LIFE CYCLE [Fig. 3.24(b)]

Adult : By piercing one end of the cocoon the adult moths come out. Adults are very large, with colourful wings spanning about 15 cm. Female has yellowish grey wings and males possess yellowish red wings with prominant eye spots on each wing (Fig. 3.24a) . The males mate the females just after emergence.

Eggs : After copulation, the female starts laying the eggs. A single female can lay 100-150 eggs.

Larvae : The eggs hatch within 9-10 days during summer and 15-20 days in winter. Following hatching the larvae start taking food leaves from the host plants. The worms are generally green in colour with hairs on cuticle [Fig. 3.24b(ii)]. It moults 4 times during the whole larval period that continues for 30-55 days during summer and 50-60 days during winter. The mature larvae spin cocoons which are hard with different colours according to the types of plant leaves they fed. Some are light green, others are yellow, and some are grey and others almost white (Fig. 3.24c). Unlike mulberry cocoons, the tasar cocoon has a stalk (*peduncle*) which helps in fixing the cocoon with the twig of host plant [Fig. 3.24b(v)]. The tasar cocoon attains a size like that of hen's egg.

Pupa : The worms grow into pupae within the cocoon. **Cultivation of tasar worm** : The tasar rearer collects the female moth from the wild and

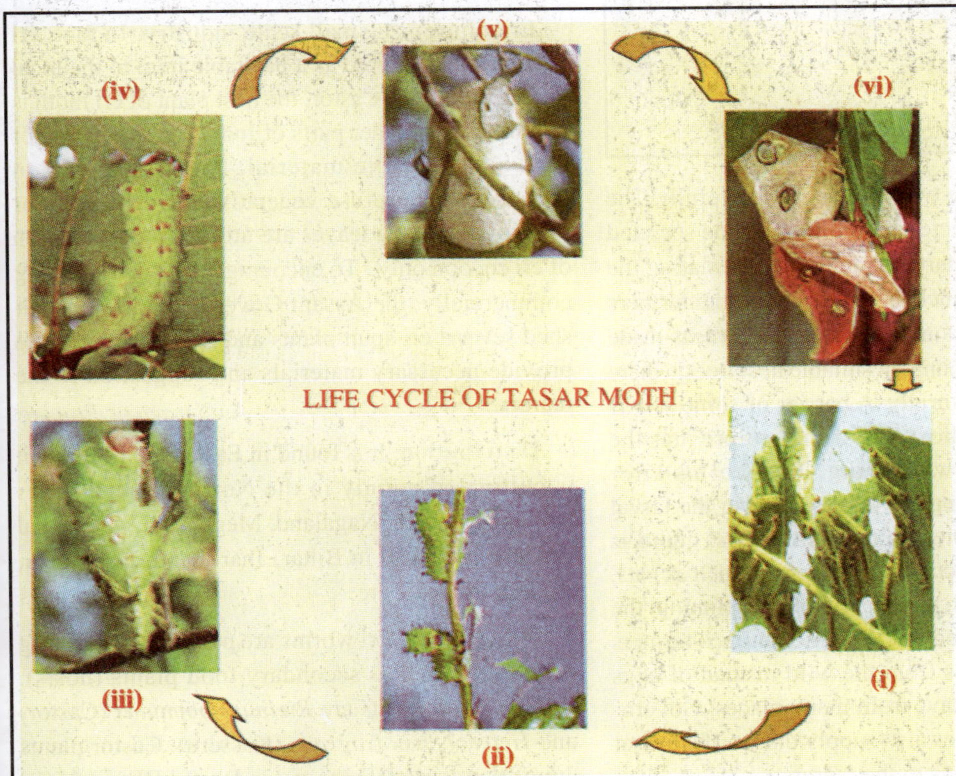

Fig. 3.24(b). Life cycle of tasar moth : (i) newly hatched larva; (ii) young larva; (iii) moulting larva; (iv) ripe larva; (v) cocoon; (vi) copulating adults

Fig. 3.24(c). Collected tasar cocoons from wild

Fig. 3.24(d). Colourful tasar cocoons

keeps it tied with the twig of host plant carefully. The males are then attracted by the pheromone secreted by the female and start copulation. After mating the female lays eggs which are collected by the farmers and are gummed on long strips of paper- or leaf-made cup which are then hung on suitable host trees, where they hatch quite naturally. A barrier of some sort is coiled around the host tree trunks to prevent the caterpillars from wandering. The worms after hatching start feeding the leaves. During the larval stage, which occupies up to eight weeks, guard is kept against the insect's natural enemies such as bats, birds and beetles. On completion of the spinning, the cocoons are harvested like fruit. Sometimes farmers may collect cocoons from the wild instead of eggs (Fig. 3.24c). Nowadays from many places, Government breeding centres also supply disease-free tasar eggs of desired breeds to the rearers.

Tasar silk : By two methods, reeling and spinning, yarn can be produced from the cocoons. Tasar filaments show the greatest length among the non-mulberry silks, 700 m. The spun yarn is generally coarse with a denier of 270-280, while the reeled yarn is fine and thin with a denier value of 80-100. Tasar silk has natural shades of pale gold, pinkish honey, creamy copperish, etc. depending on the colour of cocoons (Fig. 3.24d). It is less lustrous but coarser than mulberry silk. The fabric is light, airy, and stiff, and has its own feel and appeal.

Oak Tasar silk moth : A finer variety of tasar silk is generated by the silkworm *Antheraea proyeli* and *A. pernyi*. China is the major producer of this tasars but in India it is available in Manipur, Assam, Meghalaya, Himachal Pradesh, Jammu & Kashmir. These moths feed on leaves of oak, are found in abundance in sub-Himalayan belt. The oak tasar is finer than the common tasar silk.

Eri silk moth : The silk worms of the silk moths *Philosamia ricini* and *Attacus ricini* generate a different kind of white silk, called eri silk.

Ericulture is a small scale industry. In fact, eri cocoon is considered as a palatable item to the local people of Assam. Castor, the host plant of eri moth is mainly used in other parts of India for production of seed as a valuable material; but in N.E. region because of its profuse vegetative growth and poor yield of seeds, the leaves are utilised for production of eri cocoon only. To encourage the eri silk rearing commercially, the Assam Government has established several eri spun plants and research centres to provide necessary materials and suggestion to the rearers.

Distribution: It is found in East Asia. In India, it is cultivated mainly in the North-Eastern states including Assam, Nagaland, Meghalaya, Arunachal Pradesh, and also in Bihar, Jharkhand, Karnataka, Andhra Pradesh, etc.

Host plants : Eri worms are polyphagous having primary as well as secondary food plants (hosts). Primary food plants are *Ricinus communis* (Castor) and *Heteropenax fragrans* (Kasseru). Castor plants are of two varieties; the green leaved [Fig. 3.25(i)]

(i) Green castor plant

(ii) Violet castor plant

Fig. 3.25. Host plants of eri-worms : (i) Green castor; (ii) Violet castor

and violet leaved [Fig. 3.25(ii)]. Both are equally suitable for feeding the eri silkworm.

The secondary food plants are *Manihot utilissima* (Tapioca), *Evodia flaxinifola* (Payam), *Plumeria acutifolia* (Plum) and *Carica papaya* (Papaya).

Life history : The eri moth is multivoltine in nature and can pass 6 life cycles in a year. These worms are totally domesticated and are reared indoors.

Life cycle : Its life cycle consists of 4 stages as follows.

Adult : Adult moths are large with wings spanning about 10 cm. The wings are greyish brown in colour with prominent eye spot [Fig. 3.26(a)]. When the adult moth emerges from a cocoon it makes a hole to get out. Like other silk moths, eri moths also start mating activities following their emergence from pupal stage.

Fig. 3.26 (a). Adult moth

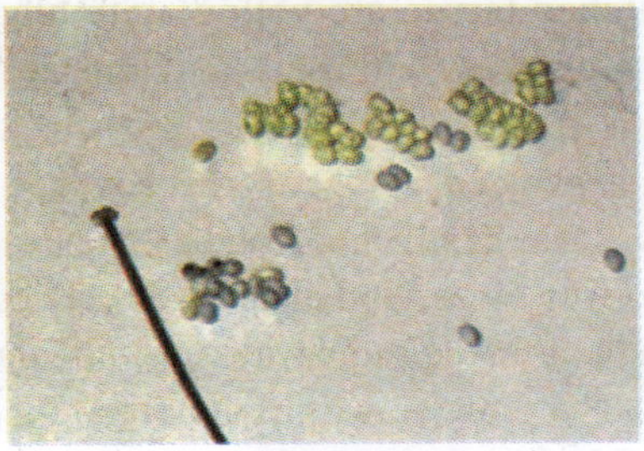

Fig. 3.26 (b). Egg

Fig. 3.26 (c). 1st instar

Fig. 3.26 (d). 2nd instar

Fig. 3.26. Life cycle of eri moth

Fig. 3.26 (e). 3rd instar

Fig. 3.26 (f). 4th instar

Fig. 3.26. Life cycle of eri moth

Fig. 3.26 (g). 5th instar

Fig. 3.26 (h). Cocoon

Egg : Following mating, the female starts egg laying that may continue up to 3 days. Each female can lay about 350-500 eggs [Fig. 3.26(b)]. Hatching depends on prevailing environmental conditions and may occur from 8-20 days.

Larvae : The hair like newly hatched larvae are yellow in colour [Fig. 3.26(c)]. Initially they feed on soft castor leaves but at later stages, worms can feed on mature leaves. After moulting for 4 times by 30-32 days [Figs. 3.26(a-g)] they spin the cocoon among the leaves. Caterpillars in their final stages start spinning the cocoons. The spinning is completed in 2-3 days. The cocoons are open mouthed, tapering at one end and flat rounded at the open end. Eri cocoons are stalkless, flossy, white or brick red in colour [Fig. 3.26(h)] 5 cm long in case of female and 4.6 cm long in male.

Pupa : Pupal stages last for 15-37 days during when larvae change to pupae.

Rearing of eri silkworm

Disease-free seed cocoons are obtained from grainages or agencies and reared fully indoors. Healthy cocoons are spread on bamboo trays in cool dark room. On hatching, active males are separated from passive females and are then allowed to mate in quiet dark room. Fertilised females are then tied to 'kharikas' (see pg. 89 for details) by passing a thread around the thoracic joint of the right wings. Kharikas are then suspended from a string. Eggs are laid within 25 hours on Kharika and are normally selected for rearing. The eggs are white, oval and covered with a gummy substance, which makes them adhere to one another. The eggs are disinfected with 2% formalin solution and then washed thoroughly with water. Eggs are incubated at 26°C, the colour changes to blue on the day prior to hatching. Hatching takes place in the morning after ten days of incubation. The newly hatched larvae are yellow with black segments. These larvae are brushed to rearing trays over which few tender food leaves are spread, and crowding is avoided. As the worms advance in age, older leaves can be given as food at 2-hour interval for four to five times. Bed cleaning is carried out at regular interval in the same way as for the mulberry silkworm. The growing worms undergo four moults and have five instar stages. Total larval period lasts for 30-35 days. The 5th instar, mature larvae stop feeding and start searching for a proper place to spin the cocoon. At this stage, the mature worms are picked up and transferred to mountages (*Chandrikas*). In wild, cocoons are spun between folds of leaves.

Post cocoon processing : Stifling is done by spreading and exposing the cocoons to the sun for 1-2 days. For degumming, cocoons are tied in a cloth sac and dipped in boiling soda solution. After sufficient boiling, the cocoons are taken out, washed with water several times to remove soda, squeezed to remove water and then spread on mats to dry. Being open mouthed, the thread of the cocoons is discontinuous. So, the thread can only be spun and not reeled. Traditionally, spinning is done in wet condition on takli and in semidried condition on a *charkha*. Improved spinning machines like N.R. Das type *charkha* and Chaudhury type *charkha* are also available for spinning of silk from eri cocoons.

Eri Silk : The silk produced by eri moths is referred to as eri and endi or errandi silk by local people. This silk is collected from pierced cocoons, so it is spun silk. The fibres in the cocoon are like a tiny bale of cotton, all wound together and tangled. So when eri silk is spun, it forms a lower grade of silk of unusual quality. Depending on how it is spun and woven it can give a very woolly result and as with poorly made wools, eri can feel like cheap acrylic. But it can also give a tight, strong fibre-like linen. This is what makes eri textiles amazing — they can have the drape and weight of linen but are warm and insulating like cotton or wool. Eri silk is less glossy and wrinkleless.

Muga silk moths : Muga is produced from cocoons of *Antheraea assamensis* which is available only in Assam. The most expensive of silks, *muga*, is intrinsically woven into the cultural traditions of the people of Assam. The vibrant Sualkuchi sarees and *mekhla-chaddars* are the traditional items made from *muga* silk. The speciality of this silk is its golden bright colour. "Muga" is derived from the Assamese word 'muga' meaning yellowish. It is superior in

(i)

(ii)

Fig. 3.27. Host plants of muga worm — (i) Som tree and (ii) Soalu tree

every respect from the normal white silk available all over the world.

Distribution : The native place of this moth is Assam. Its production was confined to Assam, border areas of neighboring north-eastern states and Cooch Bihar in West Bengal. Now it is reared in Nagaland, Meghalaya and Andhra Pradesh also.

Host plants : The muga worm feeds on aromatic leaves of Som [Fig. 3.27(i)] and Soalu [Fig. 3.27(ii)]. It can also be reared on host plants similar to that of tasar worms.

Life history : The moth is multivoltine, the entire life cycle lasts for about 50 days in summer and 120 days in winter. This moth is semidomesticated and can be raised outdoor. Muga moth ('Muga Polu' in Assamese) also has the same life cycle as other silkworms, i.e., egg, larva, pupa and adult.

Life cycle [Fig. 3.28(a-j)] — **Adult** : The wings and body of the male moth are copper brown to dark brown, while those of female is yellowish to brown. Besides colouration, the male moth can be distinguished from the female by its slightly smaller size, slender abdomen, bushy antennae and sharply curved forewing tips. *Antheraea assamensis* can be identified by the orange eye-spots, the pale leading edge of the forewing, and a black spot in the rear wing eye-spot located towards the body [Fig. 3.28(a)]. Typically, the males find the females upon emer-gence and copulate immediately.

Egg : The female moth's eggs (popularly known as seeds) are laid on the Som and Soalu leaves.

Larvae : Eggs are hatched into larvae of about 2 mm long. They grow rapidly, eat voraciously and end up about 30 mm long after 4–5 weeks. During this

Fig. 3.28. Life cycle of muga moth : (a) Adult muga moth; (b) egg; (c) 1st instar; (d) 2nd instar; (e) 3rd instar; (f) 4th instar; (g) 5th instar; (h) pupating larva; (i) cocoon and (j) pupa

time, they moult four times. At the end, they search suitable place for cocooning.

Pupa : Within the cocoon the larva pupates.

Rearing of muga moth

The seed cocoons intended for preparation of eggs are obtained from commercial rearers or from Government grainages. These are then laid in a single layer in trays to facilitate the emergence of moths. Emergence starts from dusk and continues till morning. The emerging adults are allowed to mate and in the coupled state, the pair is tied with a piece of cotton thread to 1.5-2 feet long stick made of dried straw which is known as *Kharika*. After overnight mating, the couples separate in the morning and if they do not decouple naturally they are made to do so by heat of fire lighted some distance away. The female moth lays about 150-250 eggs on *Kharika* (Fig. 3.29). During the rearing period, farmers restrict entry of people to the rearing plot as they believe that the evil sight of outsider may cause Mukhloga disease (Flacherie, a bacterial disease of muga). During summer, the worms hatch out in the morning in about 8 days. The Kharikas with the hatched worms are hanged on the host plants. The larvae immediately crawl and start feeding on leaves. When the leaves are exhausted, the larvae crawl down and are collected on triangular bamboo sieves with long handles **(Chaloni)** Fig. 3.30**,** which are again hanged on a fresh tree. A band of straw with a little sand or ash is tied around the tree trunk 1-1.2 m above the ground to prevent the worms from crawling down the ground. The larvae feed voraciously, pass through 4 moults and reach the mature stage. In the final stage, larvae become greenish blue with prominent tubercles. Larval period lasts for 30-35 days. The ripeworms come down the trees searching for a suitable place for spinning of cocoon. They are then collected by rearers and put in baskets containing mango twigs and leaves, which are set as cocoonages (Jali) for the spinning of cocoons. The jalies are then hung (Fig. 3.32) and left undisturbed in separate rooms or at some shady place till cocoons are formed. Dry leaves of singari, bhomloti, azar, etc. are utilised for preparation of jali by the farmers. It is believed that cocooning in singari leaves produces shining and compact cocoons. Spinning takes about

Fig. 3.29. Female muga moth laying eggs on Kharika

Fig. 3.30. Muga larvae are collected on chaloni

Fig. 3.31. Muga larvae crawling down the tree trunk

Fig. 3.32. Jali cocoonage for muga larvae

2-3 days in summer and 7 days in winter. Muga cocoon is golden or light brown in colour, 4-6 cm long and 2-3 cm broad with a rudimentary peduncle without ring [Fig. 3.28(j)].

Post-Cocoon Processing

The muga cocoon is compact and leathery in structure. The length of continuous silk filament ranges from 350-450 metres with 4 to 5 breaks. Immediately after removal from the mountages, cocoons are spread on bamboo mats in the sun during hot hours of the day that partially kills the chrysalis. These are then subjected to heating in oven that kills the chrysalis completely, and thus the cocoons are stifled. For degumming of cocoon, local people use alkali (*khar*) made by burning banana peel/pseudostem or paddy straw/husk. Cocoons are boiled in such mild alkaline solutions for about 15-20 minutes.

Fig. 3.33. Muga reeling by Bhir

Almost entire reeling of muga is done with a primitive machine, called Bhir or Bhawri, operated by two persons (Fig. 3.33). The cocoons are kept in basin with warm water. Reeling requires two persons: one person releases the filaments from cocoons while the other twists the filament into one thread and wind it on Bhir. Two persons can reel around 100 gm raw silk per day on an average. Only 40-45% silk filament is reeled and rest is rejected as waste.

Muga silk : The silk produced is called 'muga silk'. It is golden yellow is colour, more strong, durable and lustrous. It needs no bleaching or dying and is stain resistant. *Muga* farmers also produce ghisha or spun muga yarn known as *jotha-muga* from pierced/flimsy cocoons.

Ahimsa silk/piece silk : This silk is woven from cocoons after the moth has flown away from the cocoon, so that the worm escapes the dip in the fatal boiling hot pail of water. Traditionally eri silk is the best example of ahimsa silk. Mr. Kusuma Rajaiah of A. P. discovered this silk .Unlike the traditional method, where pupae are killed before reeling the yarn from the cocoon, in the production of non-violent silk, the adult moth is allowed to emerge alive from the cocoon. Then the silk thread is spun from open ended eri or *tasar* cocoons and from pierced *muga* cocoons that have been used for breeding or those found in jungles (Fig. 3.34). However, in

Fig. 3.34. Peace silk is made from cocoons from which the moth has freely flown away

Chattisgarh, crushed seeds of *Mahua* are used to force the worm to come out of the cocoons. The silk fibres are then extracted from those abandoned cocoons. These silk is hand-spun and hand-woven. So it takes much time and costly than the traditional silk. **Non-violent or piece or ahimsa silk is made essentially from non-mulberry silk cocoons.**

STRUCTURE OF SILK GLAND OF SILKWORMS

The silk of silkworms is secreted by a pair of modified labial gland, known as **silk glands.** The glands are well-developed in IV and V instars.

Location : The glands lie ventral to the alimentary canal. In full grown larvae, these occupy most of the body cavity.

Structure : The glands are tubular in shape with different diameters in different regions. Each gland has 3 distinct regions [Fig. 3.35(a&b)]—(1) **Posterior region** : Blunt, highly folded tubular posterior regions of both glands remain attached to tracheal bushes. This part secretes fibroin as fibrinogen which converted to fibroin upon extrusion. (2) **Middle region** : Most prominent and widest part of silk gland. It remains folded in a W-shaped structure and thus has 3 limbs — posterior, middle and anterior

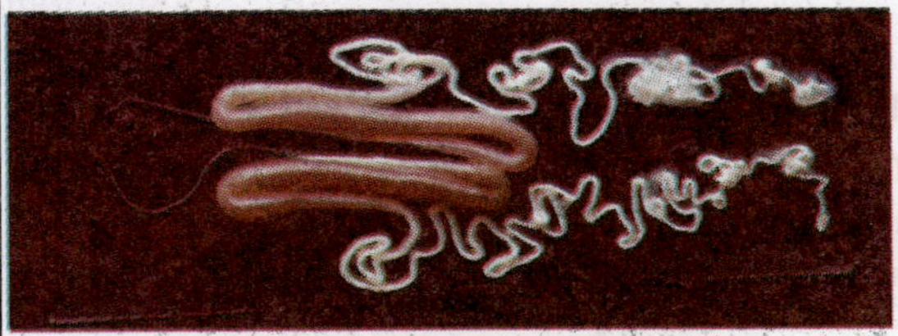

Fig. 3.35(a). Photograph of silk gland

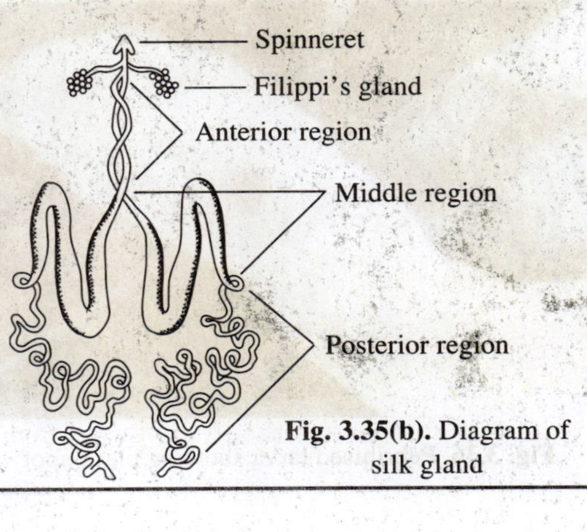

Fig. 3.35(b). Diagram of silk gland

limbs. The posterior arm secretes sericin-I. It gets surrounded by serecin-II secreted from the middle limb. This sericin again gets surrounded by sericin-III secreted from the anterior limb. The middle region of silk gland also acts as the reservoir of fibroin where the later gets mature during the storage period. (3) **Anterior region** : The thin anterior region of silk gland has no secretory role and only transports the assembled silk to the spinneret.

Spinneret : It is a projection of the median part of the labium, which draws the silk out in the form of fine filament. The secreted silk comes out as a thread or filament as it passes through silk press which resembles a typical salivary pump. The two filaments coming out of two sides are called *brins.* The sericin (gum) layer of the two brins then bind together into a single filament or *bave.*

Histology : Histologically the entire gland has 3 layers : (1) **The outer tunica propia** with uniform thickness; (2) **The middle glandular layer** with gland cells which increase in size during later instar stages of larval development and (3) **The inner tunica intima** : It has varying thickness. In the anterior region of the gland, this layer is very thick and is shed at each moult. In other regions of silk gland, it is thin and not shed at each moult.

Filippi's gland or Lyonnet's gland : In the head region of the larvae, a pair of glands is situated which open into the anterior part of silk gland near its opening into the spinneret. It is thought that these glands contribute some waxy materials to the silk thread or lubricate the passage of silk while coming out.

Disease Management : Common Diseases of Silk moth/ Silkworm

In South East Asian countries, the whole sericulture industry, especially silkworm rearing system faces severe problems because of diseases that attack the valuable silkworms. Due to prolonged domestication, silkworms developed neither any immunity nor any morphological or behavioural adaptations against the organisms causing various deadly diseases. In spite of continuous efforts by different researchers for healthy and hygienic practices of silkworm rearing, till date only a partial success has been obtained to evolve disease-resistant varieties of silk moth.

Silk moth larvae or silkworms are the most vulnerable stage for diseases. They are, in general, infected by the following groups of pathogens : (1) Protozoa, (2) Bacteria, (3) Viruses and (4) Fungi.

(1) **Protozoan diseases** : **Pebrine** : The most virulent protozoan disease of silkworm is *Pebrine*, so named as the infected worms show some pepper-like black spots.

Causative organisms : Pebrine is caused by the protozoan *Nosema bombycis* that has several strains of which most virulents are NIK-2r, NIK-3h and NIK-4m in India. The protozoa passes its life cycle through 2 stages.The infective *spore stage* and *vegetative stage*.

Modes of transmission : The disease is transmitted to silkworm in three different ways :

(a) **Oral** : Spores liberated through the faeces of infected worms or through dead larvae, contaminate the mulberry leaves given in rearing bed. Such leaves when taken by other larvae, they get infected. Again, eggs may be contaminated by such faecal matter or dead tissues containing the spores during the time of oviposition or after oviposition. The newly hatched larvae who take part of chorion of such contaminated eggs may get infection.

(b) **Contact** : Larvae may get infection through their skin from the rearing bed, contaminated with *Nosema* spore containing faecal matter or dead tissues.

(c) **Transovarial** : If infection occurs in 5th instar stage, the adult moth normally emerges through pupal stage. However, the spores of *Nosema* ingested then sporulate within the oocyte and pass on to the eggs. Thus, the egg itself acquires infection from the layers (mother) and hatches into primary infected larvae. However, these infected eggs hatch in very low percentage and the hatched out larvae usually die before reaching 3rd instar. These larvae act as the source of *secondary infection* for other larvae through their faeces or dead tissues. These larvae, if spin, form weak cocoons and exhibit pebrine symptoms and may die during or after spinning. Thus, they serve as source of *tertiary infection* to other 5th instar larvae which may spin and may even live up to the adult stage.

Symptoms : **Egg** : Infected eggs do not show firm attachment with the egg card due to improper deposition of glue. These eggs are pale yellow in colour and may fail to hatch.

Larvae : Primary infected larvae usually die before exhibiting any typical symptoms. Larvae having secondary or tertiary infections may show different symptoms like loss of appetite, *'unequals'* appearance due to slow and irregular growth; *clean worm* symptom that arises due to irregular and incomplete moulting, black pepper-like spots on the body (Fig. 3.36), irregular brown patches due to dead hypodermal cells, spitting and wasting silk instead of spinning the cocoon, etc. Infected larvae also pass soft faeces and often die after spinning without pupating.

Pupa : Live infected pupa if present inside the cocoon, may show black, swollen body with black spots on sides of abdomen.

Adult moth : Pebrine infected adult moths show black spot on abdomen, deformed antennae, unstretched, and discoloured wings. Females usually lay eggs in irregular, loose heaps. Scales fall off easily from the body of pebrine infected moths.

Fig. 3.36. Pebrinised larva showing black spot

Detection of pebrine : Pebrine at any stage of silkworm life cycle can be detected by observing the above mentioned symptoms. Besides, microscopic observation of homogenate / fluid from dead tissues, faecal pellets can indicate the presence of infective spores of *Nosema*. Again, by advanced immuno-enzymatic methods, presence of pebrine spores can be detected quickly.

Control measures : (i) In grainages only disease-free layings (DFLs) will be allowed for rearing. (ii) Diseased larvae if detected in rearing tray/bed should immediately be removed and burnt. All rearing appliances including the rearing room then should be disinfected with 4-5% formalin solution or by spreading bleaching powder. (iii) Adult moths showing symptoms of pebrine should not be undertaken for any sericultural processes. (iv) For disinfection of rearing accessories, other agents like Benomyl, Bavistin, Bengard, etc. can be used instead of routine formalin/bleaching powder.

(v) Rearing of disease resistant races of *Bombyx*, e.g., Nistari can be considered as an important preventive measure against pebrine.

(2) **Bacterial diseases** : In sericulture, silkworm can be attacked by various bacteria resulting in different diseases. Most common bacterial diseases are :

(A) **Bacterial Flacherie :** The diseases having symptoms of diarrhoea and vomiting are collectively called **Flacherie.** Again, symptoms of flacherie may not be same in all types of flacherie. The disease is also known as shrinking disease /clear head / faecal chain disease.

Causative organisms : Bacterial flacherie is caused by *Bacillus bombycis*, *B. sotto*, *Streptococcus bombycis*, *Pseudomonas aeroginosa*, etc. In general, the pathogenic bacteria may attack their alimentary canal and haemolymph.

Modes of transmission

(a) **Oral** : Larvae when take contaminated leaves can get infection directly. The entered bacteria attack and grow in their alimentary canal. Due to bacterial growth and consequent accumulation of lactate, acetate, etc., the permeability of gut is lowered due to lowering of pH. Then the bacteria become able to invade the haemocoel and spread the infection in whole body of larvae. The infected larvae pass bacterial spore containing stool which acts as new source of infection.

(b) **Bad rearing condition** : Temperature and humidity if not maintained properly in the rearing room, may enhance the growth and spreading of flacherie bacteria.

Symptoms : The diseased silkworms show following general symptoms — (i) Lack of appetite, (ii) Stunted growth, (iii) Sluggish movements, (iv) Soft and loose skin (Fig. 3.37a), (v) Shrinkage of body after moulting and these larvae usually refuse to take any food, (vi) the body become dull coloured to black-brown, (vii) the larvae show signs of pain, convulsions and die, (viii) the carcass turns black and emits foul smell (Fig. 3.37b), (ix) Diarrhoea and vomiting.

Fig. 3.37(a). Flacherie attacked larvae showing soft and loose skin

Fig. 3.37(b). Dead larva with black carcass after flacherie

Detection of bacterial flacherie : Above mentioned symptoms help to detect about flacherie - infected larvae. Besides, histopathological studies can reveal affected cylindrical cells in gut wall in case of *streptococcus* infection and affected goblet cell in case of *Diplococcus* infection.

Control measures : (i) Optimum environmental conditions with required temperature and humidity should be maintained in the rearing place. (ii) Healthy and quality leaves should be supplied to rearing larvae. (iii) Unhealthy larvae should be discarded from rearing trays. (iv) Regular disinfection of rearing room and appliances should be done with 2% formalin.

(B) **Septicemia** : It is resulted following infection of haemolymph by bacteria.

Causative organism : Septicemia mainly caused by *Bacillus, Streptococcus* and *Staphylococcus.*

Modes of transmission : (i) **Contact** through wounds on skin, and (ii) **Oral route** or intake of contaminated food can also transmit the infection.

Symptoms : (i) Softening of body, (ii) Change of body colour to brown, (iii) Liquid excreta, (iv) Loss of clasping power of prolegs and (iv) Death of larvae.

Control measures : (i) Good sanitation and healthy environment should be provided. (ii) Infected larvae should be removed/discarded. (iii) Larvae should be handled carefully to reduce skin wounds.

(C) **Sotto disease** : This disease is caused by the toxins released by some bioinsecticidal bacteria.

Causative organism : *Bacillus thuringiensis*, bacteria widely used as a bioinsecticide, release some endotoxin *thuricide*, that remains stored in a parasporal crystalline structure during the time of sporulation of the bacteria. This toxin, if taken by the larvae orally causes two types of responses : in type I response, the larvae of *B. mori* stop feeding, haemolymph becomes alkaline and paralysis occurs within 60-80 minutes of intake. In type II response, other lepidopteran larvae only show rapid inhibition of feeding.

Mode of transmission : Oral intake of mulberry leaves contaminated with toxin-crystals can cause toxicity to the larvae. But if infected by injection to the larvae, the same bacterial toxins show no effects to the silk worm larvae.

Symptoms : (i) Lack of appetite, (ii) Sluggish movement, (iii) Shrinkage of skin, (iv) Diarrhoea, (v) Loss of clasping power of prolegs, (vi) Death. After death the body becomes black in colour, internal organs become liquid and emit foul odour.

Control measures : (i) Good sanitation, specially cleaning the room with hot water to inactivate the toxin. (ii) Removal of dead / infected larvae from rearing tray, (iii) Bacterial spores if detected can be destroyed by exposing to 2% formaldehyde for 3 hours or to 100°C for 5 minutes.

(D) **Court disease** : This bacterial disease is also known as *Rangi* in India because of the colours shown by dead larvae.

Causative organism : In this disease, both larvae and pupae get infected by the bacteria *Serratia marcescans* and also by *S. piscavotar* following the primary infection with *Streptococcus faecalis.*

Mode of transmission : (i) Through skin wounds and (ii) oral route.

Symptoms : (i) Affected larvae are flaccid (ii) Death (iii) Following death, the body colour of larvae changes to crimson red.

Control measures : (i) Hygienic and clean rearing, (ii) Isolation and killing of diseased larvae and pupae and (iv) Disinfection of rearing place and appliances with 2% formalin.

(3) **Viral diseases** — (A) **Grasserie** : The complex of diseases caused by viruses showing jaundice-like symptoms are collectively referred to as grasserie or polyhedrosis. According to the kind of causative viruses, location of the latter in host's tissue and cell, polyhedrosis again are of 2 types : (1) *Nuclear polyhedrosis* which according to recent concept is considered as grasserie and (2) *Midgut cytoplasmic polyhedrosis* or *Midgut nuclear polyhedrosis.*

Causative organisms : The nuclear polyhedrosis of *B. mori* is caused by Nuclear Polyhedrosis Virus strain Bm (NPV Bm). This DNA virus multiplies only in the nucleus of the host cells, where it forms hexagonal polyhedra which remain embedded in a proteinaceous matrix, called Polyhedral Inclusion Bodies (PIB). The inclusions of PIB are synthesised by the silkworm (host) machinery, independent of viral components and so, may or may not contain viral particles.

In midgut polyhedrosis, the RNA *Smithia* virus forms polyhedra in cytoplasm in case of midgut cytoplasmic polyhedrosis or in nucleus in case of midgut nuclear polyhedrosis.

Modes of transmission : Nuclear polyhedrosis can be transmitted to healthy worms from infected worms by the following routes: (i) **Oral** : Ingestion of leaves contaminated with PIB from the dead tissue or faeces from infected worms can transmit the disease to fresh worms. (ii) **Environment** : High temperature, humidity, etc. may induce the spreading of this disease by transforming latent stage to virulent stage of the pathogen. (iii) **Blockage of spiracles** : Contaminated leaves with formalin, exposure to camphor during rearing and blocking of spiracles by dust, Uzi fly, etc. may induce the disease. (iv) **Skin wound** : Skin wound may get infection from contaminated rearing bed.

Midgut polyhedrosis, on the other hand, can be transmitted through oral route or by induction due to bad environment.

Symptoms : In case of nuclear polyhedrosis, the late stage of infection shows the following symptoms : (i) Loss of appetite. (ii) Loose and shiny white skin with elevated protuberances in intersegmental zones. (iii) Ruptured skin exuding milkywhite or yellowish fluid containing polyhedra (Fig. 3.38).

Fig. 3.38. Grasserie attacked larva with ruptured skin

(iv) Restless larvae usually try to leave the rearing tray and die due to exuding body fluid. (v) Infected larvae usually do not moult, (vi) Body colour turns to yellow and faeces are white in colour because of accumulated polyhedra, so the name 'jaundice' is given to the disease. For midgut viral infection, the worms show initially no symptoms. The symptoms appear only in late stage and include : (i) Loss of appetite, (ii) Translucent cephalothorax and shrinkage of body, (iii) Reduced growth, (iv) Opaque midgut due to accumulation of polyhedra in cytoplasm (v) White, loose faeces due to accumulation of polyhedra. (vi) Irregular moults etc.

Control measures : (i) Rearing environment should be hygienic with proper ventilation, (ii) Sterilisation of rearing room with steam or 2% formalin or bleaching powder or Resham Keed Ouzhad (RKO) or Labex, (iii) Removal of dead or infected worms from rearing tray (iv) Feeding the larvae with healthy leaves, (v) Oral administration of nalidixic acid, P-aminobutyric acid, etc. or topical application of imanine can control NPV to some extent.

(B) **Infectious Flacherie** : Besides bacteria, Infectious Flacherie Virus (IFV), an RNA virus can also cause flacherie in the host silkworm where it forms irregular inclusions mostly in the midgut goblet cells.

Modes of transmission : The causative virus enter the host midgut through oral route via ingested contaminated leaves.

Symptoms : (i) Shrinkage of body with transparent appearance, (ii) Vomiting of digestive juice.

Control measures : (i) Feeding the larvae with healthy leaves, (ii) Hygienic rearing and (iii) Disinfection of rearing room and appliances.

(C) **Gattine** : This disease is also known as clear head disease because of the transparent appearance of affected worms.

Causative agents : Primary agent is some virus while *Streptococcus bombycis* acts as the secondary invader.

Symptoms : Following secondary infections by bacteria the larvae show : (i) Lack of appetite,

(ii) vomiting of an alkaline, clear fluid, (iii) the anterior part of the alimentary canal appears transparent due to absence of any mulberry leaves and hence given the name of "Clear Head" to this disease.

Control measures : (1) Rearing of larvae in hygienic conditions (2) Disinfection of rearing place and appliances between each stage of whole rearing process.

(4) **Fungal Diseases — (A) Muscardine** : Fungal diseases of silkworm are called muscardine where the body of larvae get mummified due to deposition of calcium oxalate and hence the disease is also called 'calcino'.

Causative organisms : Different fungi cause this disease but the name varies according to the colour of conidiospores, produced by them on the body of worms. Name of different muscardines and their causative fungi are given below.

White muscardine is caused by *Beauveria bassiana* where the conidia are spherical/oval and spores are like white powdery specks. Black muscardine is spread by *Metarrhizium arisopliae*. The conidia are cylindrical and the spores are dark greenish black in colour.The fungus *Poecilomyces farinosus* causes yellow muscardine, where oval to spherical spores collectively look yellow in colour. *Aspergillus flavus* and *A. oryzel* cause brown muscardine with producing spores dirty brown in colour. Red muscardine may be caused by *Sorosporella uvella* and *Spicaria fumosorosea* with their red spores.

Modes of transmission : **Contact** : All types of muscardine infections are transmitted through skin. However, oral transmission through ingestion of contaminated leaves and transmission through spiracles are also noticed. The fungal spores upon falling on the body surface of silkworm germinate and gradually penetrate the cuticle by mechanical and enzymatic forces. The growing hyphae then draw nourishment from host's tissue and ultimately kill them. The fungi then take saprophytic nutrition from the dead larvae tissue and grow mycelia as white flakes over the dead tissue. Reproductive conidiophores with characteristic colour then grow from the mycelia. Due to continuous growth of fungi-colony, and their release of calcium oxalate as metabolite, the dead tissue of host larvae becomes hard and mummified.

Yellow muscardine fungi affect the gut of larvae, while red muscardine fungi attack the body wall. In case of white, green and black muscardines, the pathogenic fungi reach the haemocoel where they spread their hyphae.

Symptoms : (i) Sluggish movement showing "unequals" in rearing bed, (ii) Loss of appetite, (iii) Coloured patches may appear around spiracles or at leg bases. (iv) Body shrinks with loose cuticles, (v) Vomiting, (vi) Early infected larvae do not spin, but late infected instars may spin flimsy cocoon within which the pupae die, (vii) Dead larvae may be covered with coloured spores of causative fungi or may be mummified into chalky white structure (Fig. 3.39).

Control measures — (i) Proper rearing : Rearing room and bed should be free from humidity and darkness as these can stimulate the growth of fungi. Free aeration by keeping windows and doors open or by using fan for continuous flow of air should be maintained. Temperature should be kept above 22°C. Any kind of dampness in and around

Fig. 3.39. Shrinked body of muscardine-larva

rearing place is to be avoided. For this purpose, anhydrated lime can be kept in the corner of rearing room. All rearing appliances along with leaf-food should be kept free from extra moisture. During moulting, new bed should be kept thin and old news paper can be kept below the bed to prevent muscardine infection during wet season.

(ii) **Disinfection** : For preparing disinfected rearing beds, 0.4% formalinised husk can be used for the beds of I and II instars; 0.5% for II and III instars, 0.6% for IV and 0.8% for V instar. The ratio of diluted formalin and burnt husk is usually kept 1:10 (v/v). After each bed cleaning and during moulting, a mixture of dithane-M45 and kaoline can be applied on silkworm to prevent the germination of fungal spores if present on body surface of larvae. Again, for every 100 layings, 3-4 kg of Resham Keed Ouzhad (RKO) can be spread on the worms after each moulting and 30 minutes before the larvae resume feeding.

Labex, a mixture of lime and bleaching powder has anti-muscardine and grasserie effects and can improve larval growth and commercial characters as well. This mixture taken in muslin cloth can be spread directly on silkworm in each instar at a recommended dosage of 4 gm/ 0.1 m^2 of rearing tray. Other chemicals used as contact disinfectants for muscardine include benzoic acid, aliband, chemichlon, benzoalkonium chloride, mixture of ceresan and lime, etc. Burning of 20-30 gm of sulphur/m^3 area of rearing space is very effective to prevent growth of muscardine fungi.

(iii) **Removal of infected larvae** : Infected larvae if detected should be removed immediately from rearing tray before the formation of conidiospores on them. Infected larvae may be kept in 2% formalin to kill the spores or buried in deep pits.

(iv) **Cleaning the rearing bed** : Litter from the infected bed should be removed immediately after detection of infection and be dumped in deep pits and burnt.

(v) **Care in between rearing** : The rearing room and appliances should be disinfected immediately after muscardine infection before the next rearing. These should also be kept dry during the period between rearing.

PARASITES OF SILKWORM

Uzi fly larvae : Uzi flies prefer to deposit their eggs on the lepidopteran larvae. *B. mori* larvae are the most favourable oviposition site for the Indian Uzi flies, *Tricholyga bombycis* and *Exorista sorbillans*.

Modes of transmission : The female Uzi fly oviposits the eggs singly on the intersegmental membrane or among the bristles of body surface of III or IV or V instar (Fig. 3.40). Following 2-3 days of oviposition, the fusiform, legless maggot penetrates into the host body by making a hole on the cuticle. It forms a conical tube-like chamber, where it resides and grows by taking nearby fat tissues of the host and respiring aerobically. Later, it leaves that chamber and starts to take internal tissues of the host and respires anaerobically. After III maggot stages, the full grown maggot comes out of host body and pupates in nearby soil or crevices and cracks.

Symptoms : (i) Uzi maggot-infected silkworm shows black spots on their body surface at the place of maggot entry; (ii) loss of weight due to continuous loss of fat bodies and other tissues; (iii) stop feeding; (iv) proneness to other bacterial or viral infections and (v) sluggish movement.

Fig. 3.40. Uzi fly infestation on silkworm

Control measures : (i) **Prevention of Uzi fly entry in the rearing room** : Use of uzinet (nylon net of fine mesh) all around the rearing shelf stand and on windows of rearing room. Fly leaf paper coated with vaseline or insecticide on window can also be used to trap the flies.

(ii) **Destruction of Uzi-eggs** : 'Uzicide' (formulatd by CSR & TI, Mysore) or Vijeta (formulated by CSR & TI, Berhampore), can be spread on rearing tray that kills the Uzi-eggs without affecting the silkworm growth. Besides, bleaching powder is also found to deter Uzi fly from egg laying.

(iii) **Control of Uzi fly population** : Reproductive potential of adult Uzi fly can be reduced by releasing radiation-exposed sterile fly in the field, which would produce less viable eggs following mating with normal or other such fly. Chemosterilants like 10% dimilin, if applied topically on Uzi fly larvae can reduce the number of eggs laid by that adult flies.

(iv) **Biological control** of Uzi fly population includes release of natural predators like spiders in the field and some hymenopteral pupal parasites of Uzi fly like *Nesolynx thymus, Dirhinus himalayanus, Spilomicrus karnatakensi,* etc. Hymenopteran if cultured and released in Uzi affected area, they pupate inside the Uzi puparia and kill it while the adult parasites emerge. To prevent Uzi larvae to complete metamorphosis, Uzi larvae, pupae found in rearing room should be collected and killed by putting in hot water. No cracks and crevices should be left in rearing room that may allow pupation of the parasitic fly. Besides, transportation of larvae, cocoon of silk moth from Uzi infested area should be avoided.

OTHER ENEMIES OF SILK MOTHS

(1) **Beetles** : Many species of dermestid beetles can cause a great damage to sericulture. Some of them include *Dermestes vorax, D. valpinus, D. coar, Antherenus verbasi, A. piceus,* etc. Their larvae as well as adults feed on the silk-pupae, thus they pierce the freshly spun, stored or stifled cocoon and make them unreelable.

Control measures : (i) Frequent cleaning of rearing house and storerooms, (ii) Repeated turning of stored cocoons to check pest infestation, (iii) Fumigation of cocoon storage room with methyl bromide can destroy the beetles' larvae and (iv) Spraying of BHC in the storeroom.

(2) **Ants** : Many varieties of ants can attack the silkworms in the rearing tray, the spinning larvae as well as cocoons on the mountages.

Use of ant well below the rearing stand and *Raksha Rekha* can prevent such ant attack.

(3) **Straw mite** : Bamboos used for making rearing house, sometimes are contaminated with straw mites (*Pedicolus ventricosus*), which may be transported accidentally to rearing trays of silkworms. The female mites attach themselves to the larvae or pupae and draw nourishment from their body, leading to their death. The mites also secrete toxin which can kill the hosts.

The mite infested larvae lose appetite and become inactive. They vomit green fluid and excrete black fluid. Ultimately their dorsal vessel shows slow and irregular pulsation and then they die.

Control measures : (i) Materials taken for making rearing room should be checked carefully, (ii) Rice or wheat straw should not be stored near the rearing room and (iii) Acaricide may be applied in rearing room if mites are noticed there.

(4) **Vertebrates predators** : Some larger predatory vertebrate animals like, rats, squirrels, birds may enter the rearing room to feed the larvae or pupae. Their entry may be prevented by putting wire mesh on the windows and holes of rearing rooms.

MORICULTURE

Bombyx mori is a monophagous insect, feeding only mulberry leaves. Mulberry is a perennial plant belonging to the family *Moraceae* and includes a number of species and varieties. The nutritive value and palatability of different varieties are variable. The returns from sericulture, mainly the quality of cocoons, depend primarily on the nutritive quality of the mulberry leaves given to the larvae during rearing. In different countries, having sericulture practice, attempts were made to find out suitable substitute for mulberry as the sole important diet of silkworms. In the first approach, feeding of larvae with leaves of some other plants was tried. Their larval life was prolonged, cocoons spun were smaller and metamorphosis often failed. Therefore, other plants could not replace mulberry for commercial silkworm rearing.

In another attempt, polyphagous races of silkmoth have been raised, e.g., Asagiri (1990), Shin (1992) in Japan, but these were not suitable in Indian environment. In Japan, synthetic or semisynthetic diet using no or little mulberry leaves was formulated and was found to be very useful. However, in India artificial diet was not successful. Therefore, in Indian context, successful sericulture solely depends upon mulberry cultivation.

Moriculture : **The effortful cultivation of mulberry plants for harvesting the leaves that to be used as food for the silkworms in sericulture is referred to as *moriculture*.** In every successful sericulture practice, moriculture is run along with other processes of sericulture and about two-third of the cost of production incurred is spent on moriculture.

Common Indian Mulberry plants and their morphological characteristics

In India, several species of mulberry plants are available. But for sericultural practice, excessive hybridisation was done between recognised, useful species, to evolve new varieties with characters favourable for sericulture. Most common Indian mulberry species include : (i) *Morus nigra* (Fig. 3.41a), *M. alba* (Fig. 3.41b), *M. latifolia* (Fig. 3.41c), *M. indica* (Fig. 3.41d) and *M laevigata.*

Geographical distribution : Mulberry is a native plant of lower ranges of the Himalayas either of India or China. In China, the earliest record of mulberry cultivation was noted in 2800 B. C. The other important mulberry growing countries are Japan, India, former USSR, Myanmar, Bangladesh, Vietnam, Egypt, Greece, France, Hungary, Indonesia, Lebanon, Korea and Syria, etc.

Fig. 3.41(a). *Morus nigra*

Fig. 3.41(b). *Morus alba*

Fig. 3.41(c). *Morus latifolia*

Fig. 3.41(d). *Morus indica*

Fig. 3.41. Different species of mulberry plant (a, b, c and d)

In India, it is cultivated in West Bengal, Karnataka, Tamil Nadu, Himachal Pradesh, Jammu and Kashmir, Uttar Pradesh and Assam, etc.

Factors required for mulberry cultivation

Mulberry can grow in a wide range of environment. However, the optimum physical conditions required for its growth are as follows :

(1) **Temperature** : The temperature range required for mulberry growth is 24°-29°C. The plant cannot sprout below 13°C or above 38°C. So cultivation should not be done beyond that range.

(2) **Rainfall** : Rainfall ranging from 600-2,500 mm per year is ideal for mulberry growth. However, low yield and growth under low rainfall can be improved by supplemental irrigation.

(3) **Humidity** : The ideal atmospheric humidity for mulberry growth varies from 65 to 80%.

(4) **Daylight** : In temperate regions 5 to 10 hours of daylight and in tropical zone 9 to 13 hours of daylight are sufficient for normal growth of mulberry plant.

(5) **Elevation** : In nature, mulberry can grow in higher altitudes. But for cultivation purpose, an elevation of 300 to 900 m above the sea level is favourable. Above that elevation, harvesting and other cultural practices will be difficult.

(6) **Soil** : Being deep-rooted plant, soil should be deep, fertile, porous, well-drained with good water-holding capacity. Thus, loamy, clayey-loamy or sandy-loamy soils are best suitable for mulberry plantation. The expected pH of soil should be 6.2 to 6.8.

(7) **Other desirable factors** : Location of mulberry field should be near to silkworm rearing place. It should not be close to any chemical factories, tobacco fields, chilli fields and industrial belts.

Recommended varieties of mulberry for Moriculture

In different states of India like West Bengal, Karnataka, Tamil Nadu, etc., varieties of *Morus indica* are found to grow. However, because of their poor yield, many improved high yielding varieties have been evolved through hybridisation between local and exotic varieties by scientists of different Central and State Sericulture Research Institutes. The varieties that can be cultivated throughout India include *Mysore Local, Kanva-2, MR-2, Jatinuni, Maulainuni,* C-763, S-799, etc. After a great trial, scientists of CSB and different State Boards have suggested some varieties specific for different conditions which are as follows :

(1) **According to soil type** : For alluvial soil of Gangetic plains, S-1, S-1635 and S-799 are suitable. On the other hand, C-776, S-1 for saline or alkaline soil; C-776 and C-763 have been recommended for laterite soil.

(2) **According to water availability** : In heavy rainfall regions of Assam and West Bengal, most suitable varieties are S-1, C-763 and S-795. Drought tolerant varieties include C-776, C-763, S-13, S-14, etc.

(3) **According to plantation** : For tree plantation, recommended varieties are S-1301, S-36, S-523, TR-8, *M. laevigata,* Ber-S-1, etc. Kanra 2, and local are ideal for bush type, while there is no specific variety for middling plantation.

Recommended practice schedule of moriculture

In India, since ancient period, mulberry has been cultivated for its multipurpose uses than for sericulture. After the establishment of Central Silk Board, emphasis has been put on mulberry culture or moriculture along with sericulture. Research workers from Central Silk Research and Training Institute, Mysore and Berhampore have carried out intensive work on different aspects of moriculture in order to improve mulberry yield per unit area at reduced cost. They provided some guidelines for economic mulberry cultivation and distributed that to commercial rearers through state governments and their extension centres. However, rearer adopted some modifications, suitable for specific varieties, ecoclimatic conditions and silkmoth species of their choice. Here the basic guidelines for ideal moriculture will be discussed in brief.

Land preparation for mulberry cultivation

Before starting moriculture, the land taken for mulberry cultivation will be prepared properly. After removal of gravels and weeds, the land is levelled and then ploughed up to a depth of 12" to 15". Organic manures like compost or cattle yard manure is added with the soil by repeated ploughing. Recommended dose of manure is 10 tonnes per hectare for rainfed land or 20 tonnes per hectare for irrigated land.

Season and direction of plantation

In Tamil Nadu and Karnataka, ideal plantation time is July-August when south-west monsoon begins. In West Bengal, the recommended time is November when north-east monsoon is over. No plantation should be made during summer or winter and rainy season which may result in rotting of cuttings or seedlings. In tropical places, plantation can be done in any direction due to plenty sunlight. In temperate regions, plantation should be made in a direction facing sunlight and parallel to wind direction.

Plantation system

(1) **Pit system** : This system is adopted for rainfed crop. Here, instead of ploughing the entire field, pits (40 × 40 × 40 cm) are dug with an inter row and inter pit space of 90 × 90 cm for bush type of plantation. In case of high bush type, the distance should be 180 × 90 cm and 270 × 270 cm for tree plantation. Where the pit size will also be larger like 45 × 45 × 45 cm. After planting the cutting or sample, each pit will be filled with a mixture of organic manure, red soil and sand. Till root develops, they should be watered daily.

(2) **Row system** : This system is followed in case of irrigated cultivation, e.g., in South India. In the land, following thorough ploughing, ridges and furrows are made keeping a distance of 45-60 cm between two ridges (Fig. 3.42). Then from one end of a ridge to the opposite end, a rope with knots at equal distance of about 45-60 cm is tied. On either side of that ridge, cuttings are then planted at the point of each knot. Irrigated water flows through the furrows between the rows and feed the plants.

(3) **Kolar system** : It is a modified row system that found in Kolar district of Karnataka. Here the distance between the rows is kept 30-45 cm and that between two plants is about 10-15 cm (Fig. 3.43).

(4) **Strip system** : It is adopted in West Bengal. This is a very close system of cultivation where mulberry is grown in strips. Each strip has either two (*Dothaki*) or three (*Thethaki*) rows. In such strips, cuttings/samples are planted at a distance of 15 cm between the rows and also between two plants, within a row. Each strip is separated from the adjacent strip by a wide distance.

Fig. 3.42. Row system of mulberry plantation

Fig. 3.43. Kolar system of mulberry plantation

(5) **Angular system** : For better soil and moisture conservation, increased leaf production and economic water management, a new system of mulberry cultivation was developed for the slopes for Nilgiri. In this system, the cuttings are planted in the rows in such a way that they form a triangular arrangement. It allows more plants in a given space.

Propagation of mulberry plants

(1) **Seedling propagation** : Because of a wind pollinated plant, purity of mulberry cannot be maintained by seedling propagation in commercial practices. Moreover, it takes long time to reach the leaf at harvestable stage if propagated by seedlings. However, in research centres, through manipulation of pollination process, this propagation is carried out to produce desired varieties of mulberry plants.

Freshly collected viable seeds are sown in well ploughed seed bed at a depth of less than 25 mm. The seed beds are watered through a can and kept protected from sunlight by covering with coconut leaves or mats of bamboo strips. Seeds germinate in 10 days. After three months the seedlings are first transplanted to give a wider space of 23 cm between them. They can be transplanted once again to their final field after two years. Thereafter, these plants can be used for making cuttings or grafting.

(2) **Vegetative propagation** : In all commercial sericulture practices, vegetative propagation of mulberry plant is adopted by following methods :

(a) **Cutting** : Pencil thick branches from 8-10 months old plants of desired variety are cut into 7"-8" long pieces. For irrigated plantation, each cutting should have minimum 3 internodes with well-developed buds and for rain-fed plantation, there will be 5-6 internodes in one cutting. These cuttings are then planted in already prepared nursery, keeping about 2.5 cm with one node above the soil. Following 10 days of watering, roots develop in the internode under the soil and leaves from the aerial bud.

For quicker or easy root development, various growth regulators or hormones like indole acetic acid (IAA), indole butyric acid (IBA), L naphthaline acetic acid (NAA), etc. can be applied to the lower cut end of cuttings prior to plantation. Healthy cuttings are supplied by Central and State Sericulture Departments to the farmer at nominal cost. Cutting method of propagation is followed in South India.

(b) **Grafting** : In grafting, the scion is selected from desired variety of mulberry while the stock is taken from hardy, local variety. Cut end of scion is placed on the cut end of stock in such a way that they unite and grow as one plant. Grafting may be root grafting, shoot grafting or bud grafting depending upon the scion material used. But in all cases scion and stock should be compatible.

(c) **Layering** : Layering involves development of roots from a stem (mature) while it is still attached to the mother plant. Rooted stem is called layer. To make the layer, the branch is bent so that it contacts the soil and then get covered with soil. After roots develop, the branch is cut from mother plant and planted again. In *trench layering* as adopted in Japan, considerable length of the middle part with few nodded buds is covered under soil. New shoots and roots develop from each bud. Thus, a number of plants can be grown from a single branch.

(d) **Gooting/Air layering** : In this method of vegetative propagation, 12 cm circular bark is removed from the middle part of a healthy branch. Peat mass or decomposed organic manure with little root hormone is then placed around that region and tied with polythene covering. Water should be sprinkled in that region till the roots develop in the next 1-2 months. Then the rooted branch is cut from the parent and planted elsewhere.

(3) **Micro-propagation** :

Traditional methods of propagation like all vegetative processes including seedling propagation need locally adapted varieties of mulberry plant. So these methods require long time to establish. Grafting though allowed generation of new varieties, take much labour, time, and yield is also less. Therefore, with the development of current research in plant biotechnology, several techniques have evolved to propagate in short time and generate new varieties of any plant with desired characters. These techniques have been tried for mulberry plant as well. Some of such in vitro techniques are as follows :

(a) **Tissue culture** : Meristem from axillary or apical buds are taken from desired variety of mulberry plant. These are then grown to form callus in defined nutrient medium. From the callus derived from a single bud, 8-9 new plants can be grown by 3 months.

(b) **Endosperm culture** : Tissues of triploid (3n) endosperm is collected from triploid variety of mulberry plant and grown to form callus in tissue culture medium. Plants generated from the callus triploid are high yielder and rich in nutrients.

(c) **Synthetic seeds** : In this method, somatic embryos/axillary buds or meristem are encapsulated in a gel, that mimics the seed coat. These artificial seeds can be germinated in vitro in nutrient medium or in the soil directly. Propagation by artificial seeds save time and labour spent for pruning, cutting, transportation, etc.

(d) **Somoclonal variation** : Somoclonal variation can be induced externally with mutagens. Callus with desired character can be generated from such somoclonal variated tissues. SV1 is one of mulberry variety that developed from callus with somoclonal variation.

(e) **Somatic hybridisation** : Somatic hybridisation can be done through fusion between protoplasts taken from two varieties having desired characters (e.g., high yield, hardy nature, disease resistant). Hybrid plants produced by this protoplast culture are of better quality than produced by routine grafting.

(f) **Genetic engineering** : Genes of desired characters can be isolated and cloned. The cloned gene then can be introduced into the plants by direct methods (e.g., electroporation, microinjection, microprojectile bombardment) or by indirect method (e.g., by plasmid or other vectors).

All such laboratory-based techniques for propagation are now being practised in different research centres of India and the resultant products are then handed over to the farmers.

Irrigation : The survivality of mulberry plant does not depend on soil water content but the quantity as well as quality of leaves harvested from the plants absolutely depend on availability of soil water. Irrigation almost doubles the yield and improves the protein and moisture content of leaves than those from rain-fed crops. So mulberry cultivation needs irrigation at least once in 8-10 days for sandy soil to once in 15 days for clayey soil.

The mulberry field may be irrigated by various methods like furrow method, basin method, flat bed method, overhead sprinkler method, micro-irrigation method, etc. However, in all cases, appropriate drainage system must be in the field to remove the excess water. This is essential (i) to prevent root rotten by water logging, (ii) to increase the microfauna population of the soil and (iii) for leaching of excess salts.

Manuring : Mulberry is a fast growing plant and thus consumes the soil nutrients quickly. Regular manuring the mulberry field is required for the normal growth and maximum leaf-yield from the plants. Manures may be applied as *organic* or *synthetic chemical forms* or as *biofertilisers*.

Most frequently used organic manures include compost, sericulture wastes, poultry manure, green manure, oil cakes, biogas slurry, farm yard manure, etc. Organic manures increase the nutrient content of the soil, conserve soil moisture and enhance biological activities of the soil. Chemical fertilisers used for mulberry growth and yield are ammonium sulfate, ammonium chloride and nitrate, sodium nitrate, urea, potassium ammonium nitrate, bone dust, etc. Chemical fertiliser may be supplied to the plants along with irrigation water or directly as foliar sprays.

Biofertilisers suitable for mulberry cultivation are different nitrogen fixing bacteria like *Azotobactor*, *Azospirillum* and aquatic fern-like *Azolla*. The Vesiculoarborescent mycorrhiza (VAM), *Glomus fasciculatum* are also grown in the mulberry field to enhance the PO_4 utilisation by the plants. Biofertilisers should not be mixed with chemical fertilizers unless specified.

Pruning : *Pruning is periodical* and *methodical cutting of branches from mulberry plants*. It is a very important operation in moriculture. It ensures – (i) good yield of quality leaves, (ii) sprouting of young shoots from auxillary buds, (iii) good vegetative growth, (iv) removal of dead and defunct wood, (v) better exposure of the plant to sunlight, etc.

Prunings are of four types, viz., (a) **Bottom pruning** or low cut pruning where branches are cut to the base, (b) **Middle pruning** which allows cutting of branches around mid-height, (c) **Top pruning or high cut pruning** in which branches are cut at the top and (d) **Kolar pruning** – here pruning is done at the ground level. During pruning plant sap oozes out from cut ends leading to fluid loss and consequent growth problems. To minimise these problems, several methods are adopted for pruning like step-up and step-down pruning, rejuvenation pruning, alternate row harvest pruning, etc. All such prunings allow the plant to maintain their optimum growth and leaf-yield as required by the farmers. Pruning also keeps the height of mulberry plants within reach during harvesting.

Training : *Regular pruning gives a specific shape to the mulberry plant which is referred to as* ***training***. Two types of training are given to mulberry plants :

(i) **Fist form training** : In this type, the mulberry plant looks like closed fist when freshly pruned. Due to repeated pruning of branches at the same place every year, the height does not increase, but the diameter of the top part gradually increases. The plants may have single, two or even three fists.

(ii) **Non-fist form training** : In this case, the pruning is made at a level higher than the branching point every year. As a result, branching points of shoot increase in height every year not giving fist-shape to the growing plant.

Intercultivation : In all types of mulberry cultivation, the inter-row spaces are ploughed periodically either manually or by tillers. It helps to remove the weeds, loosen the soil with better aeration and percolation of water.

Mulching : Sometimes, the inter row spaces of mulberry field are covered with cut-straw mats, or mulberry twigs or stubbles or leaf moulds, etc. It allows conservation of soil moisture, loose and fragile form of soil and also adds organic manure following their slow decomposition.

These above mentioned steps, as formulated by research workers of different sericulture institutes for economic mulberry cultivation are generally followed by mulberry farmers. But some modifications in all steps according to local environment, types of silk moth and varieties of mulberry plant have been done by the field workers. However, following the above package system of moriculture, yield of mulberry leaves has been raised to remarkable quality in both irrigated and rain-fed conditions.

Harvesting mulberry leaves : *Harvesting is the collection of mulberry leaves from the plants in a rational method.* It is an important part of moriculture that controls the growth of the silkworms in one hand, and quality as well as quantity of cocoons spun by them on the other. Like moriculture, harvesting also follows some general practices.

Time of harvesting : Recommended time for harvesting the leaves is early hours of the morning. If harvested during afternoon, the leaves may contain less water and more carbohydrate due to active transpiration and photosynthesis that occurred in the day.

Choice of leaves : It was been found that the nutritive requirements of larvae vary with their age. Again the nutritive quality and quantity of the leaves vary with their position, age and physical appearances. In general, early instars are given young, large, glossy, leaves with more water content that are available in the top part of branches. Late staged instars can be given mature leaves harvested from the branch just below the top part specified for early larvae. The position of leaves in the branches that are usually taken for different instar larvae is shown below (Fig. 3.44).

However, in every sericulture practice, following leaves are to be avoided for harvesting or if harvested, not to be fed to the larvae: decline leaves, buds, too tender or too mature leaves, pest infested or diseased leaves, insecticide sprayed leaves and too coarse leaves.

Methods of Harvesting : (1) **Leaf picking** : In this case, individual leaf with petiole is picked from the main branch. The terminal buds are nipped off at the same time. It helps the lateral shoots to grow rapidly so that sufficient leaves are available during next harvesting. After 10 weeks of bottom pruning, leaf picking can be started with intervals of 7-8 weeks for subsequent pickings. By leaf-picking desirable leaves can be collected but this method requires more labour, time and further, the leaves may get withered too quickly. This method is suitable for small scale rearing operation.

Fig. 3.44. Mulberry leaves for warm-feed

(2) **Whole-shoot harvest** : In this method, branches /shoots are cut to the ground level so the sprouting shoots can reach uniform maturity by the next harvest. The entire shoots collected are given to mature larvae (after IV moult). Whole shoot harvesting is done either manually or by machine at an interval of 10-12 weeks. It is practised in Karnataka, West Bengal in India, Japan and Russia also.

(3) **Branch cutting** : Here, the entire branch is harvested and used to feed worms after III moult directly. It is applicable for tree and middling plantations. This harvesting method saves time and labour in harvesting and bed cleaning operations also. Further, leaves of entire branch remain fresh for a longer period. It is practised in Kashmir, West Bengal, parts of Karnataka in India and also in Japan and Russia.

Preservation of harvested leaves

Harvested leaves are usually not given immediately for feeding to the larvae because of greater starch content which is not palatable. Moreover, the larvae feed 4-5 times a day while leaves are harvested 1-2 times per day. A certain time is needed to reach the harvested leaves to rearing place. For these reasons, mulberry leaves are needed to be preserved after each harvest.

During storage, the leaf loses water by transpiration and hydrolysis. Water loss is directly proportional to storage time. During storage, nutritive value of the leaves also deteriorates. To overcome these problems, harvested leaves may be stored or preserved by following ways :

(i) Leaves to be fed to young larvae, may be kept in baskets but for grown-up larvae can be stored in wooden or bamboo leaf chambers covered on all sides with wet gunny cloth.

(ii) For better preservation, leaves may be kept in gunny cloth bags of 2' × 3' size, which are double layered and easy to handle.

(iii) Leaves should be kept in cool place.

(iv) Exposure to air current should be avoided.

(v) Heap storage must not be done.

(vi) The storage time must not exceed 24 hours.

(vii) If leaves get rotten or dry, must be discarded.

Diseases or enemies of mulberry plant

For each healthy moriculture practice, it is desired that the mulberry cultivation should be kept absolutely free from (i) diseases caused by infectious or non-infectious pathogens (ii) nutrient deficiency-diseases or (iii) attack by any pests or parasites. Because all such cases, if happen, would result in low and inferior yield of food leaves.

Diseases caused by pathogens : Important parts of the mulberry plants like stem, leaf and root may be attacked by various kinds of pathogens. The diseases, their causative agents (CA), symptoms and control measures are tabulated below (see Table 3.9).

Mulberry pests : For sericulture-practice, mulberry plants are pruned periodically and leaves are harvested regularly. Therefore, most of the mulberry pests are polyphagous and attack the plants sporadically and seasonally. Following main pests are found to attack mulberry leaf, stem and also known to ingest plant sap.

Table 3.9 : Diseases of mulberry plants caused by pathogens

Name of the disease and causative agent (CA)	Affected part and symptoms	Control measures
Wilt Disease CA : *Sclerotium rolfsii* (fungus)	Stem. Black patches of fungal mycelia appear on cuttings. The sprouted shoots suddenly wilt and the cuttings ultimately die.	Use of chemical fungicide Bavistin or antagonistic fungus *Trichoderma* as biofungicide.
Stem Canker Disease CA : *Botryodiplodia theobromae* (fungus)	Stem. Greenish black to dark black lesions appear on affected stem.	Before plantation, soaking of cuttings in 10 ppm solution of Bavistin or 0.1% carbendazin can prevent canker fungus attack.
Twig Blight CA : *Fusarium pallidorosum* (fungus)	Stem. The affected stems become feeble, fragile and show irregular black lesions on surface. Gradually the twigs get splitted and dry.	Drenching of soil with Captafol or dithane M-45 at a concentration of 0.5% can prevent primary inoculation
Collar Rot CA : *Phoma mororum.* (fungus)	Stem. The affected stem turns brown at ground level that gradually encircles the shoot as a collar. Leaves on it show withering.	It is a soil-borne disease. So pruning the stem at a height of 80 cm above the ground and subsequently at lower level can check the contamination.
Pythium Damping off Disease CA : *Pythium aphaanidermatum* (fungus)	Stem. Affected stem shows water soaked patches at first that become brown to black patches ultimately. Terminal leaves wilt and fall off.	This deadly epidemic disease of mulberry plant can be checked by soaking the cuttings in fungicides prior to plantation.
Powdery Mildew Disease CA : *Phyllactinia corylea* (fungus)	Leaf. Initially white powdery patches appear on ventral surface of affected leaves (Fig. 3.45) that later become brown to black and cover whole leaf. Affected leaves become dry, leathery and fall off.	Physical : Picking and burning of diseased leaves. Harvesting leaves at proper time. Proper pruning and weeding increase air circulation and reduce humidity.

Table 3.9 cont. to next page

Table 3.9 : Diseases of mulberry plants caused by pathogens (continued from previous page)

Name of the disease and causative agent (CA)	Affected part and symptoms	Control measures
Powdery Mildew Disease CA : *Phyllactinia corylea* (fungus)		Chemical : Spraying of Morestan, Karathana, sulpher power to entire field. Spreading of lime-sulphur mixture on the underside of leaves. Biological : Release of predatory spotted lady beetle in the field.
Leaf Spot disease CA : *Cercospora moricola* (fungus)	Leaf. Affected leaves show small white circular/irregular spots with a yellow halo (Fig. 3.46); leaves ultimately fall off.	Removal of weeds (alternate host) from field. Picking and burning of affected leaves. Spraying of Bavistin or Difolatan, etc.
Leaf Rust CA : *Aecidium mori* (fungus)	Leaf. Affected leaves including green apical shoot show small irregular reddish borwn spots (Fig. 3.47). Leaves loss their shape.	Affected leaves should be picked up and burned. Spraying of lime sulphur mixture, Antracol, dithane, etc.
Leaf Blight disease CA : *Pseudomonas mori, Xanthomonas campestris* (bacteria)	Leaf. Affected leaves show water soaked irregular spots at lower side to brown spots with yellow margin at upper side. Leaves become wrinkled and curled outwards. In wet weather, the spotted area fall off but in dry weather, leaf-spots remain as dry on intact leaf.	Disease resistant variety may be planted. Diseased plants should be removed and destroyed immediately. Spraying of foliar sprays like phytobacteriomycin may be effective.
Leaf Mosaic disease CA : Mulberry mosaic (virus)	Leaf. Affected leaves show mosaic dark green patches without chlorophyll around the veins. Leaves wrinkle, with chloric lesions. The shoot tips shorten and leaf size also reduced. Total leaf yield reduces.	Only preventive measures can be taken against such diseases. Removing and burning of affected plants, removing all old crops from field. Exposure of plant-cultivating apparatus to high temperature, washing the hands and tools with disinfectants. Plantation of virus-resistant variety.
Yellow Net Vein Disease CA : Mulberry yellow Net Vein virus (virus)	Leaf. Virus affected leaves turn yellow with net-vein appearance. The leaves become distorted and twined. The plant growth also retarded.	Only preventive measures can be taken against such diseases. Removing and burning of affected plants, removing all old crops from field. Exposure of plant-cultivating apparatuses to high temperature, washing the hands and tools with disinfectants. Plantation of virus-resistant variety.

Table 3.9 cont. to next page

Table 3.9 : Diseases of mulberry plants caused by pathogens (continued from previous page)

Name of the disease and causative agent (CA)	Affected part and symptoms	Control measures
Dwarf disease CA : Mycoplasma	Leaf and branches. Affected parts show short, fragile branches with irregularly arranged leaves.	Only preventive measures can be taken against such diseases. Removing and burning of affected plants, removing all old crops from field. Exposure of plant-cultivating apparatuses to high temperature, washing the hands and tools with disinfectants. Plantation of virus-resistant variety.
Nematode disease CA : *Meloidogyne incognita, M. Javanica* (nematodes)	Root. Stunted plant growth and reduced leaf-yield. Galls are present on rootlets and on feeder.	Fallowing the land, mulching the soil, rearing resistant variety. Application of organic fertilizers like neem oil cake, and compost can check nematode attack.
Root Rot diseases : (i) White Root Rot CA : *Rosellinia necatrix* (fungus)	Root. Weak growth of plant, shedding of leaves, feeble growth of buds, rotting of roots showing white patches of mycelium.	For different root rot diseases of mulberry plant, following physical, chemical and biological control measures are useful. (a) *Physical method* : The diseased plant should be uprooted, burnt; the soil around that plant will be dug out and removed to far-off places. The soil should be sterilised by putting calcium cyanamide (2400 gm/m^2) in one metre deep holes.
(ii) Violet Root Rot CA : *Helicobasidium mompa* (fungus)	Same as white root rot except the rotten root and stump show violet patches of fungal mycelia.	(b) *Chemical method* : Before plantation, soil can be mixed with 6.1% Terrazole or 23% PCNB (Pentachloro-nitrobenzene). The pits can be dusted with 2-3% of dithane 45 before planting. The cuttings or samples can be dipped in 0.2%. Bavistin solution or 20% lime water and should be washed in water before plantation.
(iii) Fusarium Root Rot CA : *Fusarium oxysporum* (fungus)	Root. The surface of root beneath the bark and cortex becomes gummy. The plants die ultimately.	(c) *Biological control* : Some biofungicides like *Trichoderma, Bacillus, Pseudomonas,* etc. can be used to check the growth of root rotting fungi.

Table 3.9 cont. to next page

Table 3.9 : Diseases of mulberry plants caused by pathogens (continued from previous page)		
Name of the disease and causative agent (CA)	**Affected part and symptoms**	**Control measures**
Root Rot diseases : (iv) Stem Rot/Heart Rot CA : *Polyporus hispidus* and *Ganoderma applantum* (fungi)	Stem. The heart wood of branches/ shoots are attacked resulting in drying, rotting of twig and branches. The attacked plants ultimately die.	The branch/twig should be free from wound or any cut. The branches of affected tree should be detached and burnt immediately.

Fig. 3.45. Powdery mildew disease of mulberry

Fig. 3.46. Leafspot of mulberry

Leaf-eater pests

(1) ***Spilosoma obliqua* (Lepidoptera)** : The females moth lay eggs under surface of mulberry leaves. Larvae feed voraciously on mulberry leaves, defoliate the plants, leaving only mid-vein.

Control : (i) Collection and destruction of leaves with eggs. (ii) Deep ploughing and flooding the soil to kill the pupa therein. (iii) Spraying of 0.2% DDVP (Dimethyl Dichloro Vinge Phosphate) on leaves. (iv) Release of larval parasite *Apantales obliqua* as biocontrol measure.

(2) Caterpillars of other lepidopterans like *Ceryx godarti, Amata passalis, Margaronia pyloalis,* etc. also lay eggs underside of mulberry leaves. The emerged larvae feed on and destroy the leaves. Control measures are same as described above except the biopredator.

(3) The most destructive leaf-eating coleopteran pests include *Mimastra cyanura* and *Episomus fugulus*. The adult beetle feeds on mulberry leaves. Control measures include digging and flooding the soil to kill either egg or grub living there. Soil may be sprayed with Bilcyp, and mixed with neem oil cakes.

(4) Besides above insects, some grasshoppers like *Letana inflata, Ailopus sinulatrix* may cause defo-

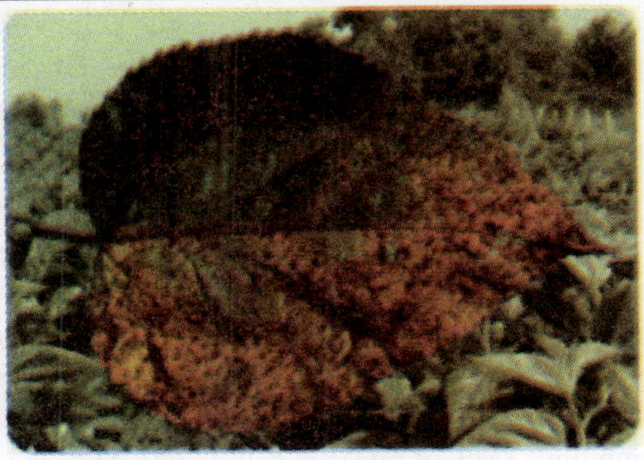

Fig. 3.47. Leaf rust of mulberry

liation of mulberry plants in different parts of India. Again some terrestrial *snails*, e.g., *Chochiopa* sp, *Cryptozona* are also found to cause defoliation of mulberry in India.

Stem borer pests

Important shoot or stem borers of mulberry plants includes beetle, viz., *Sthenias grisator* (Stem girdler), *Simoxylon pubens* (powder pest beetle), etc. They either girdle or bore the young shoots of mulberry plant and take stem tissues resulting in poor leaf-yield, drying of shoots, etc.

Removal or cutting the affected shoots, cleaning of weeds from mulberry fields can check the beetle infestation.

Sap feeders and other pests

Besides leaf eaters and stem borers, mulberry cultivation often faces attack by various sap feeders also. Nymph as well as adult of Jassid (leaf hopper) *Emposca* suck sap from leaf veins causing Hopper burn. The brown hopper *Eusarcocoris ventralis* suck sap from tender buds and leaves; the bug *Nezara viridula* is known to suck sap from both leaves and tender stems. The mites *Tetranychus equitorius* and *T. telarius* spin nets and feed on leaves.

To control the sap feeders, cutting and burning of affected shoots are suggested. Spraying of chemical pesticide and use of light traps can also provide better result in other cases.

Nutrient-Deficiency diseases

Like other plants, mulberry also suffers from various nutrient-deficiency diseases due to improper constituents of soil. Presence of excessive nutrients, on the other hand, is of same harmful for their growth and health. A brief discussion follows :

(1) **Nitrogen** : Nitrogen is a macronutrient being constituent of protein, chlorophyll, enzymes, hormones and nucleic acids. It controls utilisation of phosphorus and potassium from the soil. For mulberry, the critical level of nitrogen below 8 kg/acre (nitrate) or 30 kg/acre (ammonia) causes deficiency symptoms. Lack of nitrogen causes retarded growth with yellowing of leaves. Application of nitrogenous fertilizers like urea, ammonium nitrate, potassium nitrate can recover the deficiency.

(2) **Phosphorus** : Development, metabolic activity, maturation and differentiation of tissue require phosphorus. Besides, this macronutrient increases vigour and quality of crop, and prevents lodging of the plant. The minimum level of phosphorus in soil for mulberry cultivation should be 50-65 lb /acre. Deficiency symptoms include stunted and slender stem, chlorosis with marginal necrosis, decolouration of leaves and stems, defoliation, etc. Application of triple or superphosphate corrects the deficiency. Excess of phosphorus results in mottling of leaves and burning in the mottled parts.

(3) **Calcium** : Calcium promotes cell division, formation of cell wall; activates enzyme activity and nitrogen fixing bacteria. Calcium deficiency results in retarded growth of mulberry plants with necrosis and abscission of leaves. Roots become dry and stubby. Application of lime in recommended dose can recover the deficiency symptoms. Excess calcium, on the other hand, makes the soil alkaline and precipitates iron from soil water.

(4) **Magnesium** : Magnesium is a structural component of chlorophyll and helps in photosynthesis and phosphorus uptake. Lack of magnesium causes yellowing of interveinal regions on mature leaves and premature defoliation. To correct the deficiency, magnesium sulphate or oxide can be applied to the soil or sprayed to leaf as 2% solution.

(5) **Potassium** : This very important macronutrient regulates enzyme activity, photosynthesis, synthesis of carbohydrate and protein, stomatal movement, etc. Deficiency of potassium leads to necrosis and intra-veinal chlorosis of leaves, retarded photosynthesis and stem, root growth, scorched and coarse margin of leaves, etc. Fertilizers with potassium can correct all deficiency symptoms.

Besides the above macronutrients, some important micronutrients are also required for healthy mulberry cultivation. *Iron* being a component of chlorophyll leads to chlorosis when present in low amount in soil. Iron, manganese, zinc, molybdenum are very essential for the activities of many enzymes. Their deficiencies may cause chlorosis (Mn, Mo) yellowing (Zn) of leaves.

RECENT ADVANCES IN SERICULTURE

Biotechnology in Sericulture : The recent research works in the field of silkworm genetics and breeding were directed to studies on the interaction between the genotype and environment, heterosis expression in F1, positive transgressions expression in F2, combining ability, inheritance of the important qualitative and quantitative characters, implementation of the intensive inbred-lines breeding, breeding of sex-limited races, use of partheno- and androgenesis in the silkworm breeding, breeding of silkworm races and hybrids for summer-autumn rearing, and having higher tolerance to adverse rearing conditions, etc. Few outcomes of such research activities are mentioned below :

(A) Transgenic/genetically modified silk moth

With the help of different genetic techniques like cloning, RNA interference, etc., scientists have developed many transgenic silk moths with desired traits as follows :

(1) Transgenic silk moths (Fig. 3.48) expressing fluorescent proteins can produce large amounts of fluorescence-coloured recombinant silks. Tape-stries, other textiles and lamp shades were produced from the recombinant silks and evaluated for practical use.

(2) By introducing a gene for an acidic peptide into silkworm, scientists developed thin silk. This silk thus produced by transgenic silkworm is thinner than that of the original race.

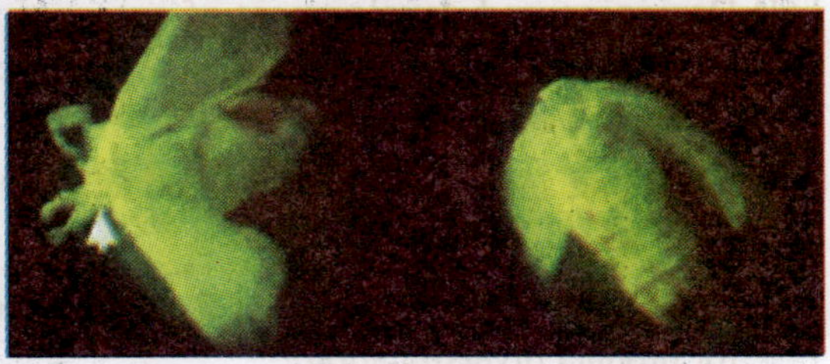

Fig. 3.48. Transgenic silk moth with green fluorescent protein

(3) Spider silk produced by transgenic silk moth has more strength and elasticity and is much more stronger than normal silk, so could possibly be used to make sutures, artificial limbs and parachutes.

(4) High yielding silkworm hybrids were developed using the DNA marker technology and then released to farmers of many places like Andhra Pradesh, Karnataka and Tamil Nadu (*viz.*, Swarnandhra, Hemavathy and Kalpava-tharavu). Again, many tropical bivoltine breeds, viz., CSR hybrid series (CSR6 × CSR26), (CSR2 × CSR27), (CSR48 × CSR5), (CSR2A × CSR4A), (CSR2 × CSR4), (CSR2 × CSR5), (CSR18 × CSR19), etc. were also developed and are being introduced in Indian sericulture successfully.

(5) Transgenic *B. mori* was produced by RNA interference (RNAi) for viral resistance against nucleopolyhydra virus BmNPV.

(B) Transgenic/genetically modified Mulberry plant

(1) Transgenic mulberry lines with HVA-1 gene for abiotic stress tolerance have been developed. Total area under mulberry cultivation for sericulture in India is about, 79, 065 ha of which 45 to 50% area are prone to abiotic stresses (drought, alkalinity and salinity). In order to develop suitable mulberry varieties, tolerant to drought and salinity, the DBT launched a programme at University of Delhi South Campus (UDSC), New Delhi.

(2) Evolution of high yielding mulberry varieties like V I (victory I) and S36 giving phenomenally higher leaf yields.

(C) Silk as Bioreactor

The silkworm, *Bombyx mori,* has been used for silk production for centuries but recently also for protein production, as a bioreactor. Since Maeda *et al.* (1985) reported the production of **human α-interferon** in silkworm larvae, using recombinant *B. mori* nucleopolyhedrovirus (BmNPV), the production of many other proteins has been achieved.

(1) TORAY Ind. Inc. (Tokyo, Japan) used silkworms to produce two recombinant proteins for veterinary use. (i) **Intercat**, a drug principally com-

posed of feline interferon, for feline calicivirus infections (commonly known as "cat flu") was introduced in February 1994 as the world's first antiviral drug for animal use. (ii) **Interdog**, a drug principally composed of canine interferon-γ for canine atopic dermatitis, was launched in December 2005. Nowadays, recombinant proteins produced from silkworms are being investigated for commercial purposes.

(2) Luciferase and green fluorescent proteins were successfully expressed in silk moth utilising silkworm as a bioreactor (Indian Institute of Science, Bangalore).

(D) Other progresses

(1) Characterisation of anti-bacterial proteins in silkworm has led to the discovery of unique insect specific lysozyme (similar to C-type lysozyme and alpha-lactalbumins).

(2) Process for degumming of silk with fungal protease enzyme has been standardised and found economically viable without chemical hazard.

(3) Development of integrated nutrient management measures comprising of various bio-agents for mulberry cultivation.

(4) Development of biocontrol agents against root rot, root knot and nursery diseases of mulberry plant and bed disinfectants against various silkworm diseases.

(5) Development of integrated pest management measures against various pest of mulberry and silkworms.

DATA BASE

With advances in genomic technologies, there is a rapid increase in silkworm related data including genome sequences and genetic markers. It is necessary to upload those in internet. Several scientific organisations have developed database system to provide integrated information about various gene resources of important qualitative characters. Such characters include egg serosa and chorion colour, larval markings, cocoon shape and colour, high hatchability, optimum larval duration (27-29 days) and fifth instar duration (7-9 days), high pupation rate, maximum number of normal eggs in the laying (more than 550), good quality cocoons, good fresh cocoon weight and cocoon shell weight, optimum filament length, good raw silk percentage, etc.

Now these average data are translated in English and converted to electronic files. The main purposes of such data bank was for promotion of global sharing of silkworm germplasm information and sharing of information about the silkworm genetic resources. However, there are many problems in sharing this information like, language, lack of enough techniques for fast and easy communicaiton, lack of internationally accepted methodology for silkworm qualitative and quantitative characterisation and evaluation. Again, in some countries this information is considered as "secret" and in others, the genetic resources belong to private companies and the government organisations do not have enough information about them. Some countries are in competition between each other in selling silkworm eggs to third markets, hence they are not interested to share this information. There are political problems between some countries and they also do not want to share the information. Nowadays many of these problems are tried to be solved by mutual discussion.

CENTRAL SILK BOARD AND RECENT ACTIVITIES

In India, the Central Silk Board is continuously putting efforts for total improvement of Indian Sericulture. Some examples include :

A. General Development

(1) Designing and development of automatic conveyor cooking machine for medium scale reeling establishments.

(2) Designing and development in production of superior grade dupion silk.

(3) Designing and development of Eri spun yarn for product diversification.

(4) Designing and development of improved silk handloom.

(5) Development of tasar reeling technology package to produce superior quality tasar silk, etc.

B. Special Projects

The Central Silk Board is implementing integrated sericulture development projects in different states in collaboration with the State Sericulture Departments and the NGOs as follows :

1. M/s. Arpitha Grameen Vikas Sansthan, Kanpur (Eri Project), UP.
2. M/s. Appropriate Technology of India (ATI) Guptkashi, Rudraprayag Dist. (Uttarkhand).
3. M/s. Grameen Evam Krishi Vikas Samithi Haldwani, Nainital (Uttarkhand).

C. Application of Remote Sensing and Geographical Information System (GIS) in Sericulture

A project was implemented in collaboration with the NESAC, during the XI Plan (2007-12) covering all the states, involving a total cost of 2.50 crores. Their main objectives were — (i) to map and identify additional potential areas for development of food plants for mulberry and vanya sericulture in different states, (ii) to carry out an appraisal survey to evaluate the progress of sericulture development in few selected areas using remote sensing technology.

D. International Collaboration

CSB being a member of the International Sericulture Commission (ISC), providing consultancy support to third world countries on sericulture development. CSB is exporting silkworm eggs to African and South Asian countries like Ghana, Ethiopia, Bangladesh, Srilanka, Vietnam, etc. Since 1989, many aided projects were implemented in India, viz, Swiss aided Seri-2000 project for mulberry in Karnataka, Tamilnadu, AP and West Bengal (1997-2004), UNDP supported project for development of Vanya silks (1999-2003), World Bank assisted National Sericulture Project for mulberry in 17 states (1989-1996), etc.

E. Training on Bivoltine Sericulture Technology under the 3rd Country Training Programme

As a part of technical cooperation programme between the JICA and the Central Silk Board (CSB), CSR&TI, Mysore is imparting training for many Afro-Asian countries like Bangladesh, Cambodia, Ethiopia, Ghana, Indonesia, Kenya, Laos, Madagascar, Nepal, Nigeria, Pakistan, Philippines, Sri Lanka, Vietnam and Uganda on bivoltine sericulture technologies with the expertise developed by CSB.

F. Cluster Development Programme

Some special sericulture projects are being implemented through cluster approaches. CSB in coordination with State Sericulture Departments, have jointly developed 51 clusters so far (45 precocoon and 6 postcocoon) covering 16 States. During the year 2009-10, CSB has released Rs. 19.11 crores and during the year 2010-11, Rs. 15.27 crores of central share of funds under CDP.

G. Establishment of Automatic Reeling Units in Andhra Pradesh, Tamil Nadu and Karnataka

Three Automatic Reeling Machines (ARM) have been established, one each in Karnataka, Andhra Pradesh and Tamil Nadu till 2009-10. Under XI Plan, five more Units are being established, two each in Karnataka/AP and one in TN.

H. Eri Spun Silk Mills at Andhra Pradesh and Assam

Three Eri Spun Silk Mills have been established at Hindupur (AP), Kokrajhar and Guwahati (Assam). These Units have started commercial production of eri silk.

I. Eri Silk Knitwear Products

As a part of Vanya silk development effort, the Vanya Silk Market Promotion Cell (VSMPC) of CSB has developed eri silk knitwear production, in collaboration with NIFT-TEA Knitwear Fashion Institute, Tirupur in Tamilnadu with a view to get the attention of both export and domestic markets.

J. Policy Initiatives

Important policy initiatives taken recently for the development of silk industry are : (i) Sericulture is included as agriculture allied activity under RKVY (Rashtriya Krishi Vikas Yojana) . This will enable the sericulturists to avail the benefits of the scheme for the entire sericulture activities up to reeling. (ii) The Govt. of India has taken necessary steps to import 2500 metric tonne of raw silk through NHDC (duty free) for distribution among small weavers. (iii) The CSB (Amendment) Act, Rules and Regulations have been notified by the Govt. of India to bring quality standards in silkworm seed production. (iv) Forest Conservation Act has been amended to treat non-mulberry sericulture as forest based activity. It will enable the farmers to undertake Vanya silkworm rearing in the natural host plantation in the forests.

Seri Products for Value Addition

Mulberry and silkworms are excellent bio-resources for various medicinal and neutraceutical products. In China, Japan and Korea, the mulberry plant, silkworms as well as silk are utilised for preparation of several products, food additives, cosmetics and drugs.

Mulberry

(1) *Neutraceutical compounds in leaves* : Tea made from mulberry leaves is popular in China, Japan and Korea; it is beneficial for diabetic and hypertensive patients. DNJ, GBBA present in mulberry leaves reduces glucose level and blood pressure.

(2) *Antioxidants in mulberry fruits* : Ripe mulberry fruits are used to make jam, jelly, pickle and wine.

(3) *Fuel and compost making from mulberry* : Mulberry waste wood is a good source of cellulose for paper and chipboard industries.

(4) Mulberry shoots can be used as staking material for climbers of vegetable, ornamental or fruit crops and mushroom cultivation.

(5) Hard woods from mulberry are used for making hockey sticks and cricket bats.

Silkworm Larva

(1) *Products from silkworm faeces* : In China, chlorophyll, phytol, carotene, and pectin are extraccted from silkworm faeces and used in food, chemical and pharmaceutical industries.

(2) Silkworm litter is also used as fodder and compost material.

Pupa

Oil extracted from silk moth pupa has numerous industrial applications. After oil extraction, oil cake can be used as moulding material in bakelite industry. Silkworm pupa can also be used as feed for fowl, fish and pig as it is rich in protein. A fungus *Cordyceps sinensis* cultured on pupae/moths, by use of recombinants have anti-tumour and anti-ageing properties.

Waste Silk

Waste silk which cannot be reeled or spun is used to make silk film or silk powder which is used as feed additive for cosmetic products. Sericin can be used for the preparation of skin care items as it can provide excellent protection against ultraviolet rays.

Silk Moths

Silk moths are used as feed for livestock. Moths are also used for preparation of fungal medicine.

Chapter Four : Lac Culture

INTRODUCTION

Lac is a resinous exudation from the body of a typical female scale insect, which it secretes for its own protection. Lac is the only known commercial resin of animal origin and thus, is nature's gift to mankind. Since Vedic period, Indians, Greeks, Romans, Chinese were familiar with various uses of lac. Lac, being an eco-friendly biodegradable and self-sustaining material has also assumed profound relevance in the current universal environment awareness.

Lac culture is the scientific propagation of lac insects through the cultivation of their host plants, in order to obtain lac in large scale. It also includes propagation of host plants, collection of lac from these plants, subsequent processing of commercial lac and prevention of enemies to both lac insects and host plants.

DISTRIBUTION

Lac insects thrive best on some tropical plants which are distributed in South East Asian countries. At present lac is produced in India, Myanmar, Thailand, Malaya, China and some other countries like Nepal, Japan, Srilanka, Africa, Australia, Brazil, France, West Germany, Turkey, USA, etc. In India, few years ago, cultivation of lac was carried out practically throughout the country. But changing socio-economic conditions have recently restricted lac cultivation in some states only. Current major lac producing states are Jharkhand, Chattisgarh, West Bengal, Bihar, Madhya Pradesh, Eastern Maharashtra and Northern Orissa (Table 4.1). Some pockets of lac cultivation also exist in Andhra Pradesh, Punjab, Rajasthan, Karnataka, Gujarat and Uttar Pradesh.

Table 4.1 : Lac production in different states of India during last few years

State	Year-wise production of raw lac (tonnes)						
	2003-04	2004-05	2005-06	2006-07	2007-08	2008-09	2009-10
Jharkhand	9,860	10,780	6,414	7,490	6,385	4,000	6,925
Chattisgarh	5,140	4,270	6,015	8,875	7,160	7,198	4,995
Madhya Pradesh	1,905	2,870	2,515	3,164	3,755	2,970	2,390
Maharashtra	633	875	1,160	1,300	1,070	1,050	520
West Bengal	1,270	1,395	926	1,240	1,140	830	855
Total	18,808	20,190	17,030	22,069	19,510	16,048	15,685

Source : **Indian Institute for Natural Resins and Gums, Ranchi**

Taxonomy of Lac Insect

(after E.E.Ruppert and R.D. Barnes, 1994)

Phylum : Arthropoda;

Sub-phylum : Uniramia;

Class : Hexapoda/Insecta; Subclass : Ectognatha

Scientific name : *Kerria (Tachardia) lacca*

The first scientific account about lac insect was given by J. Kerr in 1782, published in Philosophical Transaction of Royal Society of London. The first scientific name given to it was *Tachardia lacca* after the name of French Missionary Father Tachardia. Later it was changed to *Kerria lacca.* Out of nine genera and 87 species of lac insects reported from the world, two genera and 19 species are found in India. Of these 19 species, only *K. lacca* is exploited for commercial production of lac. However, *K. chinensis* in northeastern states and *K. sharda* in coastal regions of Orissa and West Bengal are also cultivated to certain extent.

IMPORTANCE OF LAC CULTIVATION

(1) Lac cultivation is a good source of livelihood for poor tribals inhabiting sub-hilly tracts with meagre investment.

(2) It is a good scope for marginal and degraded lands and has no competition for land operation with agriculture/horticulture crops.

(3) Highly remunerative cultivation. For example, one hectare of Ber plantation with Kusmi lac cultivation can produce net return of Rs. 3-5 lakh/year.

(4) It is like an insurance crop especially during drought year as the crop is very good during such adverse climate.

(5) A good number of host trees like Ber, Palas, etc. occurring naturally in forests and sub-forests in Chattisgarh are available for commercial exploitation.

(6) It is a good source of employment generation. One hectare of Kusmi lac cultivation on Ber generates 620 man-days of employment/year.

(7) Lac is the most superior natural resin. Besides resins, lac insects also yield lac dye and lac wax, all of which find extensive uses in Ayurveda and Siddha systems of medicines.

(8) Lac culture can help in ecosystem development. Cultivation of lac not only provides livelihood to millions of lac growers but also helps in conserving vast stretches of forests, lac insects and associated biota. Most of the lac host plants grow in forest areas and farmers resist felling of these trees, thereby protecting them for lac cultivation. Thus, lac culture plays a vital role in the protection of our bioresources. Cultivation of lac host plants for timber and fuel yields revenue in cycles of long years, whereas cultivation of lac on these trees gives a return almost every year.

(9) It has high export potential about 75%. In the world, India is the leader in lac production and export.

(10) Lac cultivation involves significant women participation.

STRAINS OF LAC INSECT AND LAC CROPS

Two strains of the lac insects are recognised in India, RANGEENI and KUSMI. The lac insects that thrive on the host plant Kusum is referred to as Kusmi whereas the Rangeeni strain generally grows on host plants other than Kusum. Each strain completes its life cycle twice a year but the seasons of maturity differ considerably. There are four lac crops in a year that are named after the Hindi months (Table 4.2). Lac insects under the genus *Kerria* are generally bi-voltine with two broods in a year. But few species like *K. lacca mysorensis* (host plant–Sal), and *K. sharda* (host plant–Kusum) are tri-voltine having three broods in a year. Again, species belonging to the genus *Paratachardia* (host plants-Tea, Sandal, etc) are all univoltine.

LAC INSECT

The lac insect is a minute, crawling, scaly insect with suctorial proboscis through which it sucks

Table 4.2 : Strains of lac insect and their characteristics

Characteristics	Kusmi	Rangeeni
Host plant	Kusum is main, others include Ber, Siris, Semialata, etc.	Rangeeni strain grows well mainly on Palas and also on a few other trees, but not on Kusum
Types of crop	(a) Jethwi (b) Aghani	(a) Kartiki (b) Baisakhi
Inoculation time	(a) Jethwi : Jan/Feb. (b) Aghani : June/July	(a) Kartiki : June/July (b) Baisakhi : Oct/Nov
Duration of life cycle	(a) Jethwi : Jan/Feb to June/July (b) Aghani : June/July to Jan/Feb	(a) Kartiki : June/July to Oct/Nov. (b) Baisakhi : Oct/Nov to May/June
Harvesting month of crop-yield	(a) Jethwi : June/July (b) Aghani : Jan/Feb	(a) Kartiki : Oct/Nov (b) Baisakhi : May/June
Quality of lac	Superior	Inferior to Kusmi lac

phloem sap of the host plant twigs. During growth, the insects secrete resinous lac from their body that gradually encases the body of the insect totally in a structure referred to as '**CELL**' (Fig. 4.1).

Life Cycle of Lac Insect : The life cycle of lac insect is completed within 6 months and it consists of stages like adult, egg and nymph instars (Fig. 4.2).

Adults : (i) **Female** : The female lac insect has a pyriform body measuring about 4-5 mm in length (Fig. 4.3a & 4.3b). The body is indistinctly divided into head, thorax and abdomen. Head bears a pair of degenerated antennae. Eyes are absent. Mouthparts are of piercing and sucking type. Posterior to mouth lie a pair of spiracles which ensures smooth

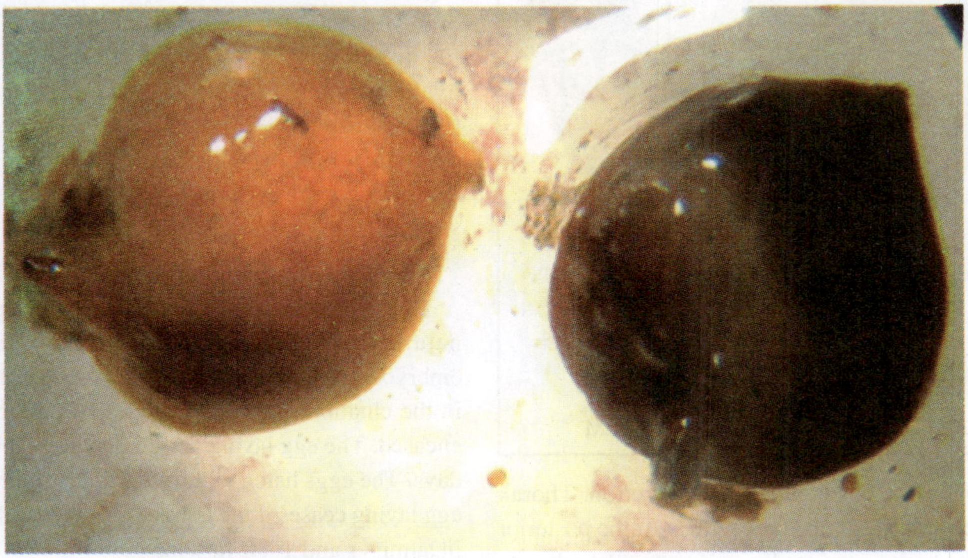

Fig. 4.1. Female lac cell encasing the insect body

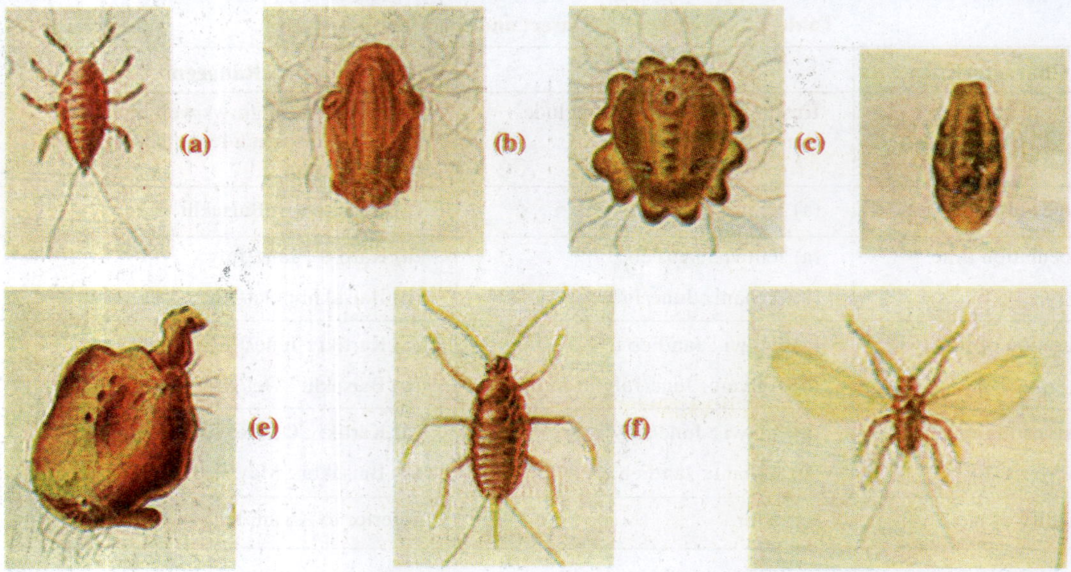

Fig. 4.2. Life cycle stages of lac insect : (a) First imter; (b) Female 4 weeks after inoculation; (c) Female, 13 weeks after inoculation; (d) Male cell, 13 weeks after inoculation; (e) Dead female cell with young emerging; (f) Wingless male; (g) Wingled male.

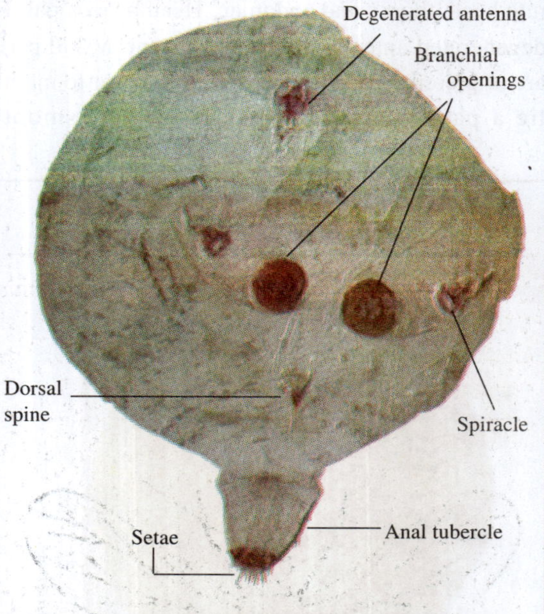

Fig. 4.3(a). Female lac insect under LM

breathing of the insect during lac secretion. Thorax has degenerated legs and lack wings. At the posterior part of the body, a triangular area – the anal tubercle encircling some setae is present. Near to it are present 2 branchial openings and one small chitinous spine called dorsal spine.

(ii) **Male** : The adult male is red in colour and smaller in size than the female insect. The length is about 1.2-1.5 mm. Head bears reduced eyes and ten segmented antennae. Mouth parts are similar to that of the females. Thorax has 3 pairs of legs. The male lac insect may be either winged with one pair of hyaline wing on its thorax or wingless (apterous) [Fig. 4.4 and 4.2(f)]. The eight-segmented abdomen ends into a short chitinous prominent sheath containing penis. A pair of white elongated caudal seta or filament is present on either side of this sheath.

Egg : The female lac insect is ovoviviparous in nature. So the laid eggs contain fully developed embryos within it. About 300-1000 such eggs are laid in the chambers (cell) in which the female remains encased. The egg laying period may last from 7 to 10 days. The eggs hatch within few hours of laying. But egg laying ceases if the temperature falls below 17°C in summer and 15°C in winter.

Nymphs : Following hatching, the first instar

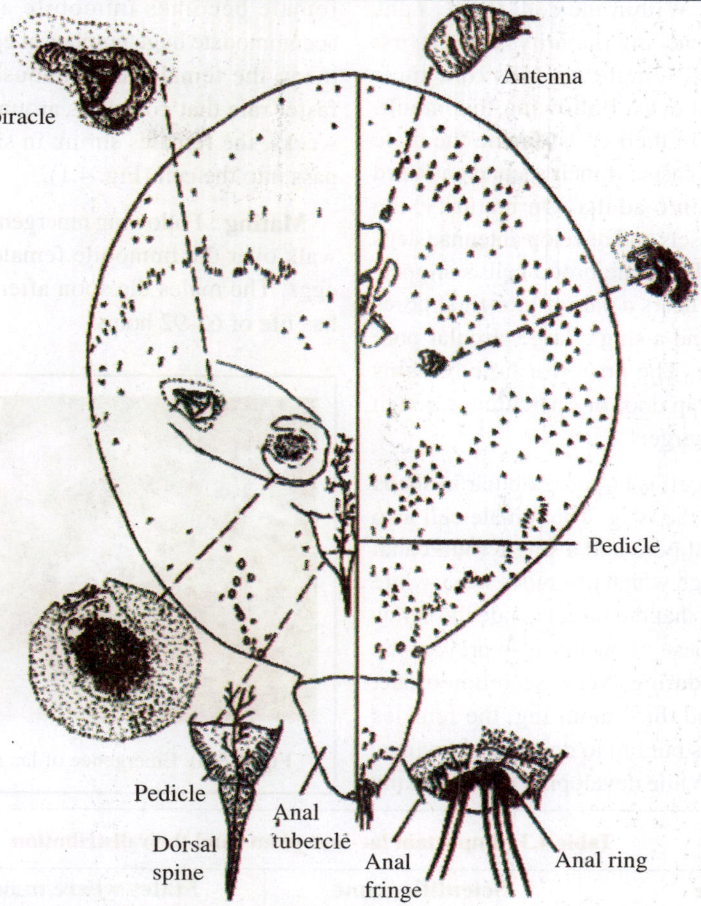

Fig. 4.3(b). Diagram of a female lac insect (Left half : dorsal view; Right half : ventral view)

nymph (Fig. 4.2a) stay within the cell for a brief period. Then the crimson red coloured nymphs, referred to as **'crawlers'**, come out of the cell in search of suitable host plant branch for settlement. The emergence of lac insect nymphs in huge number is commonly called **swarming** (Figs. 4.5a&b) that continues for several weeks. Boat-shaped nymphs are very small in size (0.5 mm) and divisible into head, thorax and abdomen. Head bears antennae, ocelli and mouth. Thorax has 3 segments, each with one pair of leg and caudal setae are found at the end of abdomen. On reaching soft succulent twigs, the nymphs settle down close together and start to suck phloem sap through their suctorial proboscis. After one day or so of settling, the nymphs start secreting lac from the hypodermal glands lying under their cuticle keeping open their mouthparts, breathing spiracles and anus. The secreted semisolid lac hardens on exposure to air and the nymph gets fully covered by the lac encasement, called as **lac cell** (Fig. 4.1).

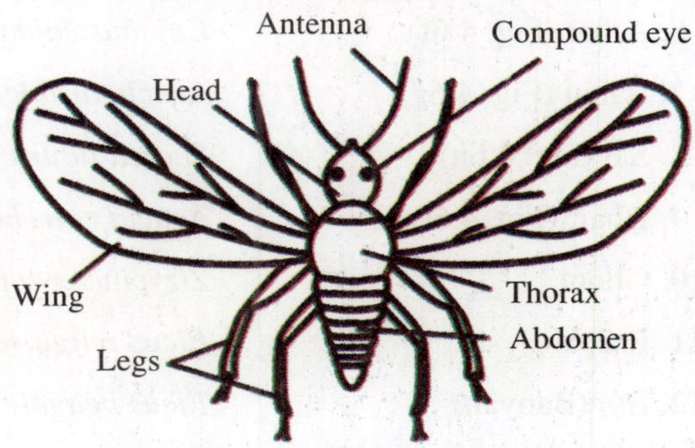

Fig. 4.4. Diagram of a male lac insect

Metamorphosis : Within the cell, the nymphs moult thrice before reaching maturity. During first moult both male and female nymphs lose their appendages, legs and eyes. Following this moult, dimorphism appears in their cells. Inside the male cells, the male nymph casts off their second and third moults and matures into adults. On maturity, the males lose their proboscis and develop antennae, legs and a pair of wings. The male brood cell is slipper-shaped (Fig. 4.2d). It bears a pair of branchial pores on the anterior side and a single large circular pore on the posterior side. The posterior hole remains covered by a round trap door or operculum through which adult males emerge.

The female brood cell is larger, globular in shape that remains fixed to the twig. The female cell also has a pair of branchial pore and a single round anal tubular opening through which protrudes waxy white filaments (it indicates that the insect inside the cell is alive and healthy). These filaments, also prevent the blocking of the pore during excess secretion of lac. Following second and third moulting, the females retain only mouthparts but fail to develop any wings, eyes or appendages. While developing into adult, the female becomes immobile and large in size to accommodate huge number of eggs. During development, the females continuously secrete resin at a faster rate that coalesces around its body. After 14 weeks, the females shrink in size allowing light to pass into the cell (Fig. 4.1).

Mating : Following emergence, the winged males walk over the immobile females and fertilize their eggs. The males die soon after copulation. A male has life of 62-92 hours.

Fig. 4.5(a). Emergence of lac nymphs from cells

Table 4.3 : Important lac-host plants and their distribution

Local name	Scientific name	States where mainly cultivated
1. Palas (Fig. 4.6a)	*Butea monosperma*	Jharkhand, MP, and UP
2. Ber (Kul) (Fig. 4.6b)	*Zizyphus mauritiana*	Jharkhand, West Bengal, Punjab, UP
3. Kusum (Fig. 4.6c)	*Schleichera oleosa*	Jharkhand, MP, Orissa, Tamil Nadu
4. Akashmoni (Fig. 4.6d)	*Acacia auriculiformis*	Jharkhand, West Bengal
5. Siris/Gulwang (Fig. 4.6e)	*Albizzia lucida*	Jharkhand, West Bengal
6. Arhar (Fig. 4.6f)	*Cajanus cajan*	Assam
7. Babul (Fig. 4.6g)	*Acacia arabica*	Jharkhand, West Bengal
8. Sal (Fig. 4.6h)	*Shorea tulura*	Mysore
9. Khair (Fig. 4.6i)	*Acacia catechu*	Jharkhand
10. Ghont	*Zizyphus xylopyrus*	MP, UP
11. Peepal	*Ficus religiosa*	All states
12. Bot (Banyan)	*Ficus bengalensis*	West Bengal
13. Semialata (Fig. 4.6j)	*Flemingia semialata*	West Bengal, Jharkhand

Fig. 4.5(b). Swarming of lac insect nymphs (crawlers)

Egg laying : About the time of egg laying, two yellow spots appear at the rear end of the female cell which gradually enlarge and become orange in colour (Fig. 4.9). At this time, the females contract at one end of the cell and thus vacates a space, called 'ovisac', inside the cell. Then the anal tubercle is drawn in the ovisac and the females start oviposit large number of eggs in the ovisac. The ovisac looks like orange due to presence of crimson lac dye which indicates the time of egg hatching.

Following hatching the nymphs emerge and the whole process begins all over again. After one cycle has been completed, and around the time when the next generation begins to emerge, the resin encrusted branches are harvested. From each crop, some encrusted twigs are retained for inoculation to the new host plants.

Parthenogenesis : In the life history of lac insect, parthenogenesis is known to occur when unfertilised eggs are directly hatched into nymphs. It is common in Kartiki crop of Rangeeni Strain.

Lac Host Plant

The plants on which lac insects thrive by sucking phloem sap from succulent twigs are called lac host plants. Till date, more than 120 plants have been recognised as host plants, but the lac insect is a natural pest on these plants. However, they are not treated as pests as such, because (1) they yield most useful product, (2) the host plants are economically not so important and (3) the insects cause temporary and recoverable damage to the host plants. What factors distinguish host plants from non-host plants are not exactly determined. It is assumed that phloem saps of all known host plants have pH ranging from 5.9–6.2 and density from 0.14-0.17. Among the host plants, some serve as hosts only in certain areas but not in others. For example, 'Arhar' in Assam and 'Ghont' in M P. The name of major host plants of lac insect found in India are given in Table 4.3 (Fig. 4.6).

It is important to mention that the quality and quantity of lac are directly related to the host plant. Recently, instead of only Kusum plant, Ber, Siris and Semialata have also been identified as good hosts for Kusmi brood lac.

General practice for lac cultivation

The process of lac cultivation or propagation of lac insects means the spread of lac insects on the same or different host plants. It involves few successive stages as follows :

(1) **Inoculation** : Inoculation is the first stage of lac cultivation. It is the process by which newly hatched (brood) nymphs get associated with new branches of host plants. It may occur by two ways : (a) Natural/Self-/Auto- inoculation and (b) Artificial inoculation.

(A) **Natural/Self-/Auto- Inoculation** : It occurs naturally and routinely. In this process of inoculation, the swarmed nymphs of one plant infect the same host again and start to inhabit (encrust) on it. However, such inoculation process is discouraged for certain reasons.

Drawbacks of Natural Inoculation

(i) **Nutrition deficiency** : If successive inoculations take place in the same host plant, it results in

Fig. 4.6. Host plants of lac insect : **(a)** Palas; **(b)** Ber; **(c)** Kusum; **(d)** Akashmoni; **(e)** Siris; **(f)** Arhar

Fig. 4.6. Host plants of lac insect : **(g)** Babul; **(h)** Sal; **(i)** Khair and **(j)** Semialata

stunted growth of the infested plant. Consequently, the insects may not get sufficient nutrition from the host plant. This in turn, will adversely affect the insect's growth and lac secretion.

(ii) **Non-uniform inoculation** : In natural system, inoculation may not take place in uniform sequence. In such case one/two crops may be absent, resulting in decreased lac production.

(iii) **Adverse climate** : The swarming of nymph depends on several climatic factors like rainfall, intensity of sunlight, direction of wind, humidity, etc. These factors, if not favourable, may affect the inoculation of nymphs and growth of the host plant at the same time.

(iv) **Parasites and predators** : If the lac is not harvested in time, the insects may invite their predators and parasites to multiply in large numbers on them. Thus, the ultimate growth of the insect, as well as lac production may be hampered.

Fig. 4.7. Brood lac sticks

(B) **Artificial Inoculation** : Due to several problems in natural system of inoculation, as mentioned above, nowadays artificial inoculation is being practised. It is brought about by the agency of man. The main idea behind this system of inoculation is to overcome the drawbacks of natural system. After the brood lac has been cut from the plants, it is necessary to subject it to proper examination, so that only healthy lac with minimum signs of predator and parasite damage is selected for use as brood lac. This is necessary to minimize the chances of propagation of the lac insect enemies. Artificial inoculation involves the following steps :

(1) **Transfer/inoculation of Seed twig/Brood twig** : The encrustation on a branch of host plant, if on squeezing is seen to exude red liquid, then this part of the branch can be considered as seed twig/ brood twig/brood stick. It indicates that living eggs are present within the encrustation that can be used for inoculation as seed.

In artificial inoculation, such seed twigs are cut in length of 20-30 cm (Fig. 4.7). Then these cut pieces are tied to fresh succulent twigs of the host plant in such a way that each brood twig gets touched at several places through which migration of nymphs from brood twig to host twig can take place (Fig. 4.8a). Brood twigs can be tied to host plant branch by 3 ways : (i) Longitudinal fashion where brood twig is tied longitudinally with host branch, (ii) Lateral

Fig. 4.8(a). Brood twig tied with host plant branch

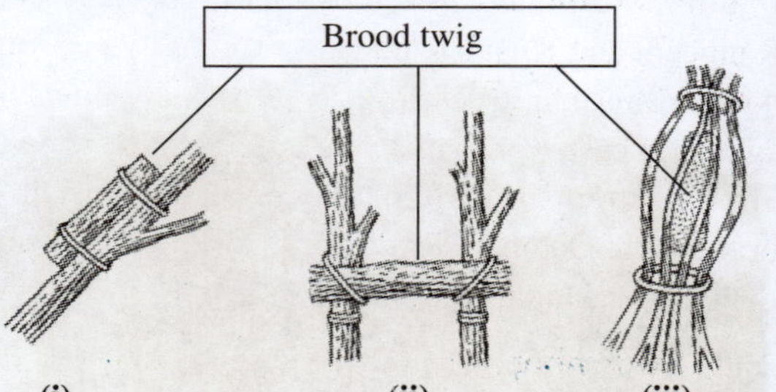

Fig. 4.8(b). Tying of brood twig with host plant branch. **(i)** Longitudinal; **(ii)** Lateral and **(iii)** Interlaced

fashion where brood twig is placed in between 2 host branches and tied with them and (iii) Interlaced fashion where singled brood twig is tied with several host branches together (Fig. 4.8b).

Precautions to be taken during Artificial Inoculation :

(a) Fully mature and healthy brood twigs, free from pest infestation and disease should be taken.

(b) Brood twig or seed twig meant for inoculation should be used immediately after crop cutting and should not be kept for a long time.

(c) To prevent falling of seed twig, tying of seed twig should be done securely on the upper surface of branches. It will ensure full contact for quick and easy crawling of nymphs from seed twigs to host twigs.

(d) Generally, for better inoculation, 3-4 seed twigs should be tied at one place of host twig.

(e) If the brood twig is taken from a particular host plant year after year, it may deteriorate the quality of lac production. Therefore, alteration of both brood and host is preferred for better quality of lac and proper nutrition of the nymphs.

(f) Humid weather may prevent swarming of nymph from brood. Hence, room-stored seed lac twigs may be heated to 200°C to induce swarming prior to tying.

(g) Cultivation of Kusmi in place of Rangeeni and vice versa should be avoided.

Time of inoculation : Each strain of lac insect (Rangeeni and Kusmi) yields two crops a year: Kartiki and Baisakhi in case of Rangeeni, and Jethai and Aghani in case of Kusmi. So the inoculation time of different strains is different. For Kartiki June/July; for Baisakhi October/November; for Aghani July and for Jethai is January/February.

(2) **Swarming** : It is most important stage of lac cultivation. Following tying, the nymphs swarm from brood and migrate to tender and succulent twigs of host plant to infest them (Fig. 4.5b). The time of swarming can be noted by the colour change of the eggs. At the time of hatching, the eggs become orange coloured. Following swarming of nymphs, the hollow cavities left in the cell get covered with-wax. Thus it is an indication that swarming has taken place.

(3) **Phunki Removal** : The operation pertains to the removal of brood lac twigs used for inoculation purposes, and the used up brood lac after complete emergence of lac nymphs from female cells is called ***phunki removal***. Ordinarily, the emergence of lac nymphs from the brood lac ceases after three weeks. The phunki lac is then scrapped off from the brood lac to avoid emergence of enemy insects. Phunki bundles are pulled down from the host trees with the help of pole mounted phunki hook or by climbing on trees.

(4) **Harvesting of lac** : Harvesting is the process of collection of lac from the host trees. It is done by cutting the lac encrusted twigs when it is matured. Harvesting can be done in two ways :

(i) **Immature Harvesting**: In this case lac is colleted before swarming. Lac, thus obtained, is called 'ARILAC'. The lac insect may be damaged at the time of immature harvesting which would cause population destruction of lac insects and may result in great economic loss to cultivators. However, in case of Palas lac (Rangeeni lac), this type of lac was found to give better quality of production. So immature harvesting is encouraged only for this lac.

(ii) **Mature Harvesting**: In this case, lac is collected after swarming. The lac, thus obtained is called 'Mature Lac'. For mature lac harvesting, it is very important to know the exact date of emergence and swarming of nymphs. A simple visual method is adopted for this purpose. A yellow spot develops on the posterior side of the female lac cell toward crop maturity. Gradually the spots spread forward until it covers half of the cell. Cutting of twigs for harvesting can be done at any time between the stages while the yellow spot occupies one-third to one-half of the cell area (Fig. 4.9). Sometimes, it is desirable to wait till the emergence of first new nymphs.

(iii) **Harvesting Time**: In accordance with different inoculation time, harvesting time for different crops are also different. The Kartiki crop is harvested in October/November; Baisakhi in May/June; Aghani in January/February and Jethwi in June/July (Table 4.2).

Some special measures to be taken during harvesting : Following considerations are recommended for harvesting :

(1) Only mature crops should be harvested. Immature/arilac cutting should be avoided except in Palas lac.

(2) A moderate crop is said to be the one from which nymph will emerge in 7-10 days. So the crop should be harvested within the above said days prior to nymphal emergence.

Fig. 4.9. Mature lac crop ready for harvesting. Note the yellow spot on female cells.

(3) If self inoculation is not required, then the entire crop may be reaped. In case of Rangeeni crop, only the lac encrusted twigs are cut, while in case of Kusum crop, reaping should be accompanied with pruning.

(4) The brood sticks harvested, should be utilised for inoculation as soon as possible. If needed, storage should be done in shaded, well ventilated room.

(5) Harvesting of lac crop at maturity can prevent the crisis of brood lac to a large extent without affecting the quality of lac (obtained as stick lac).

LAC CULTURE PRACTICE IN INDIA

(A) **Local practice** : As such, lac cultivation is simple and does not require any large investment. In India, local lac cultivation is carried out casually, as it is being regarded as subsidiary crop. Only natural inoculation occurs in such cases. However, such local practices have some disadvantages like :

(a) The same host plants are exploited year after year without giving rest for recoupment. So the trees lose vigour and are incapable to grow new succulent shoots. The trees gradually become weak and die subsequently.

(b) No pruning is done. Partial harvest is done keeping few branches untouched for self inoculation of next crop.

(c) The self-inoculation may invite heterogeneous infestation of nymphs leading to morbidity of total brood during hot season. Such situation may force the cultivators to abandon lac cultivation.

(B) **Improved practice** : For sustained production of lac as well as steady returns, some improved method of cultivation can be adopted. The underlying principle of such improved method is to provide sufficient rest to the host plants after each harvest. For this purpose : (1) **Coupe** and (2) Alteration of plant systems of cultivation are adopted.

(1) **Coupe System** : In this system, the host plants of a definite area are divided into coupe, i.e., groups, each containing certain number of plants. Trees of only one coupe are inoculated at one time. Following harvesting, these trees are given rest when they gain back the lost vigour, while other trees (which were in rest) with succulent twigs become ready for inoculation. Thus, in this coupe system, alternate groups of trees are used for lac cultivation. Full inoculation and full cropping can be done in this improved system.

(a) **Rangeeni Coupe** : In Rangeeni farms, *two coupe systems* with equal number of Palas trees in two coupes having six months rest is adopted for raising *Baisakhi-cum-Kartiki* crops, in alternate seasons. The trees are inoculated with about 500 gm of *Rangeeni* brood lac per tree, in the month of October-November. Harvesting is done after a year, after allowing self-inoculation in June-July by partial harvesting and then harvesting the combined *Baisaki-cum-Kartiki* crop in next October-November.

(b) **Kusmi Coupe** : In the *Kusmi* farms, Kusum is the major lac host plant species of *Kusmi* strain of làc insect. *Five-coupe systems* with equal number of trees in each coupe having 18 months rest in between pruning and inoculation is adopted. The trees are pruned 18 months prior to inoculation. Thereafter, in the subsequent crops, harvesting will serve the purpose of pruning. The harvesting of crop is done after 6 months of inoculation.

(2) **Alteration of Plant** : Another system adopted in improved lac practice is alteration of plant. In this system, the variety of host plant is changed after one crop. During inoculation, the seed twigs are transferred on the host plant of another variety. In this way, every type of host plant gets sufficient rest to regain its vigour and growth.

PREPARATION OF FEEDING GROUND FOR LAC INSECTS

To get good quality lac through cultivation, it is necessary to ensure proper type of feeding ground to the lac insects. The insects need to be provided with succulent shoots, as it cannot drive its slender proboscis through thick bark. For getting a good number of requisite succulent shoots, the most important method is pruning.

Pruning : Pruning means cutting away old, weak and diseased twigs from the host plants. It is done in January or June. It is very important for cultivation as it induces the host plants to throw out new succulent twigs. Pruning should be done with a sharp instrument to give a short and neat cut. If trees are old and have lost their capacity to produce vigorous shoots of new flush, heavier pruning is carried out to produce the new wood at the expense of the old. Such operation will bring the tree to a better shape, so that subsequent pruning will give the desired flush. Proper pruning should result in a good shape and give plenty of chances for the development of new shoots.

Objectives of Pruning

(1) To ensure new, good, healthy and succulent shoots.

(2) To ensure availability of large number of shoots (larger area for lac insect settlement).

(3) To provide rest to host plant for maintaining its vigour.

(4) To remove dead, diseased and broken branches.

Types of pruning in lac host plants

Two types of pruning have been recommended for lac culture.

(i) **Apical/ light pruning** : Branches less than 2.5 cm diameter should be cut from base and branches more than 2.5 cm diameter should be sharply cut leaving a stump of 30-45 cm from the base. Diseased and dead portion of branches should be removed completely. Light pruning is recommended for slow growing conventional host tree species like Palas, Kusum and Ber.

(ii) **Basal / heavy pruning** : Branches having less than 7 cm thickness should be removed from the base, whereas thicker branches should be cut at a place where it has a diameter of 7 cm. In quick growing bushy host, pruning should be done at a height of 10-15 cm from the ground level, e.g., *Flemingia macrophylla, F. semialata.*

Pruning time : After several years of experiment done at Indian Lac Research Institute (presently I.I.N.R.G.), Ranchi, Jharkhand, it has been found that the best results are obtained by pruning in February for raising the Kartiki crop and in April for raising the Baisakhi crop of Rangeeni for host plants Ber and Palas. Pruning in these months will give shoots of four and six months old respectively, for the lac larvae to feed on. In case of Kusum, pruning is best done in the month of June-July and January-February. These months coincide with those in which the crops mature, and so, harvesting of the mature crop serves the purpose of pruning also. Pruning time will, however, need to be adjusted to suit local conditions.

GENERAL MEASURES FOR IMPROVED LAC CULTURE

During lac culture, some important measures are also considered for improving lac cultivation.

(1) Use of superior breed of lac insects.

(2) Use of good quality brood lac in appropriate quantity.

(3) Post harvest management of lac crop, host plant management and lac pest management.

(4) Quality brood lac ensures high fecundity of insects and fewer requirements of inoculums.

(5) Timely harvesting of crops and proper inoculation may reduce the risk of loss of lac insect to a large extent.

(6) Involvement of forest and forest fringe dwellers in the Joint Forest Management (JFM) programme in different lac cultivating states to enhance lac production.

(7) If forest lac cultivators and state forest departments work jointly, the host plants under the custody of state forest can be used properly. It will result in a boom for lac cultivation and production as well.

PROCESSING OF LAC

The quantity of lac depends on the brood cell type. The female insect has long life span than the male. So the female secretes resin throughout her life yielding more lac than the males. The quality, on the other hand, depends upon the type of host plant. Depending on the purity, lac may be of different types which are obtained at different stages of whole lac processing system as described below :

(1) **First Stage** — **Production of stick lac** : During harvesting, the lac encrusted twigs are cut from the host plants. Then the lacs are removed from the twigs by scrapping. The lac thus obtained is

Fig. 4.10. Machine for crushing stick lac

Fig. 4.11. Machine for washing crushed seed lac

known as **stick lac** or **scrapped lac** or **crude lac** or **raw lac,** that contains insect body, lac dye, sand and twig debris, in addition to resins. If harvesting is done before emergence of larvae, then the lac obtained is referred to as **Arilac.** The stick lac, if required, can be stored on the floor in layers, less than 30 cm in height and racked frequently.

(2) **Second Stage** — **Production of seed lac** : Stick lac, if stored for long time, may form lump due to its high moisture content and may invite other storage pests, and thereby causes substantial loss and deterioration in quality of desired industrial demands. So seed lac is produced from the stick lac by further processing.

Stick lac is first crushed in a machine (Fig. 4.10) and sieved to remove sand and dust. Once the sieving is done the crushed stick lac is fed by basket into a Barrel Washing Machine (Fig. 4.11). The loading time taken is generally 15-20 minutes. Capacity of loading at a time of the barrel is its (diameter × length)/2. This washing removes the dye from the sick lac and also removes other materials like sand, wood impurities and insect body. The process usually takes around two to three hours to complete. It uses a lot of water and the rotary device inside the barrel, helps the churning of the stick lac and helps in the cleaning process. The water that comes out of the barrel is often taken onto a field and is allowed to dry down. The contents are sometimes used as pesticides and fertilisers. However, the use is much restricted nowadays. The washing time and the methods adopted to do this improves the quality of the seed lac.

Different types of washing are done for different grades single wash for low quality, double wash for medium quality and triple wash for superior quality. (a) **Single Wash** : This involves simple washing without or with negligible amount of soda for some 45 minutes. (b) **Double Wash** : Soda cleaning (750 gm for 500 kg) with low quantity of water for 20 minutes. Water is poured into the barrel from the top through pipe for about 25 minutes and stopped when colour of water passing through a 60 no. net at the back side of the barrel appears white. (c) **Triple**

Wash : Water cleaning for 20 minutes. Soda washing (1 kg for 500 kg material) for 1 hour (superior quality) or 45 minutes (medium quality). Then again water cleaning for another 30 to 35 minutes. This is done till the water coming out of the barrel shows no pigment.

After the completion of washing, the bottom part of the barrel is opened and the total mass is poured into a circular cemented basin (size of the basin depends upon the capacity of the washer). The masses thereafter are filtered through cloth (different quality of cloth is used for different grade). After removing the lighter impurities ('Khari' or 'Pati') from the mass, seed lac is collected.

The seed lac is then spread into the open cement floor for drying (Fig. 4.12). The layer of seed lac is continuously rolled over the floor from one side to the other side with the help of a 'Patri' (a wooden flat strip having a long handle) for facilitating the drying. The process of drying would take between an hour or two. Usually the washing should not go beyond 2 p.m. so that the finished product is dried by the end of the day.

After drying, the material is sieved through a number of '**Chaluni**' (circular hand operated sieving tools (Fig. 4.13). Each of them are of different size and helps sieve of the seed lac based on the size of the sieve. The residue so left out at this stage is called the '**Kuni**' and '**Molamma**'. Then all materials so sorted out are gathered and remaining impurities are cleaned using the hand operated winnowing baskets (bamboo made instrument called 'Kulo') usually done by women.

Fig. 4.12. Sun drying of washed seed lac

Fig. 4.13. Sieving and winnowing of seed lac

Fig. 4.14. Seed lac

The seed lac is obtained in the form of grain of 10 meshes or smaller, and yellow or reddish brown in colour (Fig. 4.14). Different grades of hand-made seed lacs are available in market like genuine Baisakhi, Fine baisakhi, Golden baisakhi, Golden Kusmi, Golden Kusmi-bold grain, Manbhum fine seedlac, etc.

(3) **Third Stage — Production of shellac from seed lac** : The lac most commonly used across the world is shellac which is produced by any of the following processes :

(a) **Indigenous process/country process** : In this traditional method, the seed lac is first filled into long sausage-shaped cloth bag of about 2 inches diameter and 30 feet length (Fig. 4.15). The bag is then gradually passed in front of a charcoal-fired hearth to melt the lac. By twisting the bag, the molted lac is then squeezed out through the sieve of the cloth. This molten filtered mass is then either stretched into sheets of about 0.5 cm thick with the help of glazed ceramic cylinder, or allowed to solidify in form of dice, called **button lac** (Fig. 4.16a). The residue left inside the cloth bag contains apart from lac, sand, insect debris, and is referred to as **kirrilac** (Fig. 4.16b).

Fig. 4.15. Processing of shellac by indigenous process

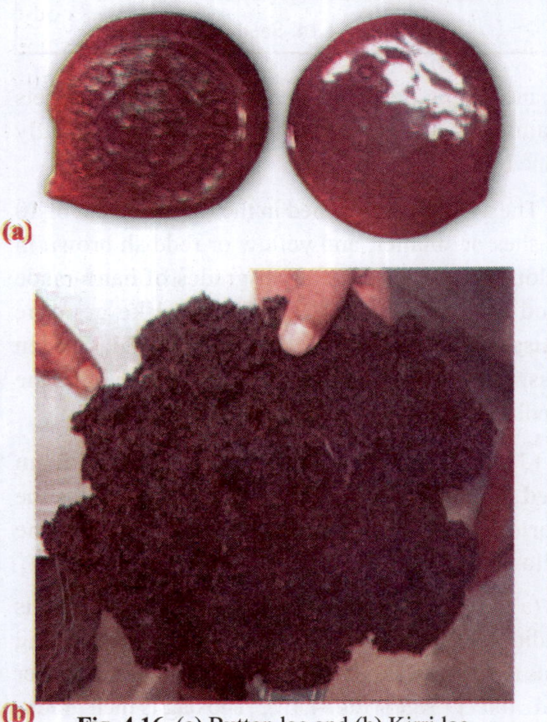

(a)

(b)

Fig. 4.16. (a) Button lac and (b) Kirri lac

The shellac obtained by such method are available in various commercial forms like Lemon one shellac, Lemon two shellac, Superior shellac, Kusmi button lac, Light pure button lac, etc.

(b) **Heat process (Fig. 4.17)** : In this process, the seed lac is melted by steam heat. Oxalic acid and titanium dioxide are mixed with the melted lac at this stage. The molten lac is then squeezed through filter by means of hydraulic pressures. The filtered molten lac is drawn into long and continuous sheets with the help of roller (Fig. 4.17a). These sheets are broken into pieces, called **flakes.** Various grades of machine made shellac flakes are available commercially, that include Orange shellac, Black T.N. shellac, Orange fine shellac, (Fig. 4.17b), etc.

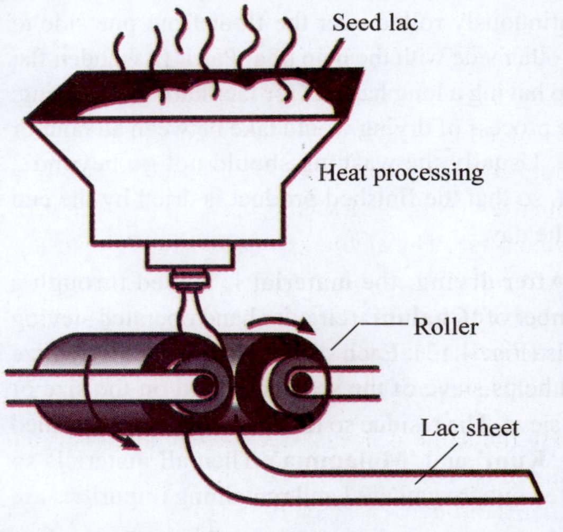

Fig. 4.17(a). Heat process of making shellac *(diagrammatic)*

Fig. 4.17(b). Different varieties of shellac flakes

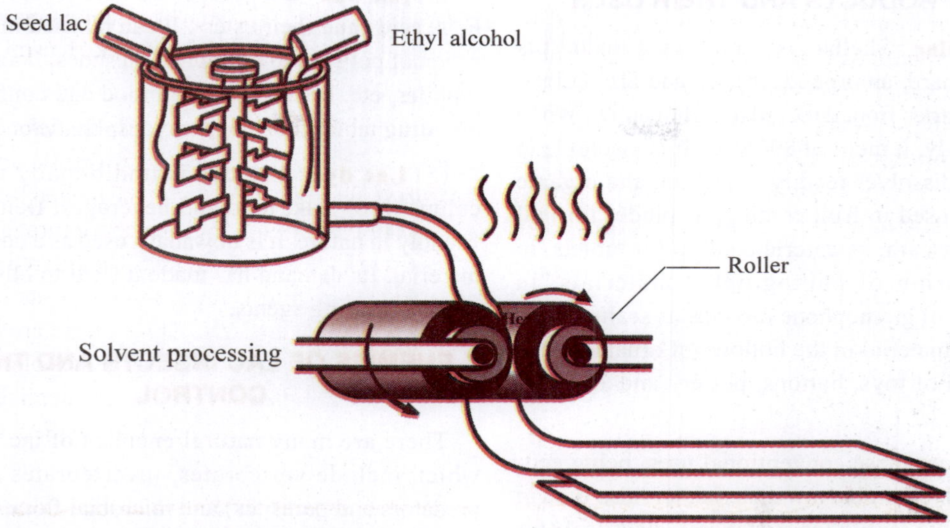

Fig. 4.18. Solvent process of making dewaxed lac (*diagrammatic*)

(c) **Solvent process** : In this process, the seed lac is dissolved in refrigerated alcohol and then filtered through filter press to remove the wax and other impurities (Fig. 4.18). The lac thus obtained is **dewaxed lac**. The alcohol is then removed with the help of roller; the molten shellac is then stretched. The solvent process may yield various grades of commercial shellac, like Dewaxed platina, Dewaxed blonde, Dewaxed lemon, Dewaxed garnet, etc.

By the same process, decolourised lac can be obtained by removing the colour (to any required standard) through charging of the dissolved lac with activated carbon. The basic procedure of bleached shellac manufacture includes following steps : (1) Dissolving the shellac in aqueous sodium carbonate solution at 90-100□C; (2) stirring with sodium hypochlorite and (3) filtering after cooling. The bleached shellac, thus obtained, is then reclaimed from the filtered solution with sulphuric acid. This bleached shellac is then again filtered and finally dried. The bleached lac is white in colour and commercially available in two grades, dewaxed bleached shellac and waxy bleached shellac.

COMPOSITION OF LAC

Major constituent of lac is resin (68%). Lac resin is thought to be a polyester complex of straight chain hydroxy fatty acid of C14-C18 carbon chain (e.g., Aleuritic acid, butolic acid), mono- and di-hydroxy acid. Other components include dye (1-2%), wax (6%), sugar, protein, soluble salts, sand, woody matter, insect body debris, etc. Lac dye is a mixture of anthroquinoid derivatives and contains erythrolaccin and laccaic acid.

PROPERTIES OF LAC

(i) Lac is soluble in alcohol and weak alkalis but insoluble in water.

(ii) It has capacity of forming uniform durable film.

(iii) It possesses high scratch hardness.

(iv) Heat soluble, at 80□C it melts.

(v) It is good adhesive in nature.

(vi) Ability to form good sealers, undercoat primers.

(vii) Lac has the capacity to allow quick rubbing with sandpaper without slicking or gumming.

(viii) It is non-conductive and non-toxic.

No other single resin, both natural and synthetic, possesses so many desirable properties, and so lac is also termed as Multi Purpose Resin.

LAC PRODUCTS AND THEIR USES

(1) **Shellac** : Shellac is natural gum resin. It is non-toxic, hard, amorphous, brittle and also edible. Its colour varies from dark red to light yellow. When heated slowly, it melts at 89-90°C. It is insoluble in water, but dissolves readily in alcohol and organic acids. It is used in fruit coating, as binder for nail varnish, mascara, as enteric coating for tablets, in manufacturing of photographic materials, in preparation of gramophone records, as sealing wax, as a filling material in the hollows of ornaments, in preparation of toys, buttons, pottery and artificial leather.

Apart from these conventional uses, being eco-friendly and renewable raw material, it is also used in industries like textile (as stiffeners), electrical (for insulation, capping, etc), paint, aluminium, pharmaceuticals, confectionary, food processing, cosmetics, antique, etc.

(2) **Dewaxed bleached shellac** : This non-toxic shellac is a bit harder and brighter than other shellac. It is completely free from wax and has high gloss and excellent adhesive quality to various substrates. It can safely be used in cosmetics (hair sprays, shampoo, perfumes, lipsticks, nail polish, eyeliners, etc.), confectionary (coating of cakes, eggs, chewing gums, cheese, fruit coating), etc. Shellac aleuritic acid is now being used in synthesis of glucose mono-aleuritate (a non-toxic, non-hemolytic water soluble compound), an isocaloric substitute for dietary tripalmitin. Aleuritic acid esters are used in plastic, fibre materials and perfume industries also.

(3) **Bleached shellac** : It is also non-toxic and physiologically harmless and thus is widely used in food industries, food packaging and allied industries. It is also used in binding and adhesive drying-related works. Besides, it can also be used in industries related with flexographic printing, confectionary, stamping ink, coating for processed food and dry flowers, textiles; cosmetics, wood finishing, fire works and pyrotechnics (as binder for fireworks, match sticks etc.), grinding wheels, rubber and plastic, electronics, etc.

(4) **Lac wax** : Lac wax is a mixture of higher alcohols, acids and their esters. It can be used in preparation of polishes to be applied on shoes, floor, automobiles, etc. It is also used in food and confectionary, drug tablet finishing, crayons, lipsticks, etc.

(5) **Lac dye** : Lac dye is traditionally used to colour wool, silk, food and beverages. Being eco-friendly in nature, it is nowadays used as a colouring material. Its demand has made it rival to other synthetic colouring agents.

ENEMIES OF LAC INSECTS AND THEIR CONTROL

There are many natural enemies of lac insects which include vertebrates, invertebrates (insect predators and parasites) and microbial flora.

(1) **Vertebrate enemies of lac insects**

The important vertebrate enemies are squirrels and rats. In worst conditions, the damage caused by these enemies can be as serious as 50% of brood sticks. Squirrels are active during the daytime and the damage by them is more common under forest conditions. Rats are active at night-time and the damage usually occurs near about the villages. Towards the crop maturity, these pests gnaw the mature lac encrustation on the tree, damage the brood lac tied to trees for inoculation and consume the full grown gravid female lac insects. The damage to brood lac tied to trees interferes with the inoculation, as the brood bundles and the lac encrustations drop to the ground where the larval emergence is taking place. Besides squirrels and rats, monkeys also cause some damage to lac encrustations and to the newly developing shoots from pruned host trees by breaking them.

Control : It is difficult to control the squirrels and rats under the open field conditions where lac is cultivated. However, scaring away of these animals or poisoning them may be adopted to keep the rodents under attack.

(2) **Insect enemies of lac insect**

It has been estimated that on an average, up to 30-40% of the lac cells are destroyed by insect enemies of lac crop. At times, the enemy attack can be so serious as to result in total crop failure. There are two kinds of enemy insects : (i) Parasites and (ii) Predators

(i) **Parasites** : All parasites causing damage to lac insect belong to the Order Hymenoptera of class Insecta. A list of parasites associated with lac insect, *Kerria lacca* is presented in Table 4.4.

Among the parasites listed, *Tachardiaephagus tachardiae* and *Tetrastichus purpureus* are the most abundant lac associated parasites. They lay their eggs in the lac cells and the grubs (larvae) after hatching start to feed on the lac insect within its cell.

(ii) **Predators** : The predators, on the other hand, are more serious and may cause damage up to 30-35 per cent to the cells in a crop. The important predators of lac insects are listed in Table 4.4. Of the predators, *Eublemma amabilis* and *Pseudohypatopa pulverea* are the most destructive to lac insects and are in regular occurrence but their incidence may vary from season to season, place to place and crop to crop. A brief description of these two predators are given below :

(1) **Eublemma amabilis** : This white moth is the most destructive predator of lac insect (Fig. 4.19) and causes most damage during *Kartiki* and *Aghani* lac crops, i.e., during the rainy season in comparison to the other two crops.

Life history : A single female moth lays greyish, flat and rounded eggs singly on the test of lac insect. The newly hatched larvae, 0.51 to 0.54 mm long, get at the lac insect either through the opening of the test or by making a hole through encrustation. A single larva can destroy 40-60 lac insect cells in its whole larval period. It has six generations in a year and the duration of the generations are about 37, 45, 42, 125, 80 and 40 days respectively. Attacked lac cells can easily be identified because of its pinkish colouration due to presence of pink coloured discs of excreta inside the hollow lac cells.

(2) **Pseudohypatopa pulverea** : It is also destructive predator of lac insects and found in all lac

Table 4.4 : Some important Parasites and Predators of lac insect

Parasite of lac insects	Family	Predators of lac insects	Family
Anicetus dodonia	Encyrtidae	*Eublemma amabilis* (Fig. 4.19)	Noctuidae
Atropates hautefeuilli	Encyrtidae	*E. coccidiphaga*	Noctuidae
Aphrastobracon flavipennis	Encyrtidae	*E. cretacea*	Noctuidae
Bracon greeni (Fig. 4.20)	Encyrtidae	*E. scitula*	Noctuidae
Campyloneurus indicus	Encyrtidae	*Pseudohypatopa pulverea*	Blastobasidae
Coccophaqus tchirchii (Fig. 4.21)	Aphelinidae	*Catablemma sumbavensis* (Fig. 4.22)	Blastobasidae
Erencyrtus dewitzi	Encyrtidae	*Cryptoblabes ephestialis*	Blastobasidae
Eupelmus tachardiae (Fig. 4.23)	Eupelmidae	*Phroderces falcatella*	Cosmopterygidae
Eurymyiocnema aphelinoides	Aphelinidae	*Lacciferophaga yunnanea*	Momphidae
Lyka lacca	Encyrtidae	*Chrysopa madestes*	Chrysopidae
Marietta javensis	Aphelinidae	*C. lacciperda*	Chrysopidae
Parageniaspis indicus	Encyrtidae	*Berginus maindroni*	Mycetophagidae
Parechthrodryinus clavicornis	Encyrtidae	*Silvanus iyeri*	Cucujidae
Protyndarichus submettalicus	Encyrtidae	*Tribolium ferrugineum*	Tenebrionidae
Tachardiaephagus tachardiae	Encyrtidae	*Phyllodromia humbertiana*	Blattellidae
Teachardiobius nigricans	Encyrtidae	*Ishonoptera fulvastrata*	Blattellidae
Aprostocetus (Tetrastichus) purpureus (Fig. 4.24)	Eulophidae	*Dolichoderus thoracicus*	Formicidae

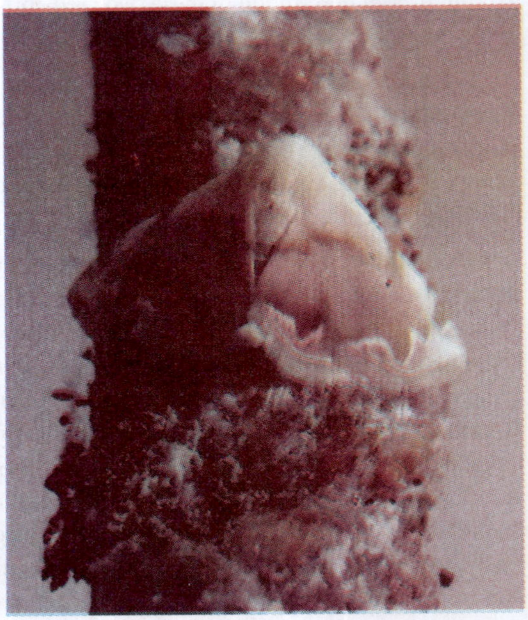

Fig. 4.19. *Eublemma amabilis*, an important predator of lac insect

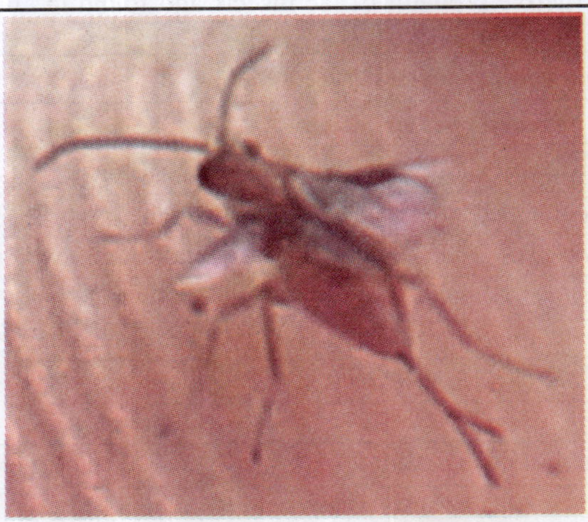

Fig. 4.20. *Bracon greeni*, an important predator of lac insect

growing areas in the country. It feeds on the live and dead lac insects and is found in large numbers in stored lac and so it is responsible for the qualitative and quantitative deterioration of stored lac.

Life history : It lays oval (0.5 mm □ 0.3 mm), colourless eggs, singly on the test of lac insects. Larvae pass normally through 5 instars but the

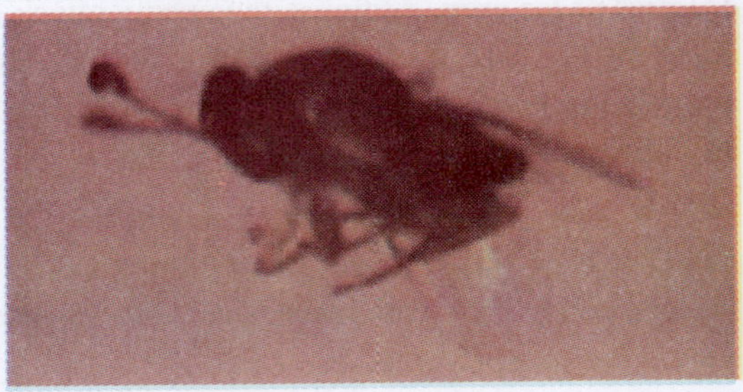

Fig. 4.21. *Coccophaqus tchirchii*, a parasite of lac insect

Fig. 4.22. *Catablemma sumbavensis*, a parasite of lac insect

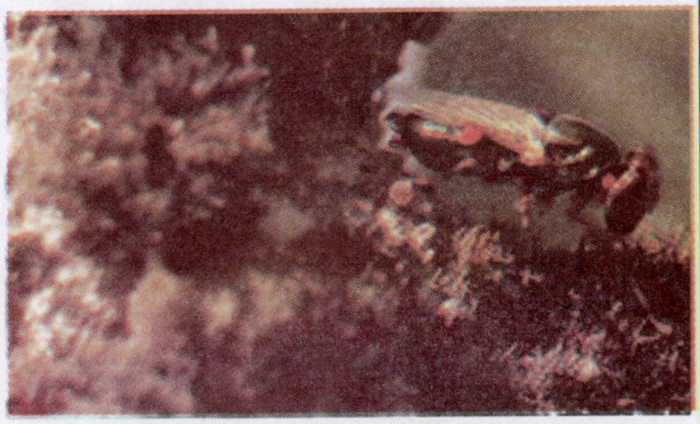

Fig. 4.23. *Euplemus tachardiae* a parasite of lac insect

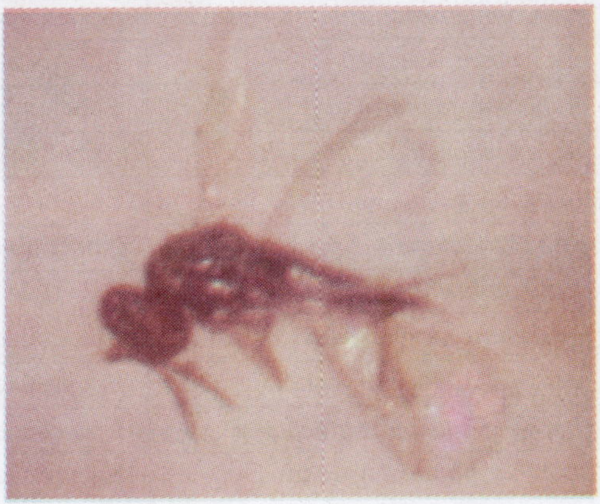

Fig. 4.24. *Aprostocetus purpureus,* a parasite of lac insect

hibernating larvae have nine instars. The newly hatched larva is about 1.35 mm long whereas a mature larva is 10-12 mm in length and 2 mm in breadth. Larval stages feed on the lac larvae and spins a loose web. A single larval predator is capable of destroying 45-60 mature lac cells.

PREVENTION AND CONTROL OF INSECT ENEMIES

Preventive measures :

(a) Parasite- and predator- free brood lac should be used for inoculation.

(b) Self-inoculation of lac crops should be avoided as far as possible.

(c) Inoculated brood bundles should be kept on the host tree for a minimum period only.

(d) Phunki (empty brood lac sticks) should be removed from the inoculated trees in 2-3 weeks time.

(e) All lac cut from the tree and all phunki brood lac (after use as brood lac) not required for brood purpose should be scraped or fumigated at once.

(f) Cultivation of Kusmi strain of lac should be avoided in predominantly Rangeeni area and vice versa.

Mechanical control : Use of 60 mesh synthetic netting (brood bag) to enclose brood lac for inoculation purposes can reduce infestation of enemy insects of lac. The emerging lac larvae easily crawl out from the minute pores of the net and settle on the twigs of the lac host plants, whereas the emerging adult predator enemies cannot move out of the brood bags and get entrapped within the net. This can check the egg laying by the predator moths on the new crop.

Chemical control : Application of 0.05% endosulfan at 30-35 days stage of crop has been identified as the most effective dose of insecticide without any adverse effect on the economic attributes of the lac insect.

Microbial control : Use of bio-pesticide, Thuricide *(Bacillus thuringiensis)* at 30-35 days stage of crop is the effective microbial control measure for important enemy insects of lac in field condition.

Biological control : Two ant predators, viz. *Camponotus compresus* and *solenopsis geminate rufa,* are the most important and promising for biological control of predator enemies of lac in field condition. Egg parasitoids, viz. *Trichogramma pretiosum, T. chilonis, T. poliae, Trichogrammatoidea bactrae* and *Telenomus remus,* have been found to be effective in management of many lac predators like *P. pulverea.* Again, hyperparasitism is found to happen in some lac cultivation areas where parasites of lac insects could also be controlled biologically by hyperparasitic insects, viz., *Aprostocetus (Tetrastichus) purpureus* (Fig. 4.24) is secondary parasite of *Coccophaqus tchirchii* (Fig. 4.21), and *Eupelmus tachardiae* (Fig. 4.23) is a secondary parasite of *Eublemma amabilis.*

Microbial flora associated with lac insects :

Two types of microflora, viz. bacteria and fungi, are associated with the lac insects. Bacteria could be symbiotic or pathogenic. Microbial studies revealed that four species, viz. *Micrococcus varians. M. conglomerates, Clostridium sp.* and *Bacillus subtilis,* are found in permanent association with various stages of lac insects. Presence of various symbiotic microflora is considered beneficial for good yield of lac, particularly during rainy season crop. However, on the other hand, association of fungi with lac insect is not always beneficial. Fungal infection in lac culture causes severe losses of lac yield by :

(1) Killing the lac insects by inhibiting respiration.

(2) Hindering mating process.

(3) Blocking larval emergence.

(4) Affecting lac host efficiency.

Lac culture during rainy season is prone to fungal attack particularly when grown on Ber and Kusum trees due to their steady and spreading crown. Three species of fungi belonging to family Eurotiaceae and Aspergillaceae, viz. *Aspergillus awamori, Aspergillus terricola* and *Penicillium citrinum,* are reported to cause maximum loss in lac crop. *Aspergillus awa-*

Table 4.5 : Name of some important pests of lac host plant and their nature of damage

Name of the pest	Host plant	Nature of damage
Myllocerus cardoni	Kusum, Palas, Ber	Leaf defoliator
Diacrisia oblique	Kusum	Infest young seedlings of Kusum
Serinetha augar	Kusum	Nymphs and adults damage fresh dry fruit and seedlings to kusum
Megalothrips distalis	Palas and Kusum	Adult and nymphs feed on young shoots and leaves
Tetatodus monticollis	Palas, Ghont and Ber	Pest of leaves
Cyclopelta obscura	Palas	Damage leaf and tender shoots
Halys dentatus	Kusum	Feed on leaves
Ericeia umbricola	Ber and Khair	Leaf pest
Tarucus theophrastus	Ber and Khair	Pest of soft tissues of leaves
Selepa celtis	Ficus and Kusum	Caterpillars are defoliators
Oocardia pudibunda	Ber	Both adult and larvae feed on leaves
Semiothisa fidoniata	Khair	Feed on different parts of tree
Parasa lepida	Sal, Palas and Kusum	Caterpillar acts as defoliator

mori (Fig. 4.25) and *Penicillium citrinum* are black and greenish in colour respectively, and make a continuous cover on lac insect cells and thereby blocking their breathing pores which ultimately lead to mortality of lac insects. A pathogenic fungus, *Pythium sp.* in female tests, causes a heavy mortality of the larvae which fail to enclose satisfactorily and lie dead in clusters within the famale resinous cells.

Fig. 4.25. Fungal-cover on lac encrustation

Prevention and control : Application of fungicides, Bavistin (carbendazim 0.05%) and Dithane M-45 (mancozeb, 0.18%) by both dipping of brood lac before inoculation and spraying on standing crop gives significantly better yield of lac. Significant reduction (84% to 75%) in mortality of 2nd inster lac nymphs/larvae can be done by the application of different concentrations of carbendazim and aureofungin on Kusmi stain of lac insects.

PESTS OF LAC HOST PLANTS

Lac insects require plant species for survival. More than 400 lac host plants have been observed to carry lac insects throughout the world. Besides lac insects which thrive on the sap of known host plants (Table 4.5) , a number of other insects also feed on them and directly consume various vegetative and floral parts from those plants. These insects act as pests and cause not only damage to the lac hosts but also offer direct competition to the lac insects by devitalising the host trees resulting into decreased production of lac of commerce. Some pests of lac hosts have been known to inflict very serious damage in seasons of serious outbreaks. Some common and important pests of lac host plants, their nature of damage are listed in Table 4.5.

Control measures : Control of host plant pests is a complicated problem because of close association of lac insect with its host and its sedentary and colonial habit. Again, lac insect, its own pest as well as the pest of host plants live very close to each other. So the convenient pest control technologies available till date are not applicable to suppress the pest population either on the host or on lac insect, because the common insecticides in generally recommended doses while killing the pests, may also prove extremely detrimental to the lac insect. So the selection of insecticides, their application and doses should be very judicious. After exhaustive experiments on control aspects of pests of lac host plants, following are recommended : Aqueous solution of 0.01 to 0.3% Endrex-20 EC, Aldrex-30 EC, Dieldrex-18EC, BHC 50% WP, Chlordane-65% dust, DDT-25 and 50 EC, DDT-50% WP and Endosulfan-35 EC have been used for control of most destructive pests of host plants.

LAC ECOSYSTEM

In India, most of the lac host trees are confined in forest areas where the lac insect along with its host plants form an ecosystem. It is a complex multi-trophic web of flora and fauna, representing a rich biodiversity. It includes besides the lac insects, lac-host plants, several predators and parasites of lac insects, secondary parasites, microboes and a variety of pests of host plants. In this ecosystem, **the lac host plants constitute the first trophic level, lac insects and pests of host plants make the second, predators along with primary parasites of lac insect form the third and hyperparasites constitute the fourth trophic level.**

Biodiversity in lac ecosystem

Host plants : **First trophic level** — Lac insects require plant species for survival. More than 400 lac hosts have been observed to carry lac insects throughout the world. Important host plants available in our country were listed in Table 4.3.

Lac insects and pests of host plants : **Second trophic level** — Nine genera and 87 species of lac insects are reported from the world. Most important and common lac insect species found in India include *Laccifer (Kerria) lacca, L. albizziae, L. fici, L. indicola, L. longispina* and *L. sindica* etc. In addition to lac insects, many other insects and organisms live on lac host plants. These include pests (see Table 4.5), birds and mammals.

Predators and parasites of lac insect : **Third trophic level** — In spite of the protective lac covering, the lac insect due to sedentary habit is always a sitting prey to predators and parasites. Twenty-two lac insect predators, 30 primary parasites, and several fungal as well as bacterial pathogens are known to present in lac ecosystem.

Hyperparasites : **Fourth trophic level** — A number of arthropods (insects) often visit the lac encrustation which prey on or parasitise many important lac predators, thus cmprising the fourth trophic level of lac ecosystem. These secondary parasites or hyperparasites are indirectly beneficial to lac cultivation in a way that they keep off the predator enemy insects. Beneficial insects mostly belong to Chalcidae (*Brachymeria tachardiae*) (Fig. 4.26), Braconidae (*Apanteles tachardiae, Chelonella cyclopyra*), Elasmidae (*Elasmus claripennis*), Chneumonidae (*Pristomerus sulci*) (Fig. 4.27) and Bethylids (*Perisierola pulveriae*). They parasitise the eggs or larvae of *Eublemma amabalis* and *Pseudohypatopa pulverea* (see also biological control).

Fig. 4.26. *B. tachardiae*, a secondary parasite

Fig. 4.27. *Pristomerus sulci*, a secondary parasite

Fig. 4.28. Visiting ant to lac encrustration

In lac ecosystem, some ants are always found which are the body guards of lac insects (Fig. 4.28). Again, few vertebrates (rat, squirrel etc.) attack lac cells during maturation time. Moreover, natural lac insect ecosystem maintains a variety of tree flora, macrofauna and soil micro-organisms.

PROBLEMS OF LAC CULTURE IN INDIA

Lac cultivation is simple, free from high technology and does not need large investment. It requires only part-time attention of cultivators, thus, eminently suited to the tribal/rural people living in the vicinity of natural and man-made forests. However, sustained attention on the lac cultivation will lead to the sustained production. Despite good unrealised demand for lac (both domestic as well as overseas), the lac production growth had not been very healthy in India. Since 2008, Indian lac cultivators are facing several problems as stated below :

(1) **Change in climatic conditions** : In many parts of India, entire colonies of lac insect, *Kerria lacca*, have been dying during the reproductive period. The reason for such mass death is uncertain. According to one school of thought, unbearable heat during March and April may hinder the insects to infest the next crop. Again, the excessive heat makes the resin melt, killing the newly produced young insects.

(2) **Parasites** : Some parasites grow within the lac insect and thus insecticides cannot be used because that will kill the parasite along with the lac insect.

(3) **Wrong cultivation** : Delay in harvesting of brood lac, crop inoculation and wrong method of transportation of brood can cause huge loss in production of lac.

(4) **Wrong pest management** : Untimely and incorrect pest management operations may result in substantial loss to lac crop or even complete failure.

(5) **Lack of adequate processing facilities** : The absence of adequate processing facilities, makes lac cultivators unable to realise the full value of their product. This is because the market value of crude lac is rather limited in comparison to the final processed products, such as shellac, which commands a premium in the market.

(6) **Inefficient extraction method** : The lac dye isolated during the processing is a valuable product for the textile industry as a colouring agent. The pigment content of the stick lac can be as high as 10% but the yield of isolated lac dye usually is less than 1% due to inefficient extraction method as followed in country system.

(7) Failure of protection mechanism and uncontrolled illicit cutting has resulted in degradation of forest as well as lac host plants.

(8) The lack of access to credit and insurance services.

Measures taken for advancement in lac culture

(1) **Conservation** : A conservation package including protection of lac host plants, their management and propagation for the purpose of lac cultivation is now being started for each region to ensure viable income for rural people. The National Forest Policy (1988) has given due importance to people's participation in creation and conservation of resources.

(2) **Women's involvement** : Women can do a majority in the operations in lac cultivation and post-harvest operations, in addition to the household works. Thus, women are now given training and their skills and competencies around various operations are utilised for lac production.

(3) Use of alternate host plants, specially for areas where traditional host plants are not available.

(4) Development of multi-locational field trials for testing new breeds.

(5) Development of integrated lac based farming system.

(6) Establishment of nursery for distribution of saplings of lac host plants in large number.

(7) Establishment of information hub for lac cultivators.

(8) Improvement of existing training facility at ILRI.

(9) Demonstration of lac cultivation process via models to encourage the lac cultivators.

Recent activities for improvement of lac culture in West Bengal

In West Bengal following districts are practising lac cultivation, viz., Purulia, Bankura, Midnapur, Murshidabad and Malda. However, during last few years, due to various problems lac production as well as lac cultivation has undergone many sufferings.

So the Government of West Bengal has undertaken an active role through Directorate of Micro and Small Scale Enterprises (M and SSE). Their main aim is to develop lac cultivation for the improvement of livelihood and socio-economic condition of poor tribal and backward families. Therefore, government has taken up following programmes :

(i) In order to educate lac growers and to make them aware of improved method of lac cultivation, demonstration centres have been set up by Directorate of M and SSE.

(ii) In order to cover the demand of the lac cultivators, the Directorate of M & SSE has established supplementary scheme for purchase of brood lac from local market and distribution through local Panchayets and Panchayet Samities to the poor lac growers at free of cost.

(iii) Organising cultivation training camp in collaboration with Panchayet Samity/Gram Panchayets.

(iv) In order to encourage the development of lac industry within the state, West Bengal Government has been promoting and assisting the lac artisans' Cooperative Societies (LACS) for producing quality lac products. The existing 3 LACS (Purulia-1, Bankura-1 and Murshidabad-1) are manufacturing quality sealing wax as per I.S. specifications and supplying them to the districts during election.

(v) Establishment of Training Cum Servicing Centre (TCSC) for industrial training to lac cultivators and lac artisans. These centres educate many youth regarding row lac processing and lac based handicrafts.

(vi) The TCSC of Chalna, Bankura has undertaken a special Lac Training Scheme for manufacturing lacquered dolls and jewelleries.

(vii) Imparting training to Self Help Groups in the cultivation of lac.

(viii) Apart from above, another programme under R.K.V.Y has been continuing since 2003-04 with a view to increase lac crop production, specially at Purulia and Bankura.

Recent activities for improvement of lac culture in Jharkhand

In India, Jharkhand produces maximum quantity of lac per year. Palamau, Latehar, Garhwa, Ranchi, Lohardaga, Gumla, Simdega, Saraikela, Hazaribagh, Chatra, East and West Singhbhum are main districts of Jharkhand where lac host plants are available and thus lac culture is being practised. Due to lack of proper knowledge, rural people of this state are facing several problems and losing interest in it. To overcome these problems, The Indian Institute of Natural Resins and Gums (IINRG), Ranchi organises several training programmes, scientific knowledge and training on lac cultivation for all stakeholders. Some of such programmes are mentioned below :

(1) **Training on modern method of lac cultivation** : It is mainly meant for unemployed rural youth, forest guards and supervisory staff.

(2) **Education programme** : This is one week duration programme, meant for agricultural and non-agricultural science graduates.

(3) **Farmers training** : This one week duration programme is on scientific lac cultivation, primary processing and uses.

(4) **Trainers training programme** : The participant includes community workers, SDO, Forest Rangers, District Rural Agriculture Extension Officers, members of Agriculture Development Officer, lac facilitator, primary society managers, Deputy Managing Directors of Districts (Forest), Lac executives, and others. In this programme, they all are given special training.

(5) **Field out-reach training programme** : This is of 1-2 day(s) and a very effective programme. This programme helps to enhance visibility of the Institute and help in development of linkages with NGOs and state government organisations.

(6) **Field visits for lac crop monitoring** : The IINRG also undertakes monitoring of lac crop for possible incidence of disease and pest, and then suggests on-spot remedial measures.

(7) **In- campus orientation programme** : It also organises one-day exposure visit to institute's farm, museum. Interaction session is also being organised round the year.

SPACE FOR NOTES

Chapter Five : Poultry

INTRODUCTION

Several bird species, like fowls, ducks, geese, turkeys, guinea-fowls, pigeons, are recognised as **poultry**. But it is more often used for fowls by common people. Non-vegetarians today make chickens an integral part of their life. Chickens are one of the major sources of animal protein, aside from meat and fish. Keeping fowls for eggs, raising broilers for meat, poultry breeding and hatcheries are the common poultry enterprises. As a business, poultry farming offers an opportunity for people to earn a considerable amount of money, that is why many young men become interested to poultry farms in our country. One of the major customers for poultry farms are the major fast-food companies, although some of them have their own chicken farms to supply their needs.

India has made considerable progress in broiler production in the last two decades. High quality chicks, equipments for poultry farming, vaccines and medicines are easily available in market. With an annual output of 41.06 billion eggs and 1000 million broilers, India ranks fourth largest producer of eggs and fifth largest producer of poultry broiler in the world. The broiler production has also sky rocketed at an annual growth rate of about 15 per cent at present. Broiler farming has been given considerable importance in the national policy and has a good scope for further development in the years to come.

In this chapter we will mainly discuss on the farming of **chick, duck, quail** and **emu** as poultry.

(A) CHICK FARMING

Poultry farming is the practice of raising poultry birds, such as chickens, turkeys, ducks, and geese, as a subcategory of animal husbandry, for the purpose of farming meat or eggs for food.

Usually chickens are farmed in most of the poultry farms. More than 50 billion chickens are reared annually as a source of food, for both their meat and their eggs. Chickens farmed for meat are called **broilers**, while those farmed for eggs are called **egg-laying hens**. Some hens can produce over 300 eggs a year. Chickens naturally live for 6 or more years. After 12 months, the hen's productivity starts to decline. When productivity declines considerably, most commercial laying hens are then slaughtered.

Advantages of chicken farming

- Initial investment is a little lower than layer farming.
- Rearing period is 6-7 weeks only.
- More number of flocks can be taken in the same shed.
- Broilers have high feed conversion efficiency, i.e., least amount of feed is required for unit body weight gain in comparison to other livestock.

- Faster return from investment.
- Demand for poultry meat is more compared to sheep/goat meat.
- Loan from banks with refinance facility from NABARD is available for starting poultry farming.

Note : For obtaining bank loan the farms should apply to the nearest branch of a Commercial or Co-operative or Regional Rural Banks in their area in the prescribed application forms which is available in the branches of financing banks. The technical officers or the manager of the bank can help/give guidance to the farmers in preparing the project report to obtain bank loan.

COMMON POULTRY BREEDS (FOWL)

There are numerous breeds of fowl exist in the world. Here some of the important breeds are mentioned under the following heads (see Box 1).

A. Indian Breeds

1. **Aseel (Fig. 5.1)** : The literal meaning of ***Aseel*** is real or pure. Aseel is well-known for its pugnacity, high stamina, majestic gait and dogged fighting qualities. The name Aseel appears to have been given to this indigenous breed because of its inherent qualities of fighting. Andhra Pradesh is said to be the

Fig. 5.1. Aseel

BOX – 1 : SOME COMMON BREEDS OF CHICKEN USED FOR FARMING IN INDIA

BREEDS

- **Indian**: Aseel, Ghagus, Chitagong, Karacknath, Bursa
- **Asiatic**: Brahma, Cochin, Langshan
- **Mediterranian**: Leghorn, Minorca, Ancona, White-faced black spanish, Blue andalusian
- **English**: Orpington, Corning, Sussex, Dorking, Red cap, Australop
- **American**: Plymouth rock, Wyandotte, Rhode island red, New hampshire

home of this important breed. The best specimen of this breed, although rare, is encountered with the fanciers and the people engaged in cock-fighting show throughout the country. Aseel is larger in built with noble looking and dignified appearance. The standard weight varies from sexual maturity, i.e., 196 days. Number of annual egg production is 92.

2. **Ghagus** : It is a pure Indian breed having feathers on the legs. It is a big and hardy breed found in Andhra Pradesh and Karnataka. This is a good table bird. The hen is a fair layer, good sitter and a real mother.

3. **Chitagong or Malay** : This is a native bird of Malay Peninsula and breed in Chitagong district of Bangladesh. Now it is also found in some parts of Eastern India. The colour of this bird is white with splash of golden markings on the wing. This bird grows faster and is ideal for table purposes. Its flesh is excellent in taste.

4. **Karacknath** (Flesh is black in colour) : The original name of the breed seems to be Kalamasi, meaning a fowl with black flesh. However, it is popularly known as Karaknath (Fig. 5.2). Jhabua and Dhar districts of Madhya Pradesh and the adjoining districts of Rajasthan and Gujarat spreading over an area of about 800 sq. miles, is considered to be its home tract. These are mostly reared by tribals, adivasis and rural poor. It is considered to be a sacred bird and offered as sacrifice to Goddess after Diwali. The eggs are light brown in colour. The day-old chicks are bluish to black with irregular dark stripes over the back. The adult plumage varies from silver and gold-spangled to bluish-black without any spangling. The skin, beak, shanks, toes and soles of feet are slate-like in colour.

Fig. 5.2. Karacknath

The comb, wattles and tongue are purple. Most of the internal organs show intense black colouration, which is pronounced in trachea, thoracic and abdominal air-sacs, gonads and at the base of the heart and mesentery. Varying degrees of black colouration are also seen in the skeletal muscles, tendons, nerves, meninges, brain, etc. The blood is darker than normal. The black colour has been due to deposition of melanin pigment. The flesh of this breed though black and repulsive to look at, is considered not only a delicacy but also of medicinal value. The tribal uses Karacknath blood in the treatment of chronic disease in human beings and its meat as aphrodisiac*. The meat and eggs are reckoned to be a rich source of protein (25.47% in flesh) and iron. A medium layer lays about 80 eggs per year. The bird is resistant to diseases in its natural habitat in free range, but is more susceptible to Mareks disease under intensive rearing conditions.

Table 5.1
Some Indian breeds of chickens and their distribution

Breed	Home Tract
Ankaleshwar	Gujarat
Aseel	Andhra Pradesh & Madhya Pradesh
Busra	Gujarat & Maharashtra
Chittagong	Meghalaya & Tripura
4Danki	Andhra Pradesh
Daothigir	Assam
Ghagus	Andhra Pradesh & Karnataka
Harringhata Black	West Bengal
Karacknath	Madhya Pradesh
Kalasthi	Andhra Pradesh
Kashmir Faverolla	Jammu and Kashmir
Miri	Assam
Nicobari	Andaman & Nicobar
Punjab Brown	Punjab & Haryana
Tellichery	Kerala

* Aphrodisiac — Food, drink or drug that stimulates sexual desire

5. **Bustra or Busra** : It is a minor breed of Indian class found in small number in Gujarat and Maharashtra particularly around Mumbai. These birds are majestic, alert, deep bodied and light feathered. They have body conformation of a typical layer, but the hen is a poor layer.

B. Asiatic breeds

1. **Brahma** : Brahma (Fig. 5.3) is an Asiatic breed of fowl, previously called Chittagongs, Gray Shanghais, and Brahma Pootras, later shortened to Brahma. It is thought to be a product of a cross of the Malay and Cochin breeds in India. Imported from Shanghai, China, in the early 1840's, they landed in New England much later. American poultry fanciers refined the original parent stock into the large stately and useful Light and Dark Brahma varieties. The Buff Brahmas are of more recent origin. The head and skull are important breed characteristics. Texture of the feathers is also of great importance, for the plumage should be smooth fitting and not loose-feathered and soft as in the Cochin variety. It is a general purpose fowl cultured for heavy meat production. The skin colour is yellow and the egg shells are light to dark brown.

2. **Cochin** : The Chinese Shanghai fowl came to England and America in 1845. The name of this Asiatic breed was later changed to Cochin. The earliest Cochins were more or less buff in colour. Its striking appearance, due to great size and profuse soft feathering, distinguished it from all other known breeds at that time. Cochins created a sensation in England, resulting in a great boom for the "Cochin China," as it was called in those days of the "Cochin craze." Currently, the American Poultry Association recognizes Buff, Partridge, White, Black, Silver-laced, Golden-laced, Blue, Brown and Barred varieties of the Cochin. Both male and female are massive in appearance, with an extraordinary profusion of long, soft plumage and a great abundance of down fibre in the underfluff, producing a rather bulky appearance and conveying the idea of even greater weight than actually exists. Primarily bred for exhibition, the Cochin is capable of being bred as a meat type fowl. The skin colour is yellow and the egg shells are brown.

Fig. 5.3. Brahma

3. **Langshan** : The Black Langshan originated in China, where it has been bred over a long period of years. Its predominant reproductive qualities indicate it to be a pure race of domesticated poultry. Black Langshans were imported from China into England by the late Major Croad, which resulted in a Class of "Croad Langshans" in the English Standard. Langshans are dual purpose fowls of the Asiatic Class and are smaller than the Brahma and Cochin and more active. The male develops a large well-spread tail with feathers of great length, the sickles often attaining a length of sixteen or seventeen inches. The closely-fitting saddle feathers, full hackled neck and upright carriage give the effect of a short back. The surface plumage throughout is close and smooth. The body in both sexes should be evenly balanced on firm, straight legs, with very little backward bend at the hocks. The height of the Langshan should be gained by depth of body and erectness of carriage and not from what may be described as stiltiness of legs. This is a general purpose fowl for production of meat and eggs. The skin colour is white and the egg shell is very dark brown.

C. Mediterranean breeds

1. **Leghorn** : The original Leghorn (Fig. 5.4) breed came from Italy, but its many sub-varieties originated or were developed in England, Denmark and America. They comprise a group characterized by great activity, hardiness and prolific egg-laying qualities. The females are non-sitters, very few of them exhibiting a tendency to broodiness. Aside from the manifold points of beauty in type and colour, found in all varieties of Leghorns, their excellent productive qualities

Fig. 5.4. Leghorn

are valuable assets of the breed. They are specially noted for their production of eggs. The skin is yellow and the egg shell is white in colour.

2. **Minorca** : Minorcas, originally called Red-Faced Black Spanish, are the largest and heaviest of the Mediterranean breed. The distinct characteristics of the Minorca breed are their long strong bodies, large combs, long wattles, large white ear-lobes, large and full tail moderately elevated, with firm muscular legs set squarely under the powerful looking body. They are excellent producer of large white eggs. The skin is white and the egg shell is a chalk-white in colour.

3. **Ancona** : The Ancona takes its name from the City of Ancona, a port in Italy, from which the first Anconas were shipped to England. They originated from crosses of Italian fowl common in central Italy, which is evident by their close resemblance in type and size to the Leghorn breed. Anconas came to America from England in 1888. This is a non-sitting fowl. They are excellent layers of white shelled eggs and have yellow skin.

4. **White-faced black Spanish** : This is probably the oldest breed in the Mediterranean Class. The large, smooth, white face and ear-lobes are distinct characteristics of this Spanish breed, but these features should not be subjected to abnormal development. A non-sitting fowl noted for the reproduction of large chalk-white eggs. The skin colour is white.

5. **Blue Andalusian** : Blue Andalusians are credited with being natives of Andalusia, a province in Spain. They originated from crossing a black fowl with one of its white sports, these two colours producing a bluish-slate fowl. In Cornwall and Devon, England, similar blue fowls were produced by crossing black and white sports. This was before Andalusians were imported into England. They resembled the earlier Andalusians in type and colour. The modern Andalusian should be symmetrical, graceful, compact, medium in size and stately in carriage. The dull and uneven blue coloured fowl of the past has been transformed into the attractive, laced breed of today, by years of scientific breeding. This is a non-sitting fowl, laying large chalk-white eggs. The skin colour is white.

D. English breeds

1. **Orpington** : The Black Orpington (Fig. 5.5) produced in 1886 at Orpington, County Kent, England, from a Black, Langshan-Black, Minorca-Black, Plymouth Rock cross, is the original Orpington. The Buff and White varieties were produced from crosses other than those which were used to make up the Black Orpington. Cochin blood was introduced into some of the earlier strains of Orpingtons as evidenced by some of the more loosely feathered specimens. The original Black Orpington came to America in 1890. The Blue Orpingtons were produced from crossing the Black and White varieties in 1923. This is a general purpose fowl for heavy meat production and for eggs. The skin colour is white and the egg shell is light to dark brown.

2. **Cornish** : The Cornish fowl originated in Cornwall, England. This was the Dark Cornish, the parent breed generally known abroad as "Indian Game," although the name Cornish is more correct. It is a composite breed of several different blood lines – Aseel (or Asil), Black-Red, Old English, and

Fig. 5.5. Orpington

Malay. A distinguishing characteristic of the Cornish is that the body of both male and female is of the same conformation. White Cornish were produced from White Malay, Dark Cornish crosses in 1890. White-Laced Red Cornish were produced in America in 1893, from Shamo Japanese-Dark Cornish cross. This is a super-heavy meat producing fowl, valuable also for crossing with other breeds for the production of market poultry. The skin is yellow and the egg shell is brown in colour .

3. **Sussex** : The Sussex is a very old English breed, which originated in the County of Sussex more than a century ago. It was primarily bred for market purposes; Sussex being famed for its production of table fowls. This is a dual purpose fowl for production of meat and eggs. The skin is white and the egg shell is brown in colour.

4. **Dorking** : The Dorking is one of the most ancient of all domesticated races of poultry. It was brought to Great Britain by the Romans with Julius Caesar, but was known and described by the Roman writer Columella long before it became a popular breed in England. He speaks of the hens as being "square-framed, large and broad-breasted, with big heads and small upright combs," adding that "the purest breeds are five-clawed." This is a dual purpose fowl for meat and egg production. The skin and egg shell is white in colour.

5. **Redcap** : Redcaps originated in Derbyshire, England, probably from Hamburg crossed with other breeds. The very large rose-comb is the most striking characteristic of the breed. They are good producers of eggs. The skin and egg shell are white in colour.

6. **Australorp** : The Australorp (Fig. 5.6) is a production breed of Australian Black Orpington, noted for its splendid egg production. The breed was developed in Australia, where it has been bred principally for egg production rather than meat, as has been the case with its predecessor, the Orpington. It is a medium weight active bird, laying tinted eggs. It is a valuable fowl for those who desire abundance of eggs without sacrificing too much value in meat quality.

E. American breeds

1. **Plymouth Rock** : The Plymouth Rock (Fig. 5.7) was first exhibited at America's first poultry show in 1829 in Boston. It is believed that these original fowls lost their identity and that the progenitors of our present Barred Plymouth Rocks were first exhibited in 1869 at Worcester, Massachusetts. These were composites of several blood lines. The first and most prominent cross was that of a Dominique male with Black Cochin or Black Java female, which was originally made at Putnam, Connecticut. The Dominique male used was not the American or Rose-Comb Dominique, but rather a single-combed, hawk-coloured fowl commonly found in that locality. This Plymouth Rock was recognized as a distinct breed and was admitted to the first American Standard of Excellence in 1874. Whether the Plymouth Rock originated from a Dominique-Black Cochin or from a Dominique-Black Java cross is a debatable question. Both have been used synonymously in classifications. White

Fig. 5.6. Australorp

Fig. 5.7. Plymouth Rock

Plymouth Rocks were white sports of the Barred variety. Other strains, trace their ancestry to other white fowls. Buff Plymouth Rocks originated in Rhode Island, not far from Fall River, Massachusetts and were first exhibited as Golden Buffs. Buff Cochin blood was introduced in some strains in New York State. Silver Penciled Plymouth Rocks originated in the state of New York. Dark Brahma and Silver Penciled Wyandotte blood were used to produce this variety. Partridge Plymouth Rocks are a result of Partridge Cochin, Dark Cornish, Single-comb Golden Wyandotte males, Brown Leghorns, Golden Laced Wyandotte and Barred Plymouth Rocks being amalgamated to form the new variety. Columbian Plymouth Rocks originated in Ohio as the result of crosses between Light Brahmas, Barred Plymouth Rocks, White Plymouth Rocks and Columbian Wyandotte. The Plymouth Rock is a dual purpose fowl for the production of eggs and meat. The colour of the skin is yellow, the egg shell is brown. There may be considerable differences in the shades or tints of shell which vary from a very light to a dark brown, depending on the strain and the stage of production.

2. **Wyandotte** : There is no reliable data on the origin of the American Dominique. It is probably a selection from the many hawk coloured or grey fowls kept in the New England States. The type of the breed indicates composite blood lines, possibly of the Asiatic and Hamburg breeds. The Dominique is a dual purpose, medium weight fowl, for the production of meat and eggs. The skin is yellow and the egg shell is brown in colour.

The Silver Laced Wyandotte (Fig. 5.8A) is the parent variety of the Wyandotte family. It originated in New York State. It was first known as "American Sebright" and "Sebright Cochin". The Dark Brahma and Spangled Hamburgs used by the originator of the "American Sebright" were evidenced by the cropping out of the Hamburg comb and Dark Brahma colour markings in some of the earlier Eastern strains. Golden Laced Wyandotte originated in Wisconsin. They were produced by mating Silver Laced Wyandotte females with a cross-bred Partridge Cochin-Brown Leghorn cockerel. White Wyandotte originated from sports of the Silver Laced variety in New York State. The texture of feather is important in Wyandotte, especially in the White variety, where broad feathers and smooth fitting plumage are necessary to preserve the characteristic curvilinear breed type of the true Wyandotte. Buff Wyandotte are light coloured Rhode Island Reds. Golden and White Wyandotte crosses and Buff Cochin-White and Golden Wyandotte crosses dominated in other strains. Black Wyandotte originated from black sports of the Silver Laced variety. Partridge Wyandotte originated in the East and West. The Partridge Cochin furnished the colour patterns for both the Eastern and Western strains. The Eastern strain was the result of a Partridge Cochin-Golden Wyandotte cross; the Western strain was produced from Partridge Cochin-Cornish-Golden Wyandotte crosses. Silver Penciled Wyandottes were produced by blending of two strains — one a Partridge Wyandotte-Dark Brahma crosses, the other, a cross of Silver Laced Wyandotte and Silver Penciled Hamburg cross. Columbian Wyandottes (Fig. 5.8B) were named for the Columbian Exposition held in Chicago in 1893. This variety was produced by crossing a White Wyandotte and a Barred Plymouth Rock. Wyandottes are of medium weight, dual purpose fowl for the production of meat and eggs. The skin is yellow and the egg shells vary in colour from very light to a rich brown, depending on individual, strain, and the stage of production.

Fig. 5.8A. Silver Laced Wyandotte

Fig. 5.8B. Columbian Wyandotte

3. **Rhode Island Red** : The origin of the Rhode Island Red (Fig. 5.9) dates to a fowl bred in the section of New England that is located between Narragansett Bay and Buzzard's Bay. The name Rhode Island Red was given to the breed in honour of the state where it originated from crossing the Red Malay Game Leghorn and Asiatic native stock.

Fig. 5.9. Rhode Island Red

The earlier Rhode Island Reds sported both single and rose combs, some even having pea combs, due to their mixed ancestry and as because they were being bred primarily for market purposes. This is an important dual purpose breed, capable of excellent egg production. Rhode Island Whites originated in Rhode Island from which they took their name. They are a result of crosses of Partridge Cochins, White Wyandotte and Rose-Comb White Leghorns. The distinct shape characteristic of both Rhode Island Reds and Rhode Island Whites is the horizontal, oblong body. Rhode Island Reds and Whites are general purpose fowls, bred for the production of meat and eggs. The skin is yellow and the egg shell is brown to dark brown in colour.

4. **New Hampshire** : New Hampshires developed over a period of years beginning around 1915 from a foundation of Rhode Island Reds, first brought into New Hampshire from Rhode Island and Southern Massachusetts. There is no record of any outside blood having been introduced and the breed has been developed by farm poultry men of New Hampshire by the continual selection of breeding stock for early maturity, large brown shelled eggs, quick feathering, strength and vigour during its evolution. Special meat type strains are now developed featuring extreme rapid growth and increased weights. This type has a very light shade of surface colour for better dress, but extreme meat type strains do not have the egg production qualities of the dual purpose Standard New Hampshire. A very popular general purpose utility fowl for eggs and meat production. The skin colour is yellow and the egg shell is brown.

COMMERCIAL STRAINS OF CHICKENS

Usually easily available breeds in a country are selected for commercial culture practice. Among the breeds, birds are classified as following strains and reared for specific purpose.

BROILER : Broilers (Fig. 5.10 and 5.11) are young chicken of either sex, which are reared primarily for meat purposes and marketed at an age of 6-8 weeks. In India following breeds are usually cultured for meat : Caribro-Vishal (Caribro-91), Cari-Rainbro (B-77), Caribro-Dhanraja (Multi-coloured), Caribro-Mritunjai (Cari-naked neck). Commercial culture of broilers is more popular in India.

LAYER : The management of birds during 21-72 weeks of age is for the purpose of laying eggs (egg

Fig. 5.10. Caribro-Vishal (Caribro-91)

Fig. 5.11. Cari-naked neck

production). Indian layers are Cari Priya, Cari Sonali, Cari Devendra, etc. Layers are also cultured in India for commercial purposes, because eggs have huge market and easy to manoeuver.

GROWER : The management of birds during 9-20 weeks or to the point of egg laying is referred to as grower period. Hitcari (Naked neck cross), Aseel, Karacknath, Upcari (Frizzle cross), etc. are generally cultured in India.

HOUSING SYSTEMS OF POULTRY

Poultry rearings are made in two ways in India — (1) **Free range** or **Extensive system** and (2) **Indoor** or **Intensive method**. Intensive or indoor system includes two types — (A) **Battery cage system** and (B) **Deep litter system**.

Sometimes other two types are also regarded — (i) Semi-intensive and (ii) Folding unit system. These systems are used for very few numbers of chickens by farmers in their courtyards. In folding unit system, the position of the portable folding units is changed each day, giving the birds fresh ground. The birds find a considerable proportion of food from the herbage. The farmers are benefitted from the scratching and raking of the land, and the excreta of the birds act as fertilizer.

(1) Free-range or Extensive system

This is the oldest system of farming used by general farmers, where there is no shortage of land and monetary involvement. Free range poultry farming consists of poultry birds permitted to roam freely instead of being contained in any manner. It is practised in rural area in small scale. A free range chicken must have daytime access to open-air runs during at least half of their life. In free-range broiler systems, the chickens are given continuous access to an outdoor range during the daytime and sheds where they are housed at night. Free-range chickens grow more slowly than intensive chickens. Free range poultry production requires that the poultry have access to the outside. In some cases this means, the poultry are raised on pasture, enabling the poultry to move around, forage for their natural diet and live in cleaner conditions than those in batteries. In some farms, the manure from free range poultry can be used to benefit crops. The benefits are also a reduced growth rate and opportunities for natural behaviour such as pecking, scratching, foraging and exercise outdoors, as well as fresh air and daylight. As they grow slower and have opportunities for exercise, free-range chickens have better leg and heart health and a much higher quality of life. Finding suitable land with adequate drainage to minimize worms and coccidian oocysts, suitable protection from prevailing winds, good ventilation, access and protection from predators can be difficult. Excess heat, cold or damp can have a harmful effect on the animals and their productivity. Unlike battery farms, free range farmers have little control over the food their animals come across which can lead to unreliable productivity. Diseases are common and the animals are vulnerable to predators. In South-East Asia, a lack of disease control in free range farming has been associated with outbreaks of Avian influenza. Therefore, for commercial purposes free range culture method is not suitable.

(2) Indoor or Intensive method

In this method chickens are kept indoors but with more space (around 12 to 14 birds per square metre). They have a richer environment, for example, with natural light or straw bales that encourage foraging and perching. The benefits of indoor systems are higher welfare with less crowding and more opportunities for natural behaviour. Intensive or indoor system includes two types — (A) Battery cage system and (B) Deep litter system.

(A) **Battery cage system** : In this system batteries of cages are placed one upon another providing all amenities to the birds. This is the most intensive type of poultry farming and is useful to those with limited floor space. This type of farming is suitable for commercial egg farming (layer farming). In this system hen is confined to a cage just large enough to permit very limited movement and allow her to stand and sit comfortably.

(B) **Deep litter system** : In this system poultry birds are kept in large pens up to 250 birds each on floor covered with litters like straw, sawdust or leaves up to a depth of 8-12 inches. This system is used mainly for farming broilers. Farmers with sufficient space opts for this type of culture, because no cage and floor system is possible here.

Modern and well-established *Indoor with higher welfare scientific practices* should be used to obtain maximum economic benefits from poultry farming. It is necessary to obtain training/experience in layer farming before starting a farm. One has to stay on the farm for constant supervision. Some of the major norms and recommended practices are given below :

(i) Preparation for Poultry Housing

At the onset one has to decide whether he will start poultry farming for broiler or layer. Accordingly pre-arations will have to be made. In both the cases some preparations are common.

(ii) Selection of poultry site

Well-raised land should be selected for poultry sheds. Land with hard rock or murram is more suitable, so that water logging and flooding near the sheds can be avoided. Separate sheds for growers and layers is recommended. Adequate facility for water, electricity, approach to road, supply of chicks, feed, veterinary aid and nearness to market for sale of cull birds and eggs should be ensured.

(iii) Construction of the sheds

Construction of the sheds should be in such a way that the end walls face East-West direction and the side walls face North-South direction, so that rain water cannot enter the sheds. 3 to 4 feet overhanging of the roof is recommended to avoid entry of rain-water inside the shed. The plinth of the shed should be at least one foot above the ground level. At least 50 feet distance between two sheds in the same sector and about 150 ft between growing and laying sector is recommended. Adequate light, ventilation and comfortable housing conditions during all seasons (cool in summer and warm in winter) are prime objectives. Construction of sheds should be in such a way that predators (cats/dogs/snakes) will not get entry into the shed. Rat proof civil structures may be constructed to avoid entry of rats. The shed should be kept clean and free from flies/mosquitoes, etc. Adequate floor space per bird is to be provided.

Construction of Battery cages for egg-laying chickens

In egg-producing farms, birds are typically housed in rows of battery cages (Fig. 5.12). The usual floor space is 14×16 feet and the height is 1.5 feet for each cage. Many small cages can be assembled together. Usually these are multistoried. One cage could be accommodated with approximately 100 birds. In recent years the recommendation is 67 to 86 square inch space per bird, which is about 9 inches by 9 inches. The whole set of cages are placed under a well-built roof. The floor is made up of strong galvanized wire set at a slope from back to front, so that the laid eggs roll out of the cage to a receiving gutter kept at the front side. Underneath is a tray for droppings. Food and water containers are kept outside the cage. Environmental conditions are automatically controlled, including light duration, which mimics summer day length. This stimulates the birds to continue to lay eggs all year round. Normally, significant egg production occurs only in the warmer months. In egg-producing farms, cages allow for more birds per unit area, and this allows for greater productivity and lower space and food costs, with more efforts put into egg-laying. Critics argue that year-round egg production stresses the birds more than normal seasonal production.

Fig. 5.12. Egg-laying chickens in battery cages

Fig. 5.13. Broilers in a deep litter house

Construction of deep litter cages for Broilers

The term **deep litter** may be defined as *the accumulation of the material used for litter with poultry manure until it reaches a depth of 8 to 12 inches.* In this system poultry birds are kept in large pens (Fig. 5.13). On an average 100 birds can be accommodated in a pen of 25 × 20 feet size, i.e., 5 square feet per bird in climate-controlled housing. In larger pens accommodation of maximum 250 birds is recommended. The floor is covered with dry litters like straw, sawdust, rice hulls or leaves up to a depth of 8-12 inches. This system is used mainly for farming meat chickens, commonly called broilers. Farmers with sufficient space opts for this type of culture, because cage and floor system is not possible here.

Maintenance of poultry

Poultry producers routinely use nationally approved medications, such as antibiotics, in feed or drinking water to treat disease or to prevent disease outbreaks arising from overcrowded or unsanitary conditions. Some approved medications for improved feed utilization may be applied. In advanced countries law prohibits the use of hormones or steroids in poultry production.

After every batch of growers or when culled birds are disposed of, the dirty litter material and manure should be removed, walls and floors should be cleaned, whitewashed with lime and disinfected with 0.5% malathion or other recommended insecticide spray. If deep litter system is followed, always use dry and clean litter material (sawdust, paddy husk, etc.). Spread 4-inch layer of litter on the floor, keep clean, disinfect brooding, feeding and watering equipments and then introduce chicks in the house. The litter material should always be kept loose and dry. Stir the litter twice a week. Any wet litter, droppings, etc. should be removed and replaced with fresh/clean dry litter. If cage system is followed, ensure that droppings are spread with lime powder or 10% malathion spray twice a month to prevent menace of flies. The droppings under the cage can be removed after 6 months.

Modern poultry farming is very efficient and allows meat and eggs to be available to the consumer in all seasons at a lower cost than free range production and the poultry have no exposure to predators.

Debeaking

The cage environment of egg producing does not permit birds to roam. The closeness of chickens to one another frequently causes cannibalism. Cannibalism is controlled by debeaking (removing a portion of the bird's beak with a hot blade so the bird cannot effectively peck).

Osteoporosis

Another condition that can occur in prolific egg laying breeds is osteoporosis. This is caused from year-round rather than seasonal egg production and results in chickens whose legs cannot support the body and so can no longer walk. During egg production, large amounts of calcium are transferred from bones to create egg shell. Although dietary calcium levels are adequate, absorption of dietary calcium is not always sufficient to fully replenish bone calcium.

In intensive broiler sheds, the air can become highly polluted with ammonia from the droppings. This can damage the chickens' eyes and respiratory systems and can cause painful burns on their legs (called **hock burns**) and feet. Chickens bred for fast growth have a high rate of leg deformities because they cannot support their increased body weight. As they cannot move easily, the chickens are unable to adjust their environment to avoid heat, cold or dirt as they would have in natural conditions. The added weight and overcrowding also puts pressure on their hearts and lungs. In the U.K., up to 19 million chickens die each year in their sheds from heart failure.

COMMON POULTRY MANAGEMENT PRACTICES

Modern and well-established scientific practices should be used to obtain maximum economic benefits from poultry farming. Some of the major norms and recommended practices are given below.

(a) Selection of well-raised land for poultry sheds is of prime importance. Land with hard rock or murram is more suitable. Water logging and flooding near the sheds should not be allowed. Separate sheds for growers and layers will yield good result.

(b) It should be ensured that there is adequate facility for water, electricity, approach road, supply of chicks, feed, veterinary aid and nearer to market for sale of cull birds and eggs.

(c) A farmer should obtain training/experience in layer farming before starting a farm. Dedicated farmers should be mentally prepared to stay on the farm and have constant supervision.

(d) Adequate floor space per bird as per specifications for construction of poultry sheds should be provided.

(e) Sheds should be constructed in such a way that the end walls face East-West direction and the side walls face North-South direction, so that rain water will not enter the sheds.

Fig. 5.14. Modern poultry housing for large scale farming

(f) Strong roof and hard flooring is recommended. Plinth of the shed will be raised at least one foot above the outside ground level (Fig. 5.14).

(g) Provide 3 to 4 feet overhanging of the roof to avoid entry of rainwater inside the shed.

(h) Provide at least 50 feet distance between two sheds in the same sector and about 150 feet between growing and laying sector.

(i) Egg producing chicks should be kept in cages with egg collecting trays (Fig. 5.15).

(j) Provision should be there for adequate light and ventilation and comfortable housing conditions during all seasons (cool in summer and warm in winter).

(k) During dry and hot season, if required, water sprinkler should be used for spraying water uniformly (Fig. 5.16). This will keep the poultry cool.

(l) During cool period heater should be used to keep the cage hot and dry (Fig. 5.17).

Fig. 5.15. Modern well ventilated cage with egg collecting tray and feeding tray for layers

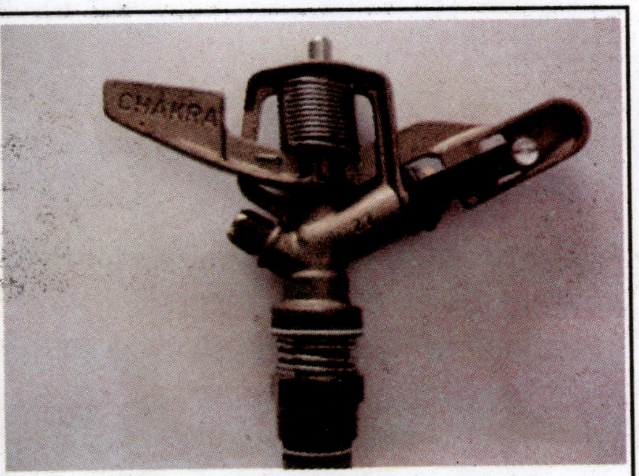

Fig. 5.16. Water sprinkler for cooling

Fig. 5.17. Poultry cage heater

(m) Construction of the sheds should be in such a way that predators (cats/dogs/snakes) will not enter the shed. Entry of rats can be avoided by constructing rat-proof civil structures.

(n) Shed should be kept clean and free from flies, mosquitoes, etc.

(o) After every batch of growers/culled birds is disposed of, the dirty litter material and manure should be removed. Walls and floors should be cleaned, whitewashed with lime and disinfected with 0.5% Malathion or DDT insecticide spray.

(p) If deep litter system is followed, always dry and clean litter material (sawdust, paddy husk, etc.) should be used. 4-inch layer of litter should be spread on the floor. It is necessary to keep the litter clean and disinfect brooding, feeding and watering equipments before introducing the chicks into the house.

(q) The litter material should always be kept loose and dry. The litter should be stirred twice a week. Any wet litter/droppings etc. should be removed and replaced with fresh/clean dry litter.

(r) If cage system is followed, droppings should be spread with lime powder or 10% Malathion spray twice a month to prevent menace of flies. The droppings under the cage should be removed after 6 months.

POULTRY EQUIPMENT

Scientifically designed cages and equipments for brooding, feeding and watering purposes should be used. BIS (Bureau of Indian Standards) specifications for equipments are available. A good design can be shown and manufactured locally, so that cost can be reduced. Some of the equipments are discussed here, which are readily available in market.

Fig. 5.18. Egg collectors — **(a)** for temporary storing and **(b)** for long term storing

A. **Egg collector** : Several types of egg collectors are available. Some are for temporary collecting and storing (Fig.5.18a) and some for long time storing with cooling devices (Fig. 5.18b).

B. **Incubator** : Fertilized eggs are kept in this instrument for successful hatching. This instrument is provided with temperature and humidity regulator devices (Fig. 5.19).

Fig. 5.19. Incubator

C. **Chick cage** : These are PVC cages for keeping and transporting chicks to market or other destinations safely (Fig. 5.20).

Fig. 5.20. Chick cage

CHICKS

Improved strain of one-day old healthy eggar type chicks or broiler chicks from a reputed hatchery should be purchased. These chicks should be bright and fresh looking (Fig.5.21). Usually 2-5% extra chicks are recommended for initial phase. If cages are used for housing of birds, proper cage space should be ensured. These chicks should be supplied

Fig. 5.21. Healthy broiler chicks

Fig. 5.22. Water-nipple for chicks

with proper recommended food with feeder (Fig. 5.23) and water in recommended water-nipple (Fig. 5.22). All equipments should be cleaned by washing and disinfecting with 0.5% Malathion spray after every batch of bird disposal.

Feeding

High quality balanced feeds should be provided to the chicks through purchasing or farm preparation. **Starter feed** (up to 8 weeks of age), **grower feed** (9 to 16 weeks of age) and **layer feed** (17 to 72 weeks of age) manufactured by reputed institutions/ companies should be used. BIS (Bureau of Indian Standards) recommended feed formulae and specifications are available. With proper knowledge/ experience, the feed can be prepared on the farm itself. The feed should be stored in clean, dry and

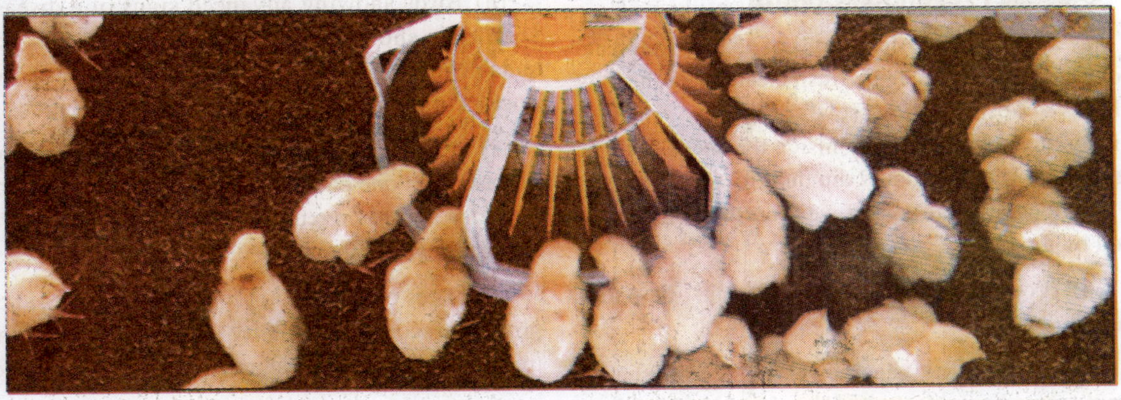

Fig. 5.23. Chick feeder

well ventilated room. If the feed remains wet it may cause fungus infection. Properly designed feeders (Fig. 5.24) should be used. Now various well-constructed feeders are available in market. Careful observation should be made before selecting the appropriate feeder so that the rats do not waste feed.

Adequate feeding space should be provided per bird. More space is required as the bird grows in age. Proper records should be kept on feed consumption per bird for each batch. About 7 kg feed up to 20 weeks and 38 kg feed from 21 to 72 weeks of age is required. Excess utilization may be due to feed wastage by rats, low temperature of shed or poor feed quality (low energy feed). Too low feed consumption may be due to disease condition, low quality/ unpalatability feed and high temperature in poultry shed.

Water for Birds

Always fresh and clean drinking water should be provided. Water should always be available to birds. Use of properly designed watering equipment (Fig. 5.25) is helpful. Several beautifully designed water pot for chicks are available in the market. Auto-regulated water supply will provide regular and continuous supply to every bird. Provide adequate watering space per bird. Water-pots should always be kept clean. Drinking pots should be such that the birds do not fall into it. Cool water to be provided during summer. Store the water in tanks that are not exposed to hot sun in summer.

Fig. 5.24. Modern feeder for chicks

Fig. 5.25. Modern watering device for chicks

Disease Prevention/Control

(a) Clean sanitary conditions of poultry sheds and equipment, balanced feed, fresh clean water, healthy chicks are essential to prevent diseases. Entry of visitors to farm, especially inside the sheds must be restricted. Visitors should dip their feet and hands in a disinfectant solution and wear apron/boots provided by the farm.

(b) Proper vaccination schedule must be maintained.

(c) High quality vaccines should be purchased from reputed manufacturers. Vaccines should be kept in cool, dry conditions away from sunlight. Any left-over vaccine should be properly disposed. Vaccines should not be used after their expiry date is over.

(d) Vaccination will be done with proper sterile vaccinators (Fig. 5.26).

(e) Any dead bird should be immediately removed from the shed and sent to laboratory for post-mortem examination or buried/burnt suitably away from the poultry sheds.

(f) The waste of farm should be suitably disposed. Different workers should be employed in brooding and laying sheds.

(g) Any bird showing advanced signs of a disease should be removed from the shed and culled. It can be sent to laboratory for diagnosis.

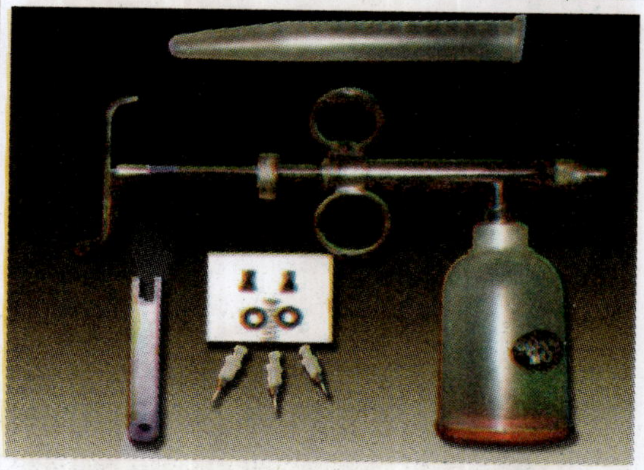

Fig. 5.26. Different types of vaccinators with accessories used in poultry farming

(h) Birds showing advanced signs of a disease should be shown to a qualified veterinarian and suitable medication/treatment should be given as per doctors or drug manufacturer's recommendations.

(i) Poultry manure, if infections, can spread disease from one batch to another. The litter is to be kept dry and removed after the flock is sold. The infected manure should be disposed off quickly.

(j) Dates of vaccination for each flock should be properly recorded. Mortality and its causes and the treatment given to birds should be properly recorded.

(k) Rats are important carriers of poultry disease. To avoid rats suitable rat poison/rat trap is to be used.

(l) Many poultry medicines can be given along with drinking water. When medication is to be given, removal of the waterers in poultry sheds on the previous evening is necessary. Next morning medicine in measured quantity of water should be employed, so that entire medicine is quickly consumed and there is no wastage.

(m) Mild infection of disease may not cause mortality but it will reduce growth. Proper recording of body weight for growers, mortality rate and egg production is necessary. Identifying the possible causes, if weight is low or egg production is low and taking steps to improve the management of the subsequent batches would prevent losses. A constant vigil and analysis of records/results are necessary to keep up the efficiency in farming.

Processing/Marketing of eggs and meat

1. Constant and steady demand for eggs should be ensured and the market should be nearer to the farm.

2. Study of the market demand for particular egg weight/size is helpful. One nest box for every 5 birds should be provided. Eggs from the shed can be collected 4 times a day. Storing them in a cool dry place and marketing them quickly is necessary.

3. Birds should not be kept on the farm beyond 18 months of age, as their egg production will go down considerably and their efficiency of feed conversion will reduce progressively as they grow older.

4. If live culled birds are sold after dressing (processing), clean dressing hall and processing equipments should be used. Dressed birds should be chilled with ice-cold water for 3-4 hours and excess water is removed. Birds should then be packed in clean plastic bags and the mouth of bag is to be sealed.

5. Processed birds should be marketed as early as possible. If they have to be preserved, deep freezing equipment (–10 to –20°C) be used. Refrigerated vans may be required for long distance transportation.

NECESSITIES FOR HIGHER EGG PRODUCTION

Quality birds : The strain that will perform best and is known to have good livability under reasonable environmental conditions is to be chosen. Good chicks cost more but they perform better.

Crowding : Overcrowding increases mortality, stress as well as production cost.

Feeding : Adequate fresh feed is to be given. Birds that are without feed for six hours will record a drop in production and a 12 hour starvation will result in moult of wing feathers. There should be adequate feeder space for the birds. Feed wastaging should be avoided. Maintaining records of daily feed-consumption is essential. It will enable to determine feed utilization and bird's performance.

Watering : Plentiful and clean disinfected water is to be provided. This management factor is commonly violated. Water restriction is a quick way to accidentally forcing the flock to moult. The waters are to be so placed such that they are easily accessible to birds.

Vaccination : All birds are to be vaccinated for Marek's Disease and Ranikhet Disease. Birds not vaccinated are highly susceptible to these diseases.

Debeaking : Proper debeaking programmes are to be undertaken, as poor debeaking can adversely affect egg production.

Culling : Unsuitable and uneconomic birds should be timely culled.

Health : Early signs of disease should be spotted for its timely treatment before it flares up in a big way. Some of the symptoms that indicate the onset of disease are : (1) Drop in egg production and feed consumption; (2) Increased morbidity and mortality; (3) Inactivity and lack of vigour; (4) Droopy ruffled appearance and respiratory distress and (5) Sudden changes in egg quality.

Sanitation : Sanitary measures are of vital importance in poultry operation. Roundworms, tapeworms and caecal worms are to be kept under control. External parasites are a serious farm hazard and can reduce production, if unchecked. Deworming at regular intervals should be practised.

Egg Quality : Respiratory and intestinal diseases should be kept under control for keeping of the quality of egg shells. Indiscriminate use of sulpha drugs can affect the egg shell quality. The use of tetracycline can, however, improve it.

Records : A daily record of feed consumption, egg production, mortality, income and expenditure is essential to help improve farming efficiency and pinpoint area of troubles and their solutions.

Routine Checking : Critical items of management should be listed on a daily, weekly or seasonal check list. Every item must be checked. It helps to locate the cause of trouble when it occurs. Routine checks are—Cleaning and refilling of waterers and feeders; cleaning the house and spraying insecticide; stirring the litter; dusting; culling of birds; egg collection, etc.

Why poultry farming for eggs?

Poultry eggs and meat are important sources of high quality proteins, minerals and vitamins to balance the human diet. Specially developed breeds of egg type chicken are now available with an ability of quick growth and high feed conversion efficiency. Depending on the farm-size, layer (for eggs) farming can be main source of family income or can provide subsidiary income and gainful employment to farmers throughout the year. Poultry manure has high fertilizer value and can be used for increasing yield of all crops.

Managing financial support for agencies

For poultry farming schemes with very large outlays, detailed reports will have to be prepared. Banks provide financial assistance for the following purposes :

- For construction of brooder/grower and layer sheds, feed store, quarters, etc.
- For purchase of poultry equipment such as feeders, waterers.
- For creating infrastructure items for supply of electricity, feed, water, etc.
- For purchase of day-old chicks or ready to lay pullets.
- For meeting working capital requirement in respect of feed, medicines, veterinary aid, etc. for the first 5 to 6 months (i.e., till the stage of income generation).

The cost of land is not considered for financial assistance. However, if land is purchased for establishing a poultry farm, land cost can be treated as party's margin up to a maximum of 10% of total cost of project. The rest can be obtained as loan.

DISEASES OF POULTRY

BACTERIAL DISEASES

***Escherichia coli* infections** : **Omphalitis (navel infection)**. It is characterized with reddening and tissue oedema in the umbilical region. *Escherichia coli* infections are widely distributed among poultry of all ages and categories. They are primarily related to poor hygienic conditions, neglected technological requirements or to respiratory and immunosuppressive diseases. A common sequel of navel infections is local or diffuse peritonitis. Preventive measures should aim at minimizing the probability of faecal contamination of eggs. This implies the maintenance of clean nests, discarding floored eggs and removal of eggs that are cracked or contaminated with faeces. Breeder eggs should be fumigated or disinfected in the farm prior to their transportation in the storage premise. The treatment is effective if initiated soon after testing the antibacterial sensitivity of isolates.

Salmonelloses : **Pullorum** disease is an acute systemic disease in chickens and turkey. The infection is transmitted with eggs and is commonly characterized by white diarrhoea and high death rate, whereas adult birds are asymptomatic earners. The morbidity and the mortality rates increase about the 7th - 10th day after hatching. The affected chickens appear sleepy, depressed and their growth is retarded. The feathers around the vent in many chickens is stained with diarrhoeic faeces or pasted with dry faeces. The edema of tibiotarsal joints is a frequent associated sign. Pullorum disease is widely distributed among all age groups of chickens and turkeys. The highest losses are in birds under the age of 4 weeks.

Paratyphoid infections : Fowl paratyphoid is an acute or chronic disease in domestic fowl and many other avian or mammalian species, caused by some motile Salmonella serotypes that are not host-specific. The highest morbidity and death rates are usually observed during the first 2 weeks after hatching. The chickens are drowsy with eyes closed, ruffled feathers and grouped near the sources of heat. Diarrhoea, dehydration and pasted down appearance around the vent are observed. Arthritis and periarthritis are encountered in broilers as well as in growing birds. Tibiotarsal joints are most commonly affected. The skin and the mucous coats are the entrance door of the infection. The prevention is based upon detection and elimination of the causative agent. Strict hygiene in hatcheries and throughout the injection of birds is essential for the prevention of *Pseudomonas* infection. The test of sensitivity is of most importance for the treatment, because the micro-organism is resistant to a high number of antimicrobial drugs.

Fowl cholera : Fowl cholera is an infectious disease in domestic fowl, waterfowl and other avian species. It is manifested either in acute septicaemic form with a high morbidity and death rates or as chronic local forms (independently or secondary to acute ones).

Acute fowl cholera : The sudden and unexpected death could be the first sign of the disease. In this form, the lesions are predominantly related to vascular injuries. Commonly observed signs are anorexia, ruffled feathers, oral and nasal mucus discharge, cyanosis and white or greenish watery mucoid diarrhoea. Frequently, subserous petechial or ecchymosed haemorrhages in the anterior part of the small intestine, the gizzard or the abdominal fat are discovered. Congestion (overfilling of blood vessels with red blood cells) of the liver is seen as an initial manifestation of *E. coli* septicaemia in a broiler chicken.

Chronic fowl cholera : It is characterised by local inflammations. The periorbital sinuses are frequently affected by a serofibrinous inflammation. It is possibly spread from sinuses to adjacent air-filled skull bones with subsequent necrosis and onset of neurological signs (opisthotonus and torticolis). The diagnosis is made on the basis of disease history, clinical signs, the lesions and the results of bacteriological studies. The immunization of birds at the age of 8 -12 weeks gives very promising results. Many antibiotics and sulfonamides could lower death rate, but at discontinuation of the treatment, the disease could recur. Sulfonamides are appropriate for treatment, but they inhibit egg-laying.

***Riemerella anatipestifer* infections** : *Riemerella anatipestifer* (RA) infection is a contagious disease in domestic ducks, turkeys and other fowl. It is encountered as acute or chronic septicaemia and is characterised with serous fibrinous polyserosites. The respiratory tract could also be affected. The ducklings at the age of 18 weeks are especially susceptible. Clinically, sneezing, cough, trembling of the head and neck, ataxia and greenish diarrhoea could be present. The most characteristic gross lesion is the deposit of fibrinous exudate on the pericardium, the liver capsule or air sacs. The chronic lesions affect the skin and the joints. Although a tentative diagnosis could be made on the basis of observed clinical symptoms and lesions, it is confirmed upon the isolation and identification of RA. The RA infection should be distinguished from septicaemia due to *P. multocida, E. coli, Salmonella,* etc. The treatment with antibiotics (Flumequine) and sulfonamides I Jrimetoprim, Sulfadiazine has a varying success. *Staphylococcus aureus* infections are commonly seen in poultry. Usually, bones, tendon sheaths and joints, particularly coxofemoral and tibiotarsal joints are affected. The clinical signs include unilateral or bilateral lameness, reluctance to move and lying down. When the tibiotarsal joints are affected, swellings, fever and sometimes necroses of overlying tissues and purulent exudation are observed. Other lesions related to endocardites are heart, liver and spleen infarctions. Liver infarctions are usually peripherally located on margins, have a pale creamy colour and are sharply demarcated. The penetration of the infection occurs mainly via the oral or aerogenic route, but could also enter through the injured skin, especially in battery cage layer hens. The differential diagnosis includes other types of bacterial septicaemia as staphylococcosis, fowl cholera, *E. coli* infections, etc.

Mycoplasma : The aetiological agent is *Mycoplasmosis gallisepticum (MG)*. In many cases, however, the pathogenicity of the micro-organism is enhanced because of its association with any or some of the following agents : *E. coli, Pasleurella multocida, Haemophilus paragallinarum* and Infectious Bronchitis (IB) or Newcastle Disease (ND) viruses. MG is characterized by respiratory symptoms and a prolonged course of the disease. Particularly susceptible are hens and turkeys at all ages. The most characteristic signs in adult flocks are tracheal rales, nasal discharge, and coughing, decreased egg production. Most outbreaks are in broiler chickens older than 4 weeks. The course of the disease is more severe during the winter and in cases of associated infections. Often, conjunctivitis, facial skin oedema and profuse tear secretion could be observed. Most common gross finding is aerosaccultes, the air sacs being filled with fibrinous gaseous exudate. The majority of routine chemical disinfectants are effective against *M. gallisepticum* that rarely survives longer than a few days away from the host. The birds could carry the micro-organism and be asymptomatic until the disease is triggered by stress factors such as change of the premise, the diet or weather, vaccinations against or infections with IB or ND, increased levels of dust or ammonia. Affected birds get progressively exhausted. When the joints and tendon sheaths are open, a serofibrinous exudate is most commonly observed. The aetiological agent is *M. synoviae*. The micro-organism shows a certain tropism to synovial structures as joints and tendon sheaths. An important route of dissemination of the agent is the transovarial transmission. The distribution by a horizontal route via the respiratory tract is also possible. The commonly used means of diagnostics is ELISA.

Necrotic enteritis : Necrotic enteritis (NE) is an acute *Clostridium* infection characterized by severe necrosis of intestinal mucosa. The disease begins suddenly, with a sharp increase in death rate. A strong dehydration is observed. The skin is sticked on or adhered to body musculature and is hardly removed. Chickens at the age of 25 weeks are usually affected, NE is also encountered in hens particularly near the period of the beginning of egg laying or peak egg laying, most commonly associated with coccidiosis. In acute cases, marked congestion of liver, responsible for its dark red to black appearance. The aetiological agent is *Clostridium perfringens*, which mainly forms type A and more rarely type C. They produce "a and p" toxins from *C. perfringens* type A and type C respectively, which are responsible for the necrosis of intestinal mucosa. The small intestine is often distended with gases and the necrotic mucosa is visible through the wall. *C. perfringens* is ubiquitous and normally reside into the intestinal tract. The alterations are particularly in the jejunum and the ileum because of their higher pH and the lower oxygen content in these areas. Sometimes, haemorrhages are seen through the intestinal wall. Throughout the simultaneous occurrence of NE and coccidioses, the content of the lumen is bloody, mixed with necrotic detritus and gas bubbles. The diagnosis is based on the distinctive gross lesions. When necessary, a histological investigation is performed or attempts for isolation of the causative agent. NE should be distinguished from ulcerative enteritis and some small intestinal coccidioses. The control should be aimed at predisposing factors. An appropriate medication of feeds is recommended. A good effect is obtained with oxytetra-cyclinedihydrate (OTC 50% premix). NE could be effectively treated with doxy-cycline hydrochloride, amoxicillin, etc.

Cholangiohepatitis in broiler chickens : Cholangiohepatitis (CAH) in broiler chickens is characterized by inflammatory proliferative and dystrophic necrobiotic alterations in bile ducts and the liver parenchyma. Usually, no clinical signs are observed. The increased daily mortality is insignificant, although in some chickens, a retarded growth and dehydration could be present. Pathoanatomically, the liver is enlarged and with paler yellow colour. In some cases, its surface has a characteristic acinous appearance and in others is mottled with multiple small greyish-white or greenish foci. *Clostridium perfringens* is the aetiological agent. CAH is experimentally reproduced in broilers by ligation of bile ducts and inoculation with *C. perfringens*. The gall bladder is filled with a thick bile secretion or a dense matter with a creamy colour. Some cases, haemorrhages with various intensities are detected in the mucous coat of the gizzard. Prevention– separate housing of the different age groups of birds, avoiding the contact with other avian species. The premedication of forages with some antibiotics and their rotation would prevent the reproduction of *C. colinum*. A good effect is achieved with oxytetra-cyclinedihydrate (OTC 50% premix). UE could be effectively treated with doxycycline hydrochloride, amoxycillin, etc.

Gangrenous dermatitis : The gangrenous dermatitis (GD) is a disease affecting fattening and growing birds, characterized with necrotization of different skin areas and a severe cellulitis of the subcutaneous tissue. The sudden and quick increase in death rates is often the first signal for the incidence of GD. The affected birds die after less than 24 h. The death rate is from 1% to 60%. The lesions are dark red to blue green macerated skin areas, usually featherless, beginning generally from wings and the adjacent areas. Most outbreaks are encountered in broilers at the age of 4–8 weeks. The disease is also observed in stock layer hens at the age of 6–20 weeks and broiler parents at the age of 20 weeks. The outbreaks are frequently observed in extremely wet and warm premises. The gangrenous dermatitis affects birds while still alive. In more severe cases, the gangrene could begin and involve the skin of the head, neck and the breast. Affected skin is macerated or totally necrotic, resulting in exposure of underlying tissues in a number of cases. Microscopic lesions are characterized by oedema, emphysema, hyperaemia, haemorrhages and necroses in the subcutaneous tissues. GD could be effectively treated by tetracyclines: doxyxycline hydrochloride, erythromycin thiocyanate, chlortetracycline and copper sulfate in the water.

Botulism : Botulism is an intoxication caused by the toxins of *Clostridium botulinum*. With the exceptions of vultures, most birds are susceptible. The clinical signs appear within a few hours to several days. Flaccid paresis of legs, wings, necks and eyelids is observed. The paresis is rapidly progressing to paralysis and the birds fall into a deep coma with neck and head typically extended forward. There are no typical gross lesions. The treatment with selenium, vitamins A, D and E as well as with some antibiotics as chlortetracycline, bacitracin, etc. could reduce mortality rate. Spirochaetosis is a septicaemic disease characterized by depression, progressive paresis, paralysis and inflammatory necrobiotic changes in parenchymal organs and the gastrointestinal tract. The birds are depressed, cyanotic; greenish diarrhoea with considerable amounts of urates is observed. After removal of plumage, the ticks, adhered to the skin, are discovered. The aetiological agent is Borrelia anserina, a highly motile spirochaete with 5-8 coils. The organism is not resistant outside the host and could exist only in some vector. The liver is often enlarged, mottled with necroses of various sizes and marginal infarctions. Usually, bile-coloured mucoid enteritis is observed. The arsenics as well as some antibiotics including tylosin, tetracycline and penicillin are effective for treatment of infected birds. The application of vaccines is successful, but the acquired immunity is short and revaccinations are needed.

Avian tuberculosis : Avian tuberculosis is a chronic infectious disease characterized by the formation of granulomatous lesions in viscera, a progressive weight loss and death. It is usually encountered sporadically in birds reared in small yards, zoos and is a problem among caged exotic birds. The aetiological agent is *Mycobacterium avium*, a very resilient and acid-resistant microorganism. It is resistant to temperature changes, drying, pH changes, to many disinfectants and survives in the soil for years. Along the small intestine, single or multiple sub-serously prominating tubercles are detected. In advanced cases, tubercles could be observed along the bone marrow of the femur or the tibia. Characteristic features of avian tuberculosis are conglomerate tubercles (gross and histological appearance). The diagnosis is based upon the complex evaluation of history, persisting lethality in adult flocks and the patho-anatomical findings. Avian tuberculosis should be differentiated from neoplastic diseases, coligranuloma (Hjarre's disease), pullorum disease, etc. The treatment is not advised, as the disease is contagious for men.

VIRAL DISEASES

Viral inclusion body hepatitis :The viral inclusion body hepatitis (IBH) is an adenovirus infection characterized by haemorrhages and dystrophic necrobiotic changes in the liver and kidneys, accompanied by intranuclear inclusion bodies. A characteristic macroscopic lesion is the enlarged, dystrophic liver with yellowish colour and crumbly texture. IBH outbreaks are encountered primarily in meat type chickens, most commonly at the age of 3-8 weeks. IBH often occurs as a secondary infection to immunodeficiency resulting from other diseases (IBD, CIA). On the background of dystrophic liver changes, haemorrhages of various intensity and size are outlined, thus creating a variety of liver lesions. IBH is characterized by a sudden onset and a sharply increased death rate that reaches peak values by the 3rd - 4th day and returns back within the normal range by the 6th–7th day. The total death rate is usually under 10% but sometimes could attain 30%. More rarely, macroscopically visible necrotic foci could be detected in the liver. Clinical signs could be observed only several hours prior to death occurrence. They consist in pale comb and wattles, depression and apathy. Sometimes, the skin is icteric. Often, ecchymoses and striated haemorrhages in skeletal muscles are observed. Microscopically, extensive dystrophic changes and necroses of liver parenchyma are detected. In the nuclei of hepatocytes, basophilic or eosinophilic inclusion bodies are detected. The basophilic inclusion bodies are usually dense and occupy the entire nuclear inner space, whereas the eosinophilic ones are round or irregularly shaped and surrounded by a light halo. The diagnosis is based upon the typical gross lesions and the history records. A principal approach in IBH diagnostics is the histological investigation that helps to detect the intranuclear inclusion bodies. IBH should be distinguished mainly from IBD and chicken infectious anaemia (CIA). With regard to IBH prevention and control, the eggs of broiler parent flocks, where the disease is consecutively appearing in the progeny, should not be used for hatching. The access of wild birds should be prevented as they are potential carriers and distributors of the virus. The most important steps in IBH prevention are the control of IBD and CIA. There are neither vaccines, nor an effective treatment.

Haemorrhagic enteritis of turkeys : The haemorrhagic enteritis (HE) is an acute viral disease in young turkeys, characterized by a sudden onset of bloody faeces causing high death rate. Blood discharge from the vent (187), fresh blood in faeces or melena (188) could be observed. HE is observed in turkeys at the age of 6–11 weeks but most commonly at the age of 7–9 weeks. As an exception, HE could be observed in turkeys under 4 weeks of age, presumably due to maternal antibodies. The small intestine, especially the duodenum, has a dark red colour and ramiform blood vessels prominating under the serous coat, and sometimes, haemorrhages are seen through the intestinal wall. The staining of peripheral blood smears reveals a picture, characteristic for septicaemia. The typical gross lesions and the history of the disease allow making a tentative diagnosis for HE. The histological detection of intranuclear inclusion bodies in the reticuloendothelial cells of the spleen or intestines confirms the diagnosis. For identification of the HEV, the agar gel precipitation method could be used. The viral antigen could be detected in fresh or frozen spleen tissue, diluted 1:1 with saline or in sera obtained from diseased turkeys. HE should be distinguished from other cases of severe enteritis in turkeys, acute *E. coli* septicaemia and some other septicaemic states (streptococcosis, fowl cholera, etc.). Within weeks after the beginning of the disease, an appropriate antibiotic therapy should be initiated for prevention of secondary E. coli septicaemia. Practically, there is no effective therapy for HE in turkeys.

Egg drop syndrome-1976 : The egg drop syndrome - 1976 (EDS 76) is an infectious disease in layer hens manifested by a quick drop in egg production, failure to reach peak production, irregularly shaped eggs, soft-shelled or shell-less eggs and depigmentation. The aetiological agent is an adenovirus of group III. The horizontal transmission occurs slowly in battery systems and rapidly in floor housing systems. The first sign is the loss of egg pigmentation, rapidly followed by the appearance of soft-shelled, shell-less of deformed-shell eggs. If defective eggs are discarded, the remaining ones have no problem with fertilization and hatching. The drop could be sudden or prolonged. Usually, it lasts for 4-10 weeks and the egg production is reduced by about 40%. Apart from the inactive ovaries and oviduct atrophy, other lesions are not discovered. The replication of the virus in epithelial cells of oviduct glands results in severe inflammatory and dystrophic changes in the mucous coat. The appearance of eggs with impaired quality and the dropped egg production are suggestive for EDS 76. The diagnosis is supported by some serological studies and is confirmed after isolation and identification of the virus. In many instances, no antibodies are detected in infected flocks until egg production approaches levels between 50% and peak production.

Adenovirus group I — associated infections : Necrotic pancreatitis and gizzard erosions in broilers. It is characterized by focal necroses in the pancreas. They are manifested by erosions affecting the cuticle and the underlying tissues of the gizzard. Avian adenovirus splenomegaly is characterized by enlargement of the spleen, pulmonary oedema and congestion. It is observed in broiler breeders at the age of 20-45 weeks. Its course is peracute or acute. The mortality could reach 8-9%. The most typical lesions are splenomegaly, mottled or marble-like appearance of the spleen, oedema or hyperaemia of lungs.

Infectious bursal disease (Gumboro) : Infectious bursal disease (IBD, Gumboro) is an acute, highly contagious viral infection in chickens manifested by inflammation and subsequent atrophy of the bursa of Fabricius, various degrees of nephroso-nephritis and immunosuppression. Clinically the disease is seen only in chickens older than 3 weeks. The feathers around the vent are usually stained with faeces containing plenty of urates. The period of most apparent clinical symptoms and high death rate is at the age of 3-6 weeks. IBD could, however, be observed as long as chickens have a functioning bursa (up to the age of 16 weeks). In chickens younger than 3 weeks, IBD could be subclinical, but injured bursa leads to immunosuppression. Also, diarrhoea, anorexia, depression, ruffled feathers, especially in the region of the head and the neck is present. A natural IBD infection is mostly observed in chickens. In turkeys and ducks it could occur subclinical, without immunosuppression. Most isolates of the IBD virus in turkeys are serologically different from those in chickens. In premises, once contaminated with the IBD virus, the disease tends to recur, usually as subclinical infection. The dead bodies are dehydrated, often with haemorrhages in the pectoral, thigh and abdominal muscles. The IBD virus belongs to the Birnaviridae family of RNA viruses. Two serotypes are known to exist, but only serotype 1 is pathogenic. The virus is highly resistant to most disinfectants and environmental conditions. In contaminated premises, it could persist for months and in water, forage and faeces for weeks. The incubation period is short and the first symptoms appear 2-3 days after infection. The lesions in the bursa of Fabricius are progressive. In the beginning, the bursa is enlarged, oedematous and covered with a gelatinous transudate. IBD virus has a lymphocidic-effect and the most severe injuries are in the lymph follicles of the bursa of Fabricius. Most commonly, IBD begins as a serous bursitis. The kidneys are affected by a severe urate diathesis. In an acute outbreak and manifestation of the typical clinical signs, the diagnostics is not difficult. The diagnosis could be confirmed by detection of typical gross lesions throughout a patho-anatomical study. IBD should be differentiated from IBH (inclusion body hepatitis). The application of live vaccines in chickens is a key point in the prevention of IBD and should be related to the levels of maternal antibodies.

Infectious bronchitis (IB) : In chickens up to the age of 4 weeks, IB is manifested by severe respiratory signs (sneezing, coughing, and rales). Rhinites and conjunctivitis, depression and crowding around heat sources are observed. The death rate could reach 100%. The mortality in young chickens is usually insignificant provided that a secondary infection with a different agent is not occurring. In such cases, there is a moderate to severe inflammatory cell infiltration of upper respiratory tract mucosa, resulting in thickened and more compact mucosa. In layer hens infected with the IB virus, oophorites and dystrophic necrobiotic lesions affecting primarily the middle and the last thirds of oviduct's mucous coat are observed. The consequences are drop in egg production, appearance and increase in the number of deformed and pigmentless eggs or eggs with soft shells and watery egg white. The nephrotropic strains of the IB virus cause severe inflammatory and dystrophic necrobiotic damages of kidneys: urolithiasis (229), interstitial nephritis (230), and haemorrhages (231) that considerably increase the death rate. Under natural conditions, only hens are infected. Non-immune birds of all ages are susceptible. The disease is even seen in vaccinated flocks. The serological methods (VN, ELISA, etc.) are widely used in the diagnostics. At present, PCR is used for rapid identification of IB virus serotypes. IB should be distinguished from other acute respiratory diseases as ND, laryngotracheitis and infectious coryza. The vaccination with live or killed vaccines is effective only if they contain the respective serotype of the virus for the given region.

Laryngotracheitis : Laryngotracheitis (LT) is a viral infection in hens, pheasants and peacocks characterized by catarrhal haemorrhagic to fibrinous inflammation of the respiratory tract. It is manifested in laryngotracheal and conjunctival form. In the laryngotracheal form, suffocation, rales and cough are observed. The head and the neck are strongly extended forward and upward during inspiration. mucous coats of the larynx and the trachea are catarrhally, haemorrhagically to fibrinously inflamed. Most outbreaks are encountered between the age of 4 and 14 weeks although the disease affects fowl of any age. LT is caused by a herpesvirus that is rela-tively resistant. In the conjunctival form of LT, wet eyes, tear secretion and oedema of infra-orbital sinuses are observed, especially in a complicated infection. The typical clinical and morphological signs are sufficient to assume the presence of LT. The diagnosis is confirmed with the detection of intranuclear inclusion bodies in the trachea throughout the histological study in the early stages of the disease, serological studies (VN, ELISA), etc. LT should be differentiated from IB, SHS, *M. synoviae* infections, etc. Premises, contaminated with the LT virus, should be freed, cleansed, disinfected and occupied again after 5-6 weeks. The vaccination of unaffected birds and these from other premises of the infected farm could protect and stop subsequent outbreaks.

Swollen head syndrome : Swollen head syndrome (SHS) is a complicated infection in broilers and broiler breeders, where the primary aetiological agent is an avian pneumovirus (APV) and the secondary — usually E. coli; it is characterized by respiratory and nervous signs. In broilers chickens, SHS is generally seen after the 4th week of life. The first clinical signs are sneezing, coughing, rales and conjunctivitis. A profuse tear secretion, reddened conjunctivas and a characteristic oblong almond-like shape of eyes are observed. The inflammatory exudate is initially transparent, but afterwards becomes opaque. Subcutaneous oedemas in the head region, involving unilaterally or bilaterally the periorbital sinuses and the mandibular space, are emerging. A seasonal pattern in the prevalence with peaks during spring and summer is observed. SHS in broiler breeders is usually encountered around or after the peak egg laying period only in female birds. Unilateral or bilateral swellings of the head, affecting the periorbital sinuses, the mandibular space and the wattles are seen. The conjunctiva and the mucous membranes of sinuses are considered to be the entrance door of the infection. A unique cytotoxin has been identified in many E. coli isolates in SHS that could be involved in the pathogenesis of the disease. The inflamed by the virus infection conjunctiva-associated lymphoid tissue is the site where the bacterium enters the subcutaneous tissue. In laying hens, the ovaries are affected in many instances (serofibrinousoophoritis), resulting in reduced egg production. The diagnosis is based upon the distinctive clinico-morphological signs. SHS should be differentiated from Mycoplasma and Pasteurella infections and the skin form of aspergillosis. Prevention — improvement of the microclimate of premises, use of live and killed vaccines.

Infectious encephalomyeliti : The infectious encephalomyelitis (IEM) is characterized by signs of ataxia, progressing to paralysis, prostration and marked tremor of the head and the neck, and because of that, is also called epidemic tremor. The chickens with prostration are usually in lateral recumbency. The expression of the eye is dull. IEM outbreaks are generally observed in chickens at the age between 8–9 and 20 days. The morbidity rate could reach 40–60%. The average death rate is about 25%, but could be more than 50%. The history of the disease, the age of onset and the typical nervous signs, especially the head tremor are indicative for the diagnosis that could be finally confirmed by histology. IEM should be differentiated from other diseases with nervous symptoms as encephalomalacia, mycoticencephalites, toxicoses (salt, pesticides). The chickens from naturally infected flocks obtain an adequate maternal immunity that preserves them from the disease.

Newcastle disease : The Newcastle disease (ND) is characterized by marked variations in morbidity, death rate, symptoms and lesions. The clinico-morphological signs possess a distinct viscerotropic or neurotropic character. In the viscerotropic form, haemorrhagic diphtheritic lesions of the entire alimentary tract, from the beak to the vent, are present. The haemorrhages of gizzard epithelium are remarkable. The mucous coat is oedematous, covered with thick mucus and mottled with haemorrhages varying from single to multiple, sometimes gathered at the boundaries with the gizzard or the oesophagus. Typical for this form are the haemorrhagic, necrotic and focal diphtheroid lesions affecting the mucosa of the buccal cavity (259, 260), the stomach and the intestines. The disease is generally prevalent in hens, more rarely in turkeys, exotic or wild birds. Birds at any age are susceptible.

It is caused by a **paramyxovirus**. Depending on their pathogenicity, the numerous known strains are classified as lentogenic, mesogenic and velogenic. The vaccines made of lentogenic strains provoke a shorter immunity that requires a revaccination. The vaccines from mesogenic strains result in a lasting immunity, but could provoke a lethal issue especially in birds without a primary immunity created on the basis of lentogenic vaccinal strains.

Fowl pox : Fowl pox (FP) is a viral disease in hens, turkeys and many other birds, characterized by cutaneous lesions on the feather-less skin and/or diphtheritic lesions of mucous coats of the upper alimentary and respiratory tract. FP is encountered in either cutaneous or diphtheritic form or in both. In most outbreaks, the cutaneous form is prevailing. The lesions vary according to the stage of development : **papules, vesicles, pustules or crusts**. The lesions are usually in the region of the head. Lesions around the vent in a pigeon. FP is caused by an epitheliotropic DNA virus from the Avipox genus, the Poxviridae family. Some virus types (strains) exist : fowl pox virus, turkey pox virus, pigeon pox virus, canary pox virus etc., different in pathogenic and immunogenic aspects. The viruses are very resistant to environmental factors and could persist for several months. Diphtheritic lesions look like whitish or yellowish plaques that are deposited and grown on the mucous coats of the buccal and nasal cavities, the sinuses, the larynx, the pharynx, the trachea or the oesophagus. The diagnosis is made upon the typical cutaneous and diphtheritic lesions. The prevention is performed by vaccinations that could be made at any age, if necessary.

Reovirus infections :**Viral arthritis/tenosynovitis**. The most frequent reovirus-associated disease in poultry is viral arthritis. Clinically it is manifested by lameness and swellings affecting primarily tarsometatarsal joints and the feet. Many infected birds are in a good general condition, but some could be lethargic and exhausted. The mortality is very low. The infections affect predominantly meat type poultry. In some cases, joint cavities or tendon sheaths contain a small amount of straw-yellow exudate whereas in other the exudate is haemorrhagic or fibrinous.

Fibrinous tenosynovitis : The inflammation of the tendon progresses to a chronic type lesion characterized by tissue fibrosis in the affected area. The aetiological agent is a reovirus, member of the Orthoreovirus genus, Reoviridae family. Several serotypes are identified. For their identification, the agar gel precipitation test could be used. Reoviruses are highly resistant to a number of environmental factors (temperature, pH, etc.).

Stunting syndrome in broilers : The stunting syndrome in broilers is associated with a reovirus infection but according to some studies, the role of the reovirus is probably secondary. It is characterized by a considerably reduced live weight in affected birds and a various degree of ununiformity in the flock varying from 5-10% to 40-50%, usually seen after the age of 14 days. The small intestine is pale, dilated and contain indigested forage. The tentative diagnosis of viral arthritis and stunting syndrome could be made on the basis of symptoms and lesions. The detection of reovirus antibodies via ELISA supports the diagnosis. Viral arthritis should be differentiated from *M. synoviae* induced, staphylococcal arthritis and the spontaneous rupture of the tendon of the gastrocnemius muscle. The vaccination of broiler breeder flocks with live, or inactivated vaccines protects one-day-old chicks. Infected birds should be removed from the premises. The iodine disinfectants are considered effective for inactivation of agents.

NEOPLASTIC DISEASES IN POULTRY

Virus-induced neoplastic diseases Marek's disease (MD) : **Acute (visceral) form** : It is characterized by diffuse or nodular lymphomatous lesions in various viscera (liver, spleen, heart, kidneys, lungs, gonads, proventriculus, pancreas, etc.), the skeletal muscles and the skin. MD affects mainly hens, and is rarely observed in turkeys. It is most commonly encountered in birds at the age of 89 weeks and in layer hens. The cases at the age of 16-20 and 24-30 weeks are predominant. MD is prevalent all around the world and, in fact, all flocks are exposed to the effect of the aetiological factor. Diffuse lymphomatous growths in the heart result in its transformation into an amorphous tumorous mass. Lymphoid cell proliferations occur in the iris and the ciliary muscles in the ocular form of MD. Three classes of viruses are able to protect fowl from MD: attenuated serotype 1 of MDV-cell associated vaccines; HVT could be used for preparation of cell-free lyophilized vaccines; naturally apathogenic isolates of serotype 2 cell associated vaccines. The vaccines against MD provide over 90% protection. HVT gives excellent results but in case of failure, a bivalent vaccine could be used. **Transient paralysis** : They are observed in chickens and hens, especially non-vaccinated against MD. Most cases present the classical form manifested by flaccid paralysis of the neck and legs for 143 days followed by complete recovery. The syndrome has to be differentiated from the neural form of MD on the basis of its transient nature and the flaccid, but not spastic paralysis.

Lymphoid leukosis : It is characterized by a gradual beginning, persistent low mortality in the flock and diffuse or focal neoplastic growths of lymphoblasts in viscera. The neoplastic changes begin always from the bursa of Fabricius, where various-sized lymphomas are detected. Clinically, pale comb and wattles, sometimes swelling of the abdomen because of the highly enlarged liver are observed. Diffuse or nodular neoplastic growths could be detected in many organs, but they are more common in liver, spleen, kidneys, heart and ovary. Usually, it is not seen in birds younger than 14 weeks. The lethal tissues occur mostly at the age between 24 and 40 weeks. Distinct nodular tumours. Tumours in the bursa of Fabricius.

Myelocytomatosis (MC) : MC is caused by viral strains of ALSVs from subgroups A, B and J (MC29, MC31, CMII, OK10, HRPS 103, and ADOL HC1). It is encountered relatively infrequently. Its occurrence is sporadic or enzootic. Susceptible birds are hens, pheasants, guinea hens and quails. In most cases, the liver is enlarged, thick and mottled with dark red spots or fat-like nodules. Sclerotic changes in the liver are possible because of regression of neoplastic lesions. The spleen is usually enlarged, but sometimes, could be atrophied. Carcinosarcoma of the pancreas. Tubulous glandular epithelial formations of the carcinoma component among the liposarcoma part of the parenchyma. The diagnosis is based upon the entity of data about the history, the gross appearance and location of the tumours and the specific histological lesions. From a differential diagnostic point of view, myeloblastosis and erythroblastosis should be considered.

Erythroblastosis : Erythroblastosis (ER) is characterized by intravascular proliferations of immature precursors of erythrocytes. ER has a leukaemic character and is manifested with signs of severe anaemia. The liver and the kidneys are moderately enlarged with a characteristic dark red to mahogany colour, sometimes with haemorrhages. ER is caused by the avian eryhtroblastosis virus (AEV); the most frequently encountered strains are E-26, ES4, R, etc. The spleen is unusually enlarged or atrophied in cases of severe anaemia. Osteopetrosis is a neoplastic disease. It is characterized by a significant thickening of bone periosteum. The diaphyses of the tibia and/or tarsometatarsal bones are most commonly affected.

Adenocarcinomatosis : The intestinal and reproductive tracts are primarily affected. Quite often, the neoplasms invade the peritoneum and other serous coats. Macroscopically they appear as numerous disseminated thick nodes of various sizes (with a diameter from several mm to 1 cm). Sometimes, the tumours are of cystic formations (cystadeno-carcinoma). The leiomyoma of the mesosalpinx is a common tumour in hens. Usually, it is located in the ventral ligament of the oviduct. Its size varies from small to large (several cm in diameter), smooth, thick, sometimes highly vascularized nodes.

PARASITIC DISEASES

Coccidiosis : Coccidiosis is a common protozoan disease in domestic birds and other fowl, characterized by enteritis and bloody diarrhoea. The intestinal tract is affected, with the exception of the renal coccidiosis in geese. Clinically, bloody faeces, ruffled feathers, anaemia, reduced head size and somnolence are observed. The area around the vent is stained with blood. The infection is realized by a faecal-oral route. After ingestion of sporulated (infective) oocysts, sporozoites are released that enter asexual and sexual cycles of development resulting in the emergence of thousands of new oocysts in the intestines. Oocysts are distributed by faeces. Soon, they sporulate and become infective for chickens. The intestinal lesions provoked by coccidia, are due to injury of the epithelial cells of the mucous coat where the parasites are developed and multiplied. The oocysts exist in the litter in premises and are distributed by clothes, shoes, dust, insects, etc. Pathoanatomically, dehydration and a high degree of anaemia of the body and viscera are discovered. The microscopic examination of a native preparation of intestinal content or superficial mucosal layer reveals a significant number of oocysts in one observation field.

Prevention. The use of coccidiostatics with forages on a rotation basis is the most extensively used means. The immunization against coccidiosis with commercial vaccines is used in broiler breeder flocks. If the chickens are exposed to the natural effect of a moderate number of oocysts in their environment, they develop immunity to the respective parasitic species.

Histomonosis : The histomonosis is a protozoan disease, caused by *Histomonas meleagridis*, and characterized by necrotizing lesions affecting the liver and the caeca. Clinically, sulfur-yellow coloured faeces and depression are observed. A characteristic feature is the blackening of the skin of the head (blackhead), due to cyanosis. Pathoanatomically, bilateral enlargement of caeca with thickening of walls is observed. The aetiological agent is *Histomonas meleagridis*, a polymorphic flagellate that is present as flagellate in caeca and as amoeba in tissues. The trophozoites survive for several hours in the environment but in eggs, they remain infective for more than a year. Some wild birds serve as vectors. The caecal content is often mixed with blood. The main sources of infection are hens and chickens. The morbidity rate amounts to 90% and the mortality rate to 70%. In the liver, irregularly outlined coagulation necroses with various size and colour, are observed. **Diagnosis** : It is made on the basis of the typical macroscopic lesions. When necessary, a histological study and phase-contrast microscopy of native preparations could be performed.

Ascaridiosis : Ascaridiosis is one of the most prevalent helminthoses in fowl. It is caused by various species from the Ascaridia genus. The ascarids have a direct life cycle. Sometimes, it could involve paratenic hosts (earthworms). Infected birds are progressively emaciated, anaemic and sometimes diarrhoeic. Pathoanatomically, haemorrhages of various intensities are found out in intestinal mucosa, catarrhal haemorrhagic enteritis, and the parasites are also observed. In cases of extensive invasion, the ascarids block the intestinal lumen and could cause a complete obstruction. The treatment and the control are realized through regular pathoanatomical and copro-ovoscopic studies and performance of therapeutic and protective dehelminthizations.

Raillietinosis : Raillietinosis is a cestodosis characterized by diarrhoea (sometimes bloody) during the acute stage and emaciation to cachexia and anaemic during the chronic stage. It is caused by some representatives of the Raillietina genus that parasitize in various areas of the small intestine. The usual intermediate hosts are ants or other insects. Pathoanatomically, haemorrhages with various intensities in the intestinal mucosa, catarrhal haemorrhagic enteritis and the parasites are found out throughout the gross examination. The treatment and the control are done by dehelminthization of all birds in the affected farm.

Knemidokoptosis : Knemidokoptosis, known also as scaly or chalky legs, is characterized by the appearance of extensive, rough and hard crusts on the featherless part of the legs. Adult birds are generally affected (hens, turkeys, pheasants, exotic birds). It is caused by *Knemido koptesmutans*. The lesions are secondary to an inflammatory reaction, in which the thickening of the cutaneous epidermis together with released exudate forms the crusts on legs. The infection occurs by contact between birds. The control is done by isolation of affected birds, leg bathing with warm acaricide solution and warm vegetable oil that helps the decrustation.

MYCOSES AND MYCOTOXICOSES

Aspergillosis : Aspergillosis is an acute or chronic respiratory disease. In rare instances, peritoneal, visceral or systemic lesions could be observed. It is caused by *Aspergillus fumigatus*. Dyspnea, enhanced tense and heavy breathing are observed. Sometimes, rales and cyanosis could occur. In the acute form, a serous fibrinous pneumonia is observed. Within the trachea and the main bronchi, usually near the bifurcation, obturation masses of coagulated fibrinous exudate are detected. In the nodular form, a characteristic *Aspergillus granulomatous* structure of tissues is discovered. The control includes : collection of clean eggs for hatching, disinfection and fumigation of eggs, incubators and hatcheries, regular checking of ventilation systems and change of air filters in hatcheries, monitoring of hatcheries and the environment for mould contamination, use of dry and clean litter and non-contaminated forages, optimisation of aeration and the humidity in premises where birds are housed. The treatment with fungicide antibiotics and different antimycotic drugs is not always efficient. Nistatin and copper sulfate (1:2000 :: vol : vol) with drinking water could be used. Mould is developed after inoculation of material on an appropriate nutrient media.

***Aspergillus granulomatous* dermatitis** : The *Aspergillus granulomatous* dermatitis as a postvaccinalcomplication is observed in growing broiler breeders. Following subcutaneous application of oil adjuvant vaccine in the lower part of the neck, a marked local reaction with extensive swelling of the head is observed. The swellings involve the subcutaneous tissue of periorbital sinuses, the mandibular space, even the entire skull without the skin adnexa.

Candidiasis : Candidiasis is an alimentary tract disease, caused by the fungal yeast *Candida albicans* and usually occurs as a secondary infection. Predisposing factors are the lack of good hygiene, prolonged antibiotic treatment, vitamin deficiency, severe parasitic infections and immunodeficiencies. The production of a soluble endotoxin is also contributing to the pathogenicity of the agent. The lesions are usually detected in the crop, lips, oesophagus, but could affect the proventriculus and more rarely, the intestines. The affected mucosa is diffusely or focally thickened, raised and wrinkled, white, resembling a towel. Possibly, the lesions could look as pseudomembranous or diphtheroid coatings. The histological study confirms the diagnosis by showing the fungal hyphae in the affected mucosa. Usually, a copper sulfate solution is used with drinking water for prevention and control. The application of nistatin in water or forage is efficient against candidiasis in turkeys.

Fusariotoxicoses : The Fusarium genus produces numerous mycotoxins, out of which the most important for poultry pathology are trichothecenes, fumosinins, moniliformin, fusaro-chromanone and zearalenone. The Fusarium toxins possess a pronounced caustic effect, resulting in necroses and crusts of the buccal mucosa. The caustic effect of some fusariotoxins is the cause of the commonly detected erosions and ulcers in gizzard cuticulum. Note the thickened wall of the proventriculus. Frequently, the proventriculus is dilated at a various extent, its wall is thickened and in its mucous coat, fibrinous necrotic and haemorrhagic lesions are detected. The prevention from mycotoxicoses requires the detection and control of mycotoxin-contaminated forage components, avoidance of forage moulding. The application of commercial mycotoxin-binding agents could reduce the effects of some of them. In some mycotoxicoses, the systemic requirements for vitamins, minerals and proteins are increased and they could be balanced by supplementation with forage or water.

DEFICIENCY DISEASES

Vitamin B_1 deficiency : In birds, vitamin B_1 (thiamine) deficiency is clinically and morphologically manifested with paralysis of limbs and muscle atrophy, beginning from the flexors of toes and ascending towards the extensors of legs, wings and the neck. The chickens acquire a specific posture with flexed legs and the head drawn back (stargazing). As hypovitaminosis B_1 causes a severe anorexia, thiamine supplementation to the drinking water is advised until the recovery of acute deficiency and thereafter, the vitamin could be added to the forage.

Vitamin B_2 deficiency : The hypovitaminosis B_2 (riboflavin deficiency) is characterized with impaired oxidation processes and dystrophic changes in the peripheral nerves. The typical clinical sign is curled toes due to paralysis. In the beginning, the toes are slightly flexed and chickens tend to stand on their hocks. In moderate cases, a marked leg weakness and toe flexion in a different extent are observed. In severe cases, the toes are completely curled downward and inward and complete weakness of legs is present. A considerable improvement and relief could be expected if the treatment is initiated in the initial stage of the disease. Water soluble vitamins, that are easily utilized, are recommended.

Vitamin E deficiency : The deficiency of vitamin E in poultry is manifested in three different forms : encephalomalacia, muscular dystrophy and exudative diathesis. Each of them is usually encountered independently, although sometimes they occur at a time. Encephalomalacia is a crazy chick disease — an abnormal softening of the brain tissues. The signs are associated to the central nervous system lesions. They include ataxia, disequilibrium, falling on the back, frequent movements with the wings, strongly stretched legs (clonic spasms) and twisting of the head. Rarely, torticollis or opisthotonus could be observed. Encephalomalacia is usually observed at the age of 15-30 days but could also be present as early as after the 7th day as well as after the 56th day of life. Pathoanatomically, oedema, haemorrhages and colliquativenecroses are detected in the cerebellum. In the majority of cases, the haemorrhages vary from hardly perceptible to petechial. Muscular dystrophy signs are usually unapparent, but locomotor problems could occur. The pectoral and thigh muscles are most commonly affected, and in them, white or yellowish muscle fibres are found, responsible for the striated appearance of skeletal muscles, due to Zenker's necrosis. Gelatinous subcutaneous oedemas are discovered in the ventral body parts : breast, abdomen, mandibular space. The skin of legs is often cyanotic. The diagnosis is based upon the typical clinical signs and the lesions. The results of the histological examination have a particular value for confirmation of the diagnosis, especially of encephalomalacia and muscular dystrophy.

Fatty liver haemorrhagic syndrome : The fatty liver haemorrhagic syndrome (FLHS) is a widely prevalent sporadic disease mainly among commercial layers. The FLHS outbreaks are often associated with hot weather and a period of extensive egg-laying. The hens in the flock are overweight (on the average by 20% or more) and a sudden drop in egg production is observed. The birds are discovered suddenly dead, with pale head skin. In the abdomen, large blood clots are detected. Clinically healthy birds in the flock could also exhibit liver haematomas, dark red (fresh) or green to brown (old). Considerable amounts of fat are detected in the abdominal cavity. The only successful approach for prevention is the reduction of obesity in layers. The use of lipotropic agents such as vitamin E, vitamin B_{12} and choline chloride gives conflicting results. The avoidance of heat stress and moulded forages is helpful.

Slipped tendon (perosis) : Perosis or chondrodystrophy is encountered in young birds whose diet is deficient in manganese (Mn) or some of the following vitamins: choline, nicotic acid, pyridoxine, biotin or folic acid. This is an anatomic deformation of leg bones in young chickens, turkey poults, pheasant poults, etc. It is characterized by retarded growth of long bones, widening of the tibiometatarsal joint, twisting or bending of the distal end of tibia and the proximal end of metatarsus and finally, slipping of the gastrocnemius muscle tendon from its condyles. Clinically, it is manifested by impaired locomotion. The vitamin D_3 deficiency or lack, or the impaired ratio of calcium and phosphorus result in rachitis in young birds. It is encountered more frequently in industrial poultry breeding where the forages are often inadequately balanced. The deficiency of vitamin D_3 and phosphorus are more common in growing birds whereas the calcium deficiency in young or adult layer hens. The newly hatched chickens have depleted calcium stores. The calcium deficiency is rapidly manifested if they do not receive an adequate diet supplement. Clinically and morphologically, soft bones or various degree of bone deformation are determined. The birds usually lie down and their growth is retarded. The diagnosis is based upon the complex evaluation of their age, signs and lesions. The supplementation of vitamin D_3 with water or forage and the balancing of Ca/P ratio in the diet contribute to the favourable outcome of the disease.

OTHERS

Pulmonary hypertension (ascitis) syndrome in broiler chickens : The pulmonary hypertension syndrome (PHS) or ascitis is related to the rapid growth and enhanced metabolic processes in broilers. Affected chickens are with a strongly distended abdomen, reluctance to move, respiratory troubles and cyanosis. The pleuroperitoneal cavity of affected chickens is filled with straw-yellow fluid. The rapid growth in contemporary broilers is related to higher needs for oxygen, and the lung remains relatively small in comparison to body dimensions. The main factor for PHS is hypoxaemia that results in enhanced cardiac activity. The consequences are pulmonary hypertension, right cardiac failure and ascitis. Immediate causes for hypoxaemia are the inadequate ventilation, the high ammonia or dust levels in premises, low temperatures, stimulating the metabolic processes. Generalised passive venous hyperaemia and stasis in the liver and other viscera are observed. The right ventricle at an early stage is hyper-trophied whereas at a later-dilated with extremely thinned wall. The diagnosis is based on gross lesions. The reduction of oxygen requirements by slowing down the metabolism protects from and decreases the incidence of ascitis. Restriction of forage consumption and various light patterns are recommended.

Amyloidosis : Amyloidosis is seen mainly in adult birds. It is characterized by extracellular build-up of the protein amyloid in different viscera and the joints. Amyloid arthropathy is primarily associated with *Enterococcus faecalis* and *Mycoplasma synoviae*. Brown layers are particularly susceptible. Some genetic factors could be involved in the incidence of amyloidosis too. The affected joints are enlarged, swollen and contain orange-yellowish matter. Amyloidosis of the liver, the spleen and the kidneys are observed. The affected organ is multifold enlarged with stretched capsule, rounded margins and pale colour. The state is usually seen following severe disturbances in protein metabolisms, prolonged and exhausting diseases (tuberculosis, etc.). The cannibalism in birds is usually consequence of feather pecking or pulling out, or vent pecking, that are behavioural reactions. Feather pecking or pulling out are observed in birds, reared in closed, often overcrowded premises. This happens frequently in growing broiler breeder flocks during the period of restricted feeding.

Deep pectoral myopathy : The deep pectoral myopathy (DPM), called also green muscle disease or Oregon disease, is observed in heavy meat types of turkeys or chickens. The disease occurs because of ischaemic necrosis due to inadequate blood supply of variously sized deep pectoral muscle groups. The lesion is uni- or bilateral and is detected as a slaughterhouse finding. Affected muscles have an unusual green colour.

Rupture of the gastrocnemius tendon in broiler breeders : This state could cause significant economical losses in broiler breeder flocks. It is usually seen in birds older than 12 weeks but is also observed in breeders at the age of 7 weeks or older than 24 weeks. The rupture could be uni- or bilateral. Clinically, lameness is observed. In affected birds, swelling of the posterior surface of the leg, just above the tibiotarsal joint could be palpated. In acute lesions, haemorrhages are visible through the skin. In older lesions, a blue-greenish discolouration is present whereas in chronic lesions, very hard masses as abnormal subcutaneous tissue are palpated. In part of birds, the muscles of affected legs undergo a various degree of atrophy. In acute lesions, swellings and haematomas under the skin of the posterior surface of the leg are detected. In older or chronic lesions, the blood is partially or completely reabsorbed. The end of ruptured tendon and the adjacent tissue are involved at a various extent by a newly grown fibrous tissue. RGT should be differentiated from reovirus-and MS infections, where the gross and histological lesions are with a marked inflammatory character.

Cage layer fatigue : Cage layer fatigue syndrome in birds is characterized by an inability to stand on their feet and fragile bones. It is mainly observed in young layer hens reared in batteries in the period of maximum egg-laying. Affected birds lie down and stopped eating. Egg shells become thin. The calcium deficiency in layer hens results in initial removal of calcium from bones, to complete depletion of the medullary bone and thereafter, of the bone wall. The bones are strongly thinned and spontaneous fractures, especially of the tibia and the femur, could occur. Although the severe calcium deficiency is often a triggering factor, the aetiology of the syndrome seems to involve other, yet unknown factors. The supplementation of calcium, phosphate and multivitamin preparations in the diet and rinking water, the regulation of avian population density into cages and ensuring adequate nutritional and drinking fronts are also contributing for the favourable outcome of the condition.

Hyperkeratosis : A local form of hyperkeratosis in an ostrich. Increased amount of keratin in the horny tissues — down feathers and beak — is present, resulting in their coarse appearance. The state is linked to impaired metabolism of sulfur-containing amino acids, vitamin A deficiency, etc.

Gout : The gout is characterized by retention and build-up of urates in tissues. It is seen as two separate syndromes: visceral gout and articular gout. The visceral gout is manifested with deposition of urates in renal tubules and the serous coats of the heart, the liver, the mesentery, the air sacs or the peritoneum. The urate deposits on serous coats resemble a chalky white dust. Visceral urate deposits are generally due to renal failure. Possible causes for this could be obstruction of ureters, renal damage or dehydration. Most commonly, visceral gout following dehydration is observed in newly hatched chickens after overheating or a more prolonged stay in the hatchery. Visceral gout outbreaks are related to vitamin A deficiency, treatment with sodium bicarbonate, mycotoxicoses, etc. Urolithiasis is an aetiologically unknown state, occurring primarily in cage layer hens, characterized by obstruction of one or both ureters with urates, atrophy of one or more renal lobes and a various degree of renal and visceral gout. A number of aetiological factors are related to this condition : protein excess, calcium excess (3% or more), sodium bicarbonate toxicity, mycotoxins (ochretoxin, etc.), vitamin A deficiency and nephrotropic strains of the infectious bronchitis virus. The lower phosphorus levels (under 0.6%) are probably helping the manifestation of the disease. There are no specific clinical signs except for the depression and the weight loss. The death rate could increase and persist around

2-4% monthly during the productive period. The total mortality is heavily affected flocks could reach 50%.

Hyperandrogenism in broiler chickens : Hyperandrogenism in broilers is a condition, manifested with marked signs of masculinization in chickens from both genders. The first signs could start very early after the age of 10-12 days. A very noticeable reddening of the comb and wattles, coarse feathering of the face, strong growth of nails and a highly aggressive behaviour in about 100% of birds are observed. The peak of clinically manifested masculinization traits is between the age of 20 and 30 weeks in all chickens. The serum testosterone is manifold higher. In 23-day-old broilers, average testosterone concentrations were 709 ng/dl vs. 36 ng/dl in intact chickens at the same age. The mycotoxins following contamination of some of forage components with moulds are assumed to be a possible cause for this state, but yet, it is not clear if this or another factor is causing hyperandrogenism.

Effect of cleavers (*Galiumaparine*) seeds on healthy status and production traits in broiler chickens : *Galiumaparine*, cleavers or sticky-willy, is an annual weed in cereals and earthed-up crops, and an herbal plant belonging to the *Galium* genus (herbarium). The seeds (fruit) are dry, round or kidney-shaped, with smooth, corrugated or granular surface, sometimes covered with hooked prickles, brown or grey colour and are notched on one side (nuts). On the left — seeds with coat, on the right — without coat. In the gizzard of chickens, the seeds become white with a contrasting black-coloured unipolar groove, similar to an eye. Most probably, that is why they are called "snake eye". Purulent necrotic processes in the gizzard of a chicken. The health is influenced by consequences of mechanical obstruction because of seeds' accumulation in the gizzard. A direct toxic effect of *Galiumaparine* is also supposed (primarily hepato- and nephrotoxic), due to release of toxic glycosides and other toxic substances from the seeds.

Gizzard impaction in turkey : The gizzard impaction could result in high mortality in turkey poults during the first three weeks of life. Affected turkey poults are dehydrated with empty gut, but the gizzard is filled with hard fibrous masses. In some cases, the fibrous masses enter the first part of the duodernum or the lower parts of the small intestines. The impaction is resulting from eating litter that gizzards cannot ingest. The prevention is aimed at control of litter eating.

Roundheart in turkeys (dilated cardiomyopathy) : The aetiology is not known, but genetic factors or early viral myocarditis is supposed. The state is lethal in turkeys at the age of 1-4 weeks. Pathoanatomically, a severe dilated cardiomyopathy, often accompanied with ascitis, hydropericardium and congestion of other organs is detected.

Acute selenium intoxication : Acute selenium intoxication in birds is usually seen after multifold overdosing of preparations containing inorganic selenium with the aim of prevention or therapy. In the early hours after intake of toxic doses of selenium, a high death rate (up to 100%) and massive haemorrhages in the liver are seen. In birds that survived the later stage of the selenium toxicosis, liver haemorrhages are outlined on the background of dystrophic necrobiotic changes.

Acute propane-butane intoxication : In cases of acute intoxication with propane butane (due to damage in heating appliances where propane-butane is used as energy source for heating), asphyxia, cyanosis of the featherless skin, pulmonary edema and sub-capsular haemorrhages in the liver are observed. Pasted vent is usually observed in diarrhoea, when the rectal masses are sticked to adjacent feathers. The observed plugs prevent the evacuation of the next faecal masses. This results in dilatation and obstipation of the rectum.

Spontaneous rupture of the caudal renal artery in turkeys : A case of spontaneous rupture of the caudal renal artery in broiler turkeys is observed at the age of 8–12 weeks. Some birds in an excellent body condition die. The carcasses are anaemic. The necropsy shows that in some birds, the entire body cavity is filled with clotted blood and in others — massive sub-capsular coagula from the side of the affected kidney are detected. In all cases, unilateral rupture was observed, primarily of the left renal artery.

Subcutaneous emphysema : Subcutaneous emphysema in the right inguinal region occurs after penetration of air or gases in the subcutaneous connective tissue. Usually, it results after air-sac tearing or skin injury. Subcutaneous emphysema also occurs in the region of the head.

Gastrointestinal impaction : A sacciform blind pouch occurs after attachment of the intestinal wall to an unabsorbed yolk sac causing impaction.

Intestinal invagination : Involvement of anterior parts of the small intestine, the mesentery and the tops of the caecum into the ileum. Usually, it occurs with a strong intestinal peristalsis, often following ingestion of feed after a restrictive feeding regimen. Rotation of the small intestine on the longitudinal axis of the mesentery resulting in venous stasis and necrosis of the intestinal wall.

Crop impaction : It is caused by hard, fibrous feed or litter, whose accumulation results in impaction. The retained content into the crop sometimes causes putrefactive necrotic processes, affecting the crop wall and the covering skin.

MALFORMATIONS

Persistent right oviduct : If the regression of the right Mullerian duct, that is embryonally present in female chickens as oviduct, does not occur, a cystic dilatation of this structure is seen, whose size varies from a long cystic formation to a big sac, filled with fluid. Absense of eyes (anophthalmia) and cranial malformations are frequent malformations in newly hatched chickens.

Open body cavities : An additional skin formation (pouch) to the left laterocaudal abdominal wall, with part of the abdominal organs found within.

Polypodia : The occurrence of more legs than normal is often seen in broilers. Such birds frequently survive up to the end of the fattening period.

SPACE FOR NOTES

(B) DUCK FARMING

INTRODUCTION

Duckery is a very good income generating occupation for small and marginal farmers. This can be even reared by landless classes and women. Duck rearing seems to be a new enterprise to farmers in dry areas, but it is also a major source of income for farmers in southern and eastern coastal areas, north-eastern India and Jammu and Kashmir. Duck rearing is prevalent among weaker sections of rural population which provides them supplementary and steady income on daily basis. Duck eggs provide nutrition to the family of the farmer. Duck rearing, engaging family labour in their leisure hours, generates rural employment.

Duck domestication dates back in Roman literature around 2000 years ago. Today among the duck producing countries of the world, Vietnam, Poland, Indonesia, Thailand, United States, Brazil, China, Bangladesh are rated among the best in decreasing order.

In India, Gujarat State has higher land under inland water (ponds or water holes), particularly in South Gujarat region. In this area tribal communities keep duck for production of eggs and meat, which are considered to be poor man's poultry productions available at affordable prices. West Bengal has the highest duck population followed by Assam, Bihar, Manipur, Kerala, Andhra Pradesh, Tamil Nadu, Orissa and Tripura.

Duck rearing can be done in all Indian states. Marshy riversides, wetlands, ponds, tanks, barren moors, etc. are good for ducks rearing. Land based duck farming is being recently started on artificial water stores, near big cities on small and large commercial scale.

Weaker and disorganised sections of people are mostly involved in duck rearing. Therefore, considering duck rearing as an effective tool for socio-economic development of rural masses, a scientific approach in rearing ducks have been initiated recently by Government of India. A central duck breeding farm at Hassarghata, Bangalore, supplies improved duck birds of "**Khaki Campbell**" and "**Indian Runner**" (produce around 250-300 eggs per year) at subsidised rates to the farmers. It also imparts training in duck rearing. Indian Council of Agriculture Research (ICAR) has launched many programmes on integrated farming systems, such as **duck cum fish farming**, where the droppings of duck serve as feed to fish and some trash fish can be used as duck feed. **Duck-cum-crop farming**, where weeds in the field, by-products of the crops, are broken, and shrivelled grains can be used as duck feed. Although duck is an aquatic bird, yet swimming in water is not essential for them. However, they need to wet or clean their heads in water at times. Their eyes and bills need cleaning. For this, waterers should be sufficiently deep enough to dip their heads.

Systematic position [After : J. Z. Young (1981)]

Phylum : Chordata
Sub-phylum : Vertebrata/Craniata
Super-Class : Gnathostomata
Class : Aves
Sub-class : Neornithes
Super-order : Neognathae
Order : Anseriformes
Scientific name : *Anas* sp. (Ducks)
Cygnus sp. (Swans)

REASONS IN FAVOUR OF DUCK CULTURE

Duck rearing is profitable and simple in management because (1) ducks are less hazardous bird, (2) ducks are disease resistant, (3) ducks have longer egg-production life, (4) their eggs are heavier and fetch a better price, (5) their meat is also a delicacy and relished by the people, (6) they do not need elaborate housing, (7) they lay 95% to 98% eggs in early morning before 9 : 00 am after which the flock can be taken for foraging to nearby ponds, by the duck farmer or his family members and (8) they kill snails, slugs and other crop pests.

ADVANTAGES OF DUCK REARING OVER POULTRY CHICKENS

There are a number of advantages of rearing ducks over chicken or other poultry birds which are as follows :

1. Ducks lay on an average 40-50 more eggs than the layers of hen.
2. Duck's eggs are about 15 to 20 grams heavier than hen's egg.
3. They require less care and attention compared to other birds.
4. They are good foragers and are able to meet part of their feed requirements from the wild.
5. Ducks have longer, economically laying period than chicken.
6. They are hardy and more easily resistant to many avian diseases.
7. Ducks flourish well in marshy and wetland where most other domestic animals cannot survive.
8. Ducks are free from cannibalism and other objectionable tendencies.
9. Ducks lay their eggs before 9:00 am and hence egg collection is easy for the keeper.
10. They are easily trained to manage themselves.
11. They live longer than chicken.
12. They are suitable for integrated farming systems such as duck-cum-fish farming, duck-cum-pig-cum-fish and duck-cum-kitchen garden.
13. They are resistant to a number of diseases that are usually found among chicken.
14. Ducklings become sexually mature earlier compared to chicks.
15. They are good predators of insect pests.
16. Ducks do not need elaborate housing.

Fig. 5.27. Drake and duck of Khaki Campbell

BREEDS OF DUCK

Breeds of the duck can be classified into three main groups :

(A) Egg types consisting of **Khaki Campbell** and **Indian Runner** breeds.

(B) Meat type consisting of **White Pekin**, **Muscovy** and **Aylesbury** breeds.

(C) Ornamental type consisting of **Crested White** breed.

(A) DUCK REARED FOR EGG PRODUCTION

These breeds are reared for production of eggs.

(1) Khaki Campbell

The khaki campbell breed (Figs. 5.27 and 5.28) has been developed in England from a cross of Fawn and White Runner with Mallard ducks. Ducks have brownish bronze lower part of back; tail coverts, head and necks rest on the body and plumage is khaki. They have green bills and dark orange legs and toes. The adult drake weighs between 2.2 to 2.4 kg, while the duck weighs between 2.0 to 2.2 kg and that of an egg is about 70 gm. Egg production average per duck is close to 300 eggs per year.

Fig. 5.28. Khaki Campbell duckling

Fig. 5.29. Fawn drake

Fig. 5.30. The white runner

(2) **Indian Runner**

There are three standard varieties of this breed, namely the **Fawn**, the **White Runner** and the **Penciled variety**. Indian Runner is a good layer having an average record of 250 eggs per year.

(a) **The Fawn** : They are white with fawn coloured back and shoulders (Fig. 5.29). The upper part of its breast and wings are fawn coloured, while the lower part is white. Body is long and narrow, slopping gradually into the neck and resembling somewhat a penguin in shape. The shanks and toes are orange red. The bill of the young drake is yellow, later becoming greenish yellow while a young duck has a yellow bill spotted with green, which later becomes a dull green.

(b) **The White Runner** : This variety is pure white in colour all throughout the body (Fig. 5.30). The bill is yellow and the shanks and toes are orange coloured.

(c) **The Penciled variety** : The head of the male is dull bronze-green and white in colour. The back has a soft, fawn ground, finely stippled with a slightly darker shade of fawn (Fig. 5.31 A&B). The body and the upper section of the breast are medium fawn and the tail of the female is medium fawn and white in colour. In female a white marking in the plumage resembles those of the male. The coloured markings are a medium fawn throughout, with a light line of fawn colour running around the edge of each feather, the border being a darker shade.

(B) DUCK REARED FOR MEAT PRODUCTION

Duck meat is highly acceptable among the non-vegetarians. Though not popular due to non-availability in the market. The meat is rich in fat (14.5%) and total energy (190 kcal/100 gm). The protein content averages about 13 per cent. White Pekin is the most popular meat type duck in the world. Some of the meat type ducks are described here.

1. **White Pekin** : This breed originated in China. Birds are large white-feathered. They have orange-yellow bills, reddish yellow shanks and feet, and yellow skin (Fig. 5.32 A&B). Their eggs are tinted white in colour and its production is considered less important than its capability to produce excellent quality meat. Birds of this breed are bad sitters and very sensitive to outside interferences. Hence they should be treated gently.

Fig. 5.31A. Penciled runner (Male)

Fig. 5.31B. Penciled runner (Ducklings)

Fig. 5.32A. White Pekin (Duck)

Fig. 5.32B. White Pekin (Ducklings)

2. **Muscovy** : It is very much questionable whether Muscovy is really a duck or not (Fig. 5.33). It is more like a goose in many ways than one; for instance, it is a grazer and eats grasses in the same way as a goose. They are very strong and powerful in flight, yet very tame and friendly. The ducklings are late in feather formation by four weeks; it takes about 16 weeks, whereas in other breeds the ducklings are in full feather at 12 weeks. The drakes are big in size and weigh above 4.5 kg, whereas ducks weigh about 3.0 kg. The breed is native to Brazil and very popular in Australia. When crossed with other breeds, it produces sterile ducks, known as **mule ducks** which produce higher meat yield.

There are two varieties of Muscovy ducks, the **white** and the **dark**. The head and face of the Muscovy are partly bare, with red, rough skin. It has a long, broad body, with greater breadth. The **white variety** has pure white plumage, pale orange or yellow legs and a pinkish, flesh-coloured beak. The **dark variety** has got a lustrous blue-black body and back. Both the varieties provide meat of excellent quality and taste, if marketed before 17 weeks of age. They are also good sitters and will hatch and care for approximately 30 ducklings in one batch. They produce 40-45 eggs, annually. Muscovy male has no curl feathers in his tail. Ducklings emerge from eggs after 36 days of incubation period.

3. **Aylesbury** : This breed originated in Buckingham in U.K. The plumages are white; legs are short but sturdy and orange in colour (Fig. 5.34). Due to its light bone and high percentage of creamy white flesh, the breed is regarded as a delicious table bird. It reaches market weight in 2 weeks. Eggs are tinted white in colour.

Fig. 5.33. Muscovy duck with ducklings

Fig. 5.34. Aylesbury duck

Fig. 5.35. (A) White crested duck (Adult) and (B) White creste baby

4. **Broiler duck** : Another short broiler type of duck, found in Kalluru lake area of Andhra Pradesh is famous for meat production. They are similar to broiler chicken and are marketed at the age of 6-8 months. The meat of such ducks is said to be tasty and nutritive.

(C) ORNAMENTAL TYPE DUCK

Crested White breed (Fig. 5.35 A&B) : This variety is very popular for duck lovers in European and American countries. They are beautiful looking and delicate. Adults are with rosy crest. They have very low economic value but have recreational importance.

EGG PRODUCTION IN DUCKS

Annual rate of egg production varies with breed. Ducks normally begin to lay at about 6 months of age. Indigenous ducks produce 90–130 eggs per year. Muscovy breed known for meat production in the world lays only 40–45 eggs per bird per year. Indian Runner breed is a good layer having an average record of more than 250 eggs per annum. Khakhi Campbell breed is highly prolific layer and its ducklings start laying eggs when they are 120 days of age and produce about 300 eggs per bird per year. About 95% of eggs are laid in the morning. Ducks can live for about ten years but the effective laying period is about 4 years. For commercial egg producers, it is advantageous that eggs are laid during morning between 6–9 a.m. can be collected on paddy straw spread in the field within a temporary fence.

A. Eggs for hatching

Egg for hatching must be fertilised. A duck normally starts laying eggs at the age of 16 weeks. For the purpose of producing fertilised eggs, one drake (male) is used for mating with 3-4 ducks. During early periods of breeding this ratio can be one drake (male) to four females and similarly at older age the number of males is increased, since during these times the drakes are less active.

Drakes should be kept with ducks all the time. One month later (the time from which they were kept together), the eggs produced are collected for hatching. It is a mistaken belief that effective copulation can only take place in water. Mating is also successful on dry ground. The role of water in the vicinity is that ducks keep themselves clean and they keep their eggs at a place having correct humidity and temperature.

B. Incubation of eggs

Ducks of the improved laying breeds are generally non-sitters. On a small scale, eggs can be hatched by a brood hen; duck breeds like Muscovy and East Indian breeds are good sitters. A hen can cover about 12 eggs and a brood duck about 15 eggs.

The most common method of getting eggs hatched is by the use of incubators. The incubation period of duck egg is 28 days. The optimum temperature is 37.2 to 37.5°C and the relative humidity is about 65 to 70 per cent. In the incubator, eggs are turned at 2 to 4 hours intervals up to the 25th day. On 26th day, they can be transported to hatching tray and no turning is needed afterwards. Ducklings are kept for a day in these trays.

The main problem that needs attention, while incubating artificially, is that shells of duck eggs have much larger pores. For this reason, they should not be stored longer before incubating. Stored eggs dry up and get infection of bacteria. Eggs laid in low temperature hatch better than those laid in high temperature. More important is that hatched eggs should be kept in cooler condition prior to incubation at temperature below 20°C. Eggs laid during cold season show about 80% fertility; while eggs laid during hot season the fertility goes down to as low as 12 per cent. Eggs of medium size show more fertility than eggs of small and large size. Shells that are small, large, pointed, round, greasy, chalky and thin should be avoided.

BROODING DUCKLINGS

In commercial duck farming, brooding of duckling is very important. Brooding management determines the growth, mortality rate and eventually the productive life of the ducks.

Under natural conditions, a duck or a broody hen can take up 10–15 ducklings for brooding. For the first few days after hatching, feed and water should be provided in the house. After 5–7 days, the broody hen or duck along with the ducklings may be allowed to move within the restricted area. Young ducklings fall easy prey to wild birds. The entire brood should be protected from rain, cold winds, predatory birds, and rodents. After 2–3 weeks no mothering is necessary.

Artificial brooding is necessary for large number of ducklings. Any equipment that broods chicks efficiently can be used for brooding ducklings. The brooding period of ducklings varies with breed. Khaki Campbell duckling requires 3–4 weeks of brooding. On an average brooding period of ducklings is about 4 weeks. A temperature of 30°C is enough for the first week and it is reduced to about 3°C per week, till it reaches the room temperature. Then ducklings are transferred to shed or other place where they grow and mature. Ducklings can be reared in intensive, semi-intensive and free range system. In free range system, they need a little high shelter.

Table 5.2 : Space allowances for ducklings

Sl. No	Age in weeks	Floor space per bird (sq. ft.)
1	1-2	1.0
2	2-3	1.4
3	3-4	2.0
4	4-5	2.5
5	5-6	4.0

Ducklings may be brooded on wire floor, litter or in batteries. A wire floor space of 0.5 square feet per bird or solid floor space of an average of 1 square feet per bird would be sufficient up to 2 weeks of age. The space requirement from first to eighth weeks of age is given in Table 5.2.

DUCK REARING

After brooding ducklings for 4 weeks, birds are transferred to sheds of convenient size for their comfortable stay and growth. Ducklings can be reared in (1) **Intensive system,** (2) **Semi-intensive system,** and (3) **Free range system**.

(1) Intensive system

Under intensive system ducks are reared on deep litter till they attain the age of 16 weeks. A confined space of 3 sq. ft. per bird is allowed. As the ducklings become 17 weeks old, they are vaccinated against duck plague and given more space, about 5 sq. ft per duck.

Ducklings grow at a very fast rate and, therefore, require a ration, rich in all nutrients. Khaki Campbell duck consumes about 12.5 kg of feed up to 20 weeks of age. Afterwards the consumption varies from 120–170 gm per bird per day depending upon the rate of production and availability of greens. The starter, growers and layers ration should contain a protein percentage of 21, 18 and 18 respectively with a metabolisable energy (ME) of 2850, 2900 and 2800 kcal per kg of feed, respectively.

(2) Semi-intensive system

Under semi-intensive system (Fig. 5.36) birds are grown on deep litter with floor space of 2 sq. ft in high shelter and 10 sq. ft as outside run till they attain an age of 16 weeks. For feeding of wet mash, 'V' shaped feeders can be used. A 4-inch space on the hopper for feeding is suitable.

Under semi-intensive system local feeds are also made use off. One third of the meal ration may be replaced by cheaper vegetable feeds, household scrapes and fodders as available under local conditions.

(3) Free range system

In free range systems, ducklings are provided ample space for run and night shelter. Under this system a flock of 2000 ducks can be reared per acre (0.0456 hectare). Ducks on free range system obtain most of their protein by foraging small fishes, crustaceans and insects.

LAND BASED DUCK REARING

In land based duck rearing arrangement, artificial water stores are made at the site. It is a recent innovation, suited for places where natural water bodies are not available and ducks can be reared economically due to vicinity of the market. Moreover, this is a unique system for even large scale commercial farming in minimum space. Normally more than 120 ducks are reared along with pisciculture (fish farming) on an area of 1200 sq. feet including night shelter, water channel, feeding floor and open run space.

The water channel of 25 feet long, 1.5 feet wide and one foot deep is sufficient for 100 ducks, which is filled with freshwater everyday. Duck farming in such intensive system is always advantageous from the point of its management, feed utilisation and maintenance of hygiene. A night shelter may be constructed using appropriate materials with sufficient provision for ventilation.

DUCK FEEDING

Ducks are the most efficient type of birds to convert fallen grains of the fields, insects, plant materials and pond materials into edible meat and eggs. Ducks of various age groups are fed on different feeds viz. starters 0–2 weeks, growers (3-8 weeks) and for adults above 9 to 20 weeks till they start laying eggs. Commercial feeds are available as mash pellets and crumbs. Ducks should never have access to feed without drinking water because they need wet feed. Pellet feeding though costly has distinct advantage of saving the amount of feed, labour convenience, no scouring and improvement in sanitary conditions.

(1) Feeding ducklings

Ducklings should be fed as soon as they are removed from the incubator to the brooder or immediately after they are received from the hatchery. Since ducklings do not readily learn to eat, it is necessary that extreme care should be given to them for the first few days to save them from dying or starvation. If it is possible to put several ducklings that have already learnt how to eat with the new hatch in the brooder, then the newcomers will quickly learn how to eat from their older mates. If it is not possible, it might be necessary to hand-feed some of them for the first day or two. Day-old ducklings should be given coarse milled cereals moistened with milk or water as a first feed and then one with an approximate composition given in Table 5.3.

The mash should be damped just sufficiently to make it "crumble'. If it is too wet, much of it is lost through the sieving process to which it is subjected in the ducks bill. No more feed that can be eaten in about 10 minutes should be fed at any time. Grit or sand and water should be available. Ducklings normally consume 12.5 kg of feed in 20 weeks.

Table 5.3 : Composition of feed for ducklings

Milled cereal	35 parts
Fine cereal bran	30 parts
Fish or meat meal	20 parts
Extracted oil-cake meal	10 parts
Fine grit and minerals	5 parts

(2) Feeding growers and layers

Feeding ducks require much care. Duck feed stored under comparatively high relative humidity, get poisoned due to fungal growth. In order to avoid problem of feed toxicity, feed may be formulated by eliminating maize and groundnut cakes as far as possible. Broken rice, damaged wheat, soybean meal, etc. can be included in formulating the feed. Fish meal should be of fresh and good quality. Strong feed for longer time should be avoided. All other precautions must be observed before selection or making feed.

A generalised formula for different categories of duck feed is given in Table 5.4. Depending upon the availability of different feed ingredients at different situations the formula may be altered.

Table 5.4 : Diet of duck of different categories

Sl. No.	Ingredients Kg/100 kg	Starters	Growers	Layers/ breeders
1	Broken wheat	38	40	35
2	Broken rice	20	20	20
3	Soybean meal	25	20	25
4	Fish meal	10	8	10
5	Deoiled rice bran	5	10	5
6	Mineral mixture	2	2	2
7	Shell grit	–	–	3
8	D. L. Methionin (g)	50	50	50
9	Manganese sulphate (g)	10	10	10
10	Vitamin A,B_2,D,K (g)	25	25	25
11	Copper sulphate (g)	1	1	1
12	Gentian violet (g)	30	30	30
13	Niftin 200 (g)	25	–	–
14	T. M-5 (g)	100	–	–

COMMON DUCK DISEASES

Ducks appear to excel all other domestic poultry in their resistance to common avian diseases, but they suffer from duck plague, duck virus hepatitis and some other diseases. A few, for example, are presented here. In case of need, veterinarians must be consulted.

(1) Duck plague

Cause : Serious outbreak of duck plagues, also known as duck virus enteritis, can cause 80–90% mortality in flocks of all ages. It is a highly contagious disease and strikes swiftly without warning.

Symptoms : Birds become restless with drooping wings and ruffled feather. Their eyes become swollen and moist with sticky discharge. They stop feeding, drink water frequently and feel difficulty in breathing. Sometimes watery yellow diarrahoea with blood is seen. Occasionally penis is swollen and protruding. In laying females, haemorrhages can be observed in the deformed and discoloured ovarian follicles.

Prevention and control : The birds can be protected by duck plague vaccine, available in the country, given at the age of 8 weeks. The affected ducks should be isolated and killed, if necessary.

(2) Duck cholera

Cause : Highly infectious disease caused by bacteria, *Pasteurella multocida,* in ducks over 4 weeks age.

Symptoms : Birds lose appetite and their body temperature becomes high. Initial diarrhoea is followed by mucoid droppings.

Prevention and control : Birds must be vaccinated first at the age of 4 weeks and again at 16 weeks age with duck cholera vaccine. Sulphonamides and antibiotics are effective in controlling the disease and reducing mortality. The dead birds should be burnt.

(3) Aflatoxicosis

Cause : Ducks are very susceptible to aflatoxin content of the feed. The minimum toxic dose for ducks is 0.03 mg per kilo of feed. Ducklings are more susceptible. Aflatoxin is produced by the mould, *Aspergillus flavus,* in the feedstuffs such as groundnut, maize, rice polish, etc.

Symptoms : Birds first show poor feed intake, poor growth, falling of feathers, lethargy, unthriftliness and lameness. At later stages, birds encounter liver lesions, ataxia followed by convulsion and death.

Prevention and control : Avoid feeding mouldy feeds. Feeds should be checked for aflatoxin, particularly during and after rainy season.

(4) Botulism

Cause : The disease occurs both in young and adult stocks. It is caused by the bacteria *Clostridium botulinum,* which grows in decaying plant and animal materials.

Symptoms : Birds lose control of their neck muscle and usually drown during swimming in water. Birds show dullness, ruffled feathers, lameness, drooping wings, labour breathing, coma and death.

Prevention and control : The affected birds can be given C-type antitoxin. Maintenance of cleanliness including removal of rotting vegetation and dead birds, can prevent the disease.

(5) Internal parasites

Ducks are resistant to internal parasites. These are more when ducks are kept in a range or when they have access to ponds. The parasites comprises of flukes, tapeworms and roundworms. There are suitable medicines available in the market which can kill or expel these from the body. Periodical examination of faecal material is very useful to identify and to treat ducks.

(6) External parasites

External parasites are comparatively less in ducks as compared to chicken. They produce annoyance to them leading to reduction in egg yields. Some of these do transmit a few disease producing organisms. So, it is essential to keep them free from parasites.

SOURCE OF DUCKLINGS AVAILABILITY

Various programmes by the Central and State Governments are being launched to popularise duck production. A beginner in duck rearing can obtain required facilities from any one of the sources listed below :

(1) Hatching eggs facilities and Khaki Campbell ducklings are available at the Central Duck Breeding Farm, Hessarghatta, Bangalore.

(2) Khaki Campbell ducks can also be procured from the Animal Husbandary Department, Agartala, Tripura.

(3) Ducklings are available for sale from the State Poultry, Gobardanga, North 24-Parganas and State Poultry Farm, Krishnanagar, Nadia, W. Bengal.

(4) Hatching eggs of khaki Campbell duck are available for sale from the Divyodaya Krishi Vigyan Keoorn, Ramshai, Jalpaiguri (North Bengal).

Fig. 5.36. Integrated farming — Duck-cum-fish farming

Table 5.5 : Cost of duck rearing unit of 100 Ducks + 15 Drakes maintained under semi-intensive conditions — NABARD Report - 2010

1.	Shed/enclosures for 100 + 15 duck for 500 sq. ft @ Rs. 20/- per square feet	Rs. 10,000
2.	Duck feeding, watering and egg laying boxes, buckets etc lumpsum	Rs. 1,500
3.	Duck lings 3 months old 115 + 15 i.e. 130 @ Rs. 130/- per duck including transport cost	Rs. 16,900
4.	Feed cost for first 2 months during growing of one month and one month of initial laying period @ 1.5 kg + 2 kg = 3.5 kg per bird @ Rs. 7/- per kg	Rs. 3,185
5.	Cost of medicines and vaccination	Rs. 1,500
6.	Other misc cost including insurance cost etc (lump-sum)	Rs. 675
	Total cost	Rs. 33,760

SPACE FOR NOTES

(C) QUAIL FARMING

INTRODUCTION

The Quail, also called *"bater"* in Hindi, is a small type of bird known as backyard chicken. The bird originated from a wild environment just as any other domesticated animal and was first domesticated in Japan in 1595. There are 45 species of quail worldwide. However, only two species of quail are popular in India; the **black-breasted jungle** or **rain quail** (*Coturnix coromandelica*) (Fig. 5.37) found in jungle and the **brown-coloured Japanese quail** (*Coturnix coturnix japonica*) (Fig. 5. 38) which is bred for meat or used for commercial purposes. While Indian jungle quail weighs up to 100 gm and lays about 100 eggs a year, the Japanese quail weighs up to 250 gm and lays about 250 eggs a year. Although the Japanese quail is the largest species, it is much smaller than pigeon. Because of prolific egg production and meat yield, it attained the status of viable commercial poultry enterprise. It is very ideal for creation of rural employment, solving the gender issue in employment and also provides supplemental income and protein requirement to rural farmers.

DISTRIBUTION : The Japanese quail is found in East Asia. Being migratory in nature, the wild Japanese quail is available in Manchuria, south-eastern Siberia, northern Japan, and the Korean Peninsula. It is also found in some parts of the north eastern regions of India like Sikkim, Bihar and parts of Assam. They dwell in grasslands and cultivated fields.

QUAIL FARMING : **Quail farming means rearing of different breeds of quails under controlled and hygienic environment in order to get their eggs, meat and also litters for commercial purposes**. Though the Government of India is encouraging entrepreneurs to start quail farming, a government licence is required to sell commercial Japanese quail, considering the jungle variety of the bird as a protected species. The Ministry of Environment and Forest delegated the power to grant such licence to the department of Animal Husbandry.

Fig. 5.37. Black-breasted quail (*Coturnix coromandelica*)

(a) Female

(b) Male

Fig. 5.38. Brown-coloured Japanese quail (*Coturnix coturnix japonica*)

Brief history : The reference to quails can be traced back in the old testament of Bible. Egyptians were known to catch quail for meat. In China, these birds were raised as pets as singing birds. In late eleventh century, quail was brought to Japan from China. It is known that a Japanese Emperor got cured from tuberculosis after eating quail meat. Afterwards, Japanese people started raising quail extensively for meat and egg. Since the wild quails were tamed and bred in captivity by Japanese people, the birds came to known as Japanese quail.

In India, quail keeping started in 1974 at Ijatnagar, when Central Avian Research Institute, introduced improved germ plasm of domesticated quail varieties from Japan, and then in 1983 at Tamil Nadu. Thereafter, agricultural Universities of Andhra Pradesh, West Bengal; veterinary colleges as well as animal husbandry department played a vital role in the promotion of quail breeding as a farm enterprise.

ADVANTAGES OF QUAIL FARMING

1. Quail farming requires minimum floor space (3-4 adult quails can be reared in one sq. ft).
2. Needs low investment. Quail can be bred up to commercial numbers very quickly.
3. Low cost of production.
4. Low feed consumption (550–600 gm of feed/ bird up to 5th week).
5. Quails are comparatively sturdy birds.
6. Can be marketed at an early age because of fast growth, i.e., 170-190 gm body weight by 5th week.
7. Early sexual maturity — starts laying eggs in about six to seven weeks of age.
8. High rate of egg laying (250-300 eggs in a year / bird) and short generation interval (3-4 generations in a year).
9. Quail meat is tastier than chicken and has less fat content. It promotes body and brain development in children.
10. Nutritionally, the quail eggs are on a par with that of chicken eggs. Moreover, they contain less cholesterol.
11. Quails are resistant to common diseases of chicken.
12. Quail can be sexed by anyone from an early age.
13. Quails have the shortest incubation time of all poultry and game birds.

Taxonomy of quails : Quail is a collective name for several genera of mid-sized birds generally belong to the order Galliformes. Old World quails are found in the family Phasianidae that includes mainly *Coturnix* Quail, the Chinese Blue Breasted "Button" quail and the African Harlequin Quail. The New World quails belong to the family Odontophoridae and include Marbled Wood Quail, Bobwhite, Spot-winged Wood Quail, Black-earnbed Wood Quail, Rufous-fronted Wood Quail, Black-fronted Wood Quail, etc.

In recent years, Japanese quail, *Coturnix coturnix japonica,* is the most productive commercial strain in India. But there is a great deal of confusion about the origin of the *Coturnix* quail as we know them today. No one can know how the various breeds and sub-species developed, but it is generally acknowledged that all the *Coturnix* types are originally based on the common quail, *Coturnix coturnix*, the wild migratory bird of Europe, Asia and Africa. Various sub-species then developed from this, including the European quail, *Coturnix coturnix coturnix;* the Eurasian quail, *Coturnix coturnix communis,* and the Japanese quail, *Coturnix coturnix japonica.*

Systemic position of quail found in India

Kingdom : Animalia
Phylum : Chordata
Class : Aves
Order : Galliformes
Family : Phasianidae
Genus : Coturnix

Some popular breeds of quail

There are many breeds of *Coturnix* quail are available. According to colour varieties some popular breeds are mentioned here.

Fig. 5.39. Thee Eurasian or Pharaoh quail (*Coturnix coturnix communis*)

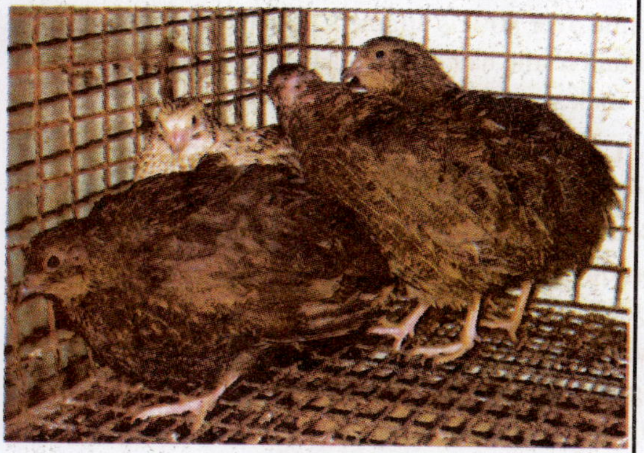

Fig. 5.40. Range quail

(1) **The Eurasian or Pharaoh quail** (*Coturnix coturnix communis*) : It is a wild type quail breed, having a mix of feather colour with black and various shades, brown colour being predominant. In male, the face and throat is rusty brown and the upper breast is cinnamon in colour. In female, the face, throat and upper breast are light cinnamon with black, while the lower breast is tan in colour (Fig. 5.39). This breed is a good layer.

(2) **Range quail** : The overall colouring is dark brown, so that some people refer to them as Brown quail. The markings of the range are essentially a lighter brown body colour overlaid with darker-brown, along with a certain amount of dark grey feathering on the back and as a faint patch on the throat (Fig. 5.40). All range quails are good layers.

(3) **Fawn quail** : Essentially like all other *Coturnix* breeds, the overall impression of the fawn is a lovely warm pinkish-brown. The fawn feathers are pencilled with white and the white eyebrow lines are present. Beak and legs are light pinkish-brown and eyes are dark brown (Fig. 5.41).

(4) **White quail** : Sometimes called English White; good specimens are completely white, with no discernible markings, other than on the head. Beak and legs are pinkish brown and eyes are dark brown (Fig. 5.42). English white quail is a layer but American white breeds are good broiler.

Fig. 5.41. Fawn quail

Fig. 5.42. English white quail

Fig. 5.43. The golden quail

Fig. 5.44. Tuxedo quail

(5) **Golden quail** : These quails are golden to white straw coloured. The male is lighter while the female is darker and more striated. She also has distinctive brown stripes around the face. In the USA, golden strains known as Manchurian Gold. In Britain and Europe, Italian or Golden Italian is the name given to the golden variety (Fig. 5.43).

(6) **Tuxedo quail** : This is an apt name for a bird with a smart white waistcoat to contrast with its dark brown overcoat. The two-colour pied pattern is the result of a crossing between the dominant dark brown and recessive white genes. The colour of back feathers is identical with those of the Range, while the white plumage is like that of the English White. The markings are often variable. The ideal ones are clear while face, chest and belly, with dark brown back, tail and crown (Fig. 5.44).

LIFE HISTORY OF JAPANESE QUAIL

Adults : The male Japanese quail has an even coloured rusty breast feathers with only few speckles of white. The adult females have grey and black speckled breast feathers (Figs. 5.38 a & b). The males possess a cloacal gland that secretes a white foamy material. Male quails usually make a different sound which is usually disturbing to the human. Sex differences in plumage colour appear about 3 weeks of age.The body weight of adult male quail is about 140 gm, while the females are slightly heavier, weighing 150 to 180 gm. Growth rate is three times faster than domestic fowl. They attain very early sexual maturity. By about 50 days of age, the females become capable of full egg production.

Rate of egg production in domesticated quail is controlled by the availability of light. Balanced diet is another factor which influences the rate of egg production. Animal protein in the diet improves egg production better than vegetable protein, as the latter lack essential amino acids like methionine and lysine.

Mating and Fertility : In animal, inbreeding reduces the rate of egg production in the females and poor semen production in males. The rate of egg production is reduced with the advance in age of the bird. Hence, the young stock should be preferred. The optimum period for good egg production is 10-30 weeks of age. Fifty per cent level of egg production is generally achieved at the age of 7-8 weeks, while over 80% is attained after 10 weeks of age. To obtain fertile hatching eggs from quail, equal number of males and females are not required.

Laying and Hatching : The female lays eggs every 16-24 hours for 8-12 months during first year. Second year egg production is 48% of the first year's production and falls drastically in subsequent years. Quail eggs are multi-coloured and heavily mottled with black, brown and blue (Fig. 5.45). Weight of egg is 6.7-13.8 gm. Quails require about 16-18 hours light for optimum production of eggs. The proportion of yolk (the yellow inside part) to albumen (the white part) is 39:61 and is higher compared to chicken eggs. The composition of quail eggs is given in Table 5.6).

Fig. 5.45. Quail eggs are coloured variously

Table 5.6 : Composition of quail egg (Whole, Raw)

Water	Protein	Fat	Carbo-hydrate	Total ash	Calorific value
74%	13%	11%	1%	1%	649 kcal 100 gm liquid

For natural incubation of eggs quails, mini hens, doves, pigeons, etc. can be utilised. First, unfertilised eggs are given to the birds for 2-3 days to check their acceptability. Thereafter, 2-10 fertilised eggs are given for incubation. Broody birds are provided with straw for nest, light, food, water, etc. Artificial incubation is a more controlled process. Incubation period for quail eggs is 18 days. After that the young chicks are hatched out.

Fig. 5.46. Japanese quail chicks. Note two golden colour stripes running down their back and face

Japanese quail chicks are dark brown in colour with two golden stripes running down their back and face, looking like pheasant chicks (Fig. 5.46). Newly hatched chicks require desirable temperature of about 35°C during the first week of brooding, which is gradually reduced at the rate of 3.5°C per day until it reaches 21°C. Floor, feeder and water space should be increased with advancement in age. At the age of four weeks, chicks are fully feathered. Feather picking or other forms of cannibalism may occur when Japanese quails are kept in wire-cage. Beak trimming may be necessary as early as two weeks of age and may be done by using nail-cutters. Young birds can be transferred from brooder to cages or floor around the fourth week.

REARING OF QUAILS IN FARM

For commercial purposes, quails are reared in farm (quailery) with various systems. Here, some general systems are being described.

(1) **Breeding**

Two main methods of breeding are normally practised in quail farms.

(a) In one method, seven quails are kept in a cage/housing with the male-female ratio of 2 : 5. It is time saving and can be adopted for large scale breeding but has some disadvantages as well, like frequent fights between the males, which causes general nervousness in the cage and the fertilisation and egg laying rate can become irregular.

(b) In the other method, individual males and females are kept separately and the males/females are placed into the cages periodically for mating. It's a good practice on a small scale with high fertility rate where desired colours and patterns of the quails can be produced. The only disadvantage of this method is that it is time consuming.

(2) **Egg laying** : Under ideal conditions, females produce about 250 to 300 eggs per year. In farms, females usually lay 75% eggs between 3 to 6 p.m. and 20% of the eggs in total darkness. Quails require about 16-18 hours light for optimum egg production.

(3) **Incubation**

The domesticated quail has lost the instinct of nesting. So for breeding purposes in quail farms, the eggs are hatched either with the help of incubator or by placing the eggs under a broody hen (ideally Bantam*). Healthy eggs with strong shell are collected, checked and then placed on wire mesh or grid of 1.5 x 1 inch, fitted on the setter trays, to keep very small number of quail eggs in the incubator. It is then adjusted with a temperature of 37.8°C and humidity level of 55-60%. If the incubator is not equipped with an automatic egg turning system, the eggs should be turned manually 5-6 times a day. Egg turning is very

* Bantam — A hen of a small breed. It is named after the province of Bantam in Java.

important to keep the little chick embryos in the middle of the egg and prevent them from sticking to the inside of the shell. On the 15th day, turning the eggs should be stopped and the temperature is slightly lowered to 36.9°C. At this stage, humidity level is raised to above 80% which will soften the shell and help the little quail chicks to break out easily. Then quail eggs are put on the hatching tray. Japanese quail chicks usually begin to hatch by the 16th day of incubation, but most of them should hatch on the 17th day. Any quail chicks that hatch after the 18th day will probably die. Hatching rate of 75% or above, indicates a good success. Incubation requirements in quail farm are as follows (Table 5.7).

Table 5.7 : Incubation requirement in quail farm

Duration (days)	Temperature (°C)	Humidity (%)	Turning of eggs
0–14	37.8°C	60	Through 45°. 5-6 times a day
15–18	36.9°C	80	Not required

Precautions to be taken for incubation

Successful quail farming depends on proper incubation and hatching. So following precautions are to be taken prior/during the process of incubation :

(1) Prior to incubation, eggs should be stored in a dry, cool place on egg trays with the broad end up. The storing temperature should be between 12–17°C and the relative humidity around 70–80 %. Eggs should never be stored in fridge.

(2) Washing the quail eggs is not allowed which can remove their porous protective coating and making them vulnerable to bacterial infections.

(3) For hatching, quail eggs should not be older than 7 days.

(4) It is also advisable to perform candling check of the eggs for the presence of any possible cracks, health of the yolks and the size of the airspace.

(5) Before incubation, eggs can be fumigated with a mixture of potassium permanganate and formalin. To prepare the mixture, potassium permanganate is placed in a low walled, large enamel or ceramic pan, and then to it formalin is added. Following fumigation, eggs are kept in a tray or dish which is then placed in the incubator with a temperature between 20-30°C and relative humidity of 80-90%.

(6) The room where the incubator is set up should be at constant temperature.

(7) The correct temperature level is very important. Under-incubating (temperature too low) will result in late hatching and over-incubating (temperature too high) will result in early hatching. Both cause death of quail chicks.

(8) The quail chicks will be allowed to come out from the shell on their own. No external forces should be put on the eggs during hatching.

(9) The chicks should be kept in the incubator until they become dry and fluffy. They are then placed into the pre-heated brooder. Sufficient time should be given before placing them, or else they will catch a cold, get curled up feet and die.

(10) The quail chicks should be left in the incubator for 24-48 hours after hatching.

(4) **Brooding** : Brooding is a technique to keep the quail chicks in a safe and controlled environment with proper air, space, heat, water, and food in order to make them healthy as they can be, until they feather out. In quails, the brooding period extends between 0-3 weeks. However, in winter it can extend up to the 4th week. The average mortality is 6-10% during brooding period. So quail chicks need some tender care throughout the first week of life.

The Brooder : Brooder is the place where quail chicks are kept just after hatching with special care (Fig. 5.47). It is of suitable size depending upon the number of quail chicks that will be kept into it. The chicks should have enough space to walk away from the heater lamp, if they want to cool down a bit.

Fig. 5.47. A typical quail brooder

The brooder is built before the hatching time and placed in a draught-free area. A bedding of about 2 to 4 inches, usually made of fine wood shavings, is put on the floor. For the first week, it is kept covered with paper or even better, with some textile, like an old heavy curtain or a bed sheet. It is important to tack the sides as well, making it tight and secure to prevent the quail chicks from getting stuck in the gap or even in worse circumstances, to get under the cover. In the brooder the heating lamp is placed about 12 inches high from the chicks for fire safety reasons. The heat level should be 35–36°C for the first week, followed by a decrease of 3.5°C for a week until the chicks are ready to move to their permanent place. The brooding of chicks can be done using infrared bulbs or gas brooders (Fig. 5.48). Reducing of heat will depend on what type of a heater is used. In case of a heat lamp there are generally two ways. One is to keep lifting the heat lamp by a bit each time. The other is to start with a 250 W bulb, then change it to a 100 W bulb, then to a 60 W bulb. Thus, the chicks will gradually be used to day and night conditions.

Fig. 5.48. Brooding of quail chicks using infrared bulbs

Precautions to be taken in Brooder

(a) Prior hatching, the heater lamp should be kept on in the brooder to make sure that the bedding and the walls are sufficiently warm for them.

(b) Drinking area should not be too deep to prevent young quail chicks from drowning (They can drown in $\frac{3}{4}$ inch or less deep water).

(c) The quail chicks increase their size significantly by the time they reach 3-4 weeks of age and get fully feathered. The size of brooder should be made accordingly.

(d) Predators should be prevented from getting to the chicks.

(e) Quail chicks should have constant access to feed and cold water.

(f) Constant and strong light can cause aggressive behaviour in the chicks. So the light should be adjusted properly.

(g) After the first week, the bedding cover and the protection from drowning should be removed after proper checking.

(5) Rearing of Adult Birds

Quails can be reared in both deep litter system and cage (battery) system.

Deep litter system : Space requirement for different purposes in deep litter system is mentioned below (Table 5.8).

Table 5.8 : Rearing space required in deep litter system

Age (in weeks)	Space requirement (sq.cm/bird)	Feeding space (sq.cm/bird)	Watering space (sq.cm/bird)
0-4	75	2.0	1.0
5-6	200	2.5	1.5
6 onwards	250	2.5-3.0	1.5

Fig. 5.49. Rearing of quails in deep litter system

Fig. 5.50. An ideal modern quail cage

Fig. 5.51. Quail colony in cage

BEDDING : Litter materials, like rice husk, saw-dusts, crushed groundnut shells, wood shavings, ground corn, cobs, etc. are used in combination as bedding materials. The litter should be of organic origin but non-proteinaceous in nature. It should be hygroscopic, dry, free from pathogenic bacteria, moulds or fungi and should be locally available at cheap rate. A five centimetre thick layer of suitable litter material may be spread on a clean floor at the beginning (Fig. 5.49). Gradually this thickness may be increased to 10 cm by 4-5 weeks.

Faeces and evaporation from the body and drinkers add moisture to the litter. The moisture content of the litter material should be between 25-30% and the excess amount of moisture should be removed. Frequent stirring of the material and proper ventilation, generally help to maintain the litter in healthy condition. To ensure the healthy condition of the birds, the freshly spread litter material should be sprayed with any disinfectant, about 96 hours before the birds are expected to use it.

When humidity is high in the atmosphere, it may be necessary to add slaked lime (4-5 kg/10 square metres of floor area) to the litter. The wet patches around the waterers should also be replaced with fresh material. During very dry conditions some moisture may be added to the litter material by fine spraying of water, to continue its proper conditioning and maturation process. A quail of 1 to 6 weeks of age may add about 360 gm of manure to the litter during the rest period of its life.

CAGE (BATTERY) SYSTEM : Quails can be reared in cages of various types. Modern quail cage possess built-in systems for manure removal, feeding, drinking and egg collection (Fig. 5.50). Typically vertical battery cages with 4-6 tiers are used to save space and in each row, there may be more than one cage. Cages are made up of weld mesh of galvanised wire. The bottom of the cage is fixed with removable plates to clean the bird-droppings. Commercial egg layers are usually housed in colonies of 10-12 birds per cage (Fig. 5.51). Rearing space as followed in most quail farms for cage system is mentioned below (Table 5.9).

Table 5.9 : Rearing space required in cage system

Age	Cage Size	No. of birds
First 2 weeks	3 x 2.5 x 1.5 ft	100
3- 6 weeks	4 x 2 .5 x 1.5 ft	50

FEEDING AND DRINKING : Feeding in deep litter system is done by quail feeder which is small pet bowl or small chicken feeder. Again, the quail drinker can be a small pet bowl or small chicken drinker that can be hanged onto the wire mesh from the outside. Obviously quails should have always dry litter and access to feed and drink at all times. In cage system, long narrow feed troughs are placed in front of the cages (Fig. 5.52). Water troughs are usually placed at the back side of the cages. For feeding, quails are classified as starter (3 weeks old), grower or finisher (4-5 weeks old), and layer or breeder (6 weeks onward). Food requirements for different stages of quails are shown in Table 5.10 and their growth performance after food consumption is listed in Table 5.11.

For first two days ground maize can be offered to quails. After that, feed can be prepared by using commonly available ingredients like maize, rice polishing, soybean extraction, groundnut extraction, sunflower extraction, etc. Commercial poultry feed that is available in the market is not suitable as it contains less protein than what is required for quails. However, it can be modified by adding more soybean extraction or any other suitable protein source that are acceptable for quails. In India, at present commercial feed for quail is not available, one has to specially get it prepared.

Fig. 5.52. Quail cage with trough feeder

Table 5.11 : Growth performance of quail

Age in weeks	Average body weight (gm)	Total feed consumption (gm)	Cumulative feed efficiency
1	25.6	45.7	1.78
2	52.2	136.9	2.62
3	87.9	237.2	2.69
4	136.8	391.8	2.86
5	192.1	561.6	2.92

CARING OF ADULT BIRDS : When rearing the male and female quails together, the male quails peck other quails and may make them blind. At times, death of the quails is also noticed. So usually adult males and females are reared separately and kept together only during breeding.

Keeping birds in continuous dark results in complete inhibition of sexual behaviour and reduces egg production, whereas providing continuous light results in a very high rate of egg production during the first year, and production coming to a halt by the end of the second year. Again, colour of lights have varying effects on the egg production rate. Red light gives much greater stimulus than blue light on

Table 5.10 : Food requirements for different stages of quails

Nutrients	Broiler quail		Layer quail		
	Starter (0-3 weeks)	Finisher (4-5 weeks)	Starter (0-3 weeks)	Grower (4-5 weeks)	Breeder (> 6 weeks)
Energy (kcal/kg)	2800	2800	2750	2750	2700
Protein %	27	24	24	20	19
Calcium %	0.8	0.6	0.8	0.6	3.0
Phosphorus %	0.3	0.3	0.3	0.3	0.45

Note : Brolier quail : Quails reared for meat. Layer quail : Quails reared for eggs.

growth of testes in male. Red light and incandescent light increase egg production in female. Availability of oxygen less than 21% in the air reduces the rate of egg production. Therefore, quailary at higher altitude in hilly areas is not successful.

Males not required for breeding, or those grown for meat production, can be given only about 8 hours of low intensity light a day, for suppressing their sexual maturity and increasing faster fattening growth.

QUAIL DISEASES

Quails, though suffer from various diseases are quite resistant to Ranikhet disease, fowl-pox, ascariasis, etc. No vaccination programme for quails yet been developed. Some important diseases, their causative agents, symptoms and control measures are described here.

(1) **Ulcerative enteritis** : It is an enteric disease, particularly occurring in the early age. The causal organisms identified are—(1) *Corynebacterium perdicum* and (2) anaerobic *Bacillus*. The natural mode of transmission of disease is through ingestion of contaminated material. The spread of disease is very fast from one flock to other through insects.

Symptoms : The disease may occur in acute or chronic form. In acute form, the mortality may be as high as 100% in young quails. The incidence is more common in birds raised on litter as compared to those raised on wire floors (cages). In certain cases, the birds may die without showing any apparent sign and such birds are usually healthy. The characteristic symptom of the disease is watery droppings containing urates. The birds become dull, humped up, have ruffled plumage and partially opened eyes. The chronic cases linger for months before their death. The most prominent lesions are the extensive ulcerations in the intestine and caecum. The birds dying suddenly may show haemorrhagic enteritis in the upper part of the intestine. The lesions on the liver may vary from yellow spots to yellowish areas of necrosis along the edge of the liver. Spleen may be enlarged haemorrhagic.

Treatment : The disease can be effectively treated by drugs like bacitracin, streptomycin and chloromycetin. Streptomycin at the rate of 2 gm/4.5 litres of drinking water for 25 days or streptomycin sulphate at the rate of 1 gm/litre of drinking water for 10 days gives complete protection against infection. The disease outbreak can be prevented by administering streptomycin sulphate at the rate of 1 gm/4.5 litres of drinking water in chicks during the first four weeks of their rearing period. Bacitracin at the rate of 100 gm/tonne and chloromycetin at the rate of 500 gm/tonne of feed give good protection from the disease.

(2) **Quail bronchitis** : It is caused by a filterable virus, serologically distinct from infectious bronchitis virus of chicken.

Symptoms : The disease usually occurs in an acute form in chicks and affects their respiratory system. It is characterised by tracheal disturbances, sneezing and coughing but without nasal discharge. Some cases show nervous symptoms. The disease may linger for about three weeks and cause mortality up to 80% in young chicks.

Treatment : There is no satisfactory treatment for this disease at present. However, maintaining hygienic conditions and protecting birds against bad weather are certain measures to be suggested.

(3) **Aspergillosis** : This disease generally affects young chicks. It is caused by the spores of *Aspergillus fumigatus*. The birds get affected through contact of contaminated feed and litter material or through inhalation of the infected material.

Symptoms : The affected bird may have difficult and increased rate of breathing. Reduced feed consumption, increased thirst, fever and emaciation may also be observed. There may be bulging of the eyelids and ulceration of cornea in older chicks. The lungs, bronchi and other parts of the respiratory system may show pinhead-like or even larger nodules. In occular cases, there may be cheesy material under the nictitating membrane.

Treatment : The feed and litter material used should be free from fungi. The materials and equipments from the farm and hatchery premises should be kept free of infection. The humidity in the brooder house should be kept low. Adding calcium propionate at the rate of 2 kg/tonne of feed arrests the growth of fungi.

GENERAL HEALTH CARE OF FARM QUAILS

In farms, the quails need extra care and proper hygiene to keep them away from diseases. The important points for consideration are mentioned here :

(1) The adult and young quail stocks should be kept separately.

(2) The breeder stock should be free from egg borne diseases like Salmonellosis, and healthy stocks should be added.

(3) The wild birds should be kept away since they might be the carriers of some transmissible diseases.

(4) Birds of different species should be kept separately.

(5) Efforts should be made to break the cycle of infection by opting clean sanitation measures from hatchery to marketing.

(6) The feed and water should always be obtained from a clean source.

(7) The houses and equipments used should be disinfected while starting a fresh lot and, thereafter, cleaned at regular intervals.

(8) The litter materials or the scraping from the dropping trays should be buried in a pit. In the event of any disease outbreak, such materials must be burnt.

(9) The dead birds should be disposed of either by burning them to ashes or burying deep into a pit under a layer of lime and earth.

Care of Quailery

To keep the quailery healthy and disease-free, frequent disinfection of equipments and other articles is often required. Thorough physical cleaning is necessary before application of any disinfectant. After the cleaning, any of the following methods can be used :

(1) Boiling water can be used to disinfect clothes, instruments, etc. Addition of 2-5% lye to hot water helps cleaning by washing.

(2) Sunlight is good disinfectant. The equipments can be exposed to it, after proper cleaning.

(3) Cresols and phenols are quite active against many viruses but are not much useful against spore forming bacteria. They have strong odour and low solubility in water.

(4) Chlorine is generally used as sodium hypochlorite. Organic matter inactivates chlorine and at 300 ppm concentration it is harmful.

(5) Formaldehyde fumigation or formalin spray is a potent disinfectant for quailery in warm and humid conditions. Potassium permanganate 20 gm and formalin 40% in 40 ml solution are required for each 2.8 cubic metres of air space. In a large shed the chemicals should be placed in divided quantities at 3-4 places in the house. The room should be kept closed for a day and then opened for 2-3 days for proper aeration before placing the birds. Since the chemical is very much irritating and harmful to skin, gas mask, gloves and close fitting clothes should be worn by the operator while working.

(6) Quaternary ammonium compounds are less effective against viruses and fungi, and do not kill spores. They are quick acting and non-toxic and, therefore, useful for disinfecting hatcheries and food processing equipments. They are also used as water sanitizers at 65 ppm.

(7) Copper sulphate is a low grade general disinfectant, but is a strong fungicide and algicide. It can be used at 0.5% level to clean the feeders and waterers. However, the utensils must be rinsed thoroughly after the treatment. It may be used in the ratio of 1 : 2000 with water for spraying the house.

ECONOMICS OF QUAIL

Egg : Quail eggs are excellent source of vitamins and proteins and are consumed variously. Raw eggs can be stored for one month at 13°C with 70% relative humidity. Boiled eggs can be stored for more than three months at room temperature in the form of egg pickles.

Meat : Quail meat is delicious and nutritious. The ready to cook quail meat can be stored in refrigerator for six months at 12°C or for one week at 5°C without much loss in nutritional qualities. In addition to its use as a dish in the kitchen, quail meat can be further stored for two months in vinegar based pickles or for five months in oil based pickles.

Quail Chicks : Quail chicks produced by breeders can be supplied to quail farmers.

SPACE FOR NOTES

(D) EMU FARMING

INTRODUCTION

Emus (*Dromicieus* sp.) belong to **ratite** group (flightless birds) and have high economic value for their meat, eggs, oil, skin and feathers. These birds are adaptable to varied climatic conditions. Although emu (Fig. 5.53) and ostrich were introduced in India, emu farming has gained much importance. Ratite birds have poorly developed wings and include **emu, ostrich, rhea, cassowary and kiwi**. The anatomical and physiological features of these birds appear to be suitable for temperate and tropical climatic conditions. These birds can be well-maintained in extensive (ranches) and semi-intensive rearing systems with reasonably high fibrous diets. United States, Australia and China are leading countries in emu farming. Emu birds are well adapted to Indian climatic conditions.

Rajasthan University of Veterinary and Animal Sciences (RAJUVAS) in Bikaner has started emu breeding centre for providing way to subsidiary income of farmers through diversification of farming system. Initially 30 pairs of emu birds have been brought from Australia; they are being reared and acclimatised in Indian atmosphere.

Fig. 5.53. An adult Emu

Systematic position : [After: J. Z. Young (1981)]

Phylum : Chordata

Sub-phylum : Vertebrata/Craniata

Super-class : Gnathostomata

Class : Aves

Sub-class : Neornithes

Super-order : Neognathae

Order : Casuariformes

Scientific name : *Dromicieus* sp.

Reason for choosing emu for farming

- These are second largest bird after ostrich by height and suitable for arid and semi-arid climate. They can thrive well in temperature ranging from 0ºC to 50ºC.
- The bird is omnivorous and can easily be maintained in modified poultry feed.

Biology : These birds reach their full size by the end of the first year. They grow up to 5.5 to 6 feet with a weight of 40 to 70 kg (Fig. 5.53). In India laying period of emu, starts after 18 to 24 months and eggs are laid during September to February. Eggs are dark bluish green in colour with a weight of 450 to 700 gm (Fig. 5.54). One bird can lay 20-60 eggs in a season. Incubating period is 50-52 days. The life span is 30-35 years while the productive economic life is of 20-25 years.

Potentiality of emu culture

- One boiled egg or omelet can provide for entire family.
- Emu meat is healthy with less cholesterol (98% fat-free). The meat is lower in fat than chicken, turkey, pork or mutton.

Fig. 5.54. Emu eggs

- Good quantity of oil can be obtained from the bird. It is reported that 4-6 litres of oil devoid of any colour, taste or odour is available from this bird.
- The oil has penetrating properties and is effective as an analgesic, anti-allergic and antiseptic.
- This oil is used in beauty creams and lotions, soaps, hair oils, shampoos, perfumes and massage oils.
- Traditionally emu oil has been used for treatment of muscle and joint pains.

CULTURE PROCEDURE

Emu Pen

Emus will not survive in a small enclosure let alone kept inside a barn or roost all year round. They need wide open space to run, kick, sun bathe, and wallow in the dust (Figs. 5.55 and 5.56). Minimum requirement for a pair of emus is at least half an acre of properly fenced pen. Therefore, starting with two males and two females, one acre of land is required. The fencing will be of wire net bordered by strong iron pipes (Fig. 5.55). The fence should not be made with ordinary wood because a single kick from the emu is enough to send a wood plank flying over few metres. The height of the fencing should be at least 8 feet. Although the emus are flightless, they can jump over a fence that is shorter than 7 feet.

Emu Shelter

Within the pen there should be a small cottage with proper roofing (Figs. 5.57 and 5.58). Emus need a roof over their head and walls to protect them during adverse seasonal conditions. A pair of emus can be sheltered in 16 feet by 16 feet housing with roof high enough for good ventilation and walls that are good to protect the birds from wind chills. There should be hay or straw to provide cushion and bedding for the emus. Emus make a pile of mess which should be cleaned every day to keep their shelter from smelling bad. They cannot be kept a day without cleaning the shelter.

Fig. 5.55. Emu roaming in a large sized pen

Fig. 5.56. An ideal emu pen

Fig. 5.57. Emu shelter within the emu pen

Fig. 5.58. Emu shelter within the pen

Baby pen

Within the shelter a specially designed baby pen should be prepared (Fig. 5.59). The baby pen is like an iron cage with soft floor, where chicks are kept for

Fig. 5.59. Baby pen with chicks

a period of at least one month of age. The size of the cage will be 3 feet length × 2 feet breadth × 3 feet height. The cage will be placed on the table of 3 feet height and in well-ventilated place. It should be regularly cleaned. It is advisable to keep the baby pen out of reach of the adults.

Laying of Emu eggs

Breeder ration with sufficient calcium (2.7%) for ensuring proper calcification of egg with strength should be fed. Feeding excess calcium to the breeding bird before laying will upset egg production and it also impairs fertility in male. Extra calcium in

Fig. 5.60. Sketch of an emu pen

the form of grit or calcite powder is provided to the birds by placing it in a separate trough. Eggs should be collected frequently from the pen. If eggs are soiled, it is cleaned with sand paper and mopped up with cotton. Eggs should be stored in a cooler room with a temperature of 20°C. Eggs should not be stored for more than 10 days so as to ensure better hatchability.

Management of Chicks

Generally emu chicks weigh about 370 to 450 grams (about 67% of egg weight) (Fig. 5.61). At the onset (48-72 hours) of their life, emu chicks are kept in incubator for quick absorption of the yolk and proper drying. During this period the brooding shed should be cleaned and disinfected thoroughly for receiving chicks. The floor will be covered with litter (paddy husk) and then new gunny bags or burlap will be laid over the litter. In a brooder about 25-40 chicks may be kept to provide 4 sq. ft per chick, for the first 3 weeks. For successful brooding, a temperature of 32°C is provided for the first 10 days followed by 25°C for 3-4 weeks. Sufficient water mugs of one litre capacity and equal number of feeder troughs should be provided in the brooder. A chick guard must be 2.5 ft high to avoid jumping and straying of chicks. A 40 watt bulb should burn in the brooder shed throughout the day for every 100 sq. ft area. After 3 weeks of age, slow extension of the brooder

Fig. 5.61. Emu chicks of 24 hours in a brooder

area is done by widening the chick guard circle and later the chick guard circle is removed by the time chicks attain 6 weeks.

Starter mash should be fed for the first 14 weeks or till attaining standard body weight of 10 kg. Proper floor space should be ensured for the birds house, as these birds require running space for their healthy life. 30 feet run space is required; hence floor space of 40 ft × 30 ft is required for about 40 chicks, if outdoor space is provided. Floor must have proper drainage and free from dampness.

Grower management

As emu chicks grow, they require bigger size waterers and feeders and increased floor space.

Different sexes should be reared separately. If necessary, sufficient paddy husk should be added in the pen to manage the litter in good and dry condition. Birds are well-fed on grower mash till they attain 34 weeks age or 25 kg body weight. Greens should comprise about 10% of diet particularly different kinds of leaf meals, for making the birds adapted to fibrous diets. Sufficient amount of clean water is provided at the time of feeding. Ensured dry litter condition throughout the grower stage is recommended. Provision of 40 ft × 100 ft space for 40 birds in outdoor space is recommended. Floor must be easily drained and should be void of dampness. At the time of capturing, it is required to hold the adult birds firmly by grabbing both the wings and preventing them from kicking. It is essential that the birds are not harmed during the period of capture.

Breeder management

Emu birds attain sexual maturity by 18-24 months of age. In a pen, a sex ratio of 1 male : 1 female is maintained. In case of pen mating, pairing should be done based on the compatibility. During mating, about 2500 sq. ft floor space per pair, is required. Trees and shrubs may be provided for privacy and inducing them for mating. Breeder diet should be offered well in advance, i.e., 3-4 weeks prior to breeding programme, enriched with minerals and vitamins to ensure better fertility and hatchability in birds. Normally, adult bird consumes 1 kg feed /day. But during breeding season, feed intake will be drastically reduced. Hence intake of nutrients must be ensured.

First egg is laid at around two and half years of age. Eggs are laid during September to February, particularly in cooler days of the year. The egg is laid in evening. Eggs should be collected daily to avoid damage in the pen. Normally, a hen lays about 15 eggs during first year cycle. In subsequent years, the egg production increases to about 30-40 eggs. On an average, a hen lays 25 eggs per year. Egg weighs about 450-700 gm with an average weight of 560 gm. Egg appears greenish and looks like tough marble (Fig. 5.54). The intensity of colour varies from light, medium to dark green. The surface of the egg varies from rough to smooth. Majority of eggs are medium green with rough surface.

Incubation and Hatching

Fertilised eggs after adjusting to room temperature should be kept carefully in horizontally or slantingly arranged rows in a tray. The eggs are kept in the incubator after thoroughly cleaning and disinfecting them. The machine should be set for the correct incubating temperature, i.e., dry bulb temperature of about 35°C–36°C and wet bulb temperature of about 25°C–28°C (about 30-40% RH). Eggs are then carefully placed on the tray in a setter, after setting the required temperature and relative humidity. Identification slip, for date of set and pedigree may be pasted on the eggs, if required. Prior fumigation of the incubator with 20 gm potassium permanganate + 40 ml formalin for every 100 cu. ft of incubator space is essential. The eggs should be turned regularly till the 48th day of incubation. From 49th day onwards, turning of the eggs should be stopped. By 52nd day, the incubation period ends. The chicks need drying. The chicks should be kept for at least 24 to 72 hours in the hatching compartment, for reducing the chance of producing weak chicks.

Feeding

Emus need balanced diet for their proper growth and reproduction. Based on the literature, certain nutrient requirements for different age groups are given in Table 5.12. Feed can be prepared by using common poultry feed ingredients (Table 5.13). Feed alone accounts for 60-70% of the production cost, hence cheap rations will economically improve the margin of returns over feeding. In commercial farms, feed intake per emu breeding pair per annum varies from 394-632 kg with a mean of 527 kg.

Healthcare and management

Ratite birds are generally sturdy and live longer (80% livability). Mortality and health problems in emus are mainly in chicks and juveniles. These include starvation, malnutrition, intestinal obstruction, leg abnormalities, coli and clostridial infections. The main causes being improper brooding or nutrition, stress, improper handling and genetic disorders. Other diseases reported are rhinitis, candidiasis, salmonella, aspergillosis, coccidiosis, lice and ascarid infestations. Ivermectin is given every month to prevent external and internal worms beginning from 1 month of age. In emu, enteritis and viral eastern equine encephalomyelitis (EEE) are reported. In India, so far few outbreaks of Ranikhet disease have been recorded based on gross lesions but are not confirmed. However, the birds vaccinated for Ranikhet disease at the age of 1week (lasota), 4 week (lasota booster), 8, 15 and 40 weeks (by Mukteswar strain) gave better immunity

Economics

Emu farm economic survey indicated that cost involved in purchase of breeding stock are expensive (68%). The rest of the investments are on farm (13%) and hatchery management (19%). Feeding cost during 2010-2011 per breeding pair per annum was estimated to be Rs. 3600.00. Cost of production of hatching egg and day-old chick was Rs. 793.00 and Rs. 1232.00 respectively. Annual feed intake per pair was recorded as 524 kg, costing Rs. 3578.00. The cost of salable chick at day-old age was Rs. 2500.00 – Rs. 3000.00. Better returns from emu are possible with good hatchability (more than 80%), lower feeding cost and minimised chick mortality (less than 10%).

Table 5.12 : Feeding ration for Emus of different ages

Food ingredients	Starter 10-14 weeks age or up to 10 kg body weight	Grower 15-34 weeks age or up to 10-25 kg body weight	Breeder
Crude Protein%	20	18	20
Lysine %	1.0	0.8	0.9
Methionine%	0.45	0.4	0.40
Tryptophan %	0.17	0.15	0.18
Threonine %	0.50	0.48	0.60
Calcium %	1.5	1.5	2.50
Total phosphorus %	0.80	0.7	0.6
Sodium chloride %	0.40	0.3	0.4
Crude fibre %	9	10	10
Vitamin A (IU/kg)	15000	8800	15000
Vitamin D3 (ICU/kg)	4500	3300	4500
Vitamin E (IU/kg)	100	44	100
Vitamin B12 (IU/kg)	45	22	45
Choline (mg/kg)	2200	2200	2200
Copper (mg/kg)	30	33	30
Zinc (mg/kg)	110	110	110
Manganese (mg/kg)	150	154	150
Iodine (mg/kg)	1.1	1.1	1.1

Table 5.13 : Recommended ingredients for Emu feeds (kg/100 kg feed)

Ingredients of feed formulation	Starter	Grower	Breeder
Maize	50	45	50
Soybean meal	30	25	25
DORB	10	16.25	15.50
Sunflower	6.15	10	0
Di-calcium phosphate	1.5	1.5	1.5
Calcite powder	1.5	1.5	1.5
Shell grit	0	0	6
Salt	0.3	0.3	0.3
Trace minerals	0.1	0.1	0.1
Vitamins	0.1	0.1	0.1
Cocciodiostat	0.05	0.05	0
Methionine	0.25	0.15	0.25
Choline chloride	0.05	0.05	0.05

Fig. 5.62. Adult Emu with chicks

Recent information

In India, the hope balloon of Emu farming has been punctured as farmers are not getting customers, as was expected. Emu meat is neither tasty nor healthy nor of medicinal value. Moreover, farming is space consuming and not at all cost effective. As farmers are not getting customers they fail to recover bank loan. They have to feed the emus from their own pocket. The project has becomes totally failed and abandoned.

SPACE FOR NOTES

Chapter Six : Fisheries and Aquaculture

FISHERIES

Fish has high nutritive and medicinal values. Its flesh provides an excellent source of protein for human consumption, which is relatively of high digestibility, biological and growth promoting. Nutritionally fish proteins are superior to milk, mutton and egg albumen. Fish protein comprises **all the ten essential amino acids**, namely *lysine* (present in high concentration), *leucine*, *isoleucine*, *arginine*, *histidine*, *valine*, *threonine*, *methionine*, *phenylalanine* and *tryptophan*. Fish flesh also provides us minerals, iodine, vitamins and high content of polyunsaturated fatty acids. *Eicosapentaenoic acid* (polyunsaturated fatty acid) of fish has been found to reduce the cholesterol level of blood and save humans from coronary diseases.

The major components of fish flesh are as follows :

Water	— 53-89%
Protein	— 20%
Carbohydrate	— Negligible (only as Glycogen in liver)
Fat	— Variable (1-8% or more)
Minerals	— 1-2%
Vitamins	— Good sources of vitamins A, D, E and B-complex.

The scientific study of fishes is termed as **Ichthyology** (Greek : *Ikhthus,* a fish; *logos*, a discourse). Ichthyology encompasses all the pure and applied aspects of the fishes. Applied aspects of Ichthyology, relates to the human welfare by way of exploring the fields of economic value. In a broader sense it is referred to as fisheries.

MEANING OF FISHERY

The term 'fishery' ordinarily means culture and propagation of edible and marketable fishes. This definition, however, should be restricted to **pisciculture**. Fishery is a vast domain. In a broader sense it includes the judicious exploitation of natural resources of water for human consumption and benefit. The natural resources of water include all the organisms such as fishes, prawn, shrimps, crabs, sharks and rays, various molluscs, dolphins, whales, seaweeds, algae, etc.

The science of fisheries is a very complex subject. It is multidisciplinary and includes physical, chemical, biological, microbiological, hydrobiological, limnological, and meteorological sciences interacting with the biology of fish, ecology of the resources, production management, marketing, etc.

Fishery science has expanded to such an extent that today various colleges and universities are running MSc courses in Zoology with specialisation in fish and fisheries or BSc, MSc courses exclusively in Fishery Science (BFSc and MFSc). Moreover, under the provision of U.G.C (University Grants Commission), a vocational course in Fishery Science or Industrial Fisheries is being run by a number of colleges and universities.

Apart from the above, a number of technological and Research Institutes, under the Indian Council of Agricultural Research (ICAR) are actively engaged in conducting research and evolving technological improvements for the betterment of fisheries. Some important Fishery Research Institutes are listed below :

(1) Central Inland Capture Fisheries Research Institute (CICFRI) with headquarter at Barrackpore (West Bengal).

(2) Central Institute of Freshwater Aquaculture (CIFA) with headquarter at Bhubaneswar (Orissa).

(3) Central Institute of Brackish water Aquaculture (CIBA) with headquarter at Chennai (Tamil Nadu).

(4) Central Institute of Fisheries Technology (CIFT) with headquarter at Cochin (Kerala).

(5) National Bureau of Fish Genetic Resources (NBFGR) with headquarter at Lucknow (Uttar Pradesh).

(6) National Research Centre on Cold Water Fisheries with headquarter at Nainital (Uttaranchal).

(7) Central Marine Fisheries Research Institute (CMFRI) with headquarter at Cochin (Kerala).

(8) Taraporevala Aquarium and Marine Research Station at Mumbai (Maharastra).

WHY CULTURE FISHES ?

Fishes are advantageous for culture because :

(1) Fishes do not spend much energy on temperature regulation as they are poikilothermic in nature.

(2) Fishes can convert food into body tissue more efficiently than any other farm animals. The conversion factor to flesh is high in fishes (80.9%) than chicken (64.7%) and beef (51%).

(3) There is always a certainty of getting a rich haul from culturable waters than from natural resources.

(4) Fish is a prolific breeder. It has high fecundity.

(5) Fish production can be organised according to the market demand in respect of quality, quantity, size, colour, preservation and processing.

(6) Low trophic feeders (herbivores) can be raised at comparatively lower costs than those which are higher in the food chain.

Indian Ichthyodiversity

India globally ranks 3rd in fisheries and 2nd in aquaculture, contributing about 4.4% of the global fish production. This is mainly due to its vast diverse aquatic resources ranging from deep seas to lakes in the mountains and more than 10% of the global biodiversity in terms of fish and shellfish. India's marine wealth comprises 8129 km of costline and 2.02 million sq. km of Exclusive Economic Zone (EEZ). Its brackishwater and estuaries comprise 1.24 and 0.29 million ha, respectively. India's freshwater resources comprise Rivers and Canals (1,97,024 km), Reservoirs (3.15 million ha), Ponds and Tanks (2.35 million ha) and Oxbow lakes and derelict waters (1.3 million ha).

In such varied aquatic ecosystems, India has vast and varied fish germ plasm resources comprising about 2,200 fish species. The total number of species reported worldwide is more than 24,618. The National Bureau of Fish Genetic Resources, Lucknow has identified fish germ plasm resources of India and their status (Table 6.1), from different ecosystems of the Indian waters.

Table 6.1 : Indian fish germ plasm resources and their status (NBFGR)

Eco-systems	Total species	Endan-gered	Vulner-able	Rare	Indeter-minate
Cold water	73	01	04	–	12
Warm water	544	03	13	02	28
Brackish water	143	–	02	–	04
Marine	1440	–	02	–	08
Total	**2200**	**04**	**21**	**02**	**52**

Of 2200 species recorded, about 400 species are commercially important.

Types of Fishery

I. Depending upon the management, fishery can be classified as under :

(a) **Capture Fishery** — Capture fishery aims at the management and capture of fishes from the hill stream, rivers, lakes, estuaries and seas. Capture does not mean, over-harvesting. It is done with an eye to conservation of the fishes and other aquatic organisms, so that sustained productivity is maintained. This can be done by restricting the mesh size of the nets, so that smaller size fishes are not caught.

(b) **Culture Fishery** — It includes the practice of rearing fishes in small impounded water such as ponds, jheels, small tanks, etc. Here important culturable fishes (endemic or exotic) are reared and bred on scientific lines and when they reach the desired length and weight, they are captured and marketed.

Nowadays, culture fisheries are also practised in large impoundments through pen or cage culture, mariculture, ranching (culture based capture fisheries), raft, rope, rack, tray or pole culture, etc.

II. Depending upon the habitat of fishes, fishery may be divided into following :

(A) **Inland Fishery** — As has been discussed earlier, India has vast inland fishery resources. With increase in culture practices, the inland fish production (3.3 mmt) has overtaken the marine production (3.0 mmt) as shown in Fig. 6.1. Inland fishery can be broadly subdivided into following :

(i) **Freshwater fishery** — Freshwater fishery comprises both capture and culture fishery. The aim is to culture fishes in small impoundments and to exploit fish resources from the great river systems of the country and from the vast networks of irrigation canals, lakes, tanks and reservoirs.

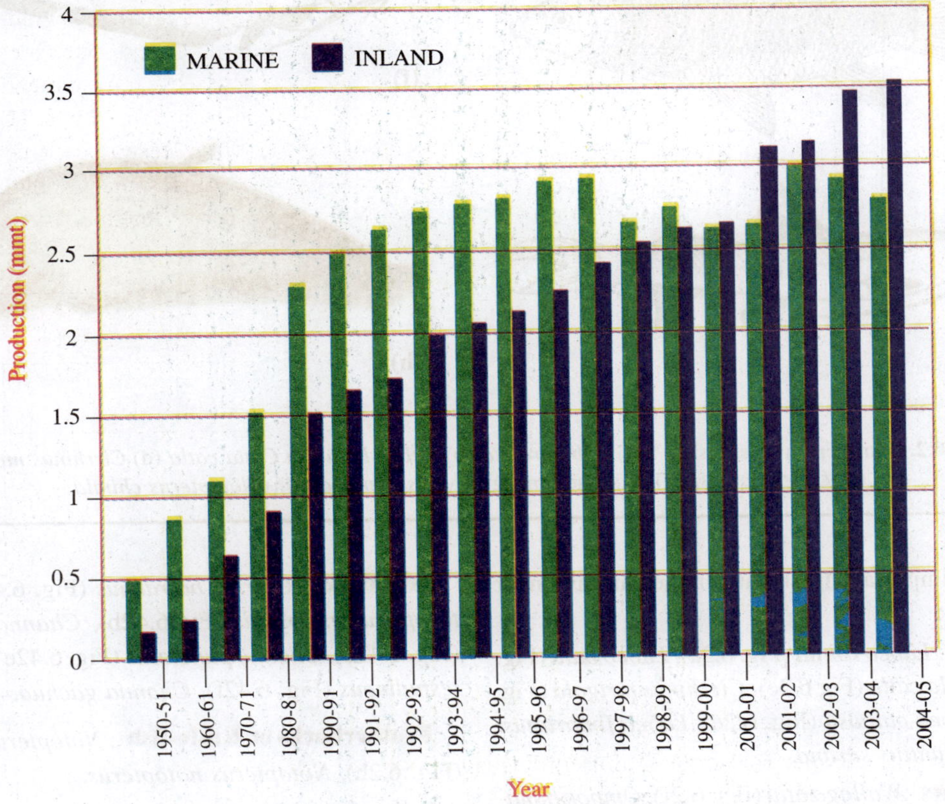

Fig. 6.1. Marine and Inland fish production trend

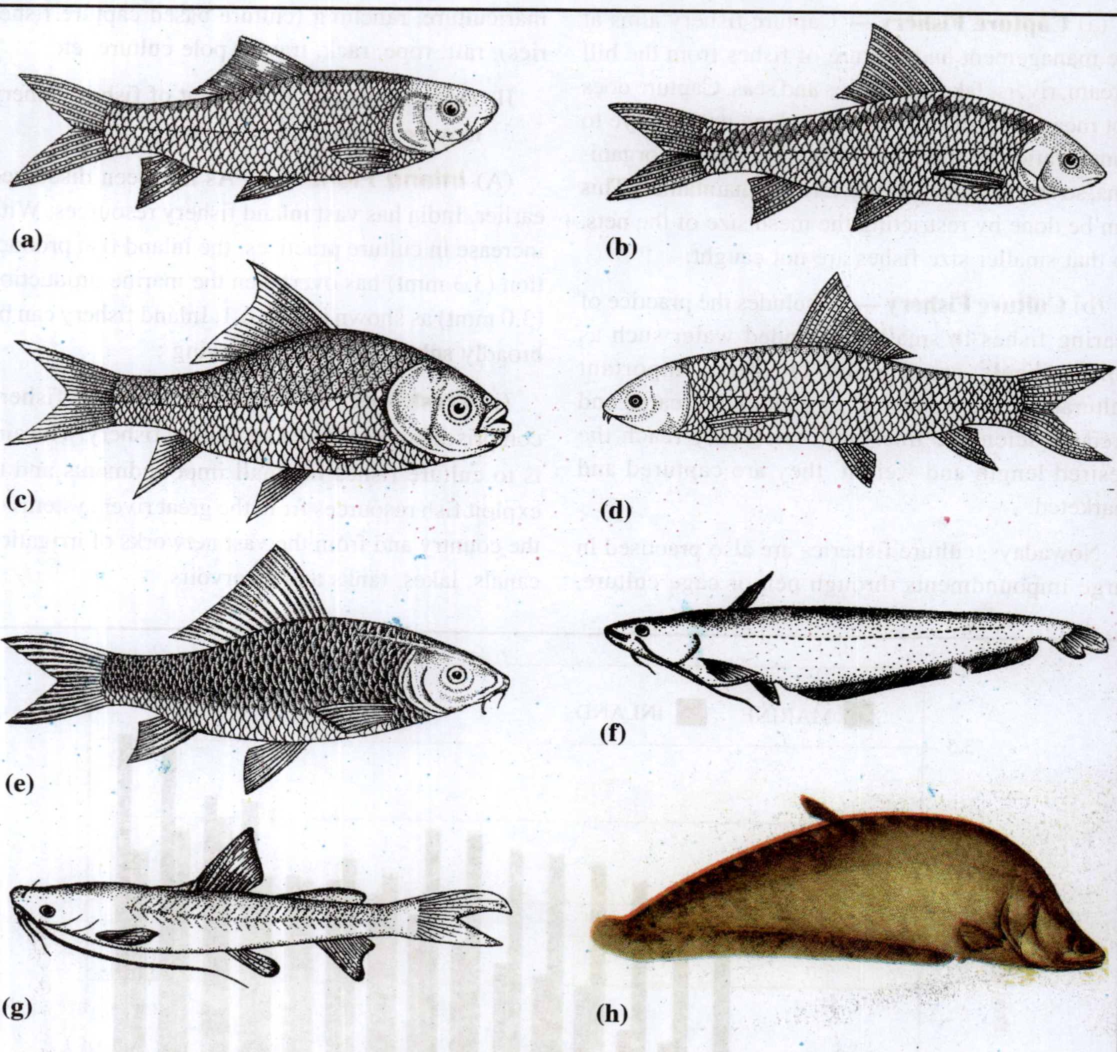

Fig. 6.2. Some Freshwater fishes : (a) *Labeo rohita* (b) *Labeo bata* (c) *Catla catla* (d) *Cirrhinus mrigala* (e) *Labeo calbasu* (f) *Wallago attu* (g) *Mystus seenghala* (h) *Notopterus chitala*

Some important freshwater fishes are as under (Fig. 6.2) :

Carps : *Labeo rohita* (Fig. 6.2a), *Labeo bata* (Fig. 6.2b), *Catla catla* (Fig.6.2c), *Cirrhinus mrigala* (Fig. 6.2d), *Labeo calbasu* (Fig. 6.2e), *Labeo fimbriatus, Tor tor, Puntius sarana.*

Catfishes : *Wallago attu* (Fig. 6.2f), *Ompok bimaculatus, Pangasius pangasius, Mystus seenghala* (Fig. 6.2g) *Mystus aor, Eutropicthyes vacha.*

Jeol fishes: *Clarias batrachus* (Fig. 6.42a), *Heteropneustes fossilis* (Fig.6.42b), *Channa striatus* (Fig. 6.42d), *Channa punctatus* (Fig. 6.42e), *Anabas testudineus* (Fig. 6.42f), *Channa gachua.*

Featherbacks or Knife-fish : *Notopterus chitala* (Fig. 6.2h), *Notopterus notopterus.*

Mullets : *Mugil* sp.

Freshwater eel : *Anguilla bengalensis.*

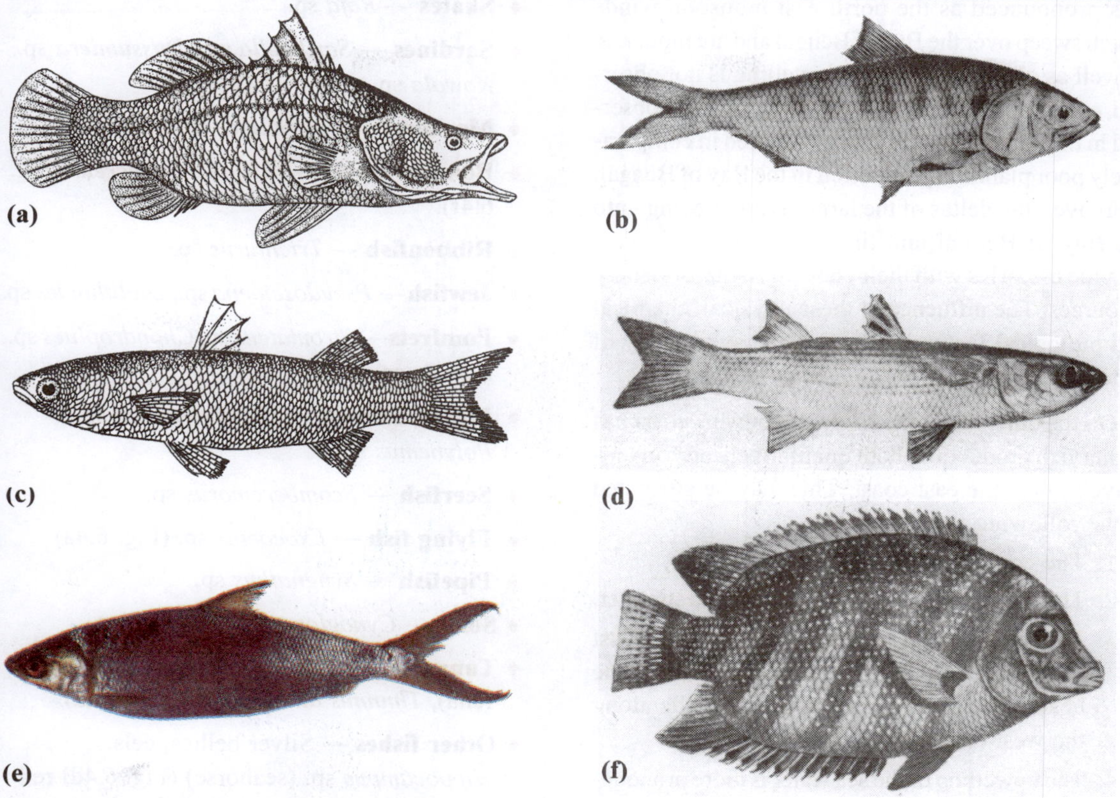

Fig. 6.3. Some Brackish water fishes (a) *Lates calcarifer* (b) *Hilsa (Tenualosa) ilisha* (c) *Mugil parsia* (d) *Mugil cephalus* (e) *Chanos chanos* (f) *Etroplus suratensis* (Pearl spot)

(ii) **Brackish water fishery** — It includes culture and exploitation of fishes from different bheris in the coastal regions, estuaries and brackish water ponds and jheels of the deltaic regions. Some of the important fishes found here are — *Lates calcarifer* (Fig. 6.3a), *Hilsa (Tenualosa) ilisha* (Fig. 6.3b), *Pangasius* sp., *Mugil parsia* (Fig. 6.3c), *Liza corsula, Mugil cephalus* (Fig. 6.3d), *Anchoviella* sp., *Chanos chanos* (Fig. 6.3e), *Etroplus suratensis* (Fig. 6.3f), *Osphronemus* sp.

(B) **Marine Fishery** — Exploitation of the fishes and other organisms of the sea comes under the preview of marine fishery.

(i) **Coastal fishery** : Marine fishery resources of a country is dependent upon a number of factors. The oceanographical features of the sea have a direct bearing upon the fish fauna and the nature of fisheries. The chemical nature of the sea water, nutrients and salt-contents determine the production of plankton which forms the food of fishes. The currents, winds and upwelling of oceans also somewhat influence plankton production and exercise a great influence on the movements of fishes and their young ones. Moreover, the continental shelf (the land beneath the sea), its area and slope, also to some extent determine the facilities of fishing.

India has two main seas, the Bay of Bengal on the east and the Arabian sea on the west. Both the seas are radically different from each other.

The **Coromandel** or eastern coast has a narrow continental shelf, that declines sharply into the deep sea waters. The circulation of water in this coast is less pronounced as the north-east monsoon winds which sweep over the Bay of Bengal and are moderate as well as of short duration. Upwelling is not observed, except for a certain degree which has been observed in the Andhra zone. This has resulted in comparatively poor plankton production in the Bay of Bengal. Moreover, the deltas of the large rivers flowing into the Bay of Bengal and the several coastal lakes provide estuaries with their characteristics of fishery resources. The influence of these physical, chemical and biological factors, reflect on the availability of different varieties of fishes and their quantity.

On the other hand, the **Malabar** or west coast has higher fish production, both qualitatively and quantitatively, than the east coast. This may be attributed to the following facts :

1. The continental shelf is broader.
2. The south-west monsoon is longer and stronger.
3. The somali ocean current from the eastern coast of Africa moves northwards, sweeps around the head of the Arabian sea and then drifts along the west coast of India.
4. The upwelling of the sea water is more pronounced and brings about a better circulation in the sea.
5. The seasonal cycles of the sea water of Arabian sea are more pronounced.

All these leads to higher plankton (food of fishes) production and thus, richer fish fauna. It is estimated that about 75% of the total marine landing comes from the west coast.

Some of the fishes that are captured from the coastal area are listed below :

- **Sharks** — *Scoliodon* sp. (dogfish), *Sphyrna* sp. (Hammer-headed shark) (Fig. 6.4b), *Carcharhinus* sp. (Blue shark), *Rhincodon* sp. (Whale shark), *Galeocerdo* sp. and *Stegastoma* sp. (Tiger or Zebra or Carpet or Nurse shark).
- **Rays** — *Pristis* (Sawfish) (Fig. 6.4c), *Rhinobatus* sp. (Guitarfish), *Torpedo* sp. (Electric Rays), *Aetobatus* sp. (Eagle rays), *Rhinoptera* sp. (Cow-nosed rays), *Dasyatis* sp. (String or Whip or Butterfly rays).
- **Skates** — *Raja* sp.
- **Sardines** — *Sardinella* sp., *Dussumiera* sp., *Kowala* sp.
- **Mackerel** — *Rastrelliger* sp.
- **Bombay-Duck** — *Harpodon nehereus* (Fig. 6.4f).
- **Ribbonfish** — *Trichiurus* sp.
- **Jewfish** — *Pseudosciaena* sp., *Otolithoides* sp.
- **Pomfrets** — *Stromateus* sp, *Chondroplites* sp., *Formio* sp.
- **Indian Salmon** — *Eleutheronema* sp., *Polynemus* sp.
- **Seerfish** — *Scomberemorus* sp.
- **Flying fish** — *Exocoetus* sp. (Fig. 6.4a)
- **Pipefish** — *Syngnathus* sp.
- **Soles** — *Cynoglossus* sp. (Fig. 6.4g)
- **Tunas** — *Gymnosarda unicolor* (dog tooth tuna), *Thunnus alalunga* (big eye tuna).
- **Other fishes** — Silver bellies, eels, *Hippocampus* sp. (seahorse) (Fig. 6.4d) red mullets.

(ii) **Offshore and deep sea fishery** — The area beyond the continental shelf is the expansive sea which is rich in various kind of fishes. These fishes dwell either at the surface, mid-water or at the bottom of the sea and their exploitation constitutes offshore and deep sea fisheries. The most profitable depth is 20 fathoms (37 metres) and the catches are better during the day than night and also during neap-tide than spring-tide. Various types of trawlers and trawling nets are in operation.

The important commercial fishes found in this area are :

Sharks — *Carcharhinus* sp., *Pristis* sp.
Indian salmon — *Eleutheronema tetradactylum*
Giant headfin — *Polydachylus indicus*
Jewfish — *Pseudosciaena diacanthus*
Sea eel — *Muraenosox* sp.
Perches — *Epinephelus* sp., *Lutianus* sp.

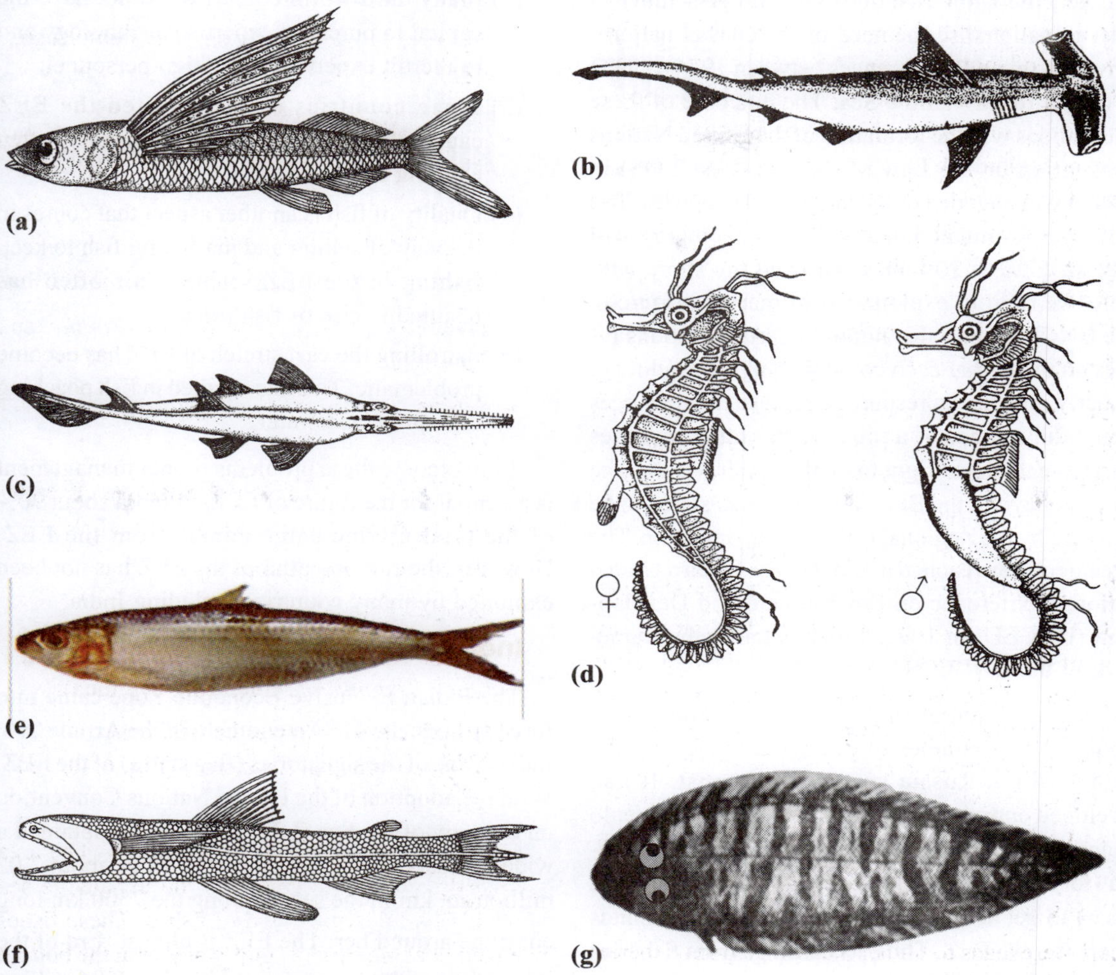

Fig. 6.4. Some marine fishes (a) *Exocoetus* sp. (b) *Sphyrna* sp. (c) *Pristis* sp. (d) *Hippocampus* sp. (e) *Sardinella* sp. (f) *Harpodon* sp. (g) *Cynoglossus* sp.

EXCLUSIVE ECONOMIC ZONE (EEZ)

Exclusive Economic Zone (better known as EEZ), was the outcome due to a conflict on the fishing rights among coastal nations. It resulted due to the fact that some nations found that their waters were being exploited by foreign countries equipped with better and more powerful fishing gear and craft, more sophisticated navigational and fish detection systems than their own. Several such exploited countries joined hands in 1950s and together unilaterally established fishing limits in their coastal waters for exclusive fishing. Initially, the limit was only six to twelve miles. Subsequently many more countries joined hands and by 1970s established an extended limit of 200 nautical miles (1 nm = ~ 1.85 km).

Later, the United Nations General Assembly in order to nationalise the need of the coastal nations, convened several conferences between 1974 to 1982 on the Third Law of the Sea. The outcome of these delebrations was the formation of the United Nations Convention on the Law of the Sea (UNCLOS) in 1994. It was attended by as many as 60 countries. The UNCLOS formulated that each coastal country will have an EEZ of 200 nm zone from the shore with exclusive right to exploitation of marine resources. The UNCLOS has also imposed certain responsibilities on them, that each country shall have a duty to properly manage its resources. In case the resources are not fully tapped by a country, then other countries shall have the legal right to exploit such waters. The fisheries beyond the EEZ (high sea) was stated to be regulated by the Regional Fisheries Organization. The agreement was reached at the Rio De Janeiro United Nations Conference on Environment and Development (UNCED) in 1992, which is under the framework of the UNCLOS.

The imposition of EEZ to 200 nm posed a lot of problems to a number of ocean-fishing countries, among which Russia suffered the most. It had developed enormous technology and fishing fleet to explore fisheries around the globe including Antarctica (for krill). The implimentation of EEZ deprived Russia to get access to these fertile fishing grounds except its own EEZ. Thus, its earnings from fisheries have been badly hit. The same problem was also faced by Japan, Spain, etc.

The EEZ comprises a stretch of water from 12 to 200 nautical miles. The inshore coastal water fisheries up to 12 nm remained with the traditional and artisanal fishermen for exploration.

EEZ had changed the fishing pattern. However, many countries including India faced a number of problems such as :

(1) Many countries found it difficult to explore new waters of EEZ that were not only too far out but deep also, and they lacked proper gear, vessel and technical know-how. The only solution to this problem was to have joint ventures with foreign countries who have the expertise and experience.

(2) Many third world countries did not have the capital to purchase expertise, technology and to recruit experienced foreign personnel.

(3) Some countries over-exploited the EEZ causing damage to the inshore fishery, often evoking unrest.

(4) Quality of fish is another aspect that comes in the way of selling and marketing fish to keep fishing in the EEZ viable. This often has resulted in rise of fish price.

(5) Patrolling the vast stretch of EEZ has become problematic. This has resulted in fish poaching in unprotected water.

To overcome these problems proper management is essential for the future of EEZ. In India about 90% of the total marine catch comes from the EEZ. However, the full potential of the EEZ has not been exploited by many countries including India.

Indian Exclusive Economic Zone

The Indian Exclusive Economic Zone came into force through the 41st Amendment of the Article 298. India is one of the signatories (flag states) of the EEZ. With the adoption of the United Nations Convention on the Law of the Sea (UNCLOS), India obtained a wide exclusive economic zone (EEZ) of about 2.02 million sq. km in the sea all along the 7500 km long coastline around her. The EEZ is almost $\frac{2}{3}$rd of the inland area of the country and comprises 0.86 million sq. km on the west coast, 0.56 million sq. km on the east coast and 0.60 million sq. km around the Andaman and Nicobar islands. The species composition of EEZ includes tuna, anchovies, perches, ribbonfish, lizardfish, catfish, shark, rays and skates, squid, cuttlefish, bivalves, shrimps, lobsters, crabs, etc. Of these, shrimps are the most exploited.

India presently has been able to actually exploit only inshore water (up to 50 m depth), while the offshore water (beyond 50 m depth) has largely remained unexploited. This has resulted in India being unable to tap the full potential of this region which has been estimated to be 3.9 to 4.5 million tonnes. Fishing beyond the 50 metres depth is negligibly exploited due to lack of infrastructure. Of the total estimated potential in this zone, 70% is constituted by pelagic* resources, 19% by demersal** resources and the rest by variety of low-value fishes, crabs, etc. Deep sea holds enormous scope, and fishing in this region is, therefore, a matter of great concern and attention. Moreover, to check poaching of unauthorised intrusion into the Indian EEZ, fast and well-equipped patrol vessels should undertake survellance as well as search and rescue works in the EEZ.

The Indian EEZ resources to the tune of 3 million tonnes per year has remained unexpolited. Strategies have been planned for exploration and exploitation of the EEZ resources. Deep sea fishing policy was brought into force in 1991. Joint ventures with foreign countries were tried up to 1983. The deep sea fishing policy in India was not welcomed by the traditional and artisanal fishermen and their association who opposed strongly, particularly the joint ventures with foreign countries.

Several million people living along the coastline are directly influenced by oceanography of the EEZ, various environmental hazards and related social issues. Thus, it is essential to comprehensively map the entire EEZ so as to understand the natural processes that have been active in EEZ. The National Institute of Oceanography (NIO) has initiated the process to explore the Indian EEZ. This enormous task has been undertaken in collaboration with other national institutes, like NIOT (Chennai), NCAOR (Vasco), GSI and some universities.

AQUACULTURE

Fish production was previously heavily dependent upon capture fishery and in particular the marine resources. However, the capture fishery cannot be expected to be a perennial protein donar. Moreover, a substantial portion of the marine catch is being utilised for making industrial products which are not directly consumed by man. Therefore, a viable alternative by which fish production could be increased through a popularized biotechnique, called *aquaculture*. Aquaculture may be defined as the "**farming and husbandary of economically important aquatic animals and plants under controlled conditions**".

The term 'aquaculture', though over a decade has been widely used, still has a misconception by many users. Aquaculture is sufficiently expressive and multifaceted. It includes the following aspects :

(1) The type of organism cultured. For example, the culture of fish, prawn, oyster, mussel, seaweed, etc.

(2) The type of culture technique, such as pond culture, raceway culture, cage culture, pen culture, raft culture, etc.

(3) The aquatic environment in which the culture is done such as freshwater, brackish water or salt water.

(4) A specific factor of the environment used for culture, such as cold water or warm water aquaculture.

Although aquaculture is considered to be a branch of fisheries science, still there is a tendency to denote the distinction between the two because of some of the basic differences in development and management.

Salient characters of aquaculture

(1) **High productivity** : In comparison to agriculture or veterinary practices, aquaculture has been found to be more productive.

(2) **Integrated farming** : Aquaculture with agriculture, horticulture or animal husbandary is found to be more profitable.

(3) **Recycling** : Aquaculture gives efficient means of recycling domestic and/or agriculture wastes and thus, helps in protecting our environment.

* **Pelagic fishes** are those species that inhabit the water column, such as mackerel, anchovy, herring and tuna.

** **Demersal fishes** are those species that inhabit the bottom such as cod and haddock. These species tend to concentrate on the broad continental shelves.

(4) **Rural development** : Aquaculture helps to integrate rural development by generating employment opportunities and would thus help to arrest the exodus of population from rural to urban areas.

(5) **Intensive fish culture** : In aquaculture practice, owing to the fishes three-dimensional utilization of the water body can be crowded more closely (200/m^3) and grown through water recirculation system. This gives high yield of about 25 t/ha/year.

(6) **Earning foreign exchange** : Commercially important items, such as ornamental fishes, *Artemia* cysts, prawns, lobsters, crabs, frog legs, etc. produced through aquaculture are highly valued and can earn good foreign exchange.

(7) **Ranching** : Fish seeds and larvae of economically important fishes of capture fisheries are artificially recruited in fish hatcheries through aquaculture, called ranching or *aquarange farming*. It is commonly known as "culture based capture fishery" and involves restocking of the wild stock for harvesting. This process has replenished the dwindling stock of rivers and seas.

(8) **Replenishment of coral reefs** : Coral reefs that have been destroyed naturally or anthropogenically are being replenished through construction of artificial coral reef.

(9) **Creating leasure-time activities** : Leasure-time activities can be maintained through sport fishing and creating home and public aquaria.

(10) **Mariculture and fisheries enhancement** : In many reef areas the use of mariculture is fast growing. It provides an alternative source of income and employment in coastal areas. Replenishment schemes of species that has been greatly diminished has been undertaken in certain areas. This has been useful for returning species such as giant clams and trochus to reefs where they have been largely or completely exterminated by overfishing.

(11) **Promoting agro-industrial development** : Aquaculture can promote agro-industrial development through — (i) processing and marketing of fishery products, feeds and equipments for aquaculture, (ii) Seaweed culture for the production of marine celluloids, (iii) pearl oyster culture, etc.

TYPES OF AQUACULTURE PRACTICES

I. Depending on Hydrobiological Features

Depending on hydrobiological features (particularly salinity) aquaculture is divided into following:

(a) freshwater, (b) brackish water, (c) metahaline and (d) mariculture. Various types of culture practices are carried out in each of these divisions (Fig. 6.5).

(a) **Freshwater Aquaculture** — Freshwater aquaculture includes culture of various economically important aquatic organisms in small impoundments. They are as follows :

(i) **Composite fish culture or polyculture or mixed fish farming**. Fast growing compatible species of fishes with different feeding habits are cultured together.

(ii) **Monosex culture** — In this culture, separate sexes of individual species are cultured. By this method the yield can be enhanced.

(iii) **Monospecies culture** — This is concerned with the culturing of individual species of fish.

(iv) **Jeol fish culture** — Shallow water bodies with poor oxygen content is used for culturing air breathing fish species such as *Channa, Clarias, Heteropneustes,* etc.

(v) **Predator-prey culture** — Predator-prey (murrel-tilapia) culture is also undertaken in shallow waters and in swampy areas. Here, predator fish like murrels are cultured along with their prey-fish, tilapia.

(vi) **Culture of fishes in temple tanks** — The quality of the water in the unutilised temple tanks are poor owing to stagnation. Such tanks can be utilised for the culture of fishes such as common carp, murrel, tilapia, catfishes, etc.

(vii) **Culture of fishes in irrigation tank** — Irrigation tanks are fitted with inlet and outlet water sources, which can be profitably utilised for culture of Indian major carps and exotic carps.

(viii) **Raceway culture** — Raceways are series of rectangular or circular earthen or cemented tanks which is constructed in such a way that a running water system like that of a stream is maintained (Fig. 6.6).

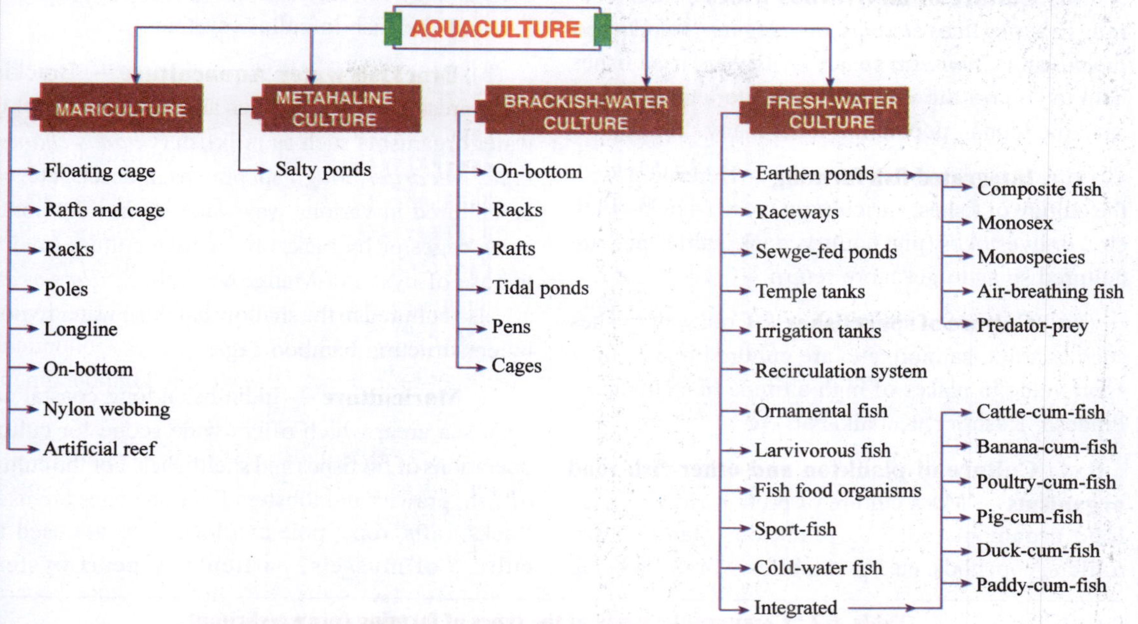

Fig. 6.5. Different types of aquaculture practices

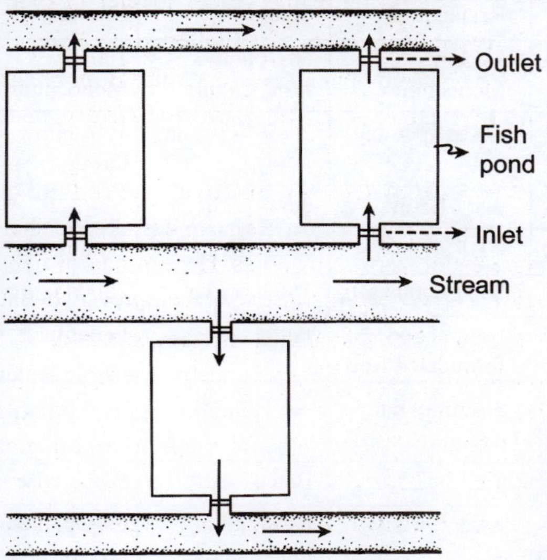

Fig. 6.6. Raceway

(ix) **Sewage-fed fish culture** — In such a culture sewage water after treatment in oxidation ponds are utilised. Such water contains high nutrients which are responsible for production of plankton bloom, the food of fishes.

(x) **Fish culture in recirculating system** — In areas of water scarcity, the recirculatory system of fish culture is practised. Hence, water is circulated continuously through a biological filter.

(xi) **Ornamental fish culture** — Ornamental or aquarium fishes (Fig. 6.7) such as guppy (*Lebistes reticulatus*), gourami (*Trichogaster trichopterus*), goldfish *(Carassius auratus*), molly (*Mollienisia latipinna),* siamese fighter (*Betta splendeus*), sword tail (*Xiphophorus helleri*), etc. are cultured due to their high demand in international market.

Fig. 6.7. Ornamental fish culture

(xii) **Culture of larvivorous fishes** — Larvivorous fishes such as *Oryzias melastigma, Aplocheilus blochii* and *Gambusia* sp. are cultured. These fishes help to control the vectors (mosquito, etc.) and can save the human population from many diseases.

(xiii) **Integrated fish farming**—Here along with the culture of fishes, agricultural crops (rice, banana, etc.) or livestocks (pig, poultry, duck, cattle, etc.) are cultured so as to get more return.

(xiv) **Culture of sport fishes**—Cold water fishes such as trout, salmon, etc. are cultured due to their sport value in places of high altitude like Himachal Pradesh, Kashmir, Kodaikanal, etc.

(xv) **Culture of plankton and other fish food organisms** — Mass culture of protein-rich live fish food organisms such as plankton (cladocerans, rotifers, copepods, etc.) and tubifex worms, is being done as they are essential for growing larvae of fin fish and shellfish in culture systems.

(b) **Brackish water Aquaculture** — Brackish water areas are rich in oxygen and plankton. Brackish water organisms such as milkfish (*Chanos chanos), Lates calcarifer, Mugil* sp. prawns, mussels, etc. can be cultured in various ways such as in tidal ponds, pens, cages or by rack, raft or rope culture (mainly in case of oysters). Mangrove crab, *Scylla serrata,* are also cultured in the shallow brackish water regions by constructing bamboo cages.

(c) **Mariculture** — India has a long coastal and open sea area, which offers wide scope for culture operations of fin fishes and shellfishes. For the culture of fish, prawns and lobsters floating cages are used. Racks, rafts, rope, pole and long lines are used for culture of mussels, particularly pearl oysters.

Table 6.2 : Comparative study of the types of farming (prawn/shrimp)

	Extensive/Traditional	Modified Extensive	Semi-intensive	Intensive	Super-intensive
Pound size	5 hectares	1 – 2 hectares	1 hectare	0.2 – 0.4 hectare	0.05 – 0.1 ha
Shape	Any	Any	Rectangular	Square	Circular
Stocking density	Mixed culture uncontrolled	3/m^2 Mixed culture	8-15/m^2 Monoculture	50-100/m^2 Monoculture	100/m^2 Monoculture
Water exchange	Sometimes with the help of pump	5%/day	5–10% per day	20-50% per day	100% flow through
Water depth	1.2–5.0 m	1.0 – 3.0 m	0.8 – 1.5 m	0.6 – 1.5 m	
Aeration	Not needed	Very less, if needed	2 HP	8–12 HP	Continuous
Production	500 kg/hectares	1000-3000 kg/ha	3000-6000 kg/ha	15000 kg/ha	25000 kg/ha
Feed used	Natural feed	Natural and formulated feed	Natural and formulated feed	Pellet feed	Pellet feed
Rate of feeding	—	1-4 times daily	3-6 times daily	—	—
Crops/year	1	2	2	Batch-wise	Batch-wise
Engineering	Not needed	Very less needed	Essential	Essential of high quality	Essential of high quality
Investment	Very less	Less	High	Very high	Very high
Care	Not much needed	Needed	Essential	Essential	Essential
Generator and current	Not needed	If needed	Necessary	Compulsory	Compulsory
Prawn/shrimp larva source	Wild	Wild	Hatchery	Hatchery	Hatchery

Sea-weeds also are widely cultured with the help of nets or webbings.

(d) **Metahaline culture** — In off seasons of salt manufacture, the salt pans are utilised for fish culture. Culture of brine shrimp (*Artemia salina*) can be undertaken in these supersaline salt pan (salinity more than 200%) area. The nauplii of *Artemia* serve as a protein-rich live food for the fin fishes and shell-fishes. The dormant eggs (cysts) of *Artemia* formed under unfavourable conditions also can earn good foreign exchange.

II. Depending on Motive of Farming

Depending upon the motive of farming, based on economic and commercial considerations, fish culture practices may be classified as under :

(i) **Extensive fish culture** — Large ponds, beels, etc. are brought under this culture (Table 6.2) where little care is taken with regard to its improvement. So it is the least managed fish farming. Here, the yield is modest and the expenditure is less as it is raised on natural food.

(ii) **Intensive fish culture** — In this fish culture (Table 6.2) an all out attempt is made to achieve maximum production of fish from a minimum quantity of water. It is the best managed form of fish farming and the fishes are fed on artificial food in addition to the natural feed. Here the yield is very high (over 6000 kg/ha/year). Although the cost of investment is high, the earnings from this culture far exceed the cost, so as to ensure high profit.

(iii) **Semi-intensive fish culture** — Intensive culture possesses certain hazards, for which a culture between the first two, called semi-intensive culture (Table 6.2), is generally practised. Here certain amount of management is required and the net profit is in between the above two.

III. Depending on Special Operational Techniques

Based on special operational techniques, fish culture practices may be classified as below :

(i) **Cage culture**: In this culture the fishes are held in a section of water, either in a flowing river or in a big impoundment. The fishes are imprisoned in a cage made of galvanised steel wire frames with nylon meshes or simply of bamboo frames with split bamboo mats. The cages are of various sizes and the commonly used size is about 10 sq. metres.

Cages are of the following types : (a) Floating cages (Fig. 6.8), (b) Fixed cages, (c) Submerged cages and (d) Moveable cages.

Fig. 6.8. Diagramatic sketch of a typical floating cage

The principal aim of culturing fishes in cages is to practice monoculture of selected species. Cage culture was first started in Cambodia. Since 1980 cage culture has been in practice in India. Cage culture has proved a boon for reclaiming extensive waste waters, i.e., swamps and derelict waters of Assam, Bihar and Karnataka. Air breathing fishes (*Clarias, Heteropneustes, Anabas, Channa, etc.)* have responded well to cage-culture practices. The **advantages** are :

1. Minimum use of all available water resources.
2. Economic use of water.
3. Easy observation and handling of the fish population on daily basis.

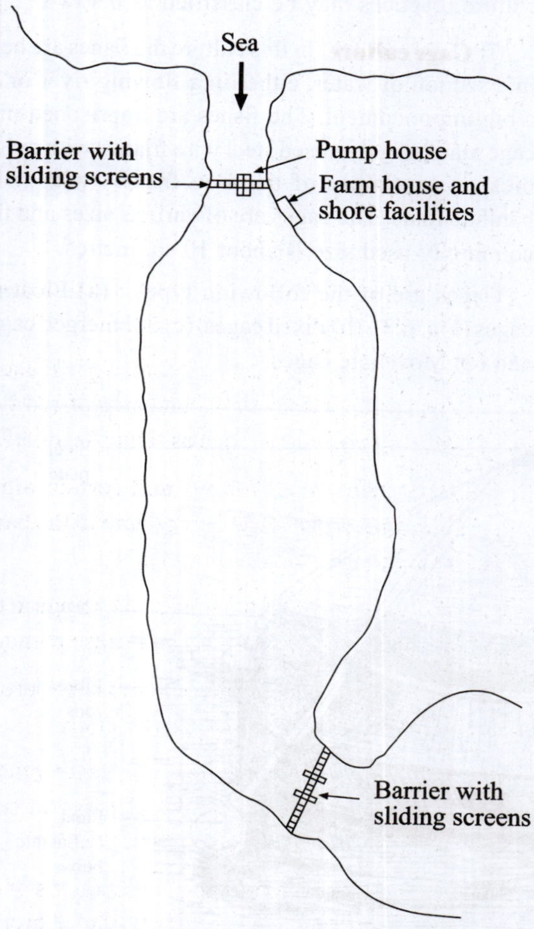

Fig. 6.9. Diagram of a pen constructed on an enclosed fjord

4. It reduces fish mortality.
5. Easiness in the control of competitors, predators and parasites.
6. Easy and complete harvesting of fishes possible.
7. Control of diseases and isolation of the diseased fishes is easy in floating cages.
8. Initial investment is relatively less.

However, there are certain **limitations** :

1. Risk of theft is high.
2. Adequate renewal of water in the cages for eliminating metabolites is not possible.
3. There is a need for maintenance of high dissolved oxygen content.
4. Fouling agents like snails, etc. accumulate at the bottom of the cage (mesh) which needs periodic removal.
5. The release of uneaten food and faeces adversely affect the water quality.

(ii) **Pen culture** : Pens generally refer to small enclosures used for confinement or safe keeping of domestic animals. In fisheries pens are formed by damming a bay, cove, fjord (an arm of a sea), estuary, river, lake or reservoir. Sites are selected where the barriers can be constructed across narrow sections or channels (Fig. 6.9) in order to reduce costs and increase the ease of operation. The enclosures should be relatively small (about 2.0-7.0 ha). However, in Japan large enclosures measuring up to 120 ha or more are also seen. The depth of the pens should be more than 1 metre, even in low tide condition.

Most of the perimeter of the enclosure (pen) is formed by the natural shoreline. When the blind end of a water area is enclosed, there may be only one or one series of barriers. In the enclosures where there are continuous flow of water, there may be two or two series of barriers : one upstream and another downstream. The barriers are made of stones, sand, soil or concrete and are provided with 'SCREENS'. These 'screens' prevent the escape of the fish stock (Fig. 6.10).

Fig. 6.10. Bamboo scaffolding enclosures of different sizes

The types of enclosures generally seen are :

(a) Bamboo scaffolding enclosures used in the bays of Philipines and in the lakes of China (Fig. 6.10).

(b) Floating net-enclosures useful for the culture of tilapia and milkfish in lakes.

(c) Single-layered pens of nylon webbing.

(d) Double-layered pens suitable for use as nurseries for fish and prawn seeds.

The introduction of pen culture in many countries have not met with much success. This probably is due to the difficulties in the use of intensive techniques and in some cases due to the high costs of embankments and water management (for example, through pumping).

The **merits** of pen culture are :

(1) As it is a continuous process due to continuous supply of water, production (yield) is greater in a limited space with rich food and oxygen supply.

(2) Higher growth rate is possible as energy is saved towards locomotion, feeding, etc.

(3) It generally generates employment opportunities for the coastal people.

(4) There is less danger of fish mortality as toxic metabolites (ammonia, etc.) are flushed regularly.

The **demerits** of this culture are :

(1) Unfavourable weather conditions may damage the pen or flood the culture sites.

(2) Pen culture sites may be polluted particularly during the occasional abundance of red tide causing organisms, such as 'Dinoflagellates'.

(3) Organisms like *Balanus* and certain algae (*Enteromorpha,* etc.) may adhere to the bamboo poles causing biofouling.

(4) Terrestrial insects sometimes take refuge in the exposed portions of pens and cause damage.

(5) The nylon-webbing enclosures may be cut and damaged by certain species of crabs.

(6) Predatory fishes sometimes may enter into the pen and cause considerable damage to seed and growing fishes or prawns.

(7) Abundance of sea weeds in the outskirts of the pen, may bring down the oxygen level through release of hydrogen sulphide on death and decay.

Role of Biotechnology in Aquaculture

A considerable advancement in aquaculture through the use of biotechnology has resulted in an increase in production in terms of quality and value addition. The areas of approach and improvement through use of biotechnology is given below :

(A) **Nutrition** : Indian aquaculture industry lags behind in the science of nutrition. A good approach would be to improve the quality and utilisation of various agricultural and industrial by-products for efficiency. Various foodstuffs having potential values as fish feed ingredients have not been nutritionally utilised as they contain toxic or anti-nutritional factors. The need of biotechnology is to develop new varieties of foodstuff with totally free or low content of such intoxicants.

Other developments of improved feeds are :

(1) Development of food additives, such as anabolic steroids and thyroid hormones, have resulted in increased growth of salmonid fishes.

(2) Development of new artificial diets, such as micro-encapsulations for filter-feeders (oysters and mussels).

(3) Improved culture of single-celled feed organisms (algae, yeast, bacteria).

(B) **Feeding stimuli and chemical signal** : To obtain the full sequence of feeding behaviour from initial recognition of feeding 'strike' to ingestion of food particles, it is necessary to identify the feeding stimuli or chemical signal. It is thus required to genetically design or engineer the bacterial strain with characteristics of having better cellulose degradation capacity.

(C) **Reduce the in-pond chemical oxygen demand** — Microbial processing of livestock manure through anaerobic digestion in biogas plants reduces the in-pond COD. It also improves the fertilisation quality of the inputs. Biotechnology has attained great significance in aquaculture development by increasing the nitrogen fixation levels in ponds through bio-fertilisation with blue-green algae (azolla) and green-manuring and through genetic manipulation of bacteria to improve their nitrogen fixing capacity. This has definitely increased fish production.

(D) **Administration of mammalian hormones**— Mammalian gonadotropic hormones have been used extensively for spawning of carps and catfishes. Hormonal manipulation of sex is being practised to control unwanted reproduction in prolific breeders such as common carp and *Tilapia*. Administration of mammalian growth hormones through feed or injection has enhanced fish growth. Manipulated monosex population improves nutritive value of fish flesh and growth.

(E) **Cryopreservation** — Cryopreservation refers to preservation of gametes and embryo generally at below freezing temperature for ready availability of genetic material for breeding and research purposes. Cryopreservation of milt has been successfully developed for several fishes. It is essential since the males of some fish species mature earlier than the females and there would be dearth of oozing males when fully mature female fishes would be available. Cryopreservation of zygotes and embryos have facilitated large scale fish seed production programme.

(F) **Use of allomones and pheromones** — In large scale induced breeding, use of allomones and pheromones is of great significance. Under biotechnological programme it is necessary to study the role of pheromones in alarm and social behaviour, species and sex recognition, sex behaviour and territorial and space recognition.

(G) **Genetic manipulation** — Change in genetic make-up of any organism by application of improved bio-techniques is referred to as genetic manipulation. Fishes, generally, can tolerate a wide range of genetic manipulation than terrestrial farm animals.

(a) **Chromosome manipulation** — Through chromosome manipulation fish stock can be improved for breeding and culture by way of gynogenesis, androgenesis and polyploidy.

(i) **Gynogenesis (all maternal inheritance)**— Gynogenesis in fishes can be induced by stimulating parthenogenetic development of fish eggs by artificially inactivated spermatozoa. Through X-ray or UN-rays, the genetic contents (DNA) in spermatozoa can be destroyed. With such spermatozoon fertilisation would occur without any contribution from them. The diploid parthenogenetic individuals would give rise to offsprins that would be females having maternal inheritance only Ex. Grass carp, Salmon, Rainbow trout, etc.

(ii) **Androgenesis (all paternal inheritence)** — Similarly with X-rays or UN-rays the genome of ovums can be destroyed and fertilisation with normal spermatozoon would result in homogametic males having paternal inheritance only. Ex. Common carp.

(iii) **Polyploidy** — Polyploidy (individuals having extra sets of chromosomes) can be achieved by subjecting fertilised eggs to heat, cold, pressure or chemical shocks. Polyploidy produces (1) sterile individuals, (2) enhanced growth and survival, (3) improved quality of flesh in some fishes, (4) more viable and ideal for producing new hybrids.

(b) **Gene transfer or Transgenesis** — Transgenesis results in formation of 'Transgennic fish'. The techniques involved is through micro-injection and electroporation, genes are introduced into one-celled embryo or oocyte. The potentialities of transgenesis are :

(1) Tolerance to physical factors. Ex. tolerance to cold using antifreeze gene.

(2) To accelerate growth by using growth hormone genes.

(3) Efficient use of food by fishes through manipulation of biochemical pathways.

(4) To increase disease resistance by using specific disease resistance genes (Ex. T-cell receptor, immunoglobulin, lymphokines, etc.).

(5) Behavioural modification (maturation, reproduction, sex control, etc.) by regulation of endocrine function.

(c) **Application of cell culture biotechnology to marine macro-algae** — Seaweeds can be easily genetically engineered. In many maritime countries, tons of seaweeds are produced through mariculture, which are used as human food (green seaweeds) and for commercial production of phyco-colloids (agar, algins, carrageenas, etc.) for industrial purposes.

(H) **Hormonal manipulation for sex control** — Sex hormones (steroids) are used to produce monosex for culture of tilapia in particular. Tilapia being a prolific breeder, creates problem in pond culture as it results in overcrowding with small size fishes and thus, in decreased production. Therefore, monosex culture controls reproduction and also helps in increased production.

To obtain only males by sex reversal method, the early fry of tilapia are fed with androgen 17-a methyl testosterone for induction of sex reversal from genetic females to phenotypic males. Similarly, in species where females are bigger and grow faster than males, only females can be produced for monosex culture by feeding fry with estrogen 17b-Estradiol benzoate or diethyl stibestrol.

(I) **Fish disease management** — Fish diseases can be controlled through development of biotechnologically produced vaccines. Vaccines against vibriosis and furunculosis have been developed and are available in the market. However, they are for limited fish species. There is an increased demand to use biotechnology for developing vaccines against bacterial, viral and other diseases affecting several commercially important culturable fish species.

(J) **Current breakthrough and essentials for future research** — Technological advancement has led to freshwater and marine pearl culture product development such as chitin, chitosan, etc. Further studies for even more efficient and large scale production technologies is essential. Attention should be drawn towards genetic improvement of fish stock through selection and genetic engineering, *ex-situ* conservation methodologies, gene banking both with live and gamete level approaches, which would subsequently lead to future research. Social impact of fisheries and aquaculture development has to be studied and the need is to develop parameters to undertake Social Impact Assessment (SIA).

Thus, for the development of biotechnology it is urgently required to upgrade laboratory facilities and attract young blood for research and training. It should be noted that water resources in the country are not exclusively available for aquaculture as there is excessive pressure on their use from several other sectors. Therefore, any further growth in aquaculture for quantum jump would have to be vertical and technology based. The transfer of technology from the laboratory to the field has had its own causilities. Today only about 30% of the proven technology available in various fields of aquaculture has been manifested in the field. There is thus, urgent need to narrow the gap between the results of research and those in the farmers field.

QUALITIES OF CULTURABLE AQUATIC SPECIES

In India about 465 cultivatable aquatic organisms, belonging to 107 families of animals and 28 families of plants were catalogued by Jhingran and Gopalakrishnan (1974). All these aquatic organisms can probably be used for culture and this would be an advantage to choose from such a large number of aquacultured species that are adapted to different environmental conditions. This would definitely eliminate the thought of introduction of exotic species for aquaculture.

However, the main question is whether it would be worth to culture all these 465 aquatic organisms and how can all of them contribute to the main objectives of aquaculture. Another fact is that, in such a case aquaculture would certainly miss the advantages that agriculture have had in concentrating research on a few species that had led to the development of advanced technologies of production of selected high-yielding strains and hybrids. However, development of such technologies had required a lot of effort and time. Aquaculture science, on the other hand, is relatively new and would probably require much longer time to reach that level of advancement if efforts have to be shared among so many species.

Aquaculture till today, through long traditional experience and scientific research, have actually succeeded in domesticating only a few number of species. This, therefore, indicates that there is a tendency towards limiting the number of species in large scale commercial aquaculture. At the same time there is a real need to have species suited for different environmental conditions and economic circumstances. Species for aquaculture, therefore, has to be selected according to the objectives of culture.

Characteristics necessary for cultivable species in aquaculture

The characteristics that determine the suitability of a species for aquaculture are as follows :

(1) Should be palatable, nutritive and economically important.

(2) Should be able to live in captivity in impoundments with other fishes without any disturbances.

(3) **Rate of growth and production** — The rate of growth and production under culture conditions is a major characteristic that determines the suitability of a species for aquaculture. Certain slow-growing species may be candidates for culture because of their high market value, but it is generally difficult to make their culture economical. The use of heated water can definitely improve the growth rates of many species, but use of such a method has not yet proved commercially successful.

(4) **Size and age at first maturity** — It is preferable to select those fishes which will reach marketable size before attaining first sexual maturity, as most of the feed would be used by them for somatic growth. Early maturity before reaching marketable size would obviously be a great handicap (example : *Tilapia*). However, early maturity would ensure easier availability of breeders for hatchery operations.

(5) **Breeding** — If breeding criteria are taken into account then species preferable for culture should consider the following :

(a) **Seeds available from the wild** : In species where controlled breeding techniques under captive conditions have not been perfected, the aquaculturist would have to depend upon seed availability from the wild. This, however, proves to be an unreliable source in large scale farming because their abundance in nature depends upon a number of unpredictable factors. Moreover, large scale collection of spawn and fry has resulted in conflict with commercial fishermen, who blame the aquaculturist for the decline in the catch. However, there is no scientific evidence for this point of view. So, it is advisable to select species that can be propagated in hatcheries.

(b) **To culture a species that can breed under captive conditions** : It is advisable to select species for aquaculture that can breed under captive conditions. This would enable production of adequate quantities of seed through hatchery production. If the selected species matures more than once a year, then it would be possible to have several crops of seed, if other conditions remain suitable.

(c) **High fecundity and frequency of spawning** : It would be advantageous if the selected species have high fecundity and greater frequency of spawning. However, small sized eggs and small sized larvae make hatching operations difficult. If the incubation period and larval cycle are shorter, then mortality of larva would be lower with greater hatching survivability.

(d) **Artificial feed acceptance of larvae** : Rearing of live food is difficult and often expensive. Therefore, larvae that accepts artificial feed is easier to rare in hatcheries.

(6) **Adult feed** — Feeding may account for more than 50% of the total cost of production, and is, therefore, of prime concern in modern aquaculture.

(a) **Culture of herbivorous and omnivorous species** : Herbivorous and omnivorous species have been preferred for culture as they feed on natural food organisms in water. The growth of natural food can be increased through fertilisation and water management. The cost of feeding in such cases will be relatively low. Shorter the food chain, faster is the growth rate. So, herbivorous and omnivorous species are preferred for the production of low-priced products.

(b) **Species adopted to feeding with artificial food stuff** : Feed efficiency in relation to growth and productivity is an important criteria. Some low trophic level feeders can sometimes be highly selective in their feeding habits. For example, some filter-feeders may require plankton of a particular size and shape, which sometimes may not be available in the impoundment. So, the need to grow these species to market size within a limited season or period often makes it necessary to resort to artificial feeding. Also sometimes the natural feed may dwindle due to grazing. So, in such cases, species adopted to supplementary feeding with artificial foodstuffs is generally preferred.

(c) **Culture of carnivorous species** : Carnivorous fishes require high protein diet which is more expensive and largely depends on local availability. However, the high feeding cost is compensated as most carnivorous species command high market price. Such species have greater export market and thus, attract substantial investments.

(7) **Preference of hardy and disease resistant species** — In relatively poor and partially polluted water bodies, the temperature and oxygen concentrations can fluctuate, and deterioration of the water quality may occur unavoidably . In such unfavourable conditions, species that are hardy will obviously fare better. At the same time they should be resistant to diseases.

(8) **Influence of species on environment**— Some species may have certain influence on the environment. For example, soil erosion may be caused due to the feeding habit of carps. Such species which upsets the ecological balance would need special protective measures, leading to higher cost of maintenance and environmental concern.

(9) **Intensive and semi-intensive cultures** — Intensive and semi-intensive cultures involve high stocking density, confined in a limited space, with water recirculation system and intensive feeding. Overcrowding may lead to an increase in transmission of diseases, cannibalism in the early stages and accumulation of waste products. Species that can withstand such environmental hazards are better candidates for aquaculture.

In the selection of species for culture, the economic considerations are more important to an aquaculturist than biological factors. The presence of proven technologies for culture along with economic viability acts as a guideline for investors and aquaculturists, in the selection of a species for culture.

SEED PRODUCTION IN CARPS

Procurement of quality fish seeds is the prerequisite for fish culture. Seed production of important cultured fishes in India varies according to their breeding habits — (1) Some fishes such as *Cyprinus, Tilapia,* etc., breed in confined waters (ponds); (2) Some fishes breed in flowing waters (rivers), such as catla, rohu, mrigal, calbasu, silver carp, grass carp, etc; (3) Fishes which inhabit salt water conditions (sea or brackish water) do not breed in freshwater, such as *Chanos chanos, Lates calcarifer,* etc. and (4) Fishes which migrates from salt water to freshwater or vice versa for breeding, such as *Tenualosa ilisha,* etc.

The Indian Major Carps (IMC) generally breed in running water conditions of rivers. Natural spawning of carps occur in flooded sections of rivers during the monsoon months. In early seventies, the collection of fish seed* from rivers formed the bulk and accounted for 90% of the total spawn collected. However, fishes collected from natural sources has the following disadvantages :

(1) The seeds collected is a mixture of economic and uneconomic fish species, along with seeds of predatory species. It is difficult to separate them.

(2) Due to mixed quality seed the yield or production will be low and, therefore, the economic return will also be low.

(3) The diseased seeds may get mixed up with good quality seeds.

(4) The resource for seeds has become limited as collection spots remain localised in certain pockets.

(5) Transportation of live fish seed requires expertise and is expensive, and during transportation may result in mortality.

* Fish seed = Spawns, Hatchlings, Fry, Fingerlings, Juveniles, Yearlings, etc.

Procurement of stocking material

The initial phase of fish farming requires stocking of the required fish seed intended for farming. In India fish seed requirement is estimated at 80.20 crores fingerlings, of which 38.71 crores are collected from natural resources and rest 41.49 crores are produced at Fish Seed Farms. Fish seed can be procured by the following three ways :

I. **Riverine source**

II. **Bundh breeding**

III. **Induced breeding (Hypophysation)**

I. Procurement of seeds from riverine source

Rivers in India form a major source of collection of fish seeds during the monsoon. At one time (1964-65) the riverine seed collection contributed about 91.67% of India's total fish seed production. At present with the advent of aquaculture techniques and the rise of pollution in the rivers, the fish seed farms has taken the leading role in fish seed production.

The rivers acquiring a serpentine course, show bends and curves at various places (Fig. 6.11). The flowing water current of the river particularly during the monsoon, results in the erosin of the bank on one side, called the **Erosion zone**. Just opposite to it is the flat, gentle sloping bank, called the **Shadow zone**. Both these sites are unsuitable for fish seed collection. Good collection areas are located on the side of the sloping bank, at a place where the water current just diverges, casting fish seed to the sides. From such areas a large number of fish seeds are collected with the help of a **shooting net** (Fig. 6.12) which is a funnel-shaped net of finely woven netting. It is operated in shallow margins of the flooded river with its mouth facing the water current. At the cod end a ring of split bamboo or cane is stitched. To this a

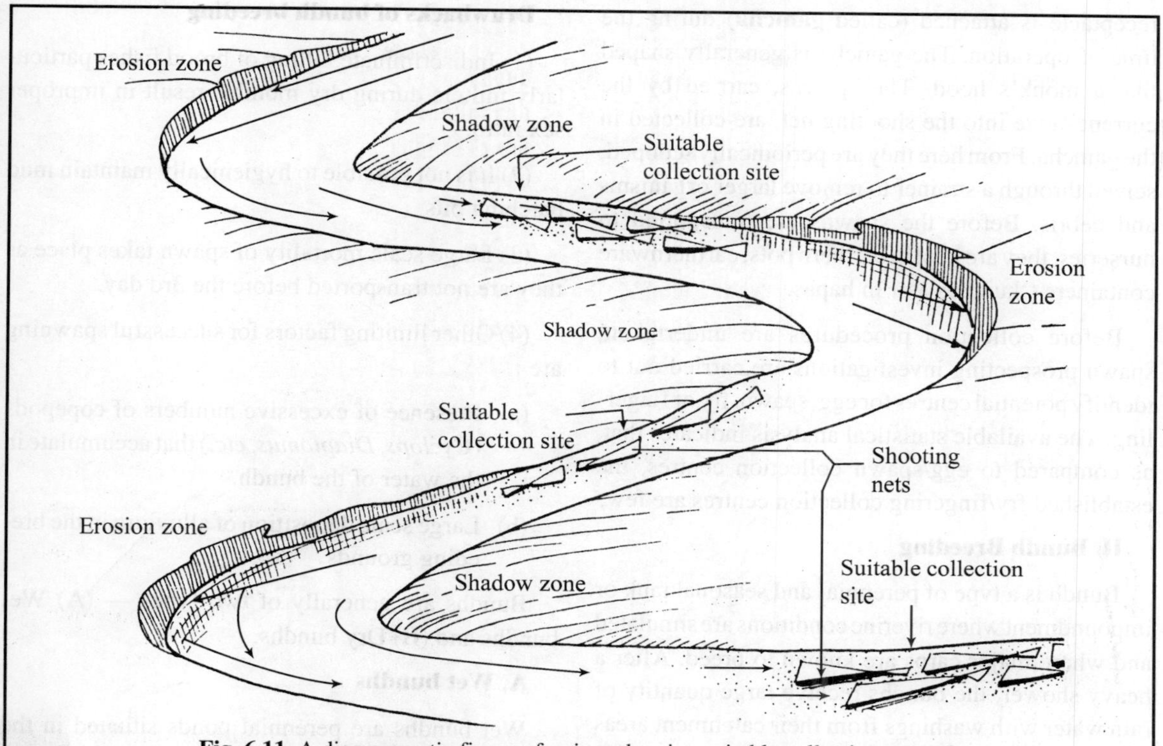

Fig. 6.11. A diagrammatic figure of a river showing suitable collection grounds

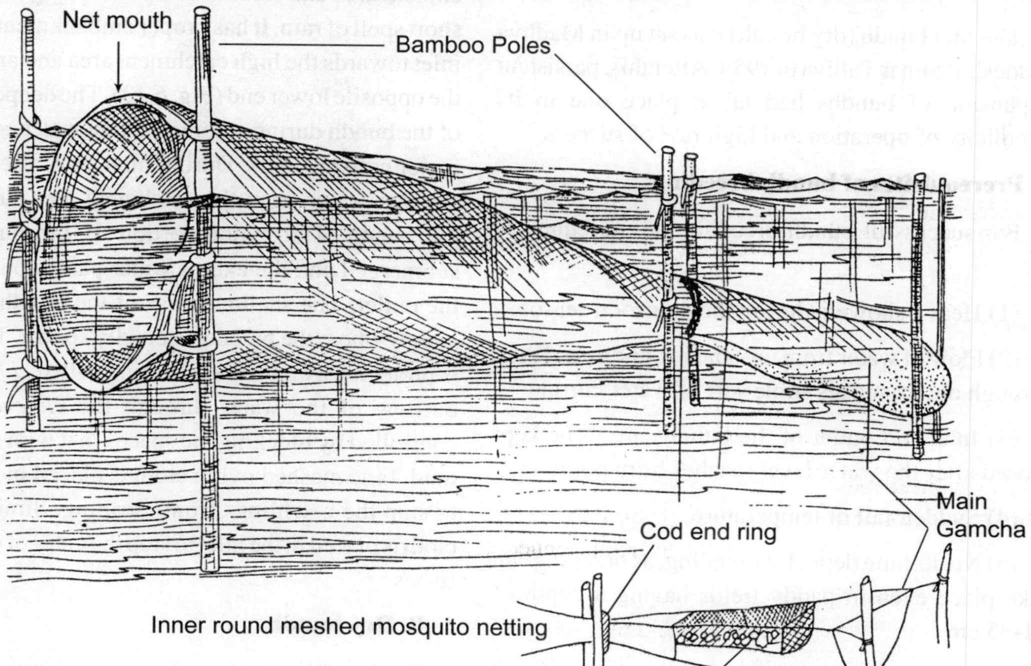

Fig. 6.12. Schematic diagram of a shooting net

receptacle is attached (called **gamcha**) during the time of operation. The gamcha is generally shaped like a monk's hood. The spawns, carried by the current move into the shooting net, are collected in the gamcha. From here they are periodically scooped, seived through a strainer to remove larger organisms and debris. Before the spawns are transported to nurseries, they are kept in earthern pots, earthernware containers ('hundies') or in hapas.

Before collection procedures are undertaken, spawn prospecting investigations are carried out to identify potential centres for egg, spawn, fry or fingerling. The available statistical analysis indicates that, as compared to egg/spawn collection centres, the established fry/fingering collection centres are few.

II. Bundh Breeding

Bundh is a type of perennial and seasonal tank or impoundment where riverine conditions are simulated and where major carps are known to breed. After a heavy shower, the bundhs receive large quantity of rain water with washings from their catchment areas and provide large shallow areas that serve as spawning grounds for the fishes.

The first bundh (dry bundh) was set up in Madhya Pradesh at Sonar Talliya in 1958. After this, persistent expansion of bundhs had taken place due to its simplicity of operation and high rate of success.

Prerequisites of bundh breeding

For successful bundh breeding the prerequisites are :

(1) Heavy showers during the monsoon season.

(2) Flow of water from up-land to bundhs and then through outlet, makes some sort of water current.

(3) Inlet and outlet of the bundhs are to be kept closed after the water level reaches brim point.

(4) Sudden fall of temperature.

(5) No definite depth for breeding, as breeding can take place even in paddy fields having a depth of 31-35 cm.

(6) Smaller fishes first get inducement to breed and the bigger ones are attracted later.

Drawbacks of bundh breeding

(1) Indiscriminate killing of brood fishes particularly milters during dry months, result in improper sex ratio.

(2) It is not possible to hygienically maintain mud hatching pits.

(3) Large scale mortality of spawn takes place as they are not transported before the 3rd day.

(4) Other limiting factors for successful spawning are :

(a) Presence of excessive numbers of copepods (*Cyclops, Diaptomus, etc.*) that accumulate in the water of the bundh.

(b) Large scale deposition of alluvium in the breeding grounds.

Bundhs are generally of two types — (A) Wet bundhs and (B) Dry bundhs.

A. Wet bundhs

Wet bundhs are perennial ponds situated in the slope of an undulating terrain. It provides a vast catchment area and fascilitates quick filling even with a short spell of rain. It has proper embankments with an inlet towards the high catchment area and an outlet at the opposite lower end (Fig. 6.13). The deeper portion of the bundh during summer, retain water containing major carp breeders. During monsoon, after a heavy shower, water from the catchment area rushes into the bundh. The major portion of the bundh gets submerged and the excess water passes out through the outlet. The shallow areas of the bundh is called *moans* where the breeders actually spawn. The outlet is protected by a bamboo fencing called *chheva*. The outflow of the water through the *chheva* can be controlled by blocking the spaces in it with straw and mud. Fine meshed nylon cloth is placed in the outlet to stop the hatchlings from escaping. Similar nylon cloth is placed in the inlet to stop the entry of unwanted fishes.

B. Dry bundhs

Dry bundh is a seasonal one, which remains more or less dry during greater part of the year. It is a

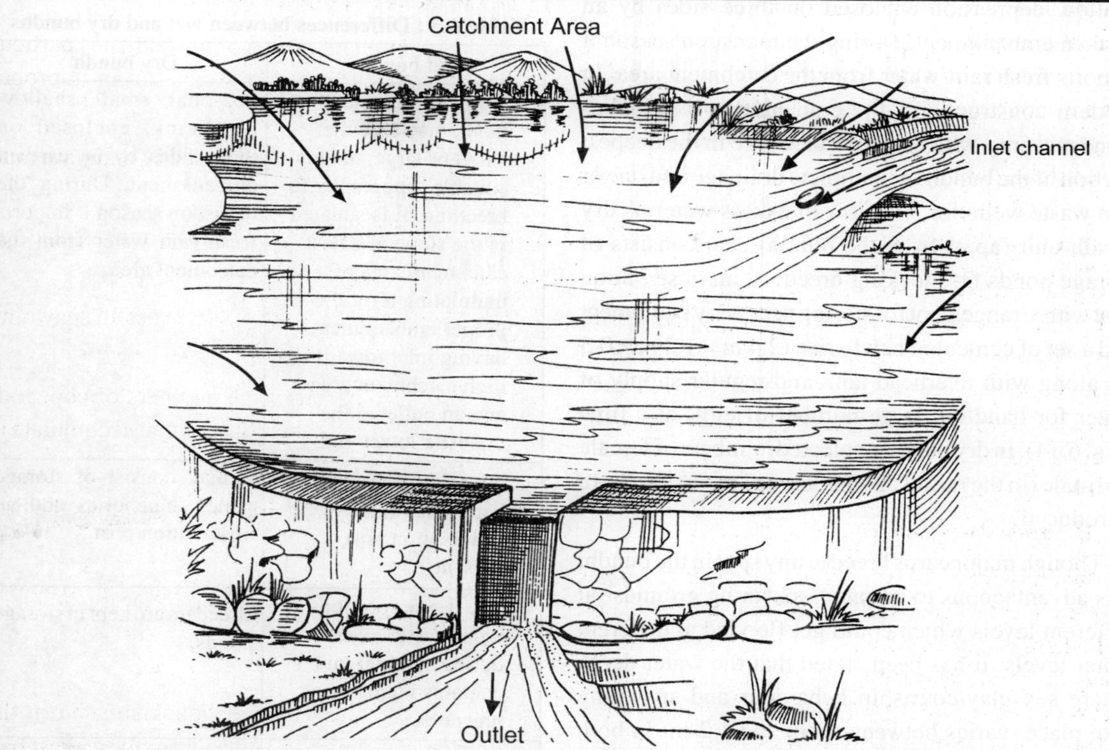

Fig. 6.13. A schematic representation of a wet bundh (arrows indicate catchment area)

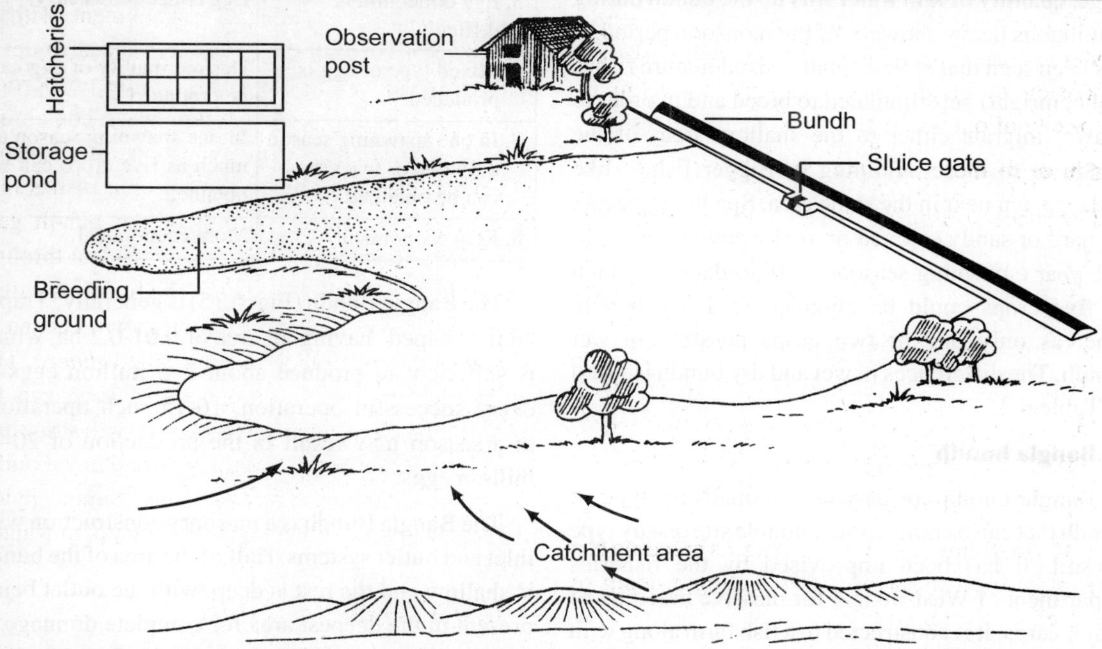

Fig. 6.14. A schematic representation of a farm with dry bundh and having provisions for observation post, hatcheries and storage pond

shallow depression enclosed on three sides by an earthen embankment. During the monsoon season it imports fresh rain-water from the catchment area. In modern constructions, the embankment is a pucca stone masonry with a small sluice gate in the deepest portion of the bundh for complete drainage and one or two waste weirs for overflow of excess water. A dry bundh unit (apart from the bundh) also consists of storage ponds for stocking breeders, an observation post with arrangement for storing necessary equipment and a set of cemented hatcheries (2.4 m × 1.2 m × 0.3 m) along with overhead tank and regular supply of water for handling large number of eggs at a time (Fig. 6.14). In dry bundhs, selected numbers of female and male (in the ratio 1 : 2) breeders of major carps are introduced.

Though major carps breed at any spot in the bundh, it is advantageous to prepare 'spawning grounds' at different levels which could get flooded at different water levels. It has been stated that the water depth where sex play/courtship behaviour and spawning take place, varies between 8 cm and 1.2 m. In both wet and dry bundhs, spawning usually occurs after a large quantity of rain water fills up the bundh during continuous heavy showers of the monsoon period. It has been seen that at first smaller sized mature fishes (rohu, mrigal) get stimulated to breed and in order to spawn, migrate either to the shallow areas of the bundh or to those adjoining it. Bigger fishes, like catla, spawn next in the same area. Spawning occurs on hard or sandy soil and on rocky embankment. In one year (spawning season of 30-40 days) as much as five crops could be obtained in a dry bundh, whereas only one or two crops possible in wet bundh. The differences in wet and dry bundh is listed in Table 6.3.

Table 6.3 : Differences between wet and dry bundhs

Wet bundh	Dry bundh
1. Perennial impoundment, may be either small or large, having suitable topography for breeding. It is situated in the slope of a vast catchment area of undulating terrain with proper embankments having inlet towards the high catchment area and an outlet at the opposite lower end.	Seasonal, small, shallow pond/tank, enclosed on three sides by an earthen embankment. During the monsoon season it imports fresh rain water from the catchment area.
2. No storage ponds, hatcheries and observation post present.	It may consist of storage ponds, hatcheries and an observation post.
3. The deeper portions retain water even during summer and contains major carp breeders.	Breeders are kept in storage ponds.
4. It is difficult to manage.	It can be easily managed.
5. Egg collection is difficult.	Egg collection is easy.
6. Mixed type of egg is produced.	Desired quality of egg can be produced.
7. In one spawning season only one or two crops can be obtained.	In one spawning season as much as five crops can be obtained.
8. Less economical.	More economical.

Bangla bundh

Bangla bundh are basically a small dry type of bundh that can be built up at a suitable site on any type of soil. It has been improvised by the fisheries department of West Bengal for massive breeding of major carps. It is constructed in a fish farm along with breeding pool and chinese hatchery, where massive induced breeding is undertaken.

The Bangla Bundh (Fig. 6.15) is generally 'Trapezoidal shaped' having an area of 0.01-0.2 ha, which is sufficient to produce about 2–5 million eggs in every successful operation. 10-15 such operations in a season may result in the production of 20-40 million eggs.

The Bangla Bundh is a masonry construction with inlet and outlet systems. Half of the area of the bundh is shallow and the rest is deep, with the outlet being present in the deepest area for complete drainage of water. The water depth of the shallow zone is maintained at 30-50 cm and that of the deeper zone at

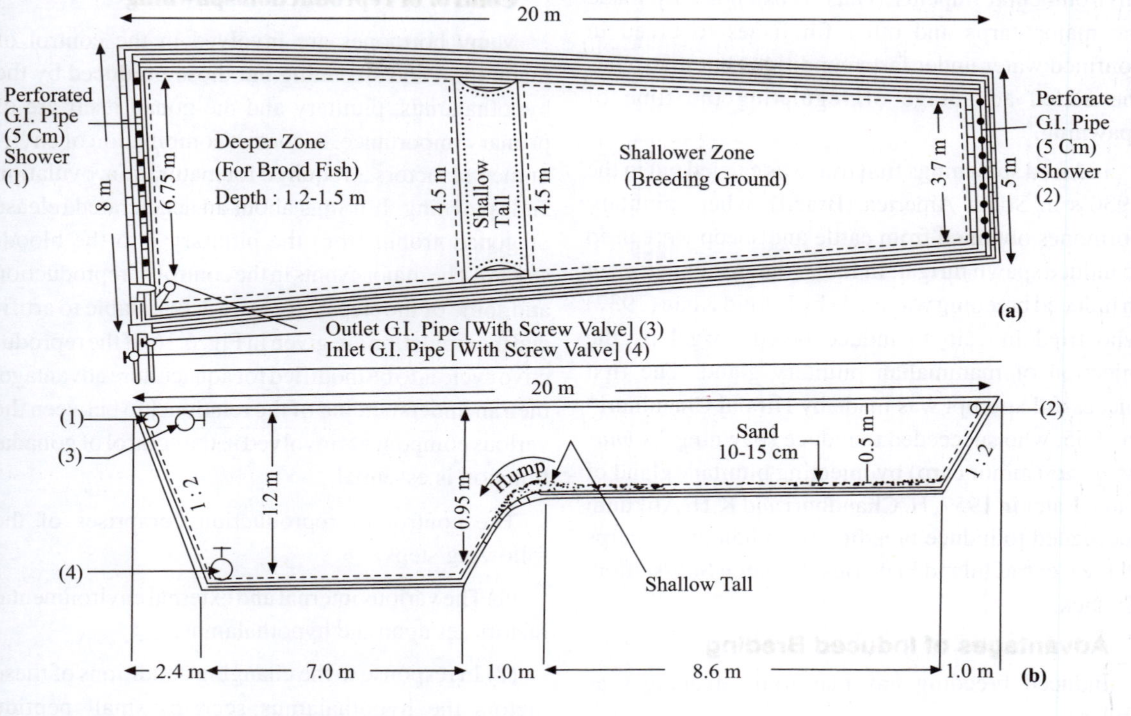

Fig. 6.15. A schematic diagram of a typical **Bangla Bandh** — (a) Top view and (b) Transverse sectional view

80-100 cm. The transverse and longitudinal slopes of the tank floor are so constructed so as to create and maintain a feeble water current just like that of a river. The water supply to the bundh is through a tubewell which is pumped into an overhead tank. A showering system is present for oxygenation and cooling the water.

Prior to the introduction of water into the bundh, a thin layer of fresh sand is uniformly spread over the shallower zone to facilitate natural breeding by simulating riverine conditions. A bump separates shallower zone from the deeper one and prevents escape of sand from the breeding ground to the deeper zone. 50 kg of spawner can be introduced in the bundh at a time. The deeper zone of the bundh gets filled first with clean water of the overhead tank. The brooders after the 2nd injection of the spawners (female) and the only injection of the mitters (males), are released into this deeper zone. The running water condition and the showering system are then maintained for about 3-4 hours. After this the males start chasing the females and they move into the shallower areas of the bundh. The male and female are seen to bend their bodies around each other during their courtship behaviour. This act exerts pressure on the abdomen of the mating pair resulting in the extrusion of ova and exudation of milt. The spent fishes then move to the deeper areas of the bundh. Fertilisation takes place and the fertilised eggs are left for 4 hours for water hardening. They are then collected with the help of a *chot jal* into a hundi and then transferred to the chinese hatcheries.

III. Induced Breeding

Many cultural species of fin fishes (particularly the major corps) under farm culture conditions, do not get the required environmental impetus for spontaneous maturation. This has led to the development and standardisation of a technique, called **induced breeding** or **hypophysation**. Here, through the injection of pituitary homogenates or extracts, the natural gonadotropin surge is stimulated, disregarding the environmental impetus. Thus, it has not only made the major carps and other fin fishes to breed in confined water under farm conditions but it also has the added advantage of regulating the time of spawning.

Induced spawnings first trial was carried out in the 1930's in South America (Brazil), where pituitary hormones obtained from cattle and sheep were used to induce spawn mrigal. In India, the first experiment in induced breeding was made by Hamid Khan (1937) who tried in vain to induce breed mrigal by the injection of mammalian pituitary gland. The first successful attempt was made by Hiralal Chaudhuri* in 1955, who succeeded to induce spawning *Esomus dauricus* (minor carp) by injecting pituitary gland of catla. Later in 1957, H. Chaudhuri and K.H. Alikunhi succeeded to induce breeding of Indian major carps at the Central Inland Fisheries Research Sub-station, Cuttack.

Advantages of Induced Breding

Induced breeding has manifold advantages as under :

(1) Pure and disease-free spawn of a desired variety of fish under cultivation can be obtained through induced breeding.

(2) Induced breeding ensures availability of seeds for fish culture at any time (except during the winter months) and do not have to depend upon the monsoon season only for its collection, like that of natural sources.

(3) The demand for any specified quality of pure fish seed of a particular species can be met through induced breeding.

(4) The same fish can be made to breed twice in one year.

(5) The technique of induced breeding is very simple and can be easily handled by a layman without much training.

(6) The cost incurred for production of a spawn from induced breeding is comparatively lower than that of obtaining from natural sources.

* Hiralal Chaudhuri is recognised as the "Father of Induced Breeding". As a mark to honour the great scientist, the Government of India declared July 10 (his birthday) as **'Fish Farmers Day'**.

Control of reproduction/spawning

Many hormones are involved in the control of maturation and spawning but those produced by the hypothalamus, pituitary and the gonad itself are of primary importance. A correct combination of environmental factors is required for maturation, ovulation and spawning. It brings about an accelerated release of gonadotropin from the pituitary into the bloodstream. The major events in the control of reproduction and some of the steps which are accountable to artificial manipulation are given in Fig. 6.16. If the reproductive cycle is to be modified for aquaculture advantage, then an understanding of the relationship between the various components involved in the control of gonadal function is essential.

The control of reproduction comprises of the following steps :

(1) The various internal and external environmental factors act upon the hypothalamus.

(2) In response to the changing conditions of these factors the hypothalamus secretes small peptide hormones, called releasing hormones.

(3) These peptides pass a short distance to the ventral surface of the brain, where they control the activity of specific gonadotropic cells in the pituitary gland.

(4) These cells then secrete gonadotropic hormones which are carried by the blood to the gonad of the fish.

(5) Here, these cells control all the structural and functional changes in the testes and ovary.

(6) One important influence of gonadotropic hormones is the induction of hormone secretions from the gonads, which either act independently and/or in concert with the gonadotropic hormones which directly control all the aspect of gonadal growth and maturation.

(7) In female fishes two main ovarian hormones are produced — (i) *Oestrogens,* which include oestradiol – 17 β and oestrone, and (ii) *progestagens,* 17α-20β-dihydroprogesterone and 17α-hydroxyprogesteron.

(8) In males, the major testicular hormones produ-

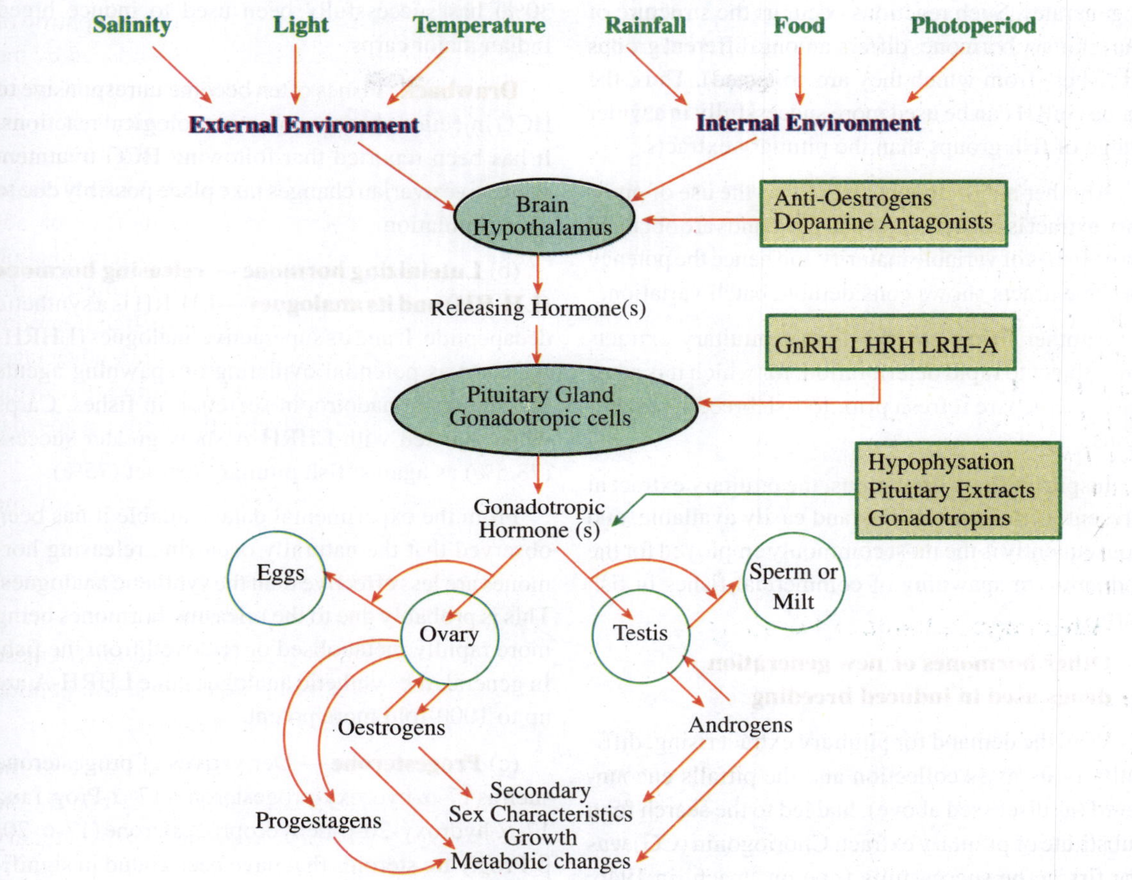

Fig. 6.16. The different factors involved in reproduction and the hormones used to induce breed artificially

ced, referred to as *androgens*, are testosterone and 11-ketotestosterone.

(9) All of the above gonadal hormones have a steroid structure which is same or similar to that of the corresponding hormones of higher vertebrates.

Fishes of commercial importance vary in their reproduction strategies and tactics. However, despite their diversity, all of them have their reproductive cycles controlled by the same cascade of endocrine controls (as discussed above). By hormonal administration and environmental manipulation, fish culturists have been able to alter the times of spawning of a wide range of fish species for commercial advantage.

Drawback in the use of pituitary hormones over the gonadotropin releasing hormones

The use of gonadotropin releasing hormones (GnRH) to induce spawning has gathered interest, particularly as these peptide molecules can be easily synthesised artificially. The relatively small size of the GnRH means that these materials do not provoke the immunological reactions which generally occur in recepient fishes in response to the injection of pituitary extracts or gonadotropic hormones. Such reactions involve the development of antibodies against the gonadotropic principles injected. It considerably reduces the effectiveness of subsequent hypophysations. In some recepient fishes the injected extract is totally inactivated by the antibody reaction it generates. Such reactions occur as the structure of the pituitary hormones differs among different groups of fishes (from which they are collected). Thus, the same GnRH can be used more successfully in a wider range of fish groups than the pituitary extracts.

Another major drawback as far as the use of pituitary extract is concerned, is that the glands are obtained from fishes of variable maturity and hence the potency of the extracts shows considerable batch variation.

Another inconvenience is that pituitary extracts are subject to rapid deterioration, for which it is advisable to prepare it fresh prior to fish breeding operations.

Inspite of these limitations, the pituitary extract at present, is the most widely and easily available, and consequently is the most commonly employed for the induction of spawning of commercial fishes in fish farms.

Other hormones or new generation drugs used in induced breeding

With the demand for pituitary extract rising, difficulty in its mass collection and the pitfalls encountered (as discussed above), had led to the search for a substitute of pituitary extract. Choriogonin (CG) was the first to be successfully tried on 'loach' in 1946. Other hormone preparations have also been used to induce spawning fishes, which includes mammalian hormones like human chorionic gonadotropin (HCG) or the hypothalamic gonadotropin releasing hormones (LHRH, GnRH) and their synthetic analogues (LRH-A/LHRH-A).

Fishermen in India have found *Synahorin* (a mixture of mammalian hormones and mammalian pituitary extracts in combination with fish pituitary extracts (FPE) to give positive results. In Indian market HCG is sold under the name *'Sumaach'*.

(a) **Human Chorionic Gonadotropin (HCG)**— Among all the mammalian gonadotropins, HCG has given the best result. HCG is widely used as it has an edge over pituitary gland (PG) mainly because of its easy to acquire, relatively cheap, long shelf-life and of known and consistent potency. HCG alone or in combination with fish pituitary (HCG 70% + PG 30%) has successfully been used to induce breed Indian major carps.

Drawback: Fishes often become unresponsive to HCG injections because of immunological reactions. It has been reported that following HCG treatment regressive ovarian changes take place possibly due to overstimulation.

(b) **Luteinizing hormone — releasing hormone (LH-RH) and its analogues**— LH-RH is a synthetic decapeptide. It and its superactive analogues (LHRH-A) serve as potential ovulating or spawning agents that induce gonadotropin secretion in fishes. Carps when injected with LHRH-A show greater success (78.5%) as against fish pituitary extract (75%).

From the experimental data available it has been observed that the naturally occurring releasing hormones are less effective than the synthetic analogues. This is probably due to the releasing hormones being more rapidly metabolised or removed from the fish. In general, the synthetic analogues like LHRH-A are up to 1000-fold more potent.

(c) **Progesterone** — Derivatives of progesterone such as 17-α-hydroxy-progesterone (17-α-Prog.) and 17-α-hydroxy-20-β-dehydroprogesterone (17-α-20-β-Prog.) are steroids that have been found in significant quantities in the blood plasma of maturing fishes. Injection with these steroids can bring about final maturation and ovulation in fishes. Therefore, progesterone can partially replace gonadotropin.

(d) **Antiestrogens** — These are synthetic, non-steroidal compounds (such as Clomiphene citrate and Tamoxifen) that are capable of competing with estrogen for binding sites on estrogen receptors. They are used for inducing ovulation. Antiestrogens inhibit gonadotropin secretion, when used in excessive dosage. It may work better in those fishes where vitellogenesis is complete and blood estrogens are high.

(e) **Ovaprim (LINPE method)** — The Chinese researchers have developed a technique, called the 'Linpe' method, where the female fish is induced to ovulate by injecting them with a combination of LHRH-A and the drug **domperidone**. The drug inhibits the action of dopamine (a substance produced by the fish that inhibits ovulation). Therefore, stimulating the sex organs of the fish.

Syndel International, a Canadian company, developed a single product, known commercially as *Ovaprim*, that contains both the hormone and the dopamine antagonist. Ovaprim has been marketed in India by 'Glaxo India' since 1990. Ovaprim has been tested successfully on a wide range of fresh, salt and brackish water species. It has been found to be very fruitful in the aquarium fish industry in India.

Advantages of ovaprim

(1) Although ovaprim is costly (almost double the cost of pituitary), yet it has the advantage of known potency and assured breeding response. Therefore, it reduces the loss incurred due to spawning failure resulting from poor quality of glands.

(2) Ovaprim being in ready-to-use form, appears to be more convenient and can be easily adopted in carp seed production.

(3) Rates of spawning, fertilization and hatching and survival has been reported to be much higher than pituitary extract.

(4) Refrigeration of the solution is not required and it has a long shelf life.

Induced Breeding Techniques

A. Fish pituitary gland—collection and preservation

(a) **Fish pituitary gland** — Pituitary gland of a fish is small, soft bodied and whitish in colour. In carps it is more or less round in shape. It lies on the ventral side of the brain behind the *optic chiasma*. It is placed in a concavity on the floor of the brain box, called *sella turcica*, and is covered by a thin membrane, called *duramater*. In some fishes the pituitary is attached to the brain by a thin stalk, the *infundibular stalk* and is called **Leptobasic**. In other fishes the gland is without stalk and is called **Platybasic**.

(b) **Donor fish** — Donar fishes are those from whom pituitary glands are collected. Pituitary glands should be collected from fully ripe gravid fishes, as the gland is most potent at the time of breeding or just before spawning. Recently killed or live fishes are supposed to be the best donors for the collection of pituitary glands. However, glands can also be collected from donor fishes which are well preserved in ice or kept in cold storage.

Certain points should be noted at the time of pituitary gland collection :

(1) The potency of the gland decreases after spawning. So glands should not be collected from immediately spawned fishes. However, glands from induced-bred carps immediately after spawning are found to be fully potent and can be used for further hypophysation work.

(2) Glands collected from immature or spent up fishes usually do not give satisfactory results.

(3) Glands should not be collected from rotten or semi-rotten fishes.

(4) At the time of collection, care should be taken not to damage the gland as the torn gland loses its potency.

(5) Another big problem is that pituitary glands are generally collected from fishes of variable maturity and hence the potency of the extracts shows considerable batch variation.

(6) There is no qualitative difference in the pituitary gland of male and female donor fishes.

(c) **Collection of pituitary gland**—Pituitary gland is located in the floor of the brain box and can be collected by one of the following ways :

(i) **By cutting and removing a portion of the brain-case**—The skull or brain-case above the brain is cut with the help of a sharp butcher's knife. The cut scalp is then removed. The grey matter and fatty substances over the brain are removed. Then, with the help of a tweezer, the olfactory and optic nerves at the interior part of the brain are detached. The entire brain is then lifted up and laid back. The pituitary gland (Fig. 6.17) is then located at the floor of the brain, ventral to the optic chiasma. The gland is then carefully picked up (without damaging it) by a pair of fine tweezer and kept in a dark bottle containing absolute alcohol.

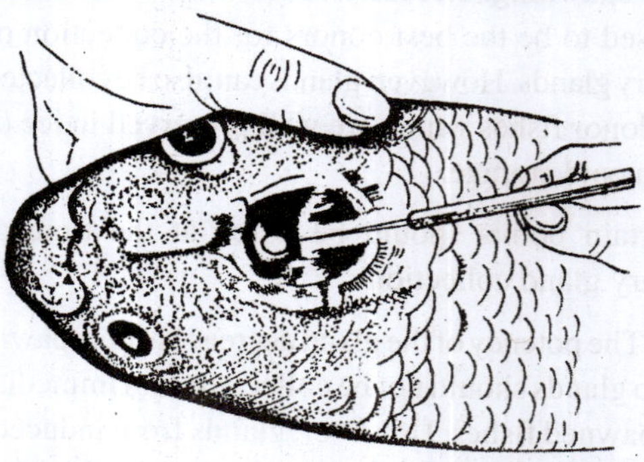

Fig. 6.17. Carp head with pituitary gland fully exposed after removal of cranial bone and brain

Collection of the gland by this method results in damaging of the fish head which would then be a problem of being sold and thus, the cost of collection of a single gland would be very high.

(ii) **Collection of gland through the foramen magnum** — By this method pituitary gland can be collected from the cut fish head. In the fish market where cut fish heads are available, the foramen magnum is first exposed by removing vertebral parts adhering to the skull by the help of a sharp knife. The brain lies on the ventral side of the foramen. The grey matter and fatty substances over the brain is then removed by the help of blunt end of the forcep. Care is taken so that the brain is not detached or damaged. With the help of the forceps, the anterior end of the brain is then detached. The brain is carefully lifted out of the foramen magnum. The pituitary gland is then located and picked up carefully by the help of tweezers.

This method is easier and less time-consuming. It is also more economical as the cut head can be sold after the removal of the gland, as much damage has not been incurred.

(d) **Preservation of pituitary gland**—For induced spawning, pituitary glands can be used in the fresh state, but it is generally preserved for future use. The gland may be immediately frozen and stored in frozen condition in a refrigerator. Otherwise, it can be preserved by one of the following ways :

(i) **Preservation in absolute alcohol**—Immediately after collection the glands are put in absolute alcohol in a dark coloured bottle. After 24 hours, the glands are washed with absolute alcohol and kept in a phial with fresh absolute alcohol. Pure absolute alcohol must be used because the hormones are soluble in water. Absolute alcohol dehydrates and defattens the glands. Occasional changing of the alcohol helps in proper dehydration and defattening of the glands and keep them in good condition for a long period. The phial is made air-tight and kept in a cool, shady place at room temperature. It may be kept inside a dessicator containing anhydrous calcium chloride.

Preservation of pituitary gland in absolute alcohol is preferred in India as very few fish-culturists can afford to have a refrigerator. Moreover, alcohol-preserved glands have given positive results than acetone-dried glands when used on Indian carps.

(ii) **Preservation in acetone** — The 'Acetone-Dried Pituitary' is largely in use in Russia and U.S.A. where refrigeration facilities are common. It is available from a number of reliable suppliers in the U.S.A, Israel and Germany. Immediately after collection of the pituitary gland, it is put in acetone containing, phials. The phials are then kept in a refrigerator or placed on dry ice. The acetone should be changed several times over a 24-hour period to ensure that the glands have been properly dehydrated and defattened. After 36 hours of storage in the refrigerator, the phials are taken out and kept at room temperature. After some time the glands are taken out of acetone and dried on a filter paper at room temperature. It is then stored either intact or after being grounded to a powder, in sealed air tight sterile phials, in a refrigerator or preferably in a –20°C freezer. Glands stored under these conditions retain their potency even after 15 years.

B. Preparation and preservation of pituitary extract

The pituitary extract is then prepared from the gland for injecting it into the recipient fish (breeders). The extract can conveniently be prepared well in advance and used when needed, thus saving time and trouble of preparing it before every induced breeding operation.

(a) **Preparation of pituitary extract for injection** — The quantity of gland required for injection is first calculated from the weight of the breeders to be injected. The required quantity of the glands are then taken out of the preserved phials and kept on a piece of filter paper (1-2 minutes) for drying or evaporation of the alcohol. On the other hand, acetone-dried glands are straight away taken from the phials for maceration.

Appropriate amount of the entire or dry powdered glands are weighed. The segregation and mixing of many glands of variable gonadotropic potency has the advantage of producing a single extract of consistent potency. The glands are then taken into a tissue homogeniser with a little quantity of distilled water or 0.3%* saline solution (for I.M.C.) and thoroughly homogenised. Further amount of the same solvent is then added to achieve a concentration of 20-50 µg of dried pituitary material per ml of the solvent. The amount of the pituitary gland per ml of the media may be increased or decreased according to the requirement of the suspension to be injected. It is suggested that a maximum of 1.0 ml and a minimum of 0.1 ml suspension should be injected to a breeder at a time.

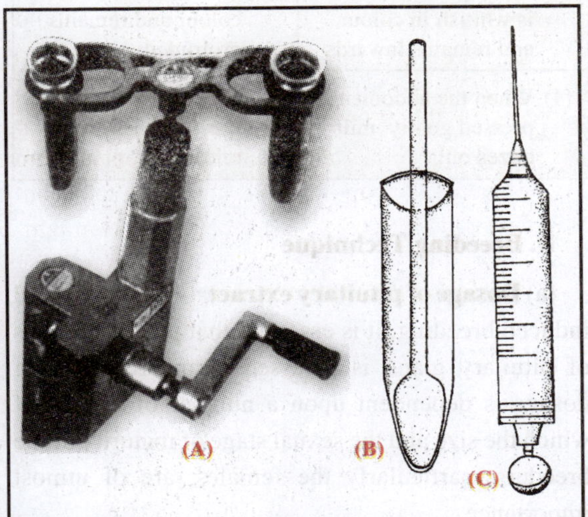

Fig. 6.18. Materials used for pituitary extract preparation and injection. (A) Homogeniser (B) Hand centrifuge (C) Hypodermic syringe

* Different strengths of common salt solution have been used in different fishes.

The macerated gland suspension is then poured into a centrifuge tube (Fig. 6.18) and centrifugation is done at 3000 revolutions per minute (rpm). The suspended particles of the crushed tissue and other solid particles settle down at the bottom of the centrifuged tube and the clear solution containing the hormones is then decanted into a small beaker to be taken into a hypodermic syringe for injection.

(b) **Preservation of pituitary extract** — Preservation of pituitary extract can be done in glycerine. In this method, a known quantity of pituitary glands, after homoginising in distilled water, is diluted to 1/3 of the total volume of the extract intended to be made with water. The suspension is kept in refrigerator for 24 hours. Pure glycerine is then added to the extract to make up the intended volume. For example, 3 ml of the extract will have 1 ml of distilled water and 2 ml of glycerine. The extract with glycerine is again kept in the refrigerator for another 24 hours, after which it is centrifuged. The supernatant fluid is then transferred into ampoules and sealed for future use. The ampoules should be properly labelled with the details of hormone concentration and date of preparation.

Glycerine preserved extracts must be prepared one or two months prior to the onset of the breeding season. It is found to retain its potency for about 1 year or more if properly refrigerated.

C. Collection, Care, Identification and Selection of Breeders

(a) **Collection and care of breeders** — The primary requirement for fish breeding work is collection and stocking of healthy recepient fishes or breeders. For successful breeding, selection of healthy breeders is essential. Generally 1.5 to 5 kg weight of Indian Major Carps are preferred for economical breeding. The breeders may be collected from any water resources, such as reservoir beel, jheel, canal, tanks, fish-farms, etc. They are stocked at the rate of 500-1000 kg of fish per acre or 1250-2500 kg per hectare water area.

If the breeders are collected during winter season, they should be stocked in a large tank and prior to the onset of breeding they should be sex-wise segregated.

If they are collected a few days or a month prior to the commencement of breeding, then they are stocked sex-wise in two sepearate tanks, called segregation tanks*. Segregation of the sexes prior to injection (although not essential) increases the affinity of maturing during the breeding procedure and gives encouraging results.

For maintaining healthy growth, the breeders are fed with artificial feed, such as powdered cotton-seed, rice bran, oilcakes (ground-nut, mustard, etc.) or a mixture of oil cakes and rice bran. They are generally fed daily for 15-30 days before injection at the rate of 10% of their body weight. During netting, proper care should be taken and rough handling should be avoided as it may cause injury to the breeders. Periodic examination of the breeders is done to see the condition of the gonads and parasitic infections, if any. In case of any infection, the breeders are dipped for half an hour in a solution of 1 ppm** potassium permanganate or acriflavine and then released back into the pond.

(b) **Identification of male and female breeders** — Although fishes are sexually dimorphic, it is difficult to differentiate their sexes morphologically in most fishes. In carps, the secondary sexual characters are not very prominent and conspicuous by which the sexes can be distinctly identified. However, some morphological characters develop during the breeding season in both male and female IMC which serve as an identification key. These characters listed in Table 6.4 are more or less the same in all the species of various genera of carps.

(c) **Selection of breeders** — Success in induced breeding largely depends upon proper selection of suitable male and female breeders. Selection of male brood fish is easy, as a fully ripe male oozes milt freely on being gently pressed at the belly. On the other hand, selection of a suitable female is a major difficulty with hypophysation. The identification characters such as tubercles or roughening of the surface of the body, changes in colouration, softening and rounding of the abdomen and reddening and protrusion of the anal papilla and vent have all been used, but each is notoriously unreliable. The most reliable character is taking the egg and examining it under the microscope. Pituitary extract injections should only be given to those female fishes in which the nucleus is on the periphery of the egg shell. Less mature females should be returned to the holding ponds or stocking tanks for re-examination at a later date. However, such practices in India is not possible and the farmers have to relay on the secondary sexual characters (Table 6.4).

Table 6.4 : Characters for identifying the sex of mature Indian Carps

Male Fish	Female Fish
(1) The inner surface of pectoral fin is rough	(1) The inner side of pectoral fin is smooth
(2) Narrow abdomen	(2) Bulging and soft abdomen
(3) Vent (genital region) is whitish in colour and remains inwards	(3) Vent is pinkish in colour and remains protruded
(4) When the abdomen is pressed gently, milt oozes out	(4) When the abdomen is pressed eggs are released

D. Breeding Technique

(a) **Dosage of pituitary extract** — For successful induced breeding, it is essential that proper dosages of pituitary gland is assessed. Standardisation of dosage is dependent upon a number of factors of which the size and the sexual stage of maturity of the breeders, particularly the females, are of utmost importance.

The dose of the pituitary injection to be given is calculated in relation to the weight of the breeders to be injected. It has been found that administration of a preliminary low dose in the female breeders followed by a higher effective dose after 6 hours has proved

* **Segregation tanks** are of regular shape with a gradual slope and a plane, flat bottom. The bottom is so made so that no problem is faced at the time of netting and all the breeders can be netted easily. It is a small tank with 25-35 metres breadth and 1.5 to 2 metres depth. It has provision for inlets and outlets so as to help in adjusting its depth of water.

** ppm means parts per million or mg per litre (mgl^{-1})

more successful than a single knock-out dose*. On the otherhand, the male spawners are far more tolerant of the farm envionment than that of female fish. For this reason male fishes require much lower doses of pituitary material for the induction of spawning by hypophysation.

The female breeders of IMC are given a preliminary dose at the rate of 2-3 mg of pituitary gland per kg body weight. After 6 hours, the females are given a second higher dose of 5-8 mg per kg body weight. During this time the males receive their only dose, at the rate of 2-3 mg per kg body weight.

A third dose of injection is generally not recommended. However, if the condition of the female breeder is found suitable and also the weather condition, then a third injection may be given. It is given 12 hours after the second injection at the rate of 4-7 mg per kg body weight to the females. In case, it is decided not to give the third injection, then the breeders are released back into the stocking pond.

The pituitary extract dose given to the exotic carps is slightly higher than that given to IMC. The females of the exotic carps are usually given an initial dose of 3-4 mg per kg body weight, followed by the second dose of 7-10 mg per kg body weight at an interval of 6 hours. At this time the males are given the only dose of 3-6 mg per kg body weight.

(b) **Injection methods** — Different methods of injection are in practice, like (i) Intra-muscular, (ii) Intra-peritoneal and (iii) Intra-cranial. However, intra-muscular injections are more commonly practised by fish-breeding workers.

In India both intra-muscular and intra-peritoneal injections have been tried but the former method has been found to be more effective and easier to follow. Intra-muscular injections are usually administered in the region of the caudal peduncle or near the shoulder region. Care must be taken to avoid the lateral line area. In case more than one injection is to be administered, it is better to give the second injection on the other side (i.e. the side which did not receive the first injection).

The most convenient hypodermic syringe used for injection purpose is a 2 ml one having 0.1 ml graduations, preferably with a locking arrangement for convenient use. The size (thickness and length) of the needle for the syringe depends upon the size of the breeders to be injected. Generally B.D.H. needle No.22 is used for 1-3 kg carp breeders, needle No. 19 for larger (above 3 kg) and No. 24 for smaller (below 1 kg) ones.

(c) **Injecion technique** — For giving injection to the breeders, the requisite quantity of gland suspension is taken in a hypodermic syringe. The brood fishes are then taken out of the hapa, where they were sex-wise segregated, and weighed. They are then brought one by one for injection and placed on a small field table provided with a cushion. The cushion is made up of wet pieces of nets, hapas, cloth, etc. Two persons are required at the time of injection; one of them holds the head of the fish pressed gently against the cushion, while the second presses the tail with one hand and with the other gives the injection on the caudal peduncle area (Fig. 6.19). For intra-muscular injection, the needle is first inserted under a scale parallel to the body of the fish and then pierced into the muscle at an angle of 45°, so that no damage is done to any of the scales.

The entire operation is done close to the water body where they would be introduced after the final injection. The breeders, after the final injection are immediately released into the breeding hapa (Fig. 6.20) or Bangla Bundh (Fig. 6.15) or breeding pool (Fig. 6.21).

(d) **Injection time** — Injection may be given at any time of the day or night, as spawning can occur at any time. But since a low temperature is helpful for spawning, injection may be given on a cool day or in the evening or night when the temperature is fairly low.

(e) **Spawning** — In case where a single high dose is administered, spawning generally takes place about 6-9 hours after the injection.

* **Knock-out Dose** —'Knock-out' Dose is a single high dose, sometimes given to the females of Indian and Chinese Carps, particularly during the later stages of breeding season when the recipient fishes are found to be in a difficult situation to breed. The 'knock-out' dose given to the females of IMC is 10-14 mg/kg body weight and that for the ripe females of silver and grass carps is 15-20 mg/kg body weight.

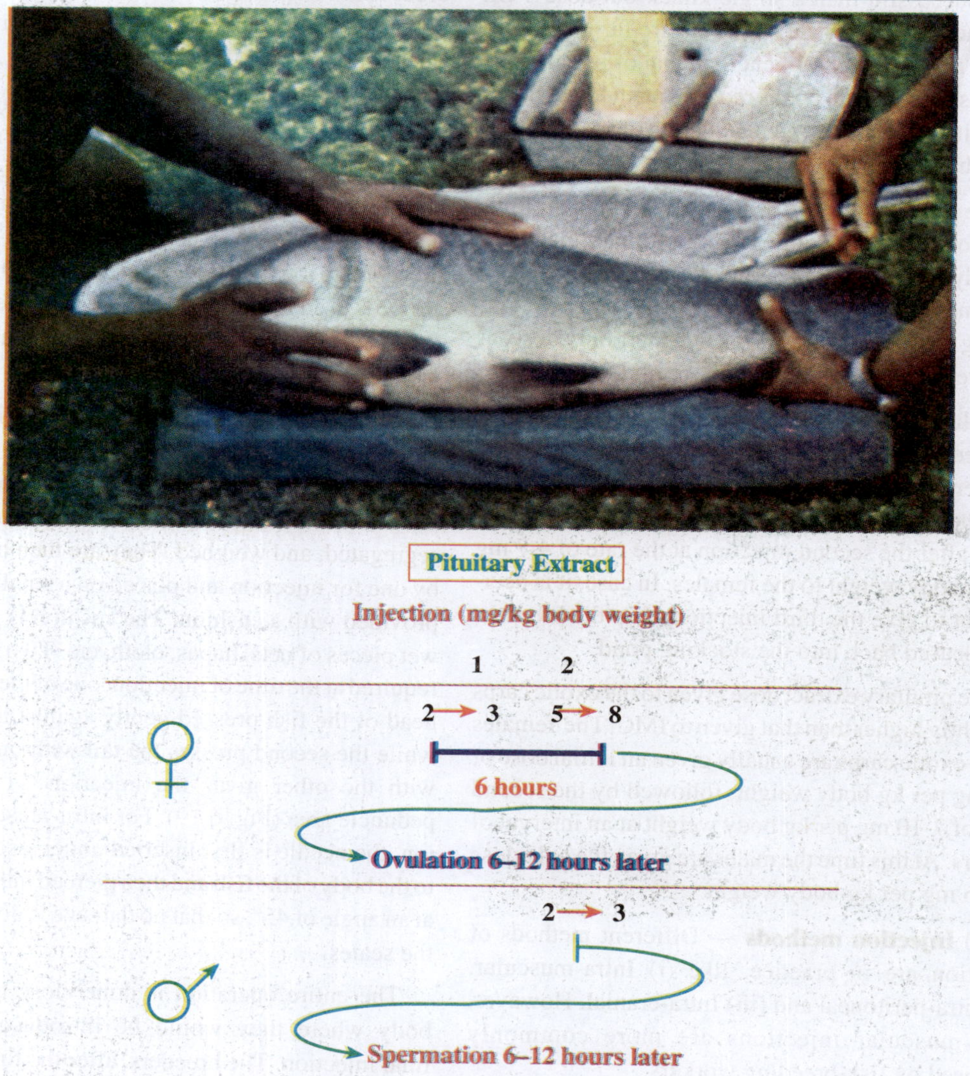

Fig. 6.19. A brood fish being injected with pituitary extract with the dose as administered above

In case of two injections (preliminary and a second dose to the female, at an interval of six hours and only a single dose to the males) spawning occurs within 3-6 hours after the second injection.

A few hours after the final injection **sex play** between the female and male is noticed. This is best observed in the Bangla Bundh as described earlier. Spawning normally is influenced by low temperature (25-30°C), rain water, slight drizzling and cool weather. The average fecundity rate of IMC has been estimated to be about 2 lakh eggs per kg body weight.

(f) **Fertilised eggs** — As the female liberates the eggs in a stream, the male oozes its milt and the eggs thus gets fertilised. After fertilisation the eggs swell up considerably within 10-15 minutes, owing to the absorption of water through the vitelline membrane. The fertilised eggs become crystaline transparent and non-adhesive with a reddish tinge in catla, pale reddish in rohu, light pale brownish in mrigal and pale bluish

Fig. 6.20. A breeding hapa constructed in a pond, showing a portion of the upper horizontal flap open for insertion of brood fishes after injection

Fig. 6.21. A breeding pool

in calbasu. The sizes of the fully swollen eggs of the four IMC's vary within 2.5 mm to 6.5 mm in diameter.

The unfertilised eggs also swell up in water but its colour becomes opaque whitish and their size is much smaller than the fertilised eggs.

(g) **Stripping or artificial insemination**—In case of Chinse carps artificial insemination is done through stripping. In stripping method the 'eggs' and 'milt' are forcibly extruded out by applying pressure on the belly of breeders when they are ready to breed.

Stripping or artificial insemination is done owing to probable reasons given below :

(1) The eggs on some occasions, when naturally released remain unfertilised or even show some initial cleavage. After which further development stops altogether and the egg turns opaque. This may be due to a sort of parthenogenetic development of eggs. Such eggs cannot be fertilised by sperms.

(2) Due to non-synchronisation spawning behaviour of the male and female in natural spawning, the eggs may remain unfertilised.

(3) The eggs may remain unfertilised due to poor or inadequate quantity and quality of the milt.

(4) The females may not be in prime stages of maturity.

(5) Adverse environmental factors (like rise in temperature) may affect spawning.

Two methods of stripping are generally practised — dry method and wet method.

(1) **Dry stripping** : The dry method of stripping has been found to be more convenient and effective and is widely practised in India. In this method after 3-4 hours of the second infection, the female fishes are examined, whether they have become fit for stripping. The female breeders become fit for stripping when the abdomen becomes very soft and eggs oozes out freely on applying slight pressure on the abdomen. The females are first stripped of eggs in an enamel tray (Fig. 6.22). The abdomen of the males are then pressed and the milt is then added over the eggs. The milt is then thoroughly mixed with the eggs, taking

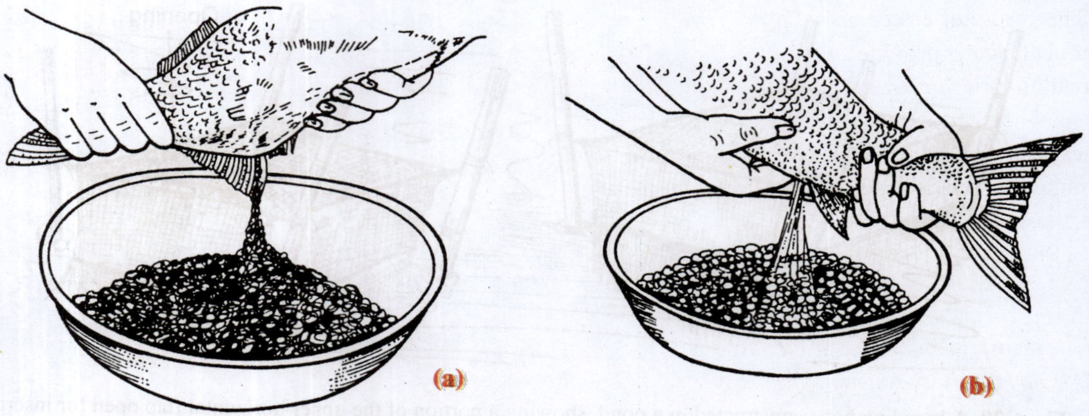

Fig. 6.22. Dry stripping — (a) Eggs are collected in an enamel tray. (b) Milt is exuded over the eggs

the help of a plastic spoon or bunch of feathers. After few minutes the fertilised eggs are washed thoroughly to remove the excess milt and blood clots. The fertilised eggs are then left immersed in water for 4-6 hours for water hardening of eggs.

(2) **Wet stripping** : In some countries wet method of stripping is also practised. In wet stripping 0.3% saline solution is used. The salt solution is taken in an enamel tray and the males are made first to exude milt. Then the female breeder is taken and the eggs are stripped into the tray. Generally the sperms of Chinese carp remain active up to 2-3 minutes in 0.3% saline, while they die much earlier in ordinary water.

E. Collection of eggs and transfer to Hatcheries

Eggs should be collected when the embryo starts the twisting movement, because by that time the eggs get properly water hardened and can withstand sufficient strain. Eggs are collected from the hapa by means of a cup or beaker and poured into a bucket with a small amount of water. In the case of Bangla Bundh, the eggs are collected with the help of a 'chot jal' and transferred to a 'hundi'. While in case of breeding pool, the eggs are directly passed on to the Chinese Hatchery through underground inter-connecting pipe system.

The breeders are then collected and the weight of the breeders after spawning is noted. The difference in weight before and after spawning gives an idea of the quantity of the eggs laid. The breeders are then transferred back into the pond.

Factors conducive for successful spawning

(1) Availability of fully matured and ripe male and female breeders.

(2) Collection of pituitary glands from fully gravid fishes.

(3) Proper assessment of dosage of pituitary hormone.

(4) Conducive physico-chemical factors of environment for spawning are :

- (i) Optimum dissolved oxygen ranges generally between 6-7 ppm.
- (ii) Optimum pH ranging between 6.0–8.0.
- (iii) Lunar periodicity.
- (iv) Optimum temperature ranging between 28-30°C.
- (v) Lightning and thunder shower.
- (vi) Stimulation effect of silt particle pressure on the body of the spawners. This pressure can be measured by conductivity meter.
- (vii) Electro-magnetic properties of water within a limited range.
- (viii) High turbidity (2000-2500 ppm).
- (ix) Stimulation on mature fish by minerals in solution or suspension in water.
- (x) Current of water.

The resultant effect of the above stated physio-chemical factors provide stimulation for sex play and spawning. Viewed from the point of physiology of the spawners, these external stimulii help in releasing necessary hormones first from the pituitary gland, for stimulating gonads as well as for the final release of ova and sperm for fertilisation.

Reasons for the failure of an induced breeding experiment

The main reasons for the failure in induce spawning of IMC by pituitary hormone injection are :

(1) Large number of failures in induced breeding often can be traced to poor condition of the brood fishes, including their health and nutrition and stage of gonadal development.

(2) Incorrect assessment of the pituitary hormone dose to be injected.

(3) Sudden rise in temperature.

(4) Accumulation of large quantity of fat in the body cavity and the fibrous tissue in the ovary of the female fish, causing obstruction in free oozing of the ova.

(5) On certain occasions, non-fertilisation of eggs occur after ovulation, or death of embryo may occur after fertilisation in the course of development. This is not due to administration of overdose of hormone, but because of the eggs not being fully viable.

HATCHERY

With increase in aquaculture activity there has been a large demand for qualitative fish seeds in quantitative term for fin and shell fishes. Thus, there is an urgent need for artificial propagation of fin and shell fishes under controlled condition. Hatchery production of seed is necessary to stabilise and ensure regular supplies and introduce breeding techniques to raise improved seed for better growth and production.

A hatchery is that part of a farm or a specialised type of farm where early incubation and hatching of fertilised eggs takes place. It is a system in which swim-up of fries, first feeding stage of development and early growth of fries take place.

Hatchery is probably the most vital component of a modern fish farm. Incorporation of hatchery in a carp farm is a relatively recent development in the history of carp culture. Hatching, care of hatchlings and larval rearing constitute the most important aspect of hatchery management and can make all the difference between success and failure of the operation.

Hatchery has the following merits :

(1) Hatching rate is very high (particularly in Chinese Hatchery — around 97%).

(2) No pollution.

(3) Well protected, well maintained and well managed.

(4) It can be installed anywhere (even in indoors).

(5) Improved and better quality of seed can be produced. Seeds bereft of diseases can be produced.

Types of Hatchery

There are different types of hatchery facilities in use, depending upon the species, locality and investment capabilities of the aquaculturist. However, all these types of hatcheries have similar basic requirement — water source, facility for holding or rearing

Fig. 6.23. A cluster of hatching hapas being used in India for hatching carp eggs

an adequate brood stock, spawning or stripping and fertilisation of ova, incubation of fertilised ova and rearing of larvae to the required stage for subsequent transfer to nurseries or transport to other places of culture.

All the different types of hatcheries has been broadly divided into the following types :

I. Traditional Method using Hatching Hapa

II. Earthen Pot Hatchery

III. Glass Jar Hatchery

IV. Chinese Hatchery

I. Traditional Method using Hatching Hapa

(a) **Hatching hapa** — It is a traditional net enclosure with the inner net being smaller than the outer one (Fig. 6.23). The outer net consists of a fine mesh (0.5 mm) seive-cloth tank about 2 × 1 × 1 m in dimension, while the inner chamber, made of the same material, has a mesh size of 2.0 – 2.5 mm. The whole device is placed in a protected water body where the water is well oxygenated. The fertilised eggs are evenly spread in the inner hapa. The hatched larvae fall or pass through the larger meshes of the inner hapa and are retained by the outer hapa as the small meshes of the outer hapa prevent them from escaping. After hatching of the eggs get completed, the inner hapa is removed together with the dead eggs, egg shells and other debris to prevent deterioration of the water quality within the hapa.

(b) **Hatching pits** — Besides the above, hatching pits are also used. Hatching pits are a series of pits located very close to the breeding ground. The size of the pit is 8 feet × 4 feet × 2 feet. Several pits are interconnected (Fig. 6.24), so as to allow the water to flow from one pit to another.

In each pit, 3 layers of cloth tanks or hapas of different dimensions are tied. The outer-most cloth tank is about 6 feet × 3 feet × $1^1/_2$ feet in size and is made up of cheap cloth. For support 2 bamboo poles are fixed to which these cloth tanks are tied. The middle tank is smaller, 5 feet × $2^1/_2$ foot × 1 feet in size and made up of mosquito cotton cloth with round meshes. The inner most tank is 4 feet × 2 feet × $^3/_4$ foot in size, having bigger mesh size.

Hatching of eggs — The eggs collected from rivers contain not only the eggs of major carps, but a mixture of eggs of several types of fishes and also certain larvae of insects, crustaceans, etc. The eggs, thus, before transferring to the hatching pits are subjected to seiving.

A sufficient number of eggs (approximately 9,00,000 to 22,00,000) should be kept in each hatching pit. Out of these, only 2.5% to 25% successfully hatch. The flowing water due to the interconnection between the hatching pits, activate the eggs to hatch. The time taken for hatching ranges from 2 to 15 hours.

After hatching, the hatchlings being slender and small pass through the meshes of the inner hapa to the outer cloth tanks, whereas the unwanted masses like unfertilised eggs, egg shells, dead hatchlings, etc., do not pass through the meshes of inner hapa. After about 20 hours of transference of eggs, the inner hapa

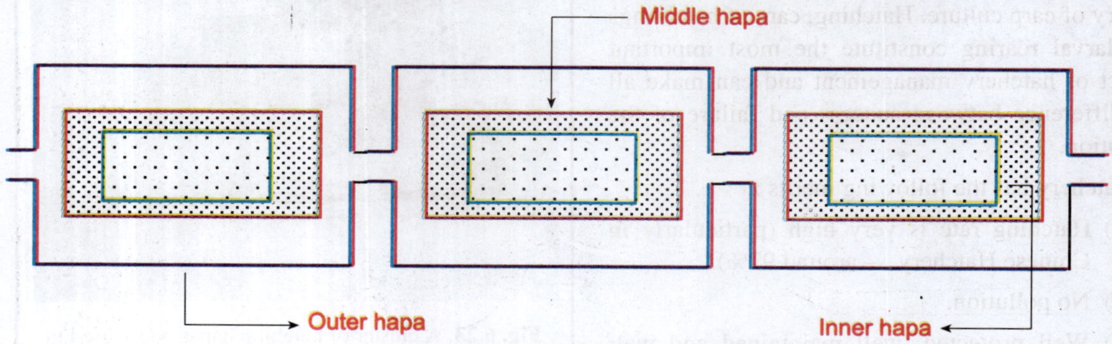

Fig. 6.24. Hatching pits

along with the unwanted masses are removed and thrown out. The hatchlings when reach the size of 4-6 mm, are transferred to nursery pond.

II. Earthen Pot Hatchery

(a) **Construction** — Here the principle of hatchery system is followed, where a overhead tank in the form of an earthen clay pot (Fig. 6.25) is installed on a brick stand about 2 feet in height. There are two different sized container. Container number I acts as incubator chamber; while container number II acts as a spawnary.

From the bottom of the earthen pot a valve and a rubber pipe are fitted. The other end of the rubber pipe contains a metal tube which is put into container number I. At the top of the container a round mesh sheet is placed and it is held in position by a bamboo rod. At the other end of the container I, near to the surface, a round spout is fitted.

This spout from container I opens into continer II which is smaller in size than the 1st one. On the top of container II, a net is placed and to its other end, nearer to its surface, an overflowing pipe is fitted. The outer end of the pipe is covered with a small net.

(b) **Working method**

(1) At first, freshwater from a tube-well is poured into the earthen pot. The valve is regulated to maintain the water pressure entering into container I.

(2) Fertilised eggs are then introduced into the container number I.

(3) The excess water flows out through the spout into container II and from it to the outside through the overflowing pipe.

(4) A running water condition is, thus, maintained.

(5) The water pressure is regulated in such a way, that the eggs do not pass out into the second container.

(6) When the eggs hatch, the hatchlings have a tendency to undergo vertical migration. So, it comes to the surfce and the overflowing water carries it into container II.

(7) The unwanted materials such as egg shells, unfertilised eggs, deformed eggs remain behind in the Ist container.

(8) The spawns are then kept in container II for 2-3 days. It is fed externally, if necessary, and later are either introduced into the nursery ponds or sold to other farmers.

Merits and demerits of Earthen Pot Hatchery

(a) **Merits of Earthen Pot Hatchery**

(1) Cool and running water condition which is essential for hatching, is maintained.

(2) Cost of making an earthen pot hatchery is very less.

(3) Very easy process to handle.

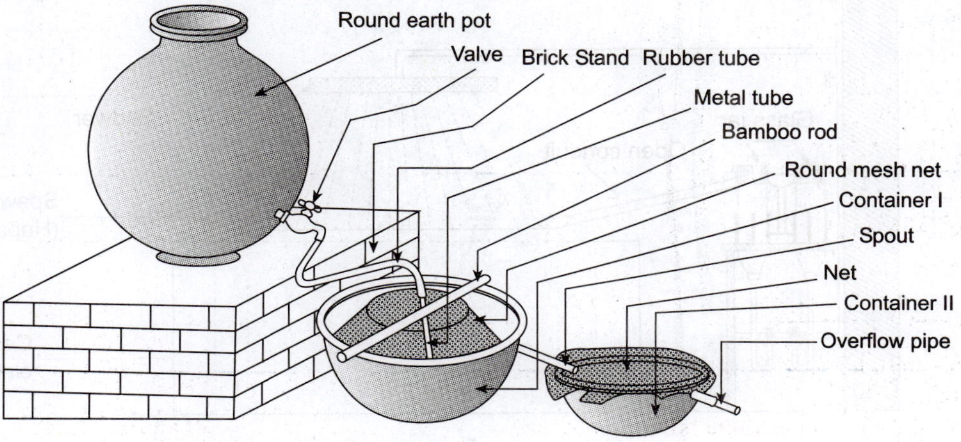

Fig. 6.25. Earthen pot hatchery used in rural India

(b) **Demerits of Earthen Pot Hatchery**

(1) Earthen pot is not capable of storing a large amount of eggs.

(2) Very laborious manual process and time consuming.

(3) Handling should be done with care as pots are liable to breaking.

III. Glass Jar Hatchery

A glass jar hatchery is a much modified set-up than the earthen pot hatchery. It is comprised of :

(a) Water supply system

(b) Breeding tanks

(c) Incubation and hatching jars

(d) Spawnery

Location : Glass jar hatchery is generally installed by the side of a water source which may be a perennial pond or a river or a water supply from a tube-well. It should be kept in a well-ventilated room with adequate light and proper drainage system.

Construction :

(a) **Overhead Tank** — Available water from the source is pumped up into an overhead tank (Fig. 6.26), which is installed above a brick wall. If water is drawn from a river, then a desilting tank should be provided. A fine meshed wire-netting is provided in the pump system. This helps to prevent the entry of organisms commonly present in the water of river and pond. A deep tube-well is of much facility as it provides clean water. The tube-well should be installed in an arsenic-free area.

(b) **Breeding Tank**—As hatchery is a self-contained unit, it generally contains a breeding tank with arrangement for overhead showers. This tank is provided for spawning the fish. An outlet is present in

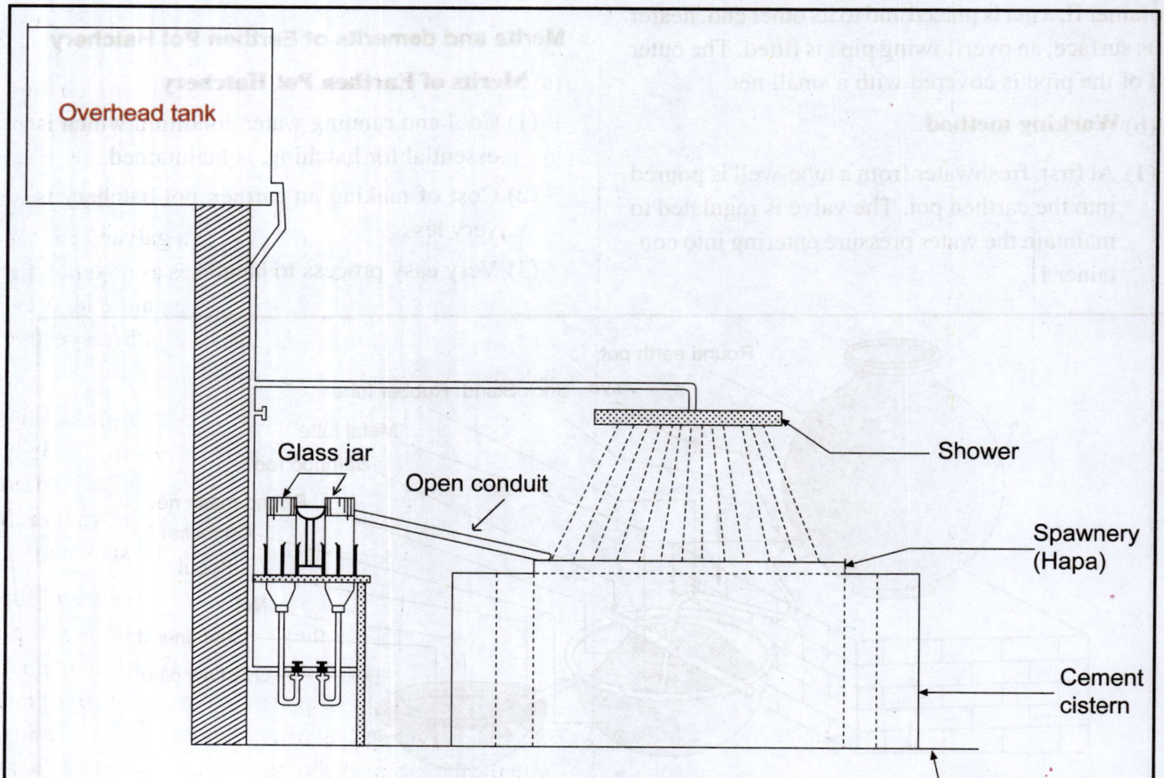

Fig. 6.26. Glass Jar Hatchery for Carps

each tank, for draining the excess water. A breeding tank is made of 1.8 m × 0.9 m × 0.9 m in dimensions. Generally, in a farm there are 2-3 breeding tanks where 2-4 breeding hapas can be fixed at a time. However, the number of breeding tanks in a hatchery depends on the spawn production target.

In a hatchery where there is no provision of breeding tanks, the operation is undertaken in hapas fixed in ponds and the fertilised eggs are released into the glass jar for incubation and hatching. Sometimes, the spawnery itself can be used as a breeding tank.

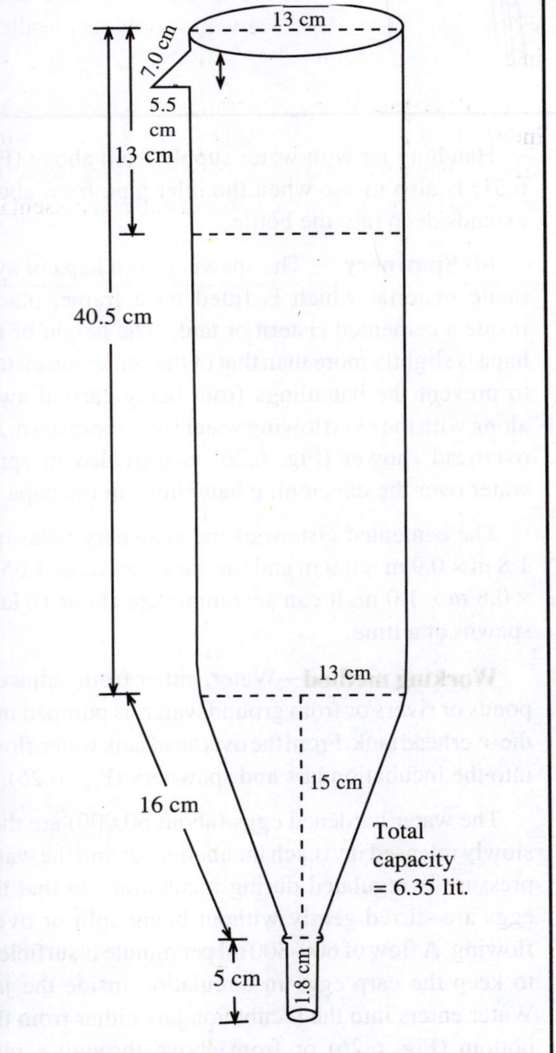

Fig. 6.27. Sketch of a hatching glass jar

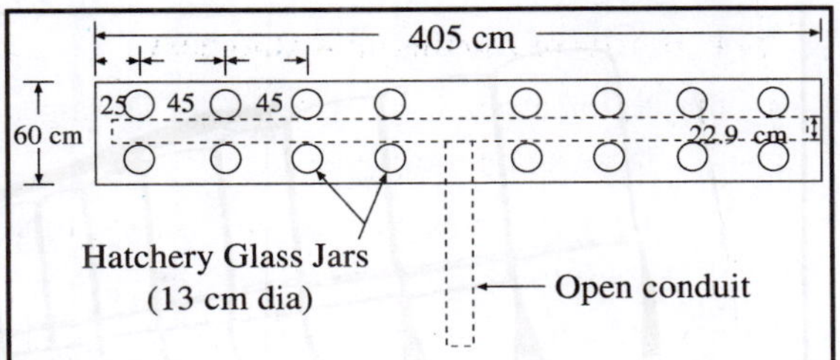

Fig. 6.28. Detailed layout of the placement of glass jars on a table

(c) **Incubation and Hatching Jars** — There are different types of incubation and hatching jars used, ranging from the glass jars, synthetic jars to Zong jars, with temperature controlled water supplies.

(i) **Glass Jars** — Glass jars are conical in shape. They are open at the top and gradually tapering towards the bottom. The various measurements of the jar are given in Fig. 6.27.

The glass jars are set up vertically through the circular holes made in a wooden table (Fig. 6.28). The table is of 4.05 m × 0.6 m with a height of 0.9 metre. The detailed plan for the sequential placement of jars on the table is given in Fig. 6.28. The jars are fitted in two rows at a gap of 22.9 cm. They are kept in a vertical position with the help of clamps (Fig. 6.26).

The top of the jar is fitted with a galvanised iron ring with a beak (Fig. 6.27). It serves as an outlet for water coming out of the jar. The jars are connected with rubber tubings at the bottom with respective taps (Fig. 6.26).

Each glass jar has a capacity of 6.35 litres and can accommodate 50,000 carp eggs at a time. A long open galvanised conduit (Fig. 6.26 and 6.28) is fitted in such a way that the overflowing water from each jar passes into it and flows down to the spawnary.

(ii) **Zug Jars** — The Zug jars, name after Zug (Switzerland) where they were first used, are large (60-70 cm in height and 15-20 cm in diameter), inverted bottles, with open tops and narrow bottoms. In a hatchery they are installed vertically in a series, with the narrow neck directed downwards attached to the water lines with suitable taps (Fig. 6.29). Water is

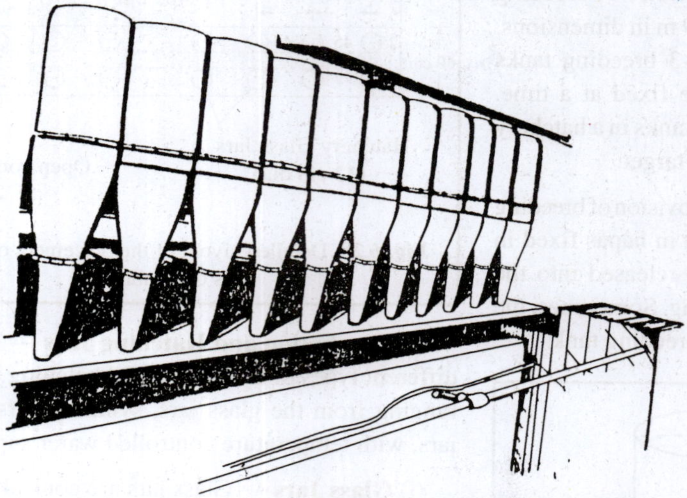

Fig. 6.29. Zug Jars

supplied from below and the outlet is at the top. The Zug jars have a capacity of 6-8 litres and can carry about 1-5 litres of eggs.

(iii) **Mac Donald or Chase Jars** — Mac Donald Jars are similar to the Zug jars. It is cylindrical in shape (40-50 cm in height) with a circular bottom for the water supply and the flow of water in the jar keeps the eggs rotating slowly.

(iv) **Farm Fabricated Incubation Funnels** — They are made of plexiglass or fibre glass, in any required dimension, and are widely used. Israeli farmers use plexiglass incubation funnels (Fig. 6.30a) of 80 cm long and 60 cm in diameter at the upper rim. Water is supplied from a bottom inlet, while water overflows from an outlet at the top. This overflowing water is used for collecting the hatchlings. Each such funnel can be used to hatch about 1,75,000 carp eggs. Woynarovich and Horrath (1980) illustrated a number of designs of incubation funnels that can easily be made with plastic and seive-cloth (Fig. 6.30b) materials, with some sprinkler-type water inlet.

(v) **Plastic Jars** — These jars are made of low density polythene. The upper part has a diameter of 440 mm, and 625 mm height, with a capacity of 40 litres. Each jar has independent control valve. For effective use, an inner egg container is used for separating the egg shells after hatching.

Hatching jar with water supply from above (Fig. 6.31) is also in use when the inlet pipe from above extends deep into the bottle.

(d) **Spawnery** — The spawnery is a hapa of synthetic material which is fitted on a frame, placed inside a cemented cistern or tank. The height of the hapa is slightly more than that of the cemented cistern to prevent the hatchlings from being carried away along with the overflowing water from the cistern. An overhead shower (Fig. 6.26) is provided to spray water over the developing hatchlings in the hapa.

The cemented cistern of the spawnery measures 1.8 m × 0.9 m × 0.9 m and the hapa measures 1.65 m × 0.8 m × 1.0 m. It can accommodate about 10 lakh spawns at a time.

Working method—Water, either from adjacent ponds or rivers or from ground water, is pumped into the overhead tank. From the overhead tank water flows into the incubation jars and spawnery (Fig. 6.26).

The water hardened eggs (about 50,000) are then slowly released into each incubation jar and the water pressure is regulated during incubation, so that the eggs are stirred gently without being split or overflowing. A flow of 600-800 ml per minute is sufficient to keep the carp eggs in circulation inside the jar. Water enters into the incubation jars either from the bottom (Fig. 6.26) or from above through a pipe which opens at the bottom (Fig. 6.31).

(a)

(b)

Fig. 6.30. Farm fabricated incubation funnels — (a) Plexiglass incubation funnels. (b) Sieve-cloth incubators

After the eggs are hatched, the hatchlings have a tendency of vertical migration. As a result they come up to the surface and pass through the outlet along with the overflowing water into the open conduit leading to the spawnery. The rate of flow of water in the jars at this time is slightly increased after the completion of hatching to facilitate speedy escape of hatchlings.

Dead eggs, unfertilised eggs, egg shells, abnormal and dead hatchlings are left behind in the jars which are subsequently removed by disconnecting the glass or incubation jar from the tap.

In the spawnery, the water is sprayed continuously for about 3 days until the fry are ready for stocking.

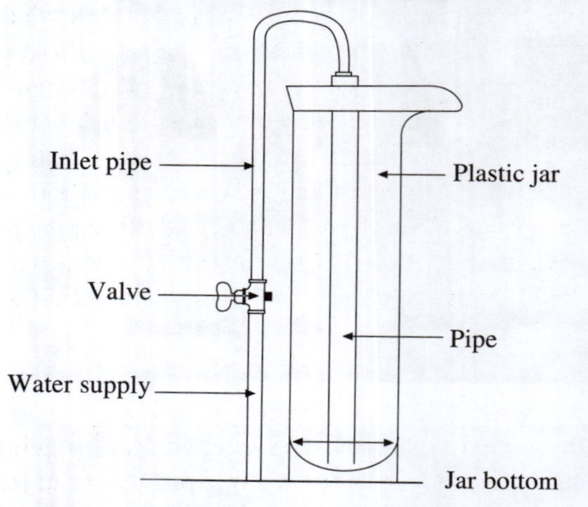

Fig. 6.31. Plastic hatching jar

Merits and Demerits of Glass Jar Hatchery

(a) **Merits**

(1) It can be operated in a compact area.

(2) As the incubation jars are transparent, the developing embryos can be seen with the naked eye and appropriate rectifications can be made.

(b) **Demerits**

(1) In case of incubation jars made of glass is liable to damage and cannot be shifted to different places.

(2) Temperature control system is not provided.

(3) Additional air circulation is not provided.

(4) Hatching success is less than that of Chinese hatchery.

(5) Damages of eggs from the glass jars during transfer may take place.

IV. Chinese Hatchery (Eco-Hatchery)

Chinese investigators have developed circular hatching tanks, in which water is in a continuous state of flow in a circular direction. This is popularly known as Chinese hatchery system. This system, within a small space, simulated some aspects of riverine environment and has proved itself a very successful method of breeding carps where commercial production of carp seed is required. The Chinese hatchery system comprises an overhead tank, spawning pool, incubation or hatching tank and spawn rearing tank.

Construction

(a) **Overhead Tank and Spawning Pool or Breeding Tank** — Description of these are given earlier.

(b) **Incubation or Hatching Tank** — Hatching tank is circular in shape and is constructed with cement and bricks. It consists of two concentric circular tanks (Fig. 6.32). The two chambers have inside diameters of 1.6 m and 5.0 m. A water of 0.92

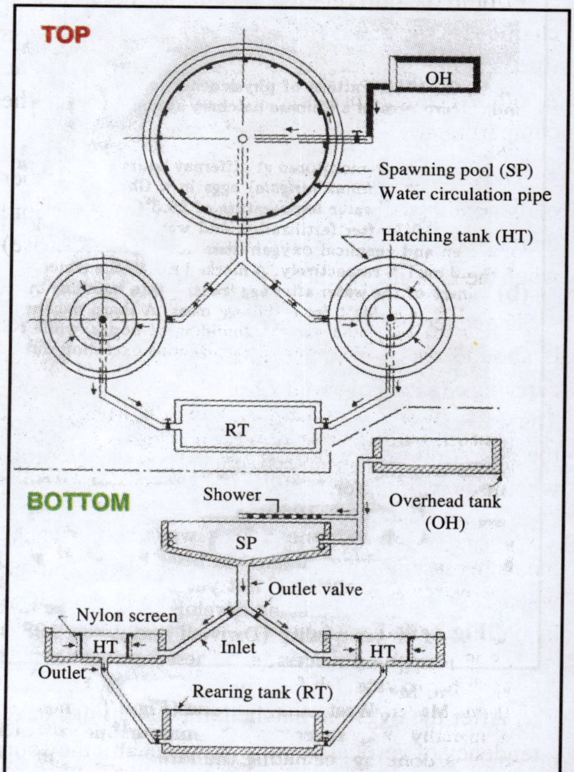

Fig. 6.32. Schematic drawing of Chinese hatchery system. Top : Cross section; Bottom : Longitudinal section of the system

Fig. 6.33. A photographic view of a Chinese hatchery

to 1.0 m depth is maintained by an outlet pipe fitted in the middle of the inner circular tank. The inner circular tank has gaps on its body, through which water passes into the inner space. This gives the shape of a double doughnut. One wall of the double doughnut lies at the periphery, while the other at the inner end that lies surrounding the outlet (Fig. 6.33).

Prior to the introduction of eggs, the inner wall or chamber is separated from the outer by a fine nylon screen which is stretched and fitted on an iron frame. A rubber belt is fastened very tightly to seal the compartment.

The floor of the outer compartment has a number of diagonally fitted pipes which face in one direction, thereby providing the circulation of water (clockwise) when the taps are opened.

The water-holding capacity of the incubation tanks is about 7 to 10 m^3 and has a relative egg density carrying capacity of about 68×10^4 m^3 to 70×10^4 m^3. The water flow is maintained at 225 l/min throughout the operation except during the period of hatching, when it is maintained at 250 l/min.

(c) **Spawning Tank** — The hatching tank acts as the spawning tank after hatching of the eggs and the water flow is maintained at an optimum level. The hatchlings are reared here for 3 days.

Working principle

After pituitary injection, the male and female brooders are released into the spawning pool. The hatchery is made ready prior to the introduction of eggs and the level of water is maintained by adjusting the outlet pipe.

The eggs may be brought mechanically from the spawning pool or may be drained directly into the outer chamber of the hatchery through underground pipe system (Fig. 6.32). The eggs in the outer chamber are kept in circulating condition due to the circulation of water made possible by the circular arrangement of pipes fitted on the floor. Continuous water circulation in the incubation tank helps to maintain a high oxygen level of the water and churning of eggs throughout. The chosen flow rate of water helps to maintain the required temperature necessary for high hatching success. The flow rate was increased during hatching because of the high oxygen demand during the process. At about 11 to 12 hours after fertilisation, hatching of the embryo starts and gets completed within 4 hours.

After all the eggs have hatched, the flow rate of water was decreased. To avoid bacterial disintegration the eggs shells, unfertilised eggs, dead hatchlings, etc. are removed periodically from the incubation chamber. This is done either by siphoning or by dipping a jute rope with a weight into the outer chamber. The circulating egg shells get attached to the jute rope, which is periodically taken out; the egg shells removed and the rope is again put into the outer chamber, till nearly all the egg shells are removed.

Some unfertilised eggs remain behind in the hatcheries and these are the focus for fungal infection caused by *Saprolegnia*. A common method of controlling fungal disease in the incubator is by the application of malachite green at a concentration of 0.02 g/litre of water for about 20-25 minutes after stopping the flow of water. When the flow is resumed, the chemical is washed out from the incubator.

Merits and Demirits of Chinese Hatchery

(a) **Merits**

(1) Chinese hatchery ensures a high hatching success (98%), as good and clean water replacement is possible.

(2) Water flow and thus the temperature can be maintained during hatching.

(3) The Chinese hatchery can be directly connected to the breeding pool on one hand and to the nursery pond on the other. Therefore, it ensures less damage of fertilised eggs and spawns at the time of transfer.

(4) It minimises the labour cost.

(5) Much larger quantity of eggs can be accommodated.

(b) **Demirits**

(1) Due to the absence of proper spawnery, total removal of shells, unhatched eggs and dead spawns and larvae are not possible.

(2) Initial cost of the set-up is very high.

(3) Proper technical application has to be maintained regarding fitting of pipes, direction and amount of water flow.

Physico-chemical nature of Water/Water quality Parameters at the time of hatching

The water quality changes are fairly evident during the development of eggs and these are regulated to ensure greater hatching success in the glass and Chinese hatchery.

(1) In case of Indian carp eggs a water temperature of 29.3°C results in mass hatching which starts about 11 to 12 hours after fertilisation and gets completed within 4 hours (Jana *et al.*, 1985). A slight decline of water temperature to 28.65°C and 28.7°C delayed mass hatching by 1 and 3 hours, respectively.

(2) The oxygen amount ranging from 6 to 7.2 mg/litre is found to be adequate to account for hatching success. It has been reported (Garside, 1959) that reduced oxygen tension retards development of fish embryo, while water rich in oxygen accelerates the process (Kinne and Kinne, 1962).

(3) The total hardness and biocarbonate of hatching water increases markedly after the transfer of eggs. This marked rise in hatchery water after egg transfer might be due to liberation of Ca^{++} and Mg^{++} from egg mass.

(4) Free carbon-dioxide of the hatchery water rises gradually as the length of the embryonic development advances, because of respiratory activity of the fertilised eggs and larvae. This, thus results in the decline of pH.

(5) The nitrogenous compounds like ammonia, nitrate and nitrite increase at the time of hatching and these compounds are released by the developing embryos.

(6) The phosphate level is greatly reduced during hatching period but tended to rise afterwards. It is due to food reserve in yolk that gets exhausted during hatching, and developing embryos utilise phosphate from the external water medium for their bone formation.

The manipulation of the above limnological conditions would thus ensure an increase in the carrying capacity of the seed rearing system.

POLYCULTURE OR COMPOSITE FISH CULTURE

Polyculture or **mixed fish farming** or **composite fish culture** is the culture of fast growing compatible species of fishes of different feeding habits (or different weight classes of the same species) in the same pond so as to utilise the various available ecological niches in order to obtain high production per hactare of water body.

A pond according to its depth can be divided into three distinct zones — upper surface zone, middle column zone and bottom zone. A particular species exploits food of a particular zone. For example — *Catla catla* is a surface feeder, *Labeo rohita* a column feeder and *Cirrhinus mrigala* is a bottom feeder. In case of single species or monospecies or monoculture only one zone will be utilised or exploited while the other zones would remain unutilised. As a result the entire ecological area would not be exploited and the yield or fish production would be less.

Principle — When different species of fast growing compatible fishes, occupying different ecological niches of a pond or any water body, are cultured together, they most efficiently utilises all the food sources available in the pond for fish production without harming each other.

Objectives of composite fish culture

(1) To obtain maximum yield or fish production.

(2) To utilise all the available niches.

(3) The fishes cultured should not cause any ecological disbalance.

(4) The fish species cultured should not have any serious competition between them but each species may have a beneficial influence on growth and production of the other. For example, grass carp by consuming aquatic vegetation, converts plant tissue into fish flesh but its excreta fertilises the pond which benefits all other species.

(5) Some species of fishes are cultured which have specific roles to play in maintaining water quality in ponds by feeding on wastes accumulated in it. For example common carp and mrigal consume the faeces of grass carp and silver carp, which contain large amounts of undigested plant matter.

(6) Recent combination of fish species cultured are based on one or two species as the main ones and the others as subsidiary compatible species which would be utilising those parts of the food resources that would have been wasted.

Origin of composite fish culture

Composite fish farming has its root in ancient China and India. In both these countries polyculture was adopted in carp farming based on mixed seed stock, collected from natural sources. It was not possible to sort out stocks of different species in their early stages. The farmers, therefore, were forced to rear the different carp species together until they reached fingerling stage. The farmers observed that there were differences in the feeding habits and different preferences for natural food organisms. This led to the thought of adjusting the stocking densities according to the occurance of natural food and the niche occupied by each species in the ponds. Thus, it could be beneficial in traditional production systems by culturing appropriate species combinations.

Such a practice could effectively lead to the utilisation of a much larger quantity of food resources and developing a symbiotic relationship between the cultured species in the system. This would lead to higher production at a lower cost and the farm environment can be maintained at required levels.

The combination of species in polyculture system cultured in China consists of :

(1) Grass carp (*Ctenopharyngodon idella*) which feeds on macro-vegetation.

(2) Silver carp (*Hypophthalmichthys molitrix*) which feeds on plankton (mainly phytoplankton).

(3) Big head (*Aristichthys nobilis*) which feeds on microplankton.

(4) Black carp (*Mylopharyngodon piceus*) which feeds mainly on snails and other molluces at the bottom.

(5) Mud carp (*Cirrhinus molitorella*) which feeds primarily on detritus.

(6) In addition sometimes the omnivorous common carp (*Cyprinus carpio*) is added that basically serves as a scavenger in farm ponds.

(7) Modern polyculture system has added a number of other species to the above traditional combination, such as tilapia, wuchang fish, crucian carp, red eye, white amur bream, snake head and mariadarin fish.

The combinations practised in Indian polyculture system comprises the following :

(1) The surface feeder, catla (*Catla catla*) which feeds on planktonic organisms particularly zooplankton. The larvae and young fry feed on planktonic unicellular algae.

(2) The column feeder, rohu (*Labeo rohita*) consuming vegetable matter including decaying aquatic plants, algae, etc. Larvae and fry feed on unicellular algae and zooplanktonic organisms.

(3) The bottom feeder, mrigal (*Cirrhinus mrigala*) prefering decayed plant and animal matter, algae, detritus, organic matter, etc. The larvae and fry feed on planktonic unicellular algae and zooplankton.

(4) The bottom feeder, calbasu (*Labeo calbasu*), feeds selectively on benthic and epiphytic organisms and organic debris.

(5) Modern culturists and scientists have developed **composite carp culture** by the addition of exotic carps such as grass carp, silver carp and common carp.

(6) New species combinations for polyculture have been developed through the introduction of benthic-feeding grey mullets (*Mugil cephalus*). The introduction of carnivorous fish, chital (*Notopterus chitala*), has been done for the control of weed fishes.

Methods adopted in composite fish culture

The methods that are adopted in composite fish culture can be broadly divided into three phases :

I. Pre-stocking management

II. Stocking management

III. Post-stocking management

I. Pre-stocking management

In pre-stocking management the pond or water body is prepared before the introduction of fry or fingerlings by the following way :

(A) **Renovation of the pond** — Generally due to rain and flood the edges or dykes of the ponds may get damaged and a lot of silt or mud may accumulate at the pond bottom, thereby decreasing the depth of water. In such cases the entire water of the pond is to be removed through the help of a pump. The pond bottom should be allowed to dry. The humus at the pond bottom should be removed through the help of manual labours. The dykes has to be repaired and plants to be planted on it so that its roots would prevent soil erosion. The pond is then filled with water.

(B) **Clearance of weeds** — One of the crucial problems encountered by Indian pisciculturists is the excessive growth of aquatic vegetation in water bodies and its control. Weeds cause the following diasters in a water body :

(1) Severely restricts plankton production as it prevents sunlight from entering into the pond water.

(2) Limits the living space for fishes and obstructs their free movement.

(3) Provides shelter for predatory and weed fishes, molluscs, aquatic insects and fish parasites.

(4) It consumes nutrient from the pond which otherwise would have been utilised for the production of plankton.

(5) Upsets the equilibrium of physico-chemical qualities of water.

(6) Causes imbalance in dissolved oxygen budget, particularly during cloudy days.

(7) Promotes accumulation of deposits leading to siltation.

(8) Obstructs netting operations.

Due to the above hazards, the control of aquatic weeds is very much essential. The control methods adopted are the following :

1. **Manual and mechanical control** — If the weed infestation is scanty, then it can be easily removed manually with the help of a long bamboo. The entire weeds are swept with the bamboo to a side and then removed manually with the hand. As the weeds grow again, it is difficult to eradicate them completely. In case of denser infestation the weeds can be removed with the help of weed cutter.

2. **Chemical control** — In case of dense weed infestation suitable weedicides may be used. A large variety of chemical herbicides are used as weedicides.

(i) 2, 4-dichlorophenoxy acetic acid (2, 4-D) is very widely and effectively used. The usual rate of application is 4.5 to 6.7 kg/ha and its application has no harmful effects on fishes.

(ii) Submerged weeds can be kept under control by the application of ammonia at the rate of 15-20 ppm.

(iii) Diuron is effectively used at the rate of 0.1 to 0.3 ppm without affecting production of plankton and fishes.

(iv) Simazine is also used at a dose of 0.5 to 1.0 ppm.

3. **Biological control** — Weeds in a water body can be controlled by means of selected varieties of herbivorous fishes. The important herbivorous fishes used are : grass carp (*Ctenopharyngodon idella*), tawes (*Puntius javanicus*); common carp (*Cyprinus carpio*); *Tilapia mosambica* and gourami (*Osphronemus goramy*).

Thus, an ideal herbivorous fish —

(i) is able to consume a wide variety of weeds as its food,

(ii) is hardy,

(iii) does not biologically interfere with other fishes,

(iv) is economical to maintain and

(v) adds to the fishery wealth.

In consideration of the above characters, grass carp has been found to be the idealist and has been used all over the world as the most effective in controlling weeds.

(C) **Eradication of weed and predatory fishes** — It is not desirable to have weed and predatory fishes in the cultural impoundment. Weed fishes by consuming the nutrients from water competes for food, space and oxygen with the fingerlings of culturable carps. They have high fecundity and ripen in summer and can breed even in the absence of rain. The predatory fishes, on the other hand, directly prey on the culturable carp spawn, fingerlings and young fishes. Therefore, their eradication from the ponds before the stocking of fingerlings is of utmost importance.

One way of eradicating these unwanted fishes is by dewatering and drying the pond. However, where dewatering is not possible then these fishes have to be killed by the application of fish poison, generally of plant origin. Although a number of organophosphatic compounds and chlorinated hydrocarbons such as aldrin, dieldrin, DDVP are tested as effective fish poisons, still the use of chemicals is not desirable as detoxification of these chemicals would require time and they would leave harmful residual effects in the pond.

The most widely used fish poison of plant origin is **mohua oil cake**. The application of *Brassica latifolia* (mohua oil cake) is of much significance in fish ponds because of its role as piscicide in the initial phase of treatment and as manure in later phase after its loss of toxicity. According to Jhingran (1982), toxicity of mohua oil cake is due to its 4-6% saponin content. The toxic effect has been found to be most acute at 16 hours of mohua treatment (Jana *et al.,* 1987). The effect of saponin lasts for 2-3 days. The recommended dose of mohua oil cake is 2500 kg/ha. After 19–25 days detoxification of mohua oil cake takes place and it takes up the role of manure.

(D) **Addition of lime** — The first step in fertilisation is the application of lime. The common form used is limestone. Lime has a number of uses such as :

(1) It corrects the acidity of soil and water.

(2) It acts as a strong pH buffer.

(3) Lime by its toxic and caustic action, kills bacteria as well as fish parasites and may render the fishes less liable to diseases.

(4) Lime counteracts the poisonous effect of excess magnesium, potassium and sodium ions.

(5) Liming is essential for successful pond manuring. A pond containing lime is found to be more fertile than the one without it.

The dose of lime depends upon the characteristics of the soil and water. A dose of 1,000 to 1,500 kg/ha is required for ponds with acid soil and water. If liming is done every year then a dose of 100 to 200 kg/ha is sufficient.

(E) **Addition of fertilisers** — The suitability of any fertiliser used in fish ponds is judged by the following :

(1) How much increase in the production of plankton it can bring about ?

(2) How cheap is it ?

(3) How easily it can be distributed over the pond?

(4) How much harmless is it to the fish under cultivation?

(5) What after-effects will it have on the conditions of the pond?

The different types of fertilisers used in fish culture ponds fall under two categories — (a) inorganic and (b) organic.

(a) **Inorganic fertilisers** : The major fertilising elements are nitrogen, phosphorous, potassium and calcium. The different inorganic fertilisers used in fish ponds are grouped as under :

1. **Phosphate fertiliser** in the form of calcium superphosphate is used at a dose of 2000 kg/ha/yr to obtain best result. Single super phosphate is used at the rate of 40 kg/ha/month.
2. **Nitrogen fertiliser** in the form of urea is widely used in India. It contains about 45 to 46% pure nitrogen. Other nitrogenous fertilisers used are sodium nitrate, ammonium nitrate, calcium ammonium nitrate, liquid ammonia, etc.
3. Potassium fertiliser.
4. Magnesium fertiliser.
5. Trace element fertiliser.

(b) **Organic fertilisers** : Organic fertilisers are either of plant or animal origin. It contains almost all the nutrient elements required in the metabolic cycle. The nutrients locked in them are either made available directly or after decomposition and transformation by the microbes. Organic fertilisers can be classified on the basis of the following :

Organic Manures

(1) Organic manures with little or no carbohydrate

(2) Organic manures with carbohydrate content only

(3) Organic manures with carbohydrate and nitrogenous matter

1. **Organic manures with little or no carbohydrate** — The manures falling under this category are :

(i) *Liquid manure from stable and byres* — This, when treated in ponds has been found to impart fertility for good fish yield. It is appplied in small doses and simulates growth of phytoplankton, filamentous algae and zooplankton. Liquid manure when applied in large doses gives profuse growth of filamentous algae and may cause depletion of oxygen.

(ii) *Guano, dried blood and slaughterhouse offal* — Through the use of dried blood and guano an increase in fish yield has also been reported. Slaughter-house offal mixed with superphosphate (P_2O_5) when used as fertiliser in fish ponds has been found to enhance fish yield.

2. **Organic manure with carbohydrate content only** — The organic manures in this category are :

(i) *Oil cakes* — Oil cakes such as mustard oil cake, mohua oil cake, cotton seed meal, soybean meal, etc. have been used for manuring fish ponds. Mustard oil cake is used either singly or in combination with cowdung. The use of mohua oil cake has already been discussed earlier.

(ii) *Green manuring* — Green manuring is used to fertilise fish ponds. In this method green plant tissues are decomposed and turned into soil. This is done by sowing a nitrogenous or other crop on dry pond bottom at the time of pond preparation or by dumping it in a heap at one end of the pond. In West Bengal green grass along with dry cow-dung, stable refuse, poultry manure and oil cake are used as green manuring. In Orissa, green manuring is practised by growing a Leguminous plant (*Sesbania* sp.) on the pond bottom and then it is crushed after 10-15 days. Generally, a dose of 1,680 kg of green manure per hectare is recommended at intervals of 3 months.

Advantages of applying green manure

- Green manure induces microbial activity in the soil resulting in increased productivity.
- After the decay of plant tissue, the humus compound thus produced increases the absorptive capacity of the soil.
- The composition also provides inorganic nutrients directly to the soil.
- The soil nitrogen is restored if leguminous plants are used.
- Conserving effect of nutrients in the soil increases, which otherwise would have been lost through drainage and erosion.

- Due to ploughing the nutrients carried by the roots are brought to the surface.
- Green manuring increases the availability and solubility of lime and phosphoric acid.
- Green manure provides a rich substratum for attached algae, zooplankton, insect larvae, worms, etc. which forms the food of fishes.

(iii) *Compost* — Another good organic manure is compost. It is formed by dumping organic wastes, vegetable debris, cut grass, cow dung, aquatic weeds, etc. into a large pit and then the pit is covered with soil. The aerobic bacteria in the presence of nitrogen converts the above mentioned wastes into humus. A dose of 5000 kg of compost/ha of water body is recommended for rich growth of plankton.

3. **Organic manures with carbohydrates and nitrogenous matter —**

(i) **Farm Yard Manure**. The farm yard manures denote refuse from all animals of the farm. The farm manures are most extensively used fertilisers in fish culture practice. Cow and pig dungs are very frequently applied, about 15-20 days before the introduction of fry and fingerlings. Ponds are manured with cow dung at the rate of 11,208 to 22,417 kg/ha. However, when applied in large quantities it may cause deoxygenation of water.

(ii) **Poultry manure**. Poultry manure is a very efficient fertiliser (Nitrogen 2%, phosphoric acid 1.25% and potash 0.75%) for fish ponds. This is why both intensive and extensive poultry production has been integrated successfully with fish culture.

(iii) **Sewage**. Domestic sewage basically contains wastes from toilets, bathrooms, kitchens and other household washings. Sewage is rich in nutrients such as phosphorus, nitrogen, etc., which are available in huge quantities. It also directly promotes primary productivity. Raw sewage is inimical to the life of fishes due to high values of BOD, O_2, NH_3, H_2S, bacterial load and negative oxygen value. Thus, treatment of sewage is a pre-requisite for waste water aquaculture. Other than its manural capacity, the potentiality to serve as an additional source of water for maintaining the deserved depth of water in the pond, particularly during the dry summer season, is an added advantage of using such waste waters.

Sewage if not properly treated prior to its application in pond water may result in profuse growth of algae, and various kinds of fish diseases are encountered.

(iv) **Sludge** — Sludge is the sediment deposited during the treatment of sewage. It is rich in total nitrogen (6.0%) and phosphorus (1.29%). Activated sludge is used as manure to fertilise pond water. It has been found to promote plankton growth and high fish yield.

II. **Stocking management**

In aquaculture practices only the common carp is cultured in monoculture system. While in polyculture system either the Indian major carps or in combination with the exotic carps are cultured in various combinations. However, prior to the introduction of fry or fingerlings the cultured water has to be tested whether it is suitable for culture or not.

(A) **Stocking criteria** — For rearing of carp the main or basic objective is the production of an optimum quantity of the desired size of fish, at minimum cost. A number of interdependent factors are present that affect productivity and cost. The factors that will influence the growth rate and production are stated below :

1. The quality and quantity of food produced by fertilisation.
2. The temperature of the water in the pond.
3. The availability of oxygen.
4. Build-up of metabolites in the pond.
5. Stocking rate.
6. The size of fish at stocking.
7. Artificial feed given to the fishes.
8. The duration of culture.
9. The size of the fish that are to be harvested.
10. The influence of productivity of natural fish food even when fertilisation and feeding are adopted in the pond.
11. The growth potential of the genetic strain used in culture.

There are two systems of stocking that the fish farmers generally adopt—(a) multi-size stocking and (b) multi-stage stocking.

(a) *Multi-size stocking* — This involves stocking, in the same pond different sizes (fry, fingerlings and yound adults) of fishes in order to utilise the food resources more efficiently. In this system additional stocking of fishes and periodic harvesting of marketable fishes are undertaken.

(b) *Multi-stage stocking* — In this system fishes are stocked in progressively larger ponds (**nursery, rearing** and **stocking**) as they grow in size. Here, the stocking rates are reduced and harvesting of marketable fishes are done after a period of approximately one year culture.

It is, thus, clear that the number of fishes to be stocked should be determined through a formula :

$$\text{No. of fishes to be stocked} = \frac{\text{Expected total production}}{\text{Expected individual growth}} + \text{Expected loss due to mortality}$$

(B) **Species combination** — Three broad combinations are practised in India, i.e. (1) Culture of Indian major carps alone, (2) Culture of exotic carps alone and (3) Culture of Indian and exotic carps together. In cases where silver carp is cultured along with catla, the silver carps introduced should be smaller in size and lesser in number. As silver carp feeds on phytoplankton, the density of phytoplankton would fall if silver carp is introduced in large numbers. This would hamper the quantity of zooplankton and ultimately the growth rate of catla, that feeds predominantly on zooplankton.

Besides the above, a number of other species are also stocked in polyculture system. The more common ones being tilapia (*Tilapia mosambica*), gouramy (*Osphronemus goramy*), grey mullet (*Mugil parsia*), tawes (*Puntius javanicus*) and a small number of carp hybrids (calbasu male × catla female). Sometimes a carnivore, chital (*Notopterus chitala*), is added to control weed-fishes. However, this carnivore should be smaller in size (2 inches to 3 inches) so that they do not prey on carp fingerlings.

Although polyculture envisages compatible combinations of species or age groups that do not compete in feeding habits, but there is often considerable overlap of the feeding habits and preferred food items. When several species are added in ponds, the concept of exclusive ecological niches and special separation are not always applied. Recent combinations are generally based on one or two species as the main ones (depending upon the market demand) and the others as subsidiary compatible species that will utilise the unutilised food resources or would help in reducing the deterioration of water quality by feeding on the wastes.

(C) **Stocking density and ratio** — Several stocking rates in different combinations is practised in different areas depending upon the market demand and the size to which it is grown. Under normal management, a stocking rate per ha of 5000–6000 fingerlings of 2.5–5 cm length is grown to a size of 600-1000 g or more in one year or less. A lower stocking density 4000-5000 is recommended if the fishes are to be grown to marketable size in a short time.

Different species of fishes are stocked in varying ratios. The commonly stocked ratio of catla, rohu and mrigal in the ponds of Bengal are 3 : 3 : 4 (Alikunhi, 1957). Recently, the most commonly stocking ratio is catla 30 per cent, rohu 60 per cent, and mrigal 10 per cent. If calbasu is included, then the percentage of catla is reduced to 50 per cent to accomodate calbasu at 10 per cent. These ratios are altered to achieve higher rates of production :

(i) depending on the market demand of the species cultured,
(ii) in accordance to the primary production in the ponds,
(iii) through more intensive stocking and supplemental feeding with locally available foodstuffs (oil cake, rice bran, etc.) and
(iv) through introduction of fast growing exotic carps (grass carp, silver carp and common carp).

III. **Post-stocking management**

Culture of carps in pond is mainly based on fertilisation and supplementary feeding.

A. **Pond fertilisation** — Fertilisation with inorganic and organic manure helps to meet the requirement of carbon, nitrogen, phosphorous and other nutrients. It is a known fact that manure increases plankton and chironomid production in fish ponds, probably due to high production of bacteria and protozoa developing on the organic matter of the manure. Detritus or the decomposing organic matter has a high protein content.

The stocking density and the environmental conditions of the pond determine the rate of application of fertilisers. Moreover, the dosage and mode of application is also important, whether the entire quantity has to be applied in one lot or sporadically. Too much of nutrients, resulting from manuring may lead to the development of algal bloom, particularly under Indian conditions. It may cause clogging of the gills and also depletion of the oxygen concentration in the ponds. Manuring, thus, should not be done during persistent cloudy weather or when algal bloom appears.

(a) **Application of organic manure** — Raw cow dung is generally used as an organic manure. Under most situations a dose of 100-120 kg (dry matter) per day can be safely used under most situations. Generally in India, the application of organic manure is 10,000-20,000 kg (wet weight/ha/year).

(b) **Application of inorganic manure** — Using chemical fertilisers is advantageous since the nutrient contents are generally of standard which would facilitate the selection of required dosage. In intensive culture, where fishes are stocked in higher densities, fertilisation at a dose of 60 kg/ha of single superphosphate and 60 kg/ha ammonium sulphate, every two weeks is considered to be beneficial. In polyculture carp ponds where supplementary feeding is adopted, nitrogen, phosphorous and potassium fertilisers are applied at a ratio of 18 : 8 : 4 at the rate of 500 kg/ha/year.

(c) **Liming**—Liming should be done monthly, 1-2 days after the application of organic manure at the rate of 25 kg/ha (i.e., 300 kg/ha/year).

(d) **Raking** — One day after liming, raking of pond bottom should be done (if possible) for proper mixing of lime and also for the release of abnoxious gases formed in the bottom soil.

B. **Supplementary feed** — The spawns after being stocked, start feeding voraciously on zooplankton. Within 2-3 days of stocking, the natural food available in the pond gets depleted. At this time artificial feed along with natural food enhances the growth and survival of spawns. Farmers in India use processed feed mixture of pellets along with fertilisers in carp ponds.

For Indian carps the commonly given feed are rice/wheat bran and oil cakes of ground-nut, coconut, mustard, etc. The husk of the rice/wheat bran are first removed by passing it through a fine-meshed sieve and also to ensure uniformity of particle sizes. Powdered oil cakes and rice/wheat bran, comprising 1 : 1 mixture are given at the rate of 3% of the body weight from the second day of stocking (Mahapatra *et al.*, 2006).

In combination of the above, feed may also comprise powdered algae and aquatic weeds, fish meal, poultry droplets, vitamins, etc. Tripathi *et al.* (1979) recommended a diet of 1 : 1 : 1 mixture of groundnut oil cake, rice bran and fish meal containing 26.7% protein and 32.5% carbohydrate.

Artificial feed may either be sprayed at a fixed place on the water surface, during specified hours or fed as a thick paste or dough in small shallow earthen vessels or bamboo tray, kept suspended in water.

C. **Periodic netting for growth and health care** — Netting operations should be undertaken regularly at least once or twice a month to enhance fish growth and at the same time to see whether they are affected with any disease or not. Healthy pond environment ensures good fish growth. If it is seen that the supplementary feed provided is not being consumed, then either the fishes are suffering from some disease or the water quality has deteriorated. The water quality should be checked and remidial measures should be undertaken.

Carps are generally affected with a number of diseases. The major diseases encountered in polyculture ponds in India are fin and tail rot, **sapro-**

legniasis (it is a fungal disease caused by *Saprolegnia*, affecting the skin and gills), Ichthyophthiriasis (it is caused by the protozoan parasite *Ichthyophthirius multifiliis*), etc. Fishes may also be infested by *Argulus*, *Lernaea* and *Ergasilus*. Affected fishes are sorted out and proper treatment is given to them.

As carp culture is done in intensively manured ponds with dense stocking of fishes, it often results in deficiency of oxygen. Oxygen deficiency in carp ponds is one of the major causes of large-scale mortality, particularly in tropical regions. The problem of anoxia generally occurs during night when photosynthesis and oxygen production ceases and the only source of dissolved oxygen is the atmospheric oxygen which gets dissolved in the water due to wind and water movements. So, proper monitoring and necessary action should be taken to rectify it. The two most common measures adopted are aeration and addition or exchange of water.

D. **Control of diseases** — Prophylactic treatments with dip or bath in 0.38% potassium permanganate are given to fishes as a precautionary measure. Therapeutic treatments are given to cure parasitic infections and bacterial diseases. In case of frequent infections by the copepod parasite (*Argulus* sp.), treatment with three consecutive applications of malathion at the rate of 0.1 mgl^{-1} is given at ten days interval.

E. **Final netting for harvesting and marketing** — Composite fish culture is generally practised for a period of 10 to 12 months, after which the entire stock is harvested in case of multi-stage stocking. During this period catla attains a weight of 800 gm–1 kg, Rohu 600-800 gm, Mrigal 400-600 gm, Silver carp 1.0-1.2 kg, Grass carp 1.0-1.5 kg and common carp 800 gm –1.0 kg. If all the above mentioned polyculture techniques are followed the yield may vary between 2000 kg/ha/year and 5000 kg/ha/year, depending upon the stocking composition and density,

In case of ponds that are undrainable, the most common means of harvesting is done by seine net (Fig. 6.34). This net is suitable for harvesting most species of fish except for common carp that may escape by burrowing into the mud bottom. Species

Fig. 6.34. Harvesting a small-sized pond by seine net

like silver carp, mullet and milkfish can escape by jumping over the net. In case of drainable ponds with proper harvesting sump or similar device, harvesting is done by draining. This is generally done when the fishes are to be sold in live condition. Some farms have **marketing ponds** where the live fishes are temporarily kept till they are marketed. After harvesting, the catches are sorted out species-wise and size-wise before marketing.

F. **Economics of polyculture** — Carp culture is undoubtedly economically viable in places where there is a market for carp. The two major factors that determine economic viability are : (1) consumer acceptance and (2) price levels in the market.

The initial cost of farm construction and equipment is high but with proper maintenance, ponds can be used almost indefinitely. Moreover, the food requirement for the fishes can be met partly or fully through pond fertilisation and the supplementary feed given is of low cost, this helps to cut down considerably one of the major costs of production. Further the cost of producing fry or fingerlings can be kept down through adoption of less expensive techniques. All these factors undoubtedly keep down the expenditure at a minimum and help to get a good profitable return.

DRAWBACKS IN POLYCULTURE SYSTEM

1. Generally the lesser number of species combinations with clearly understood specific roles are widely accepted. However, in case of larger species combinations, the question has arisen as to the value of some of the species in such combinations.

2. Although there are clear-cut differences in food preferences in natural conditions, the relevance of such feeding differences is yet to be assessed when the stocks are fed with formulated feeds.

3. It is not very easy to adopt supplementary feeding in an economical way.

4. Special skilled knowledge and efforts are required by the farmers to produce or purchase the appropriate numbers of seed stocks of the different species selected to maintain optimum balance of the species.

5. Additional labour is required for sorting out the different species after harvest.

6. In many instances, consumer acceptance of the different species varies in most areas. Difficulties have also been experienced in finding markets for certain species like the silver carp.

It has been found and proved that modern monoculture with proper feeding can be more productive than polyculture. Thus, it is opined that the value of polyculture depends very much on the situation and needs in a particular area. Polyculture probably is not widely applicable as was generally considered.

NUTRITION AND FEED DEVELOPMENT IN FISHES

The growth and activity of fishes depend upon the food it takes, not only on the quantity but also on the quality and size of the food object. Thus, fish nutrition is of great significance in fish farming, particularly in intensive culture.

Fishes do not spend energy on maintaining body temperature (poikilotherms) and in the maintenance of its position in water. The diet it takes is used in protein building material and energy requirement for metabolic activity. A deficient diet may cause nutritional diseases and other disorders that would adversely affect its growth.

Traditional aquaculturists depend totally on the production of natural food through the use of fertilisers and pond management. This practice, to some extent, is followed in extensive and semi-intensive farming. However, they resort to supplementary feeding to ensure adequate availability of food in dense stocking and for enhancing growth and production.

I. **The concept of feeding fishes**

(a) **Classification of various stages of aquaculture** — Depending on the quantity and quality of nutrients utilised by the aquaculturist, the levels of aquaculture are :

(i) *Extensive* — no nutrients added.

(ii) *Fertilisation* — feed given to enhance production of aquatic organisms.

(iii) *Supplemental feeding* — incomplete feed given.

(iv) *Intensive feeding*—nutritionally balanced feed.

(v) *Hyperintensive feeding*—High inputs of concentrated nutritionally complete feeds.

(b) **Fish production exclusively from Natural Aquatic Food** — Some fishes are planktivorous. They obtain their food exclusively from plankton. These fishes are usually grazers and have mechanism for filtering and concentrating the suspended plant and animal organisms from the water. Example of such types are catla, rohu, silver carp, etc. Tilapia, on the other hand, has the ability to feed both on plankton and on bottom materials. Mrigal is an efficient bottom feeder, while grass carp consumes large quantities of aquatic weeds. Such fishes can be cultured without artificial feeds, but usually with application of fertilisers.

(c) **Supplemental feeding** — In this case along with natural food, various feedstuffs or prepared feeds are used as a supplement to increase fish yield.

In prawn culture the impoundment is first fertilised to enhance natural productivity and in addition pelleted feeds are also given. This increases the yield of prawns. Supplemental feeding is used where natural aquatic food makes small contribution to the total protein and energy requirement of the cultured fish.

(d) **Intensive and hyperintensive feeding** — In these system of culture, natural foods are an insignificant source of nutrients. Here, primary concerns are minimum accumulation of unretained nutrients and obtaining maximum yield per unit space and effort. Thus, such a culture needs highly concentrated, nutritionally complete feed.

(e) **Nutrient requirement** — Like any other animals, fishes require nutrients for growth, reproduction and other normal physiological functions. Fishes have dietary requirements for the same 10 essential amino acids like mammals. However,

(i) the energy requirements are lower for fishes,

(ii) fishes require some lipids such as omega-3 (n-3) series of fatty acids and sterol,

(iii) fishes have the ability to absorb soluble minerals from the water and

(iv) most fishes have limited ability to synthesise ascorbic acid and thus, depend upon dietary sources.

Nutritional requirements do not vary greatly among different species of fishes. However, the nutritional requirements have been derived from experimentation upon a few species that have served adequately for the nutrient needs of others. Similarly, the ration yielding the best growth and yield has been assumed to be the most satisfactory feed for that species.

(f) **Feeding practices** — As fishes are fed in water, it may so happen that a substantial amount of feed may remain unconsumed. Such happening may not only result in economic loss but also can reduce the water quality. Therefore, the aquaculturists must consider factors such as feed allowance, feeding method and water stability.

(g) **Efficiency** — The efficiency of conversion of feed into body tissue has been found to be more efficient in fishes than in land animals. The reason for this is that the fishes are able to assimilate diets with higher percentage of protein, probably due to their lower dietary energy requirement. This lower energy requirement is due to a lower maintenance requirement and lower heat increment. Thus, fishes have a lower energy cost of protein gain, rather than superior food conversion efficiency.

The **food conversion rate** is the weight increase relative to the amount of food (which is given). It expresses the ratio between the food consumed per unit weight gained by the body.

$$\text{Food Conversion Rate} = \frac{\text{Feed intake}}{\text{Weight increase}}$$

The food conversion rate is also denoted as food quotient, food coefficient, growth coefficient, etc.

$$\text{Conversion efficiency} = \frac{\text{Weight gain}}{\text{Feed intake}} \times 100$$

II. **Feed requirements**

(a) **Bioenergetics** — Energy is not a nutrient. Energy is released during metabolic oxidation of carbohydrates, fats and proteins (amino acids). Bioenergetics or nutritional energetics is the study of the balance between energy intake (in the form of food) and energy utilisation by animals for life-sustaining processes such as tissue synthesis, various activity and self maintenance. For productive purposes, the difference between gross energy and energy available to the animal, varies widely among food materials. Digestibility accounts for most of the differences in available energy among feedstuffs for fishes.

(b) **Energy Balance** — The fate of energy balance in fishes is given in Fig. 6.35. There are two types of energy generally witnessed — heat energy that is lost for maintaining body temperature and the free energy utilised for growth and biological activities. The latter one is more important for fishes as they are poikilotherm animals. Energy metabolism in fishes, like other cold-blooded animals, is different from warm-blooded animals in that they do not expend energy to maintain their body temperature, different from its environment. In fishes, the excretion of waste nitrogen requires less energy as they can be readily released into the water through the gills.

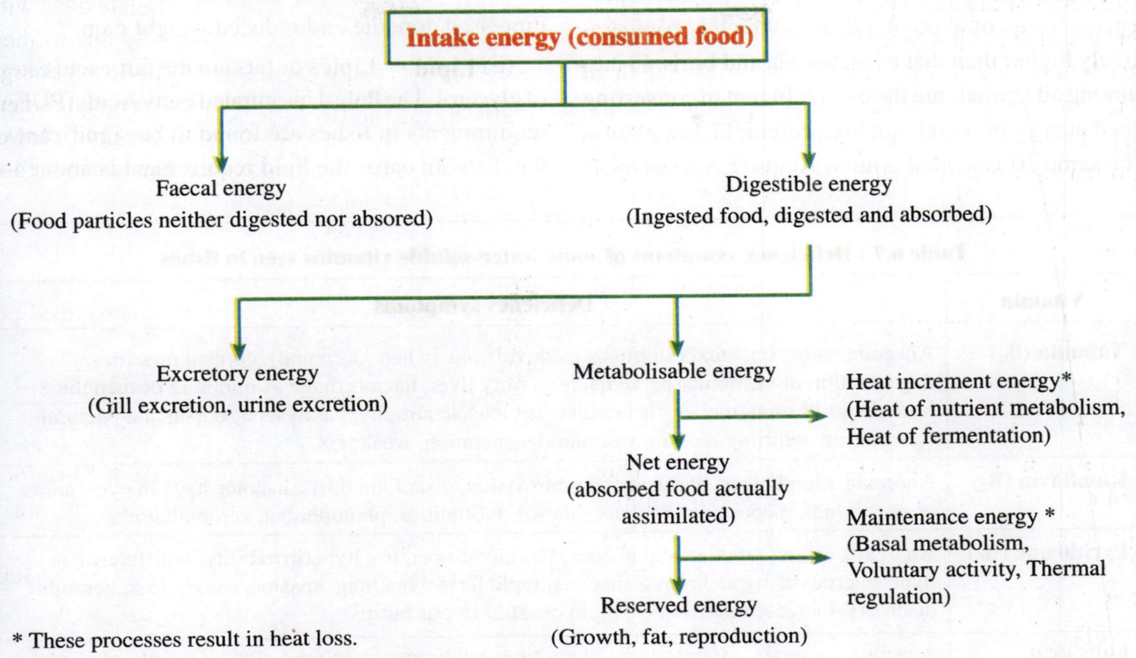

Fig. 6.35. Partitioning of feed energy consumed by fish

The metabolic rate in fish is dependent upon temperature, age or size, activity and seasonal and diurnal fluctuations of body functions. It is also influenced by oxygen or carbondioxide concentration, pH and salinity of the water. Of the total energy utilised by carps (calculated to be 25 cal/dec^2/h at 25°C), about 70 per cent is used for maintenance and growth and the remaining 30 per cent is lost to the environment.

As is shown in Fig. 6.35, heat losses occur mainly by two ways — (a) heat increment that includes energy cost of digestion, nutrient metabolism and excretion and (b) maintenance energy represented by basal metabolism (such as respiration, transport of ions and metabolites) and voluntary activity. In fishes, the maintenance energy requirement is lower, which means a greater per cent of the net energy is not dissipated as heat but retained within the body of fishes as new tissue or recovered energy.

(c) **Energy requirements and sources** — As fishes live in an aquous environment, where carbohydrates are rare, their digestive and metabolic systems are better adapted for the utilisation of protein and lipids as sources of energy than carbohydrates. Even in some fishes like grass carp, that consume macrophytes (higher aquatic plants), their digestion of native cellulose is poor.

The energy requirements of fishes are generally supplied by carbohydrates, proteins, lipids, vitamins and minerals for maintenance of growth. Most of the available information on energy requirements have been derived primarily from experimentation. On an average the various requirements of the carps are given in Table 6.6.

Table 6.6 : Nutritional requirements of adult and fingerlings of carp

Nutrients	Adult	Fingerlings
Protein	40%	45%
Lipid	20%	6%
Carbohydrate	10%	26%

(i) **Protein** — In fishes protein has a metabolisable energy value of about 4.5 kcal/g, which is comparatively higher than that of mammals and birds. Fishes among all animals are the most efficient in converting feed energy into high quality protein. Fishes require the same 10 Essential Amino Acids (EAA) as other animals. Diet deficient in any of the EAA leads to dipressed appetite and reduced weight gain.

(ii) **Lipid** — Lipids or fats are the fatty acid esters of glycerol. The Poly-Unsaturated Fatty Acids (PUFA) requirements in fishes are found to be significant in the diets. In carps the lipid requirement is about 4-6

Table 6.7 : Deficiency symptoms of some water-soluble vitamins seen in fishes

Vitamin	Deficiency symptoms
Thiamin (B_1)	Anaemia, anorexia, ataxia (terminal), convulsions (when moribund), corneal opacities, degeneration of vestibular nerve nucleus, fatty liver, haemorrhage of midbrain or medulla, loss of equilibrium, melanosis in older fish muscle atrophy, paralysis of dorsal and pectoral fins, rolling, whirling motion, vascular degeneration, weakness.
Riboflavin (B_2)	Anorexia, cloudy lens, darkned skin, dim vision, discoloured iris, haemorrhage in eyes, nares or operculum, incoordination, lens cataract, mortalities, photophobia, xerophthalmia.
Pyridoxine (B_6)	Anorexia, ascites, ataxia, convulsions, flexing of opercles, hyperirritability, indifference to light, microcytic hypochromic anaemia, rapid jerky breathing, spasms, weight loss, nervous disorders, increased mortalities, rapid onset of rigour mortis.
Folic acid	Anaemia, anorexia, ascites, dark colouration, erythropenia, exophthalmic, fragility of caudal fin, lethargy, macrocytic anaemia, pale gills, poor growth.
Pantothenic acid	Anorexia, clubbed gills, flared opercula, gill exudate, general 'mumpy' appeaance, lethargy, necrosis of jaw, barbels and fins, prostration, poor weight gain.
Inositol	Anaemia, bloated stomch, poor growth, anorexia, skin lesions.
Biotin	Anaemia, anorexia, blue slime disease, colonic lesions, contracted caudal fins, dark colouration, erythrocyte fragmentation, mortalities, muscle atrophy, poor growth, spastic convulsions.
Choline	Anaemia, poor food conversion, poor growth, vascular stasis and haemorrhage in kidney and intestine.
Nicotinic acid (niacin)	Anaemia, anorexia, colonic lesions, edema of stomach and colon, incoordination, jerky movements, muscle spasma, lethargy, photophobia, swollen gills, tetany, skin haemorrhage, high mortality.
***p*- Aminoenzoic acid**	No significant change in growth, appetite or survival.
Cobalamin (B_{12})	Anorexia, erratic haemoglobin and erythrocyte counts, fragmentation.
Ascorbic acid (Vit. C)	Anorexia, impaired collagen production, impaired wound healing, lordosis with dislocated vertebrae and focal haemorrhage, poor growth, scoliosis with haemorrhage in severe cases, twisted deformed hyaline cartilage in gill filaments and sclera of eyes.
Vitamin A	Ascites, edema, exophthalmos, haemorrhagic kidneys.
Vitamin D	Rduced conversion.
Vitamin E (tocopherol)	Anaemia, ascites, ceroid in liver, spleen, kidney, clubbed gills, epicarditis, exophthalmic, microcytic anaemia, mortalities, pericardial edema, poor growth, red blood cell fragility.
Vitamin K	Anaemia, coagulation time prolonged.

Table 6.8 : Mineral deficiency symptoms of the nine minerals needed by fin fishes

Mineral	Deficiency symptoms
Calcium	Poor growth and feed efficiency,[1,8] high mortality.
Phosphorus	Skeletal abnormalities,[5,7,8] poor growth and feed efficiency and bone mineralisation.[1,3,5,7,8]
Magnesium	Renal calcinosis,[1] loss of appetite,[1,8] poor growth,[1,8] high mortality. sluggishness, skeletal abnormalities.
Iron	Hypochromic microcytic anaemia [2,7,9]
Copper	Poor growth[7]
Manganese	Poor growth,[1] short and compact body,[1] abnormal tail growth.[7]
Iodine	Thyroid hyperplasia.[6]
Zinc	Cataract,[1] caudal fin and skin erosion,[1,7] growth depression.[1]
Selenium	Muscular dystrophy,[3] exudative diathesis.[3]

[1] *Salmo gairdneri*, [2] *Salmo fontinalis*, [3] *Salmo salar*, [4] *Oncorhynchus keta*, [5] *Ictalurus punctatus*, [6] *Oncorhynchus tshawytscha*, [7] *Cyprinus carpio*, [8] *Anguilla japonica*, [9] *Chrysophrys major*.

per cent in the diet. Most commercial aquaculture diets have less than 10 per cent of lipid. High lipid levels may lead to fatty liver disease or cause greater deposition of visceral fat.

(iii) **Carbohydrate** — Carbohydrates, in animal aquaculture, are the most abundant but relatively least expensive source of energy. Adult stages of carps utilise less carbohydrates (about 10 per cent) than fry or fingerling (about 26 per cent) stages. Although common carp utilises higher levels of carbohydrates, their diets contain over 40% dextrin that results in retarded growth. Successful fish feed contains a certain amount of carbohydrate — 20 per cent for cold-water and 30 per cent for warm-water fish feed. Carbohydrates, in addition, have the physical function of texturising manufactured feed and also help as a binder in the formulation of pellets.

(iv) **Vitamins** — Vitamin requirements of fishes are derived from the natural food. In prepared feeds adequate quantity of vitamins is generally added. However, excess of vitamin A may cause enlargement of liver and spleen, abnormal growth of bone formation and epithelial keritonisation. For water soluble vitamins, some of the deficiency symptoms observed is given in Table 6.7.

(v) **Minerals** — Fishes, like all animals, require minerals for various biological functions such as formation of skeletal tissue, blood plasma, haemoglobin formation and also for many enzymatic reactions. Of the twenty six naturally occurring essential elements required by animals, only nine have been seen to be needed by fin fishes. The deficiency symptoms of these minerals is shown in Table 6.8. Water generally contains an abundance of minerals, so supplementation in diets is generally not needed.

III. Live food of fishes

(a) **Natural sources of live food** — Most aquatic animals subsist on live food comprising of both plants and animals. These they obtain from the environment in which they live. Some, however, ingest and utilises detritus along with associated benthus organisms. All these foods are generally rich in essential nutrients.

Larval stages of many fin fishes and crustaceans depend entirely on live food, and their initial source of food is phytoplankton (probably associated with the size of the larvae at hatching). After the larvae have developed to a certain length they feed exclusively on either zooplankton or other animal species or on a combination of plant and animal matter.

For hatchery use, it is more convenient to culture algae under controlled conditions. In nursery and grow-out ponds, live food are generally produced as a result of the biological cycle initiated by mineral nutrients in solution. Light penetration is an important factor, as the sun's heat and light is used to convert the inorganic matter and carbonic acid in solution into organic matter, in the form of vegetable tissues consisting of phytoplankton and periphyton. The major nutrients required by plants are phosphorus, nitrogen and carbon. The ratio of carbon : nitrogen : phosphorus required by most species of phytoplankton is about 106 : 16 : 1, respectively.

(b) **Culture of live food** — In recent times there has been greater interest in intensive forms of phytoplankton or zooplankton cultures, which are then released into the prepared ponds for multiplication. This strategy also helps in providing the fishes with their preferred food organisms.

(i) *Algal culture* — Now-a-days algal culture facilities form an integral part of many shrimp and prawn hatcheries. The most important activity of algal culture is seen in controlled breeding of molluscs (oysters), where they are used to feed the larvae.

Various species of microalgae are cultured in laboratories for experimental uses or in special tanks for commercial purposes. In some centres of *Macrobrachium* culture, a "Green Water Method" of algal production is generally practised. Green water is a mixed phytoplankton culture, with a pre-dominance of *Chlorella* sp.

(ii) *Zooplankton culture* — Zooplanktonic organisms can be obtained either by collection from natural waters or by culture under controlled conditions. Small-scale collection of zooplankton from natural waters using standard plankton game is common, but large-scale plankton collection is practised rarely either by use of two cone type of net or by phototactic reaction using underwater lights to attract plankton. However, there are many uncertainties in the quality and quantity of plankton that can be obtained. So, for using in hatchery production of larvae, it is advisable to depend upon culturing of plankton.

Besides, the culture of Rotiferans (*Brachionus, Filinia, Keratella,* etc), Cladocerans *(Daphnia, Moina,* etc.), Copepods (*Cyclops, Diaptomus*, etc.), cultures can also be done of Oligochaetes (earthworm, *Tubifex*) and chironomid larvae. Flat trays filled with manured soil and covered with water attract chironomid flies to deposit eggs in them. A production of about 0.3 kg/m^2/week can be obtained. *Tubifex* also can be cultured by using sewage and pig or horse manure as culture medium. It develops in masses within 15 days of innoculation and can be obtained by dredging the bottom with a spade.

(c) **Fertilisation of pond for production of live foods** — Pond management has led to the practice of adding fertilisers to augment the growth of fish food organisms like phyto- and zoo- plankton. Majority of pond culture of herbivorous and omnivorous species is based on food production through fertilisation. Sometimes it is combined with supplementary feeding with easily available foodstuffs. In India, the priority is always to produce fishes to support the low income population. So, use of supplementary feed is not feasible as it would increase the cost of the fishes. Secondly, manufacture of aquaculture feeds in India is not well-developed and thirdly, the design and operation of pond farms are best suited for the growth of natural food organisms.

The fertilisers used in aquaculture are organic and inorganic in nature or a combination of both types. Details of such fertilizers and their application is given in page 254.

IV. **Supplementary feeding and processed feed**

(a) **Supplementary feed ingredients** — In traditional aquaculture, various feedstuffs have been used from time to time. It comprises oil cakes, wheat and rice bran, mill waste, brewery wastes, slaughterhouse wastes (blood, etc.), bean residues, silkworm pupae, poultry wastes, fish offal, trash fish, fish byproducts, etc. These feedstuffs in various combinations are provided as mixtures in order to improve the food conversion ratio.

Besides the above mentioned feedstuffs, processed compound feeds are also used to supplement natural food. Such formulated feeds are generally aimed at providing additional protein and fat. But the feeding cost amounts to 40-60 per cent of the overall cost of production, mostly in intensive aquaculture where complete feeds are used. However, the cost of production can be reduced if less expensive supplementary feeds are used in combination with natural food. The composition and quantity of feeds to be used is based entirely on field trials and experience of the farmer.

Various feed ingredients and their composition are as follows :

(i) **Primary plant protein sources**

(1) **Soybean meal** — After removal of the oil from soybeans, the residue is soybean meal. It is of three types — dehulled and solvent extracted, solvent extracted and expeller extracted containing 48%, 44% and 42% protein, respectively. Soybean meal is highly palatable to most warmwater fishes and is the major protein source used in aquaculture feed containing all essential amino acids.

(2) **Cottonseed meal** — It is obtined by grinding the cake, the remains after extraction of the oil. Cottonseed meal contains about 41% protein but is deficient in lysine. It is generally used in preparing catfish feeds at a level of 10 to 15% of the total feed.

(3) **Peanut meal** — After the peanuts are grinded and the oil extracted (mechanically or by solvent extraction), the peanut meal is obtained. Mechanically extracted and solvent extracted peanut meal contains about 45% and 48% protein, respectively. Although they are highly palatable to fishes they are seriously deficient in lysine.

(4) **Sunflower meal** — Sunflower meal is obtained after extraction of the oil from the sunflower seeds. It contains about 44% protein. However, its low lysine content and high level of fibre (13%), limits its usefulness as fish feed.

(5) **Canola meal** — It is prepared from a special rapeseed after the removal of oil. It contains about 38% protein, but relatively low in lysine. Canola meal is low in glucosinolates and erucic acid which is considered detrimental to fish growth. However, it can be used up to a level of 20-25% in fish feeds.

(ii) **Primary animal protein source**

(1) **Fish meal** — Fish meal contains 60 to 80% excellent quality protein and is a good source of essential amino acids. It is also rich in energy, minerals, vitamins and essential fatty acids. However, it is used sparingly for preparation of many commercial fish feed due to its high cost.

(2) **Fish soluble** — Fish soluble is obtained after extraction of fish body and liver oil. It contains about 50% dry matter, 30% protein and 4% fat.

(3) **Meat and bone meal** — It is the rendered product from beef or pork tissue, but should not contain other materials such as hoof, hair, horn, hide, blood, intestine, etc. It contains about 50% crude protein and is a good source of minerals.

(4) **Blood meal** — Blood meal is a by-product from the meat packing plants and consists of clean, fresh animal blood. The protein percentage in blood meal is 80 to 86%. It is rich in lysine but deficient in methionine.

(5) **Poultry by-product meal** — It is made up of grounded carcass of slaughtered poultry comprising head, feet, undeveloped eggs and visceral organs. However, it does not contain feathers. This product contains good quality of 59% protein.

(6) **Poultry feathers, hydrolysed** — The clean, undecomposed feathers from slaughtered poultry is subjected to high pressure treatment to produce hydrolysed poultry feathers. Although its protein percentage is high (85%), the quality of the protein, however, is poor as it is deficient in several essential amino acids, particularly lysine.

(7) **Shrimp waste meal** — Shrimp waste meal is prepared from the discarded head, limbs and exoskeleton. The exoskeleton is primarily chitin and has limited nutritional value. Shrimp waste meat contains 40% crude protein and is a good source of fatty acids, cholesterol (essential for crustaceans) and astaxanthin.

(iii) **Energy supplements** — Energy supplements are those feedstuffs that contain less than 20% protein. These includes :

(1) **Corn products** — Grounded corn grain and corn screenings are used as a relatively inexpensive source of energy to feed commercial catfishes. Cooking the grains and screenings improve the digestibility of corn for fish. Corn prducts up to 45% have been used in warm-water fish feeds without any adverse effects.

(2) **Sorghum** — In sorghum grains the nutrient composition is similar to corn but the protein content is slightly higher (11%).

(3) **Rice products** — Rice bran is high in fat and fibre. This limits its used in fish feeds to less than 7%.

(4) **Wheat products** — Wheat grain is a good pellet binder, but is generally more expensive than corn. It is a good source of energy for warm-water fishes. Wheat products when used primarily as binder, a quantity of 2-5% is gnerally used.

(5) **Animal and plant fats and oils** — They are highly concentrated source of energy and acts as a source of essential fatty acids. Animal fats and oils are generally preferred over plant oils as they are less expensive.

(iv) **Vitamins and minerals** — Although most commercial feedstuffs used in preparing fish-feeds contain vitamins and minerals, still the fish-feeds are supplemented with vitamins and trace minerals as well as phosphorus.

(v) **Binding agents** — Binding agents (Table 6.9) are generally much in use for steam-pelleted crustacean feeds, which must remain stable in water for several hours. The stability of steam-pelleted feeds in water is dependent on the amount and type of binder used. Wheat flour and high-gluten are of common use as binder for srimp feed as it is cost effective.

(b) **Feed formulation** — The nutrient content of feed may vary according to the maturity and processing of feed materials and also in relation to their protein

Table 6.9 : Binding agents used in steam pelleted feeds for shrimp and finfish (relative effectiveness are applicable for shrimp feeds)

Binder	Amount to use (%)	Nutritive value	Relative cost	Relative Effectiveness
Guar gum	1–3	No	Expensive	Good
Lignon sulfonate	2–4	No	Inexpensive	Poor
Carboxymethyl-cellulose	2-6	No	Expensive	Fair
Hemicellulose	2-3	No	Moderate	Poor
Bentonite	2-3	No	Inexpensive	Poor
Alginate	2-3	No	Expensive	Good
Agar	2-3	No	Expensive	Good
Carrageenan	0.5 - 1	No	Expensive	Good
Ethylene/vinyl acetate copolymer	2-4	No	Expensive	Good
Wheat gluten	3-5	Yes	Moderate	Good
High gluten wheat flour	20	Yes	Moderate	Good
Collagen	0.5–3	Yes	Moderate	Good
Processed milo	3-5	Yes	Moderate	Fair
Cooked starch	> 20	Yes	Inexpensivew	Fair

fibre and ash content. While formulating feed, it is essential to check the Essential Amino Acid (EAA) levels in the ingredients. *Fish meal* is widely used as one of the ingredients as it is a good source of EAA and EFA (Essential Fatty Acid). However, due to the increased cost of fish meal suitable alternatives like oil-cakes and soybean are being used, although soybean has been found to produce reduced growth rate as compared to fish meal.

The main objective of feed formulation is to provide a diet that meets the nutritional requirements at different stages of life, so as to yield optimum production at minimum cost. While trying to keep the cost at minimum it should be seen that no compromise is done with the quality of the nutrient content.

The formulation of feed is also dependent upon the type of culture system. *Semi-intensive* culture does not require vitamin and mineral supplements, as these nutrients can be obtained by the fish from the natural food organisms present in the pond water. In *intensive culture,* on the other hand, natural foods are limited and their diet should contain all the required nutrients in adequate quantities and proportion.

Computation of protein in diet formulation — In animal diet, protein is the most expensive ingredient. In diet formulation, it is first computed and is done generally by two methods — (i) Square method and (ii) Algebraic equations.

Problem 1 : In a supplementary feed 25% protein is desired using fish meal (50% protein) and rice bran (8.0% protein).

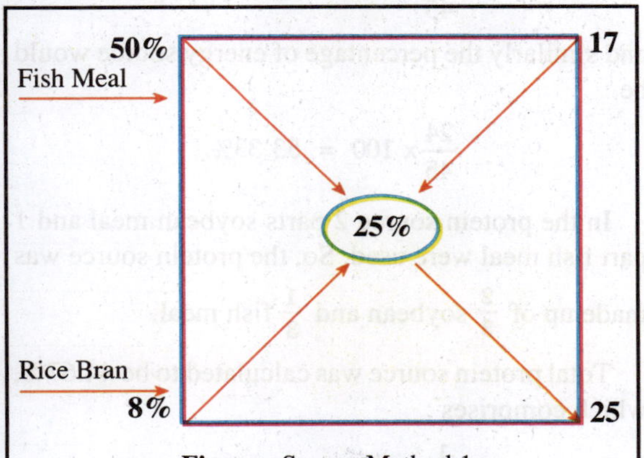

Figure : Square Method 1

Solving by the square method — Calculation can be done by both percentage basis and as ratio. For this a square is constructed as shown in **figure Square Method 1**.

In the above square the desired protein level (25%) is placed at the centre and two feedstuffs with their protein percentages (50% and 8%) are placed on the left-side corners of the square. Then, the protein levels of each feedstuff are subtracted from the desired protein level of the feed to obtain 25–50 = –25 and 25–8 = +17, respectively. The differences thus obtained (25 and 17) are placed on the corners of the square diagonally opposite to the feedstuff ignoring the '+' and '–' signs.

The proportion of fish meal needed in the feed can be computed by the difference between the percentages of protein in the rice bran and the protein required in the feed. Similarly, the proportion of rice bran required is the difference between the protein percentages of fish meal and the feed being formulated. The ratio thus obtained is 17 parts fish meal : 25 parts rice bran.

The value can also be calculated on percentage basis by the following way :

(i) The difference of percentage in the two feeds are 50–8 = 42. The figure '42' can also be obtained by adding the values on the right-hand corners of the above square : 25+17 = 42.

(ii) The percentage of fish meal required is :

$$\frac{17}{42} \times 100 = 40.48\%.$$

Similary, the percentage of rice bran required is :

$$\frac{25}{42} \times 100 = 59.52\%.$$

(iii) Thus, to prepare a balanced supplementary feed containing 25% protein, the proportions required are 40.47% fish meal and 59.52% rice bran.

Solving by the algebraic method

Suppose :

x = fish meal in kg/100 kg feed

y = rice bran in kg/100 kg feed

$\therefore$ X + Y = 100 kg feed (1)

Taking the protein into account, we get

0.50x + 0.08y = 25 kg protein/100 kg feed (2)

Multiplying equation (1) by 0.08, we get

$0.08x + 0.08y = 8.$ (3)

By substracting equation (3) from (2), we get

$0.42\,x = 17.$

$\therefore\ x = \frac{17}{0.42} = 40.48$ kg / 100 kg feed .

$\therefore$ fish meal required is 40.48%.

Similarly, for calculating the requirement of rice bran,

$y = 100 - x$

or, $y = 100 - 40.48 = 59.52$ kg/100 kg feed.

$\therefore$ rice bran required is 59.52%.

Problem 2 : In actual practice more than 2 ingrediants are used for feed formulation. As for example :

A 30% protein level is required using the following mixture :

Fish meal	60% protein
Soybean	51% protein
Rice bran	8% protein
Corn meal	10% protein

The above ingrediants are used in the following proportions :

2 parts soybean : 1 part fish meal

1 part rice bran : 1 part corn meal

Solving by the square method (2)

The desired protein level (30%) is placed at the centre of the square.

The ingradients are divided into 2 groups — protein source and energy source. The protein level of each group is calculated by the following way :

Protein source : The proportions given in the question is :

2 parts soybean : 1 part fish meal

$\therefore$ fish meal $= 1 \times 60 = 60\%$

Soybean meal $= 2 \times 51 = 102\%$

The average protein source $= \frac{162}{3} = 54\%$.

Energy source : The proportion given in the question is :

1 part rice bran : 1 part corn meal

$\therefore$ rice bran $= 1 \times 8 = 8\%$

Corn meal $= 1 \times 10 = 10\%$.

The average energy source $= \frac{18}{2} = 9\%$.

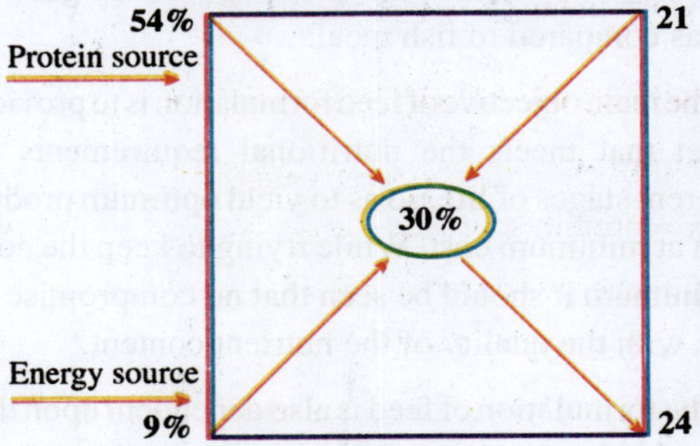

Figure : Square Method 2

The protein and energy source along with their calculated values of 54 and 9, respectively, are placed on the left-hand corners of the square. After diagonal subtraction the values 21(30-9) and 24 (30-54) are placed on the right side at the two corners of the square.

The difference of percentage of protein source and energy source are obtined by either

$$54 - 9 = 45$$

$$\text{or,}\quad 21 + 24 = 45.$$

The percentage of protein source would thus be :

$$\frac{21}{45} \times 100 = 46.67\%$$

and similarly the percentage of energy source would be :

$$\frac{24}{45} \times 100 = 53.33\% .$$

In the protein source 2 parts soybean meal and 1 part fish meal were used. So, the protein source was made up of $\frac{2}{3}$ soybean and $\frac{1}{3}$ fish meal.

Total protein source was calculated to be 44.67%, which comprises :

Fish meal $= \frac{1}{3} \times 46.67 = 15.56\%$

Soybean meal = $\frac{2}{3} \times 46.67 = 31.11\%$.

Similarly, the energy source of 53.33% is comprised of 1 part rice bran and 1 part corn meal which is calculated to be :

$$\text{Rice bran} = \frac{53.33}{2} = 26.665\%$$

$$\text{Corn meal} = \frac{53.33}{2} = 26.665\%$$

Solving by the Algebraic method

Suppose :

x = protein source in kg/100 kg feed

y = energy source in kg/100 kg feed

x + y = 100 kg feed. (1)

0.54x + 0.09y = 30 kg protein/100 kg feed. (2)

By multiplying equation (1) with 0.9, we get

0.09x + 0.09y = 9. (3)

By subtracting equation (3) from (2), we get

0.45x = 21.

∴ X = 21/6.45 = 46.67%

$$\text{Fish meal} = 46.67 \times \frac{1}{3} = 16.56\%$$

$$\text{Soybean meal} = 46.67 \times \frac{2}{3} = 31.11\%.$$

From equation (1), we have

y = 100 – x

∴ y = 100 – 46.67 = 53.33%.

$$\therefore \text{ rice bran} = \frac{53.33}{2} = 26.665\%$$

$$\text{Corn meal} = \frac{53.33}{2} = 26.665\%.$$

(c) **Types of processed feed** — Processed feed ensures that the cultured animal receives a balanced diet that meets their nutritional requirements. Processed feed also ensures the quality and availability of food. These feeds are formulated according to nutrient specifications and are usually expensive than supplementary diets.

(1) From **compositional point** of view three types of processed feeds are recognized :

(i) purified, (ii) semi-purified and (iii) practical.

(i) **Purified feed** — Such diets are made of synthetic amino acids, fatty acids, carbohydrates and chemically pure vitamins and minerals. Purified feeds are comparatively expensive. They are used only for research purpose.

(ii) **Semi-purified feed** — Such diets are made of natural ingredients in pure forms as much as are available (example : casein, fish oil, corn oil, etc.). Semi-purified feeds are commonly used in nutritional studies as test diets to determine the efficiency of different levels of dietary components in terms of food conversion or growth.

(iii) **Practical feed** — Such diets are made up of available ingredients and to a large extent natural products such as fish meals, oil seed meals, cereal grains, etc. Practical diets are aimed at keeping the cost of feed at the minimum.

(2) Based on **water content** artificial compound feeds can be classified into — (i) moist feed and (ii) dry feed.

(i) **Moist or wet feed** — Such feeds are prepared without exposure to heat such that it does not result in nutritional losses. Grinding and mixing machines are only used. Moist feeds are used for practical or test diets. Species fed on moist diet show better feed conversion ratios and growth. Such diets are to be used immediately after preparation as they are susceptible to spoilage. Otherwise freezing is required to prevent spoilage by micro-organisms and deterioration of oxygen-sensitive nutrients, such as ascorbic acid.

(ii) **Dry feed** — Dry feed contains only about 8-11 per cent water and are relatively more water stable. The water stability of such feeds can be adjusted to suit the feeding habits of the cultured species by using appropriate binders and processing techniques. Similarly, by using extrusion techniques of processing, the density of the pellets can be adjusted to make them float or sink in water. Dry feeds have the advantage of easy storage and transport.

Different sizes and shapes of pellets with different nutrient compositions are easily adjusted by the use of suitable diets. These are needed to feed successive growth stages of a species. For feeding of newly hatched larvae, the sizes vary from finely ground granules to encapsulated forms, while from small pellets to larger pellets of about 1 cm diameter and 1.5 cm length are used to feed the adult fishes. The composition of the feed is thus related to the nutrient requirement of each life stage. For example, specially made high protein diets are used to feed the brood fishes.

(3) There are various **specialised feeds** that are in use, such as —

(i) **Larval feeds** — Newly hatched larvae of fin fishes and prawn depend on live food such as unicellular algae, rotifers and nauplii of brine shrimp for better survival. Use of prepared feeds has been successful for some species. As cultured live foods (phytoplankton, zooplankton, etc.) are expensive, larval feeds have been developed to replace them. However, such larval feeds used to feed larval fishes in the hatchery should be less than 0.5 mm in size, palatable, water stable and nutritionally complete.

(ii) **Meals and crumbles** — These feeds are used for feeding fish fry and small fingerlings and have small particle size. Meal type feed is generally prepared by reducing the particle size through grinding and screening to a particle size of not more than 0.5 mm and mixing the ground ingredients. Crumbles are formed by crumbling pelleted feeds and then screening to appropriate size.

(iii) **Microencapsulated feeds** — Small particles or beadlet of feed when coated with a thin layer by a compound which will reduce the dissolving and leaching of nutrients, are called **microencapsulate feed**. Several patented processes for microencapsulation of larval feeds are in use such as nylon (N-N bonds) cross-linked proteins, calcium alginate and lipids. These materials should be non-toxic, water insoluble and easily digestible by the fish larvae.

(iv) **Flaked feeds** — For aquarium fishes feed should be nutritionally complete, palatable, should float or sink slowly and should not disintegrate quickly in water. The ingredients are first ground to fine particle size with an attrition mill and blended with water to form a sludge, that is spread evenly over the surface of a heated rotating drum, so as to dry it into a thin sheet. The dried sheet is then continuously scrapped off from the rotating drum and crumbled into flakes. Flakes contain ingradients with good hydrocolloidal properties and also good tensil strength.

(v) **Medicated feeds** — Medicated feeds contain antibiotics. There are two such antibiotics, Tetramycin (oxytetracycline) and Romet (sulfadimethoxine-ormetoprim), that are available commercially. Tetramycin is used in steam pelleted feeds as it is heat sensitive and destroyed during extrusion. On the other hand romet is heat stable and can be used in extruded feeds. Mills preparing such feeds should be licensed to add these antibiotics to their feed.

(vi) **Crustacean feeds** — As crustaceans are bottom feeders, they require diets that will remain stable in water for a longer period of time and will not float. Feedstuffs with good binding properties along with special binding agents should be used in steam pelleted feeds. Thus, binding agents are important for water stability of the feeds.

(d) **Feed preparation** — Although specialised aquaculture feed mills are increasing steadily yet there are large areas that are not yet served by commercial feed milling for aquaculture. In such cases aquaculturists will have to resort to on-farm production of feeds on a small scale. The first pelleted carp feed in India was prepared at Freshwater Aquaculture Research and Training Centre (FARTC), at Bhubaneswar in 1982.

(i) **Small scale production** — Feeds can be produced in the farm using inexpensive equipment and locally available ingredients. The equipments include weighing machines, sieves, mixer (small, of 5-10 kg capacity), meat grinder, corn meal or coffee grinder, a steamer and bamboo basket for steaming, a saucepan for gelatinising starch and a drier.

The locally available ingredients, in the relevant formula are finely ground and passed through a sieve. It is then weighed and mixed thoroughly. Oil is then added and mixed again for a few minutes. The starch, gelatinised with water, is then mixed thoroughly with the mixture, until it forms a dough.

This dough is then passed through a grinder using a 1-3 mm die. The feed that extrudes out is cut into small sizes (about 0.5 cm) and then steamed to make water-stable pellets. The steamed product is then placed on an oven and dried overnight at 60°C. Such pelleted feed is made once a week as per the need of the farm.

(ii) **Large scale production** — For requirement of feed in large-scale, the method described above is not feasible. Large-scale milling is necessary, although the basic process involved are essentially the same. It also consists of particle size reduction, premixing of micro-nutrients, mixing of all components of the diet, pelleting, cooling and then sacking (Fig. 6.36).

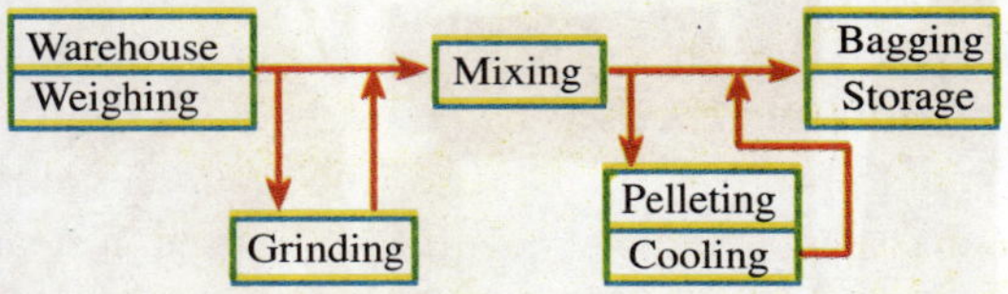

Fig. 6.36. Schematic representation of feed milling

(1) **Grinding** —Coarse ingredients are ground in a hammer mill. Grinding reduces particle size, facilitates handling of ingredients, improves feed digestibility, acceptability, mixing properties and pelletability (Fig. 6.37a).

(2) **Sieving** — Sieving is done to separate materials into sizes as per the need. Feeds, for example, are passed through 177 μm meshes. If there is excess of dust entering the grinder, then it may be controlled by sparying oil or by adding a semi-moist ingredient, such as condensed fish solubles or fermentation solubles.

(3) **Mixing** — Mixing is done to achieve uniformity of composition in the whole feed material as per the required formula. It involves scattering of particles and blending. For formula feed, the technique which is generally adopted is continuous mixing of proportions by weight or volume (Fig. 6.37b) & (Fig. 6.37c).

Various types of mixers are in use in feed mills — vertical mixers, continuous ribbon mixers and liquid mixers. To get proper mixing, addition of ingredients in a tested sequence is required, where large volume of ingredients are added first, followed by smaller amounts.

(4) **Pelleting** — Pelleting involves passing a feed mixture through a conditioning chamber where 5-6 per cent water or steam is added. Water provides lubrication for compression and extrusion. In the presence of heat, raw starch present in the ingredients of plant origin, helps in adhesion (Fig. 6.37d & e).

After entering the pellet mill, the feed is first air-dried and then given 15-16% moisture at 80-90°C. During subsequent compression and extrusion through the die, the feed temperature may increase to nearly 90°C due to friction. Pellets are then discharged from the pelletizer to a screen belt of a horizontal tunnel drier or a vertical screened hopper. The pellets are then air-cooled in about ten minutes. It is then dried to decrease its moisture content to below 13 per cent.

(5) **Binders** — When the feed mixture is difficult to pellet or is unstable in water, it is essential to add a binding agent (Table 6.9) to the formulation. Normally starch can act as a binding agent. Other successfully used binding agents include carboxymethyl cellulose, gelatin, algin (extract from marine algae, *Laminariaceae*), etc.

(6) **Granules and crumbles** — The cooled pellets are sometimes ground on corrugated rolls into various sizes of granules and crumbles (small fragments). The crumbles are used to feed small fishes, and are more easily consumed by sight-feeders. They are less expensive to manufacture than small size pellets.

(7) **Types of pellet** — Generally two qualities of the pellets are important : hardness and water stability. **Hard pellets** are special, liked by many species. However, there is always a danger of overfeeding (with hard pellets), which causes swelling and rupture of the stomach. The hard pellets are not digested thoroughly and causes fermentation and gas formation.

Feeds pelleted by the extrusion process forms **floating pellets** which are suited for certain types of culture. Advantage of this type of pellet is that the fish can be observed while feeding and the amount of feed can be regulated according to the amount eaten. Thus,

Fig. 6.37(a). Grinder

Fig. 6.37(b). Mixer

Fig. 6.37(c). Horizontal mixer

Fig. 6.37(d). Pelletiser

Fig. 6.37(e). Pellets in drier

Processing of floating pellets

Conditioning of feed containing 25-30% water
↓
Pressure cylinder
↓
Steam treatment to increase gelatinisation of starch
↓
Extrusion at atmospheric pressure through the holes of a die plate
↓
Extruding ribbon is cut by a rotating knife outside the die plate
↓
Pellets are dried at 120°C in an oven
↓
Pellets put in a cooler to lower the temperature after internal moisture becomes less than 13%

it improves feed conversion. Floating pellets are processed as follows :

(8) **Packing of the prepared feed** — The pellets are usually packed in polyethylene-lined sacks made of multilayered paper or other material.

(9) **Storage** — Raw materials and processed feeds both should be properly stored to prevent them from deterioration during storage. Loss of raw materials may occur due to insect infestation. Feed materials generally undergo chemical changes that alter their flavour and nutritive value. However, properly stored feed will keep well for 1-3 months in summer and 2-4 months in winter.

In a feed mill, care should be taken to store both feed materials and processed feeds in cool and dry conditions as much as possible. They should be raised off the ground on platforms. Warehouses or silos should be constructed in such a way that the interior remains cool and dry with adequate ventilation.

(10) **Contaminating agents** — Fats in feedstuffs often break down during storage. **Rancidity** (unpleasant smell or taste) is caused due to oxidation of lipids. Lipid oxidation can be inhibited by the addition of antioxidants such as ethoxyquin, butylated hydroxytoluene (BHT) and butylated hydroxyanisole (BHA).

Recontamination of feedstuffs by micro-organisms is another major hazards. Fungi grow at moisture contents of 15-20% and cause spoilage of feedstuffs in storage. The highly toxic and carcinogenic **aflotoxins** produced by *Aspergillus flavus* are one of the most important mycotoxins contaminating feedstuffs.

(11) **Finishing products and quality assurance** — The finished products are routinely assayed for moisture, protein, fat and fibre. Selected micronutrients are periodically checked to ensure nutritional quality. Each batch of feed should be checked for physical characteristics, such as pellet stability and floatability. Guaranteed analyses for commercial feeds is required for most states and countries.

(e) **Feeding rates** — Initially the fry (about 0.25g) requires feed as much as 10% of their weight daily. After attaining a weight of about 4 g, the feed is reduced to about 5% of the body weight per day. The total amount of feed to be given is generally estimated by multiplying the weight representing the desired growth by the conversion rate.

$$\text{Feed conversion ratio} = \frac{\text{feed intake}}{\text{weight gain}}$$

V. **Feed dispensing methods**

(a) **Hand-feeding** — Hand-feeding is the most common method of feed dispensing. Its advantage is that the farmer can observe the feeding behaviour of the stock and ensure that the feed is made available to all the fishes. 'Disc throwers' and hand-operated blowers are generally the mechanical aids used for hand feeding. In large fish ponds boats are also used for hand feeding.

(b) **Automatic feeders** — Various types of automatic feeders are used for prepared feeds in larval tanks and grow-out areas.

Electrically-activated timing devices are used to dispense set amount of feed at certain intervals. The dispersal patterns are of different types :

(1) Feed are dropped directly into the water from storage bin.

(2) Feed may be dispersed widely by falling onto a rotating disc.

In another model, electrical impulses are received by a vibrator present just inside the opening at the bottom of a conical feed hopper. A small plastic disc is present underneath the hopper. Pellets fall onto the disc, which vibrates each time impulses are received and are thrown into the water.

Mobile mechanical feeding systems such as blowers are also used to dispense feed over large areas. Mechanisms using compressed air to distribute pellets over the water has also been successfully used in raceway cultures.

(c) **Demand feeders** — Demand feeders are activated by animals themselves by means of a rod or plate hanging from the feeder into the water. The rod or plate is struck by the moving fish mistaking it as a bait. This results in the release of a small amount of pellet from the hopper. The fish soon learns to activate the feeder and it generally does so when it wishes to feed.

(d) **Computer controlled feeders** — Computer controlled feeders are used by advanced countries, where the quantity of feed and frequency of feeding are regulated through appropriate programs. It takes into account several variables such as stock density, growth rate, conversion rates, temperature conditions, mortality, harvest time, etc.

EXOTIC FISHES

Exotic fishes are those species that are non-indegenous having their origin in another country and has been introduced into the Indian waters. They generally have established culture technologies and the economics of production and marketability. For such proven exotic species, there is a strong argument for their introduction and transplantation. But, the problem often faced is whether it is necessary to introduce such exotic species and what are the procedures and precautions to be taken to prevent probable undesirable consequences.

Past practices of indiscriminate introductions and transplantations for ornamental purposes and for biological control have had detrimental effects on the local fauna and have contributed to the spread of diseases. Thus, to prevent such consequences there is a need for following appropriate procedures and effective national regulations. However, for expanding aquaculture it is very difficult to avoid the introduction of such species for experimentation or commercial production. Many such species (tilapia, common crap, chinese carp, rainbow trout, Japanese and European oysters, etc.) have already colonized outside their geographical distribution range, the majority of them had been introduced for valid reasons. It is, however, a matter of question whether such successful introduction has been preceded by detailed screening procedures.

Selection of exotic species

For introduction of exotic species Turner (1949) had suggested certain criteria. The characteristics for the selection of exotic species are as follows :

(1) Should be a fast grower, so as to compensate the dearth of protein.

(2) They should fulfil the need because of the absence of a similar desirable species in the locality.

(3) Such species should be compatible and should not compete with valuable native species leading to its decline.

(4) Should not undergo cross-breeding with native species and produce undesirable hybrids.

(5) Should be devoid of pests, parasites or diseases, so that these may not attack the native species.

(6) Should live and reproduce in equilibrium with the new environment and should not cause any adverse ecological effects.

(7) Should be hardy, easy to handle and have other usefulness (multiple benifits). For example, grass carp clears the pond of weeds and its excreta fertilises the pond water.

(8) They should have proven technologies of culture and good economic viability.

Procedures and precautions

The above criteria hold good for introduction of exotic species but critical evaluations have to be done about the consequences of introductions. Data have to be obtained through experimental studies of the rationale for introduction, advantages and disadvantages of the candidate species and the assessment of impacts on the environment. The results should be critically evaluated before the final decision is taken for general introduction or transplantation. The procedures for the introduction should be on the following lines as recommended by the Internationl Council for the Exploration of the Sea (ICES, 1979).

(1) In an approved quarantine environment, a brood stock of the species to be introduced should be established and sufficient time should be allowed for observation.

(2) The first progeny of this species should be examined thoroughly for any sign of diseases or parasites.

(3) If these exotic species are parasite- or disease-free, then they are transplanted to the natural environment. Prior to introduction, all effluents of the quarantine units are disinfected in an approved manner.

(4) This introduced species in its new environment should be kept under constant and continued study and vigilance.

Predictions of ecological effects of the newly introduced species based on controlled experiments has many limitations. Such error can be minimized if the monitoring is carried out on a long-term basis and there would be sufficient time to adopt necessary measures.

Culture of exotic fishes and their role

The first batch of exotic fishes was introduced by MacIvor in 1874. He introduced two cold-water fishes, the tench (*Tinca tinca*) and golden carp (*Carassius carassius*), into the Ooty lake. Attempts to introduce these fishes into the plains have met with less success.

Subsequently, from time to time, a number of exotic species have been transplanted into the Indian waters to serve various purposes. A list of these has been given in Table 7.10.

A. **Exotic game fishes**. Trout and salmon are the exotic game fishes that has been introduced in India. The initial introduction of trout in India was made during the latter half of nineteenth century and early twentieth century by the European residents of the country. Most of these attempts were not successful. It was in 1906 that Wilson succeeded on transplanting trout in Nilgiris and also putting its culture on a scientific footing.

The species of trout introduced are brown trout, *Salmo trutta fario* (Fig. 6.38a), rainbow trout *Oncorhynchus mykiss* (previous nomenclature; *Salmo gairdneri gairdneri*) (Fig. 6.38b), and eastern brook trout, *Salvelinus fontinalis* (Fig. 6.38c). Trout culture is also practised in the north and its seeds have been introduced in lakes, streams and reservoirs. For its propagation a number of trout hatcheries have been set up in Nilgiris, Travancore (hill regions), Garhwal, Kashmir, Himachal Pradesh, etc. Culture of trout mainly includes the following :

(i) Egg collection and artificial fertilisation by 'dry method'.

(ii) Incubation of fertilised eggs.

Table 6.10 : Some exotic fishes, their transplantation in India

Species Introduced		Source	Introduced to	Year of introduction	Purpose of introduction
Common name	**Scientific name**				
A. GAME FISHES					
1. Brown trout	*Salmo trutta fario*	England	Nilgiris (T.N)	1906	For planting in streams, lakes and reservoirs.
2. Loch Leven trout	*Salmo levensis*	England	Nilgiris	1863	For planting in streams, lakes and reservoirs.
3. Rainbow trout	*Oncorhynchus mykiss*	Sri Lanka, Germany	Nilgiris	1907	For planting in streams, lakes and reservoirs.
4. Eastern brook trout	*Salvelinus fontinalis*	Canada	—	1959	For planting in streams, lakes and reservoirs.
5. Atlantic salmon	*Salmo salar*	USA	Kashmir	1968	For planting in streams, lakes and reservoirs.
6. Sock-eye salmon	*Oncorhynchus nerka*	Japan	Nilgiris	1968	For planting in streams, lakes and reservoirs.
B. FOOD FISHES					
1. Crucian carp/ Golden carp	*Carassius carassius*	England	Nilgiris	1874	Experimental culture.
2. Tench/ doctor fish	*Tinca tinca*	England	Nilgiris	1874	Experimental culture.
3. Grass carp	*Ctenopharyn-godon idella*	Hong Kong	Cuttack (Odisha)	1959	Experimental culture and weed control.
4. Common carp (scale carp)	*Cyprinus carpio communis*	Bangkok	Cuttack	1957	Experimental culture.
5. Common carp (mirror carp)	*Cyprinus carpio specularis*	Sri Lanka	Nilgiris	1939	Experimental culture in hilly regions.
6. Silver carp	*Hypophthal-michthys molitrix*	Japan	Cuttack	1959	Experimental culture.
7. Giant Gourami	*Osphronemus goramy*	Java & Mauritius	Tamil Nadu	1916	Experimental culture.
8. Tawes	*Puntius Javanicus*	Indonesia	Kalyani (West Bengal)	1972	Experiental culture and weed control.
9. Tilapia	*Tilapia mossambica*	Bangkok, Sri Lanka	Mandapam (Tamil Nadu)	1952	Experimental culture.
C. LARVICIDAL FISHES					
1. Mosquito fish	*Gambusia affinis*	Italy		1928	Mosquito control.
2. Guppy	*Poecilia* (= *Lebistes*) *reticulata*	South America		1908	Mosquito control and for aquariums.

Table contd. to next page

Table 6.10 : (continued)

Species Introduced		Source	Introduced to	Year of intro-duction	Purpose of introduction
Common name	**Scientific name**				
C. LARVICIDAL FISHES					
3. Red tail notho	*Nothobranchus*	Africa *guentheri*	Tamil Nadu		Mosquito control.
D. ORNAMENTAL FISHES					
1. Live bearers	27 species	From various countries			Aquarium keeping.
2. Egg layers	261species	From various countries			Aquarium keeping.
E. UNAUTHORISED INTRODUCTION					
1. Big head Carp	*Aristichthys nobilis*				Aquaculture.
2. African Catfish	*Clarias gariepinus*				Aquaculture.
3. Nile Tilapia	*Oreochromis niloticus*				Aquaculture.
4. Red Tilapia	*Oreochromis sp.*				Aquaculture.
5. Red Piranha	*Serrasalmus nattereri*				Aquaculture.

(iii) Rearing of alevins and swim-up fry.

(iv) Production of fry and fingerlings.

(v) Rearing of adults for market as well as for further seed production (brood fishes).

Although a good number of hatcheries for trout propagation have been set up in Himalayan and Penin-sular states, the picture as to the production of trout is concerned, is not clear. The causes for the decline of trout fishing are :

(i) Defective setting and hatchery design, where the glacier fed streams carry a heavy load of silt, particularly during rains, which leads to mass mortality of seeds.

(ii) Nutritional imbalance caused by malnutri-tion.

(iii) Infective diseases during incumbation period of the egg and fry. If left uncared, may cause heavy mortality resulting in losses to the trout propagation programmes.

(iv) Constraints in developing commercial farms. Lack of reliable data on cost benefit, finance, trained manpower, eco-biological and engineering research.

B. **Exotic food fishes**. From time to time some exotic food fishes (Fig. 6.39) have been introduced into the Indian Waters (Table 6.10) as similar indigenous species were not available. A few of these exotic species, important from the view point of fish culture, are discussed below :

(a) **Tilapias**. Tilapias are natives of Africa. In India the first consignment of tilapia was brought by the Central Marine Fisheries Research Institute, Mandapam on 7th August, 1952 from Bangkok (Table 6.10, Fig. 6.39g). A second consignment was brought by the Madras Fisheries Department from Sri Lanka in the same year. Later in December 1953, a small stock of tilapia from Mandapam was transferred to the Pond Culture Division of CIFRI at Cuttack, to make a detailed study of its biology and culture and the probable effect of its culture on the Indian major carps.

Jhingran and Gopalakrishnan (1974) have listed 22 species of tilapias that have been used for experimental culture or for aquaculture. Tropical aquaculturists were initially excited as these fishes were handy, could breed and be reared in almost any type of water (fresh, brackish or marine), were herbivorous or omnivorous and were comparatively easy to feed. For these favourable characters, tilapias were considered to be ideal for rural fish farming.

However, some of these perceived advantages were later found to be of real constraint, particularly the prolific breeding, which resulted in overpopulation of small stagnant ponds. This created opposing reaction in countries like India, where restriction in culture was tried to prevent its possible escape into the major river systems. Moreover, their early maturation and frequent breeding, affected its growth rates, such that producing marketable sized fishes was not as easy as originally believed. Further, external feeding was found to be necessary to produce fish suitable for human consumption. There were very few profitable tilapia farms during that time and naturally interest in tilapia culture dwindle for some time.

Later, due to efforts by enterprising farmers and advanced aquaculture techniques, interest in tilapia culture again rose to the effect that it began to be described as the future 'aquatic chicken'. Despite tilapia being cultured presently on a profitable basis, due to easy workable technologies being available, still many problems have remained unsolved.

(b) **Common carps**. *Cyprinus carpio*, a native of China, is the worldwide most extensively cultivated species and is believed to be the earliest forms of fish to be cultured. It is considered to have been domesticated. Though considered to be a sport fish as well, its prime importance is as a food fish. There are three recognised varieties of common carp—(i) *Scale carp* which is orange coloured, (ii) *mirror carp* which is partially scaled and (iii) *leather carp* which is virtually scale-less. The scale carp and the mirror carp are the varieties that are preferred for culture.

The German strain of common carp (*Cyprinus carpio specularis*) (Fig. 6.39d) was introduced into the Nilgiris from Sri Lanka in 1939 (Table 6.10). It, however, did not breed successfully in the plains. Hence, in 1957 the Bangkok strain, *Cyprinus carpio communis* (scale carp) (Fig. 6.39c), was brought to India from Bangkok and introduced into the Pond Culture Division of CIFRI at Cuttack. Here, detailed investigations on various cultural aspects have been carried out. This strain breeds freely in the plains of India and has been distributed by CIFRI to the various states of the country for culture.

Although *Cyprius carpio* is exotic to India, it has already been cultivated singly and also along with

(a) (b) (c)

Fig. 6.38. Exotic Game fishes (a) *Salmo trutta fario* (Brown trout); (b) *Oncorhynchus mykiss* (Rainbow trout); (c) *Salvelinus fontinalis* (Eastern brook trout)

Fig. 6.39. Exotic Food fishes : (a) *Tinca tinca* (b) *Ctenopharyngodon idella* (c) *Cyprinus carpio communis* (d) *Cyprinus carpio specularis* (e) *Hypophthalmichthys molitrix* (f) *Osphronemus goramy* (g) *Male and female Tilapia mossambica*

Indian major carps. Its popularity in aquaculture is due to its efficiency in converting the food ingested into flesh, grows very fast and is prone to artificial feeds. It is voraciously omnivorous, feeds on a wide variety of plant and animal matters, is non-predatory when cultured along with Indian major carps (competes for food somewhat with mrigal and kalbasu) and is apparently greater disease resistant.

The culture of common carp, however, has a shortfall in the fact that its habit of sucking food organisms in the mud on the pond bottom and margins makes the water muddy and weakens the base of pond dikes (an embankment for controlling or holding back the water and also serves as a causeway — a raised road or path between low or wet grounds). This is why anglers consider common carp as a pest in sport waters. However, this habit of common carp can be greatly reduced if they are fed on artificial feedstuffs or processed feeds.

Another shortfall is the effect of common carp on the endemic cold-water fisheries of Kashmir, Himachal Pradesh, etc. The lakes of Kashmir, which once used to be full of native fauna with maximum members of snow-trouts (**Schizothoracid**) than anywhere else in India, are now full of common carp. Not only the numbers have dwindled but also the size and weight of these much prized snow-trouts have gone down considerably over the years. The reasons for the decline of snow-trouts from the lacustrine systems are competition for food, the high fecundity of common carp, auto-spawning in the lakes, better fertilisation, shorter incubation period, high growth rate, etc.

(c) **Grass carp**. Grass carp *Ctenopharyngodon idella* (Fig. 6.39b), is the natural inhabitant of the flat-land rivers of China and the middle and lower sections of river Amur in Russia. The first consignment of grass carp comprising 382 fingerlings (5.5 cm in length), which came to India from Hong Kong, was introduced into the Pond Culture Division of CIFRI, Cuttack in December, 1959. In due course they were successfully bred by hypophysation at Cuttack in 1962 and the fry were then distributed to the other states of India. Gift consignments of grass carp fry were sent for culture to other countries like Philippines in 1964, Nepal in 1966 and Burma (Myanmar) in 1968.

Although a freshwater fish, grass carp can tolerate slightly brackish water. Grass carp has a toothless mouth but has strong and specialised pharyngeal teeth (which occurs in two rows) for rasping aquatic vegetation. It is, therefore, herbivorous and feeds on macrovegetation, including grass and aquatic plants. It starts feeding on macrovegetation when about 2.5-3.0 cm in length and can ingest plants up to 50% of its weight.

Although grass carp is not of high economic value, still it has been a popular species for aquaculture due to its ability to control macrovegetation. It is a fast grower, compatible and its excreta fertilises the pond water. Grass carp, however, when cultured with prawn, makes serious inroads in the habitat of prawn and the two thus cannot be said to be compatible (Sinha, 1978).

This species has acclimatised itself in Indian conditions and can breed even in weed-infested water bodies and can restrict the growth of economically important fishes. To prevent the breeding of this species efforts have been made to produce all-male grass carp by administration of methyltestosterone, however, with little success.

(d) **Silver carp**. *Hypophthalmichthys molitrix* or silver carp (Fig. 6.39e) is the native in the rivers of China and Amur basin of Russia. A consignment of 360 fingerlings of silver carp was brought to India from Japan in 1959. It was introducd to the Pond Culture Division of CIFRI at Cuttack, Odisha. These fingerlings after attaining maturity were bred successfully by Hypophysation and their fries were distributed to other states of India.

Hypophthalmichthys is a pelagophil and planktophagous species. The larval stages of this fish subsists on unicellular algae while the fry and adult feed on Flagellata, Dinoflagellata, Myxophyceae, Bacillariophyceae, Protozoa and Rotifera along with decayed macrovegetation and detritus. It also thrives well on supplementary feeds such as bone meal, rice bran and flour.

Silver carp although a freshwater species, like grass carp, can tolerate slightly brackish water. Although silver carp is not of high economic value still it is cultured due to its ability to utilise the dense bloom of phytoplankton and its food conversion ratio is higher than that of the Indian major carps. This fish although of low-valued is suited for large-scale production to feed poor communities in rural areas.

(e) **Giant gourami**. The giant gourami, *Osphronemus goramy* (Fig. 6.39f), occurs naturally in the freshwaters of Indonesia, Thailand, Malaysia, Cambodia and Vietnam. The giant gourami was brought to Tamil Nadu from Mauritius in 1865, but it failed to establish itself. Later in 1916 it was successfully transplanted to Tamil Nadu from Java and Mauritius. They can live in swampy conditions and breathe atmospheric oxygen through accessory respiratory organs. They easily spawn under pond conditions and are very palatable. The larvae feeds on plankton, while the adults subsist mainly on aquatic vegetation. Under emergency condition the fish is known to turn omnivorous feeding on insects, frog, small fishes, etc.

The giant gourami is a highly priced species.In spite of that, its earlier interest in culturing has diminished somewhat due to its slow growth rate and is now restricted to small-scale extensive farming.

(f) **Tawes**. The tawes or *Puntius javanicus* cultivation is of little interest. It was introduced in India to determine its efficiency for weed control. In September

1972, a consignment of 500 fry was obtained from Indonesia and brought to the Kalyani Fish Farm, Kalyani (West Bengal) for observation under the **All India Coordinated Project on Composite Fish Culture and Fish Seed Production.**

Although the tawes and grass carp are used for weed control, *P. javanicus*, unlike grass carp, requires well oxygenated water for growth and survival. Therefore, tawes is unsuitale for culture in polluted waters. Thus, grass carp are better candidate for the eradication of weeds than the tawes.

C. **Larvicidal or Larvivorous exotic fishes**. The term 'larvivorous fishes' comes in connection with its importance in controlling mosquito menace, as these fishes prefer insect larvae over other food. These fishes have little food value and are of immense importance in public health. They have been found to be cheaper, more effective and more permanent than artificial methods of mosquito conrol.

Larvicidal fishes should possess the following characteristics :

(1) They should be small, so that it can move about in shallow water among weeds.

(2) They should be hardy and flourish in deep shallow and foul waters.

(3) They should be prolific breeders and breed freely in confined water areas.

(4) They have suitably designed mouth so as to pick up eggs, larvae and pupae effectively in quick succession. The mouth is obliquely directed upwards. The lips are thin making the jaws work like a pair of forceps.

(5) They are top-feeders and carnivorous.

(6) They are generally compatible with the other fishes in that environment.

(7) They are difficult to catch and can evade or escape from their natural enemies.

(8) They are absolutely worthless and insignificant as food.

(9) They should be able to stand transport and handling.

For mosquito control, larvivorous fish should be introduced into a water body after clearing dense surface vegetation and removing predatory fishes, so as to allow the larvicidal fishes to establish themselves. The breeding place of mosquitoes is generally in shallow waters which is about 22.5 cm wide adjoining the bank. There are a number of indigenous and exotic species, which at one stage or the other feed on mosquito larvae. The exotic species which have been in use in India (Table 6.10) are *Ctenopharyngodon idella, Carassius auratus, Poecilia reticulata* (Fig. 6.40b), *Gambusia affinis* (Fig. 6.40a) and *Nothobranchus guentheri* (Fig. 6.40c). The latter three has been widely and effectively used. The world famous 'guppy' *Poecilia reticulata* and 'mosquito fish' *Gambusia affinis* are popular both in mosquito control as well as for aquarium.

Nothobranchus guentheri, a native of east and central Africa, its small scale field introduction in certain parts of Tamil Nadu has been attempted. They have a very peculiar breeding habit for which they are called as **'annuals'**. This is because the habitat in which they live dry up annually and before it does so, *Nothobranchus* deposits drought resistant eggs in the mud. When the rain sets in, the mud-embedded eggs hatch and aquire maturity within a year. They are probably the fastest known grower among fishes as they grow from egg to adult in four weeks' time. Their introduction into any impoundment does not have any adverse effect on the local fauna. Once *Nothobranchus* has established itself, malaria is likely to be controlled.

MERITS AND DEMIRITS OF EXOTIC FISHES

Along with anthropogenic activities, the depletion of freshwater aquatic biodiversity has also been reported to have been caused by the introduction of exotic species. Introduced (exotic) species are those that have been intentionally or accidentally transported and released by man outside its present range. It has been reported that 168 species of fishes, representing 37 families have been worldwide introduced outside their natural distribution range. Of these, a minimum of 67 species have become established in different water bodies and of it, 27 species have turned out to be real pests.

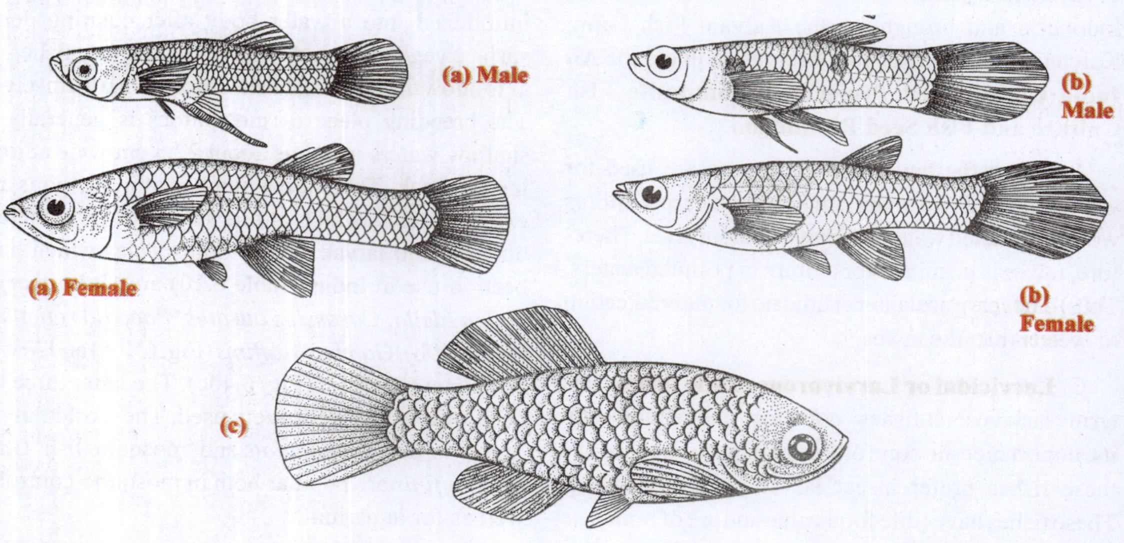

Fig. 6.40. Exotic Larvicidal fishes : (a) Male & Female *Gambusia affinis,* (b) Male and Female *Poecilia reticulata* and (c) *Nothobranchus guentheri.*

In India several exotic species of fishes have established themselves in natural water bodies. Studies on their impact in Indian waters have been carried out, which have revealed unequivocally the devastating impacts of these exotic fishes (such as Tilapia, Silver Carp, *Gambusia*, common carp, etc.) in Indian aquatic ecosystems. Some of their merits and demirits are discussed below :

I. Merits of Exotic Fishes

Although India has an array of indigenous species (about 637 fish species), yet a number of exotic species has been introduced due to their extraordinary merit wich is absent in the indigenous species or to increase the beauty of aquarium. Fishes has been introduced mainly in two ways :

(a) Scientists of the fisheries department who often investigating the pros and cons found it to be suitable for introduction which would be beneficial to fishers.

(b) Unauthorised introduction either by (i) the fishers who to gain economical milage have introduced the fishes (evading the government outline), or (ii) through any small water outlet and accidental release through the agency of fish-eating binds and mammals.

The beneficial impact of exotic fishes are summed up as follows :

(1) The exotic carps, due to their fast growth rate (high food conversion ratio), have become good candidates for the composite fish culture practice. They contribute handsomely to the yield and can solve the scarcity of protein in India. Grass carp has been reported to have attained a weight of 5 kg in a year.

(2) Common carp can be made to breed in confined waters of pond and thus, the farmers need not go to the rivers for their seed collection during the breading season.

(3) Common carp being omnivorous and a scavenger, keeps the water of an impoundment (pond) clean and pollution-free.

(4) Silver carp in Indian warmer condition, matures at the age of one, while in its native habitat it matures at the age of 5-6 years.

(5) Silver carp has the ability to utilise the dense bloom of phytoplankton and help in keeping it under check.

(6) Grass carp can feed on a wide range of aquatic vegetation and is used for controling of weeds.

(7) The excreta of grass carp contain semidigested food which serves as food for omnivorous fishes and also serves in fertilising the pond water.

(8) About 288 species of beautifully coloured exotic ornamental fishes are very popular among the aquarists.

(9) The larvicidal fishes (Guppy, *Gambusia affinis*) have been introduced for mosquito control.

(10) The game fishes (trout and salmon) have been planted in hill streams, lakes and reservoirs for leasure time angling.

(11) *Puntius javonicus* was unauthorisedly introduced in India for culturing in rice fields.

(12) Tilapia, often called as 'super-fish' and considered as 'chicken of water', breeds easily (has high fecundity), cheap to feed, can tolerate wide range of temperature and salinity. They are comparatively free from parasites and diseses. Due to their hardy nature they are ideally suited for culture in sewage-fed ponds.

(13) Tilapia needs less management for culture as they can feed on a wide range of food, attain maturity within 2 months, can breed in confined waters and have good parental care.

II. Demerits of Exotic Fishes

The indiscriminate intrdocution of exotic fishes has brought about a number of problems. Not many studies have been carried out on the impacts of exotic species in Indian waters. However, reports from elsewhere provide a bleak scenario of the damges caused by such introduction, which is true to Indian conditions. Some of the disadvantages encountered are listed below :

(1) The introduction of African Cichlid (Tilapia) in India has been claimed by many experts as a successful event. However, unanticipated impact on both freshwater and brackish water fisheries has now been felt. Being a prolific breeder and hardy in nature, Tilapia has overcrowded many water bodies, resulting in prevention of growth of IMC in pond waters.

(2) Introduction of Tilapia in Bharathapuzha (the longest river in Kerala) has resulted in replacing native fish fauna in many areas. It has also been reported that frequent gill net fishings to collect Tilapias have resulted in disappearance of freshwater turtles in certain water bodies.

(3) The introduction of *Cyprinus carpio* into Dal lake and Loktak lake has affected the indigenous fish population.

(4) Common carp has the habit of sucking food organisms in the mud on the pond bottom and margins, that makes the water muddy and weakens the base of the pond dyke, resulting in erosion. For making the pond water muddy, *cyprinus carpio*, has been considered by anglers as a pest in sport waters.

(5) In Govind Sagar reservoir the introduction of Silver carp has a negative impact on fish diversity.

(6) In India the culture of exotic carps is uneconomical due to its low cost. The flesh of these fishes are poor in quality or have undesirable grass-smell (grass carp).

(7) The introduction of Guppy and Mosquito Fish (*Gambusia*) in India has a negative impact on aquatic biodiversity. The Mosquito Fish are hardy and prolific breeders, and are capable of entering into the microhabitats of rare native species and have been reported as predators. The famous Ichthyologist Mayer (1965) labelled this species as a "fish destroyer". IUCN (1986) has also reported that the cause of extinction of a number of species worldwide has been due to introduction of Guppy.

(8) Sport fishes introduction into Indian waters has been considered previously as non-problematic. At present trouts have been reported not only to compete with native stocks but also have been found to be a major predator (Rainbow Trout) on the eggs and young ones of native species.

(9) The fishes illegally imported to India consist of the carnivorous and voracious feeders such as the Big-head Carp (*Aristichthys nobilis* Fig. 6.41a) and the African Catfish (*Clarius gariepinus* Fig. 6.41b), and the aquarium fish, the Red Piranha (*Serrasalmus*

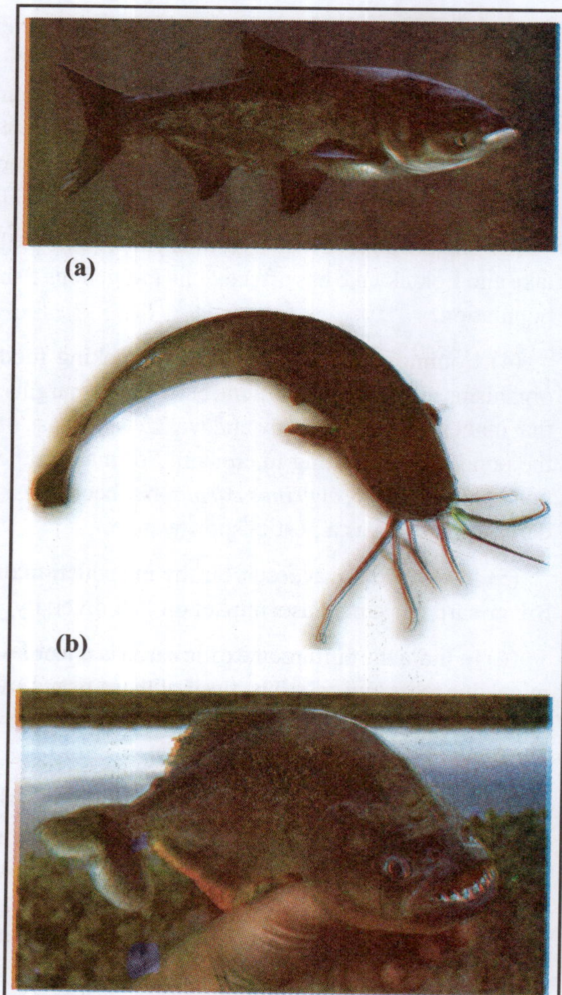

(a)

(b)

(c)

Fig. 6.41. Some unauthorised introduced exotic fish species : (a) Big-head Carp; (b) African Catfish; (c) Red Piranha

nattereri Fig. 6.41c). These exotic fishes in natural water bodies have become a very serious threat to the smaller indigenous fish species as well as invertebrates.

The threat posed by these fishes has led the Union Agriculture Ministry to order the killing of these fishes *en masse*. However, such Government order has failed to have any impact. Even today the African catfish and piranhas have been bred in the farms of West Bengal and Kerala.

(10) Along with the introduction of exotic fishes, transfer of parasites and diseases has also been reported from the Indian waters. This may lead to the formation of much dangerous mutated forms.

(11) Introduction of exotic fishes to different habitats in India may also lead to loss of biodiversity.

Thus, it can be concluded that exotic fishes not only compete with indigenous species for food and habitat, they may even be preyed upon. The exotic species introduce new parasites and diseases, resulting the production of hybrids and causing genetic 'erosion' of indigenous species and degradation of the physico-chemical nature of aquatic ecosystem. All these subsequently will lead to the loss of biodiversity. The potential risk also affects the socio economic aspects of the human community, as they depend upon the aquatic ecosystem for their sustenance.

CONSERVATON OF AQUATIC BIODIVERSITY

The need of the hour is to check the entry of exotic species as our rich freshwater biodiversity is in peril. India being one of the megadiversity countries with respect to freshwater fish diversity stands eighth in the world and third in Asia. Thus, there are plenty of culturable species and any further introduction of exotic fish species is unnecessary.

Taking into account of the numerous negative impact of exotic fishes on aquatic bodies, it is necessary to frame stringent regulations and laws for the import of non-native fishes. Generally, the introduction of exotic species should be cleared by the National Committee on Introduction of Aquatic Species in Indian Waters, New Delhi. This committee is represented by fisheries development commissioner and experts, who will study the relevance of import and the potential impact that the new non-native species is likely to produce on Indian aquatic environment. However, in spite of banning the illicit import and trade of exotic fishes (particularly in the case of piranha and African catfishes), it has continued unabatedly and there are no effective methods nor the political will to enforce the law. It is thus necessary to revise the existing laws and procedures for their stringent implementation.

Necessity thus demands the protection of the existing indigenous fish stock. The indigenous fishes should be incorporated into the value system of the society, such as sports, aesthetic, biological control, etc. It thus merits proper investigation of the utility of all indegenous fish species. A baseline data should be developed. Extreme risk areas should be identified for conservation programmes and such water bodies should be declared as **fish sanctuaries**.

CULTURE OF AIR-BREATHING FISHES

Air breathing fishes are characterised by the possession of an accessory respiratory organ. This organ enables the fishes to remain for hours out of water or for indefinite period in oxygen poor waters and even in moist mud. They are extremely hardy with respect to all environmental parameters and are suited to shallow foul waters, weedy waters of ponds and swamps. These fishes have commercial value as they are marketed in living condition and hence they are also referred to as 'live fishes' or 'jeol fishes'. The air-breathing fishes are highly priced and are well-known for their high protein, high iron, low fat-content and easy digestibility and are thus recommended by physicians as diet during convalescence.

The culture of these fishes involves low risk and simple management. Inspite of their high economic value, their cultivation has not been widely practised. Generally, due to their carnivorous feeding behaviour they are eliminated from polyculture ponds. However, if the culture of these air breathing carnivorous fishes is undertaken in shallow water pond (which is unsuitable for carp culture due to the low height of standing water), it would prove a more profitable venture. Moreover, these fishes adapt excellently to supplementary feeding with dried marine trash fish, oil cake and rice bran. Magur (*Clarius batrachus*) and Singhi (*Heteropneustes fossilis*) are known to be feeding on gastropods, microcrustaceans, tubificid worms, insects and larvae. Koi (*Anabas testudineus*) is generally micro-plankton feeder in younger stages and takes to insectivorous feeding habit in later stages. Koi also responds favourably to feed oriented culture with oil cake and rice bran.

Air-breathing fishes, at present, form the bulk and main stay of tank fisheries in India. The states of Andhra Pradesh, Assam, Bihar, Karnataka, Madhya Pradesh, Meghalaya, Maharashtra, Orissa, Tamil Nadu, Uttar Pradesh and West Bengal, support the most significant natural fishery of air-breathing fishes.

Advantage of culturing Air-breathing Fishes

Air-breathing carnivorous fishes are cultivated due to the following reasons :

(1) They have high protein, high iron and low fat content and are thus more nutritious and of high medicinal value.

(2) They have a high market demand and fetches a high price than other edible fishes.

(3) As air-breathing fishes have accessory respiratory organs, their culture can be undertaken in rejected foul water of shallow ponds, swamps, weedy waters and other kinds of oxygen depleted water bodies.

(4) They are hardy fishes and can adjust to any environmental conditions.

(5) They require little rearing.

(6) Their culture involves less risk.

(7) Their culture management is simple without much complications.

(8) They grow faster (particularly the exotic magur) when sufficiently fed.

(9) They are voracious carnivores and devours any type of flesh.

(10) In case of cage or pen culture, the yield is more and hence the income increases many-fold.

(11) They are highly recuperative.

(12) Their culture can lead to reclamation of previously unused derelicts.

Cultivable species in India

The cultivable species (Fig. 6.42) among catfishes (presence of small to very long barbels like the whiskers of cats) are Indian magur (*Clarias batrachus*) and Snighi (*Heteropneustes fossilis*). Among the murrels (snake-headed fishes) are the giant murrel (*Channa marulius*), stripped murrel (*C. striatus*) and spotted murrel (*C. punctatus*) are the cultured species. While in the case of climbing perch the species favourable for culture is Koi (*Anabas testudineus*). Among the two exotic magur cultured of commercial importance, the African catfish or Nile catfish *C. lazera*/*C.gariepinus* (Fig. 6.41) (often wrongly called as hybrid magur by common man) has received greater attention because of its voracious (carnivore) eating and faster growth rate. The other exotic catfish *C. macrocephalus*, with slower growing rate, has a higher consumer preference.

Cultural practices of air-breathing fishes

The culture of air-breathing fishes is done by the following ways :

(a) Pond culture

(b) Swamp culture

(c) Cage and pen culture

(d) High-density culture

A. **Cultivable areas** : (a) **Ponds** — Culture of air-breathing fishes is equally adaptable in carp culture ponds as well as in waters unsuitable for conventional culturable species of carps. They can be grown under poor environmental conditions. Culture is generally done in shallow water area of 0.1 ha with a depth of 2-3 feet (or 50-75 cm). Shallow water depth has an advantage as the fish has to spend less energy in travelling to surface for intake of atmospheric oxygen. Cultivation of these fishes is also done in nursery ponds during the period these ponds remain fallow (after harvesting of carp fry).

(b) **Swamps** — India has an estimated 0.6 million hactare of water that has remained unutilised for fish production. This area comprises marshes and swamps whose difficult ecology has made it derelict. West Bengal, Kerala, Bihar and Assam has such extensive areas under swamp conditions.

Fig. 6.42. Air-breathing fishes cultured in India — (a) *Clarias batrachus* (b) *Heteropneustes fossilis* (c) *Channa marulius* (d) *Channa striatus* (e) *Channa punctatus* and (f) *Anabas testudineus*.

These water bodies are potentially rich but are difficult for fish culture. Reclamation has been made possible with the introduction of cage-culture method of raising air-breathing fishes. The two main obstacles of swamp culture that can be overcome with cage culture practices are :

(1) The risk of raised fish being lost can be eliminated through cage-culture.

(2) The deoxygenated conditions of swamps which leads to mass kill, can be eliminated through culture of air-breathing fishes.

Preliminary experiments has revealed a production of over 1.2 tons/ha/7 months culture with singhi, magur and koi. Added to this would be the income through culture of '**makhana**' cash crop from the same swamp.

(c) **Cage culture** — Cage culture of air-breathing fishes is generally undertaken in swamps, running water systems like streams, raceways and channels and unmanageable water bodies like reservoirs. The cages used for culture may be *floating nylon cages* of 1 m × 1 m × 1 m, supported with wooden reapers, or *bamboo cages* of 2 m × 1 m with a depth of 0.8 m. Constant flow of water helps to bring food and wash away the waste products. Cage culture of air-breathing fishes particularly in the case of climbing perch (koi) has eliminated the risk of raised fishes being lost, thereby increasing the yield. Fry of these fishes are sometimes grown in floating baskets. The feed given in cages are the same as given in ponds.

Koi stocked at the rate of 50-100/cage and fed (at the rate of 10% of body weight) on rice bran, mustard oil cake and silkworm pupae can give a production of 1.3 kg/m^3/3 months of culture. Magur with a high stocking rate of 200/cage and fed (10% of body weight) with dried trash fish, oil cake and rice bran can yield 9.9-12 kg/m^3/1 year. As air-breathing fishes fetch a high price in the market, it may permit even enhanced expenditure to improve their culture system. This would lead to higher yield in the long run.

(d) **High density culture** — High density tank culture of air-breathing fishes in recirculating water system has been tried on an experimental scale. Although this has been found to be profitable yet it has not been adopted for commercial production.

B. **Pond management**: Pond culture of air-breathing fishes is essentially a short duration one. Pond management is very simple. Unlike carp culture, the ponds for air-breathing fish culture need not be fertilised by chemical or organic fertiliser. The only input materials are the fingerlings (6-10 gm each) and feed. In case of very heavy stocking rate and multiple cropping, replenishment of water becomes essential to obtain high yield. However, for effective management, the pond should not be more than 0.1 to 0.2 ha in size. In case, the pond (even perennial ponds) is found to be infested with predators or predatory fishes it is essential to kill them with the application of mahua oil cake at the rate of 2500 kg/ha. After 15 days of treatment with mahua oil cake, the pond bottom is raked by dragging a few bricks tied to a rope. This liberates unwanted toxic gases, etc. If the bottom silt is very heavy, then lime is added at the rate of 300 kg/ha which will not only reduce the toxicity of gases but will also enhance the fertility of the pond.

For culturing air-breathing fishes, the inner sides of the pond should be either made firm with heavy log or wood, or fenced with bamboo cane or wire screens to a height of about 50 cm. This would prevent the air-breathers from escaping the pond water through climbing or burrowing.

C. **Seed collection and transport**: Seeds of magur, singhi and murrels can be collected both by induced breeding as well as from the wild. Seeds collected from nature is dependable, as the fry of these fishes can be identified as given below :

(1) Fry of giant murrel (*Channa marulius*) is identified by its dark grey body and a lateral orange yellow band running from eye to the caudal fin.

(2) The stripped murrel (*C. striatus*) fry can be recognised by their vermillion red body with reddish golden band and a dark black band running from the eye to the caudal fin.

(3) Fry of the spotted murrel (*C. punctatus*) can be identified by their dark brown body with a golden

yellow lateral band and a mid-dorsal yellow line on the back.

(4) Fingerlings of magur (*Clarias batrachus*) can be recognised by their slate colour and long dorsal fin.

(5) Singhi (*Heteropneustes fossilis*) fingerlings can be identified by their pinkish colour and shorter dorsal fin.

(6) Koi (*Anabas testudineus*) fingerlings are known by the dark spot on the caudal peduncle and greenish hue of the dorsal body surface.

(a) **Seed collection of murrel** — Murrels attain maturity in about two years. As they are known to breed throughout the year, the fry (2-4 cm) can be collected round the year from rain-fed ditches and shallow water bodies among abundant weeds. Their peak spawning occurs, however, during April to June with the onset of monsoon. As the hatchlings move in schools, their collection in large numbers is always possible. It may be noted that when the fry reaches the fingerling stages they do not move in schools. Collection of murrel seed is generally made during monsoon.

Murrel seed can also be produced by induced breeding, but the difficulty lies in maintaining the spawn and growing them to fry stage. As the spawns after emergence from eggs do not feed on anything for about two days, their survival rate will generally be poor. For this, the hatchlings are required to be reared in suitable cemented ponds or plastic pools and have to be trained to take supplementary feed.

(b) **Seed collection of catfishes** — The matured male of magur possesses a pointed anal papilla, while that in female is oval in shape. Though catfishes breed throughout the year, the peak season for the collection of seed is pre-winter period when paddy is harvested and the low lying fields get exposed. Spawning generally takes place in fairly deep waters. The female makes a hole of 20 cm diameter and 25 cm deep on the bottom. The fertilised eggs measuring 1–1.5 mm (diameter) are yellowish-brown in colour and adhere to grasses. The males guard the eggs till it hatches within 20 hours at a temperature of about 25-30°C. About 2000 to 15,000 fries per hole can be collected from natural areas with the help of a small fine-meshed hand-net. They are then reared in nurseries till they reach a length of 5 cm.

In case of pond breeding, compartments or enclosures of 1 m × 1 m are made of wire screen in the pond having a water depth of about 25 cm. At the centre of each compartment a hole of 30 cm diameter is dug up to a depth of 25 cm. The holes are provided with some aquatic plants. The matured male and female fishes suitable for breeding are put into each compartment and they spawn within a period of 10 days. A maximum of 5,000 fry are collected from each compartment.

Seeds of catfishes can also be collected through induced breeding. The effective dosage of pituitary extract for different sized fishes is as follows :

(1) For catfishes weighing less than 100 gm is 10 mg extract.

(2) For catfishes weighing 100-150 gm is 25 mg extract.

(3) For catfishes weighing 150-200 gm is 35 mg extract.

Within 15 hours spawning takes place and the eggs thus produced is placed in shallow troughs or hatching jars. The water of the jars is changed twice a day preferably during the morning hours. After about 20 hours, the eggs hatch and the yolk sac attached to the hatchling remains for five days till they are completely absorbed.

(c) **Transport of seeds** — Fry of air-breathing fishes are generally transported without oxygen packing. The fry transported in carrier (usually polythene drum or iron drum) have enough of open space for their habitual surfacing to breath atmospheric air. If the period of transportation is less than five hours and the fingerlings are kept in open and watering tin carriers or in mud pots then a small quantity of aquatic weeds (*Hydrilla, Vallisneria, Ceratophyllum,* etc.) are placed on it. This would prevent the fingerlings from jumping out of the pot.

D. **Nursery management** : The fry rearing phase in murrels is a bit complex due to its cannibalistic habit. This can be reduced with sufficient supplementary feeding. High survival rate can be achieved through supplementary feeding and addition of micronutrients, yeast and vitamin B.

Nursery ponds of about 10-15 m^2, having a water column of 50 cm are stocked with 0.2 to 1.5 million fry per hectare. Prior to stocking, manuring is done with cow dung at the rate of 500 kg per hectare. To keep the nurseries free from insects, soap emulsion has to be applied. Advanced fry and fingerlings of catfishes collected from natural source do not require nursery management. Murrels, on the other hand, has to be trained to accept artificial feed prior to stocking in nursery ponds.

E. **Stocking of fingerlings** : Stocking of air breathing fishes is done at high density, and uniform sized fingerlings should be chosen. In case of monoculture of either magur or singhi, the stocking rate recommended in ponds is 40,000–60,000 fingerlings per hectare of water area. In case of polyculture with carps, magur may be cultured in place of common carp. Fingerlings of this catfish may be stocked at the rate of 20,000-30,000 fingerlings per hectare along with the recommended stocking density of carps.

Giant murrels, stripped murrels and spotted murrels may be stocked for monoculture at the rate of 15,000, 20,000 and 25,000 fingerlings per hectare, respectively. In the case of mixed culture, the latter two species may be stocked at the rate of 20,000 fingerlings per hectare at a ratio of 1 : 1. In case of murrel culture also, uniform sized fingerlings raised in nurseries are selected that are trained to accept artificial feed.

The selected fishes before stocking should be given a dip in 200 ppm (mg/l) formalin solution for 40 seconds or treated with 2% potassium permanganate solution for about five minutes. During periodic check on growth, the wounded fish, if any, may be treated with 0.3% acriflavin for five minutes.

Successful culture of air-breathing fishes depends upon stocking of the right size of fingerlings (6-10 gm), along with regular feeding and health checking of the fish stock. In case of mixed farming of murrels or catfishes with carps, the seeds of air-breathing fishes should be stocked only when the carps have reached a minimum size of 300 gm. This would safeguard the carps from being preyed upon by the air-breathing fishes. This practice of mixed culture not only provides an additional income through the yield of air-breathing fishes, but it also enhances the growth of carps. Such increase in growth of carps is possible as the air-breathing fishes would prey upon the trash fishes which may compete with culturable carps for food and space.

F. **Feeding and feed schedule** : To maintain an abundant food supply for the growing fingerlings of the air-breathing fishes, it is essential that the stocking pond be rich in animal food source such as tadpoles and trash fishes. In case these food sources are found to be inadequate, tilapia may also be grown, so that the young of tilapia (which is a fast breeder) may serve as a regular and staple food source for the fingerlings of cat fishes and murrels. In case of maritime states (West Bengal, Orissa, Kerala, Andhra Pradesh, etc.), abundant availbility of dried marine trash fish would make the culture of air-breathing fishes more economical. However, in ponds where the above facilities are not available, then one has to resort to supplementary feeding.

Artificial feeding for singhi and magur is undertaken using fish offal or slaughter house waste or dried silkworm pupae mixed with rice bran and oil cake in the ratio of 1:1:1. A mixture of rice bran, mustard oil cake and trash fish meal in the ratio 2:1:1; rice bran and poultry feed in 3:1; biogas slurry and rice bran in 1:2 or poultry dropping and rice bran in 1:2 ratio may also be given daily at the rate of 5-8 per cent of body weight of the fish fingerlings stocked.

Air-breathing catfishes during their six-month semi-intensive culture in stagnant ponds may be fed by the way given in Table 6.11.

On the other hand, trained fingerlings of murrel may be provided with cheaply available dried marine trash fish soaked in water, in the feeding schedule given in Table 6.12, during their eight-month semi-intensive culture.

Table 6.11 : Feeding schedule of magur stocked with 5,000 fingerlings in a 0.1 hactare pond

Period	kg/day	Feed ratio (trash fish : rice bran)
1st month	1.2	1 : 3
2nd month	2.4	1 : 1
3rd month	4.0	3 : 1
4th month	8.0	3 : 1
5th month	6.0	2 : 3
6th month	4.0	1 : 3

Table 6.12 : Feeding schedule of murrel stocked with 20,000 fingerlings in a 1 hactare pond and fed with dried marine trash fish soaked in water

Period	Feed (kg/day)
1st month	2.5
2nd month	5.0
3rd month	10.0
4th month	12.5
5th month	15.0
6th month	20.00
7th month	25.00
8th month	25.00

Feeding may be done either by broadcasting the feed in small amounts on all sides of the embankment, till the fish stops feeding or may be served in feed baskets, lowered near the banks. Feed baskets may also be given in addition to broadcasting of feed, so as to ensure feed availability to all the fishes in the pond. It is interesting to note that catfishes and murrels come in shoals when the feed baskets are lowered in the pond water. It has been observed that better feed utilisation takes place in the case of catfishes, when they are fed during dark hours.

G. **Culture duration and production** : Satisfactory response to external feeding results in good growth. Magur and singhi are expected to attain a weight of about 120 gm and 50 gm, respectively in 6 months of culture, whereas giant murrel, stripped murrel and spotted murrel grow to a size of about 400 gm, 275 gm and 160 gm, respectively in 7-8 months of culture period.

Air-breathing fishes when cultured under high input technology (semi-intensive) of both mono- and mixed- culture practices, the production potential of the following type can be obtained :

(1) A production of about 5 tonnes/hactare/6 months of magur can be obtained with supplementary feed made of dried trash fish and rice bran.

(2) An yield of 5.1 tonnes/hactare/14 months of singhi culture can be obtained with supplementary feed made of rice bran and trash fish.

(3) Murrel's production of 4 tonnes/hectare/10.6 months culture can be obtained on being fed with forage fish.

(4) Mixed culture of magur and singhi when fed with rice bran and trash fish a production of 5 tonnes/hectare/year culture can be obtained.

(5) Mixed culture of murrel and koi when fed with rice bran, mustard oil cake and trash fish may result in a production of 5.9 tonnes/hectare/6 months.

(6) Mixed culture of the three murrel species can yield about 4 tonnes/hectare/year when fed with soaked dried marine trash fish and fresh silkworm pupae.

(7) In intensive culture, the production of magur and singhi can be increased to about 7.8 tonnes and 7 tonnes respectively/hectare/5 months culture when fed with fish meal and rice bran.

In carp polyculture system magur is cultured in place of common carp. Magur, singhi and koi can also be cultured along with makhana (*Euryale ferox*), a highly priced aquatic crop, in the same shallow pond.

H. **Harvesting** : Generally harvesting of air-breathers is done in summer season from production ponds when the depth of water is less. The pond water is drained out and the fishes are harvested with the help of a hand-net or scoop-net.

I. **Diseases of air-breathing fishes** : Generally air-breathing fishes are hardy and can withstand pollution in their natural abode. Therefore, incidence of disease is less common in them. However, some of the diseases encountered during culture operation and their remedial measures are given below :

(1) Fry and fingerlings of stripped murrel have been found to be affected with fin rot and dropsy in nursery. As a remedial method the infected ones are given a bath in either 1% acriflavin or 0.3% formalin solution, for about 10 minutes or till they show sign of distress.

(2) Murrels are generally found encountered with parasites such as *Lernaea* sp. and *Argulus* sp. and can be controlled by the application of lime at the rate of 300 kg/hectare.

(3) Singhi has been found to be affeced by various bacterial diseases leading to the loss of barbel and development of reddish inflamatory patches from the base of their anal fin and running up to the caudal tip. It can be controlled by the application of terramycin at the rate of 100 mg/kg body weight in their feed, for 7 to 10 days.

(4) Pox disease has been detected in singhi from the small reddish white spot on their body. A bath in potasium permanganate at the rate of 1 mg/litre of water has been found to be effective in the prevention of the disease.

(5) Magur suffering from swollen abdomen has been found to be infected with cestoda. Such affected fishes should be removed from the pond.

(6) In intensive culture, magur has been found suffering from bacterial (generally, *Pseudomonas*) disease and the symptoms are ulcers, reddish lips and loss of barbels. Controlled methods being application of sulphadiazine in the diet at the rate of 100 mg/kg of feed for one week.

(7) Koi shows non-carcinogenic tumours caused probably by *Clostridium* sp. The tumour can be cured by the application of terramycin at the rate of 100 mg/kg of feed for seven days.

(8) In sewage fed ponds, koi has been found to be suffering from lymphocystis, detected from proliferous growth of binding tissue on the fins as well as on the body. Lymphocystis can be cured by the application of terramycin or sulphadiazine at the rate of 100 mg/kg of feed for 7 to 10 days.

INTEGRATED FISH FARMING

It has been an ancient practice to culture fishes in combination with agriculture and/or livestock. Such a venture is lucrative and provides higher farm income. There are farms where fish farming is the main activity, while in others fish farming forms the subsidiary activity. Whatever the case may be, the role such integrated farming can play has received considerable attention in recent years.

Basic principles

Integrated fish farming is based on the concept that 'there is no waste', and waste is only a misplaced resource which can become a valuable material for another product (FAO, 1977). In integrated farming, the basic principles involve *the utilisation of the synergetic effects of inter-related farm activities and the conservation, including the full utilisation of farm wastes.*

It is assumed that all the constituents of the system would benefit from such a combination. However, in most cases, the main beneficiary is the fishes which utilises the animal and agricultural wastes directly or indirectly as food. As integrated farming involves the recycling of wastes, it has been considered an economic and efficient means of environmental management.

Advantages of Integrated Fish Farming

The advantages of integrated fish farming are as follows :

(1) Full utilisation of farm wastes.

(2) Utilisation of the cooperative effects of inter-related farm activities.

(3) It increases employment opportunities.

(4) It increases nutritional source for the farmer's family.

(5) It gives higher and stable farm productivity and there is less risk (biologically and economically).

(6) It increases the income of rural population.

(7) It is a means of land reclamation in certain areas.

(8) It is an efficient and economic means of environmental management.

Disadvantages of Integrated Fish Farming

Recently, controversy has arisen among scientists on the public health aspects of integrated farming. Speculations are ripe that integrated fish farming with pigs and poultry may be a cause of influenza pandemic. This may be, as the pigs would act as 'mixing vessels' for avian and human influenza viruses, it can create new lethal strains of viruses by mutation. In such an act, fishes themselves do not play any role. However, there is no conclusive evidence that integrated farming can become a public health hazard. For safety measures, pig-poultry combinations in integrated fish culture should thus be avoided.

Types of integrated fish farming

A. Paddy-cum-fish culture

Rice fields which are water-logged for 3-8 months in a year, there is always small population of fishes that gain access to such waters. This probably had given rise to the practice of deliberate stocking of fishes and harvesting. The trapping of prawns and fishes with the help of 'gamcha or dhoti' in fallow paddy-fields has been an age old practice in India. The culture of fishes in paddy-fields has been an off-season occupation for farmers.

(a) Objectives of paddy-field aquaculture

The objectives or advantages of paddy-field aquaculture are as follows :

(1) Paddy-field aquaculture provides additional income to the farmers.

(2) In areas where rice and fish form the staple food, paddy-field aquaculture makes available an essential diet for the people.

(3) As paddy and fish can be grown either simultaneously or alternately in the same water mass, it requires very little extra input by way of additional costs, particularly in management and labour.

(4) It provides off-season employment to the farmers and farm labours.

(5) Combination of paddy and fish farming is mutually beneficial. Fish cultivation promotes better paddy production by way of exercising an effective control on unwanted weeds, molluscs, noxious insects and their larval stages.

(b) Fishes suitable for paddy-cum-fish culture

All fishes are not suitable for such a type of culture as paddy-fields provide special ecological conditions such as shallow turbid water with high temperature. Fishes having the following criteria are generally selected for paddy-cum-fish culture.

(1) Fishes that can adapt to shallow waters necessary for paddy crops.

(2) Fishes that can tolerate high temperature.

(3) Fishes that can thrive on low dissolved oxygen, which is the characteristic of paddy-fields especially in tropical countries.

(4) Fishes that can tolerate fairly high turbidity.

(5) As the duration of culture is quite short, fishes that have high growth rate is to be selected, so that it can reach marketable size within these few months.

(6) Fishes that can live in confinement and do not tend to escape from the cultivated area.

Fishes that are cultured in such waters in India are *Mugil* sp., *Mystus gulio, Haplochromis mellandi* (mollusc eating fish), *Lates calcarifer, Mugil parsia, Puntius sp., Channa* sp., prawns and shrimps. In India, limited experimental works have shown the suitability of Indian carps for such integrated farming.

(c) Management of paddy-fields

To increase the utility of paddy-fields as fish ponds, the following managements are required :

(1) A continuous flow of water in the field, with proper inlet and outlet is to be maintained.

(2) The water in the field is to be maintained at a desired level.

(3) Proper drainage of water from the field has to be made in case of flooding.

(4) At the point of entry and exit of water, some control means (such as screens) have to be provided to prevent the cultivated species from escaping and stopping the entry of wild fishes into the paddy-field.

(5) Deep pits or other devices has to be provided as shelter to the cultivated fishes at the time of distress.

(d) **Types of paddy-field aquaculture**

Three major types of paddy-field aquaculture are generally practised.

(i) Simultaneous or combined or synchronous farming.

(ii) Alternate or sequential farming or paddy-fish rotation.

(iii) Relay farming.

(i) **Simultaneous or combined or synchronous farming** — In such farming paddy and fish are grown together and harvesting the paddy and fish at the end of the rice-growing season.

Literature reveals that a variety of techniques and stocking rates have been employed in different countries on paddy-field aquaculture. In general, fishes are stocked not earlier than 5 days after transplantation of paddy seedlings, so as to enable the seedlings to form proper root. After transplantation of paddy, stocking in case of fry should be done after 10 days, while for fingerlings after 3 weeks.

The methods of culture adopted are for production of either fingerlings or fish for consumption. Farmers generally cultivate local varieties of paddy which takes up to 6 months to be harvested, so that more than one crop per year can be grown. Generally organic manures are preferred over the use of fertilisers in the preparation of the paddy-fields.

The **advantages** of simultaneous cultivation are :

(1) There is no additional cost for fish production.

(2) Fishes contribute to the enhancement of paddy production by destroying weeds, causing tillering and mineral enrichment by their digging activity, and for the fertilisation of soil by their excrement and also by the unutilised artificial feed.

(3) Due to fish cultivation paddy production gets increased by 5 to 15%.

The **disadvantages** are :

(1) The water depth being shallow hinders better fish raising.

(2) For fear of harmful effect on fishes, herbicides and insecticides uses in paddy-field is generally prevented which ultimately limits paddy production.

(3) Due to abrupt changes in temperature and dissolved oxygen, inadequate space and presence of piscivore birds, there is great loss of fishes which may be about 20-60%.

Progressive agricultural areas have slowly discarded paddy-fish cultivation in favour of the rotation of paddy and fish crops.

(ii) **Alternate or sequential farming or paddy-fish rotation** — It is the simplest form, where flooded paddy-fields after harvest are used to raise one or more crops of fish or shrimp.

In rotation system the paddy-field has to be prepared for raising fishes after the harvest of rice. To maintain the required depth of water the bunds surrounding the field have to be raised. After harvesting the rice, the stubbles are not removed. These submerged stubbles provide the substrate for the development of fish food organisms. These stubbles undergo decomposition, thereby fertilising the water and stimulating higher productivity. After fish harvesting, the residues remain in the soil which in turn serves as fertiliser for the paddy crop.

Shrimp production in paddy-fields — Shrimp production in paddy-fields is generally practised in areas (generally in the west coast and deltaic areas of eastern India) where only one crop (July to September) of a salt-resistant variety of paddy is grown. Following the harvesting of rice, the bundhs are strengthened and sluice gates are installed to control the water supply. At high tides, the fields are filled with tidal water, containing a large number of shrimp larvae. These larvae gain access to the fields, where they find food and shelter. To attract the larval shrimps, lamps are hung above the inlets. By this process, natural stocking of shrimp larvae is done at every high tide for two to three months. To prevent the escape of juvenile shrimps at low tides, conical bag-nets are attached to the sluice gates.

Shrimp harvesting from these fields starts in December, as by this time the early stock of shrimps will have attained marketable size. Regular harvesting is done which results in thinning of the stock, leading to better growth rate and harvesting of larger shrimps. The total yield of shrimp is around 780–2100 kg per hectare.

Advantages of paddy-fish rotation— The paddy-fish rotation system has the following advantages :

(1) There is no water depth limitation either to paddy or fish cultivation. Shallow water depth is maintained till the harvesting of paddy, after which the water depth is raised for the culture of fishes.

(2) Through this adequate water level management, suitable water temperature and dissolved oxygen content can be maintained.

(3) After harvesting of paddy, the submerged stubbles decompose and fertilise the water. This leads to the development of fish food organisms and ultimately stimulates higher productivity.

(4) The interval between paddy harvesting and fish stocking is sufficient enough to allow degradation of pesticides.

(5) Insect pest infestation gets reduced as their life-cycles are disrupted.

(6) Either monoculture or polyculture can be practised.

The disadvantages are : Due to presence of piscivorous birds like herons, cormorants, etc., considerable loss of fish (20-60%) takes place.

(iii) **Relay farming**— This system is much complicated and it ensures prolonged period of fish culture. It involves transferring the stock of fishes to specially prepared ditches, channels or pools at the time of paddy harvest and re-stocking them in the field for a further growing period. This system provides better growth rate than what is possible in such a short duration of one paddy crop.

For such a cultivation the farm should have additional holding or rearing facilities for the period between paddy harvesting and planting of new fish seedlings. The carps or shrimps can be grown to the required size through adequate feeding. This culture involves higher inputs including labour.

B. DUCK-CUM-FISH FARMINGS

Although duck and fish farmings have been in practice in eastern Europe and China and their compatible nature being recognised since long, the interaction and benefits of the association have been understood only a few years back. During the last decade suitable methods have been developed in various countries for raising ducks in fish ponds. Undoubtedly, such a combined culture is highly profitable as it greatly increases the production of protein in terms of fish and duck per unit area.

The combination of duck and fish farming is seen presently as a means of reducing the cost of feed for ducks, at one hand, and on the other, the excreta of ducks acts in an inexpensive way of fertilising ponds, which results in production of fish food organisms. Thus, ducks can be said to be 'living manuring machines'. The duck droppings contain 25 per cent organic and 20 per cent inorgaic substances along with a number of elements such as carbon, nitrogen, phosphorus, potassium, calcium, etc. Moreover, ducks by agitating the shore areas of the pond, help to release nutrients. Besides this, ducks feed on a variety of organisms such as weeds, snails, unwanted harmful insects and their larvae (some being vectors of fish pathogenic organisms), tadpoles, frogs and water-borne disease-causing organisms infecting man, whose eradication is one of the important aspects of farm management.

(a) **Ducks culture and their raising**

Ponds provide a clean and healthy environment for ducks. Special strains of ducks have been developed that are suited for pond raising. Such suitable strains when used, approximately 50-60 per cent of their droppings when falls into the pond, is sufficient to fertilise the water.

Reliable and timely supply of required strains of good quality ducklings are of great significance as it promotes successful farming. For supply of ducklings, small scale farmers have to depend on outside sources, while larger farms have their own breeding centres which is much convenient and profitable.

Raising of ducks is done by one of the two ways :

(i) **Extensive raising** — In extensive raising ducks are stocked at approximately 150-500 per hectare of pond surface. Here small amount of supplementary feed is provided and the number of ducks is limited due to the food they can find in the pond water. As the number of ducks are limited, they contribute little amount of manure to the pond and its effect on fish yield is also limited. Such a type of method is usually employed in Europe.

(ii) **Intensive raising** — Here ducks are stocked at a much higher density, approximately 1000-2500 per hectare of pond surface and is usually employed in Africa. The ducks are fed at the same rates as on land. As ducks are held at a much higher density per unit area, higher amounts of manure are thus loaded into the fish pond which subsequently results in higher yields.

(b) **Duck keeping**

There are two basic ways of keeping ducks in fish ponds :

(i) **Free-range system** — Here the ducks are allowed to have free access to the whole pond (Fig. 6.43). As they are allowed to swim freely around the whole pond surface, a good proportion of their droppings fall directly into the pond water and are distributed more or less uniformly. The ducks are able to forage around the whole pond. Adjacent to the

Fig. 6.43. A fish pond where the ducks have free access to the whole water body

Fig. 6.44. A fish pond where the ducks are confined to enclosures with wire fences

ponds, small duck houses are built with facilities for providing them with supplementary feed. In this system, as considerable energy is spent by the ducks in swimming activity, it is believed to affect the feed conversion ratio and growth rate of the ducks.

(ii) **Confinement in enclosures** — Here, part of the pond area and the adjacent land area are enclosed with wire fences (Fig. 6.44). About one-fourth of the enclosure will be on the land and the rest in water. Within the wire fence suitable feeding and resting areas are made. Some of the droppings of the duck falls directly into the pond water, while the rest that drops on land is washed down into the pond. Under this system fish production is reported to be almost equal to the free-range system. Keeping ducks in enclosures is preferred by most farmers, who use special strains of ducks for better growth.

(c) **Raising of fish**

For duck-fish farming the most suitable pond is the barrage type of ponds. This is made by damming shallow valleys, so that the ducks can lie on the natural slopes. There is every likelyhood that ducks may damage the earthen dykes while foraging. The problem can be solved with proper maintenance.

Commonly the fish species cultured are herbivores and omnivores. The common carp was traditionally the main species but subsequently other species of Chinese carps are used to make full use of the food resources. Other important species used are male or hybrid tilapia, grey mullet, catla, eel, asian catfish and sea perch. Fairly high stocking rates are adopted and supplementary food is also provided.

(d) **Advantages and disadvantages of fish-cum-duck culture**

(i) **Advantages :**

(1) There is practically less additional cost for fish culture, as the excreta of the ducks fertilises the pond water.

(2) Ducks by agitating the shore areas of the ponds help to release nutrients.

(3) The cost of feed for the ducks gets reduced.

(4) The ducks help to eradicate vectors of fish pathogenic organisms and water borne diseases — causing organisms infecting man, as they feed on a variety of organisms such as weeds, snails, unwanted harmful insects and their larvae, tadpole, frog, etc.

(ii) **Disadvantages :**

(1) The fingerlings released should be of more than 10 cm size, otherwise the ducks may feed on the fingerlings.

(2) While foraging for food, ducks sometimes damages the earthen dykes. This problem can be solved with proper maintenance.

C. Fish-cum-poultry farmings

Integrated fish farming with poultry is generally cultured as the poultry manure is a very efficient fertilizer for fish ponds. The poultry droppings comprises 2% nitrogen, 1.25% phosphoric acid and 0.75% potash. The low feeding cost per individual bird makes poultry farming along with fish, a common investment for poor farmers.

(a) **Poultry raising**

Both 'broilers' and 'layers' variety of chicken can be raised for fish-poultry farming. One day old chick are raised up to the **pullet*** stage after which they are put in **layer** cages.

Fish culture with both intensive and extensive poultry productions have been integrated successfully. The most intensive type of poultry production is the battery type of housing, which is installed by the side of the pond. The floor of the house is cemented and is set up at a slope so that the eggs may roll forward. For layers, the floor area required is about 30 cm^2 while for broilers, 15-20 cm^2. The usual floor space alloted for each bird is 20 × 30 × 40 cm. The birds are confined to cages which are made up of standard, stout, galvanised wire. The cages are kept on trays for collection of droppings. For further on poultry refer chapter 5.

For manuring one hectare pond water, the droppings of about 250 **layers** and four batches of 200 **broilers** each are adequate in a yerar's time.

(b) **Fish raising** — For fish raising the ponds are stocked with fingerlings of catla, silver carp, common carp, murrels, tilapia, giant freshwater prawns, etc. The stocking density of fishes is related with that of poultry and also with the period of culture.

In one hectare pond area, when stocked with 5000 giant freshwater prawns (*Macrobrachium rosenbergii*) and 1500 silver carp, and cultured for a period of four months, one can harvest 600 kg of prawns and an equal amount of fish, along with 250 culled birds.

For culturing over a period of one year, ponds may be stocked with fingerlings of catla, common carp, silver carp and grass carp at a density of 5000-6000 fingerlings per hectare. At the end of twelve months of fish-cum-poultry culture, fish yield of over 3900 kg per hectare can be normally obtained along with 42,000 eggs and 200 culled birds.

(c) **Advantages and disadvantages of fish-cum-poultry culture**

(i) **Advantages :**

(1) Chicken manure is a very efficient fertiliser, so no chemical fertiliser is needed for fertilising the pond water. This cuts down the expenditure of rearing fishes.

(2) No supplementary fish feed is required.

(3) The purchase and feeding cost per bird is low.

(4) Chicks are readily available and their productivity can be improved with simple and cheap management.

(ii) **Disadvantages :**

(1) Chicks should be examined from time to time and diseased one should be isolated, otherwise they will destroy the entire stock.

(2) Sufficient time should be given from one stocking of chicks to the next for renovation of the house and disinfectioning it.

D. Fish-cum-pig culture

In integrated fish farming with pig, the 'pig dung' is useful for conditioning the soil and providing the necessary nutrients required for fertilising the pond water. Fish-cum-pig culture is practised at large in China where pigs are considered as "costless fertiliser factories". Pig dung contains about 70 per cent diges-

* **pullet** — Chick less than one year old

tible feed for fishes. The feed while passing down the pig's alimentary canal, gets mixed up with enzymes which continue to act even after defecation. Such undigested solids serve as direct food source for tilapia and common carp.

In tropical fish ponds, weeds are a major problem in fish culture. Such vegetations are considered as valuable food resource for pigs. Thus, pigs aptly play a role in biological control of weeds.

(a) **Pig rearing** — In fish-cum-pig culture the embankments of fish ponds are made wider (over 10 m in China) to facilitate the building of pig sites and also for growing vegetables, fruit trees or other crops. In the slopes, grasses can also be grown which is used as fodder for grass carp and for other farm animals.

Various aquatic plants, such as azolla, duck weed, *Pistia, Wolffia, Lemna,* and water hyacinth (*Eichhornia*) are grown in feeder channels and irrigation ditches associated with the pond farms. These and the foliage of other terrestrial plants (vegetables, rice, corn, etc.) are used for feeding the pigs. In the water area of the pond, about 10 tons of aquatic plants can be produced, which are sufficient to feed 10 pigs. These plant materials are generally mixed with bananas, coconut meal, rice bran, soybean wastes, groundnut cakes, fish meal, etc.

Pig sites or pen or sty are generally built on nearby land or on the pond embankment. The pen enclosure is built not only for pig raising but also special consideration is given to the needs for breeding, nursing and fattening activities. Pig pen generally have a system of channels for transfering the organic matter into the pond water. Alternatively, the sty or pen may be constructed above the pond water. The structure is made of wood and provided with a lattice type of floor which permits the excreta and uneaten food to fall directly into the pond water.

Modern practices are to avoid direct washing of the wastes into the pond. The urine and dung of pigs are first allowed to the oxidation tanks (digestion chambers) where sedimentation and fermentation of the manure take place. The supernatant liquid, at regular intervals, are then discharged into the fish ponds. The sludge that remains is utilised as fertilisers in agriculture. Alternatively the pig manure may be kept in a heap on the pond embankment for later use. The chemical composition of pig wastes is depicted in Table 6.13.

The number of pigs to be raised per hectare and the manuring rates to be applied are based on years of experience. The production of manure depends upon the age and size of the pig. A piglet produces about 3.4 kg manure a day, while a one-year-old pig gives about 12.5 kg a day. The average production of faeces and urine per pig is about 7.8-8 tons per annum. A density of 60-100 pigs has been found to be sufficient to fertilise a one hectare fish pond.

Table 6.13 : Chemical composition of pig wastes

Source	Organic matter (%)	N (%)	P_2O_5 (%)	P (%)	K_2O (%)	K (%)
Fresh manure	15.00	0.60	0.40	0.18	0.44	0.37
Urine	2.00	0.30	0.12	0.05	1.00	0.83
Air-dried manure	34.32	2.12	0.98	0.43	2.45	2.03
Litter manure	34.00	0.48	0.24	0.11	0.63	0.52

(b) **Fish rearing** —Polyculture is commonly practised in such integrated farming due to the variety of food that becomes available in the pond. Herbivorous and omnivorous fishes are used for culture; generally common and Chinese carps and less frequently catfishes (*Pangasius*), Indian carps and tilapia. Due to high productivity of the ponds, fairly high rates of stocking are generally practised — 60,000 fingerlings of different species (weighing 20-30 gm) per hectare.

(c) **Production and duration of culture** — The duration of culture of fishes and pigs varies. Generally it is about one year, but culture for 6 months duration is also practised. The overall economics of combined fish and pig rearing depends on the local conditions. However, the pigs are generally sold when they have attained a weight of 90-100 kg. The production of fish generally varies between 2 and 18 tons per hectare per annum.

(d) **Advantages and disadvantages of fish-cum-pig culture**

(i) **Advantages :**

(1) Such integrated farming increases the productivity per area and thus, the farmers income becomes doubled or more.

(2) Pig dung conditions the soil of a new pond and provides ready-made organic matter, containing the necessary nutrients.

(3) Pig dung contains about 70 per cent digestible feed for fishes. The undigested solids present in the faeces of pig serves as direct feed source for tilapia and common carp.

(4) Pigs aptly playes a role in biological control of weeds, as weeds are considered as valuable food source for pigs.

(ii) **Disadvantages :**

(1) Addition of too much pig manure may lead to increased nutrient load resulting in pollution of the water body and mortality of the fishes.

(2) Considerable care and management skills are required to prevent pollution. It has been found that satisfactory fish production can be obtained with much lower manuring.

E. **Other animals in fish systems of integrated farming**

(a) **Fish-cum-cattle farming** : Cattle wastes and washings from the cattle sheds are conveyed through pipes into the ponds which acts as good fertiliser. Cattle wastes are generally collected in a pit for later use. In addition to fish yield, production of milk from cattle and beef adds to the economy.

(b) **Fish-cum-rabbit farming** : Rabbit farming has been found to be ideal for integration with small-scale fish culture. Rabbit manure have greater value as a direct food for fish compared to other livestock wastes.

(c) **Fish-cum-mulberry farming** : Mulberry plants are raised on the dikes of the fish farm and in the neighbouring fields for silkworm production. The mulberry wastes and silkworm larvae and pupae (after removal of silk) are used as feed for the fishes. It also fertilises the pond water.

FISH DISEASES : SYMPTOMS AND CONTROL

The health of aquaculture organisms, including fishes, is due to a state of physical well-being. Abnormal changes in the body arises from two factors, one for which the aquatic environment is important (dietary deficiency, attack of pathogens and other stresses) and other is due to cold-bloodedness of the fish (fish's inability to maintain its body temperature).

Nutritional imbalance results in formation of thyroid tumours, anaemia, liver degeneration, visceral granuloma, pigmentation impairment and deficiency symptoms associated with vitamin imbalance. Physical and environmental stresses play significant role in the maintenance of healthy conditions. Although, many pathogens of aquaculture species are generally found in the aquatic environment, yet their presence may not lead to the occurance of disease. Disease results due to **interaction between the species**, the **disease agent** and the **aquatic environment**. The balance between these three major factors determine the state of health of aquatic species. Disturbance in any of these three factors may lead to the disruption of the relationship, resulting in the causation of disease.

Resistance of the host

The resistance or susceptibility of the cultured species to the disease causing agents is governed by its age, exposure experience and physical barriers (such as exoskeleton, skin and mucous membranes). The physical barriers generally restrict the entry of toxic, infectious and parasitic agents. The physiological defences comprise the white blood corpuscles that engulf pathogens, avoidance mechanisms, detoxification of chemicals from water or diet by the liver, storage of certain metals by the bones and local tissue reactions. The source of fish's physiological ability to defend itself is the overall nutrition it consumes.

(a) **Immune response** — An important means of disease resistance is the immune system of fishes and its specific activity against biological agents such as viruses, bacteria and parasites. The immune system operates through antigen-antibody reactions. The antibodies of fish (teleost) are serum proteins (immunoglobulins) that belong to IgM or IgG class. These are functionally classified as agglutinating, precipitating and virus neutralising. By the stimulation from certain antigens (bacteria, virus, etc.) the agglutinating antibodies are formed. These antibodies bind the antigens rendering them less toxic and more vulnerable for easy phagocytosis. On the other hand, the precipitating antibodies are less important in fish and they generally cause precipitation of soluble antigens which are easily discarded through tests. Virus-neutralising antibodies cling onto the virus particles making them inactive.

In fishes antibody formation is temperature dependent. Lowering of temperature may delay or suppress antibody formation and would thus affect the immune response of that fish. Certain behavioural factors, such as pheromone released in crowded condition, may also suppress antibody formation.

Population with previous exposure (to a specific antigen), will be less susceptible as those on a first encounter. Moreover, young ones are more susceptible to diseases than older ones, due to their fragile defence system.

(b) **Environment**: Environment often plays a vital role in disrupting the balance between the pathogen and the host. Often the culture animals live a healthy normal life in the presence of pathogens. However, during environmental stress the pathogen gets the upper hand and the balance tips in favour of the disease.

At the time of selection of the site for aquaculture the environmental parameters required were adequately considered so that the relevant stress factors are minimum. The minimum water quality conditions necessary to maintain good fish health are :

dissolved oxygen	—	5 mg/l
range of pH	—	6.7–8.6
free CO_2	—	3 mg/l or less
ammonia	—	0.02 mg/l or less
alkalinity	—	at least 20 mg/l

Temperature plays a vital role on a number of other variable in the environment. With increase in temperature solubility of toxic compounds increases creating unfavourable conditions. As has been discussed above antibody formation is temperature dependent. Wound healing is temperature dependent as the repair of dermis is suppressed by low temperature.

Ailment from environmental factors are :

(1) Chill and cold weather results in drop of water temperature and the fish shows congestion of gill apparatus and dull appearance.

(2) Alteration in pH of water results in too acidic (pH at or below 5.5) or too alkaline (pH at or above 9) and the fish shows a sign of ailment. Mucus secretion gets enhanced. In too alkaline case, the gills appear burnt and the fins give a rot look. It may result in fatal ailment, if not duly treated.

(3) Depletion in dissolved oxygen content of water results in asphyxia and the fish shows signs of suffocation—mouth gets wide open, gill opercula gets raised and the gills spread wide apart.

(4) Constipation and indigestion results in queer movements and listlessness of fish. The fish may remain stationary in mid-water or stand on its head or lie at the bottom on its side. Stomach appears swollen and the scales get erected.

(5) "Egg binding" illness of the females occur during the breeding time. The body of the female fish become greatly swollen due to accumulation of ripe eggs that could not be released.

(c) **Genetic resistance** — As fishes are known to adapt to disease in nature, these traits of resistance can be developed to resist those infections. However, problems develop in disease-resistant strains of fish (Warren, 1983), such as :

(1) Strains of brook trout, selected for their resistance to furunculosis, were found to have acquired greater susceptibility to bacterial gill diseases during selection.

(2) Strains of steelhead trout, selected for their resistance to bacterial kidney disease (BKD), were also found to be most susceptible to *Vibrio anguillarum* infections.

Thus, it can be concluded that loss of genetic diversity in a selection process makes it difficult to develop strains of fish that are resistant to several other diseases at the same time. Even the disease-resistant fish that survives may develop a carrier state, which is harmful if the disease carrier is to be introduced into new areas where the disease would not otherwise occur. Thus, by maintaining a high level of genetic diversity in a stock, the ability increases to withstand the stress of infectious diseases.

Mode of contraction of infection

The contraction of infection is direct in most parasites and pathogens. The disease causing agents freely swim on to the host or contact it by contiguity, or through a vector (the fish leech transmits *Trypanoplasma* in cyprinids causing sleeping sickness — Trypanoplasmosis), or infection may be caused through the food medium which is already infected. For example, transmission of cestode (*Ligula*) through infected copepods (harbouring the larval stages of the parasite) which is subsequently eaten up by the fish.

Sign of sickness and effect on fish

Sickness in fishes may vary, but there are some common external symptoms, which are —

(1) fish becomes listless,

(2) loss of balance occurs and the ailing fish is unable to maintain its normal position,

(3) the diseased fish tends to lie on its side either floating at the surface or resting at the bottom,

(4) slimy, grey excretions appear on the skin,

(5) fading of normal colour occurs and the gills appear pale and

(6) the fins (including tail fin) do not function normally and with vigour.

The parasites of fish extract their nutrition from the host but do not harm it. However, they cause loss of anabolic material and render the host weak. They, thus, directly or indirectly harm the fish as described below :

(1) Cestodes (*Eubothrium*) cause mechanical damage as their large-scale multiplication blocks the lumen of the gut.

(2) Trematodes (*Diplostomum*) cause cataract in the fish eye and impairs vision.

(3) Acanthocephalans (*Pomporhynchus)* heavily perforates the intestine.

(4) *Ligula* (cestode) multiplies to such an extent that it constitutes about half the total weight of the host fish. The parasite consumes a substantial part of the fish's food and thus, retards its growth.

(5) 'White spot' and 'black spot' cause irritation in the skin and the fish scratches to bleeding.

(6) *Argulus* from its poison glands discharges toxic substances that may hasten the death of the host fish.

(7) The common aquatic fungus (*Saprolegnia parasitica*) is a secondary parasite that invariably causes death to the host fish.

Diagnosis of fish disease

The following pathological procedures are undertaken for the diagnosis of various fish diseases :

(1) *Assessment of fish environment*.

(2) *Assessment of fish health* — A sick fish behaves differently from that of a healthy one :

- Sick fish is less timid.
- When taken out of water a sick fish remains calm and sluggish.
- When a sick fish is held in hand it lets its part of the body hang down vertically with fin folded.

- When a sick fish is turned to one side, its eyeball also turns following the turning of the body (a healthy fish keeps its eyeball in fixed original position).

(3) *Assessment of probable cause of sickness* — This is done in two ways :

- Evaluation of the quality of water in which the ailing fish is present. The features of the water may be assessed by analysing such features like smell of the water, transparency, colour, oxygen content, temperature, pollutant, etc.
- Examining closely sick fishes — diagnosis of the disease and its causative factor.

(4) *Disorders by abiotic factors* —Various ingredients present in the domestic and industrial effluents may find their way into fishery resources and cause various disorders in fish. Some of these are listed below along with the disorders caused :

Substances	Disorders
(i) Phenols, synthetic detergents and alkalies	Haemorrhage on skin, head, base of fin; discolouration of gills; changes in blood corpuscles (decrease in erythrocytes).
(ii) Chlorine (> 4 mg/litre)	Damage of gills; anaemic; in carps death results after 8 hours.
(iii) Arsenic salts, manganese chloride and acids	Blindness
(iv) Iron and nitrate	Increase in erythrocytes
(v) Ammonia (> 1 mg/litre)	External bleeding; haemorrhage of internal organs.
(vi) Carbon dioxide	Calcium deposits in ureters; swelling of kidney.
(vii) Hydrogen sulphide (> 0.6 mg/litre)	Gills become reddish-violet in colour.

Table 6.14 : Some parasitic fish diseases, their causative organisms, symptoms and treatment

Sl. No.	Causative organisms	Disease	Symptoms	Treatment
(1)	**Bacteria** *Pseudomonas*	Fin and tail rot	White line on the margin of the fin; fin rays become brittle and start breaking.	1 minute dip treatment in 500 ppm copper sulphate solution.
	Aeromonas salminicida	Furunculosis	Open sores or ulcers on the body.	Dip treatment for 1 minute in 1 : 2000 copper sulphate solution for 3-4 days.
	Aeromonas punctata	Dropsy	Accumulation of fluid inside the body cavity; scale protrusion; exophthalmic condition.	Dip treatment in 5 ppm potassium permanganate solution for 2 minutes.
		Eye disease	Cornea of eye becomes vascularised and later becomes opaque; eyeball gets decayed.	Chloromycetin (8-10 mg/litre) bath for 1 hour for 2-3 days.
(2)	**Fungi** *Saprolegnia parasitica*	Water-mould disease (Saprolegniasis)	Dies after ulceration or exfoliation of skin followed by haemorrhage; blindness; tufts of white hair-like out-growth in the affected region.	Dip treatment for 3 seconds in 1 : 10000 solution of malachite green or for 5-10 minutes in 3% common salt solution or 1 : 1000 solution of potassium permanganate.

Table 6.14 contd. to next page

Table 6.14 : (continued)

Sl. No.	Causative organisms	Disease	Symptoms	Treatment
(3)	**Virus** *Rhabdovirus*	Viral haemorrhagic septicaemia (VHS)	Anaemia; pale discolouration of gill; swelling of eyes and belly; oedema of muscles; inflammation of intestine; anus becomes protruded.	No known cure for VHS. Prevention is the best approach to control like other viral diseases. Contaminated water should be thoroughly disinfected before use.
(4)	**Protozoans** *Ichthyophthirius multifiliis*	Ichthyophthiriasis (White-spot disease)	Small whitish-cysts of about 1 mm diameter on the skin, gills and fins.	5 days bathing in 2 ppm methylene blue; hourly dip treatment in 1 : 5000 formalin solution for 7-10 days.
	Myxobolus luciopercae	— —	Round or rice-shaped cysts on the gills.	All infected fish should be destroyed or burnt.
	Myxobolus exiguus	Knot disease	Small knots in the skin of common carp.	All infected fish should be destroyed or burnt.
	Myxobolus pfeifferi	Boil disease	Large boils varying from the size of a nut to that of a hen's egg on several parts of body.	Bath in 3% common salt solution or in 1 : 2500 formalin solution for 10 minutes.
	Sporozoans	Pox disease	Small milky-white spots raised 1-2 mm above the surface of skin.	Intra-peritoneal injection with 1 ml of a 1% solution of arsenic compound, Arycil.
	Costia necatrix	Costiasis	Bluish-coating on the skin; lesions as irregular patches.	Bath in 3% common salt solution or in 1 : 2500 formalin solution for 10 minutes.
	Myxosoma cerebralis	Whirling disease	Caudal bend, deformity of the oral region and blackening of tail region.	Destroy all infected fish by applying quicklime (pond disinfectant) at the rate of 2 t/ha.
(5)	**Trematodes** *Gyrodactylus* sp. *Dactylogyrus* sp.	Gyrodactylosis Dactylogyrosis	Fading of colours; drooping of scales; peeling of skin.	Dip treatment in 5% common salt solution or in 1 : 5000 formalin solution for 5 minutes.
	Diplostomulum sp. *Diplostomum* sp.	Diplostomiasis (Black-spot disease)	Small black nodules of about 1-5 mm diameter in the affected region.	Dip treatment in 3 : 100000 picric acid for 1 hour; Di-*n*-butyl tin oxide at the rate of 250 mg/kg body weight.

Table 6.14 contd. to next page

Table 6.14 : (continued)

Sl. No.	Causative organisms	Disease	Symptoms	Treatment
	Hemistomum spathaceum	Worm cataract	Whitish-eye due to the white dots, i.e. little worms situated between lens and lens capsule.	Dip treatment in potassium antimonyl tartrate (1.5 mg/litre).
(6)	Cestodes *Ligula* sp.	— —	Dull, sickly and with parts of alimentary canal swollen or completely choked by cestode cysts or worms.	Dip treatment in 3 : 100000 Picric acid for 1 hour; Di-*n*-butyl tin oxide at the rate of 250 mg/kg body weight.
(7)	Nematodes *Philometra* sp. *Camallanus* sp.	— —	Dull, sickly and with parts of alimentary canal swollen or completely choked by roundworms.	Dip treatment in 3 : 100000 Picric acid for 1 hour; Di-*n*-butyl tin oxide at the rate of 250 mg/kg body weight.
(8)	Acanthocephala *Zelanechinor-hynchus* sp.	— —	Dull and sickly.	Dip treatme·t in 3 : 100000 Picric acid for 1 hour; Di-*n*-butyl tin oxide at the rate of 250 mg/kg body weight.
(9)	Hirudinea *Hemiclepsis* sp.	— —	Abnormal movements of the fish due to irritation, as the parasites feed on the blood of host.	Dip treatment in 1 : 1000 solution of glacial acetic acid.
(10)	Copepods *Argulus* sp. *Ergasilus* sp. *Lernaea* sp. *Caligus* sp. *Pseudocycnus* sp. *Lernaenicus* sp. *Clavellisa* sp.	Argulosis Lernaeosis	Loss of scales and presence of red spots; damage of gills.	Half an hour treatment in 500 ppm formalin solution; mechanical removal by forceps followed by a bath in weak potassium permanganate solution for 5 minutes and subsequent bath in 1% common salt solution for 1 hour; pond may be disinfected by applying Lindane at the rate of 8 ml/1,000 litres of water.

(5) *Major symptoms and diagnosis of common disease and parasite infection* — For examination of affected organs certain laboratory procedures are undertaken :

- Collection of blood sample of diseased fish for haematological examination.
- Removal of fluid contents for examination, from boils, knots, vesicles, abscesses, etc.

- Culture of exudates from affected internal organs.
- Preparation of slide-smears from cultures grown from their exudates or directly from the diseased organs.

Diseases encountered by aquaculture species

Several diseases are encountered by an aquaculturist, particularly in semi-intensive and intensive systems of culture. Some of the major diseases, causative organisms, symptoms and treatment are listed in Table 6.14.

A. **Nutritional diseases**

Nutritional diseases can be said to be of three types :

(a) Those arising from malnutrition.

(b) Those arising from imbalance in the major components of food.

(c) Those arising from toxic effect of the diet.

Standard artificial diets tend to eliminate these diseases. Sub-standard diets used in supplementary feeding increases the possibility of their occurrence.

(1) *'Pin heads'* with enlarged head and slender body occur in nurseries due to starved larvae. Starvation occurs generally from inadequate supply of food or from behavioural starvation (it occurs due to overstocking).

(2) *Lipoid hepatic degeneration* disease is common among farm fishes caused due to overfeeding, where trash fish or pelleted diets have developed rancidity. The symptoms are swollen liver with rounded edge, anaemia followed by gills looking pallor.

(3) Nutritional gill disease occurs from panthonic acid deficiency in the diet. The symptoms are inappetence, respiratory troubles and anaemia.

(4) Vitamin deficiency results in diseases which are enlisted in Table 6.7.

(5) Mineral deficiencies of nine essential minerals are given in Table 6.8.

B. **Diseases caused by parasites and pathogens**

In aquaculture the known major diseases are caused through infection by pathogens of viral, bacterial, fungal or protozoan origin. Many of these have no known methods of therapy. Prophylaxis is the only control measure. Parasitic copepods and helminths also cause many diseases, but of lesser importance.

(a) **Bacterial diseases**

Bacterial diseases have worldwide distribution. It occurs both in tropical and temperate climate aquaculture. There are several bacterial diseases of cultured fin- and shell-fish. Many bacterial infections also occur in association with viral diseases (as secondary infection). However, bacterial diseases are most feared often causing heavy loss to fish farms.

(i) **Dropsy** — Abdominal dropsy of carps is caused by *Aeromonas punctata.* In temperate regions it occurs during spring. Symptoms include accumulation of fluid inside the body cavity, withering away of fins, scale protrusion and general edema. Intestines are inflamed and the liver and kidney are affected. The bacterium is rod-shaped with a flagellum, measuring 1 to 1.5 mm in length. If the infected fish is in good health then the bacteria remains dormant. The weak fishes are affected which may lead to death. The disease spreads when the liquid containing bacteria escapes the body of an infected fish after post-mortem decay.

(ii) **Furunculosis** — Furunculosis is caused by *Aeromonas salminicida.* It occurs mainly in salmonids which includes *Salmo salar, S. fario, S. salvelinus, S. irridaeus,* etc. The bacterium is rod-shaped measuring about 2 to 3 μm in length. The disease has cosmopoliton distribution and also infects various species of warm-water and cold-water fishes. The name furunculosis is obtained from the blisters or furuncules that occur on the infected fishes' skin. The furuncules (boils) are rounded swellings appearing on the sides of the body, which contains pus—a mixture of destroyed body tissues (muscles, leucocytes, etc.) and bacteria. The swelling sometimes bursts releasing the content into the water (Fig. 6.45). In case of massive infection, the intestine shows heavy inflammatory

Fig. 6.45. Trout showing furunculosis

conditions which further spread to peritoneum and muscles. Ultimately mass mortality occurs which has been reported in trouts.

(iii) **Columnaris** — Columnaris is caused by the bacterium *Flexibacter columnaris*. It occurs in chronic form in both cold-water and warm-water fishes worldwide. Fishes under stress due to crowding and increased temperatures are susceptible to infection. Generally two strains have been identified — high and low virulence. The less virulent form are primarily responsible for cutaneus infection, while the high virulent ones attack the gill tissues and causes a "gill rot" condition (Fig. 6.46).

The sign of the diseases in the less virulent cases, prevalent among most species of fishes, are :

(1) Appearance of discoloured grey patches on the dorsal fin.

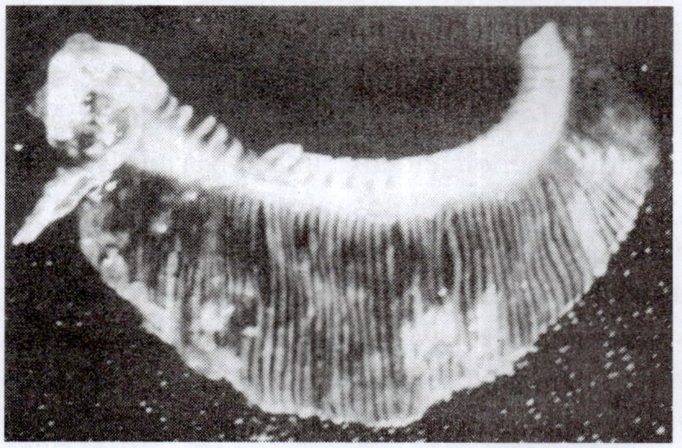

Fig. 6.46. Highly virulent form of *Flexibacter columnaris*, infecting the gill of common carp

(2) These lesions grow and expose the underlying muscle tissues.

(3) The lesions are prominent in the mouth and head regions.

(4) The lesions may become yellow and cratered subsequently.

(iv) **Gill rot** — The causative organism of gill rot, generally infecting the fingerlings of grass carp, is *Myxococcus piscicola.* It is characterised by masses of myxo-bacteria on the gills and is associated with hyperplasia of the gill epithelium. The gill covers in case of acute infection become inflammed and are eroded by the bacteria, forming small transparent patches.

(v) **Vibriosis** — The bacterium *Vibrio anguillarium* causes vibriosis, which occurs both in salt and fresh-waters. Vibriosis has become one of the most serious disease of cultivated marine species of fish such as rainbow trout, salmon, etc. *Vibrio* generally enters the fish through surface wounds and acts mostly on the skin forming lesions. The ulcers extend deep into the muscles and internal haemorrhage, kidney damage and swollen spleen occur in acute cases.

In shrimps, infection occurs in larvae, juveniles and adult. The bacterial attack may result in septi-caemia. The common sign of infection in shrimps are disoriented swimming and increasing opaqueness of abdominal muscle in juveniles and adults. Total mor-tality of the stock may occur.

(vi) **Fin and tail rot** — In fin and tail rot (Fig. 6.47), the margins of the fins become necrotic and slough away. This bacterial infection is caused by

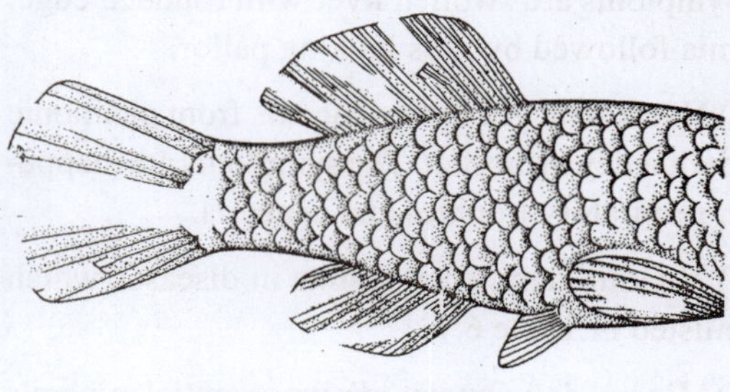

Fig. 6.47. Tail and fin rot

Pseudomonas. Fin and tail rots occur generally when unfavourable environmental conditions prevail. Improvement of water quality in ponds often results in cure of the disease.

(b) **Fungal diseases**

Fungal diseases are among the commonest diseases of freshwater. Some forms are primary invaders of fish and other aquatic organisms. Majority of them are saprophytic opportunists who take advantage of (1) necrotic tissue associated with injuries, (2) bacterial or parasitic lesions, (3) dead and decaying eggs and (4) unsanitary conditions. Some of the important fungal diseases are :

(i) **Saprolegniasis (Cotton-wool disease)** — Saprolegniasis is caused by *Saprolegnia* and some non-saprolegniaceous fungii such as *Pythium* and *Leptomitus*. This disease affects the skin and gills (in late stages) of freshwater fish and crustaceans. Small white tufts appear on skin and also on fins. They can be identified by the appearance of profusely branched, non-septate, cotton-wool-like tufts of mycelium (Fig. 6.48).

(ii) **Branchiomycosis** — Branchiomycosis, the gill rot of carps, involves the attack of the fungi, *Branchiomyces* on the gills of carps and channel cat-fishes in summer. At the onset of the infection, the gills of the fish show deep red patches. Subsequently, necrosis of the gills sets in and the colour of the gills becomes yellowish brown. Often, secondary infection of *Saprolegnia* occurs at this time.

Fig. 6.48. Cotton-like growth of the fungus causing saprolegniasis in crucian carp

Fig. 6.49. Rainbow trout — (a) Non-infected (b) Infected with IPN causing virus, showing pale pigmentation and distended vent.

(c) **Viral diseases**

(i) **Infectious pancreatic necrosis (IPN)** — IPN is one of the important viral diseases infecting trout and salmon. This virus has also been isolated from a number of other fish species (including eels) and several species of bivalve molluscs. IPN causing virus belongs to the birnavirus group and is 70 nm in diameter. It occurs as an acute disease in fry and fingerlings. Bad husbandry, unfavourable environmental conditions and overstocking help in the outbreak. The affected individuals show weak respiration and conclusive movements. Affected fishes show an overall pale pigmentation, exophthalmia, abdominal distention and haemorrhages on the ventral surface (Fig. 6.49). In the pyloric caecae, haemorrhages are also seen to occur. The liver and spleen become pale. The occurance of clear or milky mucus in the stomach and anterior intestine is a characteristic feature of IPN. When the mortality rate is considerable, infected individuals swim in a rotating manner about their long axis. This whirling behaviour is a terminal sign and the fish dies within an hour or two.

Most survivors of this infection become lifelong carriers of this virus. They intermittently shed varying quantities of this virus over a long period of time, through urine, faeces, milt and eggs. Transmission of this virus from parents to the offsprings takes place through the eggs.

(ii) **Viral haemorrhagic septicaemia (VHS)** — Viral haemorrhagic septicaemia is economically the most important viral disease seen in cultured salmonids, especially in case of hatchery-reared rainbow trout. The disease is caused by a bullet-shaped **rhabdovirus** of 50 nm in diameter.

Transmission of the disease takes place by contact, by transportation or bad handling and from fish to fish through water. Rise in water temperature leads to lesser losses and the loss ceases during spring.

Symptoms of VHS include general anaemia, pale discolouration of gills, swelling of eyes and belly, oedema of muscles, inflammation of intestines, etc. On the skin sores may appear, anus becomes protruded and kidney and liver may often be affected. Bleeding occurs in gills and muscles. Signs of excitability are seen including erratic swimming.

(iii) **Infectious haematopoietic necrosis (IHN)** — IHN is an acute viral disease caused by a bullet-shaped virus. It occurs in the fry and fingerlings of salmons and trouts and is transmitted from fish to fish and from parent to progeny through seminal fluids or infected eggs. Mortality may occur from the sac fry stage to the yearling stage.

The symptoms are weakness, dark colouration, abdominal swelling and pale gills. The kidney gets affected. It becomes translucent and speckled with pigment cells.

(iv) **Infectious dropsy of carp (IDC) and spring viremia of carp (SVC)** — Infectious dropsy of carp is the most serious disease of common carp and several other fishes. IDC comprises several aetiologically different diseases such as spring viremia of carp, carp erythrodermatitis, swim bladder inflammation, pseudomonas septicaemia and aeromonas septicaemia. The methods of treatment and prevention of these diseases are different.

SVC is caused by a rhabdovirus. It is very contagious and can cause high mortality. The symptoms include reduced movements and frequent resting, darkening of skin, protruding eyes, bleeding from eyes, fins, skin and gills, faecal casts trailing from the reddish protruding anus, accumulation of fluid in the abdomen, inflamed intestine, bleeding in the internal organs and anaemia.

The transmission of this rhabdovirus takes place from brood fish to offspring through eggs or sperm. It enters the fish through the gills.

(v) **Channel catfish virus (CCV)**—CCV is a serious virus disease caused by a herpesvirus measuring 100 nm. It occurs in fry and fingerlings of less than 4 months old channel catfishes in culture medium. Affected fish hangs vertically in the water and shows loss of balance (spiral swimming). Lesions develop at the posterior part of the kidney with increased number of lymphoid cells. Lesions later develop in the liver, spleen and the digestive tract. Haemorrhages are seen in the muscles, gills, skin and fin bases.

(vi) **Carp pox (CP)** — In cyprinids, carp pox is caused by a virus similar to herpesvirus. The disease is characterised by proliferation of the skin. This cutaneous growth reduces marketability of the fish. However, there is seldom any mortality of the infected fish. *Argulus,* the carp louse acts as a virus-carrier and transmits the virus within a pond population.

(vii) **Lymphocystis** — Lymphocystis is a viral disease that occurs as whitish nodules on the fins, head and sometimes on the whole body of the fish (Fig. 6.50). It occurs in several species of freshwater, brackish water and marine fish species. In culture conditions it is highly contagious and spreads very rapidly.

(d) **Protozoan diseases**

There are various protozoan parasites that live on fishes and other aquatic organisms. These parasites

Fig. 6.50. Flatfish showing lymphocystis nodules on whole body surface

cause both external and internal diseases, often causing havoc in hatcheries and rearing tanks. In certain cases the infected fishes may die without showing any symptoms. Some of the important protozoan diseases are given below :

(i) **Ichtyobodosis** — Ichtyobodosis is common among fishes when farmed for long periods in crowding conditions and is seen affecting various species including common carp and trout. It was previously known as **costiasis**. The causative agent being a flagellate *Ichtyobodo necator* (*Costia necatrix*). The parasite (Fig. 6.51) attaches itself to the skin of the host, with its anterior end forming finger-shaped processes at the point of attachment. The parasite asexually multiples by longitudinal fission. It dies when it falls from the hosts' body. Water acts as a medium of transmission. The parasite can tolerate temperatures varying from 2°C to 30°C.

The main symptom is the appearance of a slimy secretion on the skin, fins and the gills. Dull spots appear on the sides which subsequently becomes fused into a continuous greyish film. In advanced stages red patches may appear on the affected parts. Fins may become folded and erosion of the tissue occurs between the rays. Infected gills become pale and get partially destroyed. If unchecked the infection may prove dangerous.

(ii) **Ichthyophthiriasis (white spot itch disease)** — Ichthyophthiriasis is caused by a ciliate, *Ichthyophthirius multifiliis,* and ocurs in all species cultured in ponds including common carp, trout, etc. It causes the white spot itch disease which is fairly widespread.

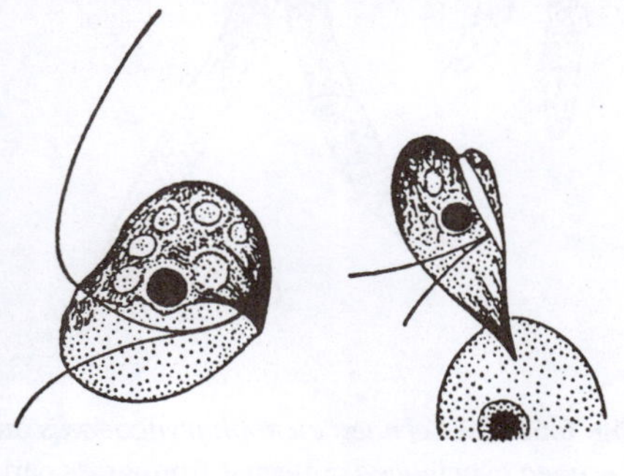

Fig. 6.51. *Ichtyobodo necator (costia necatrix)*

Ichthyophthirius multifiliis measures 0.5-1.0 mm long. It has a round or ovoid body, with longitudinal rows of cilia on the body surface. Since the parasite can withstand a high range of temperature (25°C to 26°C), outbreak of the disease usually occur in spring and summer when the water temperature is high.

The symptoms are appearance of numerous small white specks (early infection) or white spots (advanced infection) over the entire body. The spots are due to a small chamber in which the parasites are accumulated. Mature parasites leave the fish periodically and are responsible for the spread of the disease. Epidemics break resulting in mass mortality when the fish is weakened and its resistance is low.

(iii) **Knot and boil disease** — Various species of *Myxobolus* cause ulceration which appears either like a boil or knot or rice shaped. It occurs in the skin of common carp (*Cyprinus carpio*), in the gills and other body parts of tench (*Tinca*) and barbels (*Barbus*).

(iv) **Whirling disease** — The protozoan, *Myxosoma cerebralis* causes whirling disease which is prevalent among salmonids. Here the common sign of the disease is tail-chaising behaviour which is done rapidly. This takes place when the fish is frightened or trying to feed. The parasite feeds on the cartilage of young host fish. In advanced stages of infection, skeletal deformation takes place including deformed skulls and gill covers and spinal curvature. In early stages of the disease, trout may develop 'black tail'.

The infection spreads through infected fish, contaminated water and mud. The spores of *Myxosoma* can survive for 10-15 years in contaminated mud.

(e) **Copepod infection**

Copepod parsites are ectoparasites of the families Arguilidae, Ergasilidae and Lernaeidae. They generally infect several cultiviable fish species. *Argulus foliaceus* and *Lernaea cyprinacea* are the ones best known.

(i) **Fish lice** — Infection by fish lice is one of the commonests. **Argulosis** is caused by *Argulus* species, such as *A. foliaceus* (Fig. 6.52a), *A. coregoni, A. japonics* (Fig. 6.52b) and *A. giordanii*. Outbreak of argulosis is often fatal causing mass mortality. The

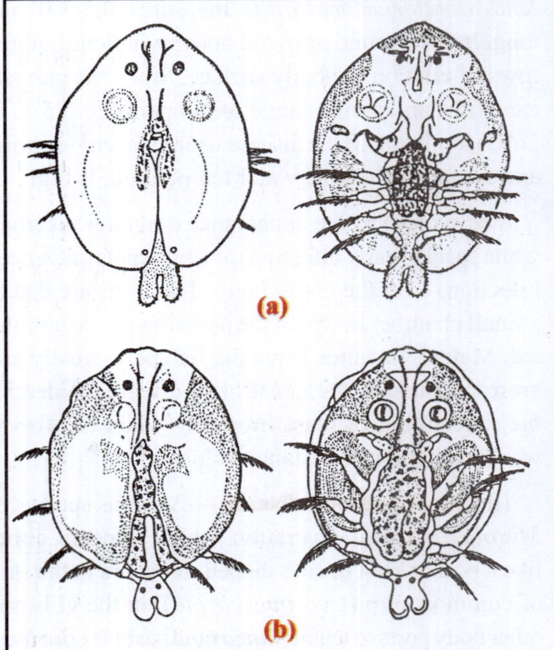

Fig. 6.52. (a) *Argulus foliaceus* and (b) *Argulus japonicus*

parasite is large, the size of adult lice ranges between 4-5 mm for males and 6-7 mm for females. They are oval with flattened greyish-green body. The parasite has large suckers and a proboscis for clinging and sucking blood from the host body surface.

Carps, trouts, tench, pike, bream, eel, etc. are known to be the common victims. The affected fish shows red blotches on the skin. The fish victim makes all effort to dislodge the parasite by way of frequent scratching. Signs of nervousness are apparent in the behaviour of the fish.

(ii) **Anchor worm disease** — Anchor worm disease, better known as **Lernaeosis,** is caused by the species of *Lernaea,* such as *L. cyprinacea* (Fig. 6.53b) and *L. ctenopharyngodonis* (Fig. 6.53a). This infection has been observed in a number of freshwater fish. The mature females of these anchor worms have a long unsegmented body. The head has branched processes, by the help of which the parasite penetrates the body of the host (Fig. 6.53). This worm has no intermediate host. The larva which is free-living, parasitises on any species of fish. However, the worm is very sensitive to salinity and can survive only on low salt concentration.

(f) Worm infections

Four groups of worms are parasitic on fishes. They are flatworms (trematodes), tapeworms (cestodes), roundworms (nematodes), thorny-headed worms (acanthocephalaus) and fish leech (Fig. 6.54).

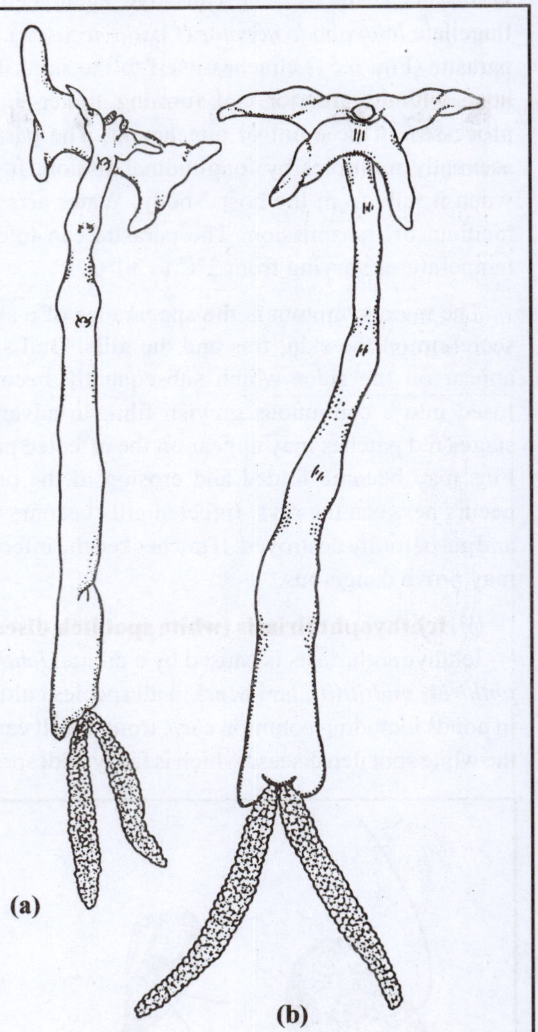

Fig. 6.53. (a) *Lernaea ctenopharyngodonis* and (b) *Lernaea cyprinacea* collected from grass carp and common carp, respectively

(i) **Flatworm infection**—Such infection is generally caused by flatworms (trematodes) *Distomum* and *Dactylogyrus* (Fig. 6.54A), who attack the gills and cause swelling and marginal discolouration. The life cycle of these parasites are completed with the help of two intermediate molluscan hosts.

(ii) **Tapeworm infection** — Such infection is caused by the tapeworm (cestode), *Ligula intestinalis* (Fig. 6.54B), which attains a length of 15 to 40 cm. Its infection is commonly witnessed more in wild fishes than in cultivated fishes. Fishes infected by this parasite show a swelling of the belly. The gonads get adversely affected with loss of fertility. Death may also occur. The parasites harbour in the fish-eating waterbirds where they attain their adulthood.

(iii) **Fish leech infection**—Fish leech infection is caused by leeches (Fig. 6.54E) of families Rhyncobdellidae and Gnathobdellidae. Small fishes are more affected. The fishes appear restless and make vain efforts to get rid of the parasites.

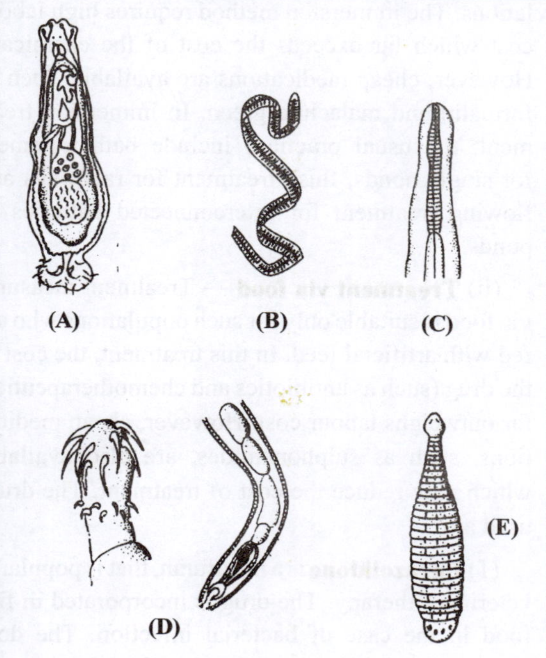

Fig. 6.54. (A) *Dactylogyrus.* (B) *Ligula.* (C) *Philometra.* (D) *Acanthocephalan* and (E) *Fish leech.*

C. **Control of fish diseases**

To cure diseased fishes on large scale is not only difficult but also too expensive to be of any practical use. For curing diseased fishes it is fruitful only at the initial stages of the outbreak of the disease and particularly when the number of such fishes is small and the damage done is less. However, to keep infections at a lower scale it is essential to take certain preventive measures, such as maintenance of hygiene and some prophylactic measures.

(a) **Hygiene maintenance**—The preventive measures to be taken are :

(i) Periodical drying of ponds and then refilling it with fresh, good quality water.

(ii) Siltation should be avoided as far as possible.

(iii) Growth of weeds to be controlled.

(iv) Pollution of the water body is to be avoided.

(v) Sustainable stocking density is to be maintained and over feeding is to be avoided.

(vi) Water bodies should not be stocked with unhealthy or poor quality eggs and fish larvae.

(vii) Frequent handling, or transport of fishes is to be avoided.

(viii) On the first detection of the disease the affected fishes (or the dead ones) should be removed and treated separately. The dead fishes is to be buried with quicklime.

(b) **Prophylaxis**—The following safety measures are to be taken :

(i) Routine disinfection of nets and other related tools are to be undertaken.

(ii) When ponds are periodically dried, they should be disinfected, each time with quicklime before water is filled.

(iii) Ponds when threatened with disease outbreak should be disinfected with either potassium permanganate (1 gm per 200 litres) or with 600 ppm benzalkonium chloride or calcium cyanamide.

(iv) Fishes before being stocked, should be given a salt bath which will not only rid them off any external parasite (Ex. *Costia*) but will also make them less vulnerable to future attacks.

(c) **Prevention of nutritional diseases** — Nutritional diseases can be kept under control if the following measures are taken :

(i) Food of good quality, rich in vitamins but containing less of fat should be used for feeding fishes.

(ii) The supplementary feed should be given according to the recommended dose and in no case should exceed 2–5% of the total weight of fish in the pond.

(iii) The supplementary food should be distributed in a controlled manner and overfeeding should be avoided.

(d) **Treatment of individual affected fish** — In the case when the disease has been identified, the affected fish is treated with medicines and quarantine baths using common salt, copper sulphate, potassium permanganate, malachite green, quicklime, formalin, lysol, etc.

(i) **Bacterial diseases** — In case of bacterial infections, antibiotics, such as chloramphenicol may either be mixed with food or injected into the body at the rate of 1 to 1.5 mg per 100 gm body weight.

(ii) **Fungal diseases** — The fungal infected fish is given a bath of potassium permanganate (1 gm in 100 litres) for 1 to $1\frac{1}{2}$ hours, or common salt (25 gm in 1 litre) for 10 minutes, or malachite green (1 gm in half a litre and then diluted to 1 ml per litre) for 1 hour, or copper sulphate (5 gm in 10 litres) until the fish gets fatigued.

(iii) **Protozoan diseases** — Fishes infected with costiasis parasite are to be given a bath of formalin (4 ml in 10 litres) for 10 minutes, while for ichthyophthiriasis, the pond water may be treated with 1 gm of malachite green per 12 square metres area thrice, at an interval of two days.

(iv) **Leech parasites** — Fishes affected with leech parasite are treated with a lysol (cresol and soap in 1 : 1 ratio) bath (1 ml in 5 litres) for 15 seconds and then rinsed with fresh clear water.

(v) **Flatworm parasite** — The fishes attacked by *Dactylogyrus* should be given a formalin bath (1 ml in 1 litre) for 15 minutes, or a salt bath (25 gm in 1 litre) for 10 minutes.

(vi) **Copepod infection** — In case of attack by *Argulus*, the fishes should be given a bath of lysol (quantity as mentioned above) or potassium permanganate (1 gm in 1 litre) for 40 seconds.

(e) **Therapeutical treatment** — Therapeutic measures for treatment of fish stocks are done through two methods — (i) treatment by immersion and (ii) treatment via food. In therapy of fish stocks, it is important to keep down the cost.

(i) **Treatment by immersion** — This is suitable in a general way for both large and small fish populations. The immersion method requires high labour cost which far exceeds the cost of the chemicals. However, cheap medications are available, such as formalin and malachite green. In immersion treatment, the usual practices include bath treatment for single ponds, flush treatment for raceways and flowing treatment for interconnected batteries of ponds.

(ii) **Treatment via food** — Treatment measures via food is suitable only for such populations who are fed with artificial feed. In this treatment, the cost of the drug (such as antibiotics and chemotherapeutics) far outweighs labour cost. However, cheap medications, such as sulphonamides, are also available which may reduce the cost of treatment. The drugs used are :

(1) **Furazolidone** is a nitrofuran, that is popular in veterinary therapy. The drug is incorporated in fish food in the case of bacterial infection. The dose recommended is 11 gm per 100 kg body weight per day for about a week.

(2) **Sulphadiazine** is a sulphonamide, found equally effective in case of bacterial diseases. It is admi-

nistered through the diet in the ratio of 5 : 1. The dose recommended is 5.5 gm per 100 kg body weight per day for about a week.

(3) **Oxytetracycline** is commonly used for bacterial therapy of vibrosis and ulcer disease in fishes. The recommended dose is 7.5 gm per 100 kg body weight per day for 1 to 2 weeks.

(4) **Di-n-butyl tin oxide** is effective against most parasites of the fish gut other than protozoans. The dose generally applied is 25 gm per 100 kg body weight per day for 3 days.

(5) **Epsom salt** or magnesium sulphate is found useful for the treatment of protozoan parasites and is mixed with the diet.

Epizootic Ulcerative Syndrome (EUS)

The outbreak of Epizootic Ulcerative Syndrome took place for the first time in India from Tripura state in 1988. It caused heavy loss to fishes in the wild as well as in farms, practically engulfing all freshwater bodies. It attacked a large number of species such as *Puntius, Clarias, Heteropneustes, Mystus, Mastacembelus,* all carps, etc.

The clinical symptoms and pathological features were :

(1) Ulceration of the skin.

(2) Erosion of epidermis and dermis with deep large ulcers.

(3) Hypodermis and muscles showed marked inflammatory haemorrhage.

(4) Raised areas of emudation and erythema were of common occurence.

(5) Necrosis of vital organs like kidney and liver.

(6) The virulence was noticed in 3 stages —

   (i) Initial stage marked with general haemorrhagic ulceration of the skin and scale pockets.

   (ii) The middle stage showed heavy loss of scales from the epidermis.

   (iii) The final stage witnessed haemorrhage and necrotic ulcers.

The causative pathogen includes all major pathogenic forms — bacteria in the skin, fungi in the liver, protozoans in the kidney and *Argulus* and *Gyrodactylus* on the gills. Viral infection was also found to be associated. All these pathogens were observed to be due to secondary infection. The primary pathogen is yet to be detected.

The control measures of this outbreak includes application of lime (500 kg per hactare), potassium permanganate at the rate of 2-4 ppm and a chemical compound that is marketed as 'CIFAX'.

FISH PRODUCTS AND BY-PRODUCTS

Fishes in addition to providing us nutrition (mainly in the forms of protein) do have other uses. A very good example is provided in case of shark carcass (Table 6.15) which not only have nutritional value but can be used to get various utility products (by-products).

Various edible fishes after landing may either be sold in fresh condition or may be preserved for long term storage, for later consumption. Non-edible, low valued trash fishes and parts of fishes which are not utilised for human consumption, do have other utility as by-products. The protein, fat and other useful body contents are processed into a number of valuable products. The products or by-products and their uses are given in Table 6.16.

Some of the by-products (mentioned in Table 6.16) are dealt below :

I. **Fish Liver Oil**— Fish liver oil is one of the most important by-products of fish. Liver, which is a part of the offal discard during dressing of landed fish, is the storehouse of glycogen and fat along with Vitamins A and D. This property of fish livers of certain fishes like Cod, Shark, Ray, Halibut, Tuna, etc. has rendered it of immense commercial value of providing highly medicinal 'liver oil'.

Table 6.15 : Chart showing various utilities of the various parts of a shark carcass

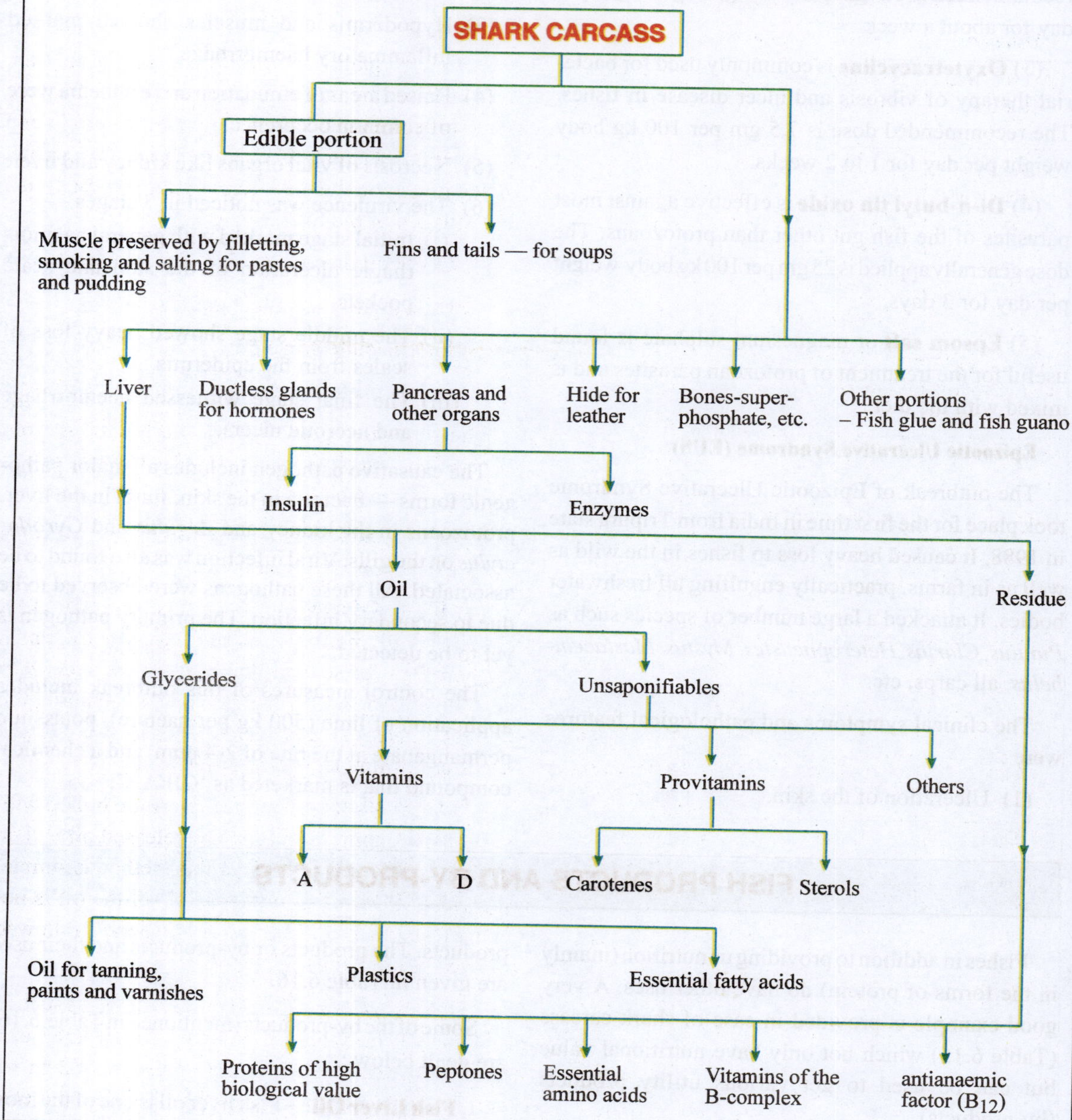

(a) **Composition of fish liver oil**

Water — 20% to 36%

Proteins — 5% to 10%

Fats — 55% to 75% (Unsaturated fatty acids higher in concentration than saturated ones).

As fat carries with it high concentration of Vitamins A and D, the fat of fish liver is the source of vitamins A and D rich liver oil. However, the concentration of vitamin A and fat in the liver varies from fish to fish and from season to season. Halibut (*Psettodes* sp.) and tuna (*Thunnus* sp.) liver have low fat but high

Table 6.16 : Parts of fishes used for production of valuable products/by-products

Fish parts	Product/By-product	Uses
(a) Fat of liver and body	1. Fish liver oil + Vitamin A 2. Fish body oil	Pharmaceutical Pharmaceutical, Industrial
(b) Protein of muscle and other tissue	3. Fish flour, sausage, ham 4. Fish meal 5. Fish silage 6. Fish soluble 7. Fish manure 8. Fish guano	Human consumption Animal feed Animal feed Animal feed Agriculture Agriculture
(c) Collagen of body tissues	9. Fish glue	Industrial
(d) Collagen of air bladder	10. Isinglass	Industrial
(e) Skin	11. Fish leather	Industrial
(f) Scales	12. Animal charcoal 13. Guanine 14. Shagreen	Industrial (purification of liquids) Industrial (artificial pearls) Industrial (abrasive)
(g) Fins	15. Dried fins	Human consumption (soup)
(h) Eggs	16. Fish roe	Human consumption
(i) Enzymes and hormones	17. Enzymes and hormones	Pharmaceutical
(j) Sterol	18. Cholesterol 19. Squalene 20. Lecithin	Pharmaceutical Industrial (mordant in dyeing). Industrial (anti-bloom agent in chocolate industry).

vitamin A content (50,000 to 3,00,000 I.U. per gram), while cod (*Gadus* sp.) liver is rich in fat but poor in vitamin A (1000-3000 I.U. per gram).

Vitamins A and D present in fish liver oil is in proportion most suitable for human needs, as it has medicinal use in the form of prophylactic and curative.

(b) **Method of extraction of fish liver oil** — For extraction of quality liver oil it is essential that the liver of fishes be in good condition (either fresh or properly preserved). This is to guard the protein and fat rather than vitamin A content which is very stable. Depending upon the method of extraction used, fish liver oils are of the following grades :

(1) Crude oil used for burning

(2) High quality technical grade oil

(3) Intermediate grade oil

(4) Edible oil for medicinal use.

The first three grades of oil are produced from diseased or discoloured livers while the last one is produced from fresh or suitably preserved livers.

(i) **Oil extraction** : Some popularly used methods of oil extraction are given below :

(1) **By Autofermentation** — Fish livers are first chopped into small pieces and exposed to sun in earthen pots for several days, so that it undergoes decomposition. As it undergoes decomposition, the oil that is released is collected. This oil is crude and used for burning lamps in curing yards.

(2) **By Boiling** — Chopped livers are boiled with sufficient quantity of water. The released oil is then decanted off easily. Although this method is simple, the yield, however, is moderate as all the oil is not extracted from liver. This method is generally followed in small scale industries.

(3) **By steaming** — Minced livers of fishes such as cod (which has high oil content) are steamed at a temperature of 85-90°C under a pressure of 2 kg/sq. cm. The liver cells disintegrate and the oil is collected from the top layer. This method is good for operation while on board of fishing vessels or trawlers, where steam is readily available.

(4) **By Chemical Digestion** — Chemical digestion is done by the following three methods :

- **Aquacide digestion** : Aquacide is a patented mixture of sodium bicarbonate and paraldehyde. When fresh liver is mixed with it, it partly denatures the protein and yields a pulp. The pulp is stirred with warm water as it is passed through a series of tanks or cylinders. This gives rise to an emulsion of water and released oil. As soon as the emulsion breaks, the oil floats on the top from where it is collected.

♦ **Alkali digestion** : Digestion of proteins in the liver with alkali has proved to be the most suc-cessful method. The livers of halibut, tuna and some sharks, which are relatively poor in oil but high in vitamin A potency, are processed in this manner. In this case, the oil is held strongly by proteins and is not totally released by steaming or boiling.

Minced liver is mixed with dilute caustic soda (1-2% by weight) or sodium bicarbonate (2-5% by weight). It is then steamed at 82-88°C with constant stirring. The pulp is then centrifuged and the supernatant oil is collected.

♦ **Enzyme-alkali digestion** : This chemical digestion is a modified version of the above, where alkali (sodium bicarbonate) is added after the chopped liver is treated with an enzyme (pepsin). As pepsin acts best in an acidic medium, HCl is added to the mixture.

At first the chopped liver is brought to pH level of 1.2 to 1.5 by the addition of HCl. Then commercial pepsin (0.5% by weight of the liver) is mixed with the pulp and digested at 43-49°C. The pH is then raised to about 9.0 by the addition of sodium bicarbonate and the temperature is increased to 80°C. After complete digestion, the oil that has gathered at the top is collected.

(5) **By Solvent Extraction** — This is an advanced and expensive process although the yield is higher. This involves solvent extraction plant/equipment and solvents, like ethylene dichloride, solvent ether, chloroform, light petroleum, etc.

In this method, at first the minced liver is dehydrated with the application of an anhydrous salt, generally sodium sulphate. When the moisture has been removed, the pulp thus formed is mixed with the solvent (preferably ethylene dichloride) and put to extraction process. After distillation the oil is separated out. Any free fatty acid present in the liver is removed during refinement of the oil. The solvent extracted oil is darker in colour and has a higher viscosity.

(ii) **Refinement of liver oil** : The oil after collection from the above methods is left to stand, so that water and any particulate matter present in it settles down. Oil is then removed from the top and is subjected to filtration, centrifuging, or treatment with Fuller's earth*. The oil that is obtained is congealing** oil. It has stearine (triglyceride of stearic acid) present in it. Destearinated or non-congealing oil is obtained by separating stearine, which is done by cooling the oil slowly to 10°C. The oil, thus obtained, is again filtered in a chill room.

(iii) **Standardisation of vitamin A potency in the extracted oil**

Due to its richness in vitamin A content, the liver oil has immense medicinal value. For commercial purposes, it is thus subjected to standardisation of vitamin A potency by the following methdos :

1. **Biological estimation** — By this method standardisation is done by feeding the refined oil to 5 weeks old albino rats which are raised on vitamin A deficient diet. The response in growth is then matched with the growth rates of rats grown on known potency of vitamin A.

2. **Colorimetric estimation with tintometer** — In this method the oil is dissolved in chloroform and then to it antimony trichloride is added. A blue colour develops and the intensity of this colour is read on a tintometer. The vitamin A potency is thus determined.

3. **Photoelectric spectrophotometric estimation**. In this method the principle of characteristic and selective absorption of a certain zone of ultraviolet region of the spectrum is used. For this, the extracted oil is first dissolved in a suitable solvent and the solution is kept in the path of ultra violet rays. The intensity of absorption is measured with photoelectric unit and can be read on its galvanometer.

* Fuller's earth — It is an absorbent clay used for removing greese.

** Congealing — To change from a soft or fluid state to a rigid or solid state by cooling or freezing.

(c) **Uses of fish liver oil** — Fish liver oil has the following uses :

(1) Crude liver oil is used for burning purposes or making lights in curing yards by fishermen.

(2) Fish liver oil because of its high vitamin A and D contents can cure or prevent occurrence of disease caused by the deficiency in the body of these two vitamins, such as rickets, xeropthalmia, impaired vision and eye defects, abnormalities of skin, mucous membranes and vertebrae.

(3) When taken orally it will ensure good growth of bone and teeth and will develop more resistance to counter bacterial attack.

(4) The **stearin** obtained by chilling the oil is used to make soaps of inferior grade and in leather curing.

(5) Inferior pharmaceutical grades are used in animal and poultry feeds.

(6) The volatile fraction of the liver oil, distilled at 250°C is toxic to bacteria and fungi.

(7) **Alkoxyglycerol** present in shark liver oil provides a unique immune supporting nutritional effect.

(8) **Squalene** present in certain shark liver oil as a constituent of the unsaponifiable fraction of fish oil, is used as a mordant in the dyeing of synthetic fibres.

(9) **Lecithin**, present in high concentration in dogfish liver oil (and in the oil of fish eggs) is used as a wetting and an anti-bloom agent in the chocolate industry.

II. Fish Body Oil

Fish body oil is the oil obtained from the entire body of fishes like Herring, Sardine, Salmon, Mackerel, Anchovy, etc. It is also extracted from offal and other wastes discarded from canneries or curing yards. Fatty fishes like Sardine yield more oil than non-fatty fishes. However, on an average about half kg of fish body oil can be obtained from five kg of fishes. Fish body oil varies according to season, sex, size, age, nature of food and the locality from which it is caught. It has been observed that the body oil content in mackerel rises to a maximum in October-November and falls thereafter.

Fish body oil substantially differ from fish liver oil. Fish body oil is poor in vitamins A and D contents and has less unsaponifiable matter as compared to fish liver oil. The presence of various proportions of the glycerides of fatty acids (both saturated and unsaturated groups) has resulted in various uses of fish body oil.

Methods of fish body oil extraction — There are two methods of extraction of fish body oil : (i) Dry method and (ii) wet method.

(i) **Dry method of extraction** — This method of extraction is used in fishes with low body oil content and the oil recovery is less than the wet process. The fishes are grounded in a grinder, cooked (with constant stirring) and pressed to recover the oil.

(ii) **Wet method of extraction** — The wet method is employed for obtaining body oil in fishes such as Oil Sardine (*Sardinella longiceps*), which are rich in oil content. In the wet method, the fishes are crushed to a pulp and cooked with steam in a continuous vertical cylindrical cooker. The cooked material is then pressed and a mixture of oil and stick-water (pressed liquid of fish soluble) are collected into setting tanks or are centrifuged to separate the oil.

Refinement of fish body oil — In both the above methods of oil extraction the residue (after pressing) is processed as a by-product — the fish meal. The oil thus collected after pressing contains protein debris and lots of other impurities such as water soluble and non-fatty acids of the body tissues. The pressed liquid is first filtered and then subjected to various refinement processes such as :

(1) To neutralise the acids it is treated with caustic soda.

(2) To remove bad odour, the liquid is treated with super-heated steam and sodium carboante.

(3) To bleach the unwanted dark colouration, it is subjected to aeration.

(4) To remove moisture, it is heated to a temperature of 105°C.

(5) To remove any stearine content, it is subjected to refrigeration.

Uses of fish body oil

(1) After various refinement processes (as above), the better grades of body oils thus formed, are used for edible purposes.

(2) The fish body oil of high iodine value is very suitable for manufacture of paints and varnishes, as it is a drying oil.

(3) It is used in dressing of leather and tanning of skin.

(4) Body oil of low iodine value are generally preferred for making solid fats, such as margarine and lard (refined fat from the abdomen of a pig) substitutes.

(5) It is used in the manufacture of laundry soap and cheaper grades of toilet soap, insecticidal soap, etc.

(6) It is used in steel and iron industry for tempering of steel, concentration of low grade iron ores, etc.

(7) Due to the presence of vitamins A and D, the fish body oil is used for medicinal purposes, for animal and poultry feed and in aquaculture practices.

(8) It is used in the manufacture of chemicals such as alkyl halides, silicons and quaternary ammonium salts.

(9) It is used in the manufacture of cosmetics, lubricants, candles and cutting oils.

(10) Fish oil formulations when sprayed on citrus trees have proved effective as fungicide.

(11) It is used in the manufacture of printing inks, waterproofer preparations, plastics and linolium (a substitute for rubber).

(12) Body oil is used for smearing the surface of boats for longer preservation.

III. Fish Meal

Fish meal is the next most important product after fish liver and body oil. It is a preparation where the body of fresh fish is ground, cooked and dried. It is a highly nutritive product and is an excellent poultry and animal feed, which increases egg and milk production.

The nutritive value of fish meal depeds upon —

(1) type of fish used (low fat and salt content to be used),

(2) the state of freshness of fish,

(3) season of landing and

(4) manner of preparation.

Chemical composition

Moisture —6-12% (Moisture content is greater in lean fish than in fat fish).

Proteins — 55-70% (The protein of fish meal is of high digestibility coefficient and contains all the essential amino acids).

Oil/fat — 2-15%.

Minerals — 10-20% (calcium 5%, phosphorous 4% and iodine).

Vitamins — Very rich in vitamins A, B, B_{12}, D, K and E.

Raw material —The raw materials include fishes such as sardines, mackerels, ribbon fish, silver bellies and other fishes. Sharks and rays yield superior quality of fish meal. Fish meal is also obtained as a by-product of the canneries and fish oil industries.

Preparation of fish meal — For manufacturing of fish meal the following methods are employed.

(a) **Preparation for small scale production** — The fish is first minced and then cooked. To remove moisture, the cooked mass is pressed in screw presses. The cake thus produced is dried in the sun, or in flame driers in which the material is exposed to high temperature or generally in steam-jacketed drums under partial vacuum.

(b) **Preparation for large scale production** — Here huge amount of raw material is used, that includes — (i) entire landed fish of poor food value, (ii) wastes from filleting plants (of canneries) and (iii) curing yards.

The raw materials, thus collected, are ground well to crush bone and flesh. The minced mass is then heated in steam, either by external application of steam (dry process) or by pressure-steaming through the minced mass (wet process). Hydraulically operated pressing is then done and after extraction of oil and water, dry cakes are ready for sac filling and marketing. Fish meal is generally packed in gunny or coir bags, which are insect and vermin proof. Storage, if done in tin containers, is done under an atmosphere of nitrogen and with soldered lids.

Use of fish meal

(1) Fish meal being a highly nutritive product (it contains all the essential amino acids) makes it an excellent poultry and animal feed which is practically good for all classes of livestock. By its use milk and egg production gets increased.

(2) As fish meal contains calcium, phosphorus, iodine and rich variety of vitamins, it is important for growing cattles, for it promotes building of tissue and bone.

(3) Trash fish meal mixed with rice bran and vitamins, forms an excellent feed in aquaculture.

IV. Fish Manure and Guano

Fish manure and guano are inferior quality of fish meal. It is not fit for animal consumption. **Fish manure** is a by-product of the curing yards, fish glue industries and oil extraction plants, where trash or spoilt fishes have been employed. Mackerel, horse mackerel, sardine, etc., which are spoilt and unfit for use as food is used for preparation of fish manure. This manure has a high content of nitrogen (5-7%), phosphates about 4-6% and lime (CaO) 1-5%. For such nutrient content, fish manure is useful in raising coffee, tea and tobacco crops.

Fish guano is the by-product of the body oil extraction plant which is the dried refuse left after the oil is pressed out. Here oil bearing species such as oil sardines are used as raw material. Fish guano contains high concentation of nitrogen (8-10%) and appreciable quantities of phosphoric acid. For this, guano is several times more effective than any animal manure.

V. Fish Flour (Hydrolised protein)

Fish flour is a superior quality fish meal, produced under strict control and care, and forms an ideal protein supplement for human diet, even for infants of 3-4 months old. On commercial scale, it is produced by a sophisticated solvent extraction process. The process is complicated and is not expensive. Fish is chopped and washed. It is then boiled with dilute acetic acid at 80°C. The mass is then washed thoroughly and the water is pressed out. This is then treated with petroleum to remove fat and also to increase its keeping quality. The mass, thus, obtained is then hydrolysed with an alkali, preferably with caustic soda (10% at 80°C). The entire mass is then liquified when it is neutralised with acetic acid (85%). The liquid is then dried with a spray, which subsequently forms a dry, cream coloured powder.

Use of fish flour

(1) It is considered as an ideal protein source to supplement diet for both adults and infants.

(2) It is used to enrich bakery products such as bread, biscuits, cakes and soup.

(3) As the product contains high protein (35%), it is very suitable for convalescing patients struck with malnutrition, anaemia, etc.

VI. Fish Silage

'Silage' generally means fodder converted in a cylindrical tank called 'silo'. Fish silage is a liquid or semisolid fish meal and is a highly nutritive animal feed. It is produced by adding 3-4% of an acid to a minced fresh fish or fish offal. Generally formic acid is used but sulphuric or propionic acid can also be used. By using these acids the pH of the mixture can be lowered to 4.0 or below. This inhibits bacterial decay. The enzymes present in the chopped fish act upon it and reduce the mixture to slurry. An antioxidant is added to prevent rancidity of fats and the liquid can be stored in a silo up to 6 months.

Fish silage is also obtained by fermentation with lactic acid bacteria in molasses. The advantage fish silage has over fish meal is that the vitamins remain unaffected to a large extent and the product is free from fishy odour. Its production is more preferred in temperate regions (Norway, Denmark, etc.) than in tropical countries.

VII. Fish solubles

The residual part of the liquid which is obtained during the extraction process of fish oil is known as fish solubles. It is valued as additives to dry feeds for animals. It is rich in protein and vitamin B-complex and has the following composition :

Water —50.0%

Protein —33.9% (all essential amino acids)

Fat —2–6%

Ash —9.4%

Vitamin B-complex and choline.

VIII. Fish sausage and ham

Fish sausages are prepared from minced fish flesh that are stuffed into a prepared intestine or similar other casings. **Fish ham,** on the other hand, has small pieces of solid fish meat (pieces of one square cm), mixed with pasted fish meat. In both the preparation, spices and additives are added to improve the taste, flavour and keeping quality. Spices include salt, sugar, chillies, onion, corriander, glutamate, egg-white, hydrogenated vegitable oil, etc. Additives comprise antiseptics and antioxidants (ascorbic acid) to prevent rancidity. Colouring agents also may be added. These products are commercially manufactured in Japan, Russia and USA, and are prepared from less valuable trash fishes.

IX. Fish Macaroni

Macaroni is a pasta (flour and egg food preparation of Italian origin), prepared from wheat flour in the form of dried, hollow tubes. Fish macaroni is a product which is prepared from *Puntius carnaticus*. The fish is first minced and then mixed with tapioca* or sorghum flour in equal parts. It is then spiced with salt, chillies and tamarind. The product (paste) is extruded and dried. The product has good keeping quality, is cheap and easy to manufacture. It is manufactured on commercial scale by the Mysore Institute of India.

X. Fish biscuits

Fish biscuits are manufactured in Chile and Morocco. It is blended with biscuit mixture prior to baking.

XI. Fish glue

Fish glue is a good adhesive obtained from trimmings, bones and skin of fishes belonging to order — Gadiformes (Cods, Pollack, Hakes, etc). The raw materials are washed, chopped and steam-heated in steam-jacketed cookers. The mass is then covered with water and to it is added small quantity of acetic acid. It is then cooked for 6-10 hours. The liquor is extracted and concentrated to form glue. The residue is dried and is used as manure. Fish glue is used for :

(1) Smearing the backs of glued stamps and labels.

(2) Glue obtained from cod is of better quality and is used in photo-engraving.

(3) Low quality fish glue is used as adhesive for paper boxes, shoes and other things like furniture where joining is required.

XII. Isinglass

Isinglass is a gelatin-like material obtained from the swim bladder or air-bladder of certain fishes. When put in water, it swells up but does not dissolve in it. At high temperature it hydrolyses in water to produce strongly adhesive gelatin.

Swim bladder is a hollow sac, the outer layer is thick and fibrous, while the inner one is thin, often with a silvery lustre. The thin, inner silvery shiny layer of the air-bladder of some fishes, such as sturgeons, carps and catfishes, is used to derive isinglass. For preparing isinglass the air-bladders are collected, washed thoroughly to remove blood and other extraneous matters. Then the outer thick and fibrous layer of the wall is separated from the inner layer, which is exclusively isinglass raw material. This is then sun-dried and marketed.

* **Tapioca.** It is a food substance prepared from cassava (a tropical euphorbiaceous plant). It is used in puddings as a thickener.

Isinglass has the following uses :

(1) Isinglass is used mainly for clarification of wine, beer and vinegar.

(2) It is used for the preparation of special grade cement and plaster.

(3) Formerly it was used as a substitute for gelatin in confectionery.

XIII. **Fish skin**

The skin of larger fishes is collected, soaked in brine and left for a day. On the next day it is salted and again put in brine containing 10% hydrocholoric acid. The skin is then taken out, drained and scraped on the surface, particularly to remove fine denticles present in the skin of sharks. They are then limed and tanned by the usual process.

Fish leather has various uses, such as :

(1) Natives of old age used the dried and spiny skins of Globe fishes or Porcupine fishes for making war helmets.

(2) In Japan the dried skin of Puffer fishes is inflated and used to make lanterns.

(3) Now-a-days the skins of larger fishes, such as cod, salmon, halibut, toadfish, sharks and rays are tanned and marketed as ornamental leather and can be dyed in different colours.

(4) Fish leather, particularly of sharks, is used in making shoes, wallets, bags and tobacco pouches.

(5) The skin of sharks, ray, skates are very hard and is used as an abrasive for polishing wood or metal.

(6) Suitably prepared and dyed skin of sharks providing 'Shagreen' is used for covering card cases, jewellery boxes, sword scabbards, etc.

XIV. **Artificial pearls**

The silvery scales of Europen cyprinid have been used for the manufacture of artificial pearls. By scrapping the scales, a glossy pigment is obtained. It is then coated on the inner surface and formed into hollow glass beads. The beads are then filled with wax.

XV. **Fish fins**

The fins (except caudal fin) of shark are cut near the root, washed in sea water, dusted with a mixture of wood ashes and lime and dried in the sun or smoked. The cured product is crisp and brittle. In China and Philippines the shark fins are used in soup.

XVI. **Fish roe and Fish caviar**

Roe (mass of eggs or spawn) of a number of fishes is considered as food. The protein of roe is tasteless and its digestibility co-efficient and biological value are 81% and 88%, respectively. Roe fat is characterised by high lecithin (59%) and cholesterol (14%). Roe is a good source of vitamin B. It also has vitamins C, E and D.

Caviar is the processed and salted form of roe of any large sized fishes. The caviar of sturgeons is held in high esteem as a very delicious food and serves as an appetiser.

XVII. **Medicinal uses**

(1) The pancreas of shark is very rich in insulin. Whales also provide a considerable quantity of insulin. Fish replaces cattles in providing raw material for the manufacture of insulin.

(2) The large otoliths* of Sciaenids are removed from the head and after rubbing and mixing them with water, are given to convalescent children suffering from rickets.

(3) *Amphipnous cuchia* has been considered by fisher folks as a highly medicinal fish. As soon as the fish is caught, gram flour (besan) or wheat flour is rubbed on its body, so as to wipe out all the mucous. The flour with mucous is then made into small balls and then dried. Such balls (tablets) are prescribed to the persons suffering from impotency. It has been claimed that such persons soon regain their vigour and strength.

(4) Live fishes such as *Clarias batrachus, Heteropneustes fossilis* and *Channa* sp. are valued for their high nutritive and medicinal properties.

(5) *Sillago sihama* is considered good and nourishing for nursing mothers.

* Fish **otoliths** are the fine concretions of calcium carbonate, useful for the proper growth of bones.

ORNAMENTAL AQUARIUM FISH CULTURE

Fish is an enterprising aquatic animal. It not only provides us protein but also entertainment. Many fishes are beautifully coloured, attractive, graceful in their shape, body-form and movement. Such facinating beauty have been attracting people from time immemorial and are termed as 'Ornamental Fish' or 'Aquarium Fish' or 'live jewels'.

Keeping such fishes has become a hobby which dates back sometimes around 800 B.C. (there is, however, no authentic document). However, the idea of keeping fishes in glass aquaria originated after 300 B.C., at the time blowing of glass was invented. Modern-style aquaria were developed in the early 19th centuary. An **Aquarium** is an antificial tank made of glass, for keeping live aquatic plants and animals.

Aquarium fish keeping, which started as a small time hobby, has now become an international trade worth millions of dollars. It has taken the shape of an industry. In India ornamental fish keeping started as a hobby through the introduction of 26 1egg layers and 27 live bearers of exotic fish species. Later the farmers took this up as an alternative income generating activity through the breeding and farming of these exotic species, as the demand of these fishes is very high within the country.

The indigenous ornamental fish species, on the other hand, has been less popular among the local hobbyists. However, at present the export trade of these species has been flourishing. India has the potential to emerge as one of the leaders in the trade if it can utilise the diverse indigenous fish resources. The export of ornamental fishes from India is dependent mainly (85%) on species caught from the wild and a few (15%) bred varieties of exotic species. India exports these fishes to USA, Singapore, U.K. Japan, Spain and Germany. Export is mainly carried out through the international Airport at Chennai, Mumbai, Thiruvananthapuram, Cochin and Kolkata. About 50% of the total export takes place from Kolkata.

Types of Aquaria

Depending upon the size of the aquaria, the place they are installed and the size of fishes kept, aquaria are of the following two types :

(a) **Home Aquaria** — These aquaria are smaller in size and made for keeping small-sized indigenous and exotic fishes. They are installed in homes, hotels, shops, offices, etc. for decorative purposes. In schools and colleges, in particular, they are used for the purpose of study and identification of ornamental fishes. In Universities and research laboratories, aquaria are used to assist in research for observing spawning, hatching and their developments and also behaviour studies or for bioassay experiments.

(b) **Public Aquaria** — They are larger in size and afford the space to keep many larger and more exotic specimens such as snakeheads, cichlids, eels, catfishes, sharks, etc., which are displayed to best effect. These aquaria are installed at the places of public exhibition, with great deal of thoughts to display the ornamental fishes to their best advantage not only for aesthetic and entertainment requirements but also for scientific interest. One such famous Public Aquarium in India is Taraporewala Aquarium in Mumbai.

THE AQUARIUM

Setting up of home aquaria is always an exciting event. It has now become a very popular hobby which appeals to young and old equally. Like philatelists, aquarists have also formed societies where ideas are exchanged. It has opened up a flourishing business in all major cities of the world.

Aquarium can always be purchased, but making it and procurring of ornamental fishes are not expensive. The raw materials for making an aquarium can be bought and assembled. Some indigenous aquarium fishes can be obtained from ponds and streams, while the fancy varieties can be bought from the market.

A. Setting up of an Aquarium

While setting up an aquarium two main factors are to be considered : (1) **Size** — Which will be determined by the type of fish to be maintained and the dimension of the intended site, and (2) **Weight** — a factor which should not be overlooked.

Different types of tanks are in use as given in Table 6.17. The most suitable size for home is an 18 gallons aquarium of size 60 × 30 × 38 cm. Standard sizes usually starts at 45 × 25 × 25 cm (18" × 10" × 10"). A tank of 60 × 30 × 30 cm (24" × 12" × 12") requires adequate support. Any flat surface can suffice to hold a fish tank, but it is preferable to use a stand. Various kinds of stands are available. However, for a standard sized aquarium a matching stand can be purchased or made. Metal stands of slotted steel angle frames may be made comprising one, two or three tiers (Fig. 6.55). The main frame should be made of sufficient thickness.

Table 6.17 : Sizes and volume of some standard aquarium

	Aquarium size (in cm)			Volume of water	
	Length	Width	Depth	In gallons	In litres
1.	47	25	25	6	27.3
2.	60	30	30	12	54.6
3.	91	30	38	20	91.0
4.	121	30	38	30	136.5
5.	152	45	45	70	318.5
6	182	45	45	80	364.0

Frames are usually of angle iron to provide the necessary rigidity and strength. For smaller tanks, pressed steel can safely be used, but care should be taken that it does not rust. To avoid rust alternative methods for constructing an aquarium can be used :

(i) **Aquaria made completely with glass** — The all glass aquaria with the bottom, sides and ends made of glass sheets cemented together with 100% silicone sealant has been very successful. Handling these glass aquaria should be done with utmost care.

(ii) **Aquaria made of plastic** — They are less expensive than glass and are made of flat sheets of clear and coloured *perspex* cemented together. These tanks cannot be made of large size as the perspex is flexible in nature. Also, in due course these tanks may become full of scratches.

(iii) **Using fibre glass frame and base for aquarium** — For the front, back and sides of the aquarium normal fibre glass is used. The fibre glass is hard. It is non-corrosive and can be obtained in a variety of colours.

While selecting any type of aquarium, it is essential to note that it is of sound construction. It is preferable to have shallow and wide tanks as they have more surface area to facilitate oxygenation through atmospheric contact. Also, such tanks can accomodate more number of fishes than tanks having lesser surface area but with equal quantity of water.

B. Requirement for construction of an aquarium

For construction of an aquarium the following primary things are required :

(a) Glass sheet (b) Silicone (c) Masking or duct tape (d) Caulking gun (e) Aquarium hood (f) Aquarium stand (g) Lighting devices (h) Heating devices (i) Thermometers (j) Thermostats (k) Composts (l) Aquarium filters (m) Aerators and diffusers (n) Planting sticks (o) Ornamental plants (p) Feeding rings (q) Hand-net (r) Scrapers or algae cleaners (s) Siphon tube (t) Water testing kits.

(a) **Glass sheet** — For construction of a rectangular shaped aquarium of size 60 × 30 × 38 cm or 60 × 30 × 30 cm, five sheets of 6 mm thick "plate" glass, also known as "sheet" or "annealed" glass are taken. The thickness of the glass sheet would vary depending upon the size of the aquarium (Table 6.18). Tempered glass should not be used. The edges of each glass sheet should be made smooth with a dry abrasive paper or the glass shop may be asked to sand or grind the edges.

(b) **Silicone** — One should be careful as to the type of silicone to be used. Aquarium silicone sealent is the same as regular household silicone, which is without the anti-mildew chemicals added to it. Silicone marketed for aquarium use is much more expensive.

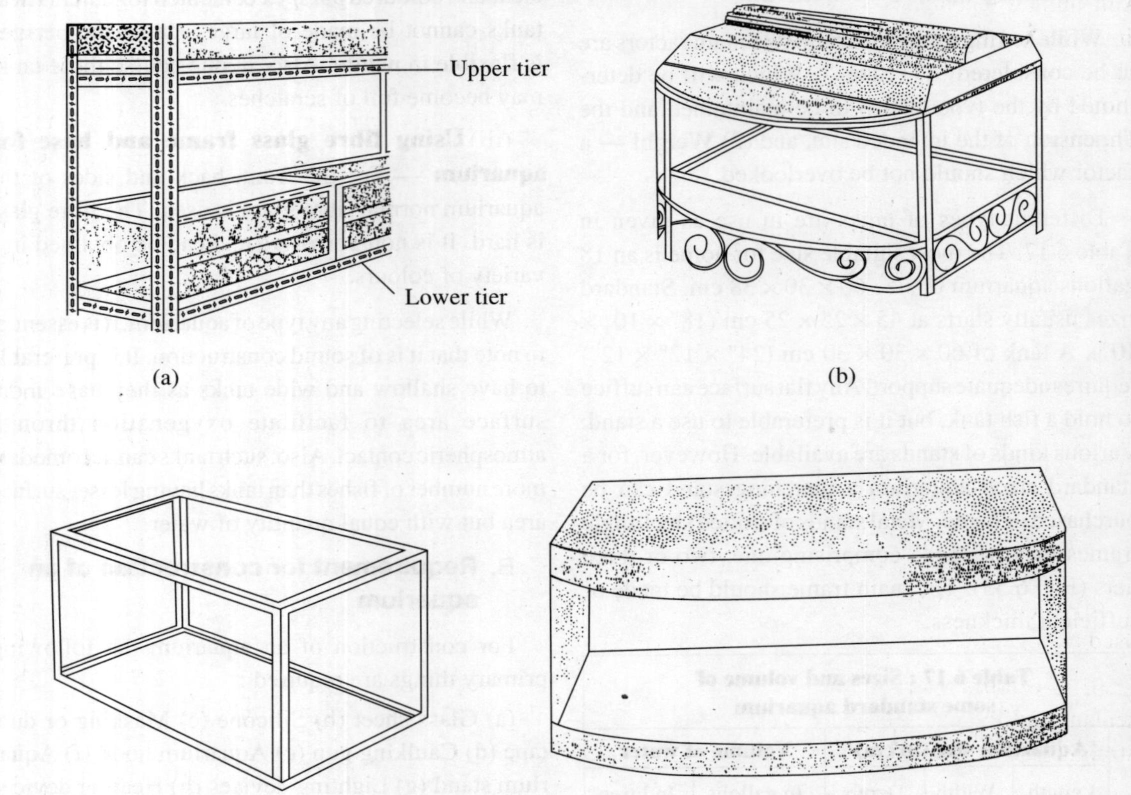

Fig. 6.55. Various designs of aquaria and supporting stand :

(a) Aquaria with slotted steel angle frame made of two tiers; (b) Bow-fronted tank with wrought-iron stand; (c) Angle iron tank; (d) All plastic tank; (e) Tank with a typical hood

Anti-mildew chemicals slowly release toxin over time which is toxic to both fresh and salt water fishes. Thus, the silicone to be used for aquarium making should not contain the anti-mildew chemicals. Therefore, the 100% silicone sealant should be used.

Table 6.18 : Thickness of glass suggested for different sized aquarium

	Aquarium size (in cm)			Thickness of glass sheet (in mm)
	Length	Width	Depth	
1.	47	25	25	4
2.	60	30	30	6
3.	91	30	38	10
4.	121	30	38	12
5.	152	45	45	> 12
6.	182	45	45	> 15

(c) **Masking or duct tape** — The masking tape is used to cover the edges of the aquarium.

(d) **Caulking gun** — Into it is put the silicone sealant and the caulking gun is used to lay it in a straight line on the glass.

(e) **Aquarium hood** — A slant hood or lid is put on the roof of the aquarium to prevent entry of any undesirable things or creatures and also to prevent fishes from jumping out of water. The hood is generally made of aluminium sheet and often painted colourfully.

(f) **Aquarium stand** —Aquarium can be placed on a table or on an aquarium stand. A variety of attractive, beautifully designed iron stands (Fig. 6.55a & b) are available, which would enhance the beauty of the room.

(g) **Lighting devices** — Direct sunlight on the aquarium must be avoided as it causes algal growth and turns the water green. However, proper illumination in the aquarium is important as it not only makes the fishes visible but also enhances the beauty of the tank. It is better to provide light artifically. Fluorescent tubes (Fig. 6.56b) are preferred to ordinary light bulbs (Fig. 6.56a) as these have a longer life and provides bright light. A special tube, 'Gro-lux', is available in the market which encourages plant growth. Decorative lights can also be used. Some aquarists use underwater lighting, which not only provides light but also helps to heat the water (Fig. 6.56c).

(h) **Heating devices** — To maintain the temperature during winter and particularly in places where the temperature goes down to about 1°C, it is essential to have a heating device. Generally an immerson heater is used for the purpose (Fig. 6.56d). The heater consists of an element wound round an insulated core and is contained in a heat-resistant glass tube. A rubber cork is fitted at the mouth of the glass tube through which the electric cable is held tightly and it acts as a water seal. Also, as mentioned above a lighting cum heating device (Fig. 6.56e) can also be used.

(i) **Thermometer** — A thermometer is installed at vertical position to record the changes of temperature of the aquarium water. The common type available in the market has a glass tube mounted on a scaled, plastic strip which is held in position in the aquarium by a rubber sucker pad (Fig. 6.56e).

(j) **Thermostat** — The temperature in a tropical aquarium should be maintained at a range of 18-29°C, depending upon the species. This temperature can be maintained by installing thermostats which are generally of two types. The internal thermostat is installed either partially or fully submerged in water, while the external thermostat is fixed outside the tank (Fig. 6.56e). Temperature, however, will always fluctuate a few degrees around the point set by the thermostat.

(k) **Composts** — For aquarium without filter, composts are used. An aquarium compost is the **gravel** which is made up of small stones and pebbles, or a mixture of these with sand. The compost is first washed thoroughly and then spread at the tank bottom, on which plants can be grown.

(l) **Aquarium filters** — The purpose of filters in aquarium is filtration of water, removal of waste food and decaying plant leaves, and providing fishes and plants with clean water. A variety of filters are in use in aquarium which can be grouped in the following three types :

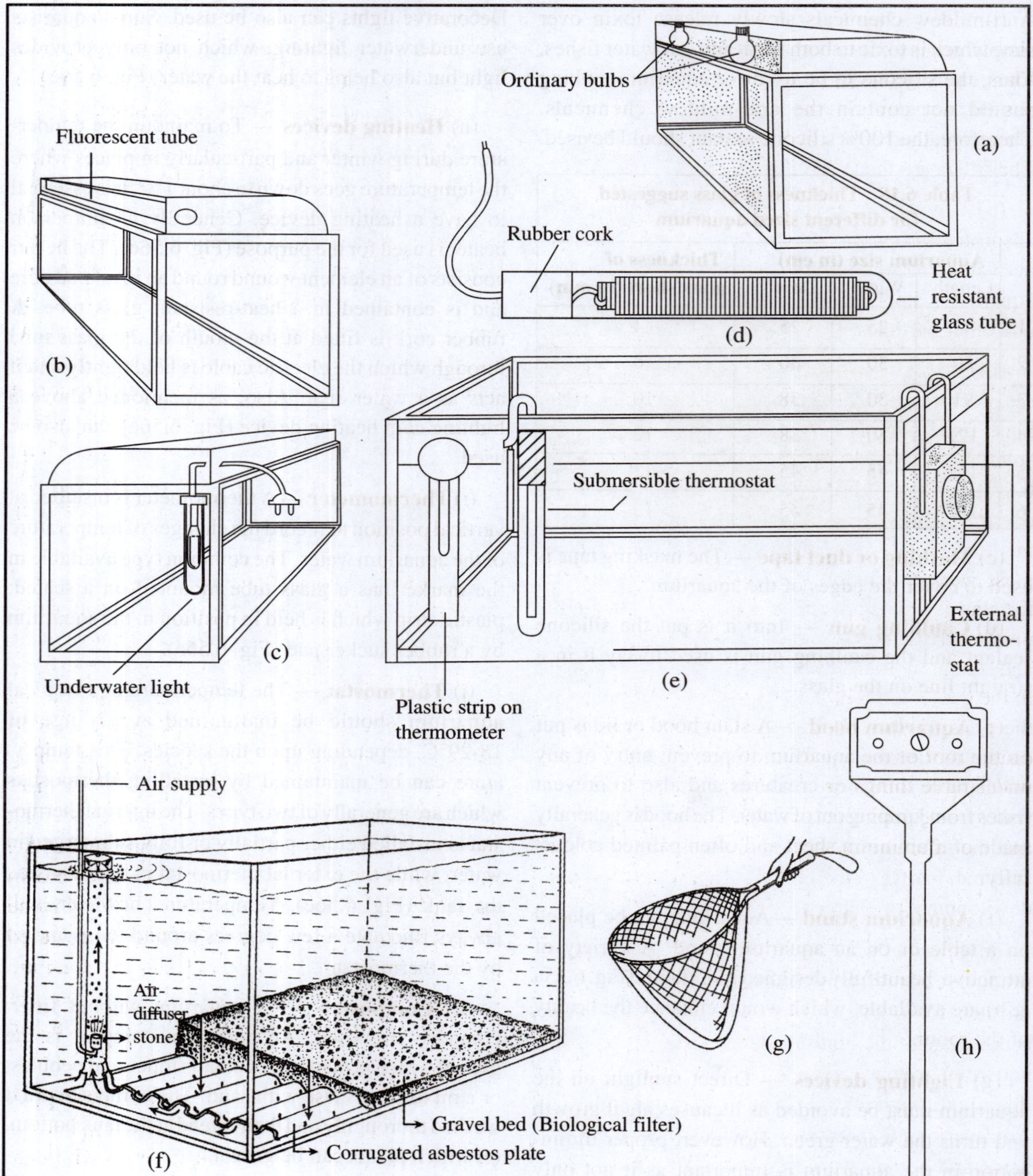

Fig. 6.56. (a) Low wattage light bulb fitted on the hood. (b) Fluorescent tube fitted on the hood. (c) A lamp is fitted which also serves to heat water. (d) Immersion heater of aquarium. (e) Aquarium fitted with thermometer, internal and external thermostat. (f) Aquarium with biological filter and aerator. (g) Hand-net. (h) Razor blade glass scraper.

(i) **Mechanical filters** — Mechanical filters are simple pumps containing a nylon filter wood or a foam cartridge. When the water of the aquarium is passed through it, the suspended perticulate matters are trapped and effectively removed from the water. The filters are then regularly cleaned.

(ii) **Chemical filter** — Chemical filter uses the absorptive properties of activated carbon. This tiny activated carbon attracts and removes from the aquarium water the dissolved organic waste products and other particles.

(iii) **Biological filters**—Biological filters comprises a corrugated and porous asbestos or acrylic plate fixed slightly above the bottom with a non-soluble adhesive on all the sides of the tank (Fig. 6.54f). A narrow glass cylinder of 8 cm diameter and open at both ends, is inserted through an opening made at one side of the plate. Through this cylindrical glass wear circulation and aeration are done. After the plate is fixed, gravel is spread over the floor of the plate to a height of 5-7 cm. Then well-washed granite and suitable colourful aquarium stones are embedded in the gravel terrace. Care should be taken such that the gravel is not coarse as food particles may get caught in the interspaces of the gravel and the food would be beyond the reach of the fish. Moreover, the water would percolate faster through the coarse gravel and the efficiency of the biological filter would be affected, owing to the limited surface area and poor development of bacteria.

Importance of Biological Filter — The excretory products of fishes, the dead plants and organisms, and uneaten food, rich in nitrogen compounds would accumulate at the bottom and cause pollution of the aquarium water. Ammonia, a toxic gas, is formed directly through the excretion of gills and indirectly from various excretory products such as urea, uric acids and amino-acids. Concentration of ammonia beyond 0.5 mg/litre is toxic to fishes which would result in the haemoglobin of blood to loose its ability to combine with oxygen or to liberate carbon dioxide. Aquarium fishes excrete ammonia in the range 0.3-4.0 g/kg of fish/day. Ammonia content also increases with an increase in pH (10 times at one unit increase in pH). Therefore, it is necessary to oxidase ammnia to nitrite (NO_2) and subsequently to nitrate (NO_3), which is non-toxic up to 400 mg/litre and it is taken up by the plants during photosynthesis.

The ammonia rich water in the aquarium passes through the gravel bed and filter plate, and accumulates at the bottom of the aquarium below the biological filter. While passing through the gravel bed, the aerobic nitrifying bacteria (*Nitrosomonas* and *Nitrobacter*) that accumulate naturally in the gravel bed due to constant circulation of water, act upon the ammonia rich water. The *Nitrosomonas* oxidises ammonia to nitrite, which is further oxidised to nitrate by *Nitrobacter*. The purified nitrate rich water which percolates into the lowest chamber of the aquarium, is lifted up to the surface by the cylindrical glass which is fitted with an aerator, thereby circulating the water again and again.

(m) **Aerators and diffusers** — Generally partial water changes in every two weeks may be adopted. Alternately, to maintain proper oxygen content in the aquarium water, particularly in the night (when there is no photosynthesis), it is essential to aerate the water. Aeration can be done by providing a small piston or diaphragm-type aerator, polythene tubing and an air-diffuser stone (Fig. 6.56f). The diffuser stone is lodged in the glass cylinder that has been set up in the porous asbestos or plate while setting the biological filter in the aquarium. When air is blown into the polythene tube, the air reaches the airstone and owing to its porous nature air bubles are released which rise to the surface and oxygenate the aquarium water. It is advisable to take precautionary measures to clean the porous stone as it often gets chocked with suspended particles.

(n) **Planting sticks** — When it is desired to plant some plants in between the rocks, planting stick is generally used, as it is difficult to reach with bare hands. The planting stick is made of thin wood or metal strip.

(o) **Ornamental plants** — Presence of plants in an aquarium serves several functions as follows : (i) it provides shelter, shade and sanctuary for fishes, (ii) it decorates the aquarium, (iii) they serve in spawning, (iv) they serve as food for certain fishes, (v) they utilise the carbon dioxide (waste of fishes) and liberate oxygen through photosynthesis and (vi) they absorb the nitrate, converted from ammonia by the biological filter.

Several types of plant are available, but hardy aquatic plants of quick growing types should be considered by an aquarium keeper. Aquarium plants (Fig. 6.57) may be of two types — rooted plants and floating plants. Among rooted plants are *Vallisneria* sp. (tape grass) (Fig. 6.57a), *Myriophyllum* sp. (Milfoil) (Fig. 6.57b) , *Amazan sored* (colourful long leaves) and *Crystocdmiyes* sp. (coloured broad leaves); whereas floating plants comprise *Salvinia* sp. (long leaves) (Fig. 6.57c) and *Crystal worth* (small leaves). Other important plants are *Ceratophyllum* sp. (Fig. 6.57d), *Clodia* sp., *Fontinalis* sp., *Cabomba* sp. (fanwort) (Fig. 6.57e), *Sagittaria* sp. (Fig. 6.57f), *Chara* sp. (stonewort), *Hydrilla* sp. (Fig. 6.57g), etc.

Before planting, the plants should be washed thoroughly so that no parasites can enter the aquarium and cause the death of fishes. These plants are disinfected by rinsing in 0.1 per cent solution of alum (potassium-aluminium sulphate) followed by washing in water. The plants are carefully installed by the planting stick. Large, wide plants like the tape grass (*Vallisneria* sp.) may be placed at the back of the aquarium, while the bushy plants may be planted on the side corners

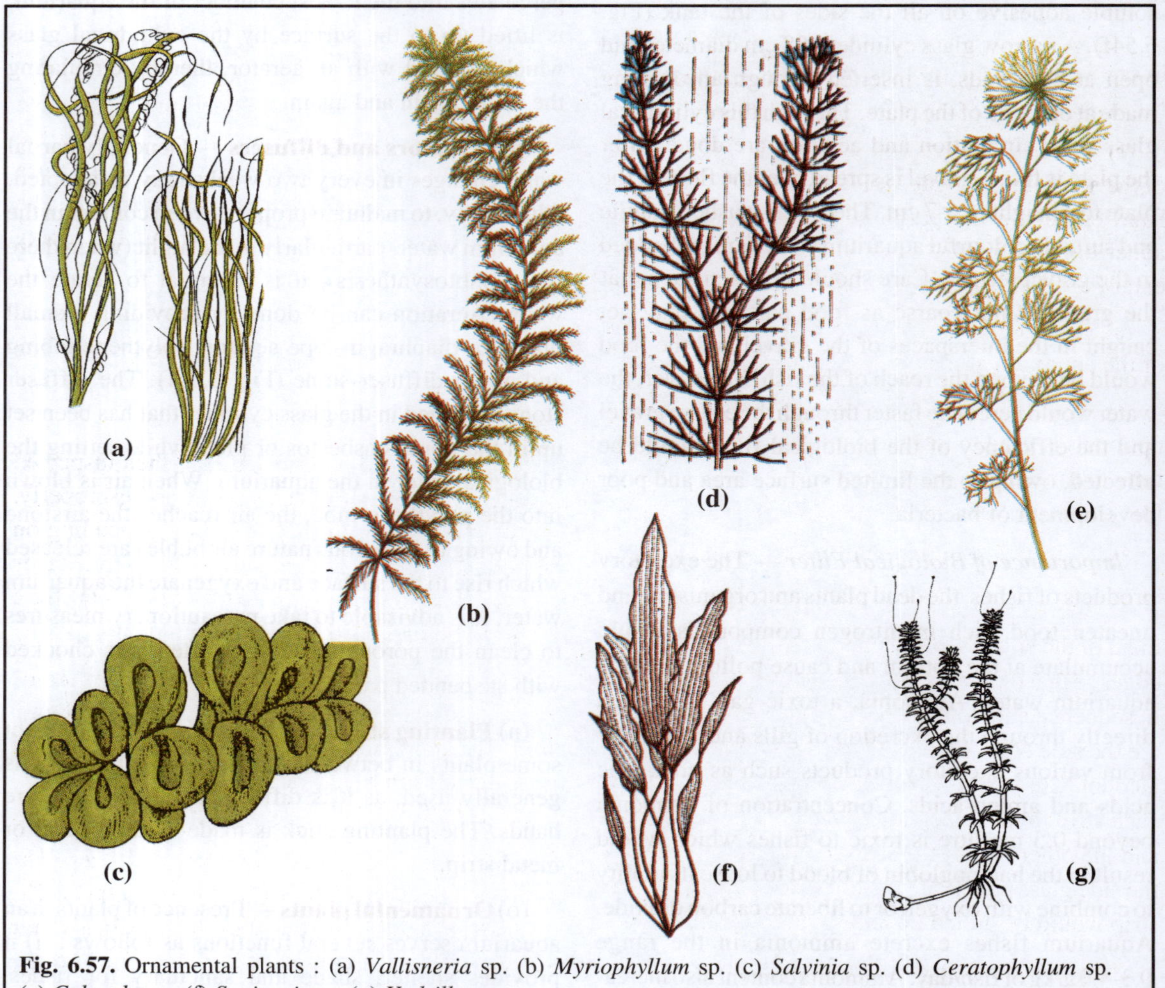

(a) (b) (c) (d) (e) (f) (g)

Fig. 6.57. Ornamental plants : (a) *Vallisneria* sp. (b) *Myriophyllum* sp. (c) *Salvinia* sp. (d) *Ceratophyllum* sp. (e) *Cabomba* sp. (f) *Sagittaria* sp. (g) *Hydrilla* sp.

and the smaller plants in the middle. The floating plants may be installed in aquarium when gouramis are bred, as these plants are used for construction of bubble-nests for breeding. After the plants are planted, the aquarium is left undisturbed for at least a week or two after which the fishes may be introduced. Artificial soft and colourful plants may be used to increae the beauty of the aquarium.

(p) **Feeding rings**— The feeding rings are made of plastic in the form of a square or a circle. They float on water and the dried food gets limited within it. Thus, it serves to limit the spread of the food. However, the disadvantage of such segregation may result in the larger species of fishes getting greater share of the food than the smaller or younger ones.

(q) **Hand-net**— A hand- net (Fig. 56g), rectangular in structure, is generally required for catching a fish for inspection and for transfer to elsewhere.

(r) **Scrapers or algae cleaner**— To clear the glass of the aquarium of algae, a fine steel wool held in the hand can serve the purpose. However, for deeper aquarium an ordinary razor blade is clipped at one end of a wooden handle (Fig. 6.56h).

(s) **Siphon tube**— This is used to siphon water out of the aquarium at the time of exchanging water.

(t) **Water-testing kit** —Water-testing kit (not optional) is used to test the nature of the water from time to time, such as pH, dissolved oxygen content, free carbon dioxide, etc.

C. Installation of a home aquarium

Step 1. Cut the glasses according to the measurement of the aquarium one intends to make. Grind the edges so that the ends are not sharp. Lay all the glasses in sequence and double check that the shorter pieces are the side ones. Clean all the edges with acetone or alcohol. Cut strips of marking tape and stick half of each strip to the underside of the bottom glass piece and the other half of the strip should be left bare. As the side glass panels are installed, the bare half of the strip would be pulled up and stuck to the glass.

Step 2. With the help of the caulk gun (or with the bare hand — it is difficult to keep it straight), apply a straight, thin (3 mm) and continuous strip of 100% silicone sealant along the top of the bottom piece, about 2 mm away from the edge. Place the front piece in place, gently press it down into place. The excess silicone should not be wiped off. The front glass piece after some time would stand straight on its own.

Step 3. Then another thin line of silicone is applied along the bottom glass pane (2 mm from the edge) for one side panel. At the same time a line of silicone (3 mm wide, 2 mm from the edge) is applied along the inside edge of the front pane which has just been installed. Then apply and press the side panel into place by applying pressure. The piece should be got in place at the first try, so help is taken, if necessary. The glass pane should not be pulled off after fixing for re-alignment. This would create bubbles in the silicone, which would create a faulty seal later on.

Step 4. For the second side panel another thin line of silicone is run, 2 mm from the edge, along the bottom pane and another line (3 mm wide, 2 mm from the edge) along the inside edge of the front pane. The second side panel is then pressed firmly, but gently into place.

Step 5. Three more thin line of silicone is run — one along the bottom pane and 2 along the inside edges of the back pane. Then the back panel is pressed into place firmly, but gently.

Step 6. Each corner and edge are then inspected thoroughly to make sure that they are lined up properly. The silicone will dry in 12-24 hours (depending on the humidity and air temperature). As silicone takes a longer time to harden, it is better to let it sit for a full week before water is poured into it.

Step 7. After a week, the aquarium is set on the stand and all the glass sheets are cleaned with a damp cloth. The aquarium is then filled with water and left for a few days to check for leakage.

Step 8. After confirming that there is no leakage, the aquarium is emptied of its water and dried. The gravel is then washed thoroughly with $KMnO_4$. The inside of the tank is also washed. The gravel is then spread on the tank bottom in a slanting manner (4 cm at the back and 2 cm at the front). This is done to allow the sediments to accumulate in front, which can be easily removed by siphoning. Colourful pieces of rocks, pebbles, etc. are then placed on the top of the gravel. Before adding water cover the bottom with newspaper so that the gravel is not disturbed. The tank initially is half-filled with water, and aerators are placed in it.

In aquariums where filters are added (instead of gravel), it is fitted in the manner as discussd earlier, along with aerators and diffusers.

Step 9. Heater and thermostat are then fixed in appropriate places, connecting each other but not to mains.

Step 10. The selected plants are then fixed at the back and sides of the aquarium with the help of the planting stick. The aquarium is then filled completely with water leaving the top 4-5 cm or so. Tap water may sometimes contain high chlorine content which is harmful to fishes. The chlorine content can be reduced if such tap water is allowed to stand for several hours in the open air. Alternatively, well water can be used to fill the aquarium, which is usually better as it contains wealth of minerals. Fresh pond water can also be considered, but it should be free of dangerous larvae, pathogens and parasites.

Step 11. The mains are then switched on. The cover is then placed on the top of the aquarium and the lights, thermostat and aerators are checked for proper functioning. The tank is left for one week. Daily monitoring of temperature, pH, etc. and checking of leakage are done.

Step 12. The ornamental fishes to be kept in the aquarium are selected and then introduced into the tank.

D. Precautions taken for an ideal aquarium

For proper maintenance of the aquarium the following precautionary measures should be taken.

(1) For an indefinite time the aquarium should not be left uncared.

(2) Aquarium should not be moved frequently from one place to another or shaken up.

(3) The aquarium should not be completely filled up with water, rather about 4-6 cm should be left vacant.

(4) An aquarium should always be provided with a cover.

(5) The aquarium once set and balanced, the water should not be changed, unless the aquarium water tends to become turbid or cloudy. If the level of water falls due to evaporation, it should be restored by addition of chlorine-free tap water.

(6) While changing the water of the aquarium, a constant flow should be maintained with the help of a siphon tube.

(7) The aquarium should be placed on a stand or a table at a convenient spot in the room, so that there is just enough light.

(8) Temperature fluctuation in the aquarium should be minimum and there should be sufficient light (at least two hours a day).

(9) The pH of water for tropical freshwater fishes should be 7.0–7.2.

(10) The ornamental fishes should be given a bath in $KMnO_4$ or copper sulphate before keeping them in the aquarium.

(11) Overcrowding of fishes in the aquarium should be avoided.

(12) Overfeeding the fishes in the aquarium should be avoided as the left over will cause pollution. It is better to feed the fishes twice a day.

(13) There should be changes in the nature of the feed given to fishes. A single type of feed should not be used always.

(14) Waste production and food left overs should be carefully siphoned out without disturbing the fishes.

(15) Aerator should be used frequently to increase the dissolved oxygen content of the aquarium water.

(16) Predators should not be allowed to get into the aquarium.

(17) Infected fishes should be removed from the aquarium as soon as they have been identified and proper treatment should be given.

ORNAMENTAL FISHES

A. Criteria for selection of aquarium fishes

While selecting suitable aquarium fishes it is essential to follow the following lines.

(1) Aquarium fishes selected should be attractive and brillantly coloured.

(2) Selected species should be compatible to each other.

(3) Fishes introduced into the aquarium should be healthy, disease-free and should be collected from reliable sources.

(4) Fishes that are sluggish should be avoided. Behaviour of the fishes should be lively with well-spread fins and should be actively swimming.

(5) The sizes of the fishes kept in the aquarium should be uniform (as far as possible), as larger fishes may harm the smaller ones.

(6) Fishes should be brought from the aquarium shop in properly packed, well-oxygenated polythene water bags kept at normal temperature.

(7) Initially only 5 to 6 fishes of choice should be introduced gently into the aquarium. Avoid adding a number of fishes within a short period and also avoid overcrowding.

B. Common aquarium fishes in use

Aquarium fishes are generally small-sized, brightly coloured with characteristic markings, such as bands, spots, etc. India has a vast treasure of ornamental fishes. About 600 freshwater species have been labelled as 'aquarium-worthy'. The north-eastern states of India are blessed with about 250 native ornamental fish species which include 187 species in Assam, 165 in Arunachal Pradesh, 159 in Meghalaya, 139 in Manipur, 103 in Tripura, 71 in Nagaland, 46 in Mizoram and 29 in Sikkim. On the other hand, West Bengal has only 52 native ornamental fish species. India exports nearly 200 ornamental fishes. Of these about 85% (155 species) are from north-eastern states. So north-eastern states plays a vital role in the ornamental fish market of India.

On the other hand, Singapore is the biggest trading centre of the world for ornamental fishes. They are also the largest exporters (22%) of the ornamental fishes.

Some of the commonly used aquarium fishes (both native and exotic) are listed below; belonging to sub-class **Actinopterygii** and Infra-class **Teleostei**.

Order Osteoglossiformes

(a) **Knife fishes**— The family Notopteridae under this order has two species in India and are known as *featherbacks*. They are easily identified by their long anal fin and a small slender dorsal fin (for which the name 'Featherback').

The aquarists, however, call these fishes as *'Knife Fishes'*— the Clown Knife fish which is larger and the Black Knife fish which is smaller (Fig. 6.58). Due to their graceful movements they are perfect for display in public aquaria. They are predatory in nature and if kept with small sized fishes in aquaria, will devour it.

Fig. 6.58. Black Knife Fish

Order Anguilliformes

(b) **Eel**— The order Anguilliformes consists of the well-known *Eels*. They are catadromous fishes. They are highly valued as food fishes. There are some species of eels which are priced by aquarists, such as the spotted Moray Eel (Fig. 6.59), Pink Worm Eel, Olive Eel, Brown Snake Eel and Bicolour Snake Eel.

Fig. 6.59. Spotted Moray Eel

Order Cypriniformes

(c) **Barbs** — This group are economically important both as food and aquarium fishes. It comprises Hi-Fin Barbs and the true barbs. The true barbs previously under the genus *Barbus* is now limited only to the African and European species. In the Asiatic region the Genus name *Puntius* is used. There is hardly any freshwater body in the Indian region which does not have this species. Some popular *Puntius* species from India are the Golden Dwarf Barb (also known as Gelius Barb), Melon Barb [Fig. 6.60(a)], Four Spot Barb, Two Spot Barb, Green Swamp Barb, Indian Rosy Barb [Fig. 6.60(b)], Long Snouted Barb, Black Spot Barb, Filament Barb, Malini's Barb, Golden Dwarf Barb, Streaked Barb, Dwarf Barb, Maharaja Barb, Sophore's Barb, One spot Barb; Tic-Tac-Toe Barb, Kooli Barb, Red Line Torpedo Fish, DMK Fish.

Other types of Barbs are under the following Genus :

(i) **Thick lips** under the genus *Catla.*

(ii) **White Carps** under the genus *Cirrhinus*

(iii) **Red Tailed Silver Shark** [Fig. 6.60(c)] under the genus *Gonoproktopterus.* This genus also consists the **Scarly Barb** and **Sombre Barb**.

(iv) **Sharks** under the genus *Labeo.* All species of *Labeo* introduced into the aquarium trade were all known popularly as sharks. They comprise of the following :

1. **Red Tail Black Shark** (*Labeo bicolour*). This is the native of Thailand. From this fish the nick name 'shark' was obtained for these group (*Labeo*) of aquarium fishes. However, now *Labeo bicolour* is known as *Epalzeorhynchus bicolour.*
2. **Red Gilled Violet Shark** (*Labeo boga*). This strikingly beautifully coloured fish [Fig. 6.60(d)] has emerged as a hot favourite with aquarists all over the world.
3. **Black Line Shark** (*Labeo bata*).
4. **Syke's Shark** (*Labeo boggut*).

Fig. 6.60. (a) Melon Barb, (b) Indian Rosy Barb, (c) Red Tailed Silver Shark, (d) Red Gilled Violet Shark, (e) Tiger Barb

5. **All Black Shark** (*Labeo calbasu*).
6. **Miniscale Shark** (*Labeo gonius*).
7. **Southern Orange Shark** (*Labeo kontius*).

(v) Other aquarium fishes under genus *Labeo* are :

(i) **Silver *Labeo*** (*Labeo kawrus*)

(ii) **Pencil Gold *Labeo*** (*Labeo nandina*)

(vi) **Tiger Barb** (*Puntius tetrazona*) is one of the most colourful active aquarium fish from Indonesia [Fig. 6.60(e)]. It is yellow with black stripes across the body.

(d) **Hatchet Fishes** — These fishes belonging to the Genus *Chela* are also commonly known as 'oriental white minnows'. There are 4 species in India — Neon Hatchet (*Chela cachius),* Burjor's Brilliance (*Chela dadiburjori,* Fig. 6.61), Indian Glass Hatchet (*Chela labuca*) and Malabar Hatchet.

Fig. 6.61. Burjor's Brilliance

(e) **Razorbelly and Streak** — They belong to the genus *Salmostoma.* Razorbelly is a silvery, elongated fish. The streak comprises Long Jawed Streak and the Silver Streak (Fig. 6.62). The Long Jawed Streak is regarded as a useful larvivorous fish.

Fig. 6.62. Silver Streak

(f) **Carplets** — The carplets are represented by the Veli Lake Carplet and the most popular (among aquarists) Brass Carplet (Fig. 6.53). These fishes are under the genus *Amblypharyngodon* and inhabit the Indus plain and the adjacent hilly areas. Glowlight carplet of the genus *Horadandia* is one of the smallest aquarium fishes found in India measuring 1 cm.

Fig. 6.63. Brass Carplet

(g) **Trouts** — These are mostly present in the hill or mountain streams but a few inhabit low land water. 17 species are found in the Indian region comprising Blue Dotted Hill Trout (Fig. 6.64), Branded Hill Trout, Metallic Hill Trout, Lineated Hill Trout, Spotted Hill Trout, Half Banded Hill Trout, the very rare Milk-fish with its silvery, milky appearance, etc.

Fig. 6.64. Blue Dotted Hill Trout

(h) **Danios** — The Danios belonging to the genus *Danio* has derived its name from the Bengali word 'dhani'. This name probably refers to their small size or to their habitat in paddy (paddy in Bengali is called 'dhan') fields, in the edges of rivers and lakes in grassy jungles. There are 11 species in India comprising Turquoise Danio (Fig. 6.65a), Zebra Danio, Giant Danio, Malabar Danio, etc. The exotic Moustached Danio (Fig. 6.65b) has a characteristic two pairs of long barbs.

Fig. 6.65a. Turquoise Danio

Fig. 6.65b. Moustached Danio (exotic species)

The Dwarf Danios are from the genus *Brachydanio* which has eight species in the Indian region.

(i)**Rasbora** — These cyprinids of the genus *Parluciosoma* (Slender Rasbora, Black Line Rasbora — Fig. 6.66, etc.) have 9 species and the genus *Rasbora* (Yellow Tail Black Tip, Yellow Tail Scissorstail, etc) comprises 4 species.

Fig. 6.66. Black Line Rasbora

(j)**Alga-eater** — The alga-eaters of genus *Crossocheilus* (Slaty Alga-eater) have two species in India. They are boring but are an excellent algae- and refuse- eater. The alga-eaters of genus *Garra* have 19 species in the Indian region. They are adapted to life in swift-flowing mountain streams. The Rinocerous Alga-eater (Fig. 6.67) is much sought after fish by the aquarists. Other known ones are cardamom Alga-eater, Kemp's Alga-eater, Khasi Alga-eater, Cauvery Alga-eater, Brown Alga-eater, etc.

Fig. 6.67. Rinocerous Alga-eater

(k)**Torrent Fish** — There are 5 hill stream species of Torrent fishes in India of the genus *Psilorhynchus.*

Fig. 6.68. Banded Torrent Fish

Notable among them are the Banded Torrent Fish (Fig. 6.68), checkered Torrent Fish and Sparkling Torrent Fish.

(l) **Loaches** — Loaches are a varied group of aquarium fishes belonging to 2 families (Balitoridae and Cobitidae), 4 sub-families and 45 genera. It comprises Sucker Loaches, Cylindrical Loaches, Wedge Tail Loaches, Puma Loach, Sidewinder Loach, Fascinating Loach, Zodiac Loach, Leopard Loach, Black Panther Loach, Air Breathing Loaches, Snake Loaches or Indian Coolie Loach (Fig. 6.69a), Ring Loach, Olivaceous Loach, Polka Dotted Loach, Banded Loaches, Victory Loach, Long Snouted Loach, Grizzled Loach, Tail Spot Loach, Goalpara Loach, Jaguar Loach, Y-Loach, Tiger Loach, Striped Loach (Fig. 6.69b), etc. The last named three Loaches are the most popular among the Indian Loaches.

Fig. 6.69a. Indian Coolie Loach

Fig. 6.69b. Striped Loach

(m)**Characins** — Characterised by the possession of either teeth or an adipose fin. A number of these fishes from South America are known as the 'tetras' such as Lemon Tetra, Rosy Tetra, Bleeding Heart Tetra, Red Tetra, Neon Tetra (Fig. 6.70), etc. Other groups of characins are the Beacon Fish or Head-and-tail-Light Fish, X-Ray Fish, etc.

(n)**Goldfishes** — These exotic cyprinids are very much in demand in the ornamental fish market. A number of fancy varieties of goldfishes are available in the market, such as the common or typical goldfish *Carassius auratus,* Fig. 6.71(a); Comet Gold-

Fig. 6.70. Neon Tetra

fish [Fig. 6.71(b)] has very long caudal fin; Veitail Goldfish has all the fins greatly elongated and is golden in colour; Shubunkin Goldfish has colours of deep blue, red and white, black and golden; Lionhead Goldfish [Fig. 6.71(c)] has expanded and warty head. Other varieties of Goldfishes are Peacock tail, Oranda, Calico Oranda, Telescope Eyes, Calico, Fringetail, etc.

Fig. 6.71(a). Typical Goldfish

Fig. 6.71(b). Comet Goldfish

Fig. 6.71(c). Lionhead Goldfish

Order Siluriformes

(o) **Catfishes** — Generally catfishes can be identified by the presence of barbels on each side of the upper jaw and some also have on the lower jaw. These fishes are priced as food item and are very numerous in the Indian sub-continent. Many catfishes are said to be venomous. With the help of their spines they can inflict severe wounds and inject a poison produced by the glandular epidermal cells. The catfishes of aquarium value comprise 13 families and 54 genera. Some notable aquarium catfishes are : (i) **Bagrid Catfishes** comprising the magnificient Goldern Red-tailed Imperial Catfish, Golden Headstander Catfish, Velvet Catfish [Fig. 6.72(a)], Shovelmouth Catfishes, Shark Catfishes, Mystus Catfishes, Hovering Catfishes, Whale Catfishes (Genus *Rita*), etc; (ii) **Eurasian Catfishes** comprising Glass Catfish, Butter Catfishes (Gulper Catfish), Killer Catfishes (Genus *Wallago*), etc; (iii) **Schilbid Catfishes** consisting of Glass Catfishes (striped Glass Catfish — Fig. 6.72(b), etc; (iv) **Scissorstail Catfishes**; (v) **Sisorid Catfishes** contain the Thread Tail Catfish endemic to India [Fig. 6.72(c)], Torrent Catfishes, Flat-Headed Catfishes, Broad-Headed Catfishes, Whiptail

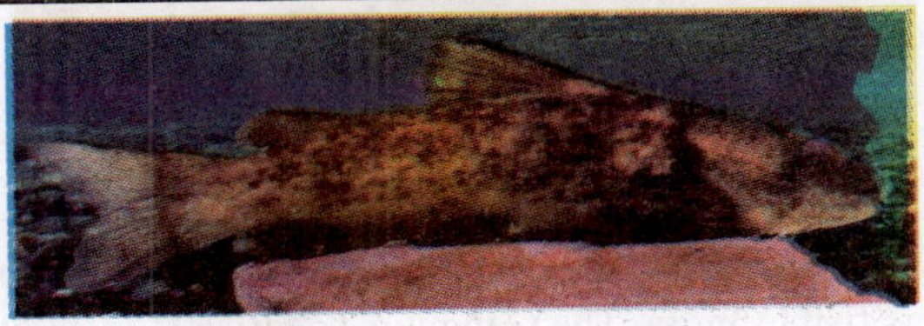

Fig. 6.72(a). Velvet Catfish

Fig. 6.72(b). Striped Glass Catfish

Fig. 6.72(c). Thread Tail Catfish

Catfishes, etc. (vi) **Air breathing Catfishes** (*Clarius*); (vii) **Stinging Catfishes** of the genus *Heteropneustes*; (viii) **Angler Catfishes** are robust but ugly to look at, such as the Devil Catfish; (ix) **Stream Catfishes**; (x) **Tiger Shark Catfishes**; (xi) **Fighting Catfishes** have elongated body and no two specimens can be kept in a single aquarium Ex. Long Fighting Catfish [Fig. 6.72(d)]; (xii) **Sea Catfishes** are primarily marine with few being confined to freshwaters. Examples are the Dusky Shark Catfish, Spotted Shark Catfish [Fig. 6.72(e)], etc; (xiii) **Eel Tail Catfishes** occur in marine, brackish and freshwaters.

Fig. 6.72d. Long Fighting Catfish

Fig. 6.72e. Spotted Shark Catfish

Order Batrachoidiformes

(p) **Toadfishes** — There are only two genera of this fish in India. The Gangetic Toadfish (Fig. 6.73) makes a grunting sound when disturbed. Handling must be done with care as its spine can inflict injury.

Fig. 6.73. Gangetic Toadfish

Order Cyprinodontiformes

(q) **Halfbeaks** — Famous among them are the Southeast Asian Halfbeak and the Long Nose Needlefish.

(r) **Killifishes or Egg-laying Tooth Carps** — They are small fishes and are excellent jumpers. They are *egg hangers* (eggs are laid on floating plants) or *egg-buriers* (eggs are burried in the bottom).

Some egg-hangers are *Rivulus cylindraeus* from Cuba and Florida; *Aplocheilus lineatus* and *A. parvus* Fig. 6.74(a) of Indian origin and *Aphyosemion australe* Fig. 6.74(b) and *A. gardneri* form West Africa have become extremely popular in recent years because of their beautiful colours. Among the egg-buriers are the Dwarf Argentine Pearlfish (*Cynolebias nigripinnis*), *Aphyosemion sjoestedti, Rachovia brevis, R oj-fia occidentails,* etc.

Fig. 6.74(a). *Aplocheilus parvus*

Fig. 6.74(b). *Aphyosemion australe*

(s) **Live-Bearing Tooth-Carps** — This family is represented in fresh- and brackish waters from North-eastern Argentina to Eastern United States. They comprise the world famous Guppy, Molly and Sword Tails :

(i) **Guppy** — *Poecilia (Lebistes) reticulatus* is the most common aquarium fish due to its brillant colours, low cost and hardiness. It is the "beginner's" species. It is the native of some of the islands of the Caribbean and the bordering countreis of South America. It has been introduced into India in 1908 from South America for mosquito control. It has been hybridised in countless permutation and combinations (Fig. 6.75). Common varieties based on caudal fin variations include round tail, top sword tail, bottom sword tail, Robson round tail, etc.

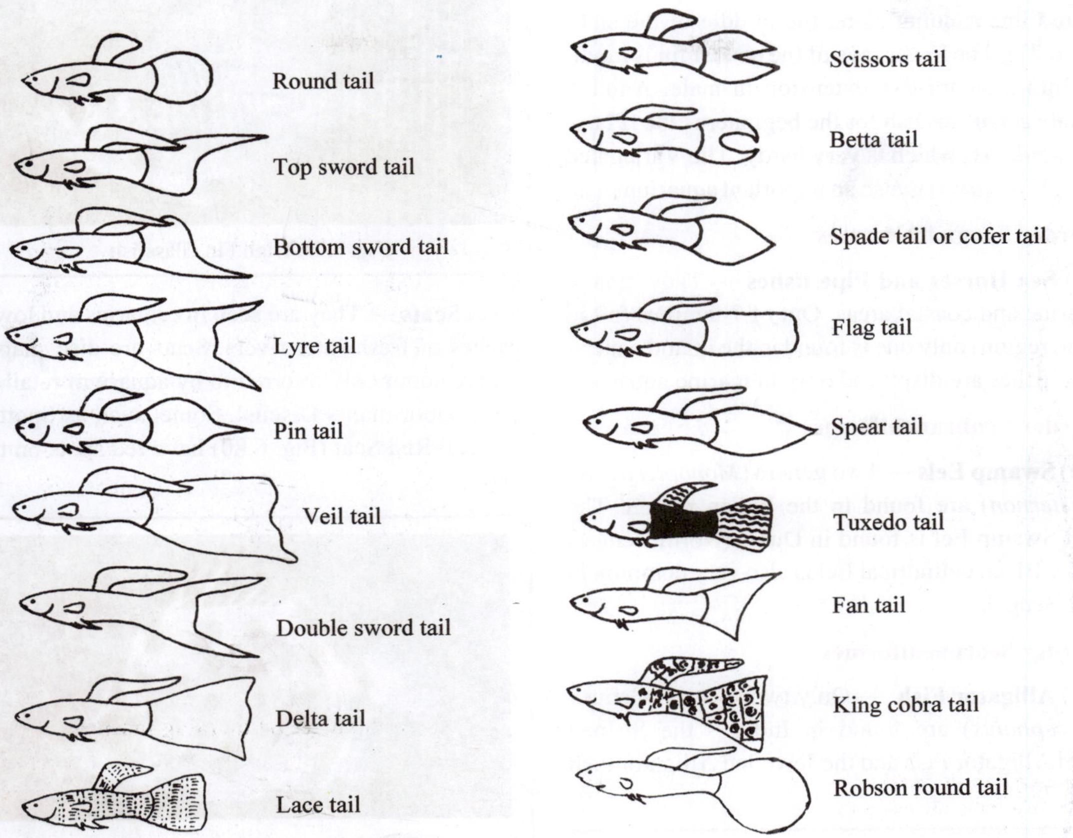

Fig. 6.75. Guppy varieties based on the variation of the caudal fin

(ii) **Molly** — It belongs to the genus *Mollienisia. M. sphenops* is the originator of Black Molly with short fin. *M. latipinna*, the Black Sail-fin Molly is by far the most attractive (Fig. 6.76). The male develops a very large sail-like dorsal fin. The Giant Sail-fin Molly (*M. velifera*) is the largest among the mollies.

Fig. 6.76. Black Sail-fin Molly

(iii) **Sword Tails and Platies** — They belong to the genus *Xiphophorus* and are distributed from Southern Mexico to Guatemala. The Sword Tail

Fig. 6.77. Sword Tail

(*X. helleri*) is the common variety with green body and red line running along the middle of the sides (Fig. 6.77). The lower rays of the caudal fin is modified into a sword-like extension in male. Another suitable aquarium fish for the beginner is the 'Platy' (*X. maculatus*) which is very hardy. The Variegated Platy (*X. variatus*) is also an important aquarium fish.

Order Syngnathiformes

(t) **Sea Horses and Pipe fishes** — They inhabit estuarine and coastal areas. Out of 7 families (all in Indian region) only one is found in the inland waters. These fishes are displayed only in marine aquaria.

Order Synbranchiformes

(u) **Swamp Eels** — Two genera (*Monopterus* and *Ophisternon*) are found in the Indian region. The Giant Swamp Eel is found in Duars (North Bengal) and the Black cylindrical Eel is also very common in West Bengal.

Order Scorpaeniformes

(v) **Alligator Fish** — Only two species (Genus–*Platycephalus*) are found in India – the Striped Tailed Alligator Fish and the Jewelled Alligator Fish Fig. 6.78.

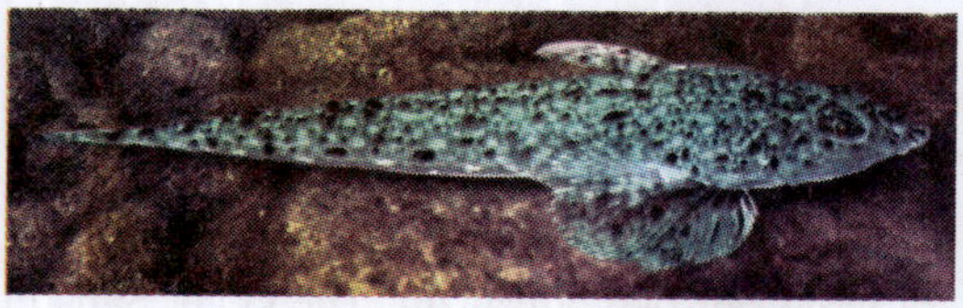

Fig. 6.78. Jewelled Alligator Fish

Order Perciformes

Perciformes is the largest order (about one-third) of fishes. Only 14% of perciformes occur in freshwater and the rest are marine shorefishes. Some of the important fishes are :

(w) **Glassfishes** — Body is semi-transparent where the vertebral column and swim bladder are easily visible. They comprise the Long Glassfishes, Giant Glassfishes, Jewel Glassfishes, etc. The Jewel Glassfishes such as Himalayan Glass Perchlet, High Fin Glassfish (Fig. 6.79), etc. are prized as attractive aquarium fishes.

Fig. 6.79. High Fin Glassfish

(x) **Scats** — They are seen in estuaries and lower reaches of freshwater rivers. Scats are disc shaped and are commonly referred to by aquarium retailers as the 'poor man's Discus'. Some, such as Spotted Scat and Red Scat (Fig. 6.80) have red spots on the body.

Fig. 6.80. Red Scat

(y) **Sunfishes** — Only two species are found in India (Southern part) of which the Malabar Sunfish (*Pristolepis malabarica*—Fig. 6.81) is beautifully coloured and could become popular with aquarists.

Fig. 6.81. Malabar Sunfish

(z) **Chameleon Fishes** — This is one of the most colourful fish. It has an amazing range of pigmentation. It can change its colour with great rapidity and vary from solid colours to intricate mottled designs.

It exists in two sub-species — *Badis badis badis* (Dwarf Chameleon — Fig. 6.82) and *Badis badis burmanicus* (Red Chameleon). They are column feeder and can feed on mosquito larvae.

Fig. 6.82. Dwarf Chameleon

(za) **Cichlids** — Cichlids are carnivorous (except *Tilapia* and *Geophagus*), hardy and colourful fishes. Their characteristic feature is the presence of a single nostril on each side. Only one genera (*Etroplus*) is present in the Indian region and two were introduced into Indian freshwaters. The well-known species are : Pearl Spot Fig. 6.83(a), Banded Chromid, Mozambique Mouth breeder (exotic), Golden-eyed Dwarf Cichlids (from Guiana), Agassiz's Dwarf Cichlid (from Amazon).

Fig. 6.83(a). Pearl Spot

Angel Fishes and the Discus — Some members of Cichlids from South America are called Angel Fishes (*Pterophyllum*) and the Discus (*Symphysodon*). Angel fishes Fig. 6.83(b) today are generally the hybrids. They are peaceful fishes. The Discus [Fig. 6.83(c)] is also a spectacular aquarium fish. It is flat and round in shape.

(zb) **Gobies** — They are sluggish fishes and are carnivorous. They are generally estuarine and coral reef species, with only a few freshwater species.

Fig. 6.83(b). Angel Fish

Fig. 6.83(c). Discus Fish

Some important ones are Bee Fishes (Bumblebee Goby, etc.), Sleeper Goby, Sailfin Gobies (Moustached Sailfin Goby, etc.), Knight Gobies, Blue Spot Dragon (Fig. 6.84), Banded Goby, Hi-Fin Gobies, Neon Gobies (Multispotted Neon Goby, etc.), Clown Gobies, Flat Headed Gobies, Bicolour Gobies, Armadillo Gobies, Eel-like Gobies, Fanged Dragon, Neon Dragon, etc. The **Flat Headed Gobies** (four species in India) has an unusual habit of using a fallen leaf as a cover under which it moves around and hunts. It also protects the goby from predators.

Fig. 6.84. Blue Spot Dragon

(zc) **Gourami** — All the four species of Gourami found in India are very beautiful and peaceful fishes. The most popular among them is the Dwarf Gourami, *Colisa lalia* (Fig. 6.85), and they thrive well in aquaria. Other famous species of *Colisa* are Striped Gourami and Honey Gourami. The Kissing Gourami is an exotic fish, native of Thailand to Malaya Arihipelago. It is a misconception that they kiss each other, rather they fight in this manner.

Fig. 6.85. Dwarf Gourami

(zd) **Paradise Fish** — Paradise Fish, *Macropodus opercularis* is a hardy exotic fish and is probably the first tropical aquarium fish. *M. cupanus* is of Indian origin.

(ze) **Siamese Fighting Fish** — Siamese Fighting Fish, *Beta splendens,* is from South East Asia. It is a beautifully coloured fish but is extremely aggressive towards other males and is not advisable to keep two males of Siamese Fighting Fishes (Fig. 6.91) in the same aquarium.

(zf) **Snakeheads** — These fishes are recognised by the shape of the head which resembles that of a snake. These fishes are air breathers. Some important colourful species of snakeheads (*Channa*) are Peacock Snakehead, Isabella Snakehead, Bengal Snakehead (Fig. 6.86), Broad Banded Snakehead, Georgette Snakehead, Cobra Snakehead, Rainbow Snakehead, etc.

Fig. 6.86. Bengal Snakehead

Order Mastacembeliformes

(zg) **Spiny Eels, Spineless Eels and Sub-cylindrical Eels** — 'Spiny Eel' is so named due to the detached, depressible spine present near the soft dorsal fin. Many of them are air breathers. Some of the noted Spiny Eels are the Peacock Eel, Tyre Track Eel (largest), Red Tailed Eel, Spiny Green Eel (Fig. 6.87), etc.

The Spineless Eels are extremely small and is represented by a single genus. The Sub-cylindrical Eel, represented by two genera is an uncommon and very inactive fish.

Fig. 6.87. Spring Green Eel

Order Pleuronectiformes

This order comprises marine fishes such as the **Flat** or **Tongue Fishes** and the **Soles**. They are benthic and carnivorous.

Order Tetraondontiformes

(zh) **Puffer Fishes** — These fishes are called 'Puffer' as it can inflate its body by swallowing water or air. They are mostly marine or estuarine and a few live only in freshwater. There are six genera of Puffer Fishes present in the Indian region. Some important Puffer Fishes are Topaz Puffer which is brilliantly coloured and not too agressive like the White Spotted Puffer. The Emerald Puffer when inflated, resembles a table-tennis ball, while the Red Green Dwarf Puffer (Fig. 6.88) is a remarkable little fish having only six vertebrae.

Fig. 6.88. Red Green Dwarf Puffer

C. Food and feeding of Ornamental Fishes

For better growth and survival of aquarium fishes it is essential to provide them with mixed and balanced diet. Algae which often grow on the pebbles, glass panes and in water of the aquarium, serve as a good source of food for sword tails, kissing gouramis and mollies. Various protein-rich live fish food such as *Cyclops, Daphnia,* Rotifers, *Tubifex,* earthworms, mosquito larvae, chironomid larvae, artemia nauplii, etc. are considered as excellent food. Freshly chopped earthworm, shrimp, fish paste, scraped boiled fish or raw liver are also considered as good food items. Of great use to aquarists are the commercially prepared dried food meant only for aquarium fishes which is easily available from requisite shops. The dried food generally constitutes a mixture of cereal, milk powder, prepared egg, dried *Daphnia,* fish or shrimp. These are dried, powdered and mixed together and formed into pellets or flakes. Liver is high in protein, rich in iron and contains vitamins. Raw liver, however, should be avoided as it will make the water cloudy. So the liver is put into boiling water for a minute to coagulate the blood. The liver is then chopped into small cubes and is shreded in a blender. Shreded shrimps can also be added to it to improve the food quality. Hard boiled egg yolk can also form useful food. Wheat-germ can also be given from time to time as it contains large amount of Vitamin B. Adequate amount of Calcium, Phosphorus and iodine should also be given to the fishes to prevent ill health and shunted growth.

A feeding ring which is in the form of a square or a circle is used for confining the dry feeds within the ring, which thus helps — (1) to avoid the spreading of the dried feed all over the water surface and (2) in allowing the excess feed to fall at the bottom in one place only. The cup-shaped feeding device (Fig. 6.89) attached to the ring, is meant only for tubifex worms.

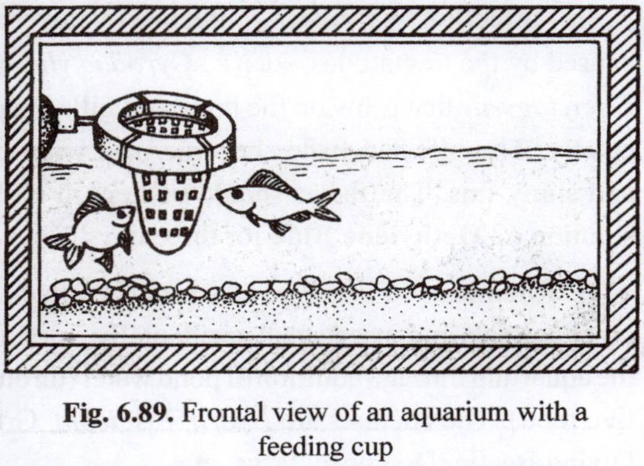

Fig. 6.89. Frontal view of an aquarium with a feeding cup

Aquarium fishes should be fed with different types of food for better health and growth. It is advisable to feed them less but often. Aquarium fishes are generally fed twice a day. However, fishes can remain without food for nearly two weeks.

D. Identification of a healthy fish

In an aquarium healthy fishes can be easily recognized through the following observations :

(1) They should be active and feed well.

(2) Their fins should be erect.

(3) Their body colour should be good and glazy.

(4) There should be no spots or blenishes on the skin.

(5) The eyes should be bright and the colour of the gills red or pink.

(6) The body cavity of the fishes should not be swollen and no skeletal deformities.

E. Fish Diseases and Treatment

In an aquarium, infectious diseases are also prevalent and can wipe out the entire stock. The common kinds of diseases or disorders are :

(a) **Degenerative diseases** — Increasing age leads to heart failure, kidney failure, skeletal deformities, development of growth and pigmented tumours, imbalance, etc. Senile changes takes place and the fishes become incapable of breeding. Air-bladder disease is also common.

(b) **Traumatic Diseases** —Injuries due to fighting or popping out of the aquarium may lead to traumatic diseases.

(c) **Congenital diseases** — Certain inborn deformities (absence of gill covers, twisted backbone, etc.) leads to congenital diseases.

(d) **Sudden shocks**—Aquarium fishes turns upside down, when transferred from acid to hard alkaline water. Similar shock symptoms are also observed in fishes when placed in water of different temperature.

(e) **Infectious Diseases**—Infectious diseases caused mainly by protozoans are common in aquarium. Such diseases are transmitted through infected fishes either directly or at the time of transfer by a net. Some common diseases witnessed are :

(i) **White spot diseases**—Caused by the protozoan *Ichthyophthirius* sp., the white spot disease is very common infection in aquarium. It causes itching and the infected fish is seen to be rubbing its body against object in the aquarium to reduce irritation. They are often caused by a drop in water temperature.

A dip of the infected fish for ten days in 5% aquous solution of **Methylene Blue** is effective but should be applied in a tank where plants are absent, as methylene blue is harmful to plants.

(ii) **Velvet disease** — Caused by the protozoan, *Oodinium* sp., velvet disease mostly affects labyrinth fishes and members of carp family such as Barbs, Minnows, Danios and live-bearing tooth-carps. Fishes affected by this parasite have golden dust-like spots on their skin as if sprayed with golden powder. Infected fry and young fishes usually succumb to the disease before it is determined. The free-swimming parasite is very difficult to irradicate as this protozoan also contains chlorophyll and can exist without fish for a long time in the presence of light.

Total treatment of the tank has to be done with **Methylene Blue** for ten days. The drug **Acriflavin** can also be used but it would cause temporary sterility in fishes. The plants in the infected tanks must be removed and washed in a solution of potassium permanganate. The gravel should be washed and preferably in boiling water.

(iii) **Chilodon disease** — Caused by *Chilodon cyprine,* the protozoan parasite attaches itself to the skin ad gills of the fish resulting in asphyxiation. The affected fish is often seen to spiral up to the surface of the water to gasp a bubble of air. Treatment with **Methylene Blue** or **Acriflavin** may obtain favourable result.

(iv) **Costiasis** — Costiasis results due to the infection of the protozoan, *costia* sp., which grows on the mucus of the fishes, skin. The affected fishes lose their appetite, becomes lethergic and breaths heavily. A dip in 2.5% **salt solution** for fifteen minutes is recommended for cure.

(v) **Fin rot** — The bacteria causing this disease enters through the damaged fin or inflamation and causes destruction of the fin tissue, resulting in death of the affected fish. Treatment with a weak solution of **Acriflavin** or **Penicillin** will be effective.

(vi) **Columnaris disease** — Caused by *Chondococcus columnaris,* this bacteria enters the fishes' body through injured areas, particularly near the mouth region. The cure recommended is 250 mg **Chloramphenicol** capsule per gallon of water.

(vii) **Pop-eye exophthalmos** — Aquarium with Siamese Fighting Fishes generally succumb to this bacterial disease, when the eyes become swollen and cloudy. The affected fish loses its eye sight if not treated in time. The affected fighter fish is put in a net and one drop of **silver eye drop** is applied four times a day.

(viii) **Cotton-wool disease** — Caused by the fungus *Saprolegnia* sp., the parasite attacks and grows at the region where damage of the skin has occured. It produces a cotton wool-like growth and the mycelia eventually invades the body tissue. The affected fishes should be removed and the wound treated daily with 50% **Methylene Blue** till its total irradication.

(ix) **Infection by Flukes** — Infection by flukes are caused by the trematodes, such as *Gyrodactylus* and *Dactylogyrus* that grow on the body and gills, respectively. The infected fishes become pale with torn and slimy fins. The fishes should be kept in a blue solution of **Methylene Blue** for three days.

(f) **Enemies** — Fishes are also prone to attack by other animals and insects that accidantally enter into the aquarium through addition of pond water (through live food). The enemies are *Hydra, Piscicola,* Great Diving Beetle, Dragonfly larva, etc.

F. Breeding of Ornamental Fishes

In nature, fishes are stimulated by changes in environmental factors which are difficult to extend in captivity. However, such conditions, to some extent, can be stimulated in the aquaria by providing specific biological requirement for spawning. Some fishes breed naturally with ease while others have to be bred by artificial techniques such as hypophysation and hand stripping.

The native aquarium fishes, which have high demand overseas, are harvested from their natural sources (i.e., from the wild). However, overharvesting of these priced native fish species may lead (in the long run) to a depletion of the stock and many such species have been pushed to the brink of extinction. Hence, captive breeding of such economic ornamental fishes is very important not only for a sustainable trade of these species but also to conserve the natural stock. Breeding of popular exotic species is also undertaken to meet the high demand of the local aquarium fish market.

The various aspects of breeding comprise selecting the breeding pair, conditioning the brood fishes, observing spawning and raising its young. According to the breeding habits of fish, the breeding tanks have to be prepared. Fishes either produce egg or live young. Egg layers comprises egg-scatterers, egg depositors, egg buriers (bubble-nest builders), mouth brooders, nest builders, etc. In live-bearing fishes live youngs are released by the females.

(a) **Identification of sexes** — In the case of live-bearers identification of sexes is essential. The mature males of egg-layers develop 'pale-tubercles' or small 'bumps'–like white spots or cysts on the head, gill cover and pectoral fin which make the male rough to touch. They ooze milt when pressure is applied on the belly. The mature female has a 'gravid spot' on the vent, and the belly is comparatively swollen.

(b) **Selection of Breeding Pair** — A breeding pair is selected that are already showing courting behaviour, when they show interest to each other. The fishes are brought to pek condition through feeding with good quality food particularly the live food. In case of egg-layers, male fishes often show too much interest in females and they need to be separated after spawning.

(c) **Egg-layers or Oviparous** — Fertilisation is external in oviparous fishes. The fry of these fishes are small and helpless whey they hatch. The egg-layers depeding on the type of eggs laid, may be of two types — (i) **Adhesive eggs** which have the ability to attach to any substratum and (ii) **Non-adhesive eggs**.

Another method of dividing the egg layers of ornamental fishes is according to their spawning habit. They are of the following types :

(i) **Egg Scatterers** — The egg scatterers, such as Zebra Danio, Goldfish, etc., show little or no parental care. Fishes like Siamese Fighter Fish have a tendency to eat up their own eggs when they are released. So the eggs or the fishes should be removed to another tank after spawning gets over. In the wild, the eggs probably get hidden in the muddy water or are swept away by the water current. In an aquarium, the eggs are provided with some egg saving measures, such as — (1) A layer of marble chips or pebbles are provided at the bottom, in which the eggs may lie concealed. (2) In case of fishes which releases adhesive eggs (Ex. common carp), thick bunches of plants or mops of nylon wool may be provided in the breeding tank. (3) For further safeguard of the eggs, a net may be kept submerged in the water, below the breeders, through which the eggs may pass through to the tank bottom.

Goldfish Breeding — Although goldfish (*Carassius auratus*) has a lifespan of 10 years, the male and female mature in the first and second years, respectively. The selected breeders are segregated sex-wise. At the time of breeding two males and one female are released into the breeding aquarium, which has been erected for the purpose. For collection of the eggs, artificial spawning nests are made of small branches of aquatic plants tied on a circular rope, as shown in Fig. 6.90. At the time of mating, the males chases the female and pre-spawning intimation takes place. The eggs laid by the females are fertilised by the milt of the males.The fertilised eggs being adhesive and demersal, stick to the spawning nest. This nest with attached eggs is then taken out of the breeding tank and transferred to the nearby hatching tank. In a period of three days hatching takes place and the hatchlings initially subsists on the availble yolk. After two days (when the yolk has become exhausted), they are fed with mashed particles of hard boiled egg yolk seived through a muslin cloth or on small sized zoo-plankton and nauplii of artemia. It has been seen that enhanced feeding takes place only at a high temperature of about 35□C.

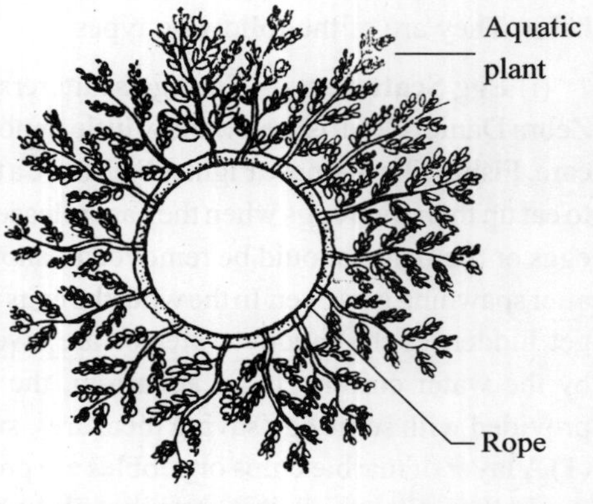

Fig. 6.90. Artificial spawning nest of goldfish

(ii) **Egg Depositors** — The egg depositors, such as *Rasbora doniconius, Hemichronis bimaculatus, Aequidens* sp. etc., show highest degree of parental care. They are the most important group among the egg-layers and unlike the egg-scatterers, these fishes do not eat their own eggs. They select sites such as filter tube, glass wall or plant leaf for laying the eggs. The Cichlid fishes deposit their eggs in a hole dug at the bottom. Before spawning the egg depositors forcibly evict any other fish from the surrounding area by chasing them away.

(iii) **Egg burriers** — The egg burriers, such as *Aplocheilus panchax,* are annual fishes, as they survive only for a year in the wild and never live to see the fate of their offsprings. In the wild, the adult fishes spawn before the onset of summer. They bury their eggs in the mud bed so that they can survive the forthcoming dry season when the bed dries up. With the onset of monsoon, the eggs hatch out.

In the aquarium, however, the condition is different as the water never dries out and most of the fishes live longer than one year. The eggs are laid in dense planted environment inside the aquarium. As the eggs are capable of remaining viable, hatching is possible even after some weeks or even months.

(iv) **Mouth Brooders** — The female tilapia, being a mouth brooder, picks up the fertilized eggs into its mouth and incubate them in its throat. Even after hatching, the female tilapia carries the spawn and young ones in its mouth. During this incubation period of two weeks, the female does not take any food.

(v) **Nest Builders** — The gouramis (*Colisa lalia, C. fasciata,* etc.) and their relatives are the most popular among the nest builders. The male generally builds the nest and also guards the eggs from any intruders including the female.

Bubble-Nest Builder — The bubble-nest is made by the male next to a floating leaf or among the floating plants. The males of Siamese fighters (*Betta splendens*), gourami, etc. forms the nest by frequently rising to the surface, gulping atmospheric air and enveloping it with mucus formed in the mouth during spawning to form a bubble (Fig. 6.91a). Several such bubbles adhere to one another to form a raft or float. The male then drives the female near to the raft, flexes around her (Fig. 6.91b) and fertilises few eggs released by the female. The fertilised eggs sink slowly to the bottom. Both fishes then drift downwards. The male picks up the eggs and blows them into the nest. The courting and mating behaviour is repeated until the female is empty of eggs. The male, at this stage, becomes aggressive even towards the female and, if not removed, will certainly kill it.

The male guards the eggs until it hatches. Some of the hatched fry (Fig. 6.91c) fall out and spiral away to the bottom. They are picked up by the male and blown back into the group. After five days when the fry starts to swin, the male fish is then removed. The fry is then fed with infusoria* or finely seived egg yolk for a few days, after which they are given freshly hatched brine shrimp, *Daphnia* and chopped worms.

---

* **Infusoria** are a collection of ciliated protozoans.

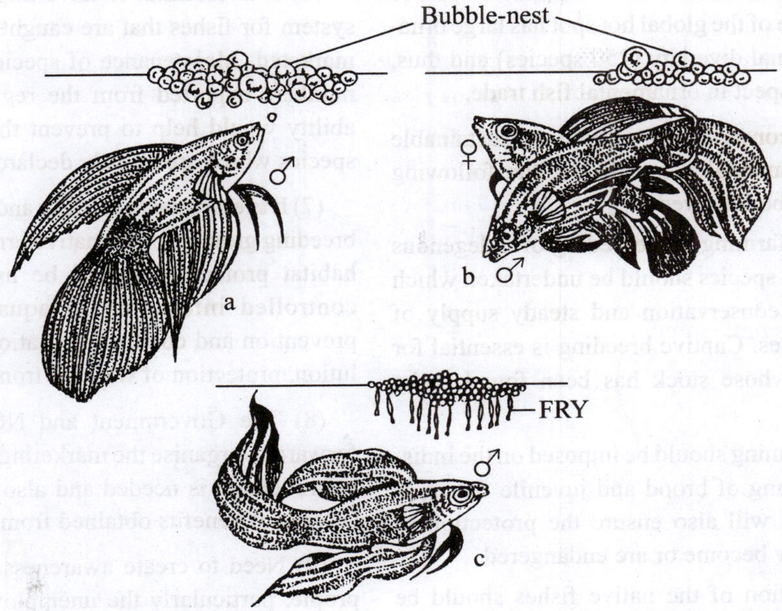

Fig. 6.91. Spawning behaviour of Siamese Fighting Fish

(d) **Live Bearers of Ovoviviparous**

The Live-Bearers due to their beautiful and bright colouration find a special place for the ornamental fish hobbyists. They also have an easy breeding habit. The live-bearers directly release the young ones. The eggs of these fishes are fertilised internally in the belly pouch of the female.

After one copulation, the female collects the sperms which is enough to fertilise 2 to 3 batches of eggs. The pregnant females are then separated in a brood cage, through which the young can escape when born. This is done because the live-bearers are known to be cannibalistic and feed on their young ones. The breeding tanks may be provided with plants for the safety of the young ones.

In India the live-bearing fishes are generally of exotic origin. Some of the common live-bearers are Guppy (*Poecilia reticulatua*), Molly (*Mollienisia latipinna*), Sword tail (*Xiphophorus helleri*), Platy (*X. maculates*), etc.

(e) **Feeding of Fry** — Food should not be given to the new brood until they are able to consume it, because the unconsumed food would decay and pollute the tank. The fry of live-bearers are able to feed immediately, but the young ones of egg-laying fishes would often remain immobile for several hours while they consume the remainder of their yolk sac.

Commercially available fry food can be had in aquarium shops and is available in liquid or powdered form. The size of the particles should be such that the fry can eat it. Generally brine shrimp, microworms and Infusoria can be fed to new fry. Infusoria are a collection of ciliated protozoans that can be produced in the laboratory. Boiling water is poured over chopped dry hay or green lettuce leaves in a jar. The mixture is kept in a warm, dark place for a few days. A healthy broth of Infusoria will be produced floating in the Sun. Small quantities of the infusoria can be added, several times a day for feed of the new fry.

Strategies for conservation and sustainable growth of Ornamental fish Trade in India

Native ornamental fish species are available in natural water bodies like rivers, lakes, swamps, streams, paddy fields, etc. The north-eastern part of India, being one of the global hot spot has large ornamental fish faunal diversity (250 species) and thus, has a large prospect in ornamental fish trade.

To ensure conservation as well as sustainable growth of ornamental fish industry, the following aspects should be followed :

(1) Captive farming and breeding of indegenous ornamental fish species should be undertaken which would help in conservation and steady supply of ornamental fishes. Captive breeding is essential for some species whose stock has been found to be declining.

(2) Legal banning should be imposed on the indiscriminate catching of brood and juvenile fishes, is important. This will also ensure the protection of species that may become or are endangered.

(3) Exploitation of the native fishes should be encouraged. Unnecessary banning of some species whose stock are plenty in the wild should be avoided to prevent natural wastage.

(4) Strict vigilance on unscrupulous fishing methods like poisoning, dynamiting, etc. should be implemented.

(5) As the ornamental fish culture is done indoor, there are vast scopes for women to be involved in this trade.

(6) It is essential to have improved information system for fishes that are caught from the wild and marketed. Maintenance of species record and their numbers exported from the region of their availability would help to prevent the harvest of those species which are already declared endangered.

(7) It is essential to identify and protect the natural breeding grounds of the native ornamental fishes. So habitat protection should be undertaken through controlled infestation of aquatic macrophytes, prevention and control of siltation and aquatic pollution, protection of wetland from conversion, etc.

(8) The Government and NGOs should come forward to organise the marketing channels, as much improvement is needed and also for fair and equal sharing of benefits obtained from the trade.

(9) Need to create awareness among the native prople, particularly the unemployed youth, to take up ornamental fishery as a means of self-employment.

(10) Transfer of technology from laboratory to the local farmers on the ornamental values of such fishes, their conservation status, methods of captive breeding, farming techniques, system designing, proper handling, disease diagnosis and prophylaxis are important. So for sustainable growth of this industry it is essential to educate all the stakeholders.

SPACE FOR NOTES

Chapter Seven : Pearl Culture

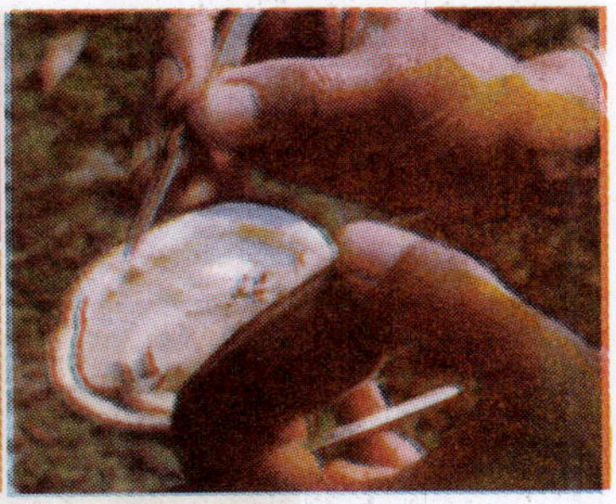

The word 'pearl' is derived from the latin word *pirula* which means pear, that is in accordance to the pear shape of the pearls. The beauty of pearl is an object of adoration and a barometer of wealth. Pearl is counted among the nine gems and needs no cutting or polishing to bring out its lustre. All over the world, pearl has been a subject of folklore and is also a subject of one of the most modern sciences — genetic engineering.

The credit for the production and development of modern pearl culture goes to Japan. The initial success was achieved in 1893 by Kokichi Mikimoto, who is considered as the 'Pearl King' and the Father of Pearl Culture industry. From the initial success the technique of pearl culture was developed and perfected.

In India the technological breakthrough in pearl culture was achieved in 1973 by the Central Marine Fisheries Research Institute (CMFRI) at Tuticorin. A second, equally significant, breakthrough was achieved in 1981, through the development of controlled production of pearl oysters in hatchery.

Pearl oysters are gregarious and they remain attached to rocks, live or dead corals, other molluscan shells or to other hard substrate. They remain at a depth of 10-12 fathoms about 20 km off the shore. The pearl banks in the Gulf of Mannar are known as 'paars' and the intertidal beds in the Gulf of Kutch as 'khaddas'.

NATURAL, CULTURED AND ARTIFICIAL PEARLS

(a) **Natural pearls** — Natural pearls may be formed within the oyster or mussels by either accidental entrance of a solid or accidental wound within the shell muscles or tissues. Pearls so produced are called natural pearls and are very rare because of their accidental origin.

(b) **Cultured pearls**— Cultured pearls are produced by human interference, when the pearls are produced through the process of culture of pearl producing oysters or mussels.

(c) **Artificial pearls**— Such pearls are cheap imitations made of plastics, glass, fish scales, etc. with an artificial lustre.

PEARL PRODUCING MOLLUSCS

Any mollusc, which has the mother-of-pearl layer on the inner surface of its shell, is generally capable of producing a pearl. However, pearl as a jewel is only produced by a few species, which are listed below.

A. Marine pearls

(a) **Pearl oysters** — *Pinctada fucta, Pinctada margaritifera* and *Pinctada maxima* are the three prime species of pearl oysters and produce superior quality pearls.

(i) ***Pinctada fucata* (Gould)** — It is commonly known as the **Indian pearl oyster**. It was formerly known as *P. vulgaris* in India and *P. martensii* in Japan. Both these species are now known as *P. fucata* (Fig. 7.1). Its distributions are in the Indian subcontinent, Japan and the Persian Gulf. In India it occurs in the Gulf of Kutch (coast of Gujarat) and in the Gulf of Mannar (off the coast of Tamil Nadu). In the Gulf of Mannar the natural pearls produced by this oyster were the famous 'Lingah' pearls or **Oriental** pearls.

Pinctada fucata occurs in the intertidal area to a depth of 12 fathoms*. It grows up to a size of 9 cm. Its maximum lifespan is about 7 years.

Fig. 7.1. *Pinctada fucata*, the Indian pearl oyster

(ii) *Pinctada margaritifera* (Linnaeus) — This oyster is commonly known as the **black-lip pearl oyster** (Fig. 7.2). It has a very wide distribution. In India it occurs at the intertidal region of Andaman-Nicobar Islands. The black-lip pearl oyster grows to about 17 cm in shell height.

Fig. 7.2. *Pinctada margaritifera*, the black-lip pearl oyster

Fig. 7.3. *Pinctada maxima*, the silver-lip pearl oyster

(iii) *Pinctada maxima* (Jameson) — This oyster is generally known as the **gold-lip** or **silver-lip pearl oyster** (Fig. 7.3). It is the largest and best, of all the pearl oysters in the world. Its distribution is restricted and is seen in Australia, Papua New Guinea, China, Philippines, Thailand, Myanmar and Indonesia. It also occurs in the Andaman and Nicobar islands.

Pinctada mxima occurs from low tide level to about 40 fathoms depth. The species grows to about 28.2 cm in diameter and the pair of shells weigh up to 5.5 kg.

(iv) Minor *Pinctada* species — The following minor *Pinctada* species (Fig. 7.4) are not considered very important as far as pearl production is concerned. Pearls are occasionally produced by these oysters, which are listed below :

Pinctada albina carcharium is present in Shark Bay (Australia). The pearls produced have strong iridescence with colours such as light yellow, light brown, light green or light pink.

* 1 fathom = 1.8 metres (6 feet)

Fig. 7.4. Minor species of *Pinctada* — *P. radiata*

Pinctada albina sugillata occurs in India (Gulf of Mannar) and Australia. It grows up to 11 cm.

Pinctada maculata is observed in Australia.

Pinctada chemnitzii is observed in India and Australia.

Pinctada imbricata is known from Venezuela and Bermudas. The pearls are translucent with colours varying among white, light pink or yellow.

In the Indian waters two other species of *Pinctada* are also recorded — *Pinctada atropurpurea* and *Pinctada anomioides.*

(b) **Winged oysters** — Winged oyster (*Pteria penguin*) rarely produces pearls.

(c) **Windowpane oysters** — *Placenta placenta,* commonly known as **windowpane oyster**, produces seed pearls of minute size. They are not considered as gems but are used for medicinal purposes.

(d) **Sea mussel** — Sea mussels such as *Mytilus edulis, Perna viridis* and *P.indica* occasionally produce seed pearls which are not considered as gem.

(e) **Giant clam** — *Tridacna,* the giant clam, produces occasionally large sized pearls, as big as a golf ball.

(f) **Gastropods** — The gastropod, *Haliotis* species, produces multicoloured irregular pearls, especially red and green in colour.

B. Freshwater pearls

Freshwater pearls are produced by freshwater mussels belonging to Unionaceae. Some of the notable species used are *Hyriopsis schlegelii* of Japan, *Cristaria plicata* of China and species of *Lamellidens.* In India and South Asia *Lamellidens marginalis* is used.

PEARL FISHERIES OF INDIA

(A) Gulf of Mannar — The Indian pearl fisheries in the Gulf of Mannar along with Sri Lanka produces the true oriental pearls of the finest quality. The species common to this area is *Pinctada fucata.*

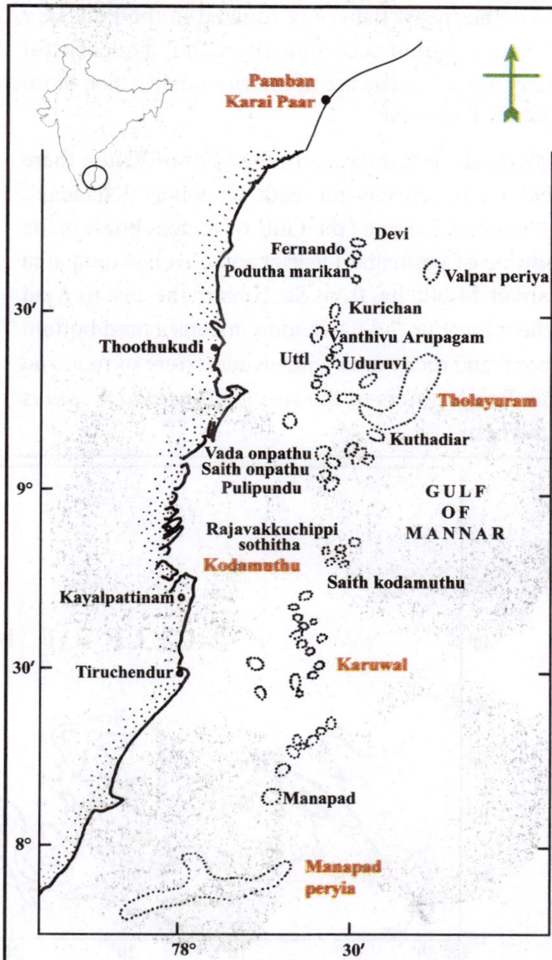

Fig. 7.5. The Gulf of Mannar with its distribution of pearl banks (paars)

In 1922, Hornell listed a total number of 72 pearl banks known as 'paars'. 'Paar' means a patch of hard ground of very low profile. These paars are located in line, parallel to and at a distance of 10-16 nautical miles* from land. This is stretched from Pamban Karai Paar in the north (head of Gulf of Mannar) to Manapad Periya Paar in the south (Fig. 7.5).

Among these 72 beds, quite a large number of them have not yielded any fisheries at any time. In general, the pearl fisheries at Tholayuram paar, Kodamuthu group and Karuwal group have been relatively consistent in the production of pearl than the others.

Another pearl fishery is situated in the Palk Bay off Toudi, which economically is not a successful fishery as the oysters are of inferior quality than those of Gulf of Mannar.

(B) **Gulf of Kutch** — In the Gulf of Kutch there are about 42 pearl oyster reefs, known as 'Khaddas'. The intertidal area of the Gulf of Kutch, borders the coastline of Jamnagar District and covers a total area of about 24,000 ha, from Sachana in the east to Ajad in the west (Fig. 7.6). 'Khadda' means a hard bottom of coral and rocky area with an admixture of mud and sand. Pearl oysters of the species *Pinctada fucata* is found here.

PEARL, ITS COMPOSITION, PROPERTIES AND FORMATION

(A) **Chemical composition of pearl** — According to Bolman (1941), the nacreous pearl of *Pinctada* has the following composition :

Water	.....	3.97%
Organic matter	.....	3.83%
Calcite and aragonite	.....	91.53%
Loss	.....	0.67%

Superior and inferior quality pearl is determined by the following way :

Quality of pearl matter	Inorganic matter	Crude
Superior quality pearl (blue)	92.60%	4.50%
Inferior quality pearl (blue)	91.30%	8.31%

The mineral calcium carbonate is secreted by the pearl-sac epithelium in two forms — aragonite crystals and calcite crystals. The third form of vaterite crystals has not been reported in the pearl oyster. The real 'nacre' substance is aragonite while calcite is the prismatic substance. Although four types of cultured pearl products are known (organic, prismatic, nacreous and compound), it is only the nacreous product that is accepted as pearl.

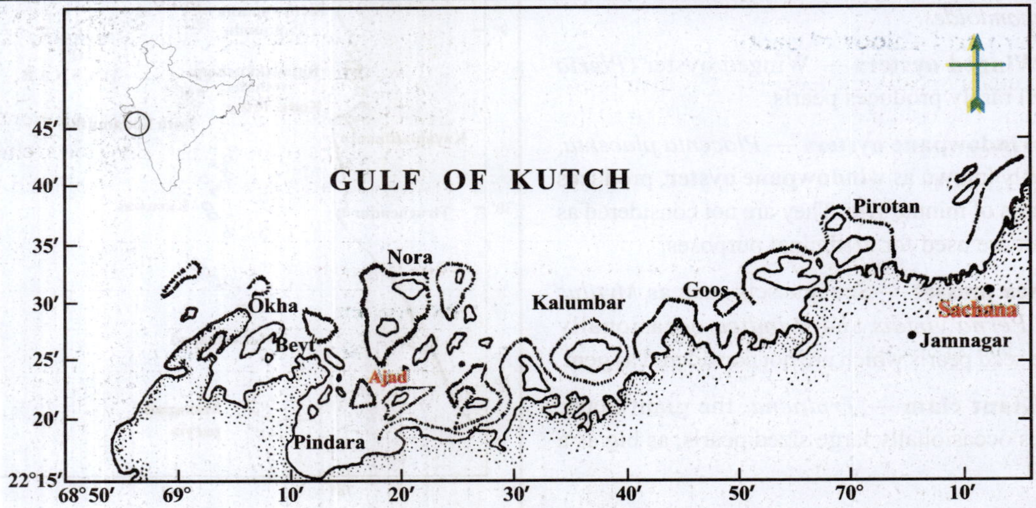

Fig. 7.6. The Gulf of Kutch with its distribution of pearl banks ('Khaddas')

* 1 nautical mile = 1852 metres or 6076 ft; 1 mile = 5280 ft.

(B) Properties of Pearl

Classification of pearls : Pearls are classified according to the size as given below :

Size (diameter in mm)	Type name
Below 2.6	Minute granular
2.6 – 4.9	Fine pearl
5.0 – 6.8	Small pearl
6.9 – 8.2	Medium pearl
Above 8.2	Large pearl

Factors responsible for the production of coloured pearls

(1) The depth of oyster culture and light penetration is related with the lustre and colour of the pearl.

(2) Lustre increases with the increase of rearing period of pearl oyster.

(3) A relation exists between the colour of the cut pieces of the pallium of the 'donor' and the colour and lustre of the pearl.

(4) Availability and quality of food phytoplankton content has distinct effect on the colour and lustre of the pearls via its role of energy supply.

(5) Amount of trace element present influences the colour of pearl.

Texture and colour of pearl

Good quality pearls are determined by their lustre and colour. The colour of the pearl is related to the metallic elements present in the oysters environment. This has been analysed through spectral analysis of different coloured pearls. It has been observed that cream-coloured and golden-coloured pearls contain good quality of copper and silver while much sodium and zinc were reported to be present in meat-coloured and pink-coloured pearls. The results of the spectral analysis is furnished in Table 7.1.

Studies revealed that the colour of the pearls is determined mainly by the colour of the cut pieces of the pallium and the part from where they are cut (Table 7.2). Further, the colours of the shell of the mussel also have marked influence on the colour and lustre of pearls (Table 7.3).

(C) Formation of Pearl

(a) Natural pearl : Pearl is formed due to secretion of the mantle tissue. The nacreous layer of the pearl is secreted by the outer epithelial layer of the mantle. This layer has the capability to rearrange and regenerate itself and remain viable when disturbed and also when transplanted in other tissues of the animal. The inner epithelium and the connective tissue, on the other hand, would disintegrate when transplanted.

A natural pearl is formed by the following way :

(1) The outer epithelial layer of the mantle, accidentally falls into the body of the pearl oyster.

(2) It regenerates a sac consisting of a single layer of cells, thus forming the pearl sac.

(3) Sometimes a foreign body accidently enters into the body of the oyster, when the shells remain open, and gets trapped between the shell and the

Table 7.1 : Spectral analysis of different coloured pearls (+* traces)

	Metallic elements									
Colour of pearl	Na	Sr	Mg	Zn	Al	Fe	Si	Ba	Cu	Ag
Meat colour	+	+	+	+	+	+	+	+	+	+
Golden colour	+	+	+	+	+	+	+	+	+	+
Pink colour	+	+	+	+		+			+	
White colour	+	+	+	+		+*			+*	
Butter colour	+	+	+	+		+*			+*	
Green colour	+	+	+							

mantle. The outer epithelium of the mantle invaginates and form the pearl sac.

(4) Inside the pearl sac, as a result of secretion of nacre, a natural pearl grows.

(5) The secretion of nacre continues till the end of life of the pearl oyster, thereby forming/producing a beautiful, natural pearl.

Table 7.2. Showing the influence of the part from where the pieces of pallium have been taken, upon the colour and lustre of the pearl

Part from where the pieces of pallium have been taken	Colour of the cultured pearl
Anterior part	Jade white, Silvery white
Middle part	Faint pink, light yellow
Posterior part	Pink, Light rosy red, golden
Dorsal part	Deep pink, golden orange, copper, rosy, violet, etc.

Table 7.3. Showing the colour and lustre of the pearl being formed by the influence of the shell colour of the mussel from whose body the pieces of pallium have been extricated

Colour of the mussel shell from whom extricated	Colour of the mussel shell up-on whom it has been implanted	Result (colour of pearls)
White	White	White 80%, Reddish yellow 20%
Red	Red	Red 55%, White 45%
Yellow	Yellow	Yellow 80%, White 20%

Figure 7.7(a) depicts the formation of a **blister pearl** or a free pearl inside the mantle. From the figure it is to be noted that every pearl has a nucleus (however tiny it may be) at its core. However, 'pearl without nucleus' may rarely be formed and would obviously be a contradication. Such possibility occurs when a few decayed blood cells or epithelial cells might provide the basis for pearl formation, which subsequently disintegrates totally. When such a pearl is cut, it reveals no nucleus and appears to be formed entirely of mother-of-pearl layers.

(b) **Cultured pearl** : The technique of the formation of natural pearl is manipulated in the cultured pearl formation process. Here for the formation of pearl, the two pre-requisites are :

(1) the outer epithelium of the mantle lobe and

(2) a core substance or nucleus.

(i) **Formation of pearl sac** — Here, a small piece of mantle from a donor oyster is grafted skilfully into the gonad of the recipient oyster along with a nucleus. The outer epithelium of the grafted mantle piece regenerates itself around the implanted nucleus and forms a pearl sac. This pearl sac secretes and deposits nacre or mother-of-pearl around the nucleus, thus resulting into a pearl (Figs. 7.7 b & c). As this pearl is produced through manipulation (surgery) in the oyster and by further cultivation of the seeded oyster, it is thus called the **cultured pearl**. The half-pearl (as shown in Fig. 7.7b) is produced by inserting the nuclei on the nacreous face of the shell at appropriate sites.

(ii) **Secretion of pearl forming matters** — Initially, after the formation of pearl sac, its internal part is alkaline and the secretion contains organic substances such as Keratin. The secretion is due to the prismatic layer containing calcites. The internal part of pearl sac later becomes acidic and subsequently neutral. The secretive matter ultimately becomes pearl forming material.

(iii) **Calcium absorption and formation of calcium carbonate** — The main ingredient of pearl is calcium carbonate (calcite and aragonite). In pearl oysters the main source of calcium is food. However, the gill, mantle and foot can directly absorb calcium from water.

In case of the gills, the calcium absorption power is strong and its rejection capacity is meagre, while, in the mantle both absorption and rejection of calcium are very strong. Thus, the gills are considered to be the most ideal organ for absorption and storing of calcium in pearl oysters.

Calcium absorbed through food and from water slowly moves into the pallium through blood. The carrying of calcium in the body depends mainly upon

(a)
(b)
(c)

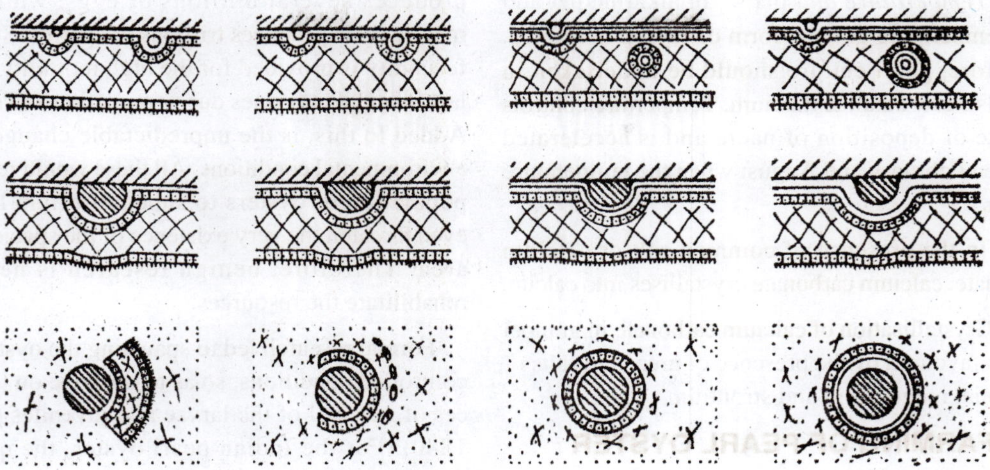

Outer mantle ephithelial cells and pearl sac epithelial cells; Inner epithelial cells; Gonad; Nucleus; Shell nacre; Connective tissuel; Foreign body.

Fig. 7.7. Diagramatic representation of the process of pearl formation. (a) Natural blister pearl on shell nacre and natural free pearl in mantle; (b) Half cultured pearl and (c) Spherical cultured pearl.

the role played by alkaline phosphatase present in the epithelial cells of the connective tissue and mantle of the pearl sac. Alkaline phosphatase combines with the calcium ion (Ca^{+2}) to form phosphate and other salts. The calcium ion in the presence of phosphatase gains energy and passes through the mantle to be absorbed by its epidermal cells. This is then carried to the connective tissue of the pearl sac. In the course of its passage the Ca^{+2} also unites with the cartilaginous sulphate (a compound protein of acidic nature). It further induces the calcium ion to become active and is transported to the cytoplasm of nacreous layer and ultimately excreted by the epidermal cells. The excreted calcium then under the activity of carbonic acid releasing hydrase combines with the carbon dioxide (released by the body of the mussel due to metabolic activity) and ultimately forms calcium carbonate. It has been reported that the concentration of carbon dioxide determines the nature of crystal formation. In the presence of increased CO_2 in the aquatic environment calcium carbonate forms sleek-like crystals, while it takes the shape of calcite when CO_2 decreases in the environment.

(iv) **Factors influencing secretion of nacre**

(1) *Morphological structure of the epidermal cells of pearl sac* — As the secretion from the epidermal cells results in the formation of pearl, the healthy condition of these cells and the quality of pearls formed by them are closely related. It has been seen that nacres, producing regular pearls, are constituted of flat epithelial cells and lacking in either granular or mucous cells. Moreover, pearl sacs responsible for the formation of hard pearls are composed of bubble-shaped epithelial tissues, while high-rising collumnar cells produce soft pearls.

Furthermore, it was observed that the morphological structure of the epidermal cells of pearl sac are not stable and undergo seasonal and environmental changes.

(2) *Calcium carbonate crystallisation* — Formation of calcium carbonate crystals is wholly determined by the environmental conditions.

(3) *Amount of* CO_2 — As has been discussed earlier the amount of CO_2 determines the nature of crystal formation.

(4) *Temperature and* pH — In alkaline pH and high temperature granpel form of crystal is formed. Therefore, pearl culture should be undertaken in neutral or low alkaline medium. Temperture controls the rate of deposition of nacre and is accelerated between February and August when the temperature is high.

(5) In the presence of common salt or calcium phosphate, calcium carbonate crystallises into calcite.

(6) Crystallisation of calcium carbonate in granpel form is favoured by the presence of urea, and salts of ammonia, magnesium and strontium.

FARMING OF PEARL OYSTER

In pearl culture, farming of oysters is a major activity. A pearl oyster farm is ideally located in a sheltered bay or lagoon which offers protection. It is also installed in coastal waters where sea conditions are not rough.

As the territorial sea is the property of the state, the sea water area has to be leased for farming. The pearl oyster farm will also have a shore establishment with facilities for all land-based works. This includes a laboratory for oyster surgery, pearl collection centre, pearl grinding and processing unit, farm stores, cage-cleaning yard and a mechanical workshop. The farm should have boats for transport of workers, oysters and farm materials. If pearl oyster breeding is to be conducted, then the farm should have hatchery establishment with facilities for supply of good sea water, aeration, maturation and spawning of oysters, live algal production, larval and spat rearing.

Pearl oyster farming comprises two phases — (I) Mother-oyster culture and (II) Post-operative culture.

I. MOTHER-OYSTER CULTURE

Mother-oyster culture deals with raising of stock by either oyster breeding or by spat collection and rearing them up to a suitable size for nucleus implantation.

A. PEARL OYSTER BREEDING

In case of pearl oyster, every mature female produces several millions of eggs, while every mature male produces trillions of sperms. Such high fecundity is provided for the high mortality of eggs, larvae and young ones due to predation and diseases. Added to this, is the unpredictable changes in the environmental conditions. All these would reduce the population of oysters to subequilibrium levels or even threaten the very existence of the species in that area. Therefore, benign research is needed to rehabilitate the resources.

Formal research led to spawning the oyster under controlled conditions, so as to improve on the chances of survival of the larvae and juveniles (Brooks, 1880). For the Indian pearl oyster, the quest for hatching technology was motivated and formulated by the Central Marine Fisheries Research Institute. It led to the development of hatchery technology for the production of pearl oyster seeds in 1978.

Hatchery Establishment—The hatchery establishment comprises facilities for good sea water supply, aeration, maturation and spawning of oysters, live algal production for food, larval and spat rearing (Fig. 7.8).

(a) **Supply of clean sea water** — Good quality sea water is to be supplied aplenty as it determines the success of hatchery production of pearl oyster seed. Larvae of pearl oyster are very sensitive to even small amount of pollutants and, therefore, sea water free from pollutants is very essential.

The water drawn from the sea is first passed through sedimentation tanks and filtration units. It is then stored and before use, it is further filtered through a series of cartridges so that all particulate matter of about 3 μm size is removed. It is further passed through germicidal ultraviolet irradiation to kill bacteria. Optimum range of temperature, salinity and pH should be maintained and any large variation of these parameters would result in failure of hatchery operation.

(b) **Aeration** — Initially larval rearing is done with change of sea water every alternate day. However, for later larval phase, for the spat and for broodstock maintenance, aeration is required to meet

(a)

(b)

(c)

Fig. 7.8. Pearl Oyster Hatchery —
(a) Live algal production; (b) Larval rearing and (c) Spat rearing

the loss of dissolved oxygen consumption by the larvae, spat and adults.

(c) **Maturation and spawning of pearl oysters** — Pearl oysters in a hatchery system has to be maintained in active reproductive phase all throughout the year. This is done by slowly raising the temperature of sea water in which the oysters are maintained from ambient condition to temperatures required for gonad maturation and then for spawning. This practice is common in oyster hatcheries of Canada, USA and Europe.

In tropical countries like India, temperature by itself is not the key factor in controlling the reproductive cycle of pearl oyster, as the annual temperature range from about 25-32°C. However, it has been observed that plentiful food supply provides the energy for reproductive growth and along with a bit of temperature manipulation (within the range of 25–32°C), maturity and spawning of pearl oysters can be obtained.

For production of larvae in the laboratory, both natural spawning and induced spawning are undertaken. Fully mature oysters are then brought from the natural beds or from the farm and placed in the hatchery along with other matured oysters. The fully mature oyster starts spawning spontaneously and in all cases the male initiate spawning. When this single oyster starts spawning, the chemical substance present in the gametes induce other oysters to follow, resulting in mass spawning in the tank.

However, through physical or chemical stimulation the mature oysters are further induced to spawn at the required time. When pearl oysters are placed in this buffer solution at pH 9.0 for 1-2 hours and then transferred to normal water, they spawn with a success rate of about 78%. Another chemical stimu-

Fig. 7.9. Live microalgal food organisms generally used as live food to feed the larvae of pearl oysters in hatchery system

lation with satisfactory spawning response is obtained when oysters are injected with 0.2 ml of N/10 Ammonium hydroxide at the base of its foot. Mature oysters can also be made to spawn on thermal stimulation when they are maintained at 24°C for a day and then the temperature is raised gradually to 32°C.

Further, it should be noted that as soon as spawning is noticed, the oysters are transferred from the culture medium to normal sea water so that the eggs and sperms are not subjected to any physiological stress, so as to remain viable at the time of fertilisation.

(d) **Production of live food** — The veliger larva of pearl oyster measures about 67.5 μm × 52.5 μm and has a mouth opening of minute size. It, thus, requires live food which is smaller than 10 μm in size. Therefore, the dietary requirement of the larvae comprises certain microalgae, particularly flagellates, which are motile, devoid of cell-wall and are not noxious. Thus, in a hatchery system mass production of live algal cells is required to feed the larvae.

Some of the microalgae, generally used as food for the larvae of pearl oysters, is given in Fig. 7.9. Among these *Isochrysis galbana* and *Pavlova lutheri* are invaluable. A concentration of 1-2 million cells per ml is given as food to the larvae at appropriate ration. The algae (Fig. 7.9) are mass cultured in 20-litre carboys or in 100-litre perspex tanks. They are cultured in appropriate nutritional medium under illumination of 1000-1500 lux and photo-period comprising 12 hours light and 12 hours darkness, in cool temperature (23-25°C) and under aseptic conditions. The inoculated cultured stocks multiply and bloom in 4-5 days and are harvested to feed the larvae.

(e) **Rearing of larvae** — The tiny veliger larvae (Fig. 7.10A) in the hatchery takes about 16-20 days to metamorphose into the adult pearl oyster. Healthy larvae are separated and reared in filtered and sterilised sea water in fibreglass tanks. They are fed once a day with adequate supplies of microalgal

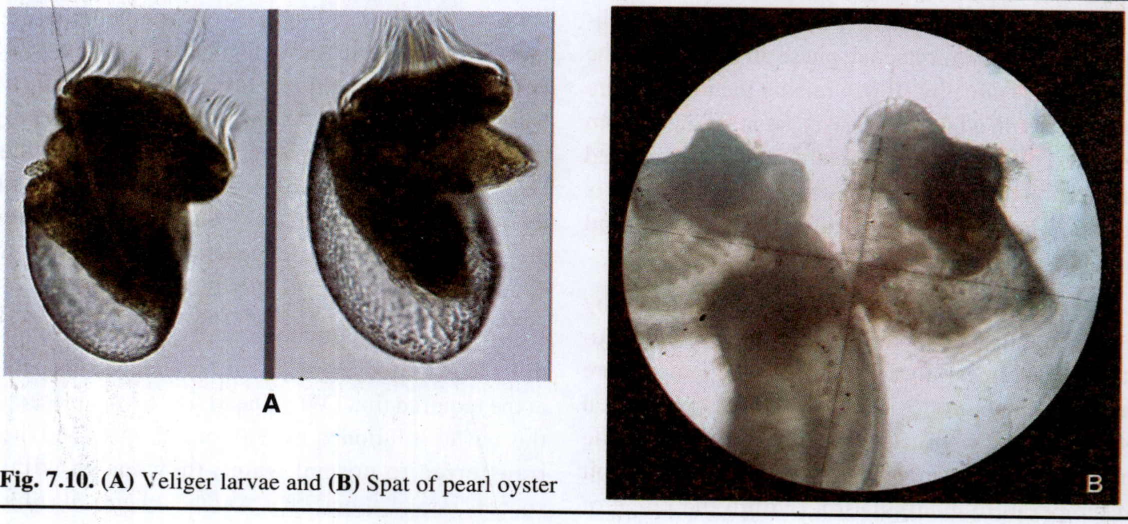

Fig. 7.10. **(A)** Veliger larvae and **(B)** Spat of pearl oyster

food. The sea water is changed every alternate day after removing the larvae on a sieve. To prevent bacterial and fungal diseases, antibiotics are added to the larval rearing medium.

The veliger larva reaches the umbo stage within 9 to 10 days. The umbo stage larva swims freely and with the development of foot, reaches the pedi-veliger stage. It soon settles down on a substratum. The brief free-swimming period of larval life comes to an abrupt halt. It then attaches itself to the substratum by secreting the adhesive byssal thread. The larva through a series of morphological changes metamorphoses and attains the shape and form of adult. It attains a size of only 300 μm. This transformed young oyster is called 'spat' (Fig. 7.10B). The spat terminology is used till it reaches a size of about 2 cm shell height. The spat and also the adults discard the bunch of byssal threads, moves a bit away from the original spot by the help of foot and refixes itself on a new spot by secreting fresh threads.

(f) **Factors influencing larval rearing** — A number of factors contribute to the success of larval rearing and spat production.

(1) Foremost among them is the water quality. Water temperature, pH, salinity and dissolved oxygen should be in the optimum range.

(2) The pipeline should be of good P.V.C. material so as to avoid rusting of metal parts. Parallel standby water line should be kept for use when the mainline needs maintenance.

(3) The material used in construction of the tank should be such that it does not leach in sea water and should remain clean.

(4) Use of detergents for cleaning the tank should be avoided.

(5) The hatchery water should be free of dust and insects.

(6) The algae food given to larvae should be disease-free.

(7) Feeding schedule and ration (quantity) should be precise to avoid decay of excess algae.

Hatching operations should be avoided in seasons when conditions are unfavourable. High mortality rates tend to occur as the larvae are delicate in nature and sensitive to environmental changes. An average of 25% spat production may be considered as satisfactory, although a rate of 50% can be achieved under ideal conditions.

(g) **Spat Rearing** : The spat initially are too small and fragile for transplantation to the farm. It needs rearing in the hatchery till it reaches a size of about 3 mm in a month. At this initial phase of spat rearing, good sea water, aeration and adequate amount of external algal food should be provided. The spat consumes a variety of algae and diatom species. When the spat reaches 3 mm, they are carefully collected from the tanks, placed in fine-meshed velon screen nets with a protective covering and transplanted to the farm. When the oyster reaches 25 mm, they are taken out and grown in regular baskets. The young oysters fall prey to a number of predators. So areas where prey are abundant should be avoided or these enemies should be periodically removed and destroyed.

B. Spat Collection from Natural Sources and Rearing

This type of pearl oyster farming deals with raising of stock from natural sources of spat collection and rearing them up to a size (about 20 g in weight) suitable for nucleus implantation.

(a) **Farm site** — The site of pearl oyster farms are selected in sheltered and protected areas of the sea, where the depth should be about 10 m and silting should be minimum. It should be seen that the salinity does not fall below 15 ppt*, as this would lead to mortality. Prolonged low salinity condition may occur at the time of heavy precipitation of rain and heavy discharge of rivers in the adjoining area of the farm. Area rich in phytoplankton should be selected as they are food for the spat. But care should be taken that this bloom is not noxious. A mild water current of 2 knots would help bring in fresh food and at the same time help in removal of metabolic wastes, faecal matters and other farm droppings.

* ppt — parts per thousand (g/l)

(b) **Farming techniques** — At the onset of culture the technique used was simple, comprising colleciton of oysters (preferably females) from the natural beds by engaging divers and broadcasting them on the seabed in a protected area near the shore. In such type of practice, the oysters suffered heavy mortality due to covering of silt and sand and also due to predation.

Subsequently, the technique developed was to grow the oysters in the water column which would be well above the bottom. This type of culture came to be known variously, such as **off-bottom culture, hanging culture, suspended culture, long-line culture** and **raft culture**. To give the oysters protection from predators, they are placed in nets, cages and baskets.

(i) **Nets, Baskets and Cages** — Modern techniqe of oyster culture has envisaged a change in design and materials of the structures without any change in principle. Pearl oysters are placed in various types of nets, cages or baskets, which are then suspended from the raft or long-line or underwater platforms, by synthetic rope at appropriate depth. Typical ones used are the frame nets (Fig. 7.11a) or the meshed cages or baskets (Fig. 7.11b). The frame net is useful to avoid crowding of oysters and is easy to follow the post-operative performance of individual oyster. The cage or basket is useful for general mother-oyster culture.

The basket or cages are of various design and made of different materials. Steel rods of 6 mm diameter form the solid frame which is given anticorrosive treatment with paint or plastic coat. The solid frame is covered with synthetic webbing of appropriate mesh size, so that the oysters do not drop out. A lid to the basket is also made of similar material. The entire structure is periodically checked and required maintenance should be done, so as to increase its life to more than 3-5 years.

The basket or cages have the following structural variations :

(1) *Size and shape* — square, rectangular or round.

(2) *Volume of space* — may be flat as in trays and pearl nets or three-dimensional as in baskets or cages.

(3) *Number of compartments* — the number of compartments in the cages or baskets may be single or multilayered (Fig. 7.11b).

(4) *Number of oysters per basket* — Oysters are segregated in diffrent size groups and the number of oysters to be kept in a basket varies according to their size. For example, a square basket of 40 cm × 40 cm × 10 cm can hold 125 oysters of 35-45 mm size, 100 oysters of 45-55 mm size, 75 of 55-60 mm size and about 50 in case of larger ones.

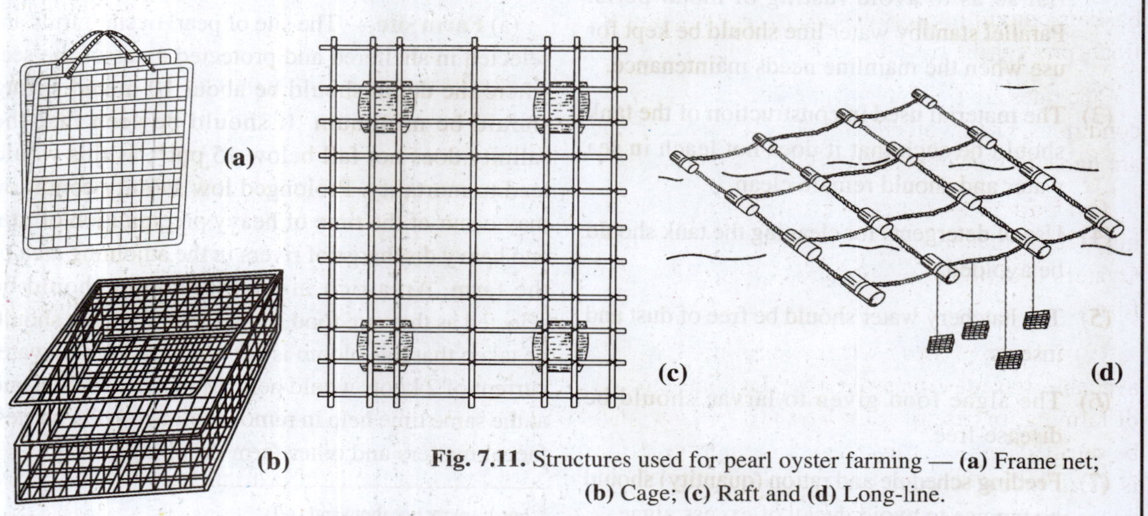

Fig. 7.11. Structures used for pearl oyster farming — **(a)** Frame net; **(b)** Cage; **(c)** Raft and **(d)** Long-line.

(ii) **Rafts** — The baskets, cages or frame nets are hung from rafts which are installed in sheltered bays or open sea. Rafts (Fig. 7.11c) are made of wooden poles placed parallel and across, and tied with nylon ropes to form a rigid frame. A standard raft is of 6 m × 5 m size and is convenient in handling. A couple of planks are fixed on the raft to enable the workers to walk on during inspection. The logs and planks are coated with coal tar. Floats (usually 4 for a standard raft) are attached on the underside of the rafts, so that when it is floated on the sea, the floats remain partially submerged and the poles of the raft remain clear of the water. Empty air-tight drums (200-litre capacity) with fibreglass coating, mild steel barrels coated with anticorrosive paints, or more modern styrofoam floats or FRP/synthetic floats are used for buoyancy. The choice of float depends on the cost and long-term economics. The raft after construction on the shore is dragged to the sea and towed by a boat to the selected site of oyster farming. The raft is then anchored to prevent it from drifting away. It is moored by two anchors (grapnel or admiralty type) on opposite sides connected to the raft with tested quality chain.

(iii) **Long-lines** — Long-line culture is best suited for open coastal conditions. Long-lines (Fig. 7.11d) are made of a series of large hollow floats in a row, connected by synthetic rope or chain and suitably anchored. From the strong rope of the line, the baskets containing oysters are suspended and the floats prevent the baskets from sinking.

The lines are flexible and can tide over rough sea conditions. On the other hand, the rafts are rigid and are liable to break under strong winds and waves.

(iv) **Underwater platforms** — For oyster farming in deeper lagoons underwater platforms are used. Oysters in strings are suspended from these platforms.

(v) **Choice of site** — Depending upon the site available, it is important to choose the right technique of farming. In case of calm bays raft culture should be selected, while for turbulent sea long-line culture is ideal. In case of moderately deep lagoons, underwater platforms are best suited.

(c) **Environmental parameters necessary** —The sea water provides the oysters with all requisites to maintain their life functions. Anthropological fouling of sea water may cause harm to the oysters. Therefore, while selecting the site for oyster farming it is essential to provide the oysters ideal environmental conditions which would ensure success in pearl oyster farming.

(i) **Topography** — It has been discussed earlier, pearl oyster farms are constructed in areas sheltered from strong winds and waves. The east and west coasts of India are effected by north-east and south-west monsoons, respectively. As a result, year-round open-sea farming becomes difficult to achieve. Pearl culture in the Gulf of Mannar is restricted to small islands such as Krusadai, Pullivasal, Hare, Karaichalli, Manauli, Nallatannitivu, Vilanguchalli and Vantivu. In Gulf of Kutch, the 'Khaddas' are open intertidal stretches where embankments are erected in the channels to shelter the oyster farm. Moreover, ideal conditions for pearl oyster farms are present in the Andaman and Nicobar Islands and in the lagoons of some islands (Bangaram) of Lakshadweep.

(ii) **Hydrographic conditions**

(1) *Depth* — For rearing pearl oyster, depth is an important factor. Quality pearls are produced in deeper waters (beyond 10 m). In the Gulf of Mannar, better quality pearls are produced by oysters in deeper beds (at 15-20 m depths) compared to those produced in intertidal areas. Fouling and boring problems do not occur at depths below 10 m.

(2) *Bottom* — Muddy bottom should be avoided. Sea bottom should preferably be gravelly or an admixture of sand.

(3) *Water current* — Water current should be mild (about 1 knot) and would be sufficient to carry dissolved oxygen and food material. Strong current should be avoided as it results in stunted growth of oysters.

(4) *Change of culture ground* — Repeated culture in the same ground often results in accumulation of biological wastes on the sea bottom, resulting in deteriorative quality of pearl production. Moreover, diseases and mortality of pearl oysters may occur.

(5) *Temperature* —The suitable range of temperature is between 18-25°C. During winter when temperature drops to about 10°C, the oysters undergo hibernation. However, in the Indian seas, temperature is high throughout the year and higher growth rate is observed in the winter months. At lower temperatures fine layers of nacre are deposited which improve the lustre of the pearls.

(6) pH—The pH of the sea water should be between 7-8. In the farm of Tuticorin (presently named Thoothukudi), the annual range of pH is 7.7-8.3.

(7) *Salinity* — Salinity in the Gulf of Mannar generally varies between 30 and 35 ppt. Heavy rain in the sea does not bring about any noticeable change in salinity. However, during monsoon heavy discharge of freshwater from rivers, considerably lowers the salinity of costal waters. Prolonged spells of low salinity cause oyster mortality.

(8) *Calcium* — Calcium is required for the formation of shell and pearl. Oysters derive calcium from sea water as well as from the body fluids *via* the mantle. The normal calcium level in sea water is 400 mg/l. In Tuticorin Bay, salinity ranges from 316 to 454 mg/l. Low calcium level would lead to formation of thin shells and poor quality pearls.

(9) *Nutrient salts* — Nutrients such as phosphates, nitrates and silicates are essential for good growth of phytoplankton which forms the food of oysters.

(10) *Trace elements* — The composition of trace elements (quality and quantity) influences the colour of pearls (Table 7.1). Pink coloured pearls contain more sodium and zinc, while copper and silver are responsible for gold- and cream-coloured pearls.

(d) **Food of pearl oysters** — Several groups of phytoplankton, particularly diatoms and flagellates, form the main food of pearl oysters. The chemical composition and amount of phytoplankton determine the quality of pearl. Blooms of certain diatoms and dinoflagellates sometime result in toxicity and may lead to death of oysters. Therefore, areas which witness such toxic blooms should be avoided for construction of pearl oyster farms.

C. Enemies of pearl oyster

In maintenance of pearl oyster stock, one of the major problems encountered are the enemies of pearl oyster. They can be classified as biofouling, boring and predator organisms.

(a) **Biofouling Organisms** — Biofouling organisms settle and grow on the shells of oysters. Several species of marine organisms get a foothold on the oyster shell and form mini-scale biological colonies. These organisms settle as larvae, metamorphose and grow vigorously (Fig. 7.12).

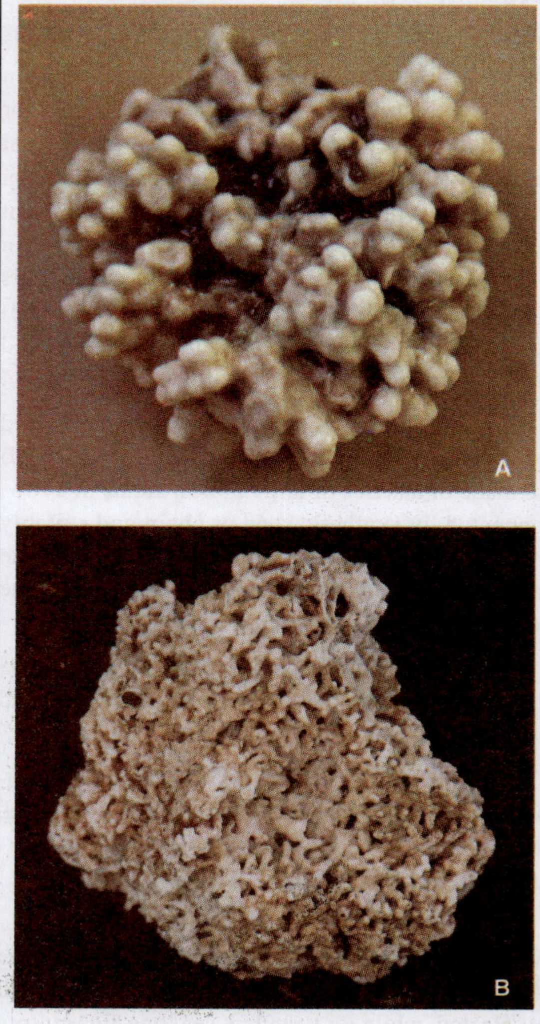

Fig. 7.12. Biofouling organism of pearl oysters. **A** : Coralline; **B.** Worm-Calcite-Cast

The various biofouling organisms are mainly the barnacles (*Balanus amphitrite,* whose number may run to a hundred and virtually occupying all the spaces on the shell), bryozoans (*Membranipora* sp., *Thalamoporella* sp. and *Lagenipora* sp.), simple and compound ascidians (*Ascidia* sp. and *Dicarpa* sp.), the spat of bivalves (*Avicula vexillum* and *Crassostrea madrasensis*) and hydroids. *Avicula vexillum* not only grows profusely on pearl oysters but also on the baskets, almost closing the meshes and disrupting waterflow through the baskets. The weaving mussel (*Modiolus* sp.) forms extensive carpet-like colonies over the pearl oyster beds but causes no serious threat. Other biofoulers are sea weeds, sponges, amphipods, isopods, polyclad worms, nematodes, opisthobranchiate mollusc, *Pinna* sp., and egg capsules of gastropods and crinoids.

Biofouling is a major problem in pearl oyster farming as it effects the growth of oyster and the quality of pearls. It necessitates intensive labour work for periodic cleaning of shells. Mechanical devices and water jets are employed to clean oysters and cages. This problem of biofouling can be minimised by choosing deeper waters and selected depths of water, where these fouling organism is relatively less.

(b) **Boring organisms**— The boring organisms riddle through the shells and makes the pearl oysters weak and friable. They include two dominant groups — the polychaetes and sponges. The boring molluscs comprising *Martesia* sp., *Lithophaga* sp. and isopod *Sphaeroma* sp. are rarely seen and are of less significance (Fig. 7.13).

Sponges such as *Cliona celata, C. vastifica* and *C. margaritiferae* form honeycomb-like ramification in the shell with numerous openings on the nacreous surface.

Polychaets such as *Polydora ciliata, Cirratulus cirratus,* members of family Syllinidae, Nareidae and Terebellidae cause havoc to the shells. *Polydora ciliata* cuts tunnels through the length and breadth of the oyster shells. The oysters, to prevent these predators from creating shell cavity, secretes excess calcium carbonate to seal the entry points. Oysters become weak in the process and in heavy infection may cause large scale mortality.

The boring organisms can be irradicated by placing the pearl oysters in freshwater for about 6 to 10 hours or by application of a thin film of 1% formaldehyde solution on the shells and after few minutes they are again put back in sea water.

(c) **Predators** — In farms the major predators of pearl oysters are the gastropods such as *Cymatium cingulatum* and *Murex virgineus,* who cause serious damage (kills an oyster per day). These gastropods are seasonal devastators. Crabs such as *Charybdis* sp., *Atergatis* sp., *Leptodius* sp. and *Thalamita* sp. also prey upon the pearl oysters. These gastropods and crabs when young (small in size), enters into the baskets through the meshes and grow by preying

Fig. 7.13. Boring organisms of pearl oysters

Fig. 7.14. Predators of pearl oysters

upon the young pearl oysters. Prevention methods have to be taken through periodic inspection, removal and killing of these predators (Fig. 7.14).

In the natural beds, coral reef perches (*Balistes* sp, *Serranus* sp. and *Lethrinus* sp.), rays (*Rhinoptera* and *Ginglymostoma*), starfish and octopus are notorious predators that predate upon pearl oysters.

D. Surgery of Pearl Oyster for Production of Cultured Pearls

For production of cultured pearl, man provides the oyster with two essential conditions — the core material or nucleus and the mantle tissue. Through skillful surgery, a mantle piece from the donor oyster is grafted into the gonad of the recipient oyster along with a nucleus. The processes involved are selection of oysters, graft tissue preparation, conditioning graft tissue and nucleus implantation. Later post-operative care is also undertaken.

(a) **Surgical tools used for surgery** — For pearl oyster surgery, special surgical tools are used (Fig. 7.15). The set comprises oyster knife, speculum, oyster stand with a clamp, graft cutting knife, incising-cum-grafting needle, nucleus implanting needle, needle hook, spatula, scalpel, a pair of scissors and fine hair brushes. The tools are maintained properly after each operation by thorough cleaning with freshwater and drying them in the sun.

(b) **Selection of oysters** — Healthy oysters are selected that are free from any boring and fouling organisms and are free of bacterial and viral diseases. Other factors that are to be considered are the weight and reproductive phase of the oysters. Oysters more than 20 gm of weight are desirable for implantation of nucleus. Fully matured oysters are not considered. Desirable sized oysters are brought from the farm for surgery and the shells are cleaned of the fouling and

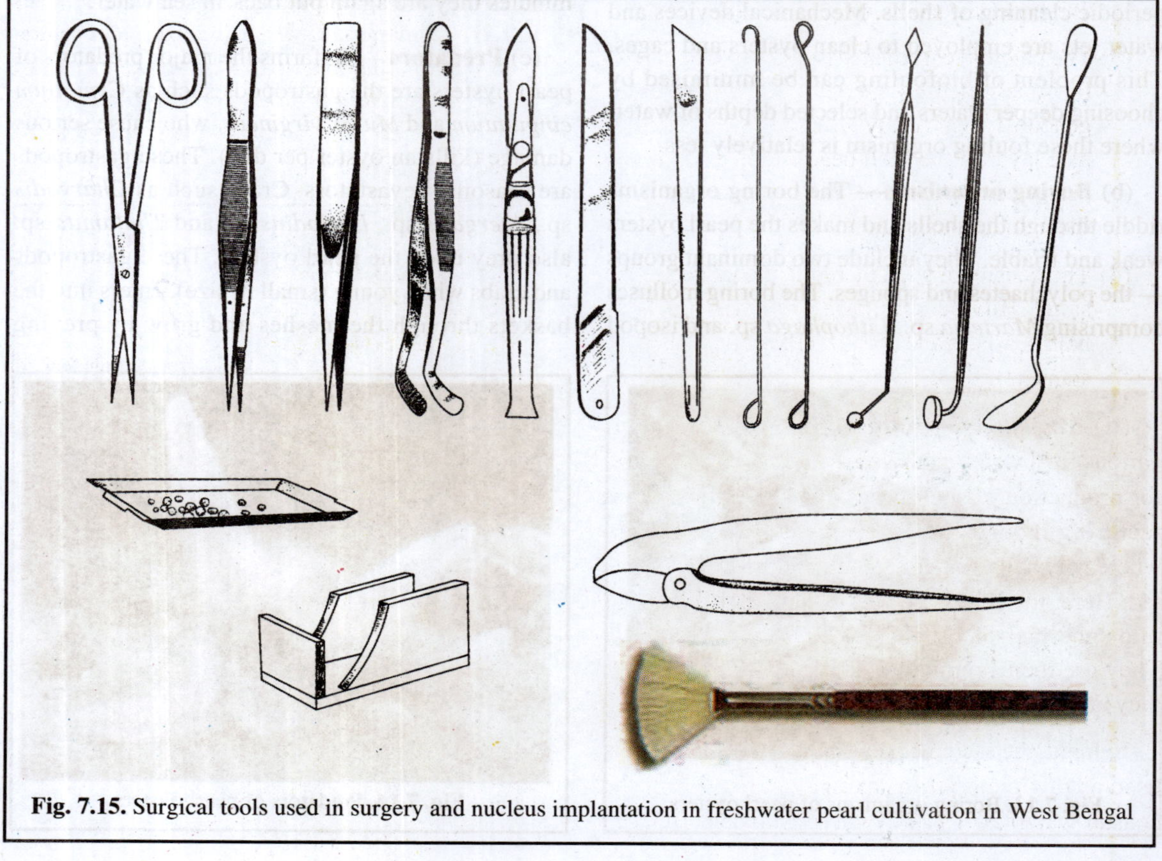

Fig. 7.15. Surgical tools used in surgery and nucleus implantation in freshwater pearl cultivation in West Bengal

boaring organisms. Initial selection is done based on the size and shell quality and the final selection is done only at the time when the oysters are opened for surgery.

(c) **Nucleus preparation** — (i) **Raw material used** —The core material used for the cultured pearl formation, can be any inert substance. Small Buddha carvings made of ivory, wood, stone or metal castings were used by the Chinese. Mikimoto, the father of pearl culture, used bits of shells, coral pieces, bits of metal, grounded scales of fishes and so on. Later, since pearl culture was commercialized, it was experimentally obtained that beads of freshwater mussel shells, were the most suitable for use as nuclei for cultural pearls. The reasons for such conclusion were :

(1) The general physical properties of shell material is similar to that of nacre.
(2) The heat resistant property of shell material and nacre are the same.
(3) The calcium carbonate (in aragonite or calcite form) secreted by the pearl sac, binds homogeneously and firmly with the calcium carbonate layer of the shell bead and, therefore, the pearly component does not get separated from the nucleus at any time.
(4) The density of shell material and nacre are the same. However, in case of blister pearls or half pearls, alabaster is used as the core substance.

(ii) **Shell bead production** — The shells of various freshwater mussels are used as raw material for production of shell bead. The shell bead manufacturing industry in Japan is highly specialised where the shells from different countries are imported. Here, the shells are cut, ground and processed into spherical shell beads of 2 to 8 mm diameter. They are then supplied to several countries where they are used as nuclei for pearl culture practices.

In India, the sacred chank shell (*Xancus pyrum*) is used as raw material for nuclei making. From the coastal waters of Tamil Nadu, Kerala and Gujarat, several thousands of *Xancus* are fished annually and sent to West Bengal, where the cottage shell industry makes bangles and rings from them. However, more than 70% of the shell mass is wasted and often used in making lime. From these shell waste, thick shell bits are sorted out and then ground into spheres. It is then polished and used as nuclei for cultured pearls. Shell beads are also produced from the shells of giant clam, *Tridacna* sp., which occur in Andaman and Nicobar Islands. However, freshwater mussels of large size with thick shells are hardly found in India.

(d) **Preparation of graft tissue** — Graft tissue preparation is essential prior to surgery. The graft tissue is obtained from the mantle of some healthy pearl oysters which are selected as donors. Donor oyster should be in prime condition and without blisters caused by boring organisms. They should not be subjected to any conditioning process (narcotisation, etc.).

The mantle of pearl oyster is divided into 3 zones (Fig. 7.16) — marginal zone, pallial zone and central zone. It is the pallial mantle that is required for graft tissue preparation as it is more productive in terms of secretion of shell material. The mantle should be removed without injury.

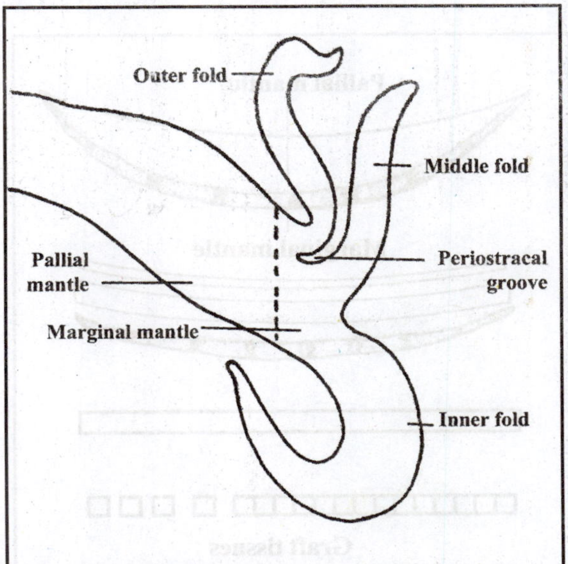

Fig. 7.16. Diagramatic transverse section showing the mantle folds of pearl oyster

The oyster shell remains tightly shut when taken out of water. It is then cut open by the following method :

(1) The oyster is held dorsal side down and the posterior side facing the person.

(2) The curved end of the sharp oyster knife is inserted between the shells on the posterior side.

(3) The knife is then pushed straight through the oyster until the knife-tip touches the anterior end.

(4) The sharp end of the knife is pressed straight downwards cutting through the adductor muscle.

(5) The valves are opened tearing the hinge, without disturbing the two mantle lobes so that the lobes do not shrink.

The mantle is then removed (taking care that it does not shrink) by the following steps :

(1) The gills are brushed gently with the tip of the spatula, exposing the mantle lobe.

(2) The mantle is then cut with the help of a graft knife, starting from the posterior margin tracing a curve up to the anterior margin.

(3) With the help of forceps, the mantle is gently lifted and the tissue is placed on a soft, clean, moist wooden block. The mantle's inner epithelium now faces the person.

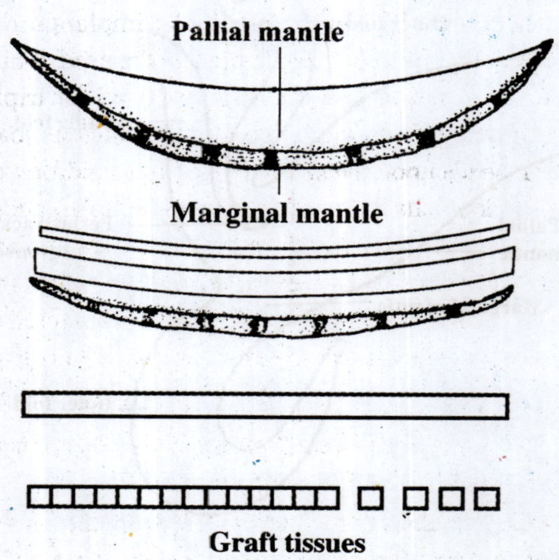

Fig. 7.17. Preparation of graft tissue from mantle

As shown in Fig. 7.17, the graft tissue is further prepared by the following way :

(1) The tissue is first gently stretched end to end.

(2) The mucus and dirt present on the mantle tissue is gently wiped off with the help of a wet sponge.

(3) With the help of a graft knife the marginal mantle (Fig. 7.16), which can be identified due to its folds and pigmentation, is cut off.

(4) Similarly, the inner muscular portion of the mantle on the opposite side is cut off.

(5) With the help of a fresh wet sponge the mucus and dirt are again wiped away.

(6) The mantle ribbon is then lifted by holding one end. The side is reversed and placed on the block such that the outer epithelium faces the person.

(7) Mucus and dirt are again wiped carefully without damaing the outer epithelial layer.

(8) The margins of the mantle are further trimmed on either side until a ribbon of about 3 mm width is obtained.

(9) All the dirt and mucus are wiped off from the wooden block and if necessary, the wooden block may be changed. Care should be taken while transferring the mantle ribbon that the sides are not changed.

(10) With the help of the graft knife, the ribbon is cut into small pieces (about 2-3 mm) in proportion to the size of nucleus.

(11) The small pieces of the graft tissue is smeared with a very dilute solution of water soluble eosin by the help of a brush.

(12) Care should be taken such that the tissues remain moist until they are used within 30 minutes.

Precautionary methods to be taken are :

(1) Care should be taken such that the oyster knife while inserting between the valves does not slip and injure the palm of the person.

(2) Clean, filtered sea water should be used during the entire operation.

(3) All the instruments should be sterilised and dried in sunlight prior using.

(4) Clean and moist sponges should be used and should be changed after each wiping of mucus and dirt.

(5) Wooden blocks must be smooth, clean and should be moist till the graft tissues are used up.

(6) The shrunk mantle lobes (due to bad handling) should not be used.

(7) The tissues, generally should be used up preferably within 10-15 minutes of preparation. Retention of tissue on the block for a long time leads to deterioration and failure in pearl sac formation.

(e) **Conditioning of oysters prior to surgery** — Conditioning of oysters are generally undertaken before surgery. It is either done naturally or by chemical.

(i) **Natural physical conditioning** — Such methods are ideal and inexpensive. It can, however, put into use in regions where there is stratificaiton of temperature in the sea (as in Japan) and sharp difference in food availability. By using the thermal difference the oysters are made to spawn in higher temperture of the surface water. The loss of stored energy thus encountered due to spawning, weakens the oysters. They are then further subjected to starvation stress in depths of low phytoplankton production. This reduces the metabolic rate.

(ii) **Chemical conditioning** — In Indian waters the above technique does not work. So the oysters are resorted to chemical conditionig. Oysters are first put in tubs placed in sea water. Over it menthol crystals are then sprinkled. Under the effect of menthol, the oysters get narcotised in about 60-90 minutes. The adductor muscle of the oysters relaxes and the valves open. The response time varies with temperature and the oysters become almost non-responsive to touch. One by one they are taken out of the tub, washed in sea water and then used in surgery. Pearl oysters should be treated with menthol in batches as prolonged exposure in this chemical causes swelling of tissues, large secretion of mucous and ultimate mortality. Safe limit of exposure in menthol is about 30-45 minutes.

(f) **Implantation of nucleus**

(i) **Nucleus implantation for round pearls**

(1) *Site of implantation* — The site of formation of natural pearls in oysters is the muscular region mainly in the pallial connective tissue, tissues around the liver and kidney, levator muscle and in the palpar region. The pearls formed are generally small in size and irregular in shape.

In case of cultured pearls the technology aims at producing pearls of large size and of smooth, round shape. For implantation of nucleus, the region of the body to be selected should have adequate space for accomodation of large sized nucleus and the organ should not be highly specialised. Thus, the gonadal region of pearl oysters has been found to be the ideal site (Fig. 7.18) and the follicles carrying the gametes, get reorganised without greatly affecting the reproductive function of the oyster. As a secondary site, the peripheral region near the liver or hepatopancreas is also used for nucleus implantation, but care should be taken so that the stomach, intestine and other vital organs such as heart, kidney, etc. are not damaged.

(2) *Number of nucleus implanted* — In oysters of sufficiently large size, a maximum of 5-6 nuclei of 4 mm diameter or less can be implanted. Single nucleus implantation is done for nucleus of large size (6 mm diameter). For double nuclei implantation, 2 nuclei of 6 mm and 5 mm diameters are used, while nuclei of 4 mm or less diameter are used for triple and higher nucleus implantation. The nucleus load also depends upon the size of oyster and condition of gonad. Generally, in case of two-nuclei implantation, the primary nucleus (larger one of the two) is implanted in the gonad at its ventral extremity, while the smaller nucleus (secondary nucleus) is implanted at the dorsal side close to the hepatopancreas (Fig. 7.18).

(3) *Oyster surgery conducted* — Pearl oyster surgery is done by the following steps :

- The conditioned oyster is first mounted on the stand and positioned between the two plates of a clamp in a manner to suit the convenience of the technician and also that it does not slip out.

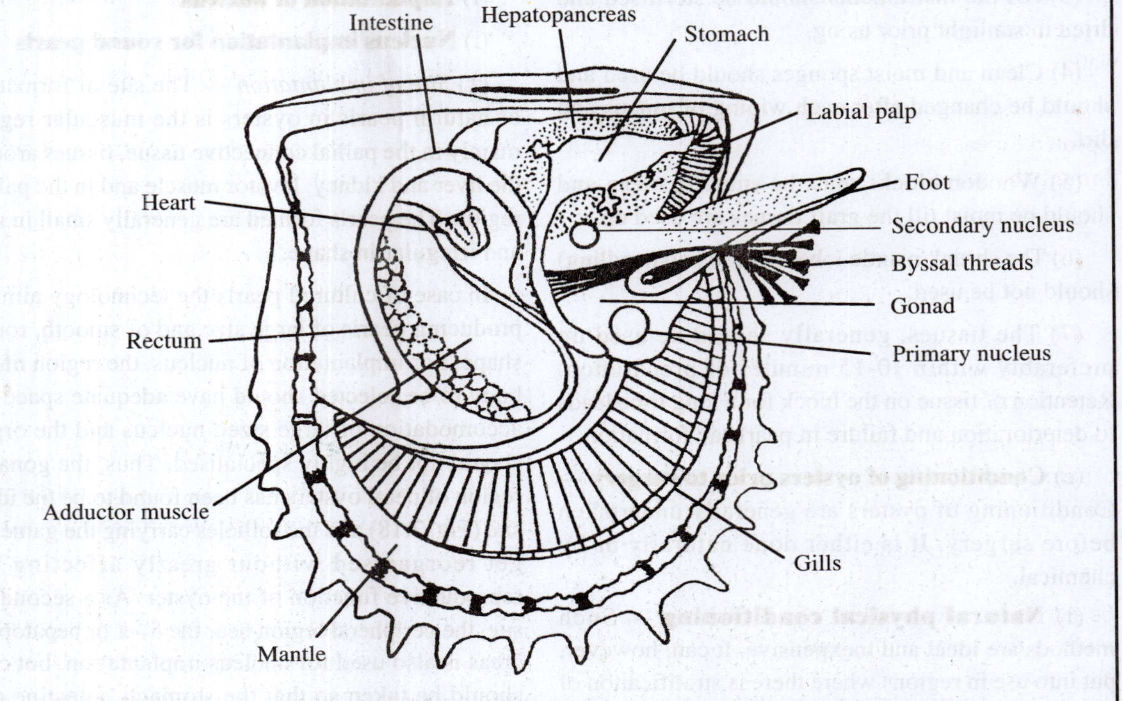

Fig. 7.18. Diagrammatic sketch of the anatomy of *P. fucata* showing the implantation of primary and secondary nuclei in the gonad.

- The end of the speculum is inserted into the posterio-ventral corner of the oyster in between the two valves, and into the opening is placed the gap-regulator ring.
- With the help of a spatula the mantle, gills and labial palps are smoothly pushed aside to get a clean view of the gonad region.
- The tip of the foot is hooked with the needle hook and is gently lifted so that the base of the foot is slightly elevated. The needle is held in position till the operation is completed.
- With the end of the oval knife, a sharp incision of length subequal to the diameter of the primary nucleus, is made in the pigmented region at the base of the foot. The muscles are gently detached with the sharp oval end of the needle and a passage is cut through the gonad up to the site of implantation. The needle is then gently withdrawn.
- A piece of the graft tissue (Fig. 17.17), present on the block is picked with the help of the reverse side of the same needle (as used before). The mantle piece is then inserted gently through the passage cut in the gonad and left at the site of implantation. The needle is then withdrawn.
- The nucleus is lifted by the appropriate nucleus-implanting needle with the help of the cup-end. The nucleus is inserted gently through the incision made previously at the foot and the passage cut through the gonad. On reaching the site, the nucleus is dropped by deflecting the needle. The nucleus comes in contact with the outer epithelium of the mantle tissue, which was previously grafted into the gonad. The outer epithelial side of the mantle piece should face the nucleus and this orientation is important for successful formation of the future pearl sac. The needle then is gently withdrawn through the passage.
- The two margins of the incision are then allowed to come in contact in the pigmented region.
- The oyster is then removed from the clamp, the

speculum is withdrawn and the gap-regulatory ring is slipped out. The oyster is then placed in fresh sea water.

◆ In case of multiple implantation, the process is repeated for each nucleus implantation which is done through the same incision.

(4) *Precautions to be undertaken*

◆ The instruments should be cleaned in sea water before and immediately after use.

◆ Care should be taken such that the valves are not opened too much, so that the adductor muscle snaps and the oyster dies.

◆ The pressure on the foot should be adjusted in a manner so that the foot does not get torn while pulling it with the needle hook.

◆ The incision made on the foot region should be a sharp cut and should be just of required length as the size of the nucleus to be implanted. If the incision made is too large and if much tear of tissues has taken place then it is advisable not to proceed further and the oyster be placed back in fresh sea water to be taken back to the farm.

◆ If there is copious flow of gametes at the time of cutting a passage through the gonad, then it is advisable not to proceed further and the oyster should be taken back to the farm with the general stock.

◆ At the time of surgery care should be taken not to damage the vital organs such as stomach, intestine and heart.

◆ The orientation of the graft tissue and nucleus should be proper, i.e., the nucleus should be in contact with the outer epithelium of the mantle tissue.

◆ Perfection in the surgery should be undertaken with skill and patience.

(ii) **Half-round pearl** — In *P. maxima*, *P. margaritifera* and the winged oyster *Pteria penguin*, half-round pearl is produced using the pearl-Buddha technique of China. In this case the nuclei implanted ranges from 10 to 20 mm in diameter and is made of alabaster. The technique adopted is as follows :

(1) The oyster is first clamped horizontally on the stand, and from the upper valve, the mantle is separated.

(2) A drop of 'aron-alfa' (a special transparent adhesive) is applied to the shell.

(3) The nucleus is pressed against the adhesive for 2 minutes by a ring.

(4) By this method, the required number of nuclei are fixed to the shell (Fig. 7.19).

(5) The oyster is then turned over and the opposite shell is treated in the same manner, taking care that the nuclei on both shells do not oppose each other.

(6) The number of nuclei fixed in each shell depends upon the size of the nuclei — 3 of 16-17 mm diameter and 5 of 13 mm diameter.

II. Post-operative Culture

The stress endured by the oysters due to surgery and conditioning makes them weak and vulnerable. Careful handling is essential until the oysters recover fully. The operated oysters, thus, should be placed in gently flowing sea water or in a static system with frequent change of sea water. In case of a bay or

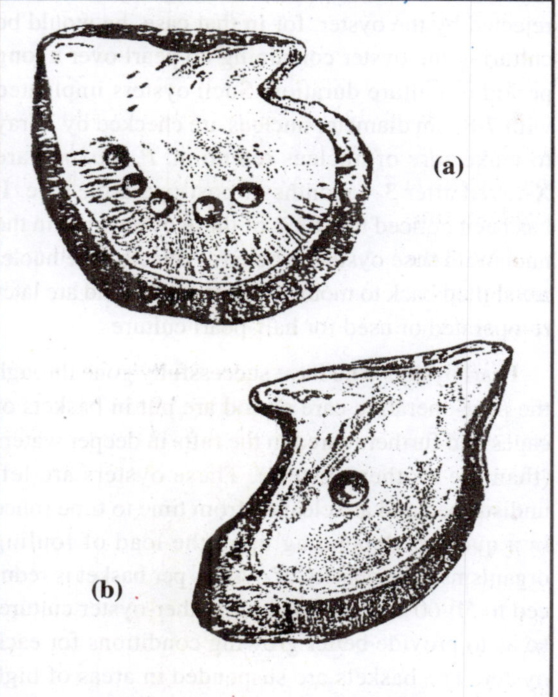

Fig. 7.19. (a) Nucleus of half-pearl implanted on the shell of *Pinctada maxima* **(b)** Blister pearl formation in *Pteria penguin*.

lagoon, where the sea water is transparent, the oysters can be caged and directly suspended from the rafts. It is, however, advisable that the operated oysters be kept for 3-4 days in the surgery facility under observation before they are returned to the farm.

The oyster, on being placed in sea water, would, start shutting the valves within a few minutes as the effect of menthol gets over slowly and would sooner resume its normal function of opening and shutting the valves. Normally the incision would heal within 2-3 days. At this time, the oyster would attempt to get rid of the nuclei. If the incision and the passage cut through the gonad are too large then the chances of the nucleus getting rejected are greater. Also, rough handling and exposure to strong stimuli (tides, current, large waves and food supply) would also result in rejection of nuclei. In case of oysters whose vital organs have been damaged during surgery, would suffer mortality.

In case of single, large nuclei implantation, the farmer has to be sure that the nucleus has not been rejected by the oyster, for in that case, he would be culturing the oyster containing no pearl over a long period of culture duration. Such oysters implanted with 7-8 mm diameter nucleus are checked by X-ray to make sure of nucleus retention. *P. maxima* are X-rayed after 3-4 months of post-operative care. It has been noticed that 83% of these oysters retain the nucleus. Those oysters who have rejected the nuclei are shifted back to mother-oyster culture and are later re-operated or used for half-pearl culture.

Pearl oysters who have successfully gone through the post-operative care period are put in baskets or cages and further reared in the rafts in deeper waters (than the mother-oysters). These oysters are left undisturbed and are cleaned from time to time (once in a quarter) depending upon the load of fouling organisms. The density of oysters per basket is reduced to 50-60% of the load of mother-oyster culture, so as to provide better growing conditions for each oyster. The baskets are suspended in areas of high phytoplankton population.

The duration of post-operative culture depends upon the time required for the pearls to attain lustre and thickness of the nacreous layers. The duration of the post-operative culture in Japan has been progressively reduced from $2\frac{1}{2}$ years in 1960 to 6-8 months in recent years. This is due to commercial compulsion. Ward (1985) has stated that a nacre thickness of 0.5 mm is acceptable as pearl and the minimum is 0.35 mm for good lustre and colour. However, the thickness obtained in 6-8 months post-operative culture is 0.2 mm.

In India, under tropical conditions, the acceptable thickness for good lustre and colour are produced within 3-4 months with nuclei of 2-3 mm diameter and in 15-18 months with nuclei of 6-7 mm diameter.

A. Influence of Temperature on nacre formation

Laying down of nacre around the nucleus is influenced by the temperature of sea water. In Japan, the temperature of waters in winter goes down to as low as 10°C. At this temperature, the metabolic activity of the oysters is very low and they enter into a state of hibernation. At this period, the nacre secretion is practically arrested. Nacre secretion recommences only when the temperature rises in spring.

In India, high temperature prevails throughout the year and hence nacre secretion is uninterrupted and hence the growth of pearl is faster.

B. Harvest of Pearls

Harvesting of pearls, also known as 'beaching' of pearls, is done during period of low temperature and pH. The cooler period of the year is selected as the lamellae of nacre layed down at this period are thin. This thin lamellae on the top are considered to be best for colour and lustre. Generally, the farmers at the end of post-operative culture, shift the oysters to grounds which favour production of pearls of superior quality. This phase is known as 'make-up' culture.

The harvested oysters are taken up individually and opened to collect the pearls. *P. fucata* oysters are cut open with a sharp knife to take out the pearls and thereby killing the oyster. However, in case of *P. maxima*, the oysters are opened carefully taking care not to kill the oysters at the time of removal of the pearls. These oysters are then reused for at least one more crop by inserting a new nucleus of appropriate size in the already formed pearl sac, without using a fresh mantle piece.

In case of old aged oysters, they may contain some natural pearls, besides the cultured ones. After removal of the cultured pearls (from these oysters), the flesh is separted from the shell and soaked in slack lime. They are then placed in a wooden vat fitted with rotating wooden blades. The flesh gets pounded into a slurry. Due to the higher density, the pearls settle at the bottom of the vats, which are then collected, washed with neutral soap water and dried.

C. **Gross production**

Some Oysters during post-operative culture period suffer mortality due to natural as well as surgical causes. It has been estimated that 20-30% of operated *P. maxima* die. It has also been seen that 10-15% of operated oysters reject the nucleus even under the most healthy post-operative conditions.

Pearl yield is calculated based on the weight of pearl produced from 10,000 invested nuclei. Generally 10,000 numbers of 6 mm nuclei produce 6.75 kg of pearls for a gross yield of 70%. While 7 mm nuclei produce 9.38 kg of pearls at a gross yield of only 40% as the mortality rate is much higher.

Recent development and future trends in Pearl Culture

Pearl culture is perceptibly changing from a monopolistic industry to a trend of diversification in several parts of the world. Success of this trend depends on development of appropriate technologies which would lead to the production of quality pearls in different species of oysters in different regions. The trends and developments are :

(a) **Science and technology trends** — The future of pearl culture would depend upon the quality rather than quantity of pearl production.

(b) **Resource management** — Oysters are sedentary organisms having restricted distributions and thrive only under certain ecological conditions. So judicious management is required as fishing pressure, and pollution is taking its toll resulting in degeneration of stock. Sea-ranching in recent years has been adopted as a technique of improving the natural stock of oysters.

(c) **Improvement of Mother-Oyster Stock** — Hatchery production of pearl oyster is widely followed to replenish the dwindling stock. Attempts now has centred in identifying the genetic characteristics of a few isolated wild stock and in building a breeding programme on this material. By selective breeding, pearls without yellow pigments can be more effectively produced. Genetic work in pearl oyster has been directed towards producing stocks yielding pearls of desirable quality, resistant to microbial diseases, adaptable to environmental stress and having high growth potential.

Mass production of *P. fucata* in India has been established through hatchery production. Selective breeding of *P. fucata* has been undertaken with regard to colour and quality of nacre.

(d) **Pathological problems** — As oysters are reared in limited areas in high density, there is every chance that diseases would infest and spread fast. This would lead to excessive mortality of oysters (80%) resulting in serious impact upon the industry. The main culprit being marine*Vibrio* bacteria, and control of the disease outbreak becomes impossible. Identification of the pathogen and prophylactic measures should be undertaken. Intensive research in this field is of top priority to safeguard the farm stock.

(e) **Environment of pearl culture**. The quality of pearls produced is influenced by the environmental factors. This has led to the establishment of farms and shifting of culture grounds to conducive environmental areas. However, further investigations are needed to understand the critical aspects of environment and their influence on the physiological function of secretion of the organic and inorganic components of the pearl. The factors that plays a vital role in the colour and texture of pearl (page 351) and better secretion of nacre (page 353) has been discussed earlier. If these factors can be artificially extrapolated in pearl farming then both better quality and higher quantity of pearls can be produced. These will be areas of future research.

(f) **Techniques of surgery** — In the past few years great improvement has been achieved in producing black pearls in *P. margaritifera*. The future challanges to be encountered are (1) to control the mortality of oysters in the post-operative phase, (2) reduce the rate of rejection of nuclei, (3) improve production of quality pearls, (4) to increase the size of the nuclei up to 16 mm diameter and (5) production of round nucleated pearls having size up to 17 mm diameter, in freshwater mussels.

(g) **Bio-mineralisation and spectral characteristics of nacre** — Mineralisation and spectral characteristics of pearl have been the subject of research to understand the formation of organic matrix and crystalline microlayers. A sound knowledge on these is requried which would help in controlling the quality of pearls and these may lead to new dimensions in the future.

(h) **Tissue culture** — Culturing of pearl oyster mantle tissue has been persued for many years. From such tissue culture experiments, it has now been possible to produce pearls by injecting around the implanted nucleus a fraction of the cell suspending liquid obtained from mantle epithelial cell culture. The cells in suspension forms the pearl sac which would secrete nacre leading to pearl formation. If the type of epithelial cells that would be secreting the aragonite crystalline matter, can be isolated, then it might be possible to culture the finest pearls *in vitro*, outside the body of pearl oyster.

(i) **Farming technology** — To make farming of pearl oyster cost-effective and less fatiguable for people working in the farms, it is essential to modify and improve the system, design and materials of farming.

(j) **Organisational set-up** — Pearl industry over the days have become a composite structure that comprises raising of mother-oyster, pearl production, processing and marketing. All these fields have become highly specialised requiring separate technology, skill, capital, manpower and equipment. The advantge is that it can mobilise resources, manage problems with competence and can have greater control over situation. However, the disadvantage is that, if there is any miscalculation in any of the areas, then it can upset the entire system.

(k) **Formation of Pearl Farmers' Co-operative Society** —Young farmers with expertise for pearl culture may organise themselves into a Pearl Farmers Co-operative Society. The society with state financial support will look after input supply as well as output managment.

(i) **Input supply** — It includes the following :

(1) Facilitation of leasing sea area and shore facility.

(2) Supply of oysters.

(3) Supply of farm materials.

(4) Supply of nuclei, chemicals and equipments.

(5) Security arrangements.

(6) Members training arrangements.

(7) Spat supply arrangements with Government organisations (in the initial stages). Collection from the natural beds to be kept as secondary option in seasons of abundance.

(8) Import of shell beads (in the beginning). The society would later set up its own unit.

(9) Society should have its own hatchery in course of time.

(ii) **Output Management** — The society will look after the following :

(1) Purchase the pearls produced by the individual units.

(2) Help in sorting, grading and processing.

(3) Sell the pearls in the domestic market and export them.

(4) The by-products to be properly utilised.

(l) **Policy formation** — Employment generating potential is present in pearl culture and it provides employment for; (i) skilled and semi-skilled persons in the coastal rural areas and (ii) managerial and technical personnel. Pearl adds to the GNP and helps in earning foreign exchange. It is a new industry and pearl culture is a long-term investment, so it will need policy support, financial assistance, subsidy and economic incentive from the Government in the early stages, till it becomes self-sufficient.

USES OF PEARL

(1) Used as an ornament and a symbol of grandeur.

(2) Used in ornamental handicraft.

(3) It is a very precious medicine for surgery, pediatrics, ENT, etc. as used by the Chinese.

BY-PRODUCTS AND THEIR USES

The by-products of pearl culture is referred to the total mass of the oysters after extraction of the cultured pearl. The by-products of pearl culture are, thus, the seed pearls, shell and the flesh.

(1) The tiny seed pearls (which are unsuitable as a gem) are used in the preparation of medicine. In India pearl powder and pearl liquid are important ingredients as follows :

(i) Pearl powder is a highly stimulant tonic and aphrodisiac. Its other medicinal values are — laxative, sedative, emetic and nutritive.

(ii) Pearl powders act as an antacid. It is also used in heart burn and bilious affections.

(2) In case of class 'C' category of pearls (not used in jewellery), the nacreous layer is removed from the nuclei and ground into a powder. It is then dissolved in phosphoric acid. The final product (pearl calcium tablet) is formed by additional chemical processes. The pearl calcium tablet, thus formed, is marketed in Japan for pregnancies, weak bodies, tooth cavities, stomach acids and allergies.

(3) Large oyster shells, for their mother-of-pearl layer, are used in shell craft.

(4) Small and broken shells are ground and used as ingredients in poultry-feed.

(5) The adductor muscle portion of the flesh of oyster is used for human consumption.

(6) After removal of the pearl, the entrails of the oyster are used for feeding fishes.

SPACE FOR NOTES

Chapter Eight : Prawn and Shrimp Culture

INTRODUCTION

The terms "shrimps" and "prawns" have been used interchangeably and the terminology varies from nation to nation. However, they are closely related belonging to crustaceans of the families *Penaeidae* and *Palaemonidae*. The standardisation of the terms was one of the agenda at the "Prawn Symposium" of the Indo-Pacific Fisheries Council held at Tokyo in 1955. The decision arrived at was that; 'prawn' should be applied to the Penaeids, Pandalids and Palemonids and the word 'shrimp' to the smaller varieties belonging to other families. However, in aquaculture literature (T.V.R. Pillay, 2005) prawns are referred to as freshwater forms of Palaemonids and shrimps for the rest, particularly for marine forms.

Prawns are the larger variety and are freshwater and brackish water inhabitants. Shrimps are smaller marine forms. Prawn and shrimps are tasty and highly nutritive, containing vitamins A and D. They are palatable due to the presence of glycogen and free amino acids, with low fat content in their muscles.

Culture of these crustaceans has been an age old practice in ponds and paddy-fields in the form of extensive culture. Intensive and semi-intensive culture of these forms has been a recent technological advancement of aquaculture. This sort of culture was triggered by increased market demand in recent years coupled with inadequacy of the capture fishery landings to meet the demand. In fact, shrimps and prawns have become high-value commodities, mainly because of their export market. Due to the prospects associated with exports, private sectors with support of the government have resorted to such culture practices.

DIFFERENCES BETWEEN PENAEID AND NON-PENAEID

In India, commercial prawns are grouped into Penaeid and non-Penaeid ones. The important distinguishing characters of these two groups are listed below :

Penaeid	Non-Penaeid
(1) The pleurae of the second abdominal segment overlap the pleurae of the first segment only. (Fig. 8.1a).	The pleurae of the second abdominal segment overlap the pleurae of the first and third segments. (Fig. 8.1b)
(2) The first three thoracic legs are chelate.	The first two thoracic legs are chelate.
(3) Male has petasma* (Fig. 8.2a) for transferring sperms. Female has thelycum‡ (Fig. 8.2b).	Such organs are absent in males and females.
(4) The females lay eggs individually in water. Fertilisation and development take place outside.	The females carry eggs between their pleopods as a cluster.
(5) The penaeid catch along the Indian coast is a little over 50%, the bulk of which comes from the coast of Kerala.	The non-penaeid catch along the Indian coast is a little less than 50%, of which 90% are hauled from Maharashtra and Gujarat coasts.

* **Petasma** : The male endopodites contain hooks and the hooks of the two sides interlock to form a rod-like structure called petasma.

‡ **Thelycum** : It is a structure in the female into which sperms are transferred through petasma of males.

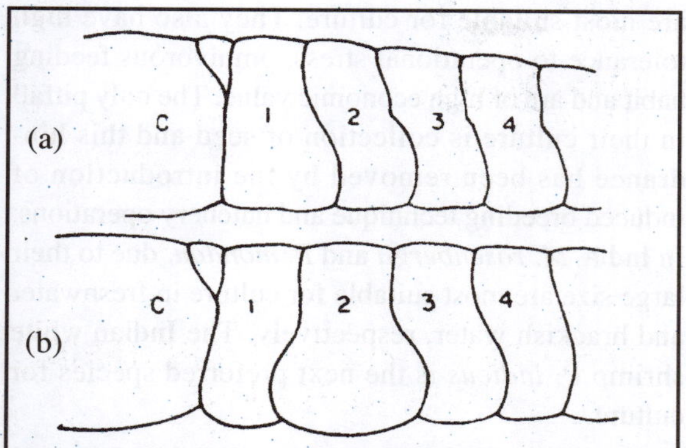

Fig. 8.1. Exoskeleton (sketch) of (a) Penaeid and (b) Non-Penaeid prawn (C, carapace; 1 to 4, first four abdominal exoskeleton)

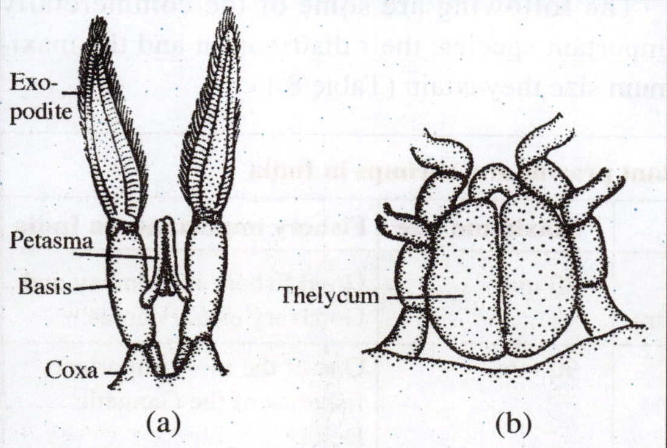

Fig. 8.2. (a) First pleopod of male penaeid showing petasma (b) Thelycum of *Penaeus indicus.*

IMPORTANT SPECIES OF PRAWNS AND SHRIMPS

There are about 150 species of freshwater prawns all over the world, of which only 40 species are found distributed in the rivers, reservoirs and estuaries of India. Some important species are *Macrobrachium rosenbergii, M. malcolmsonii, M. idae, M. mirabile, M. rude, M. lamarrei, M. scabriculum, M. villosimanum* and *M. birmanicum choprai.* Among these, *M. rosenbergii,* the giant freshwater prawn, has been found to be most suitable for culture, owing to its faster growth rate, higher tolerance to wider range of salinity and temperature, and less cannibalistic tendency.

Among the *Penaeus* species, the important ones cultured in Asia are tiger shrimp *Penaeus monodon,* the Indian or white shrimp *Penaeus indicus* [Fig. 8.3(a)], the banana shrimp *Penaeus merguiensis* [Fig. 8.3(b)], the green tiger or bear shrimp *Penaeus*

Fig. 8.3. (a) *Penaeus indicus*

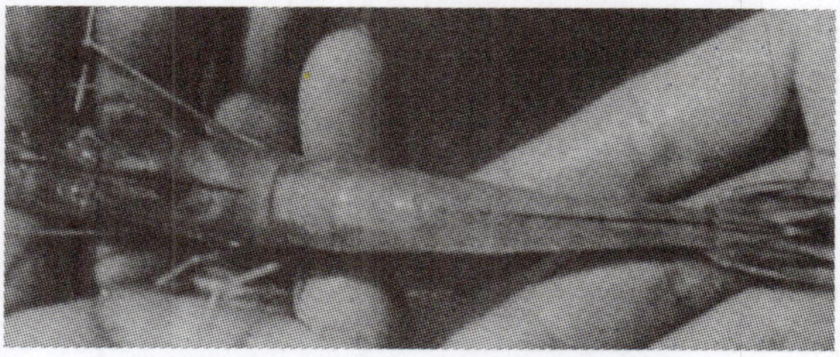

Fig. 8.3. (b) *Penaeus merguiensis*

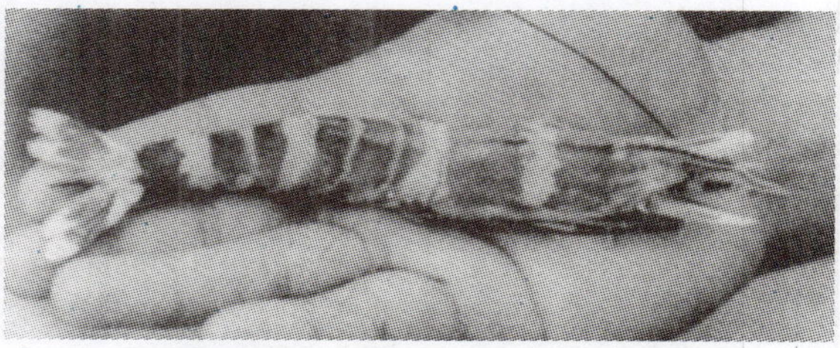

Fig. 8.3. (c) *Penaeus semisulcatus*

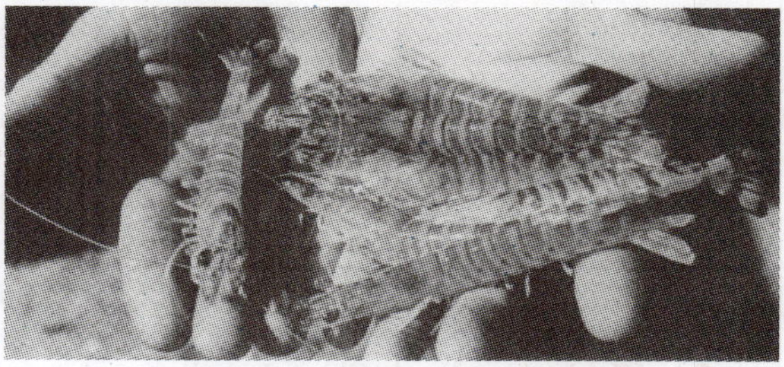

Fig. 8.3. (d) *Penaeus japonicus*

Fig. 8.3. (e) *Penaeus vannamei*

semisulcatus [Fig.8.3(c)], the oriental shrimp *Penaeus orientalis* and the Kuruma shrimp *Penaeus japonicus* [Fig.8.3(d)]. The species cultured in Taiwan is the red tailed shrimp *Penaeus penicillatus.* Among the *Metapenaeus* shrimp species cultured in several Asian countries are *M. monoceros, M. dobsoni, M. affinis, M. brevicornis* and *M. ensis.*

Some important shrimp species of economic value, also cultured outside Asia, are the Kuruma shrimp *Penaeus japonicus* [Fig.8.3(d)], the white leg shrimp *Penaeus vannamei* [Fig. 8.3(e)], the blue shrimp *Penaeus stylirostris,* the Mediterranean shrimp (triple-grooved shrimp) *Penaeus kerathurus, Penaeus notialis,* the brown shrimp *Penaeus aztecus,* pink shrimp *Penaeus duorarum*, etc.

Prawn and Shrimp, due to their high growth rate (marketable size can be attained by 3 to 6 months), are most suitable for culture. They also have high tolerance to operational stress, omnivorous feeding habit and are of high economic value. The only pitfall in their culture is collection of seed and this hindrance has been removed by the introduction of induced breeding technique and hatchery operations. In India, *M. rosenbergii* and *P. mondon,* due to their large size are most suitable for culture in freshwater and brackish water, respectively. The Indian white shrimp *P. indicus* is the next preferred species for culture.

GENERAL DISTRIBUTION, MAXIMUM SIZE AND FISHERIES IMPORTANCE OF PRAWNS AND SHRIMPS IN INDIA

The following are some of the commercially important species, their distribution and the maximum size they attain (Table 8.1).

Table 8.1 : Some commercially important prawns and shrimps in India

Name	Distribution	Maximum size	Fishery importance in India
(1) *Hippolysmata ensirostris*	India (along the coasts) Myanmar, Sri Lanka and Sumatra	80 mm	Good fishery in Bombay and Godavary estuary areas.
(2) *Palaemon styliferus* (Rushna chingri)	Pakistan to Malay Archipelago. In India, present along the northern regions of both coasts.	90 mm	One of the most important fisheries of the Gangetic Delta.
(3) *Palaemon tenuipes*	India (northern regions of both coasts) to New Zealand through Malaysia.	80 mm	Another important fisheries in Bombay and Gangetic Delta.
(4) *Macrobrachium malcomsonii*	India and Myanmar. Present in the peninsular rivers of India and migrates into brackish waters during breeding season.	Male-230 mm Female-200 mm	Fairly good fishery during monsoon in North-East Coast.
(5) *Macrobrachium rosenbergii* (Golda chingri)	Wide distribution, it extends from the Indo-Pacific zone up to Indo-China. In India common in lakes and estuaries.	320 mm	Very good fishery during the monsoon and post-monsoon months, particularly in Kerala.
(6) *Macrobrachium rude*	India (South-West Coast and East Coast), East Africa, Madagascar and Sri Lanka.	130 mm	Good seasonal fishery in Orissa, Bengal and Andhra Pradesh.
(7) *Heterocarpus wood-masoni*	In India present in Andaman and Nicobar Islands and the coast of Kerala.	130 mm	Has potential of commercial importance. It is a predominent species in deep-water trawling.

Table 8.1 contd. to next page

Table 8.1 : Some commercially important Prawns and Shrimps in India (continued from previous page)

Name	Distribution	Maximum size	Fishery importance in India
(8) *Parapandalus spinipes*	In India it is recorded only in the South-West Coast. It is also present through Red Sea, Gulf of Aden and Arabian Sea.	130 mm	Has potential of commercial importance. Present in good numbers in deep-water trawling (150 to 200 fathoms) off Kerala coast.
(9) *Plesionika martia*	In India found both in the Arabian sea and Bay of Bengal. Widely distributed through Indian seas to Japan, Australia and Hawaiian Islands.	125 mm	—do—
(10) *Aristeus semidentatus*	Soth-West Coast of India. Present in Kermadec Islands, Kei Islands and Arabian Sea.	150 mm	Has potential of commercial importance. Present in good numbers in deep-water trawling (150 to 200 fathoms) off Kerala coast.
(11) *Atypopenaeus stenodactylus*	Indian seas (West and East Coasts) through Malaysia, Hong Kong to Japan.	50 mm	Caught in large numbers all throughout the year from Mumbai coast. In East Coast caught in small numbers.
(12) *Metapenaeus affinis*	Found along the coasts of India. Indian seas to Malaysia, parts of Indonesia to Hong Kong and Japan.	180 mm	Very important fisheries along both the coasts. Juveniles are caught from estuaries.
(13) *Metapenaeus brevicornis* (Chamne chingri)	Found along the northern region of both coasts. Present from Pakistan through India, Malaysia, Thailand, Indonesia to East Borneo.	125 mm	—do—
(14) *Metapenaeus dobsoni*	Common along South-West Coast of India. Present in India through Malaysia, Indonesia to Philippine Islands.	125 mm	Important fishery along the South-West Coast. Juveniles caught from estuaries and river mouth.
(15) *Metapenaeus monoceros* (Honne chingri)	Along the entire coastline of India. Present in Mediterranean and Indian seas to Malaysia.	180 mm	Important fishery of commercial importance.
(16) *Parapenaeopsis hardwickii*	Bombay area and Godavary estuary. From Indian waters through Malaysia to South China	120 mm	Good fishery along Mumbai coast and in Godavary estuary.
(17) *Parapenaeopsis stylifera*	India (all along the coast line), Sri Lanka, Malaysia to Indonesia and Borneo.	140 mm	Commercially important species along the entire West Coast.
(18) *Penaeopsis rectacuta*	Madras, Kerala Coast and Andaman and Nicobar Islands of India. Gulf of Aden through India, Malaysia, Philippines, South China sea and Japan.	130 mm	Commercially important species, obtained in large quantities of Kerala coast (100-200 fathoms).

Table 8.1 contd. to next page

Table 8.1 : Some commercially important Prawns and Shrimps in India (continued from previous page)

Name	Distribution	Maximum size	Fishery importance in India
(19) *Penaeus indicus* (Chapda chingri)	All coastal waters of India. Found in India, Sri Lanka, Gulf of Aden, Madagascar, East Coast of Africa, Malaysia, Indonesia, Philippines, New Guinea and Northern Australia.	230 mm	Most important commercial species along all Indian coast and also in estuaries.
(20) *Penaeus monodon* (Bagda chingri)	In India more common along the East Coast (Bengal and Orissa) and all along West Coast. It is present from South Africa to Southern Japan.	320 mm	Very important commercial species particulary in Bengal and Orissa. Present in West Coast but not as dominant as in the east.
(21) *Penaeus semisulcatus* (Hende bagda)	Widely distributed in Indo-West-Pacific in India, Malaysia, Indonesia, Australia (Northern and North-Eastern), New Guinea, Philippine Islands to Southern Japan. Common on the East Coast of India.	250 mm	Small fishery in East Coast. Very common in the 'bheri fisheries' of West Bengal due to its tolerance to wide salinity range.
(22) *Solenocera hextii*	Bay of Bengal (65 to 276 fathoms) and along entire West Coast in India.	140 mm	Large size is attractive commercially and caught in varying numbers along West Coast.
(23) *Acetes erythraeus*	Found in East and South-West Coast of India. Red sea through Bay of Bengal to Malay Archipelago.	30 mm	Fairly good fishery in West Bengal, Orissa and T. Nadu. Caught in large numbers (December to April) in Thiruvanantapuram coast.
(24) *Acetes indicus*	In India, common along the northern area of West Coast and in the East-Costal areas and eastuarine regions. Present along India through Gulf of Siam to Malaya and East Indies.	40 mm	Good fishery along North-West Coast and East Coast. In Mumbai it contributes 20% of shrimp fisheries.

Note : *Names mentioned within the bracket are colloquial Bengali names*

I. CULTURE OF GIANT FRESHWATER PRAWN

The giant freshwater prawn (Fig. 8.4), *Macrobrachium rosenbergii* (De Man, 1879), is economically important, as it is a fast growing delicious species with considerable export potential. The males and females attain a size of about 25 cm and 15 cm, respectively. Although the adults of this prawn thrives in freshwater and brackish water areas, its larval rearing requires water of about 12 ppt salinity. This requirement of larvae has caused immense problem for hatchery and grow-out stages of culture. However, to some extent it has been overcome by the use of sea water or brine, trucked in from the nearest source or by the use of artificial sea water.

Macrobrachium rosenbergii is widely distributed in South East Asia besides the Indian subcontinent. It is either reared alone (monoculture) or introduced as a component in carp culture system. One of the suitability of culture of this species is that, the adults are omnivorous and feed on a variety of foods of plant and animal origin.

Fig. 8.4. Giant freshwater prawn, *Macrobrachium rosenbergii*

A major constraint of large scale commercial culture of this giant freshwater prawn is the nonavailability of seed, as this is not abundant in nature. Availability of quality seed, suitable feed and obtaining optimum sustainable yield without environmental degradation are high technological inputs that have been given priority in the past two decades, for farmers to undertake large scale commercial aquaculture ventures in the country. The semi-intensive culture technology (page 217) of this giant prawn has been standardised and encouraged for culture as farmers can obtain optimum sustainable yield without environmental degradation.

Biochemical composition : Biochemical composition (%) of adult giant fresh water prawn are as follows :

Moisture 78.29	Fat 0.27	Calcium 0.074	Sodium 0.182	Iron 0.0012
Protein 21.17	Ash 0.37	Phosphorus 0.1799	Potassium 0.2986	Dry matter 21.71

Distribution : *M. rosenbergii* is present in rivers, lakes and estuaries of India. It is found distributed from lower to upper reaches of rivers where there is tidal influence. This species prefers shallow and muddy environment. The ecological limits in distribution are water temperature of 25-34°C and salinity 0-20 ppt.

Habit and habitat : *M. rosenbergii* is nocturnal in habit. During the day time it hides in shades and shelters. It is sluggish, but shows territorial behaviour. It is pugnacious in nature, but shows less cannibalistic tendency.

Biology and Life history

The life cycle of the giant freshwater prawn *M. rosenbergii* is given in in Figure 8.5.

(a) **Differences between mature male and female prawns**

The sexual differences between matured male and female giant freshwater prawns are given in Table 8.2.

(b) **Food and feeding habits**

The giant freshwater prawn is omnivorous, bottom feeder and more importantly it responds positively to supplementary feeds. When it is extremely hungry it resorts to cannibalistic activity.

(c) **Growth and Maturity**

The growth rate of this prawn is fast, showing sexually dimorphic pattern. The males grow faster with heterogenous growth rate. In five months, it

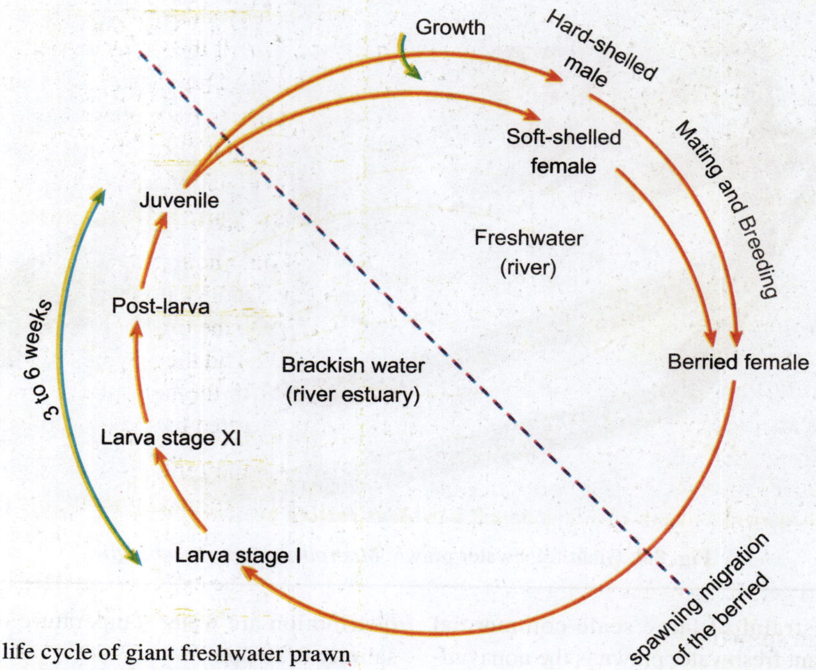

Fig. 8.5. The life cycle of giant freshwater prawn (*M. rosenbergii*)

Table 8.2 : Identification characteristics of mature male and female giant freshwater prawn (*M. rosenbergii*)

Mature male	Mature female
1. Mature male prawns are larger in size.	1. Mature female prawns are comparatively smaller, of some age group.
2. Abdominal region is narrower.	2. Abdominal region is broader than that of male.
3. The second walking legs are considerably larger and thicker (Fig. 8.6c)	3. The second walking legs are thinner and narrower (Fig. 8.6f).
4. The cephalothorax is proportionately larger.	4. Cephalothorax is comparatively smaller.
5. At the base of each fifth leg is present male genital opening (male gonopore) guarded by lid.	5. At the base of each third leg is present female genital opening (female gonopore) without lid.
6. Gonad is a narrow greyish-coloured mass.	6. A ripe 'berried' female possesses large orange coloured masses of eggs.
7. A limb-like structure of the copulatory organ (Appendix masculina) present at the base of the endopodite of the second pleopod (Fig. 8.6b).	7. Absent in females (Fig. 8.6e).
8. There is presence of a lump or hard point in the centre on the ventral side of the first segment of the abdomen (Fig. 8.6a).	8. No such structure present in female (Fig. 8.6d).
9. No modifications present in abdominal pleura.	9. The abdominal pleura become distended outwards to make space for holding eggs (popularly known as "brood-pouch").

attains average growth of 40-50 gm under scientific culture. In the size group of 40-60 gm, the tail of both the sexes constitute about 55% of the total weight.

This prawn attains first maturity when it is 136-150 mm in length, weighing about 35-40 gm, within 4-6 months under captivity. In a sexually ripe female prawn, the fully developed ovary is an orange-coloured mass.

(d) **Moulting**

M. rosenbergii stops feeding when moulting time gets nearer. At this time, it becomes sluggish, its eye colour becomes dull and the shadow of newly-formed shell appears in the rostrum. Moulting gets completed within ten minutes. The frequency of moulting is irregular and is dependent upon several factors, such as nutrition, age and stage of sexual maturity. Generally the interval of moulting is 20 to 40 days. The shell gets hardened within 2-6 hours.

(e) **Mating and spawing**

The female prawns before mating moults and this moulting is termed as **pre-mating** or **pre-spawning moult**. Mating can be observed by introducing a mature male and a sexually ripe female (3 to 6 hours after pre-mating moult) in an aquarium tank containing freshwater. The female is held strongly by the male between its chelate legs. The male then cleans the ventral part of the thoracic shell, which takes 10 to 15 minutes and then places the female ventral side up. It then presses from above and ejects the sperm through its genital pores in one gelatinous mass on the ventral thoracic region of the female. The male then releases its grip and the female swims away.

Six to twenty hours after mating, the female lays eggs and the whole batch of eggs is laid in 20 minutes. The eggs at the time of extrusion are considered to be fertilized and they are held within the pleopods in bundles with the help of extremely thin elastic membrane secreted by the ovigerous setae.

Unmated ripe females also lay eggs within 24 hours after the pre-mating moult. As the eggs are not fertilised, they drop off in 2 or 3 days.

The size of the egg is about 0.6 mm to 0.7 mm in diameter and weighs about 0.1 mg. It is slightly oval in shape. A medium sized prawn of 180 cm length and 80 gm weight lays about 60,000 eggs. Larger ones, depending upon its size, lay about 1,00,000 to 1,70,000 eggs. In tanks, mature female prawns, spawns twice in five months. In natural condition, however, 3 to 4 spawning may occur in one year. The

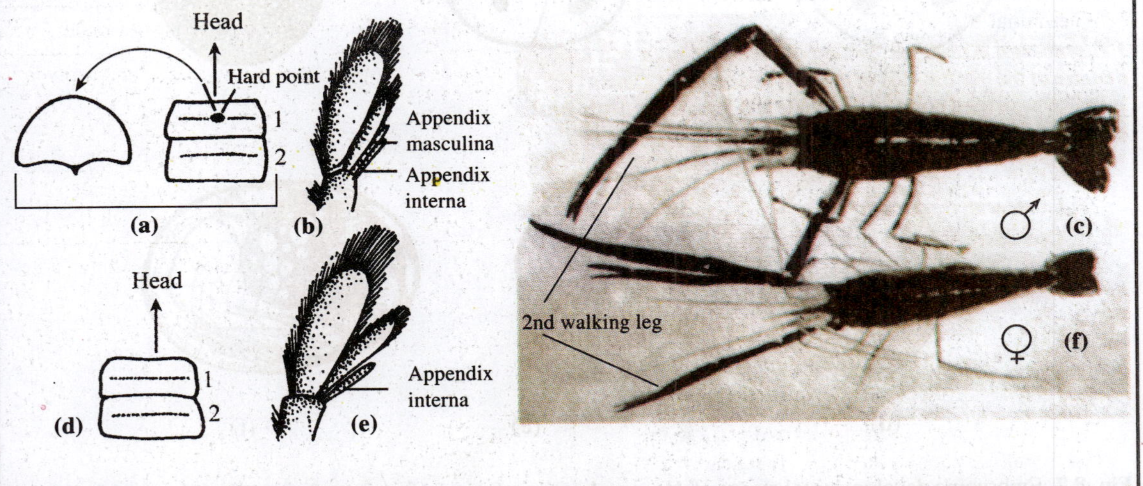

Fig. 8.6. Sexual dimorphism of male and female of *M.rosenbergii* — **(a)** Hard point in first abdominal segment of male; **(b)** Male copulatory organ (appendix masculina) present at the base of the endopodite of second pleopod; **(c)** Mature male; **(d)** Abdominal segment of female; **(e)** Sccond pleopod in female and **(f)** Mature female.

species have an average fecundity of 500 larvae/gm body weight of mother prawn.

(f) **Embryonic development**

Berried females actively migrate to estuarine regions for the purpose of hatching their young. The incubation period is 18-21 days and requires a temperature of 28-30°C. The eggs are carried by the females during this period. They aerate the eggs by beating their pleopods.

Fertilised eggs have a homogenous granulation and are bright orange in colour. Cell division begins 4 hours after fertilisation. Development continues (Fig. 8.7) and the bright orange colour of the eggs begins to fade to a pale-grey and finally darken to slate-grey prior to hatching. On the 19th or 20th day after fertilisation the larva hatches out. During the time of hatching, the mother prawn vibrates her pleopods at intervals to disperse the eggs.

(g) **Larval development**

The newly hatched larvae (Zoea) are planktonic and attracted to light (not bright illumination). They require brackish water for their survival and metamorphosis. In nature larvae hatch both in freshwater and brackish water. Those in freshwater are carried to estuaries by water currents, otherwise in freshwater they will die in 4 to 5 days.

The larvae are active swimmers. They swim tail first, ventral side upwards at an oblique angle. Early larvae are gregarious in habit. They swim together and in 10 days time the gregariousness gradually disappears. Larvae are omniovorous, feeding on zoo- and phytoplankton and other organic foods like bits of flesh, prawns, molluscs as well as grains, seeds, fruits, etc. Liquid food is not taken.

Larvae passes through eleven moults and the completion of larval cycle takes about 30 days depending upon the temperature and feeding conditions. The different larval stages are given in Fig. 8.8 and their recognised characters are depicted in Table 8.3. The total length of different larval stages range from 1.92 mm (stage I) to 7.73 mm

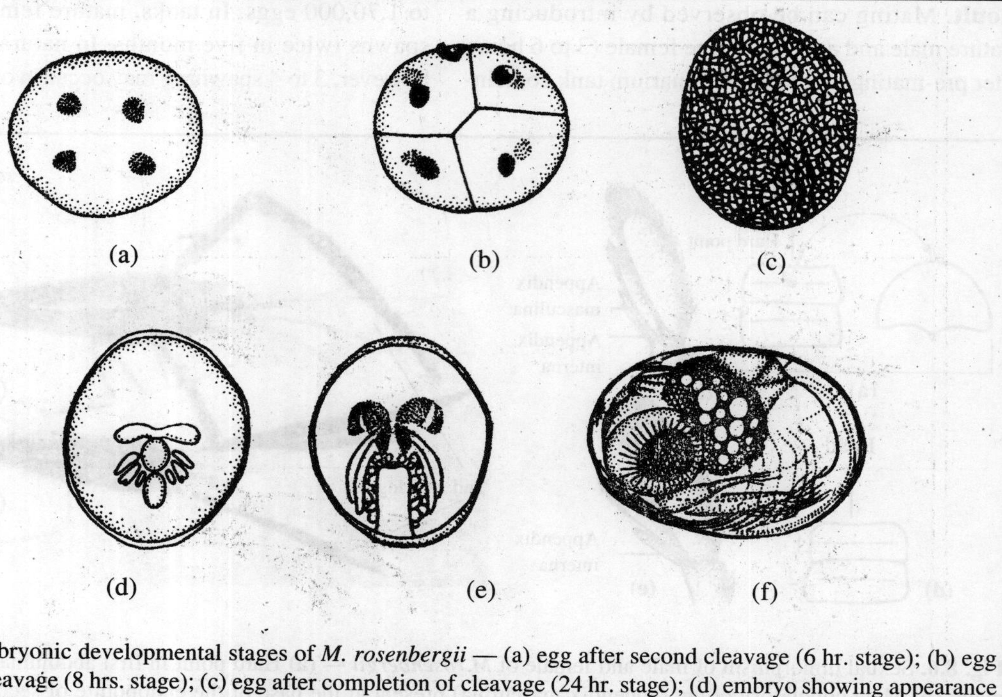

Fig. 8.7. Embryonic developmental stages of *M. rosenbergii* — (a) egg after second cleavage (6 hr. stage); (b) egg after third cleavage (8 hrs. stage); (c) egg after completion of cleavage (24 hr. stage); (d) embryo showing appearance of post-nauplius appendages (5th day); (e) embryo with optic rudiment (8th-9th day) and (f) prior to hatching (19th-20th day).

(stage XI), while the newly metamorphosed post-larvae measures 7-8 mm (Fig. 8.8).

(h) **Post-larvae, Juveniles and Adults**

Juveniles emerge when the last stage larvae metamorphose and cease their pelagic life. They sink down and settle at the bottom as crawler and clings to vegetation and submerged objects. They feed on organic matter as well as on small worms, crustaceans and insect larvae.

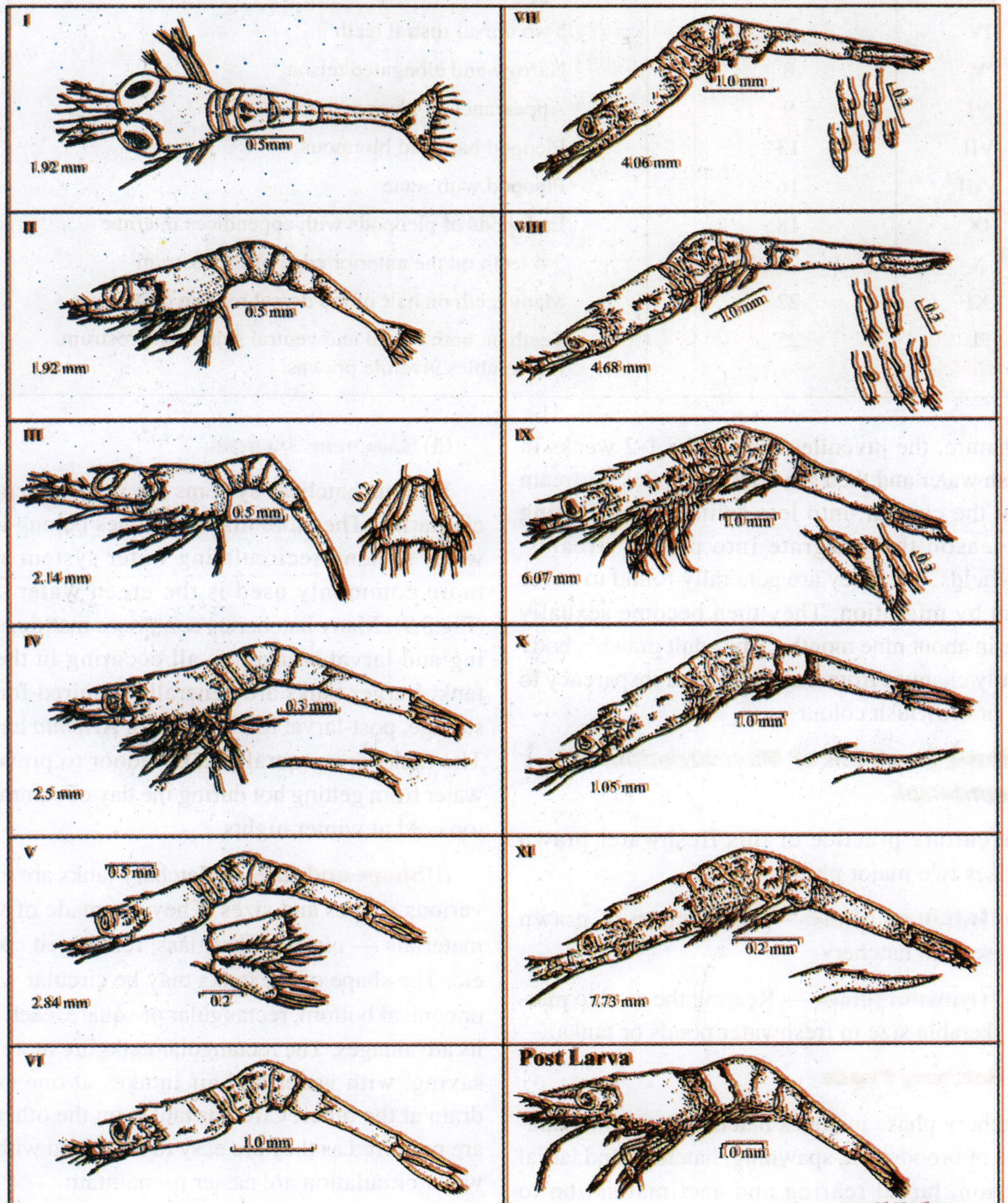

Fig. 8.8. Various larval and post-larva stages of *M. rosenbergii*

Table 8.3 : Characterisation of larval stages in *M. rosenbergii* (PL : Post-Larva)

Larval stages	Age (in days)	Recognisable characters
I	1	Sessile eyes
II	2	Stalked eyes
III	4	Appearance of uropod
IV	6	Two dorsal rostral teeth
V	8	Narrow and elongated telson
VI	9	Appearance of pleopod bud
VII	13	Pleopod bare and biramous
VIII	16	Pleopod with setae
IX	18	Endopods of pleopods with appendices internae
X	20	3-4 teeth on the anterior edge of the rostrum
XI	22	Many teeth on half of the dorsal margin of rostrum
PL	25	Teeth on both dorsal and ventral side of the rostrum. It resembles juvenile prawns.

In nature, the juveniles remain for 1-2 weeks in brackish water and then gradually migrates upstream against the current, into less saline water. During rainy season they migrate into ponds, streams, paddy-fields, etc. They are generally found to cover 200 km by migration. They then become sexually mature in about nine months. The adult prawn's body gradually changes from their original transparency to bluish or brownish colour.

Culture practices of *Macrobrachium rosenbergii*

The culture practice of this freshwater prawn comprises two major phases :

(A) **Hatchery phase** — Production of prawn seed in hatchery.

(B) **Grow-out phase** — Rearing the seed to marketable size in freshwater ponds or tanks.

(A) Hatchery Phase

Hatchery phase includes hatchery system, management of broodstock, spawning, hatching and larval collection, larval rearing and acclimatisation to freshwater under controlled conditions in the hatchery. A nursery system may also be introduced before stocking in grow-out ponds.

(a) **Hatchery systems**

Various hatchery systems are in use in commercial farms. The more important ones belong to clear water system, recirculating water system and the more commonly used is the green water system (Fig.8.9). Many hatcheries comprises mating, spawning and larval rearing — all occuring in the same tank. Larger tanks are generally required for water storage, post-larval rearing and for *Artemia* hatching. The tanks are generally kept indoor to prevent the water from getting hot during the day of summer, and too cold at winter nights.

(i) **Shape and size** — Hatchery tanks are made of various shapes and sizes. They are made of various materials — plastic, fibreglass, reinforced concrete, etc. The shape of the tanks may be circular with flat or conical bottom, rectangular or square, each having its advantages. The rectangular tanks are more space-saving, with water and air intakes at one end and drain at the other. Circular tanks, on the other hand, are preferred as they are easy to clean and with better water circulation are easier to maintain.

(ii) **Colouration** — The tanks are painted with dark colours so that the food can be well-seen by the larvae.

Fig. 8.9. The green water system of algal production in a prawn hatchery

(iii) **Aeration** — An aeration system is attached to the bottom of the tank to facilitate proper mixing of the water and even distribution of the food particles and also the larvae. The even distribution of the larvae helps in reducing cannibalism.

(iv) **Water management** — Water management procedures of larval culture refer to terms such as green water, clear water and recirculating water systems.

(1) **Green water system** — As the name specifies, the green water system (Fig. 8.9) involves the production of phytoplankton culture dominated by *Chlorella*. To maintain the plankton bloom, a fertiliser mixture of 4 parts urea and 1 part NPK (Nitrogen : Phosphorus : Potassium = 15 : 15 : 15) in tap water, is added to the tank at least once a week. Production of **rotifers** is also maintained by the additon of $CuSO_4$ (0.6 ppm) into the tank water. The **salinity** of the green water should not be more than 12 ppt. For water exchange of larval tanks, only green water of less than three days old is used.

The function of green water is to maintain the water quality in the larval tanks by recycling the waste products of larval metabolism. It provides food for live food organisms such as *Artemia*. The only drawback in this system is that only a lower stocking density is feasible.

(2) **Clear water system** — This system is adopted for different level of larval production, where circular tanks with conical bottoms (Fig. 8.10) or rectangular tanks are used. Chlorination followed by dechlorination, UV light, strong aeration and use of antibiotics are the measures taken to maintain the water quality. In this system high stocking density is feasible, but it requires high management skills.

(3) **Recirculating water system**—This system is valuable for water conservation and use of energy. The tank water is recirculated through a graded sand/ gravel filter, attached to mechanical or airlift pumps. The water quality is maintained by involving mechanical and biological filtration, chlorination and dechlorination or the use of ozone. However, com-

Fig. 8.10. Clear water system used in prawn hatchery, comprising circular tanks

mercial application of such a system has not gained much importance in prawn culture.

(b) **Management of broodstock**

One important aspect in hatchery management is the availability of berried prawns over an extended period, probably round the year. For this, it is essential to develop viable broodstock in captivity. This involves the following steps :

(1) Freshwater ponds are prepared in a manner similar to that practised under carp culture system. It is fertilised with cattle manure, 15 days before stocking of prawn seed.

(2) Post-larvae or early juveniles of prawns are obtained from the market and stocked in the ponds.

(3) The juveniles are fed with high quality ingradients in the form of pellets, which is broadcasted throughout the pond to enable all the prawns to feed. In absence of pelletised feed, supplementary feed of plant and animal origin is given.

(4) Within 5-7 months of culture, prawns attain maturity in freshwater pond depending on temperature, food and ecobiological conditions.

(5) Maturation takes place in freshwater. Berried prawns are collected by means of a cast net from the pond for hatchery purpose.

Berried females can also be obtained from other farmers initially. In such cases they are brought to the farm in open containers with fresh- or brackish water or in case of long distance transportation they are packed under oxygen pressure at the rate of 2 to 5 prawns per bag.

On arrival at the hatchery, the berried prawns are disinfected by placing them in a container containing 0.3 ppm of copper sulphate or 20 ppm of formalin for 30 minutes.

(c) **Spawning**

The berried prawns are released into a tank at the rate of 5-10/m^2 depending upon the size. The tanks are filled up with brackish water of 5 ppt salinity for efficient hatching. The hatchery tank should be cleaned every day with daily exchange of water of identical salinity.

Mating takes place between hard-shelled males and ripe soft-shelled females. The females are soft-shelled as they have just completed their pre-mating moult. The pleopods of the female are vigorously moved to keep the fertilised eggs well-aerated. The incubation period varies between 18 to 23 days at a temperature of about 28°C. The ovarian eggs of the berried females, even during the incubation period, may start maturing and a second pre-mating moult can occur within a period of about 3 weeks.

Healthy berried female prawns are selected from either natural sources, rearing ponds, brood tanks or aquaria and introduced into the larval tanks. At the time of selection it is important to ensure that all the females have black or grey eggs. This would ensure that all the eggs hatch out within two or three days and the larvae in the tanks will be of the same age.

(d) **Hatching and larval collection**

When all the eggs have hatched, the spent females are removed from the larval tank with the help of a coarse mesh dip net. Larval release from the pleopod of the female prawns occurs over a period of four days, with a peak during 24 to 72 hours. Therefore, the females have to remain in the larval tank for a period of 4 days. The salinity of the water is then raised to about 12 ppt, with an optimum temperature range of 26-31°C. The tank is aerated to maintain oxygen levels nearly to saturation point. In case of clear-water and recirculating system hatcheries, the tanks are not exposed to direct sunlight.

In farms where larvae hatch out in the hatching tank, they are removed and stocked in the larval rearing tank. The larvae after hatching in the hatching tank are attracted to one corner by means of a light and they are siphoned into a container with a known volume of water. The total number of larvae in the container is estimated and then stocked in larval rearing tanks.

(e) **Larval rearing**

The newly hatched larvae (zoea) are planktonic and require brackish water for their survival and metamorphosis. They undergo eleven stages of metamorphosis (Fig. 8.8 and Table 8.3). After 16-28 days (depending upon the temperature) the post-larval stage is achieved and the larvae are harvested for stocking in grow-out or nursery ponds.

(i) **Water intake and treatment system** — To achieve optimum larval survival and seed production, it is essential that the sea water and freshwater used should be clear and free from pollution. The sea water collected is first stored in cemented tanks and chlorinated with bleaching powder (20 gm/1000 litres of water) to kill all the micro-flora and -fauna. The water is then allowed to settle for a week and then used for preparation of larval medium. The medium is then passed through 5-10 micron filter bags and introduced into larval rearing tanks. The water from time to time should be checked for suitable hardness (50-100 ppm), iron (below 0.1 ppm) and level of heavy metals.

(ii) **Larval stocking** — Larval stocking and rearing can be done following two-phase larval rearing system utilising cylindro-conical tanks in conjunction with the large tanks having flat or U-shaped bottom. This system is suitable for obtaining high survival and production of post-larvae.

In the first phase, early larvae (stages I to V or VI) are reared in cylindro-conical tanks with high density stocking of 500 to 700 larvae per litre.

In the second phase of rearing, the advanced larvae stage of V or VI are collected from the first phase of rearing and stocked in large tanks at the rate of 50 to 80 larvae per litre.

(iii) **Larval feed**

(1) **Feeding of brine shrimp** (*Artemia* sp.) — Freshly hatched *Artemia* nauplii are supplied as food, to the prawn larvae in its early stages (stages II to VI). Feeding is done twice a day and for the later stages, once during late evening in combination with prepared feeds.

The cysts of brine shrimp are hatched in a conical tank whose capacity usually ranges between 100 and 500 litres. Hatching instruction of the cysts are normally given by the suppliers. But certain conditions are required for optimum hatching which is given below :

Salinity (ppt)	: 12-35 (depending on different *Artemia* strains).
Temperature	: 28-35°C
Light (Lux)	: 2000
Aeration	: vigorous and continuous
Dissolved oxygen (ppm)	: 2–8
Stocking density of the cysts (gm/L)	: 1 to 2

The *Artemia* nauplii after hatching are disinfected with 70 ppm of formalin after which they are fed to the prawn larvae.

(2) **Prepared feed** — A number of feed ingredients of animal and vegetable origin are combined variously to prepare diets for feeding the larvae. The feed ingredients are mixed together with a little water, in an electric blender or mixer to form a bolus which is then steamed for 30 minutes. It is then cooled and stored in a refrigerator. The refrigerated feed should be utilised within four days. The ingredients and its compositon used for formulating a diet suitable for feeding prawn larvae are given below :

Fish flesh	:	68%
Hen's egg	:	20%
Wheat flower	:	05%
Skimmed milk powder	:	05%
Vitamin and mineral mix	:	02%

(iv) **Method and feeding schedule** — For high survival and metamorphosis, proper feeding level and feed formulation is an important factor. At the 2nd or 3rd day after hatching, feeding of the prawn larvae with brine shrimp larvae is initiated. The *Artemia* nauplii is given preferably in the evning, so as to ensure the presence of food in the medium all throughout the night. At later stages of development, the prawn larvae are given prepared feed, seived through a mesh to obtain suitable particle size. This is given during the day time at an interval of 2-3 hours.

(v) **Cleaning of larval rearing tank** — Larval rearing tank should be properly cleaned to remove the excess food particles and metabolic wastes that have settled at the bottom of the tank. This is done by siphoning daily after stopping aeration and preferably in the evening hours, prior to exchange of water and introduction of brine shrimp nauplii.

(vi) **Water quality management during larval rearing** — Water quality in hatching deteriorates due to accumulation of water and unutilised feed. So, the main problem in modern hatchery management is the maintenance of good water quality. The optimum water quality parameter to be maintained is given below :

Water temperature (°C)	: 29-31
pH	: 7.0 – 8.0
Salinity (ppt)	: 10 – 12
Dissolved oxygen	: at saturation level
Nitrite (NO_2-N) (ppm)	: < 0.1
Nitrite (NO_3-N) (ppm)	: < 20
Ammonia (NH_3-N) (ppm)	: < 0.1
Ammonia (NH_4-N) (ppm)	: < 1.5

The water quality in larval rearing tank is maintained through exchange of 50% of water everyday. Water quality in the larval rearing tank is also maintained by means of recirculation through biological filter using materials such as gravel, activated charcoal, molluscan shell, etc. **Biological filtration** is a process of mineralisation of organic nitrogenous compounds, nitrification and denitrification of bacteria, suspended in the water and attached to submerged objects in the filter bed.

(f) Acclimatisation to freshwater

The post-larvae are acclimatised gradually to freshwater. During this period aeration should be continued and after acclimatisation the larval feed is stopped. The post-larvae are then fed on a variety of food such as small shrimp (*Acetes* sp.), worms, crushed maize, broken rice, molluscan meat, tapioca, etc. After a week or two, the post-larvae become suitable for stocking in grow-out ponds. If they are reared for more than two weeks, then over-crowding will take place and the survival rate will decrease.

(g) **Introduction of Nursery system of rearing**

A nursery system of rearing may be introduced before stocking in grow-out ponds, where newly metamorphosed post-larvae of *Macrobrachium rosenbergii* are reared to a juvenile stage (25 mm).

Shallow freshwater ponds of one metre depth and a water area of 200 to 500 m^2 are preferably used. The pond bottom should be made of clay or clay-loam. The pond should be prepared by liming and use of fertilisers (cattle manure, urea, single super phosphate), at least 15 days prior to stocking.

The quality of pond water should be maintained and should be free from toxic chemicals and other pollutants. Some important water quality parameters to be maintained for successful rearing are as follows :

Water temperature (°C)	:	28-32
pH	:	7.0 – 8.5
Total hardness (ppm)	:	40 – 100
Dissolved oxygen (ppm)	:	6 – 4

The post-larvae are stocked at a density of 50-100/m^2 and reared for one to two months, by which through proper feed, they attain a size of 0.5 to 1.0 gm. Suitable foods given are :

(1) Living aquatic animals—small aquatic worms, insect larvae.

(2) Small pieces of clams, snails, squids, shrimps (*Acetes* sp.).

(3) Plant material—broken rice, other grains, tapioca, sweet potato, peas, oil cakes, etc.

Formulated feed (pellets) are also recommended at the rate of 20% of post-larval body weight. The feed is given 3 to 4 times a day and is broadcasted all over the pond. With proper feeding regime the survival rate is 75 to 80%.

However, some farmers in the tropical countries bypass the nursery phase and stock the post-larvae directly into the production ponds. The outcome in such cases is unpredictable.

(B) Grow-out Phase

Grow-out phase comprises grow-out phase technology, pond preparation, water quality management, stocking density, feeding technology, survival and sustainable production.

(a) **Grow-out phase technology**

The culture of giant freshwater prawn can be undertaken either alone (monoculture) or in combination with various compatible fish species (mixed culture or polyculture). The location of the fish farm should be in a suitable area with warm climate and soil, with 85% water retention quality. Rectangular perennial ponds with pollution free freshwater is required, having an area of 0.2–0.5 ha and a depth of about 1.5 m. The pond should have a sloping bottom and a sump at a deeper end for easy harvesting of the prawns. The pond should have also an inlet pipe at the shallow end of the pond and provided with a fine seive to prevent entry of unwanted material. An outlet should also be present at the deeper end of the pond and fitted with a fine nylon mesh screen to prevent escape of prawns.

(b) **Pond preparation**

The pond should be kept dried for at least a week before filling it with water, to maintain substratum hygiene. Lime is applied at the rate of 100-200 kg/ha/crop to maintain the desirable pH and pond hygiene. The pond is fertilised with cattle manure, (5 to 10 ton/ha), urea (100 to 150 kg/ha) and single superphosphate (200 to 300kg/ha).

Freshwater prawn is a benthic dwelling organism. They exhibit territorial behaviour and is pugnacious in nature when reared in intensive culture systems. Therefore, increasing the surface area provides space for increased stocking and thus, reduces competition for space. Provision for shelters and artificial substrate such as branches, palm leaves, plants, horizontal framed nets, etc. when placed in ponds, not only reduces the number of aggressive encounters but also increases the survival rate of prawns, particularly in intensive culture systems.

(c) **Water quality management**

The water used for grow-out operation of giant freshwater prawn (post-larvae to marketable size), should be free from toxic chemicals and pollution. Water quality maintenance is one of the important aspects in management of prawn ponds and it should be maintained in the range given below :

Water temperature (°C)	: 28-31
pH	: 7.0 – 8.5
Total hardness (ppm)	: 40 – 100
Dissolved oxygen (ppm)	: above 3.0 (before dawn)
Calcium (ppm)	: 50 – 100

(d) **Stocking density**

Fifteen days after liming and fertilization, the 1-4 weeks old juveniles of prawn are stocked. In case of monoculture, the desirable stocking density is 5-8/m^2 and cultured for a period of 5 to 6 months. In polyculture systems of culture, the stocking density of prawn is reduced by 50% and along with it compatible carp fry (IMC and/exotic carps) are stocked at the rate of 300/ha.

(e) **Continuous culture**

A common grow-out practice is the continuous culture, where ponds are stocked once or several times a year. They are never completely drained, except for repairs. At regular intervals harvesting is done selectively of 30 to 45 gm size of prawn for marketing. When the whole stock is harvested, a number of small prawns will be caught which may not be acceptable in the market. So selective fishing is done with judicious management to grow undersized individuals to marketable size.

(f) **Feeding technology**

M. rosenbergii in nature feeds upon a variety of food items such as aquatic worms, insects, small molluscs, crustaceans and detritus. In intensive culture, however, nutrionally balanced feed is to be provided comprising 25-30% protein in the feed. The feed is given at the rate of 20% of their body weight per day. Pellet diets have also been found to be suitable in grow-out phase.

(g) **Growth and survival**

A number of factors, such as stocking density, food, water environment, are conducive of good growth and better survival rate. Prawns reared in ponds, usually reach marketable size (40-50 gm) within 5-6 months of culture. The optimum marketable size of *M. rosenbergii* is about 50 gm and if allowed to grow further, survival rate would decrease drastically. The growth rate would also be slower, as much of the food would be utilised not for growth but for building of reproductive structures.

The culture period varies with the size of stocking in the pond. When stocked with post-larvae the culture period is for 6 months, while when stocked with nursery reared juveniles the culture period is reduced by 1 month. The survival rate achieved in such culture is 50-75%.

(h) **Sustainable production**

Production depends on factors such as health of post-larvae, water quality, feed quality and pond management. A normal sustainable production of 1000 to 1500 kg per hectare per crop can be achieved. However, production can be considerably increased (2500-3000 kg/hectare/crop) with provisions of pond aeration during night time (preferably between 2 to 5 a.m), exchange of water and with quality diet in the form of pellet.

The farmer through such sustainable prawn culture can achieve a high profit, much higher compared to other fish culture activities.

Diseases encountered by *M. rosenbergii*

M. rosenbergii, although considered as a freshwater prawn, is not purely a freshwater species as it has to undergo a saline phase in its life cycle. This giant freshwater prawn is susceptible to various types of diseases, particularly during the intermoult period.

Table 8.4 : Some important diseases with symptoms and stages of infestation, encountered by the giant freshwater prawn

Type of disease	Name of the disease, causative agent and symptoms	Infected stages of prawn
I. Bacterial infection	(1) Chitinolytic bacteria. Erodes the surface of the exoskeleton, causes loss of appendages and lesions. "Shell disease" or "black spots".	"Black-spots" is an obvious disease of post-larvae and harvest-sized prawn
	(2) Filamentous bacteria (*Leucothrix* sp.). Clogs gills and interferes with respiration.	Post-larva
II. Fungal infection	*Saprolegnia* sp., *Fusarium* sp., *Lagenidium* sp., *Sirolpidium* sp., etc.	Egg, nauplius, protozoea, juvenile to adult.
III. Protozoan infection	*Epistylis* sp. and *Zoothamnium* sp. are very common; *Vorticella* sp. is less common. It is attached to body surface and gills.	Common cause of larval disease, Also found in adult.
IV. Hydrozoan infection		Prey upon larvae
V. Soft-shell syndrome	*Vibrio alginolutium*. Carapace brownish in colour and septicaemia.	Juvenile and adult
VI. Other diseases	(1) 'Whiteness' disease of muscular tissue which starts from the tail region. It is caused from stress condition, overcrowding, low O_2 level, temperature, shock, etc.	Juvenile and adult
	(2) Dark areas in gill chamber, caused due to precipitation of chemicals and for high level of nitrogen, wastes, etc.	Juvenile and adult

Some common diseases with symptoms and the infected stages of the prawn are given in Table 8.4.

II. CULTURE OF TIGER SHRIMP

The tiger shrimp (Fig. 8.11), *Penaeus monodon* (Fabricius, 1798), forms both capture and culture fisheries of commercial importance. It is the fastest growing species used in aquaculture with major potentiality in India, Bangladesh and a few other countries. *P. monodon* is euryhaline and can tolerate nearly freshwater conditions, although 10-25 ppt salinity is considered optimum. It can tolerate a temperature between 12°C and 37.5°C.

Apart from *P. monodon*, other commercially farmed shrimps are *P.japonicus* [Fig. 8.3(d)], *P. indicus* [Fig. 8.3(a)], *P. merguiensis* [Fig. 8.3(b)] and *P. vannamei* [Fig. 8.3(e)]. Although considerable

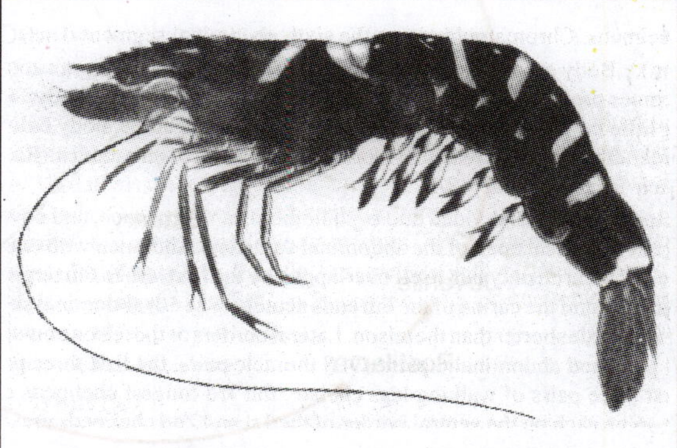

Fig. 8.11. The tiger shrimp, *Penaeus monodon*

similarities in the culture technology is present for these species, there are also a number of differences. These differences are due to environmental requirements, breeding, feeding behaviour and compatibility with other species. However, *P. monodon* is the most extensively cultivated shrimp among the 29 recorded penaeids because it is best suited for culture in confinements, attainment of large size, faster growth, hardy nature, greater tolerance to varying salinity, temperature and oxygen, natural abundance of seed and high market price with large export potentiality.

Research on shrimp culture has been in line with development in hatchery techniques for induced breeding, controlled spawning and larval rearing.

Biochemical composition

Biochemical composition (%) of adult tiger shrimp constitutes of the following :

Moisture	... 75.61 to 78.26
Protein	... 20.29 to 22.01
Fat	... 0.81 to 1.12
Ash	... 1.07 to 1.50
Calcium	... 0.0066 to 0.0081
Phosphorus	... 0.131 to 0.247

Distribution : In India, the natural distribution of tiger prawn is along the east and west coasts with high potentiality in the Sunderbans covering an area of about 16,900 km^2 at the Gangetic delta and also Chilka Lake.

Habit and habitat : Mainly marine in habitat and bottom dweller. It is present in inshore and offshore regions of varying depth (0 to 72 fathoms). Tiger prawn can invade and thrive well, both in brakish water and freshwater rivers, rivulets, paddy-fields and other impoundments for its great migratory behaviour and highly efficient hypersomatic regulatory system in adjustment with low saline water.

Biology and Life history

The life cycle of tiger shrimp, *P. monodon,* is summarised in Figure 8.12.

(a) Differences between matured male and female tiger shrimps

The differences in sexual characters between matured male and female tiger shrimps (*P. monodon*) are given in Table 8.5.

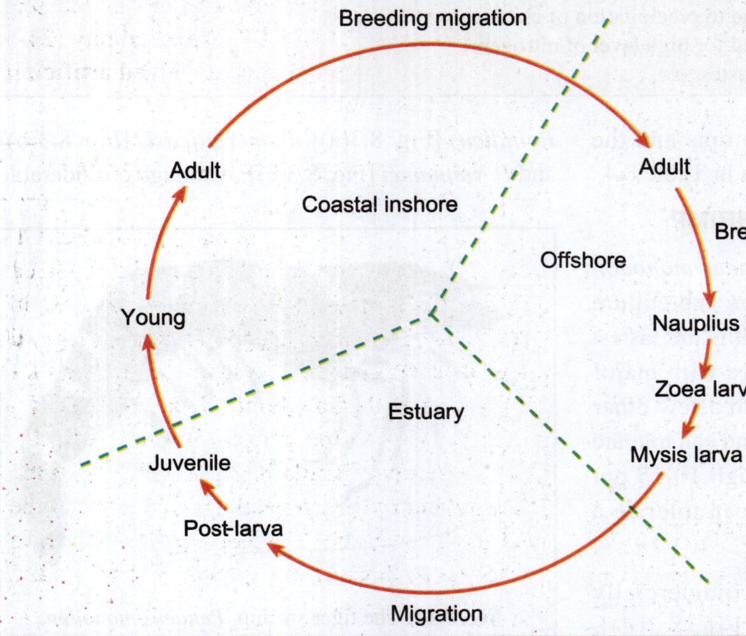

Fig. 8.12. Life cycle of tiger shrimp (*Penaeus monodon*)

Table 8.5 : Sexual characters of matured male and female tiger shrimps (*P. monodon*)

Matured Male	Matured Female
(1) Great chela is stouter	(1) Great chela not so stout.
(2) Petasma (Fig. 8.2a) present	(2) Petasma absent.
(3) Thelycum absent	(3) Thelycum (Fig. 8.2b) present and is of 'closed type'.
(4) Genital aperture opens at the base of the fourth pair of pereopods*.	(4) Genital aperture opens at the base of the coxa of third pair of pereopods.
(5) Appendix interna of the 2nd to 5th pleopods** is not interlocked.	(5) Appendix interna of the 2nd to 5th pleopods interlocked with one another during breeding season to form a basket for carrying a large number of eggs.
(6) Appendix masculina present arising from the base of the endopodite of 2nd pleopod and act as copulatory organ.	(6) Appendix masculine absent.
(7) The dactylus is as long as the protopodite, that arises from the inner side of maxillipeds. From the tip of the protopodite arises a pencil of setae lodged in the concave inner side of dactylus. The tip of dactylus is bluntly rounded.	(7) Dactylus is a tapering joint, articulating in ordinary way end-on with the protopodite.

(b) **Food and feeding habits**

P. monodon in its larval, juvenile and adult stages shows various food preferences. Zoea and mysis stages are phytoplankton feeders, mainly consuming diatoms. For post-larva and juvenile, periphyta form the main food. Adults, on the other hand, prefers animal matter like polychaetes, crustaceans, insects, small molluscs, ophiuroids, fish remains, etc. During scarcity of food and overcrowding, adult prawns show cannibalistic tendency, preying mainly upon the moulted and weaker ones.

(c) **Growth and maturity**

Tiger shrimps have faster growth rate. Males attain sexual maturity at minimum size of 31 mm or 20 gm in pond environment and 37 mm or 35 gm in wild. In case of females, sexual maturity is attained at 41 mm or 40 gm in pond and 46 mm or 63 gm in the wild environment. In deep sea, the natural abode of tiger shrimp, the likely environmental factors that stimulate sexual maturity are high water pressure, darkness, low temperature and high salinity. Under rearing condition of pond environment with nutritious rational feeding and judicial water management, maturity of tiger shrimp can also be attained. Maturation can be quickened with favourable range of temperature (27.75-30.55°C) and salinity (24-31 ppt). Maturation can also be achieved artificially through eye-stalk ablation.

* **Pereopods :** The thoracic appendages used for walking (5 pairs) are the pereopods. ****Pleopods :** The abdominal appendages used for swimming (6 pairs) are the pleopods.

INDUCED BREEDING OF SHRIMP

Sexual maturation and breeding, particularly in penaeid shrimps, have been found to be mediated by hormones produced in the endocrine system. In crustaceans (also shrimps) important related organs of the endocrine system are : (i) the Medulla Terminals Ganglionic X-organ (MTGX), associated with the ganglia in the eye-stalk; (ii) the sensory Papilla X-organ (SPX); (iii) the sinus gland of the eye-stalk located close to the cornea; and (iv) the Y-organ, a pair of glands in the anterior thoracic region in the antennary and maxillary segment.

Under captive conditions, male shrimps mature fully and spermatophores can be observed through the carapace. Female shrimps, on the other hand, do not often mature fully in captive conditions, even though maturing eggs can be seen in the ovaries. The maturation process in female shrimps is inhibited by a gonad-inhibiting hormone (GIH), that is secreted by the Medulla Terminalis Ganglionic X-organ (MTGX) and stored in the sinus gland. Also the moult hormone crustecdysone, secreted by the Y-organ, seems to have an influence on maturation.

Removal of the eye-stalks, or only the sinus glands from the eye-stalk, in non-breeding, season results in rapid increase in the weight of the ovaries. Based on the maturation inhibiting property of the sinus gland, induction of maturity under controlled conditions can be achieved by eye-stalk extirpation or ablation (surgical removal). Thus, the gland containing the inhibitory hormone when removed accelerates vitellogenesis. Also, environmental factors such as temperature, photoperiod, salinity, pH and the nature of nutrition of brood animals, seems to be important factors in the maturation and spawning of shrimps.

The technique of ablation involves the removal of either eye or total removal of the eye-stalk. Unilateral ablation may or may not give rise to maturation. Removal of both eye-stalks (bilateral ablation) has been found to be more effective, but has the risk of higher mortality in the ablated shrimp. Bilateral ablation can bring maturity within 40 days in Indian and 90 days in Philippine environments. Four different methods of ablation are practised : (1) cutting off the eye-stalk along with eyeball, (2) making an incision across the eyeball and squizing the eye-stalk tissue to destroy the sinus gland and the associated organs, (3) canterisation of eye-stalk for the destruction of ovary inhibiting factors and (4) tightly tying the eye-stalk at the base with thread.

At the time of ablation it is important to prevent excessive loss of eye fluid and infection. It is prescribed best to undertake ablation during the intermoult stage, for maturation to follow within a week. Ablation done during the premoult period may lead to immediate moulting and the latency period will get prolonged. At the period after ablation, it is recommended that a continuous flow of water be maintained in the maturation tanks with a daily exchange of 60-70 per cent of water.

(d) **Pre-spawning moult and mating**

The pre-requisite for mating in *P. monodon* is the moulting of the female, since it is a 'closed thelycum' shrimp. Mating, thus takes place between three hard-shelled males and a newly moulted female. The males insert spermatophore into the female thelycum (Fig. 8.2b) by the appendix masculina (Fig. 8.6b). During mating the male positions itself directly below the female. The pair engages in parallel swimming movements and then quickly shifts to perpendicular position (Fig. 8.13).

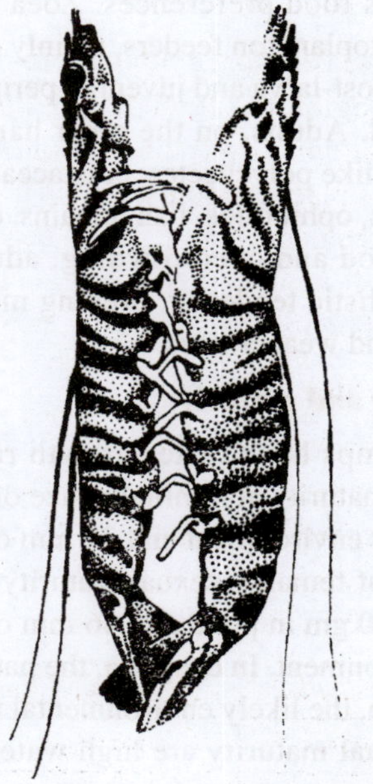

Fig. 8.13. Perpendicular position during copulation of male and female shrimps

In nature, mating of tiger shrimp takes place in the sea (in more saline areas), where the spawners and milters migrate from the estuary or brackish water areas. Complete darkness as well as bright flood light can hinder mating.

(e) **Fecundity**

In *P. monodon* fecundity range largely varies. It depends upon the varying body size, condition of the spawning female, uneven development of the ovaries in the same age group and so on. Variation in egg number is also seen in wild and captive ablated females. A wild spawner of size 150-200 gm may produce more than a million eggs, while an ablated artifically matured female may produce 3.55 lakh eggs under confinement.

Hatching rate varies from 0 to > 90% depending on maturation, sex ratio, spawner size, type of ablation (unilateral/bilateral), mating efficiency, water depth and other physiological factors.

(f) **Spawning**

Spawning in tiger shrimp takes place in late night or early morning in relatively deeper water and preferably near river mouths. High salinity of the sea is required for larval development.

Spawning season spans throughout the year and varies from region to region depending upon the environmental conditions. In Indian environment breeding takes place in almost all the months of the year, with different peaks at different coastal areas. In West Bengal it is during April/May and October/November, while in Orissa it is November-February.

(g) **Embryonic development**

Cell division of fertilised egg in *P. monodon* [Fig. 8.14 (a–f)] commences about half an hour after spawning. Incubation period is 12-16 hours at 27°C to 29°C. Embryonic development is very simple and its salient features are given below :

Post-fertilisation time	Embryonic characters
(1) 00 hr 30 minutes	2-celled stage
(2) 01 hr 00 minute	4-celled stage (Fig. 8.14b)
(3) 01 hr 50 minutes	Morula stage (Fig. 8.14c)
(4) 05 hr 00 minute	Early embryonic nauplius stage. Future antennae and body demarcated (Fig. 8.14d).
(5) 11 hr 00 minute to 12 hour 00 minute	Late embryonic nauplius stage. Antennae (3 pairs) almost differentiated, the third pair with inconspicuous setae at its tip. A pair of sharp needle-like curved outgrowth appearing at the hinder end of the body (Fig. 8.14e). Embryo exhibits faint movement inside the egg capsule.
(6) 14 hour 00 minute to 16 hour 00 minute	Embryonic nauplius about to hatch. Body shape somewhat oblong when viewed ventrally. All the 3 pairs of antennae are further developed, the third pair being formed with exo- and endo- pod and biramous setae. Hinder spines prominent and straight (Fig. 8.14f). Movement of the embryo gets vigorous before hatching and nauplii hatch out by rupturing the vitelline membrane.

(h) **Larval development**

Larval development in tiger shrimp is a complex and a much elaborate process, passing through several metamorphoses to attain the adult form. The hatchings have to pass through 6 nauplii, 3 zoea and 3 mysis stages to attain post-larval stage. The changeover from one stage to another takes place after a moult and finally results into the post-larva or megalopa larva. The four larval stages (nauplius, zoea, mysis and megalopa) take about 10-15 days. In nature, the development of egg to mysis stage (Fig. 8.14) occur in the outer littoral zone of the sea. The megalopa larva undergoes development while migrating through the inner littoral zone and estuarine area.

(i) **Nauplius larva** — The six naupliar stages [Fig. 8.14 (n_1 to n_6)] each followed by a moult, vary in length ranging from 0.32 mm to 0.53 mm and the duration of development varies between 48 and 53 hours. The naupliar stages exhibit planktonic behaviour and subsists on the small embryonic yolk. It has three pairs of antennae, the first pair uniramous and the second and third biramous.

(ii) **Zoea larva** — The three zoea stages [Fig. 8.14 (z_1 to z_3)] have size length measuring from 1.02 mm to 2.75 mm and the duration of development varies between 120 to 144 hours or 5 days. They are also planktonic in behaviour and with their relatively increased sized mouth it can accept food particles as big as 50 µm in size.

(iii) **Mysis larva** — The three mysis stages [Fig. 8.14 (m_1 to m_3)] have their length varying between 3.40 mm and 4.37 mm. Their duration of development varies from 288 to 341 hours, depending upon the temperature of water and the feeding components. Mouth large and can accept food of sizes up to 200 µm.

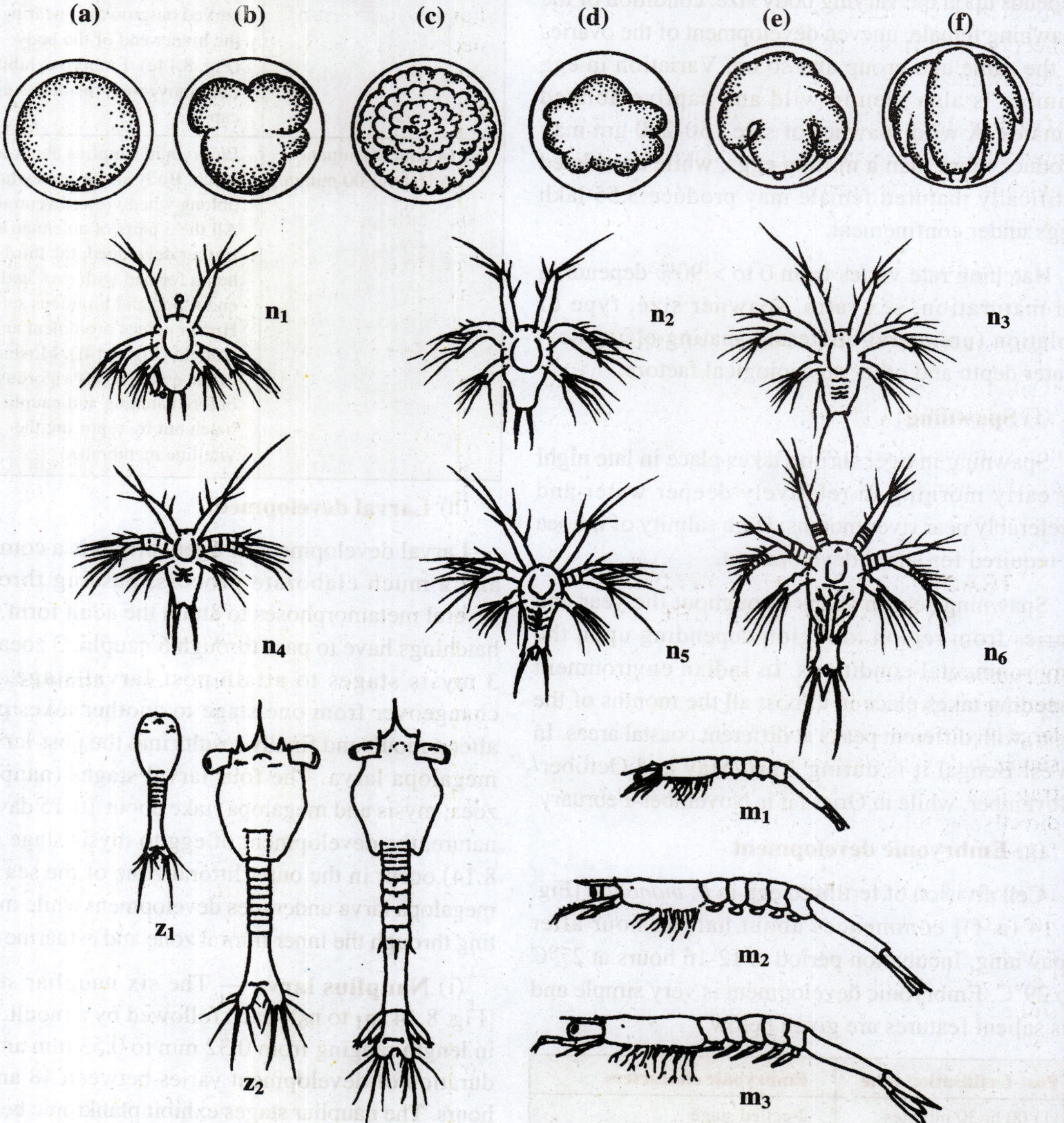

Fig. 8.14. *Penaeus monodon* showing embryonic and larval developmental stages. (a to f) developing embryo (within egg shell); (n_1 to n_6) six naupliar substages; (z_1 to z_3) three zoea substages and (m_1 to m_3) three mysis substages.

(iv) **Megalopa larva** — This is generally called as the post-larvae measuring from 5 to 20 mm in length. The first post-larvae measures about 5 mm. It has a transparent slender body and a dark brown streak on the mid-ventral side extending from the tip of the antenular flagellum to the tip of the telson which turns pink in smaller (6-14 mm) and then green in larger (15-20 mm) specimens. The body being transparent is not easily distinguishable from mixed collection in the net, unless transferred to a tray with a clean white background.

(v) **Juveniles** —They are more than 21 mm in length with complete gill system. Body is blackish or greyish in colour. It spends its life in estuarine environment.

(vi) **Adult** — Adult characters appear with sex differentiation. It lives in estuarine environment.

Culture practices of *Penaeus monodon*

Culture of shrimp is mainly carried out in ponds, in both traditional and modern type of practices. The culture can be divided into (A) seed stock production and collection. (B) Hatchery system and (C) Grow-out system.

TRADITIONAL VERSUS MODERN SYSTEMS OF CULTURE

In traditional system of culture, natural stocking was done through intake of tidal water which carried large numbers of shrimp larvae. These ponds were simple in design and were meant to serve largely as trap ponds. Shrimp larvae were also released directly into the rearing or production ponds by the farmers.

In modern culture necessary ponds have been incorporated for growing larvae to an advanced juvenile stage and later transferred to production ponds. Subsequently, with the adoption of controlled propagation, many shrimp farmers introduced hatchery units also. Specialisation in culture has also been achieved through rearing of post-larvae or juveniles for sale to farmers, for grow-out culture. Such nursery farms have incorporated brood ponds along with hatcheries and nurseries, and also facilities for growing natural (live) larval food. Farmers adopting semi-intensive systems (page 217) of culture have earthen nursery and rearing ponds with connected dykes. Various types of shelters and artificial substrates have been added to the water in ponds, such as water plants, hollow bricks, plastic hollow pipes, styrofoam sheets, framed nets, etc.

As shrimps formed only a small percentage of the harvest, the tranditional shrimp culture was necessarily a polyculture system where there was inability to control the composition of the seed stock. It is for certain that the species combinations were not always compatable. Also, the culture procedures were not favourable for high survival rates for shrimps, as the shallow depths of milkfish culture did not permit high stocking rates of shrimp.

Thus, to overcome such difficulties, modern system envisages to adopt monoculture practices. Even combinations with other shrimp species is not advisable in view of the differences in environmental requirements between species. Furthermore, intensive systems (page 217) of culture of prawn and shrimps are in practice in tank farms. In these tanks water is frequently exchanged to maintain high oxygen level as well as to remove metabolic wastes. In case of water crisis, when water exchange is low, heavy aeration is done to keep the organic metabolites in suspension, as these suspended particles serve as biological filters (the dense colonies of nitrifying bacteria grow on them).

The traditional system of paddy-cum-shrimp culture practised mainly on the west coast of India has been described earlier (page 292). In such cultures after harvesting of shrimp, the fields are stocked with carp during the rainy season when the salinity in the field is low.

(A) Seed stock production and collection

Seed stock production through hatchery technique has not gained much inportance commercially, since major shrimp-culturing countries depend mainly on the collection of post-larvae and juveniles from the natural sources.

Collection of seed stock from the wild — In traditional system of culture, post-larvae and fry are brought in by the incoming tidal water. The unwanted fish larvae and fry are eliminated from those of the shrimp by an improved system of stocking. In this improved system, the nursery ponds are filled with as many larvae and fry as possible by manipulating the tidal flow. The pond after a month is treated with tea-seed cake (this contains 10%-15% saponin) at the rate of 10-25 ppm. This kills all the fish larvae without affecting the shrimp larvae and fry. These then are transferred to the rearing or production pond.

Fry of shrimp can also be collected with different types of lure lines and nets, such as scoop nets, fine-meshed bag nets, push nets, scissor nets, etc. The post-larvae can be identified as given earlier.

PROBLEMS RELATED TO COLLECTION OF WILD SEED STOCK

The collection of wild seed stock has many difficulties as stated below :

(1) Problem is faced in sorting out the required species from the mixed collections (The mixed collection may contain larvae of slow-growing, undesirable species of shrimps and also of predatory- or weed- or trash-fish).

(2) Problem may arise due to fluctuations in their availability.

(3) Social conflicts may arise as the local shrimp fishermen ascribe for poor commercial catches of shrimp to the removal of larvae and juveniles for farming.

(4) Sorting of fry of the required species of shrimps for culture require considerable expertise. Some of the main identifying characters on which the sorting is dependent are nature and location of pigmentation, body shape and mode of locomotion.

(5) The men and women engaged in prawn seed collection from natural source, often after sorting out the required prawn juvenile, throw the rest of the juveniles on the shore, thereby killing a number of important and endangered fish and prawn species resulting in destruction of biodiversity.

(6) Similar destruction of biodiversity is also witnessed in improved system of stocking in traditional prawn culture system where a number of other prawn and fish juveniles and fry are killed by treatment of the pond with tea seed cake, containing a poison, called saponin.

(B) Hatchery systems

In hatchery system, controlled maturation, spawning and larval rearing of shrimps are undertaken. Initially such culture was directed towards mass production with gravid females being collected from the wild. But later, success was achieved through maturation and mating of shrimps in captivity. Here too, mature females that have undergone prenuptial moulting can be mated.

For large scale production in hatchery system, it is very difficult to obtain sufficient spawners of *P. monodon* from the wild. Therefore, for large scale production of adequate numbers of larvae of this tiger shrimp, maturation of captive stock, either wild-caught or pond-reared is necessary. Further, induced breeding through unilateral eye-stalk ablation is also undertaken for *P. monodon* to make them spawn freely in controlled environmental condition.

(a) **Shape and size** — Hatchery tanks of various shape and size (Fig. 8.15) are in use made of cement concrete, ferrocement, plastic, fibreglass, etc. These tanks are used for spawning and larval rearing. Cages and pens have also been used but not on **commercial basis**.

The size of the hatchery tanks ranges between 100 and 2000 ton capacity and are suitable for spawning a large number of spawners at a time. It is also used for larval rearing and fry nursing. As all is done in the same tank, it is often referred to as the **community culture method**.

Contrary to it, many hatcheries developed seperate smaller sized tanks made up of plastic or fibreglass for spawning, larval rearing and culture of live food. The capacity of spawning and larval rearing tanks varies from 100 to 250 litres and 1000 to 2000 litres, respectively. In case of *P. monodon* where availability of spawners is difficult, community culture is not commercially efficient.

(a)

(b)

Fig. 8.15. Indoor prawn hatchery system — **(a)** Circular cemented tanks and **(b)** rectangular fibreglass tanks.

A third system of hatchery tanks is also in use, which combines the advantages of the above two types. This system maximises tank utilisation, particularly for *P. monodon* culture of spawning and larval rearing. This system comprises 1000-2000 litres capacity of spawning tanks; 1000-3000 litres of larval rearing tanks and 30-100 tons of nursery tanks.

Another "ladder system" hatchery, developed by Liao (1985), is also in use, comprising four interconnected tanks that are built on a sloping ground (Fig. 8.16). At the top of the slope is the algal culture tank, followed below by the nauplii and zoea larvae rearing tank; another mysis to $PL_1 - PL_5$ stages rearing tank and finally a large post-larvae rearing tank. These four tanks are all built one below the other with descending water levels.

(b) **Larval food** — The food of larval shrimps are raised by fertilisation of the tanks to produce plankton. Among the phytoplankton, *Chaetoceros, Skeletonema* and *Tetraselmis* are preferred as food by

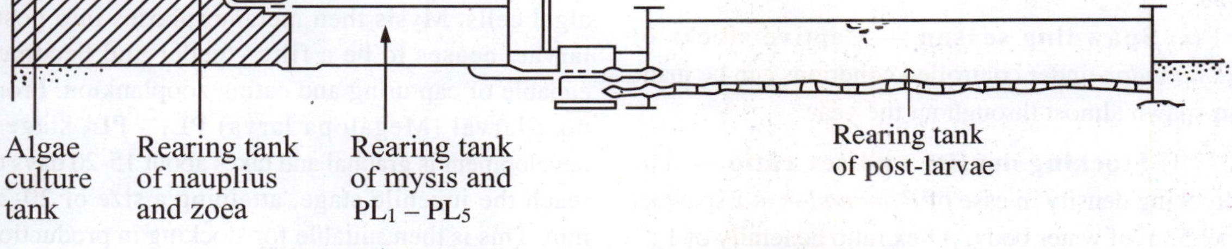

Fig. 8.16. The ladder system hatchery developed by Liao (1985)

the shrimp larvae. Among the zooplankton, *Brachionus* sp. (rotifer) is probably the most important larval food. Another important larval food is the nauplii of brine shrimp (*Artemia salina).*

In larger tank hatcheries, where community culture is being practised, the tanks are fertilised with 3 mg/l of KNO_3 and 0.3 mg/l of Na_2HPO_4. This produces phytoplankton which are used to feed the protozoea stage. The phytoplankton production is maintained, if needed, through further addition of fertilisers. Sometimes, pure cultures of diatoms are inoculated prior to the addition of fertilisers. When the mysis stage is reached, the larvae are fed on *Brachionus* or brine shrimp nauplii.

In case of hatcheries with separate small hatchery tanks or in ladder system hatchery, algae are cultured in separate algal tanks or in plastic bags. The required amount of phytoplankton are introduced daily during the protozoea stage. Zooplankton are also hatched in special tanks and are fed to mysis and early post-larval stages.

When the density of production of plankton becomes inadequate, supplementary feed is given in the form of soybean cake, soybean curd, fertilised eggs or egg yolks of oysters, etc. In the post-larval stage after the fifth day, they are fed on polychaetes, chopped mussels, clams, cockles or artificial compound diets, three or four times a day.

(c) **Spawning and larval rearing** — The minimum age of female spawners in case of *P. monodon* is 9-15 months with a weight of 32-45 gm. Spawners from the wild are bigger than those from captive brood stock. Farmers are of the opinion that wild stock spawners are superior to captive ablated spawners as the quality and quantity of their eggs are higher.

(i) **Spawning season** — Captive stocks of *P. nonodon* under controlled condtions can be made to spawn almost throughout the year.

(ii) **Stocking density and sex ratio** — The stocking density in case of *P. monodon* is 1 spawner per 5 m^3 of water body. A sex ratio generally of 1 : 1 is maintained in spawning tanks, but a ratio of one male : 2 females has produced higher spawning rates and egg production.

(iii) **Spawning** — In case of controlled spawning, advanced stages of matured gravid females (soon after their moulting) and males are stocked in spawning tanks. Spawners obtained from the wild during winter are disinfected with 3 ppm $KMnO_4$, 25 ppm formalin or with 3-5 ppm of the commercial product, Treflan[R]. In community culture method, several spawners are introduced into the large tank, while in other culture systems, individual spawners or batches of spawners are introduced in separate spawning tanks. The salinity in the tanks has to be maintained between 28 and 35 ppt and the temperature between 23 and 33°C. Spawning generally takes place in the night hours. Fertilisation is external. At the above mentioned temperature, embryonic development is rapid.

A few points are to be noted :

(1) There is generally a time lag between mating and spawning. Therefore, the eggs may not be fully matured at the time of mating.

(2) In case of *P. japonicus* and *P. indicus,* the females have a tendency to eat their own spawned eggs. So in their culture, it is advisable to install mesh trays or plates on the bottom of the spawning tanks to protect the eggs. After all the eggs have been collected, the spent females are separated.

(iv) **Larval rearing**— The nauplius subsists on its own yolk and passes through three to six sub-stages ($N_1 - N_6$). In 2-3 days time it passes through three zoea sub-stages ($Z_1 - Z_3$). During these sub-stages the larva starts feeding on unicellular algae and it lasts for 3-6 days. This is followed by the three sub-stages of Mysis ($M_1 - M_3$) which last for another 3-5 days. During these stages the larva also feeds on algal cells. Mysis then metamorphoses into post-larvae, ceases to be a filter-feeder and becomes capable of capturing and eating zooplankton. From post-larval (Megalopa larva) PL_1 - PL_5 stages, development is gradual and takes about 15-20 days to reach the juvenile stage, attaining a size of 20-25 mm. This is then suitable for stocking in production ponds.

The main difficulty faced in rearing of protozoeal stage is when they start feeding, and at this stage cultured phytoplankton is the most efficient food for larvae. Also, the protozoeal larva is most sensitive to light and so the tank should be properly covered to ensure darkness. The last sub-stage of protozoea starts feeding on zooplankton. Both mysis and post-larvae prefer zooplankton up to the fifth day, after which they consume larger food and feed at the bottom.

(v) **Nursery ponds** — Nursery ponds vary in size ranging from 500 to 2000 m^2 in area with an average depth of about 40-70 cm. Larvae are stocked at the rate of 100-150 per m^2. Prior to the introduction of larvae, the ponds are prepared by irradicating predators and then fertilising with a combination of organic and inorganic fertiliser.

Nursery cages may also be used which are rectangular in shape (1-2 m × 5 m × 1 m). They may be either floating or stationary type installed in protected bays, lagoons or ponds. They are stocked with larvae at higher densities of 1000-2000 per m^2. However, the cages are rarely used as their small mesh sizes become rapidly choked by biofouling.

(C) Grow-out System

The common grow-out system of shrimp culture is carried out in earthen ponds. Presently, extensive culture methods are giving way to semi-intensive methods of culture. The traditionally cultured coastal shrimp ponds are now stocked with sorted fry collected from the wild or bought from fry collectors and stocked at a low rate of 3000-5000 fry/ha. These ponds are prepared before stocking of fry and water exchange is maintained using tidal flow during the two months of rearing period.

(a) **Semi-intensive farming** — Modern shrimp farming is carried out following mainly semi-intensive forms of culture. Farming is done in specially designed pond farms which usually have a hatchery, nursery and rearing ponds. These ponds are generally rectangular in shape and 1–3 ha in area, having a depth of 0.8–1.2 m. Stocking rate varies considerably from 28,000 to 50,000 fry/ha. Natural food (produced by application of fertiliser) and supplementary feed or formulated feed are given. Water exchange in large scale is required in culture ponds to accelerate and synchronise the moulting cycle of the stock. In this type of culture the yield has been found to be about 1.5–2 tons/ha per crop.

(b) **Extensive Farming** — Extensive farming of shrimps in India is carried out in paddy-fields (pages 216, 292). Paddy-cum-shrimp culture has also undergone some changes due to the introduction of controlled stocking of fry and better management of water and manuring by rice stubble. Yield of about 400 kg/ha per crop has been achieved under very favourable conditions. Otherwise, the yield is around 200 kg/ha per crop.

(c) **Intensive Farming** — Intensive farming of shrimps is usually carried out in cemented tanks. Smaller earthen ponds are also used of 0.5–1 ha size and 60-150 cm depth. In such impoundments, inlets and outlets are arranged in such a manner so as to effect proper circulation of water. Aeration of the water is done with equipments such as paddle wheels and air blowers. Stocking is done at the rate of 200-250 fry/m^2. Formulated diets comprising high protein are fed to the fry. Yield of about 1.5–3 tons per crop has been achieved and in special ponds about 28 tons/ha/year.

Super-intensive culture using high stocking density and heavy feeding often results in environmental disaster and health problems and has, therefore, been seriously criticised. A comparative study of the above farmings is given in Table 6.2.

(d) **Pen culture of shrimps** — Pen culture of shrimps has been practised in lagoons in southern India in suitable intertidal areas enclosed by net fencing. These pens are constructed at a height of 1 metre for holding water during low tide. It resembles tidal ponds and the area enclosed extends to about 10,000 m^2 with a depth of 1–1.5 m. Stocking rate is between 20–30 per m^2. Average yield is about 3–4 tons/ha/year.

(e) **Polyculture vs Monoculture of shrimp** — Although it is a known fact that polyculture yields higher production, still most shrimp farmers prefer monoculture of the fastest growing species of shrimp. On the contrary, shrimp culturists do resort to polyculture but it is with other shrimp species. This is done due to shortage of seed stock of the preferred species.

From economical point of view also, monoculture is found to be more profitable due to the high market value of shrimps. It has been recorded that the gross revenue per unit area of monoculture farms growing *P. monodon* is about double than that of polyculture farms growing the tiger shrimp along with milkfish and crabs.

(f) **Food and feeding** — The dietary requirements of different species of shrimp exhibit considerable differences, particular with regard to protein diet. Some marine shrimps require much higher protein diet as given below :

Shrimp species	Percentage of protein in diet
P. japonicus	48 to 60
P. monodon	35 to 39
P. setiferus	20 to 32
P. vannamei	30
P. stylirostris	35
P. indicus	43

Shrimp feeds also require sterols and fatty acids as their synthesis does not take place in the body of crustaceans. Standard remixes of vitamins and minerals are also added in all diet formulation.

In larval and fry rearing as well as in adult grow-out, fresh and live food are given. Commercial feed are rarely used as its acceptance in commercial farming is rather slow. Farmers generally use them along with natural food and feedstuffs.

(D) Harvesting and marketing of shrimp

Efficient harvesting of shrimps are based on the habit and culture procedures. The time of harvesting is based upon the following habits :

(1) Shrimps are attracted by light.

(2) They graze at night.

(3) Shrimps actively respond to movements of water, particularly when water is let into or drained out of a pond.

(4) They are more active during new moon and full moon.

(5) Maximum activities are witnessed shortly after sunset and shortly before sunrise.

At the end of each crop total harvesting is conducted by draining out the water and placing a bag net at the sluice gate. The shrimps are caught as they swim out from the pond. When partial draining is done, partial harvesting is employed by seine nets. Electric fishing is also successfully conducted in harvesting tiger shrimps in some countries, generally during cool seasons. However, harvesting shou⊥ be avoided when most of the shrimps are undergoing moulting.

Marketing of live shrimps are done in containers placed in cold water tanks. For long-distance transport, shrimps are preserved by freezing following the shrimp processing procedures.

Diseases encountered by *P. monodon*

Tiger shrimp is hardy in nature. It is thermal and euryhaline. It can tolerate low concentration of dissolved oxygen as well as various stresses (environmental and infestational). Normally, adults are resistant to disease, but are highly susceptible to parasitic infections. Both infectious and non-infectious diseases are encountered which are listed in Table 8.6.

SHRIMP PROCESSING TECHNOLOGY

After the shrimps are caught they are first given a preliminary wash. Once on board they must be handled quickly and carefully. The shrimps are then sorted or culled. Dead or damaged shrimps, other fishes and rubbishes are removed by hand. To reduce bacterial contamination the shrimps are then washed to remove any remaining mud or sand. The shrimps are then ready for further processing such as freezing, cooking and chilling.

(A) Preservation on board

(a) **Freezing** — On sea, shrimps are frozen in cold brine or a solution of sugar and salt. This is done by air blast freezing or by plate freezing. Cooked as well as raw, whole shrimp can be frozen in a plate freezer,

Table 8.6 : Some important infectious and non-infectious diseases encountered by tiger shrimp

Type of disease	Name of the disease, causative agent and symptoms	Infected stages
I. Viral	MBV (Baculovirus)	PL_5 to adult
	IHHNV (Infectious hypodermal and haematopoetic necrosis).	Post-larva to juvenile
	HPV (Hepatopancreatic Parvo-like virus), found on hepatopancreatic tubules and duct epithelium, anterior midgut epithelium, cuticular hypodermis, haemocytes, haematopoetic organs and connective tissues.	Juvenile
II. Bacterial	Necrosis of appendages	Zoea, mysis, post-larva
	Vibro disease (*Vibrio* sp.)	Zoea
	Luminous bacterial disease (*Vibrio harveyi, V. splendidus*)	Nauplii, zoea, mysis & post-larva
	Filamentous bacterial disease (*Leucothrix mucor*)	Larva and post-larva
	Shell disease (*Vibrio alginolyticus, V. parahaemolyticus, Acromonas* sp.)	Juvenile and adult
III. Fungal	*Lagenidium* sp.	Egg, larva and post-larva
	Sirolpidium sp.	Larva
	Haliphthoras philippinessis	Mysis, zoea
	Fusarium sp.	Nauplius, protozoea, juvenile to adult
	Hyphomyces sp.	Nauplius, zoea
	Saprolegnia parasitica	Adult
	Leptoleynia marina	Adult
IV. Protozoan	*Epistylis, Vorticella, Zoothamnium, Ephelota, Gemmipara* and *Acineta*- ciliate infestations from eyes, gills, appendages, shell or body surface.	Adult and larva
Gregarine	Gregarines	Larva
Microsporidiosis	Microsporidia	Female spawner
V. Copepods	*Caligus epidemicus* (a crustacean disease)	Adult mostly
VI. Other diseases	Chronic soft shell syndrome	Juvenile, adult
	Red disease	Juvenile, adult
	Fatty infiltration of hepatopancreas	Juvenile, adult
	Blue disease	Brood stock
	Cramped tail	Juvenile, adult
	Haemocytic enteritis (secondary bacterial infection of *Vibrio alginolyticus*)	Young, juvenile and sub-adult
	Blackgill disease (for *Zoothamnium* infestation)	Juvenile to adult
	Muscle necrosis	Juvenile to adult
	Gas-bubble disease	All stages

but immersion freezing of cooked shrimp cannot be done as the thawed shrimp are difficult to peel. Improved glaze on the shrimp can be obtained by freezing in sugar and salt solution.

The frozen shrimp can be stored at sea at –30°C and transferred to shore cold storage on arrival at the port, or thawed immediately for further processing.

(b) **Cooking** — Immediately after catching, cooking helps to retain the best flavour and texture. However, if the product gets contaminated after cooking, food poisoning bacteria can grow rapidly on it. To reduce such risk of poisoning, the cooked shrimp should be frozen on board immediately.

(c) **Chilling** — The whole shrimp, after sorting and washing, are packed in ice in shallow boxes. In the box successive layers of crushed ice and shrimps are placed till it is fully packed. Subsequently, for best result the iced shrimp should be processed on shore within 2 days of capture.

(B) Shrimp handling on shore

Whole raw shrimps on landing should be processed in factories without wasting much time.

(a) **Thawing of frozen whole shrimp** — Blocks of whole shrimp frozen at sea, when brought to shore are thawed in air or water. The thawed strimps can then be processed further.

(b) **Size grading** — Whole raw shrimp in the factory are first graded for size. The bigger sized shrimps are mechanically peeled. Small shrimps are uneconomical to peel and are discarded. They are chopped and used as raw material for various products.

(c) **Peeling** — The shell of the shrimps are normally peeled by hand. Peeling machines are now available commercially which can handle deep water shrimp. The shrimps are beheaded, the shell is split along the back and then pulled out from the meats. A small incision is done on the dorsal side to remove the digestive tract.

Particular attention is to be paid to hygiene and sanitation. Regular bacteriological control is essential for this type of product. Care should be taken at the time of peeling process, particularly hand peeling. Workers should clean their hand frequently and all working surfaces must be kept clean. To keep down contamination, chlorinated water should be used and the shrimp should be kept cool throughout the process. Machine peeling, however, results in far less re-contamination.

(d) **Cooking** — Cooking of shrimp provides a product that is ready to eat. Cooking, prior to peeling, is also done to loosen the meat in the shell, so that peeling becomes easy.

Water containing 3-5 per cent salt is used with a ratio of 1 kg shrimp to 20 litres of water. Use of stronger brines can cause discolouration. The boiling time should be such so as to develop fully the flavour and texture of the shrimp meat. Overcooking would destroy the flavour and result in loss of weight.

(e) **Cooling** — The cooked shrimp should be immediatley cooled to give it a better colour. They can either be (1) immersed directly into chilled water for about 3 minutes until they are at a temperature of about 0°C, or (2) they are first put under tap water to reach its temperature and then in chilled water. The latter method is more economical. As soon as the cooked shrimps are down to chilled temperature, they should be removed from water, drained, packed in clean boxes and transferred to the chill room. Care should be taken that the water and ice used for cooling should be clean and hygenic. Preferably the water and ice is to be chlorinated.

(f) **Dipping** — Before freezing of the meat, a number of dipping treatments are undertaken in order to improve colour and flavour. Dipping in salt, monosodium glutamate, citric acid or sodium citrate, polyphosphate and dye are being used in some stage or the other, although some of these treatments are questionable. Bacterial contamination of the cooked meat should be taken care of.

Dipping the meat in 3-7 per cent salt solution is done if salt uptake during boiling is found to be insufficient. Dipping in other solutions is done for the following :

(1) Monosodium glutamate is used to enhance flavour.

(2) Polyphosphate reduces weight loss at the time of thawing.

(3) Critic acid or sodium citrate is used to reduce discolouration of meats.

(4) Dye is added sometimes to give a uniform colour.

However, most of these treatments are unnecessary for good quality shrimp meats.

(g) **Freezing of meats** — The peeled, boiled meat is then frozen individually or in blocks. Individually quick frozen (IQF) meats can be frozen in either an air blast freezer or a liquid nitrogen freezer. IQF meat requires a freezing time of about 10 minutes at –30°C. Liquid nitrogen freezers are compact and can freeze shrimp meat quickly at a freezing time of $3\frac{1}{2}-5$ minutes.

Blocks of shrimp meats are freezed in a horizontal plate freezer. The freezing time for a 25 mm block, operating at –35°C is about 40 minutes. The frozen blocks are then packed in cartons and then moved to cold store.

(h) **Packing and glazing** — IQF meats for sale are weighed into flexible film bags which are sealed and packed in fibreboard outer cartons for storage and distribution. IQF meats are also glazed (dipped in cold water so as to coat them in a layer of ice) before packing them. However, packing in sealed film bags is sufficient to protect the unglazed product under good commercial conditions. Vacuum packing is also done to give them additional protection in case it is stored in poor commercial cold storage.

Blocks of shrimp meats can also be glazed as an alternative to protective packaging. Glazed products in cold stores, however, should be inspected periodically and the glaze be renewed, if required.

(i) **Cold storage** — The packed and glazed, frozen shrimp meats should be stored in a cold storage at a temperature of –30°C. This will keep good for a period of 6 months, provided they are properly wrapped or glazed.

(j) **Canned shrimp** — The peeled shrimp meats are blanched by immersing them for 2-3 minutes in boiling water containing 6-8 per cent salt. It is then cooled and drained on a conveyor and packed by hands into cans lined with a sulphur resistant lacquer. In case of *wet pack*, hot brine containing 2-3 per cent salt is added to the can and sometimes a small amount of citric acid is added to reduce iron sulphide discolouration. The cans are then closed at a temperature of about 65°C and then heat processed in a retort. In the retort, the cans are then cooled to about 38°C. It is then removed, dried and stacked for 48 hours, after which it is labelled, put in a carton and stored. *Dry pack* is made by blanching the meats for 8-10 minutes and packing them in parchment lined cans without brine. The cans are then evacuated, sealed and heat processed for about 60 minutes.

(k) **Smoked shrimp** — Cold smoked products of shrimps can be made. Head is removed from the raw strimp and peeled. It is then boiled in a 10 per cent salt solution for about 3 minutes, drained for about 2 hours, then laid on oiled mesh trays and smoked in a mechanical kiln for $1-1\frac{1}{2}$ hours at 30°C.

PRODUCTS OF SHRIMP

Shrimp is used to make a number of commercial products such as :

(1) The meats of shrimp is used for making a number of food products such as pastes, spreads, crisps, soups, bisques, sauces, papad and other prepared dishes. Most of these can be stored for long periods after freezing or canning.

(2) Shrimp meal is prepared from shrimp waste such as severed heads and peeled shells, containing 40-45 per cent crude protein and 5 per cent moisture after cooking, drying and grinding.

(3) From shell waste, glucosamine hydrochloride can be derived which has been suggested as a possible commercial source of glucosamine.

(4) Chitin and its derivative (chitosan) have also been extracted from shrimp waste on a commercial scale.

(5) Small sized shrimps are cooked and dried and used for feeding poultry and farmed trout. These shrimps are boiled for 4 minutes and then placed on mesh trays in a suitable drying kiln for 8 hours.

Chapter Nine : Dairy

(A) COW

INTRODUCTION

Cattle (colloquially **cows**) are the most common type of large domesticated ungulates. They are the most widespread species of the genus *Bos*, and are scientifically called as *Bos primigenius*. Cattle are raised as livestock for meat (beef and veal), as dairy animals for milk and other dairy products, and as draft animals (pulling carts, ploughs and the likes). Other products include leather and dung for manure or fuel. It is estimated that there are 1.3 billion cattle in the world today. In 2009, cattle became the first livestock animal to have its genome mapped.

Species of cattle

Cattle were originally identified by Carolus Linnaeus as three separate species. These were *Bos taurus*, the European or "taurine" cattle (including similar types from Africa and Asia); *Bos indicus*, the zebu; and the extinct *Bos primigenius*, the aurochs. The aurochs is ancestral to both zebu and taurine cattle. Recently these three have increasingly been grouped as one species, with *Bos primigenius taurus*,

Systematic Position (*After The Life of Vertebrates*: *J. Z. Young*, 1981)

Phylum : Chordata
Class : Mammalia
Subclass : Theria
Infraclass : Eutheria
Cohort : Ferungulata
Super order : Paraxonia
Order : Artiodactyla
Family : Bovidae
Subfamily : Bovinae
Genus : *Bos*
Species : *B. primigenius*
Subspecies : *B. p. taurus*, *B. p. indicus*

Binomial name

Bos primigenius (Linnaeus, 1758)

Trinomial name

Bos primigenius taurus, *Bos primigenius indicus*

Synonyms

Bos taurus, *Bos indicus*

Bos primigenius indicus and *Bos primigenius primigenius* as the subspecies. Complicating the matter is the ability of cattle to interbreed with other closely related species. Hybrid individuals and even breeds exist, not only between taurine cattle and zebu but also between one or both of these and some other members of the genus *Bos*: yak (called a dzo or "yattle"), banteng and gaur. Hybrids can also occur between taurine cattle and either species of bison, which some authors consider to be in the genus *Bos* as well.

The aurochs originally ranged throughout Europe, North Africa, and much of Asia. In historical times its range became restricted to Europe, and the last known individual died in Masovia, Poland in around 1627. Breeders have attempted to recreate cattle of similar appearance as aurochs by crossing traditional types of domesticated cattle, creating the Heck cattle breed.

Cattle genome

In the April 24, 2009 edition of the journal *Science*, it was reported that a team of researchers led by the National Institutes of Health and the U.S. Department of Agriculture has mapped the bovine genome. The scientists found that cattle have approximately 22,000 genes, and 80 per cent of their genes are shared with humans. They have approximately 1,000 genes which they share with dogs and rodents that are not found in humans. Using this bovine "HapMap", researchers can track the differences between the breeds that affect the quality of meat and milk yields.

Terminology

In general, the same terminologies are used to describe the cattle in different parts of the world but with minor differences in the definitions. The terminology described here contrasts the differences in definition between the United Kingdom and other British influenced parts of the world such as Canada, Australia, New Zealand, Ireland, and the United States.

- An intact (i.e., not castrated) adult male is called a *bull*. A wild, young, unmarked bull is known as a *micky* in Australia. An unbranded bovine of either sex is called a "maverick" in the USA and Canada.
- An adult female that has had a calf (or two, depending on regional usage) is a *cow*. A young female before she has had a calf of her own and is under three years of age is called a *heifer* (pronounced as "heffer"). A young female that has had only one calf is occasionally called a *first-calf heifer*.
- Young cattle of both sexes are called *calves* until they are weaned, then they are called *weaners* until they are a year old in some areas; in other areas, particularly with male beef cattle, they may be known as *feeder-calves* or simply *feeders*. They are referred to as *yearlings* or *stirks* when they are of one and two years of age.
- A castrated male is called a *steer* in the United States, and older steers are often called *bullocks* in other parts of the world; although in North America this term refers to a young bull. *Piker bullocks* are micky bulls that were caught, castrated and then later lost. In Australia, the term "Japanese ox" is used for grain fed steers in the weight range of 500 to 650 kg that are destined for the Japanese meat trade. In North America, draft cattle under four years old are called *working steers*. Improper or late castration on a bull results in it becoming a coarse steer known as a *stag* in Australia, Canada and New Zealand. In some countries an incompletely castrated male is known as a *rig*.
- A castrated male (occasionally a female or in some areas a bull) kept for draft purposes is called an *ox* (plural *oxen*); "ox" may also be used to refer to some carcase products from any adult cattle, such as *ox-hide*, *ox-blood* or *ox-liver*.
- In all cattle species, a female that is the twin of a bull usually becomes an infertile partial intersex, and is a *freemartin*.
- *Neat* (horned oxen, from which neat foot oil is derived), *beef* (young ox) and *beefing* (young animal fit for slaughtering) are obsolete terms, although *poll*, *pollard* or *polled cattle* are still terms in use for naturally hornless animals, or in some areas also for those that have been disbudded.

- Cattle raised for human consumption are called *beef cattle*. Within the beef cattle industry in parts of the United States, the older term *beef* (plural *beeves*) is still used to refer to an animal of either gender. Some Australian, Canadian, New Zealand and British people use the term *beast*, especially for single animals when the gender is unknown.
- Cattle of certain breeds, bred specifically for milk production, are called *milking* or *dairy cattle*; a cow kept to provide milk for one family may be called a *house cow*.
- The adjective applying to cattle in general is usually *bovine*. The terms "bull", "cow" and "calf" are also used by extension to denote the gender or age of other large animals, including whales, hippopotamuses, camels, elk and elephants.

OXEN

Oxen (singular **ox**) are large and heavy set breeds of *Bos taurus* cattle trained as draft animals. Often they are adult, castrated males. Oxen are used for ploughing, transport, hauling cargo, grain-grinding by trampling or by powering machines, irrigation by powering pumps and wagon drawing. Oxen were commonly used to skid logs in forests. These are most often used in pair, for light work such as carting.

An ox is nothing more than a mature bovine with an "education" . Ox trainers favour larger breeds for their ability to do more work and for their intelligence. As they are larger animals, the typical ox is the male of a breed, rather than the smaller female. Females are potentially more useful producing calves and milk.

Ox is also used for various cattle products, irrespective of age, sex or training of the beast – for example, ox-blood, ox-liver, ox-kidney, ox-heart, ox-hide, and ox-tail.

Singular terminology dilemma

Cattle can only be used in the plural and not in the singular. Thus one may refer to "three cattle" or "some cattle", but not "one cattle". There is no universally used singular form in modern English of "cattle", other than the sex- and age-specific terms such as cow, bull, steer and heifer. Historically, "oxen" was a non-gender-specific term for adult cattle, but generally this is now used only for draft cattle, especially adult castrated males. The term is also incorporated into the names of other species such as the musk ox and "grunting ox" (yak), and is used in some areas to describe certain cattle products such as ox-hide and ox-tail.

Other terminology

Cows of certain breeds that are kept for the milk they give, are called "dairy cows" or "milking cows" (formerly "milch cows" – "milch" was pronounced as "milk"). Most young male offspring of dairy cows are sold for veal, and may be referred to as *veal calves*.

The term "dogies" is used to describe orphaned calves in the context of ranch work in the American west. Other obsolete terms for cattle include "neat" (this use survives in "neat's-foot oil", extracted from the feet and legs of cattle), and "beefing" (young animal fit for slaughter).

Word origin

Cattle did not originate as the term for bovine animals. It was borrowed from Old French *catel*, which itself came from Latin *caput*, i.e., head, and originally meant movable property, especially livestock of any kind. The word is closely related to "chattel" (a unit of personal property) and "capital" in the economic sense.

In older English sources such as the King James version of the Bible, "cattle" refers to livestock, as opposed to "deer" which refers to wildlife. "Wild cattle" may refer to **feral cattle** or to undomesticated species of the genus *Bos*. Today, the modern meaning of "cattle", without any other qualifier, is usually restricted to domesticated bovines.

SOME COMMON INFORMATION ON CATTLE

- Cattle have one stomach with four compartments, called as "**ruminant stomach**". Cattle sometimes consume metal objects which are deposited in the reticulum chamber of the stomach and irritation from the metal objects causes "hardware disease".

- The gestation period for a cow is nine months. A newborn calf weighs 25 to 45 kilograms. The world record for the heaviest bull was 1,740 kilograms, a Chianina named Donetto, when he was exhibited at the Arezzo show in Italy in 1955. The heaviest steer was eight year old 'Old Ben', a Shorthorn/Hereford cross weighing 2,140 kilograms in 1910. Steers are generally killed before reaching 750 kilograms. Breeding stock usually live about 15 years (occasionally as many as 25 years).

- A common misconception about cattle (particularly bulls) is that they are enraged by the color red (something provocative is often said to be "like a red flag to a bull"). This is incorrect, as cattle are red-green colour-blind. The myth arose from the use of red capes in the sport of bullfighting; in fact, two different capes are used. The capote is a large, flowing cape that is magenta and yellow. The more famous muleta is the smaller, red cape, used exclusively for the final, fatal segment of the fight. It is not the colour of the cape that angers the bull, but rather the movement of the fabric that irritates the bull and incites it to charge.

- Although cattle have red and green colour receptors in the cone cells of their retina, they cannot distinguish red from green. Thus, they are dichromatic, the same as most other mammals (including dogs, cats, horses and up to ten per cent of human males). They are able to distinguish some colours, particularly blue from yellow, in the same way as most other mammals.

DOMESTICATION AND HUSBANDRY

Cattle occupy a unique role in human history, domesticated since the early Neolithic period. They are raised for meat (beef cattle), dairy products and hides. They are also used as draft animals and in certain sports. Some consider cattle the oldest form of wealth, and cattle raiding consequently one of the earliest forms of theft.

Cattle are often raised by allowing herds to graze on the grasses of large tracts of rangeland. The most common interactions with cattle involve daily feeding, cleaning and milking. Many routine husbandry practices involve ear tagging, dehorning, loading, medical operations, vaccinations and hoof care, as well as training for agricultural shows and preparations. There are also some cultural differences in working with cattle — the cattle husbandry of Fulani men (*Berber people of North Africa*) rests on behavioural techniques, whereas in Europe cattle are controlled primarily by physical means like fences. Breeders use cattle husbandry to reduce *Mycobacterium bovis* infection susceptibility by selective breeding and maintaining herd health to avoid concurrent diseases.

Cattle are farmed for beef, veal, dairy, leather and they are less commonly used for conservation grazing, simply to maintain grassland for wildlife. Depending on the breed, cattle can survive on hill grazing, heaths, marshes, moors and semi-desert. Modern cows are more commercial than older breeds and having become more specialized, are less versatile. For this reason many smaller farmers still favour old breeds, like the dairy breed of cattle Jersey.

In Portugal, Spain, Southern France and some Latin American countries, bulls are used in the activity of bullfighting; a similar activity, Jallikattu, is seen in South India; in many other countries this is illegal. Other activities such as bull riding are seen as part of a rodeo, especially in North America. Bull-leaping, a central ritual in Bronze Age Minoan culture, still exists in south-western France. In modern times, cattle are also entered into agricultural competitions. These competitions can involve live cattle or cattle carcasses in hoof and hook events.

In terms of food intake by humans, consumption of cattle meat is less efficient than of grain or vegetables with regard to land use. Cattle grazing consume more area than any other agricultural production.

ENVIRONMENTAL IMPACT

In 2009, a United Nations report from the Food and Agriculture Organization (FAO) states that cattle farming are "responsible for 18% of greenhouse gases". The production of cattle to feed and other purposes, stresses ecosystems around the world, and is assessed to be one of the top three environmental problems in the world on a local to global scale.

The report of FAO, entitled *Livestock's Long Shadow*, also surveys the environmental damage from sheep, chickens, pigs and goats. But in almost every case, the world's 1.5 billion cattle are cited as the greatest adverse impact loader with respect to climate change as well as species extinction. The report concludes that, unless changes are made, the massive damage reckoned to be due to livestock may become more than double by 2050, as demand for meat increases.

Some microbes respire in the cattle gut by an anaerobic process, known as **methanogenesis** (producing the gas methane). Cattle emit a large volume of methane, 95% of it through eructation or burping, not flatulence. As the carbon in the methane comes from the digestion of vegetation produced by photosynthesis, its release into the air by this process would normally be considered harmless, because there is no net increase in carbon in the atmosphere. It is removed as carbon dioxide from the air by photosynthesis and returned to it as methane. Methane is a more potent greenhouse gas than carbon dioxide, having a warming effect 23 to 50 times greater. A little increase in methane concentration in the atmosphere exerts a potentially significant contribution to global warming. Research is underway on methods of reducing this source of methane, by the use of dietary supplements, or treatments to reduce the proportion of methanogenetic microbes, perhaps by vaccination.

Cattle are fed with a concentrated high-corn diet which produces rapid weight gain, but this has side effects which include increased acidity in the digestive system. When improperly handled, manure and other by-products of concentrated agriculture also have environmental consequences.

Grazing by cattle at low intensities can create a favourable environment for native herbs and forbs; however, in most world regions cattle are reducing biodiversity due to overgrazing driven by food demands by an expanding human population.

RELIGION, TRADITIONS AND FOLKLORE

Hindu tradition

Cows are venerated within the *Hindu* religion of India. According to Vedic scriptures they are to be treated with the same respect 'as one's mother' because of the milk they provide; "The cow is my mother", as says the Mahabharata. They appear in numerous stories from the *Puranas* and *Vedas*. The deity *Krishna* was brought up in a family of cow-herders, and given the name *Govinda* (protector of the cows). Also Shiva is traditionally said to ride on the back of a bull, named *Nandi*. In ancient rural India every household had a few cows which provided a constant supply of milk and a few bulls that helped as draft animals. Conservative *Hindus*, even though they might eat meat of other animals, almost always abstain from beef and the slaughter of cows is considered a heinous sin in mainstream *orthodox Hinduism*. Slaughter of cows (including oxen, bulls and calves) is forbidden by law in most of the states of the Indian Union. Illegal slaughter of cows in India is sometimes the reason for religious riots between Hindus and Muslims.

Other traditions

- The Evangelist St. Luke is depicted as an ox in Christian art.
- In Judaism, as described in Numbers 19:2, the ashes of a sacrificed unblemished red heifer that has never been yoked can be used for ritual purification of people who came into contact with a corpse.
- The ox is one of the 12-year cycles of animals which appear in the Chinese zodiac related to the Chinese calendar.
- The constellation Taurus represents a bull.
- An apocryphal story has it that a cow started the Great Chicago Fire by kicking over a kerosene lamp. Michael Ahern, the reporter who created the cow story, admitted in 1893 that he had fabricated it for more colourful copy.
- The first known law requiring branding in North America was enacted on February 5, 1644 in Connecticut. It said that all cattle and pigs should have a registered brand or earmark by May 1, 1644.
- The akabeko (*red cow*) is a traditional toy from the Aizu region of Japan that is thought to ward off illness.

- The Maasai tribe of East Africa traditionally believe that all cows on earth are the God-given property of the Maasai.

In heraldry : Cattle are typically represented in heraldry by the **bull**. See Fig. 9.1.

Population

The world cattle population is estimated to be about 1.3 billion as per 2010 data. India is the nation with the largest number of cattle.

Cattle population (%) in different countries of the world			
India	28.29	Brazil	18.79
China	14.03	United States	9.71
Argentina	5.13	European Union	8.80
Australia	2.93	South Africa	1.42
Canada	1.40	North Africa	4.50
Other countries	5.00		

Economy

Cattle today are the basis of a multi-billion dollar industry worldwide. The international trade in beef for 2000 was over $30 billion and represented only 23 per cent of world beef production (Clay, 2004). The production of milk, which is also made into cheese, butter, yogurt and other dairy products, is comparable in economic size to beef production. All these products provide an important part of the food supply for many of the world's people. Cattle hides, used for leather to make shoes, couches and clothing, are another wide-spread product. Cattle remain broadly used as draft animals in many developing countries like India.

BREEDS OF CATTLE

A group of animals related by descent and similar in most characters like general appearance, features, size, configuration, etc. are said to be a breed. The members of the breed may possess considerable differences; still they have as a group many common points which distinguish them from other groups. Such a common characteristic group is termed a breed.

COMMON INDIAN BREEDS OF COW

There are twenty four prominent breeds of cattle found in India. In addition a large number of non-described cattle, which are small in size and are of low productivity, can be seen in areas of heavy rainfall, coastal and hilly areas of the country, like Assam, Bengal, Bihar, Orissa and Kerala. The quality of cattle depends upon the environmental conditions, particularly on rainfall, temperature and humidity. In India good varieties of cattle are found in dry areas, like Punjab, Haryana, Rajasthan, parts of Gujrat, Maharashtra, Tamil Nadu, Karnataka and Andhra Pradesh.

The breeds are classified on some external characteristics and geographical distribution. The more important breeds of cattle and their geographical distribution are described below.

Italy Lithuania Podlaski, Poland Poland

Fig. 9.1. Heraldry representation by bull in different countries

Lyre-horned grey cattle with wide forehead flat and dished in profile

1. **Kankrej** : These are fast, powerful draught cattle (Fig. 9.2). Average milkers' yield about 1400 kg milk/year in cattle farms and less in villages. They are found in the south-east of Runn of Kutch in Gujrat, from the south-west corner of Tharparkar district (now in Pakistan) to Ahmedabad and from Dessa in the east to Radhanpur in the west.

2. **Hissar** : A powerful draught hybrid developed in course of over a century (Fig. 9.3). Intermediate between Kankrej (Fig. 9.2) and Hariana (Fig. 9.10) breed, Kankrej influence is dominant in its character. Very few animals of this breed are now available. The present government policy is to develop the Hariana breed. These are present in Hissar and Hansi districts.

3. **Kenwariya or Kenkatha** : These are small, but sturdy and powerful cows. They are grey on the barrel and dark grey on the rest of the body (Fig. 9.4). These are found along the Ken River in Banda district of Uttar Pradesh and certain adjoining areas of Madhya Pradesh.

Fig. 9.2. Kankrej

Fig. 9.3. Hissar

Fig. 9.4. Kenkatha

Fig. 9.5. Kherighar

4. **Kherigarh** : These are white cows with narrow face and thick upstanding horns (Fig. 9.5). Very active, good for light draught and trotting. These are prevalent in Kheri district of Uttar Pradesh.

5. **Malvi** : Massively built and compact bodied, grey in colour (Fig. 9.6). Good draught variety. Their size may vary in different areas. Found in Gwalior, parts of Madhya Pradesh and Andhra Pradesh.

6. **Tharparkar** : Medium-sized and compact body stature (Fig. 9.7). Good milch and draught animals. They are found in Sind (Pakistan), Kutch, Jodhpur

Fig. 9.6. Malvi

Fig. 9.7. Tharparkar

and Jaisalmer. Though this breed belongs to Pakistan, herds of high-class animals are available in India. They yield 1800 to 2750 kg of milk in a year in cattle farms and 1360 kg in average.

White or light-grey cattle with conffi-shaped skulls

1. **Bachaur** : Purely draught breed cattle and poor milker (Fig. 9.8). Present in Sitamari district of Bihar.

2. **Gaolao** : These are of medium sized, lightly built cattle. Mostly white in colour and essentially draught animals (Fig. 9.9). They are moderate milker. It is most important breed in Wardha and Chhindwara districts of Madhya Pradesh.

3. **Hariana** : This is one of the best dual-purpose breeds (Fig. 9.10). Hariana breed are powerful work animals and good milkers. Average milk yield per year is about 1360 kg in breeding tracts and 3040 kg/ individual in cattle farms. These are prevalent in Rohtak, Hissar, Karnal and Gurgaon districts of Haryana. It is a popular breed, used extensively in other parts of India.

4. **Krishna valley** : Bullocks are massive and powerful, good for slow draught and heavy ploughing. Fairly good milker, found in Maharashtra-Andhra Pradesh border.

5. **Mewati** : These are powerful animals but docile, suitable for heavy ploughing and carting. They are

Fig. 9.8. Bachaur

Fig. 9.9. Gaolao

Fig. 9.10. Hariana

similar to Haryana breed but with traces of characters of Gir variety. Present in West Alwar and Bharatpur in Rajasthan.

6. **Nagori** : Bullocks are large, powerful and prized much for fast road work (Fig. 9.11). Famous breed of trotting draught cattle. Found in north-east Jodhpur.

7. **Ongole** : Usually found in draught area of the country. Large animals are muscular and powerful (Fig. 9.12). Valued for heavy work, but not suitable for fast trotting. In cattle farms milk yield up to 2300 kg, average about 1350 kg. Yield much less in villages. This breed is distributed in Guntur district of Andhra Pradesh.

Fig. 9.11. Nagori

Fig. 9.12. Ongole

8. **Rathi** : Medium sized, powerful animals adapted for moderately heavy plough and road work (Fig. 9.13). Probably a mixed breed of Hariana, Nagori and Mewati. These are distributed along Alwar and adjacent areas of Rajasthan.

Fig. 9.13. Rathi

Cattle with heavy build and curled horns

1. **Dangi** : These are moderately active workers and poor milker (Fig. 9.14). These are perfectly suited for areas with heavy rainfall. Distributed in small areas in Ahmednagar and Nasik districts of Maharashtra.

2. **Deoni** : Bullocks are suitable for heavy cultivation. Well milker, best adapted in Andhra Pradesh. Black with white patches and red with white

Fig. 9.14. Dangi

Fig. 9.15. Deoni

Fig. 9.16. Gir

Fig. 9.17. Sahiwal

patches are common colour among the group (Fig. 9.15). They are very much related to Gir and Dangi varieties. Distributed in western Andhra Pradesh.

3. **Gir** : Bullocks are large and powerful, but slow and lethargic (Fig. 9.16). They are red or mottled in colour. One of the best milkers in India. In cattle farms the average milk yield is 1680 kg. Distributed in Kathiawar and Baroda areas of Gujarat, northern Maharashtra and adjoining areas of Rajasthan.

4. **Sahiwal and Red Sindhi** : These are mainly prevalent in Pakistan. Large sized and improved herds of these breeds are also found in India (Fig. 9.17).

Mysore breeds with prominent foreheads and long horns

1. **Hallikar** : Grey-coloured animals are good draught cattle (Fig. 9.18). They are poor milkers. Found in Hassan, Mysore, and Tumkur districts of Karnataka.

2. **Alambadi** : These are locally modified variety, i.e., not clearly marked breed (Fig. 9.19). Somewhat similar to Hallikar breed. Present in hilly areas of Salem and Coimbatore districts of Tamil Nadu.

3. **Amritmahal** : Small grey cattle (Fig. 9.20). Active and famous for power of endurance. These are poor milkers. Distributed in some parts of western India.

4. **Bargur** : Moderate, sized cattle, fiery, restive and very difficult to train. Poor milkers found in Coimbatore district of Tamil Nadu.

Fig. 9.18. Hallikar

Fig. 9.19. Alambadi

Fig. 9.20. Amritmahal

5. **Kangayam** : Poor milkers found in Coimbatore district of Tamil Nadu. These are moderate sized, active and powerful draught animals (Fig. 9.21).

6. **Khillari** : These are similar to Amritmahal breed and found in southern Maharashtra (Fig. 9.22).

Fig. 9.21. Kangayam

Fig. 9.22. Khilari

CRAZE FOR HIGH MILKING DWARF INDIAN COW

An increasing number of ultra-rich families of India are becoming crazy of keeping dwarf variety of high milking cows. These are becoming precious assets. One of these varieties is *Punganur* dwarf cow and other is *Vechur* cow. Punganur cow is originated in Chittor district of Rajasthan. It has high fat content and rich medicinal properties in its milk. While cow milk normally has fat content of 3 to 3.5 per cent, the Punganur breed's milk contains 8 per cent, similar to buffalo milk. This high quality milk is offering in the Tirumala Tirupati Devasthanam (TTD). In the TTD's cattle yard about 200 Punganur cow is reared.

Vechur is another dwarf cow and giving more milk than other Indian varieties of cow. This is the smallest of all varieties of Indian cow. In Kerala this cow has become status symbol for high income group peoples.

COMMON EXOTIC BREEDS OF COW USED IN INDIA

Jersy

The Jersey breed (Fig. 9.23) originated on the Island of Jersey, a small British island in the English Channel off the coast of France. The Jersey is one of the oldest dairy breeds and has been reported to be pure-bred for nearly six centuries. This breed was known in England as early as 1771 and was regarded very favourably because of its milk and butterfat production. At that time, the cattle of Jersey Island were commonly referred to as Alderney cattle, but later were referred to only as Jerseys. Adaptable to a wide range of climatic and geographical conditions, outstanding Jersey herds are found from Denmark to Australia and New Zealand, from Canada to South America, and from South Africa to Japan. They are excellent grazers and perform well in intensive grazing programs. They are more tolerant of heat than the larger breeds. With an average weight of 500 kg, the Jersey produces more milk per kilogram of body weight than any other breed.

Fig. 9.23. Jersey

The modern Jersey breed is widely used in dairy farms. Additional emphasis on milk production and less stress on butterfat production had resulted in general acceptance of Jersey cows. Cows show very marked refinement about their heads and shoulders, carry long, straight top lines and usually carry out long and level at the rump. There is no other appealing dairy animal than the well-balanced Jersey cow. She is usually docile and rather easy to manage. Jersey bulls, while small as compared to the other dairy breeds, are extremely masculine. They are quite muscular around their crests and shoulders and are considerably less refined throughout than are the females. They usually range in weight from 650 to 900 kg. Jersey bulls are known for having the least docile temperament of the common breeds of cattle. Modern Jerseys may be of a wide range in colour. The colour in Jerseys may vary from a very light grey or mouse colour to a very dark fawn or a shade that is almost black. Both the bulls and females are commonly darker around the hips and about the head and shoulders than on the body.

Jersey and Jersey crossbreed cows are most suitable for those parts of India where availability of green fodder is scarce and extreme climatic conditions are observed. Average production of milk for all Jersey cows reared at a Model Dairy Farm in Anand in 2007-2008 was 4000 kg. Average fat content is 4.8% and average SNF (Solid Not Fat) is 8.4%.

Holstein and Holstein Friesian

The Holstein cow (Fig. 9.24) originated in Europe. The major development of this breed occurred in the Netherlands and more specifically in the two northern provinces of North Holland and Friesland which lay on either side of the Zuider Zee. The original stock was the black animals and white animals of the Batavians and Friesians, migrant European tribes who settled in the Rhine Delta region about 2,000 years ago. For many years, Holsteins were bred and strictly culled to obtain animals which would make

Fig. 9.24. Holstein Friesian

best use of grass, the area's most abundant resource. The cross-breeding of these animals evolved into an efficient, high-producing black and white coloured dairy cow. Holsteins are quickly recognized by their distinctive colour markings and outstanding milk production. Holsteins are large, stylish animals with colour patterns of black and white or red and white. A mature Holstein cow weighs about 800 kg and stands 58 inches tall at the shoulder. Holstein heifers can be bred at 15 months of age, when they weigh about 450 kg. While some cows may live considerably longer, the normal productive life of a Holstein is 12 years. Average production for all Holsteins reared at a model dairy farm in 2007-2008 was 6000 kg. Average fat content is 3.8% and average SNF is 8.35%

DAIRY MANAGEMENT

Modern and well-established scientific principles, practices and skills should be used to obtain maximum economic benefits from dairy farming. Some of the major norms and recommended practices are as follows :

I. Housing

The following specifications are followed for construction of cattle house (Fig. 9.25) for farming :

1. Construction of shed on dry, properly raised ground should be made.
2. Water-logging in marshy and heavy rainfall areas must be avoided.
3. The walls of the sheds should be 1.5 to 2 metres high.
4. The walls should be plastered to make them damp proof.

Fig. 9.25. Cattle farm

5. The roof should be 3-4 metres high from the floor.
6. The cattle shed should be well ventilated.
7. The floor should be pucca/hard, even non-slippery impervious, well sloped (3 cm per metre) and properly drained to remain dry and clean.
8. There should be a 0.25 metre broad, pucca drain at the rear of the standing space.
9. A standing space of 2 × 1.05 metres for each animal is needed.
10. The manger space (for cleaning and providing food and water) should be 1.05 metre with front height of 0.5 metre and depth of 0.25 metre.
11. The corners in mangers, troughs, drains and walls should be rounded for easy cleaning.
12. It is recommended 5-10 sq. metres loaf space for each animal.
13. There should be proper shade and supply of cool drinking water in summer.
14. In rainy season and winter, animals should be kept indoor during night.
15. Individual bedding should be provided daily.
16. Maintenance of sanitary condition around shed is of prime importance.
17. For control of external parasites (ticks, flies, etc.), spraying the pen sheds with Malathion or Copper sulphate solution should be done regularly.
18. Draining of urine into collection pits and then to the field through irrigation channels will give additional advantage.
19. Proper disposal of dung and urine to a *gobar* gas plant will be an ideal utility of excreta. Where *gobar gas* plant is not constructed, the dung along with bedding material and other farm wastes can be converted into compost.
20. Adequate space for the animals should be provided.

II. Selection of animal

1. Healthy, high yielding animals will be selected with the help of veterinary/animal husbandry officer of State government/ Zilla Parishad, etc.
2. Freshly calved animals in their second/third lactation will be a good purchase.
3. Before purchasing, actual milk yield by milking the animal three times consecutively should be ascertained.
4. Identifying marks in the newly purchased animal should be given immediately by ear tagging or tattooing.
5. Vaccination is a must to the newly purchased animal.
6. Newly purchased animal should be under observation for a period of about two weeks and then allowed to mix with the general herd.
7. Purchase a minimum economical unit of two milch animals.
8. Purchase the second animal/second batch after 5-6 months from the purchase of first animal.
9. When the first animal is in its late stage of lactation and is about to become dry, purchase the second animal. It will maintain continuity in milk production vis-a-vis income. This will ensure availability of adequate funds for maintaining the dry animals.
10. Follow judicious culling and replacement of animals in a herd.
11. Cull the old animals after 6-7 lactations.

III. Feeding of milch animals

1. Best feeds and fodders should be provided.
2. Adequate green fodder in the ration is necessary. It will be better to grow green fodder on own land wherever available. The fodder should be chopped at the right stage of their growth and chaff roughage before feeding.
3. Crushed grains and concentrates are mixed with flaky and crumbly oil cakes. They are moistened before used as feed.

4. Adequate vitamins and salt licks besides addition of mineral mixture to the concentrate ration should be provided.
5. Aadequate and clean water should be provided.
6. Adequate exercise to the animals should be provided.
7. To estimate the daily feed requirement, remember that the animals consume about 2.5 to 3.0 per cent of their body weight, on dry matter basis.

IV. Milking of animals

1. Milking of the animals two to three times a day at fixed times is recommended.
2. As far as possible, milking should be done by the same person regularly.
3. Milking the animal in a clean place is necessary.
4. The udder and teat should be washed with antiseptic lotions/lukewarm water and dry before milking.
5. Milker should be free from any contagious diseases and should wash his hands with antiseptic lotion before each milking.
6. Milking should be done with full hands quickly.
7. Sick cows should be milked at the end to prevent spread of infection.

V. Diseases

There are several diseases which are encountered by these animals. Some of the most important and frequently occurring diseases are described :

(A) *Escherichia coli*

Escherichia coli, although considered to be a part of the normal gut flora for many mammals (including humans), has many strains. Strain E. coli 0157:H7 is associated with human illness (and sometimes death) as a food borne illness. It is reported that grass-fed cattle have much less strain of *E. coli* in their guts than their grain-fed counterparts. Also, the amount of *E. coli* they do have is much less likely to survive the first-line defense against infection, i.e., stomach acid. This is because feeding grain to cattle makes their normally pH-neutral digestive tract abnormally acidic. Over time, the pathogenic *E. coli* becomes acid-resistant. If humans ingest this acid-resistant *E. coli* via grain-feed beef, a large number of them may survive past the stomach, causing dreaded infection.

(B) *Mad cow disease*

Meat and bone meal can be a risk factor for **bovine spongiform encephalopathy (BSE)**, when healthy animals consume tainted tissues from infected animals. People concerned about **Creutzfeld-Jacob disease (CJD)**, which is also a spongiform encephalopathy, favour grass-fed cattle for this reason. The problem becomes more complicated as other feedstuffs containing animal by-products are allowed to be fed to other non-ruminants (chicken, cats, dogs, horses, pigs, etc.). Therefore, at a feed mill for instance, mixing feed for pigs, there is still the possibility of cross-contamination of feed going to cattle. Since only a tiny amount of the contaminating prion begins the cascading brain disease, any amount of mixed feed could cause many animals to become infected. This was the only traceable link among the cattle with BSE.

(C) *Campylobacter*

Campylobacter, a bacterium, contaminated through food, causes illness resulting in nausea, vomiting, fever, abdominal pain, headache and muscle pain.

(D) *Bovine Leukaemia virus (BLV)*

Bovine Leukaemia virus is insect-borne and found in grazing herds. Studies in Sweden and then Soviet Union have linked BLV outbreaks and increases in human leukaemia. BLV and HTLV-1 share a common gene, HTLV-1, which is the first human retrovirus ever known to cause cancer.

(E) *Brucellosis*

The bacterium *Brucella abortus* is the principal cause of brucellosis in cattle. The bacteria are shed from an infected animal at or around the time of calving or abortion. The most common clinical signs of cattle infected with *B. abortus* are high incidences of abortions, arthritic joints which is retained afterbirth. There are two main causes for spontaneous abortion in animals. The first is due to **erythritol**, which can promote infections in the fetus and placenta. The second is due to the lack of anti-Brucella activity in the amniotic fluid. Males can also harbour the bacteria in their reproductive tracts, namely seminal vesicles, **ampulla**e, testicles and epididymides.

(F) *Tuberculosis*

Mycobacterium bovis causes TB in cattle. It has been found that infection of herd by tuberculosis is more likely in areas where infected vector species come in contact with domestic livestock at farm/bush borders. Controlling the vectors through eradication and monitoring the level of disease in livestock herds through regular surveillance, are seen as a "two-pronged" approach to ridding of the disease.

(G) *Johne's disease*

It is a chronic, incurable bacterial infection that primarily affects the lower portion of small intestine of ruminants, although the pathology and symptoms vary among species. Infection most commonly occurs when young animals ingest the bacteria *Mycobacterium avium paratuberculosis (Map)*. After infection, the bacteria grow slowly inside the animal's intestinal cells. Over the time, the animal's immune cells multiply in response to the bacteria's presence, eventually leading to thickening of the intestine and impaired ability to absorb nutrients. Johne's disease in a herd is associated with some or all of the following costs :

- Premature culling resulting in increased replacement costs, increased cull rate, and loss of valuable animals and genetics.
- Milk production may get decreased by 5 to 25 per cent from the last lactation.
- Decreased salvage value at slaughter owing to poor condition at culling.
- Lost commercial market.
- Ineffective treatments, which require drug withholding times for further treatment.
- Cost of testing and added management to prevent spread of disease.

(H) *Mastitis*

In dairy cattle it is the persistent and potentially fatal, inflammatory reaction of the udder tissue. It is the most costly infection of the dairy industry. Milk from cows suffering from mastitis has an increased somatic cell count. *Staphylococcus aureus* is the most common etiological organism responsible, but *S. epidermidis, Pseudomonas aeruginosa* and *Strepto cocci* are occasionally isolated as well.

Lactation mastitis usually affects only one breast and the symptoms can develop quickly. The signs and symptoms usually appear suddenly and they include :

- Breast tenderness or warmth to the touch.
- General malaise or feeling ill.
- Swelling of the breast.
- Pain or a burning sensation continuously or while breast-feeding.
- Skin redness, often in a wedge-shaped pattern.
- Fever of 101°F (38.3°C) or greater.
- The affected breast can then start to appear lumpy and red.

VI. Protection against diseases

1. Regular checking and alertness for signs of illness such as reduced feed intake, fever, abnormal discharge or unusual behaviour will prevent diseases. In any unusual condition consultation with the nearest veterinary aid centre for help, is necessary. This will protect the animals against common diseases.
2. In case of outbreak of contagious disease, immediate separation of the sick animals from the healthy animals is necessary to control the disease. Periodic tests for Brucellosis, Tuberculosis, Johne's disease, Mastitis, etc. will give good result.
3. De-worming the animals regularly is a healthy practice. Examination of faeces of adult animals to detect eggs of internal parasites and subsequent treatment of the animals with suitable drugs is of prime importance.
4. Wash the animals from time to time to promote sanitation.

VII. Breeding care

Close observation of the animal and keeping specific record of its coming in heat, duration of heat, insemination, conception and calving are essential factors. This helps the animals to breed in time. The onset of oestrus will be within 60 to 80 days after calving. Timely breeding helps achieving conception within 2 to 3 months of calving. Breeding should be done when the animal is in peak heat period (i.e., 12 to 24 hours of heat). High quality of semen, preferably frozen semen, of proven sires/bulls should be used.

VIII. Care during pregnancy

Special attention should be given to pregnant cows two months before calving by providing adequate space, feed, water, medical aid, etc.

IX. Marketing of milk

Marketing of milk should be made immediately after it is drawn, keeping the time between production and marketing of the milk to the minimum. Use of clean utensils and handling the milk in hygienic way are primary requirements. Milk pails, cans, utensils are washed thoroughly with detergent and finally rinsed with chloride solution. Too much agitation of milk during transit must be avoided. Transportation of milk during cool hours of the day is recommended.

X. Care of calves

Caring of newborn calf is vital. Treatment and disinfection of the navel cord with tincture of iodine as soon as it is cut with a sharp knife is essential. Feed colostrums to calf. Assisting the calf to suckle on its own within 30 minutes of calving is a good practice. In case it is desired to wean the calf immediately after birth, then feeding the colostrums will be in bucket.

The calf should be separated from birth till two months of age in a dry clean and well-ventilated place. Protection of the calves against extreme weather conditions, particularly during the first two months is necessary. Grouping of the calves according to their size will be done. Calves should be regularly vaccinated. The calves should be dehorned around 4 to 5 days of age for easy management when they grow. Dispose off extra calves not to be reared or maintained for any specific purpose as early as possible, particularly the male calves.

PHILOSOPHY OF CATTLE FARMING

Cattle production worldwide is based on **animal genetics** and **feeding methods**, for different quality of cattle products. Cattle are basically residual claimants to crop or land resources. Those countries with excess of low-value land tend to grass-feed their cattle herds, while those countries with excess feed grains like, U.S.A. and Canada, feed cattle with a grain ration. Grain-fed cattle have more internal fat (i.e., marbling) which results in a more tender meat than forage-fed cattle of a similar age. In some Asian countries such as Japan, which is not a grain surplus country, tastes and preferences of consumers have encouraged grain feeding of cattle, but at a high cost since the grain must be imported.

CATTLE FEED

Cattle may be raised on diets primarily composed of pasture (grass) or a concentrated diet of grain, soy, corn and other supplements. Cattle raised on a primarily forage diet are termed **grass-fed** or **pasture-raised**; for example, meat or milk may be called grass-fed beef or pasture-raised dairy. Most cows have a diet that is composed of at least some forage (grass, legumes, or silage). In fact, most beef cattle are raised on pasture.

Different cattle feeding production systems have separate advantages and disadvantages. However, the composition of some of the feed for cattle is discussed here in details.

Grass fed: Chips of grass and other forage compose most or at least the great majority of the grass fed diet.

Corn-fed : Cattle called "corn-fed," "grain-fed" or "corn-finished" are typically fattened on corn, soy and other types of feed for several months before slaughter. As a high-starch, high-energy food, corn decreases the time to fatten cattle and increases yield from dairy cattle. Some corn-fed cattle are fattened in concentrated animal feeding operations.

Fatty diet : Most grass-fed cattle are leaner than grain-fed beef lacking marbling, which lowers the fat content and caloric level of the meat. Meat from grass-fed cattle has higher levels of conjugated linoleic acid (CLA) and the omega-3 fatty acids ALA, EPA, and DHA.

Flax (linseed) seeds *(Linum usitatissimum)* is one of the most biologically available botanical sources of omega-3 fatty acid which is high in protein and ALA (alpha-linolenic acid). Ground flax is used as a fat supplement in cattle rations. Being high in protein and fats, it can easily replace other energy dense foods such as corn. Flax seeds suppress inflammatory effects from bovine respiratory disease (BRD) often affecting stressed cattle due to production practices such as transport and processing. Another effect of feeding flax in cattle ration is an observed increase in the daily dry matter intake. Flax feeds are more expensive compared to other feeds, but the economical benefits from having a generally healthier herd of cattle outweigh the cost.

Antibiotics : Antibiotics are commonly used in the food production system as a way to control the growth of potentially harmful bacteria. Potential benefits from the use of antibiotics include the prevention of diseases, increase in food and water uptake and increase the digestive effectiveness of the animal. There are concerns, however, about residues of the antibiotics getting into the milk or meat of cattle. There are two distinctions between the clinical and non-clinical use of antibiotics in cows. Clinical use of antibiotics refers to the treatment of cows due to sickness. However, corn-fed cattle draw attention to the non-clinical use of antibiotics. Antibiotics are used to promote growth and treat sick cattle, yet the cattle would not get sick if they were not fed on corn-based diet, that subjects them to diseases caused by the malfunctioning of their rumen.

NECESSITY OF DAIRY FARMING

Dairy farming is an important source of subsidiary income to small or marginal farmers and agricultural labourers.

1. The manure from animals provides a good source of organic matter for improving soil fertility and crop yields.

Fig. 9.26. Gobar gas plant

2. The *gobar* gas from the dung is used as fuel for domestic purposes as also for running engines for drawing water from well (see Fig. 9.26).

3. The surplus fodder and agricultural by-products are gainfully utilized for feeding the animals.

4. Almost all draught power for farm operations and transportation is supplied by bullocks.

5. Dairy provides additional income when cultured along with fishery, aquaculture or agriculture.

6. Since agriculture is mostly seasonal, there is a possibility of finding employment throughout the year for many persons through dairy farming. Thus, dairy also provides employment throughout the year.

The main beneficiaries of dairy programmes are small or marginal farmers and landless labourers.

According to World Bank (2008) estimates, about 75 per cent of India's 940 million people are in 5.87 million villages, cultivating over 145 million hectares of cropland. Among 70 million rural households, 42 per cent operate up to 2 hectares of farmland and 37 per cent are landless households. These landless and small farmers in India have in their possession 53 per cent of the animals and produce 51 per cent of the milk. Thus, small or marginal farmers and landless agricultural labourers play a very important role in milk production of the country. Dairy farming can also be taken up as a main occupation around big urban centres where the demand for milk is high.

SCOPE FOR DAIRY FARMING AND ITS NATIONAL IMPORTANCE

The total milk production in the country was estimated at 84.6 million metric tons. At this production, the per capita availability is estimated to be 226 grams per day against the minimum requirement of 250 grams per day as recommended by ICMR. Thus, there is a tremendous scope or potential for increasing the milk production. The population of breeding cows and buffaloes over 3 years of age was 62.6 million and 42.4 million, respectively (1992 census). Central and State Governments are giving considerable financial assistance for creating infrastructure facilities for milk production.

MILK AND MILK PRODUCTS

See flow chart 9.1 on milk products and production relationships.

WHITE REVOLUTION AND OPERATION FLOOD

'Operation Flood' (OF) commonly referred to as 'White Revolution' is an activity launched in 1970 to make use of the dairy aid offered by European Economic Community (EEC), that ended in1996. It is set within an ambitious framework in terms of both time and space and in the wider context of the political economy of evolving trade and the aid policies. The main argument on this project was —Food aid is often the benign face of surplus disposal, benefitting special interests in donor countries and disrupting farming in recipient countries. In India, OF is an exception to the rule that food aid does more harm than good; here Government funded agriculture research ensures people's entitlement to food better than big commercial ventures. India's White Revolution was successful because of the creation of a small farmer-controlled net work of diary co-operatives. Initially the problem apprehended by the economists that the foreign monetary assistance would cause milk prices to crash, disincentivise milk production and ultimately make India permanently dependent on other countries for its needs of dairy products.

The challenge was to see that the flow of the dairy aid is calibrated in a way that the supplies do not flood the market and to design a consolidated national plan to mitigate disincentives to the domestic dairy farmers. The strategy was to 'monetise' the aid; invest the profits from the monetised commodities into capital-incentives through essential infrastructure in the form of National Milk Grid System (NMGS).

This holistic and multifaceted strategy paid off. Not only did the market supplies remained unaffected but also the dairy farmers got the right prices. The necessary infrastructure was established and streamlined to match the growing urban demand. In India 'Amul' is hailed as the model in Gujarat. The small farmers accounted for about 70% of milk production, and 22.5% of the rural households' income came from milk. Thus, India dispelled fears that food aid would throw it to a state of perpetual dependence on others, serving as a market for the surpluses of the rich countries.

VERGHESE KURIEN — THE MILKMAN OF INDIA

Dr. V. Kurien is considered as the father of "**White Revolution**". He was the founder of the cooperative dairy movement in the country. He was born in Kozhikode, Kerala on November 26, 1921 and at the age of 90, he expired on 9th September 2012 in Anand, Gujarat, India. Basically he was a Mechanical Engineer with a degree in Dairy Engineering. He came to Anand in 1949 at the behest of the then Union Home Minister, SardarVallabhbhai Patel, to solve some problems of the local farmers. The problem was solved but Dr. Kurien could not leave Anand. He fell in love with this place.

In 1949 Dr. Kurien formed **Anand Milk Union Limited (AMUL)** and started to solve the milk producers' problems. From one milk project to a larger one, the 'Milkman of India' saw India emerge from a country into the largest milk producing country in the world, overtaking even the once milk-abundant Netherlands. It is because of Dr. Kurien that India

FLOW CHART No. 9.1 : MILK PRODUCTS AND PRODUCTION RELATIONSHIPS

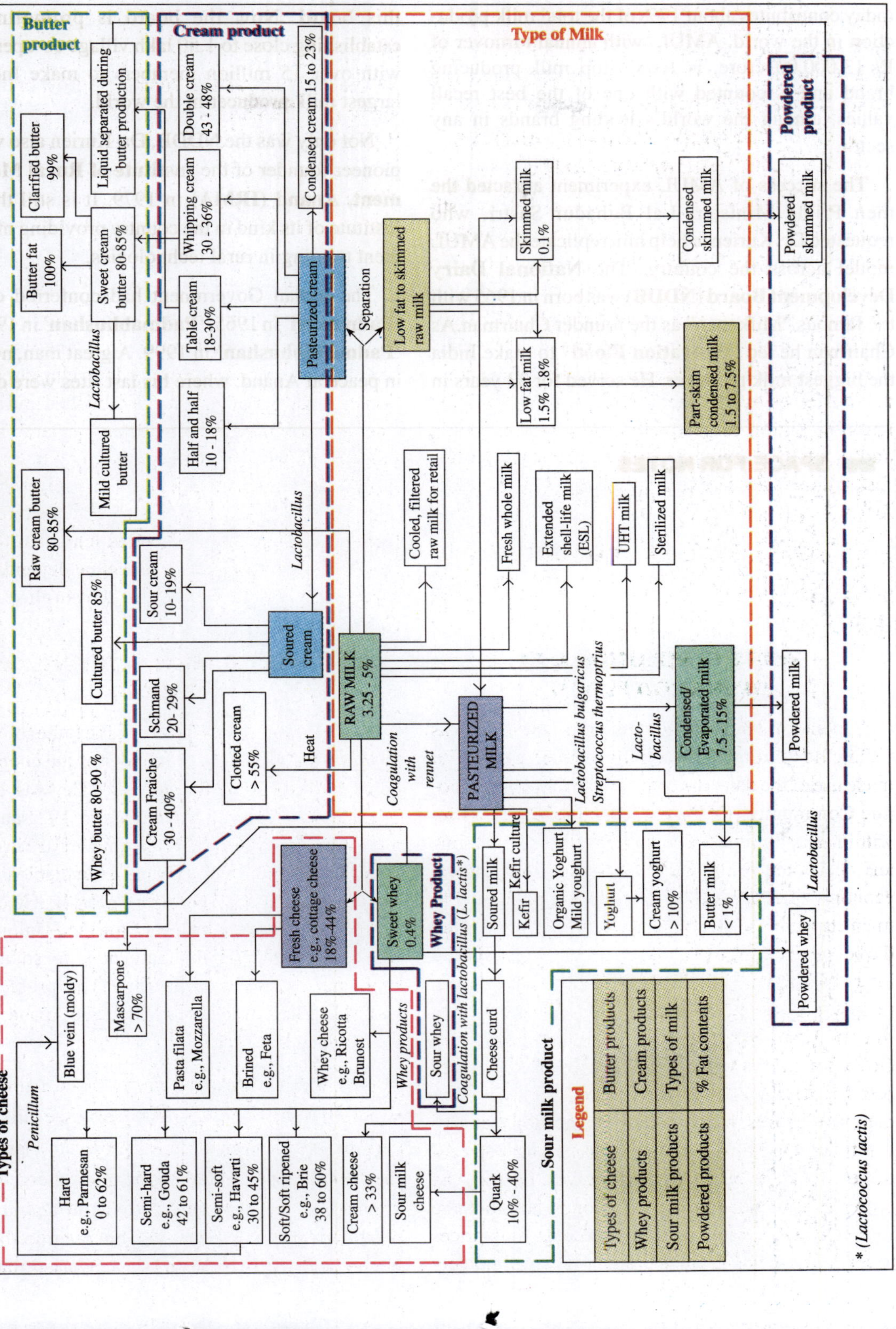

today contributes about 17% of the total milk production in the world. AMUL, with annual turnover of Rs.13,000.00 crore, is Asia's top milk-producing brand and is counted with one of the best recall values, among the world's leading brands in any sector.

The success of AMUL experiment attracted the then Prime Minister, Lal Bahadur Sastri, who requested Dr. Kurien to help him replicate the AMUL model across the country. The **National Dairy Development Board (NDDB)** was born in 1965 with the famous "Milk man" as the founder Chairman. As Chairman he led "**Operation Flood**" to make India the biggest milk producer. He served for 33 years in this board. Now the board is promoting and establishing close to 1.50 lakh village co-operatives, with over 15 million members to make India the largest milk producer in the world.

Not only was the NDDB, Dr. Kurien also was the pioneer founder of the **Institute of Rural Management, Anand (IRMA)** in 1979. It is still the only institute of its kind in the country providing management training in rural technologies.

The Indian Government had conferred on him '**Padmashri**' in 1965, '**Padmabhushan**' in 1966 and '**PadmaVibhushan**' in 1999. A great man, now lies in peace at Anand, where his last rites were done.

SPACE FOR NOTES

(B) BUFFALO

INTRODUCTION

The buffalo is a member of Bovine family and its scientific name is *Bubalus bubalis.* This is descended from the *arni* or wild Indian buffalo, and is widely dispersed throughout Southern Asia. The world buffalo population is continuously increasing and was estimated to be more than 170 million in 2003 as reported by FAO (2004). More than 95 per cent of the world buffalo population is found in Asia where they play a leading role in rural livestock production. Over the last decade buffalo farming has widely expanded in Mediterranean areas and in Latin American countries and several herds have also been introduced in Central and Northern Europe. Buffalo farming is increasing in Italy too, due to the growing market demand for buffalo milk, that is utilized exclusively for the production of "**mozzarella cheese**". Another economic benefit derived from buffalo milk production is that buffalo milk is not restricted by the specific European Union (EU) directive called "milk quotas", introduced to stop the increase of cow milk production. In fact, this regulation induced some farmers, in areas where Friesian cattle are traditionally reared, to consider the option of breeding milking buffaloes for the production of "mozzarella cheese". This led to an expansion of buffalo breeding in the north of Italy which is away from the customary area, that is southern Italy (Campania, Lazio and Puglia regions), where 95 per cent of the Italian buffalo population is reared.

Classification

The domesticated buffaloes are commonly classified into two main categories, viz., (1) the swamp buffalo and (2) the river buffalo. They belong to the same species but have contrastingly different habits. The swamp buffalo is more or less a permanent inhabitant of marshlands, where it wallows in mud, feeds on coarse marsh grass, and is found principally in Malaya, Singapore, Philippines, Thailand, Indonesia, Southern China and other countries in Asia. The river type is found throughout India, Pakistan, Bangladesh, Nepal and Sri Lanka. These are primarily milch animals.

DIFFERENT TERMS USED IN RELATION TO BUFFALO

Terms	Detail
Herd	Group of animals
Buffalo bull	Adult male
She-buffalo/buffalo cow	Adult female
Buffalo bull calf	Young male
Bellowing	Sound produced
Buffalo bullock	Castrated male
Spayed	Castrated female
Calf at foot	Female with its offspring
Caving	Act of parturition
Serving	Act of mating
Gestation	Pregnancy

Characteristics of Buffalo

Buffaloes are very docile in nature. In India, usually responsibility of care of milking buffalo is left to women and girls. The men usually take care of bullock. The buffalos are long-lived animals. It is quite common to find a reasonable milk yield or a good day's work still being achieved even at 15 years of age. Buffalos of 18 years and even more may produce calves. Up to 20 years they can be used in different works.

These animals are less heat resistant than other bovines, though they are mostly found in tropical countries. Access to water or shade is essential for the well being of buffalos. Milking animals must have liberal amount of drinking water to avoid dehydration, which can take place at any time. Sudden changes of temperature and especially the effect of cold winds and draughts may be extremely harmful to this species. These are regularly wallowing animals. The river breeds of Indian sub-continent prefer the clear water of streams and pools. Buffaloes apparently wallow for two main reasons : (i) to regulate body temperature and (ii) to control parasites.

PUBERTY

Puberty in buffalo is delayed compared with cattle. The age at puberty is difficult to establish because of difficulties in oestrus detection in this species and most estimations appear to have been extrapolated from the age at first calving. The River type puberty is earlier than the Swamp type (21 to 24 months). First conception occurs at an average body weight of 250 to 275 kg, which is usually attained at 24 to 36 months of age. This characteristic is important for buffalo culture.

BUFFALO BREEDS

Eleven Indian breeds are listed according to the breed names given in Mason (1996). The breeds are described in details following FAO data in tabular form (Table 9.1 to 9.11) .

Table 9.1 : SWAMP BREED

Habitat	Brahmpurta river banks in north-east		
Average production traits		**Phenotypic traits**	
365 days milk yield	500 kg	Horns	Straight with turn at ends
Age at first calving	55 months	Colour	Slate black/dark-tan/albino
Lactation length	313 days	Size	Small and compact
Calving interval	511 days	Forehead	Light neck and head
		Tail	Medium length

Table 9.2 : TODA BREED

Habitat	Western Tamil Nadu		
Average production traits		**Phenotypic traits**	
365 days milk yield	700 kg	Horns	Long and semicircular
Age at first calving	47 months	Colour	Light to dark grey
Lactation length	250 days	Size	Medium/long with broad & deep chest
Calving interval	480 days	Forehead	Broad and concave
		Tail	Long and slim with black switch

Table 9.3 : SURTI BREED

Habitat	South-Western Gujarat		
Average production traits		**Phenotypic traits**	
365 days milk yield	1400 kg	Horns	Sickle shaped, growing downward, tip upward
Age at first calving	56 months	Colour	Black as well as brown
Lactation length	290 days	Size	Medium
Calving interval	535 days	Forehead	Broad, long and convex between horns
		Tail	Long thin flexible, white switch common

Table 9.4: BHADAWARI BREED

Habitat	U.P. and adjoining M.P.		
Average production traits		**Phenotypic traits**	
365 days milk yield	1100 kg	Horns	Curling downward, backward — then upward
Age at first calving	50 months	Colour	Copper-coloured hair, grey skin, white ring under neck
Lactation length	270 days	Size	Medium with wedge-shaped body
Calving interval	478 days	Forehead	Small and bulging between horns
		Tail	Thick and long with brown switch

Table 9.5 : PANDHARPURI BREED

Habitat	Southern Maharashtra		
Average production traits		**Phenotypic traits**	
365 days milk yield	1400 kg	Horns	Long touching hock bone
Age at first calving	45 months	Colour	Blackish with grey
Lactation length	350 days	Size	Medium
Calving interval	465 days	Forehead	Long narrow face, prominent thick neck
		Tail	Short, white switch common

Table 9.6 : NAGPURI BREED

Habitat	Vidarbha region of Maharashtra		
Average production traits		**Phenotypic traits**	
365 days milk yield	1200 kg	Horns	Long, flat, curved backwards on sides of neck
Age at first calving	56 months	Colour	Black
Lactation length	270 days	Size	Medium with slightly deep back
Calving interval	430 days	Forehead	Long, cone-shaped. Straight nasal bone
		Tail	Short, white switch common

Table 9.7 : MARATHWADA BREED

Habitat		Marathwada region of Maharashtra	
Average production traits		**Phenotypic traits**	
365 days milk yield	1000 kg	Horns	Medium length, parallel to neck reaching shoulder
Age at first calving	57 months	Colour	Grayish black to jet black
Lactation length	300 days	Size	Light to medium, compact stature
Calving interval	435 days	Forehead	Long narow face, white mark on head is common
		Tail	Short, white switch common

Table 9.8 : MEHSANA BREED

Habitat		Northern Gujarat	
Average production traits		**Phenotypic traits**	
365 days milk yield	1700 kg	Horns	Slightly curved upward, inward
Age at first calving	42 months	Colour	Black, dark tan
Lactation length	310 days	Size	Medium
Calving interval	476 days	Forehead	Wide and slight depression in middle
		Tail	Long, switch black, brown/white also common

Table 9.9 : NILI RAVI BREED

Habitat	Western Punjab		
Average production traits		**Phenotypic traits**	
365 days milk yield	1950 kg	Horns	Small and curled
Age at first calving	45 months	Colour	Black, white spots on forehead, muzzle, feet. Walled eyes
Lactation length	300 days	Size	Medium to large with deep frame
Calving interval	487 days	Forehead	Heavy head, convex at center
		Tail	Long, white switch common

Table 9.10 : JAFFARABADI BREED

Habitat	Southern Gujarat		
Average production traits		**Phenotypic traits**	
365 days milk yield	1850 kg	Horns	Curved downward, compressed head
Age at first calving	51 months	Colour	Black, Dark tan
Lactation length	300 days	Size	Massive body with long and tender neck
Calving interval	440 days	Forehead	Heavy and wide head
		Tail	Medium

Table 9.11 : MURRAH BREED

Habitat		Central Haryana and Delhi	
Average production traits		**Phenotypic traits**	
365 days milk yield	2000 kg	Horns	Short and slightly curled
Age at first calving	44 months	Colour	Jet-black
Lactation length	300 days	Size	Long body with massive frame
Calving interval	453 days	Forehead	Light neck and head
		Tail	Long, white switch common

Husbandry

Buffaloes are traditionally managed in domestic conditions together with the calf. They are hand-milked twice a day. They are fed with different kinds of roughages (barley and wheat straw, cornstalks, sugar cane residuals). In addition, they are given concentrated mixtures (commercially available food). If grazing is available, they graze all day long. They mate naturally. Some villages also practise artificial insemination.

Breeding

Artificial breeding process is mainly restricted to insemination only. In spite of expansion in buffalo breeding, there was no improvement in milk and meat production due to slow genetic progress. The productivity increase obtained is due mainly to an improvement in management techniques rather than to genetic selection.

BUFFALO MANAGEMENT

Breeding behaviour — Buffaloes are seasonal breeders

The buffalo is poly-oestrous, breeding throughout the year, but rainfall, feed supply, ambient temperature and photoperiod influence the annual calving pattern. In tropical countries like India, she buffaloes show seasonal sex periodicity in respect of oestrous and conception. Most of the animals show periodicity in October to February with the peak around December and this period coincides with higher conception rate. Buffalo bulls are sexually least active during hot seasons. High environmental temperature upsets the normal physiological functions affecting spermatogenesis in males and ovarian activity in the females. Decreasing day length and cooler ambient temperatures favour cyclicity whereas long day length and the high summer temperatures depress cyclicity. In the Indian subcontinent, most buffaloes calve between November and March.

During this breeding period, the bulls have been found to be very active sexually and the quality and quantity of semen are very high, particularly during winter (November to February). The she-buffaloes show the maximum of ovarian activity and the largest percentage of them conceive during this period.

With judicious management practices such as heat detection, protection from thermal stress, adequate and balanced ration and optimum time of insemination, the calving in buffaloes could be evenly spread round the year to a great extent.

Silent Heat

Silent heat refers to normal follicular development and ovulations without the behavioural signs of estrus. In summer the extreme climatic stress coupled with increased daylight reduces the incidences of buffaloes coming in regular heat. During this time other reproductive problems like repeat breeding and pregnancy losses are also increased. It is generally felt that most of the buffaloes do have sub-functional ovaries causing weak or silent oestrus (in 60% cases) and such incidences are even higher than the true anestrous, especially in summer.

Corrective Measures

(1) For animals which either do not manifest oestrus or are in a state of silent oestrus or sub-oestrus firstly need proper attention rather than therapies. Two or three examinations with an interval of a week between each state will reveal that most of them are cases of apparent anestrous.

(2) In extreme summer condition these animals should be protected from direct solar radiation by maintaining them in shaded half-walled sheds. If possible animals are given showers in addition to wallowing facilities particularly at midday during April, May and June.

(3) Parading vasectomized bulls for heat detection can be gainfully employed at organized/large farms. Since in summer buffalo bulls soon lose libido, replacement by young high libido bulls may be applied.

(4) Painting of *Lugol's Iodine* : The external os is swabbed thoroughly with Lugol's solution with the help of a metallic swab holder and vaginal speculum—double application of this solution is sometimes recommended at one week's interval. It is presumed that application of Lugol's Iodine on the cervix causes local irritation and brings about reflex stimulation of anterior pituitary for secretion of gonadotrophins and thus cyclicity starts.

Repeat Breeding

A repeat breeding in buffalo is the one which has normal oestrus, oestrus cycles as well as reproductive tract and has been bred three or more times by a fertile bull. It is possible that pathology which cannot be detected by normal clinical methods may be present which include among others, failure of fertilization, anovulation and delayed ovulation, tubal obstructions, early or latent embryonic mortalities, poor breeding and management techniques including genetic, nutritional and infectious factors. Endometritis in many cases has been found to play major factor for causing repeat breeding.

Endometritis is caused by sporadic uterine infections by a variety of micro-organisms such as Coliform bacteria, *Pseudomonas aeruginosa*, *Streptococci*, etc., viral and fungal agents along with mycoplasmal agents may also cause endometritis in cattle and buffaloes. The condition follows mainly parturitions, especially abnormal ones; abortion, dystokia, retained **placenta,** genital prolapse, uterine inertia, traumatic lesions in the uterus, cervix and vagina. Daily examination of repeaters is necessary to correct the condition arising from ovarian dysfunction.

The golden rule, prevention is better than cure, applies more appropriately to reproductive problems. The incidence can be kept under control by adopting methods such as : (1) improved feeding and management, (2) providing hygienic surroundings at the time of calving, (3) aseptic precautions at the time of inseminations, and (4) educating farmers on detection of heat and maintenance of breeding records.

Weaning : Separating calf from its mother at a very early age is a bit difficult in buffaloes due to high motherly instinct. Since the method is important to assess dam's milk production and to feed calf according to live weight, a substitution technique could be followed.

A single calf is substituted for let-down of milks

in a number of buffaloes and the real calves are weaned just after birth. It can be achieved by putting blinkers (materials used to shut off the side views of buffaloes) during parturition and subsequently soiling the body of substituted calf with placental secretions. The dam starts licking the substituted calf after removal of blinkers.

Calf Mortality

Buffalo calf mortality rate under one month of age averages about 10% and varies from 3–30% in individual herds. Losses up to 50% have occurred in large dairy herds. A calf mortality rate of 20% can reduce net profit by 38%.

A number of factors considered responsible for this hazard are discussed below :

(1) Antibodies are not transferable from buffalo dam to her fetus through placental membranes and thus they are susceptible to various virulent diseases. In buffalo calves in spite of feeding colostrums, a low antibody titter exists. Immunoglobulin levels have been reported to be 29.73 mg/ml in day olds and the quantity increases to 35.66 mg/ml on the second day of life. On the other hand in one estimate it has been noted that colostrums which is the only source of immunoglobulin for the buffalo calf, contains 68.75 mg of immunoglobulin per ml on the first day, 23.75 mg/ml on the second day and 1.01 mg/ml on the fifth day of lactation

(2) Certain meteorological influences may have an effect on calf mortality rate. During the winter months, mortality may be associated with the effects of cold, wet and windy weather while during the summer it may be the hot, dry weather.

(3) Most of the deaths occur during autumn and winter months before the age of 3 months. The causes of mortality in order of priority have been found :

(a) Pneumonia, (b) Enteritis,

(c) Toxaemia/Septicaemia,

(d) Worm infestation, (e) Bloat, etc.

(4) The cause of high mortality in male calves could also be due to neglecting tendency by the management, particularly regarding feeding.

Some of the above prescribed management practices including feeding practices may be followed for minimizing the mortality rate.

(a) ***Housing*** : In commercial herds, calves after weaning may be kept in groups in large pens where individual feeding is advocated. The small herds' calves after weaning should be kept in separate pens (24 sq. ft) up to 3 months of age to avoid suckling instinct. The methods will eliminate the possible calf scour and parasitic infestation.

(b) ***Feeding colostrums*** : It is the milk secreted by the udder immediately after parturition and for the following 3-5 days. It contains 20% or more protein, a little more fat, 10 to 100 times more vitamin A, three times more of vitamin D and may be tinged pink due to blood corpuscles. It acts as a natural purgative for the calf, cleaning from its intestines the accumulated faecal matter. Of much greater importance, it is through the medium of the colostrums, antibodies which protects against various bacteria, and viruses are supplied to the newborn calves.

The calves at birth after weaning should be fed colostrums within 2 hours and its feeding should continue for a period of five days at the rate of 1 to 1.5 kg per day. The details of milk feeding schedule are given in Table 9.12.

Table 9.12 : Milk Feeding Schedule

Age (days)	Colostrums (kg)	Body weight (kg)	Milk (litre)
1—5	1 to 1.5	—	—
6—15		Up to 25 kg	1.0
11—15		26—30	2.0
15 & above		31—40	2.5
		41—45	3.0
		46—50	3.5
		51—55	4,0
		56—60	4.5
		60 and above	5.0

In case the dam does not give colostrums a substitute of equal nutritive value prepared from 2 eggs and an ounce of castor oil may be fed for building up resistance. It may be necessary to inject dam's serum to such a calf for augmenting antibody titre in the body. The feed efficiency ratio of buffalo calves is as high as 1 kg gain per 1.16 kg dry matter. Milk feeding schedule is given in tabular form. It is necessary to provide milk to weaned calves at body temperature preferably with supplements to make up the deficiencies of Fe, Cu, Mg, Zn and Mn. Green fodder containing up to 100 gram dry matter may be offered daily from 15 days of age onwards so that it stimulates early rumen development.

(c) ***Feeding Antibiotics*** : Antibiotics are necessary in calves below 3 months of age to overcome certain stress conditions developed due to clinical or sub-clinical type of scour. These drugs are usually unnecessary in calves above 3 months as by this time rumen starts functioning and antibiotics are liable to interfere in normal functioning of rumen microbes.

(d) ***Feeding Milk Replacer*** : The objectives of feeding milk replacer to calves are primarily to reduce the cost of raising buffalo calves and also to save milk for human consumption. Details of milk replacer are given in Table 9.13.

Table 9.13 : Ingredients of milk replacer

Ingredients	Quantity
Wheat	8 kg
Fish meal	12 kg
Linseed meal	40 kg
Milk	16 kg
Coconut oil	7 kg
Linseed oil	3 kg
Citric acid	1.4 kg
Molasses	8 kg
Mineral mixture	3 kg
Butyric acid	0.660 litres
Antibiotic mixture	300 g
Rovimix A, B_2, D_2	15 g

Reproductive factors in Female

Poor reproduction is a major limiting factor in buffalo production and for achieving quick genetic improvement. Buffaloes, both swamp and river, have a very high age at first calving and long calving intervals, the latter being partly due to greater lactation stress in high producing buffaloes and partly due to seasonality of breeding.

For a successful breeding to occur, a female animal must be in heat and bred at the right time after oestrous is observed. With proper nutritional management heifers could be bred successfully even as early as two years of age. Buffaloes have been reported to show less intense signs of oestrous. Problems of irregular oestrous are the most important factor obstructing attainment of full reproductive and productive potential of this species. Notable reason for this is the seasonal variation and malnutrition. In such cases the ovaries become smooth or inactive.

Estrous Cycle

The estrous cycle is about 21 days which comprises four distinct phases, viz., pro-estrous, estrous, meta-estrous and di-estrous. Overt signs of estrous in buffalo are not as pronounced as in cattle. Heterosexual behaviour or standing to be mounted by a male is the most reliable sign of estrous in the buffalo.

Mating Behaviour

Mating behaviour in many respects resembles that of cattle. On contacting a female, the male sniffs her urine, then displays the 'Flehmen' reaction and proceeds to nuzzle and lick the perineum and vulva. An oestrous female responds by standing immobile for the male to mount and ejaculate; mutiny lasts for 20-30 seconds. Rhythmic pelvic thrusts during intromission and the forward leap at ejaculation are less marked in buffalo than cattle. The male dismounts and gradually retracts penis into the sheath while the female remains with her back arched and tail elevated for a few minutes.

Gestation length

The gestation period varies according to breed and environment and appears to be slightly longer than cattle. It is generally quoted as being "6 weeks less than a year" (or 10 months and 10 days). The average appears to be 316 + 5 days.

Under the usual conditions of husbandry, a female buffalo will produce its first calf at the age of $3\frac{1}{2}$ to $4\frac{1}{2}$ years and thereafter will average 2 calves in every three years up to the age of 15. Twins are extremely rare. The average time for entire process of parturition comprising (i) dilation of the cervix, (ii) expulsion of foetus, (iii) expulsion of foetal membrane is about 1 hour. Calving seldom presents any difficulty.

Factors Affecting Libido and Semen Quality in Male

The size and weight of testes are significantly lower in buffaloes than in cattle, and are reflected in relatively poor libido and semen production and quality.

High ambient temperature and direct solar radiation adversely affect libido and semen production. Administration of exogenous gonadotrophins has produced variable results. Intramuscular injections of prostaglandins (PGF2) 1 hour before semen collection reduced the reaction time and resulted in increased ejaculate volume without affecting the total and live sperm concentration.

The major problems of male fertility are poor libido, quantitative and qualitative reduction in buffalo semen during summer, and wide range of variability in libido, semen production, and quality and freezability among bulls. The spermatogenic efficiency of the buffalo is only 43%. Although these problems are mostly related to nutrition and physical environment, especially heat stress, there are large genetic differences which could possibly be exploited in improving performance of males.

Artificial Insemination (AI) in Buffalo

In India AI has been practised in buffaloes for over 30 years but has lagged behind cattle largely because of the difficulty of detecting oestrous. Chilled semen (5°C), diluted in a variety of extenders is still widely used in buffalo AI.

Buffalo semen can be frozen in extenders used for bovine semen. One such extender is a tris-buffer containing 7% glycerol and 20% egg yolk. Extended semen is packed in 0.25 or 0.1 ml French straws to contain 30 million motile spermatozoa per dose. Straws are exposed to nitrogen vapour at –120°C to –140°C for 7 to 10 minutes before being stored in liquid nitrogen. Semen should be thawed at 37°C to 40°C. The post-thaw progressive motility of buffalo semen varies from 35 to 60%.

Nutrition : Buffaloes are mostly found in countries where feeds, fodders and pastures are limited and are of poor quality. In spite of this situation, the buffaloes have thrived reasonably well and relatively better than cattle. Energy and protein requirements have been worked out for maintenance and milk production. The energy requirement for maintenance ranges from 97.8 to 188.8 kcal in dry and lactating buffaloes.

- Reproductive efficiency is the primary factor affecting productivity and is hampered in the female buffalo by a delayed attainment of puberty, seasonality, long post-partum anestrus and subsequent calving interval, and poor oestrus expression. Regarding the latter, artificial insemination (AI) which is the normal practice in cattle breeding, is limited in buffaloes due to the weakness of oestrus symptoms and variability of oestrus length that makes oestrus detection very difficult.
- Studies have been undertaken in order to better understand the reproductive physiology of the buffalo and the factors affecting its behaviour. Considerable attention has been paid to reproductive endocrinology, with the aim of developing models to improve reproductive efficiency, particularly when controlled breeding techniques are utilized.
- There is a large variation in age at puberty of buffaloes in different countries. The delay in puberty and the consequent delay in conception is one of the problems that lead to the low reproductive efficiency of the buffalo species, thus lengthening the non-productive life.
- Many factors influence age at puberty, such as breed, season, climate, nutrition and growth. The buffalo management needs, right from birth, to

ensure a correct weight increase. In fact, the animals that show a higher daily gain (before the trials) reach puberty in a shorter time.

The age at puberty is affected by the dietary energy level. The heifers fed with a high level diet (5.56 MFU/day), had a daily gain of 562 g as against 465 g of the heifers fed with a low level diet (4.42 MFU/day), that reached puberty 30 days earlier. In fact, heifers bred with a good early management system conceived at a younger age compared with the others (543±16 and 844 ±11 days, respectively). Stimulating the hormonal changes occurring around puberty may induce sexual maturity in heifers. Trials to induce and synchronize oestrus in buffalo heifers have been undertaken using PRID (Progesterone-Releasing Intravaginal Device) plus PMSG (Pregnant Mare Serum Gonadotrophin) to induce oestrus in noncycling buffalo heifers.

ECONOMIC USES OF BUFFALO

Milk and its quality

Dairy industry in India is mainly buffalo oriented, which contribute 55% of total milk produced. The largest number of buffalo cows in India and Pakistan, calve (nearly 80%) during the period of July to December. This period becomes surplus with milk for dairymen. This surplus milk is converted into other products. Indian buffaloes on an average produce 500 kg of milk during lactation. Daily yield of the lactating buffalo in a well-managed farm is as high as 20 kg.

COMPOSITION OF BUFFALO MILK

	Constituent	% value	
	Water	83.63	
Total Solids (TS)	Fat	6.56	
Total Solids (TS)	Protein	3.88	Solid Not Fat (SNF)
Total Solids (TS)	Lactose	5.23	Solid Not Fat (SNF)
Total Solids (TS)	Ash (Minerals)	0.70	Solid Not Fat (SNF)

Uses of buffalo milk

(a) Buffalo milk is used for making Mozzarella cheese in Italy.

(b) Buffalo milk produces better yoghurt in body texture, allows easy separation of cream and churn of butter, ghee and gives better quality of desiccated milk (khoa) and also acid coagulated milk (poneer)

(c) It is a better tea/coffee whitener.

(d) Richer curd can be produced from buffalo milk.

Work

Buffaloes are classic work-animal of Asia. Buffaloes are used in conventional agricultural practices. Buffalo as a draft animal, has working capacity of 0.75 H.P. A sturdy buffalo can pull a weight of 0.9 to 1.0 ton.

Buffalo meat (Carabeef)

In Italy, Bulgaria, Czechoslovakia and Egypt buffaloes are raised mainly for meat. Buffalo meat contains white fat as the beta carotene (the precursor of vitamin A).

Buffalo horns

Processed horns are used to produce buttons, toggles, combs, spoons, forks, knife handles, napkin rings, wall decorations, shoehorns, etc.

Buffalo hide

The hide of the water buffalo is an important item for both export and local industries. Pakistan is the largest producer of good quality hides and skins.

Buffalo faeces

Tremendously used as fuel and organic fertilizer by the rural people.

Buffalo hair

Their hair is thicker than any other bovine and used for production of brush rather than felt.

SPACE FOR NOTES

Chapter Ten : Goat and Sheep Farming

(A) Goat Farming (B) Sheep Farming

(A) GOAT FARMING

INTRODUCTION

Goats are one of the earliest discovery of mankind in prehistoric times as ready and easy source of meat. Whether in cold arid up hills, or hot arid deserts, or hilly tracts of mountains or ravines constituted of leached soil, goats have survived and sustained the poor people. The present worldwide distribution of goats shows that the number of milch type goats are more in the temperate zone and dual type or meat-milk type goats are primarily located in the sub-tropical and tropical Asian and African countries. India possesses the second-largest goat population in the world. In pastoral and agricultural subsistence societies in India, goats are kept as a source of additional income and as an insurance against disaster. In addition to this, goat has religious and ritualistic importance in many societies. In the prevailing socio-economic conditions of the country, where per capita land holding is hardly 0.2 ha, goat rearing becomes an inseparable component of mixed farming system. Due to low investment, wide adaptability, high fertility and fecundity, low feed and management needs, high feed conversion efficiency, quick pay-off and low risk, goat farming has been attracting many progressive farmers, businessmen, professionals, ex-servicemen and even educated youths.

The goat belongs to the family Bovidae (hollow-horned ruminants) and is the member of the genus Capra. Domesticated goats (*Capra* hircus) are descendants of the pasang (C. aegagrus), represented in Europe by the Cretan and Cyclades races. The East was probably their original home, the earliest recorded being the Persian race.

Advantages of goat farming

(i) The initial investment needed for goat farming is low.

(ii) Due to small body size and docile nature, housing requirements and management problems with goats are less.

(iii) Goats are friendly animals and enjoy being with people.

(iv) Goats are prolific breeders and achieve sexual maturity at the age of 10-12 months. It starts giving milk at the age of 16-17 months. Twining is very common, and triplets and quadruplets are rare.

(v) In drought prone areas, risk of goat farming is very much less as compared to other livestock species.

(vi) Unlike large animals in commercial farm conditions, both male and female goats have equal value.

(vii) Goats are ideal for mixed species grazing. This animal can thrive well on wide variety of thorny bushes, weeds, crop residues, agricultural by-products unsuitable for human consumption.

(viii) Under proper management, goats can improve and maintain grazing land and reduce bush encroachment (biological control) without causing harm to the environment.

(ix) In India, there is no religious taboo against goat slaughter and meat consumption.

(x) Slaughter and dressing operation and meat disposal can be carried without much environmental problems.

(xi) The goat meat is more lean (low cholesterol) and relatively good for people who prefer low energy diet especially in summer and sometimes goat meat (chevon) is preferred over mutton because of its "chewability"

(xii) Goat milk is easy to digest than cow milk because of small fat globules and is naturally homogenised. Goat milk is said to play a role in improving appetite and digestive efficiency. Goat milk is non-allergic as compared to cow milk and it has anti-fungal and anti-bacterial properties and can be used for treating urinogenital diseases of fungal origin. A comparative chart of components of goat, cow and human milk is given in Table 10.1.

(xiii) Goats are 2.5 times more economical than sheep on free range grazing under semi-arid conditions.

Table 10.1 : Components of goat milk, cow milk and human milk

	Goat	Cow	Human
Protein %	3.0	3.0	1.1
Fat %	3.8	3.6	4.0
Calories/100 ml	70	69	68
Vitamin A (I.U./gram fat)	39	21	32
Vitamin B_1/thiamin (μg/100 ml)	68	45	17
Riboflavin (μg/100 ml)	210	159	26
Vitamin C (mg/100 ml)	2	2	3
Vitamin D (I.U./gram fat)	0.7	0.7	0.3
Calcium %	0.19	0.18	0.04
Iron %	0.07	0.06	0.2
Phosphorus %	0.27	0.23	0.06
Cholesterol (mg/100 ml)	12	15	20

(xiv) Goat creates employment to the rural poor by utilising unpaid family labour. There is ample scope for establishing cottage industries, based on goat meat and milk products and value addition to skin and fibre.

(xv) Goat is termed as walking refrigerator for the storage of milk and can be milked number of times in a day.

BREEDS OF GOATS FOUND IN INDIA

I. **Indigenous goat breeds** : In India, each goat breed is considered as a population living in a given locality with a local name and having characters distinct from other such population. At present, in India four agro-climatic regions are known to have the following goat breeds (Table 10.2).

Table 10.2 : Regional distribution of goat breeds

Region	Breeds available
1. Temperate Himalayan region (states of Jammu & Kashmir, Himachal Pradesh, Parts of U.P.	Gaddi, Chegu, Changthangi
2. Arid and semi-arid North-Western region (Punjab, Hariyana, Gujarat, Rajasthan and parts of U.P.	Jamunapari, Sirohi, Surti, Marwari, Jhakrana Beetal, Barbari, Mehsana, Gohilwadi, Zalwadi, Kutchi.
3. Southern region	Osmanabadi (Deccani), Sangamneri, Kanni-Adu, Malbari.
4. Eastern region	Bengal black, Ganjam.

Description, uses and distribution of the above mentioned breeds are presented in Table 10.3.

II. **Exotic goat breeds**

The principal exotic dairy breeds of goats are Toggenberg, Sannen, French Alpine and Nubian. They are all noted for their higher milk yield and most of these breeds were imported to India to improve milk yield of our local breeds and to upgrade our non-descript goats.

Toggenberg : It is originated in the Toggenberg valley in north Switzerland. Skin is very soft and

Fig. 10.1. Toggenberg breed

Fig. 10.2. Sannen breed

Fig. 10.3. Alpine breed

pliable. Usually both male and female are hornless (Fig. 10.1). The adult doe weights 65 kg or more and the bucks more than 80 kg. Average milk production is 5.5 kg per day. The butterfat content of its milk is 3-4 per cent. The male usually has longer hair than the female.

Sannen : It is a native of Sannen valley of Switzerland and is noted for its consistency and high production of milk. Does weigh 65 kg and the bucks 95 kg. Colour is white or light cream. The face may be slightly dished and the ears point upward and forward (Fig. 10.2). Both sexes are normally polled but sometimes horns do appear. Average milk yield is 2-5 kg per day during a lactation period of 8-10 months. Milk fat is 3-5%.

Alpine : This breed was originated in Alps mountains. It was derived from French, Swiss and Rock Alpine breeds. They are horned and have no distinct colour. Average milk yield is 2-3 kg with butterfat of 3-4% (Fig. 10.3).

Nubian : Originated in Nubia of north-eastern Africa. Also found in Ethiopia and Egypt. It is a long legged and hardy animal. The breed along with Jamunapari of India together with native breeds of U.K. formed the Anglo Nubian breed of goat.

Anglo Nubian : It is a big animal with a fine skin and glossy coat, pendulous ears and Roman nose. Anglo Nubian is known as the Jersey cow of the goat

Fig. 10.4. Anglo-Nubian breed

world. Udder is large and pendulous with bigger teats. There is no fixed colour. Bucks weigh 65-80 kg and does from 50-60 kg. Average milk yield is 3-4 kg/day. Peak yield may even go up to 6.5 kg or more (Fig. 10.4).

Angora : Originatd in Turkey or Asia minor. It produces a superior quality fibre, called **mohair**. The soft silky hairs cover the white body. If not shorn during spring the fleece drops off naturally as summer approaches. Average fleece yield is 1.2 kg. Good animals yields even up to 6 kg. The angora is small in size with shorter legs. Horns are grey, spirally twisted and inclined backward and outward. Tail is short and erect (Fig. 10.5).

Fig. 10.5. Angora breed

Table 10.3

Breeds	Distribution	Description	Uses
1. Gaddi/Himalayan	Kangra and Kulu valleys, Ladakh; Girmul and Simla and parts of Jammu.	Average body weight of male is 27 kg and female is 25 kg. Sturdily built medium sized body covered by coats which are mostly white, but black and brown combinations are also seen. Coarse, long hair measure about 8-10 inches. In both sexes, horns are large, upward and directed backwardly. Tapering muzzle and ears are long, pointed and drooping.	(1) Castrated bucks are used for transport. (2) Good for hair, average fleece yield is about 300 gm/clip. (3) Can also be used for meat.
2. Chegu	Mountaneous range of Spiti of HP, Yaksar and Kashmir of J & K, Uttar Kashi, Chamboli and Pithoragarh of U.P., etc.	Average body weight of male is 36 kg whereas that of female is 25 kg. Medium sized goat with body colour white and also mixed with greyish red. Both sexes have prominent horns, upward and directed backward with one or few twists. Hair length is about 5-9 cm.	(1) This breed is used mainly for pashmina production; average yield is 120 gm/year. (2) Small quantity of milk and good meat can also be obtained.

Table 10.3 (contd. from previous page)

3. Changthangi Pashmina (Fig. 10.6)	Ladakh, Lahaul valley of J & K, Spiti of Uttarakhand.	Small to medium sized body. Adult males weigh 20 kg and females weigh about 19 kg. These goats are mostly white, but black, grey or brown are also seen. Horns in both sexes are large, turning up and outwardly to form semi-circle with some variations both in size and shape. Ears are long and usually pendulous.	(1) This breed produces long fine pashmina on both sides of body and shoulder. Average production of pashmina is 70-500 gm/animal/year. This pashmina is used to prepare Kashmir shawl or rug. (2) In hilly areas, used for transportation.
4. Jamunapari (Fig. 10.7)	Mostly found in Etawah district of U.P, and in tracts lying between Jamuna, Ganges and Chambal rivers.	Most majestic and biggest goat breed of India. Large sized, tall body with average body weight in male 50-60 kg and in female 40-50 kg. Body colour mostly white or light yellowish tan with light brown spots on neck and face and occassion-ally black or tan patches are found on body. Most chara-cteristic feature of this breed is highly convex nose line with a tuft of hair, known as "Roman nose" or parrot mouth appear-ance. Horns are medium sized and curved outwardly. Ears long, flat and dropping. A thick patches of hair are present on buttock region.	(1) This breed is famous for dual purposes — milk and meat. Average milk yield varies from 1.5 to 2 kg/day. (2) Meat quantity is good with a bone and meat ratio of 1 : 3.9. (3) Usualy doe kids once a year. (4) This breed has been extensively used to upgrade indigenous breeds for milk and meat.
5. Barbari (Fig. 10.8)	In U.P, Etawah, Agra, Mathura and Aligarh districts. Gurgaon, Karnal, Panipat and Rohtak in Haryana.	Small goats with compact body weighing 40 kg in male and about 24 kg in female. Body colour varies, white with light brown pathces is most common. Both sexes have horns that are twisted and directed upward and backward. Ears are short, tubular and erect. Bucks bear thick long beard.	(1) This breed has dual purposes of milk and meat. Milk yield is 130-200 kg in a lactation period of 150 days. (2) Barbari is a prolific breeder, kids twice in a year and generally gives birth twin or triplets. (3) Its dwarf size and habit of stall feeding made it suitable for city rearing where grazing facilities are rare.

Table 10.3 (contd. from previous page)

6. Beetal	Throughout the state of Punjab and Hariyana	Large breed; body weight in male is 50-62 kg and in female is 35-40 kg. Body colour varies with white, red and tan spots being common. Roman nose, long dropping ears and small thin tail are the characteristics of beetal breed. Both sexes have horns which are directed backward and have few twists. Only male possesses beard.	(1) Reared mainly for milk. Yield per lactation varies from 150-190 kg. (2) Highly adaptable to various agroclimatic zones. (3) Kidding is good.
7. Surti (Fig. 10.9)	Surat and Baroda districts of Gujarat.	Average body weight is 30 kg in male and 32 kg in female. Medium sized body is white in colour. Both sexes have small horns directed backward. Females with highly developed udder.	(1) This breed is a good milk producer; yielding on an average 2.0 kg/day. (2) Very suitable for rearing under confinement as this goat can live on food waste and leaves.
8. Sirohi (Fig. 10.10)	Sirohi district of Rajasthan and Palampur in Gujarat.	Compact medium sized goat (body weight is 50 kg in male and 40 kg in female). Coat colour generally brown with light/dark brown patches. Hair short and coarse. Both sexes have small horn, curved backward. Ears flat and dropping.	(1) Sirohi is a choiceable goat breed for meat. (2) Milk yield is relatively small, about 0.5 kg/ day. (3) It is well suited to stall feeding.
9. Marwari	Marwar regions of Rajasthan	This medium size goat breed is predominantly jet black in colour. Body weight of male is 33 kg and that of female is 25 kg on an average. Both sexes have short pointed horns, directed upward and backward. Ears are small and flat. Males have thick beard.	(1) Marwari doe is a poor milk yielder, about 0.8 kg/day. (2) It is good for meat. (3) Lustrous hair can be used for different purposes.
10. Mehsana	Found in Mehsana, Gandhi Nagar and Ahmedabad districts of Gujarat.	Mehsana is larger than Marwari breed with an average body weight of 36 kg in male and 32 kg in female. Black body is covered by long coarse hair with white spot at the base of ear. Ears are always white. Horns in both sexes are twisted & curved upward & backward.	(1) Triple purpose use for milk, meat and hair. (2) Milk yield averages 1.0 kg/day. (3) Average hair yield is 200 gm/goat/year

Table 10.3 (contd. from previous page)

11. Jhakrana (Fig. 10.11)	Alwar district of Rajasthan	Large goat breed; males weigh about 55 kg and females about 45 kg. Black body with white spots on ears and muzzle. Fare head is slightly bulging.	(1) Good milk yielder. Average milk production varies 2-3 kg/day. (2) Prolific breeder. Twins and even triplets are common. (3) Its meat is also popular. (4) Skin is used in tanning industry.
12. Kutchi (Kathiawari)	Kutch district of Gujarat and few places of Rajasthan.	Large breed with medium body weight (male : 42-45 kg, female : 30-40 kg). Body colour predominantly black but may be white, brown or even spotted. Hair long, nose slightly 'Roman' type, and ears are long, broad and dropping. Cork screw horns pointed upward.	(1) Milk yield is around 1.5 kg/day. (2) About 500 gm hair can be obtained /goat if shorn twice a year. (3) Kutchi also acceptable for meat.
13. Zalwadi (Fig. 10.12)	Zalwad regions of Gujarat	Body is large; black in colour with white ears. Male weighs about 38 kg and female about 32 kg. Both sexes have long twisted horns, pointed upward. Ears are large leaf-like and dropping.	(1) This breed is reared mainly for meat. (2) Daily milk yield is about 1.5 kg. (3) Hair can be used for different purposes.
14. Gohilwadi	Bhavnagar, Amreli and Junagarh districts of Gujarat.	This is also a large goat breed. Body weight of adult male averages 37 kg and female averages 36 kg. Black coat is covered with long hair. Nose is slightly convex. Ears are tubular and dropping. Both sexes have slightly twisted horns, turned backward.	(1) Gohilwadi is useful for milk yield, average daily milk yield is 1.7 kg for a total lactation period of 228 days. (2) Their hair is used for different purposes.
15. Osmanabadi (Fig. 10.13)	Osmanabad region of Maharashtra and adjoining areas.	Average body weight in male is 34 kg and that of female is 32 kg. Coat colour varies, but mostly black; mixture of white and black or red are also found. All males and 50% of females are horned. Ears are of medium sized.	(1) Not so popular for milk; average milk yield varies from 0.5 to 1.5 kg during a lactation period of about 4 months. (2) This breed is mainly reared for meat. (3) The does can breed twice in a year and twinning is common.

Table 10.3 (contd. from previous page)			
16. Malbari/ Tellicherry (Fig. 10.14)	Northern Kearla including Calicut, Cannore and Mallapuram Districts.	Medium sized breed where males are usually of 38 kg and females vary between 30-32 kg. Coat colour highly variable from complete white to full black. Head with flat and occasional Roman nose, small twisted horns and medium sized ears. All males and some females are bearded.	(1) Malbari is mainly reared for meat. (2) They are poor milkers. (3) Skin is useful for tanning industry.
17. Sangamneri	Pune and Ahmednagar districts of Maharashtra.	It is also a medium sized breed with average body weight like that of Malbari goats. No uniform body colour; it varies from white to black/brown with spots of other colours. Both sexes are horned. Ears are dropping.	(1) Not reared for milk. (2) Meat production is good. (3) Dressing percentage is about 41 at 6 months, 45 at 9 months and 48 at 12 months of ages.
18. Bengal (Bengal Black) (Fig. 10.15)	West Bengal, Bihar, Orissa.	Goats are small sized; adult male body weight is about 30-32 kg and that of female is about 20 kg. Predominant body colour is black, but may be white, grey etc. Body is covered with soft and short hair. Shoulder and hips are of equal height, chest is wide. Both sexes have small to medium sized horns directed upward and sometimes backward. Ears are upright and pointed. Beard is found in both sexes.	(1) This breed is reared for excellent meat. Dressing percentage is around 48%. (2) Skin is also of superior quality and has great demand in footwear industry. (3) Does kid twice in a year; multiple birth is very common. (4) Not a good milker.
19. Ganjam (Fig. 10.16)	Ganjam and Koraput districts of Orissa	Tall goat breed where males are about 45 kg in body weight and females are about 31 kg. Body colour varies, black is predominant but white , brown or spotted are also available. Hairs are short and lustrous. Horns in both sexes are long and straight.	(1) Prolific breeder; litter size 1.8 kids. (2) Mainly raised for meat and is poor milker.
20. Kanni-Adu (Fig. 10.17)	Tamil Nadu	Tall goat with a body weight of 34-38 kg in male and 28 kg in female. Predominantly black body with white spots. Ears are medium in length. Males are horned and females are pooled.	(1) Very important meat goat.

Fig. 10.6. Changthangi Pashmina

Fig. 10.7. Jamunapari

Fig. 10.8. Barbari

Fig. 10.9. Surti

Fig. 10.10. Sirohi

Fig. 10.11. Jhakrana

Fig. 10.12. Zalwadi

Fig. 10.13. Osmanabadi

Fig. 10.14. Malbari/Tellicherry

Fig. 10.15. Bengal Black

Fig. 10.16. Ganjam

Fig. 10.17. Kanni Adu

GOAT MANAGEMENT

Management of goats in goat farm/gotery : Goat has been rearing since the time immemorial. Generally goat farming means rearing of goats for the purpose of harvesting milk, meat and fibre. The places where goats are reared for commercial purposes following proper scientific and modern principles, are referred to as goat farm or gotery. Common management practices that are adopted in goat farm include following aspects :

(A) Housing of goats

Goats do not like being closely confined and thus, do not need any special housing. However, in goat farm under commercial conditions, goat should be housed properly with the following general tips :

(1) Goat house/shed must be strong and comfortable with enough space for taking rest of goat.

(2) It should be made in dry weather and environment.

(3) The floor of the house must have to keep always dry.

(4) Plenty of light and air should be available in the house.

(5) There must be system for controlling temperature and moisture.

(6) The house should be kept free from being damping. Because damping condition is responsible for various diseases.

(7) Never let the rain water to enter inside the house.

(8) Goats are feared about cold and water. So extra care should be taken in rainy and winter seasons. Otherwise, they may caught by pneumonia.

(9) The place must have the facilities of cleaning well regularly.

(10) If possible the wall of the house should be made with concrete or bamboo poles.

Fig. 10.18. Litter system of goat housing

Fig. 10.19. Raised platform system of goat housing

Types of goat house :

There are no definite design of goat house or sheds and many types of houses for goat-rearing are found in India. However, following types of goat houses are most common :

(a) **Goat housing over ground (Litter system)** : Generally these types of house are made over the ground. This is the most common house for goat. Farmers make the floor of this house with brick and cement or simply with soil and spread some dry straw, paddy husk, groundnut-shell over the floor to make a thin litter (Fig. 10.18). But this should be periodically turned to remove the foul smell. This type of house must have to keep always dry and clean to reduce external parasitic infestation.

(b) **Goat housing over pole (Raised platform system)** : These types of house are made over pole. The floor of the house is about 1 to 1.5 metres high (3.5 to 5 feet) from the ground. These types of house keep the goat free from damping condition, flood water, etc. The poles and floor are generally made with bamboo or wood (Fig. 10.19). These types of house are very suitable for goat farming because it is very easy to clean. Wastes and urine of goat can easily be removed from the house. Diseases are also less in this housing system.

(c) **Concrete house** : These types of goat-houses are fully made with concrete. Concrete houses have many advantages. It is very easy to clean and always keep the goat safe from all types of enemies. The house may be built over ground or concrete poles. Both types are easily maintained. Diseases are less in this housing system. But it is very expensive method of goat housing.

Required space for goat : In accordance with increasing body size and weight of goat, they need more space. A house of 1.8 metre × 1.8 metre × 2.5 metre (5.5 feet × 5.5 feet × 8.5 feet) is suitable for staying of 10 small goats. The space can be extended or decreased according to the number of goats present in the farm. In general, the required space for proper growing of goats during different stages of life is mentioned below :

Space requirement of goats during different stages of life

Goat	Required space (Sq. metre)
Baby goat	0.3
Adult goat	1.5
Pregnant goat	1.9
Billy goat (male goat)	2.8

(B) Selection of goat breeding stock

Goats in good health and having good physical features must be purchased in consultation with veterinarian or bank's technical officer. Animals which are ready to breed and in prime stage of reproduction are generally preferred. Newly purchased goats are kept under observation for about 15 days and then mixed with the general flock.

(C) Feeding Management

Within or near the goat farmhouse, there must be bushes or shrubs for browsing of animals. Stall-fed goats are efficient converters of farm and crop residues into excellent organic manure. Goats relish the stalks and residues of most of the nutritious cereals and they do well particularly when mixed with green fodder such as grasses and subabul. So, stall feeding for goats are ideal for intensive integrated farming system (IIFS). Otherwise, as an alternative to above, supply of cultivated fodder from own farm or from surrounding farms may be ensured. As a general rule, most of the energy requirements of goats should be met through roughages*. Half of the roughages should be leguminous green fodders and rest half should be grasses or tender tree leaves. In the absence of good quality green fodders, concentrates** must be considered to replace them.

Kids should be fed colostrums*** (see below) up to 5 days of age. Later on, they can be put on kid starter rations. Green leguminous fodders should be offered sufficiently to kids from 15 days onwards. Salt and water should be provided to kids at all times. During breeding season, additional concentrates should be given to bucks and does. Concentrate feed can be prepared by mixing the ingredient in proper ratio as mentioned in Table 10.4.

Table 10.4 : Ingredients of concentrate feed required by goats during different stages of life

Ingredients (%)	Kid	Growing peroid	Lactating doe	Pregnant goat
Maize	37	15	52	35
Pulses	15	37	—	—
Oil cakes	25	10	8	20
Wheat bran	20	35	37	42
Mineral mixture	2.5	2	2	2
Common salt	0.5	1	1	1

* **Roughage** : Roughages are feed with a low density of nutrients and with a crude fibre content over 18% of dry matters including fresh and dried forages and fodders. It promotes normal bowel function.

** **Concentrate** : It is an energy rich feed with high nutrients and low crude fibre content. Usually it contains different cereals, grains, milling by-products, oil cakes, etc. It is given to livestock for proper nutrition and growth.

*** **Colostrum** : Colostrum is the first milk produced by mammary gland of any female mammals in late pregnancy or just prior to giving birth. It contains antibodies to protect the newborn from various diseases. It has lower cholesterol and higher protein content than ordinary milk.

(D) Breeding Management

In general, it is planned to obtain 3 kiddings in 2 years by adopting optimal breeding management. So, for about 25 does (female goat), one buck (male goat) should be provided in one breeding season. In goat farms, there are 3 systems for breeding, which are as follows :

1. **Pasture or pen breeding :** It occurs in normal course in the herd. This type of breeding system, thus involves very little labour and results in a large percentage of does conceiving. It also has the following disadvantages.

No control over when the does are bred and when they kid. Young kids below 5 months of age are also chased by bucks.

2. **Artificial insemination :** In this system, frozen semen is thawed and deposited into the reproductive tract of the doe by a trained technician using a plastic inseminating pipette.

This system is gaining popularity and will continue to do so as long as suppliers continue to offer a selection of goat semen.

3. **Hand mating** : In this system, the buck is kept in a lot of his place and is never allowed to run with the does. As the does come into heat, they are taken to his pen, either singly or two or three at a time, and the buck breeds them threre, the breeding is observed and as the doe is bred, she is removd from the pen.

This system has the advantages of giving precise breeding dates and it can give does a considerable period of rest if the farmers like so. The disadvantages include the need for a separate buck shed, considerable amount of time involved and unless the farmers have many does, the cost of buck per doe becomes high.

4. **Cross-breeding** : When unrelated goats are mated, the system is known as cross-breeding. First cross-progeny is usually superior to their parents. This is known as hybrid vigour.

Cross-breeding is used to improve the productivity of local animals. Using an exotic buck to breed a native doe will produce strong vigorous offspring. These offspring will have increased production potential.

Under tropical conditions, cross-breeding should be considered successful, if the offsprings are sufficiently better than the native female stock. Indian breeds of goat exhibit considerable variation in lactation milk yields, but in general, average about $1/2$ to 1 kg/day. The lactation milk yields are, however, low when compared to milk yields of about 3 to 5 kg/day of goat breeds in a temperate environment. These differences are mainly due to breed, plan of nutrition, size, age and management methods. Cross-breeding of local breeds with improved exotic breeds has brought about considerable improvement in milk production with enough capacity to withstand the direct and indirect effects of a hot climate.

(E) Care of goats during pregnancy

In advanced stage of pregnancy, the does must be transferred to either kidding pens or separately marked space for kidding within the main shed after thoroughly disinfecting it. After kidding, the does should be provided with warm bran mash for two days.

(F) Care of kids

Immediately after parturition, the naval cord will be disinfected with tincture of iodine as soon as it is cut with a sharp knife. Almost immediately after birth, the kids, if healthy and strong, are on their legs and start sucking their mother's teats. If more than one kid is born, it may be necessary especially when they are very young, to ensure that the smallest of them gets its due share of milk, because it may be prevented from doing so by the stronger kids. The kids should be protected from extreme weather conditions, particularly during the first two months.

For the first 3 to 4 days after birth, the kids must be given colostrum. The rearing of kids may be done either natural or by hand rearing and each has its advantages and disadvantages. In India, it is the natural method that is usually practised and this consists of in leaving the kid to take what amount of milk it can obtain from its mother. Hand rearing is resorted to when weaning is practised or when the goat dies. There are two methods of hand rearing, one consisting of feeding the kid with a bottle and the other is feeding it off the pail. Both methods are learned by them easily, but bottle feeding is to be preferred because the saliva that is produced during the process of suckling the milk, aids in digestion. Kids will also readily take to feeding on a foster mother when they are put on her teats.

When about two weeks old, kids begin to nibble green food or dry fodder and it would be well to see that small quantities of these are within their easy reach at this time. It is also important that kids are allowed plenty of open air and sunlight. In hot weather, it is best to keep them in an enclosure build round a tree so that they may also be provided with shade. The enclosure should be large enough to allow them plenty of exercise.

At the age of 2 to 3 months, the suckling may be practically discontinued and at four months the kids should be completely weaned because by this time they will become fit, like the older goats, to eat solid food, although they may as well be allowed to suckle a little longer.

Male kids, unless they are required for breeding purposes should be kept separated from the female kids.

At one year, a buck should receive 1.8 kg of grain mixture, the allowance being increased by 50 per cent during the breeding season. A liberal amount of fodder should be given. An average of 7 to 8 kg of green fodder per day should be adequate for a full grown Jamunapari buck, when entirely stall fed. Where the animal is unduly fat, its grain ration should be cut.

(G) Routine operations in goat farm

(1) **Castration*** : Castration of bucks, if required, should be done at the age of 2 to 4 weeks, although castration at later stage also becomes successful. It is usually done to (i) increase the palatability of meat, (ii) increase the body weight rapidly, (iii) make the skin superior in quality. Following castration, there may be swelling in the place of testis which will be treated with medicines at least for 2-4 days.

(2) **Dehorning** : Derhorning is the cutting off horns that have already grown to significant size, and disbudding is the removal of horn buds at an early age to prevent further horn-growth. In goat farms if these two methods are practised, that should be done very carefully under medical persons. These practices are done just to avoid fighting amongst the horned stocks and to keep the horned and polled animals together.

(3) **Care of feet** : Farm goats often suffer from a condition, known as overgrown feet, where deformity in feet can cause extreme discomfort, even arthritis in old age. It can be avoided by pairing the hooves when feet are overgrown.

(4) **Marking of goats** : Goats in flock or herd can be easily identified if they are marked. Marking can be done by 3 ways, e.g., ear-tatooing, ear-tagging and ear-notching. These should be carried out carefully within one week after kidding carfully.

(5) **Tethering** : Tethering is not an option for keeping goats. There are danger that tethered goats might become tangled/strangled and they cannot escape from bad weather/dogs/children, etc. However, if only one or two goats are raised, tethering can be used. According to Animal Welfare Act, 1999, a tethered goat must be provided with adequate feed, water and shelter. Following points to be noted while (a) Goats must be accustomed to being tethered and used to traffic if grazing the roadside. (b) A leather collar should be used around the goats neck to which a light chain is attached. (c) There may be one swivel at both ends of the tether. (d) The goat must be given opportunity to forage in a wide area with vegetation. (e) The foraging area to be changed regularly.

(6) **Exercise paddock for stall fed goats** : Goats when reared in stall fed system, it is of utmost importance to provide the exercise paddock.

---

* **Castration** : It is the removal or destruction of testes, epididymis and a portion of each spermatic cord from male animal. Non-breeding males and males not slaughtered at a young age are usually castrated. In goats and sheeps, it is done for market orientation production of livestocks and to keep the breeding programme in control.

(H) Marketing

The marketable products of goat farming include the fattened kids, manure, culled animals. Marketing avenues for the above products are slaughter houses, individual meat consuming customers and agriculture farms. Therefore, availability of either slaughtering facilities or traders who will purchase live animals should be ensured to convert the fatteners into wholesome meat and meat products. Further, demand for manure from nearby agriculture farms must also be ensured.

(I) Management of diseases

Though goats are robust animals and are resistant to many diseases, following diseases are however, often known to attack farm goats (see Table 10.5).

Besides the mentioned diseases, infectious abortion in a doe herd is very common in India. It is a serious problem because abortions not only lead to economic losses but can also be a public health concern because infectious agents that cause abortions in goats can also infect and cause diseases in humans and other livestock. Infectious abortions should be taken seriously by the producer and herd manager. The most common micro-organisms that cause abortions in goats are :

(1) *Chlamydia psittaci* causing Chlamydiosis.

(2) *Coxiella burnetii* causing Query or Queensland (Q) fever.

(3) *Listeria monocytogenes* causing Listeriosis.

(4) *Leptospira spp* causing Leptospirosis.

(5) *Toxoplasma gondii* causing Toxoplasmosis.

(6) *Brucella melitensis* causing Brucellosis.

Table 10.5 : Some common goat diseases, their symptoms and prevention

Disease	Symptoms	Prevention and treatment
Mastitis	Enlarged, hot, painful udder. Fever. Milk watery with flakes of blood.	Improve hygiene. Application of antibiotics.
Foot Rot	Lameness. Hoof will look as if it is rotten and with foul smell. Signs of pain is seen, if pressed.	Trimming of hooves, and soaking of feet in a bath of water with $CuSO_4$.
Brucellosis	Abortion in late pregnancy. Retention of placenta and metritis. In bucks, infertility, orchitis and swollen joints are seen.	Isolation of infected animals. Vaccination, blood testing and culling of positive animals.
Internal parasites	Loss of weight. Reduction in milk yield. Diarrhoea and anaemia.	Good quality food and clean water. Proper medication.
External parasites	Restlessness, scratching, loss of weight, reduction in milk yield.	Application of proper medicine as a dust, spray, or a dip.
Poisoning	Unsteadiness followed by dullness and unconsciousness. Great pain and vomiting. Convulsion and eventual death.	Keeping goats away from poisonous plants and chemicals. Immediate treatment is needed in such cases.
Bloat	Distended abdomen on left side, respiratory difficulty, restlessness.	Too much fresh green grass should be avoided. A cup of mineral oil may bring relief. In acute cases, removal of gas by making puncture is needed.

(J) Common preventive measures

(A) Vaccinations

(1) Kids should be given a first dose of Enterotoxaemia and Tetanus vaccination at 8 weeks of age and again at 12 weeks of age.

(2) Does should be given Enterotoxaemia and Tetanus vaccination at 4-6 weeks before breeding season and 4-6 weeks before kidding.

(3) Bucks should be vaccinated with Enterotoxaemia and Tetanus once in a year.

(B) Deworming

- Before breeding the doe should be dewormed. Those goats affected with worm infestation are lean and dull.
- Kids should be dewormed at one month of age. The life cycle of worms is three weeks, so deworming again at two months of age is recommended.
- Pregnant does should be dewormed 2-3 weeks before kidding.
- Does in early pregnancy (up to 2 months), should not be dewormed to avoid abortion.

CONSTRAINTS OF THE GOAT INDUSTRY

The followings could be considered as technical constraints for securing a thriving goat industry in the country :

(a) Non-availability of high-yielding breeding stock.

(b) Low level of nutrition and managerial efficiency.

(c) Lack of definition of the production objectives.

(d) Limited attention to application of the modern techniques for improving the reproductive efficiency such as artificial insemination (AI), synchronisation of estrous, semen freezing, etc.

(e) Limited use of outstanding exotic breeds for improved meat and milk quality.

(f) Inadequate control of diseases and parasites due to non-availability of prophylactic vaccines against important contagious diseases.

(g) Lack of knowledge on successful rearing of kids. Kid mortality is very high when weaning is practised at a very young age.

(h) Lack of knowledge on silvi-pastoral system.

(i) Housing for goats in different eco-zones requires a very elaborate and systematic study.

(j) Organised marketing is very limited. This has resulted in unscrupulous exploitation by the middle-man who is often seen moving with the goats along the migratory routes.

RESEARCH AND DEVELOPMENT OF GOATS IN INDIA

Central Institute for Research on Goats

Central Institute for Research on Goats (CIRG) is a premier research institute of Indian Council of Agricultural Research (ICAR), an autonomous organisation under Department of Agricultural Research and Education, Ministry of Agriculture, Government of India. The Institute was established on 12th July, 1979 with the vision to develop poor men's cow – the goat, as a source of livelihood, security, poverty alleviation and employment generation. The institute is located at NH-2 between Agra and Mathura. It has many working units of which two important units are working with the following objectives.

Farm based units

Objectives

(1) To estimate genetic variance of economic traits in goats.

(2) To estimate breeding value of males and females.

(3) *In situ/ex situ* conservation of elite germ plasm and its effective utilisation.

(4) To estimate production economics of goats under farm conditions.

(5) To validate farm based goat production technologies under field conditions.

Field based units

Objectives

(1) To assess the production performance of goat breeds in farmers flocks under village management system and improving the germ plasm through selection.

(2) To evaluate the socio-economic status of goat breeders and the economics of goat production in farmers flock.

(3) To disseminate the pro-poor goat based technologies under field conditions and assessment of their impact on goat production.

Some of the remarkable activities of working units :

(1) Alpine & Sannen were dairy goats; they were used for crossing with native Beetal and Malbari goats, and the cross-breeds have shown to be of high potential under average and special feeding conditions, with kids attaining a body weight of 25 kg at 6 months of age.

(2) Black Bengal and Jamunapari crosses have shown improved meat production.

(3) Pashmina goats have been successfully reared at higher and lower altitudes.

Thus, the concerted approaches adopted by the institute will help to demonstrate and implement the available improved technologies at the farmers' doorstep. Moreover, the dedicated mailing list "goat net@ cirg.res.in" and helpline are working round the clock to address the goat farmers' problems through internet.

(B) SHEEP FARMING

INTRODUCTION

Perhaps the first ruminants domesticated by man along with goats, sheep are a very valuable and important asset to mankind. Utility facets of sheep are **milk, meat, wool, skin** and **manure.** Sheep were domesticated long before the dawn of recorded history. Wool fibres have been found in remains of primitive villages of Switzerland that date back an estimated 2000 years. At present, there are over 800 breeds of sheep in the world in a variety of sizes, shapes, types and colours. They are found in tropical countries and in the arctic, in hot climates and in the cold, on the desert and in humid areas. Domesticated sheep, *Ovis aries* belongs to the phylum *Chordata,* class *Mammalia,* order *Artiodactyla,* family *Bovidae,* and genus *Ovis.*

Advantages of sheep farming

(i) Sheep do not need expensive buildings to house them and, on the other hand, require less labour than other kinds of livestock.

(ii) The foundation stock is relatively cheap and the flock can be multiplied rapidly.

(iii) Sheep are economical converter of grass into meat and wool.

(iv) Sheep can eat varied kinds of plants compared to other kinds of livestock. This makes them excellent weed destroyer.

(v) Unlike goats, sheep hardly damage any tree.

(vi) The production of wool, meat and manure provides three different sources of income to the shepherd.

(vii) The structure of their lips helps them to clean grains lost at harvest time and thus, convert waste feed into profitable products.

(viii) Mutton is one kind of meat, towards which there is no prejudice by any community in India and further development of superior breeds for mutton production will have a great scope in the developing economy of India.

(ix) The contribution of sheep through export of meat is 8 per cent of the total export value of agricultural and processed food products. Live sheep are also exported for meat purpose. Sheep skin in the form of leather and leather products is also exported.

SHEEP BREEDS

I. INDIAN BREED OF SHEEP

The Indian sheep breeds are named after their place of origin or on their prominent characteristics. In India four Agro-ecological zones are known to have the following breeds of sheep (Table 10.6)

Table 10.6 : Breeds of Indian sheep in different Agro-ecological zones

Northern-temperate		North-Western arid and semi arid		Southern		Eastern	
Breed	Distribution	Breed	Distribution	Breed	Distribution	Breed	Distribution
Gaddi	Kashmir, H. P.	Chokla	Rajasthan	Deccani	Maharastra, Andhra Pradesh (A.P.), Karnataka	Chotta-nagpuri	Bihar and West Bengal
Rampur-Bushair	H.P., U.P.	Nali	Rajasthan, Hariyana			Balangir	Orissa
Bhakarwal	Not specific	Marwari	Rajasthan	Nellore	A.P.	Ganjam	Orissa
Karnah	Kashmir	Jaisalmeri	Rajasthan	Bellary	Karnataka	Tibetan Sheep	Arunachal Pradesh
Gurez	Kashmir	Pugal	Rajasthan	Hassan	Karnataka	Bonipala	Sikkim
Kashmir-Merino	Kashmir	Mulpura	Rajasthan	Mandya	Karnataka	Shahabadi	Bihar
Changthangi	Kashmir	Sonadi	Rajasthan, Gujarat	Mecheri	Tamil Nadu	Garole	West Bengal
Poonchi	Kashmir	Patanwadi	Gujarat	Kilakarsal	Tamil Nadu		
		Muzzaffar-nagri	U.P, Delhi, Hariyana	Vembur	Tamil Nadu		
		Jalauni	U.P.	Coimbatore	Kerala, Karnataka		
		Hissardale	U.P.	Nilgiri	Tamil Nadu		
		Magra	Rajasthan	Ramnand white	Tamil Nadu		
				Madras red	Tamil Nadu		
				Tiruchi Black	Tamil Nadu		
				Kenguri	Karnataka		

Most popular Indian breeds of sheep and their uses are tabulated below (see Table 10.7).

Table 10.7 : Popular Indian breeds of sheep and their uses

Sheep breed	Description	Uses
Nali Fig. 10.20	Medium-sized sheep; face is light brown in colour; skin is pink; ears are large and leafy; both sexes are polled; tail is short to medium in length and thin; fleece is white with brownish tinge, coarse, dense and long-stapled; forehead, legs and belly are covered with wool.	This sheep are mainly reared for wool and clipped twice a year. Average cut varies between 2.5-3.5 kg per year. March clip is yellow and September clip is golden brown in colour.

Table 10.7 : Popular Indian breeds of sheep and their uses (contd.)

Sheep breed	Description	Uses
Chokla Fig. 10.21	Light to medium sized; face is generally devoid of wool and is reddish-brown/dark brown in colour; skin is pink; ears are small to medium in length and tubular; coat is dense; both sexes are polled.	Although Chokla is perhaps the finest carpet-wool breed, its wool is being diverted to the worsted* sector due to scarcity in fine apparel wool in the country. The average yield is between 1.5-2.5 kg per year.
Marwari Fig. 10.22	Medium-sized; black face with colour extending upto the lower part of the neck; ears are extremely small and tubular; both sexes are polled; tail is short, medium and thin.	This breed is clipped twice a year, yield varies from 1.5 - 2.5 kg per annum. Fleece is white and not very dense and suitable for carpet making. It is also used for meat purpose.
Magra Fig. 10.23	Medium to large in size; white face with brown patches around the eyes are characteristic; skin colour is pink; ears are small to medium and tubular; both sexes are polled; tail is medium in length and thin.	Fleece is of medium carpet quality, extremely white and lustrous and not very dense, Bikaneri magra strains have extremely white and lustrous fleeces and are suitable for good quality carpets.
Jaisalmeri Fig. 10.24	Tall, well-built, largest breed in Rajasthan; black or drak brown face with the colour extendng upon the neck; typical roman nose, long drooping ears generally with a cartilaginous appendage; both sexes are polled; tail is medium to long.	Fleece colour is white, of medium carpet quality and not very dense. Annual fleece weighs about 750 gm with staple length of 6.5 cm. It is also known as a meat breed.
Malpura Fig. 10.25	Fairly well-built with long legs; light brown face; belly and legs are devoid of wool. Ears are short and tubular, with a small cartilaginous appendage on the upper side; both sexes are polled; tail is medium to long and thin.	Fleece is white, extremely coarse and hairy. Annual fleece yield is about 1 kg which is used by carpet industry. Malpura meat is also acceptable.
Sonadi	Fairly well-built but somewhat smaller to Malpura; long legs; light brown face with colour extending to the middle of the neck; ears are large, flat and drooping and generally have a cartilaginous appendage; tail is long and thin; both sexes are polled; udder is fairly well developed.	Fleece is white, extremely coarse and hairy and yield varies from 0.6 - 1.2 kg annually. This breed also gives milk 1-1.5 kg per day.
Patanwadi Fig. 10.26	The breed includes three distinct strains : (1) Non-migratory, red-faced animals with small bodies, these are typical Patanwadis and are located in north-eastern Saurashtra. (2) Migratory type with larger body and long legs, typical Roman nose and long tubular ears. (3) Meat type, with big body and low stature.	(1) Yielding relatively finer fleeces (2) Produces coarser fleece (3) Used as a source of meat, also produces coarser fleeces.

* Worsted : Worsted is a fine, smooth yarn, spun from long strands of combed wool.

Table 10.7 : Popular Indian breeds of sheep and their uses (contd.)		
Sheep breed	**Description**	**Uses**
Muzaffarnagari Fig. 10.27	Medium to large in size; it is one of the largest breed and very well adapted to irrigated areas. Face lines are slightly convex; face and body are white with occasional patches of brown or black; both sexes are polled; males sometimes contain rudimentary horns; ears are long and drooping; tail is extremely long and reaches fetlock. Belly and legs are devoid of wool.	Fleece is white, coarse and open from which carpets are manufactured. People also take its meat.
Jalauni	Medium-sized; straight nose line; both sexes are polled; ears are large, flat and drooping; tail is thin and medium in length; belly and legs are devoid of wool.	Fleece is coarse, short-stapled and open and generally white which can be used for carpet industry. Jaalauni is also taken for meat purpose.
Hissardale	Small low-set appearance with short legs; leaf-like medium-sized ears. Most animals are polled; colour is predominantly white, although some brown or black patches can also be observed.	Its wool is fine and suitable for garments. Annual fleece yield is about 2 kg.
Deccani Fig. 10.28	Medium-sized; predominantly black with white markings; white and brown/fawn animals are also seen. Rams are horned but ewes are polled; ears are medium, long, flat and drooping; tail is short and thin. Belly and legs are devoid of wool.	Fleece is extremely coarse, hairy and open and used for rough blankets. Annual yield is about 700 gm.
Bellary Fig. 10.29	Medium-sized; body colour ranging from white to black with various combinations; one-third of the males are horned, females are generally polled; ears are medium, long, flat and drooping. Belly and legs are devoid of wool.	Fleece is extremely coarse, hairy and open. Annual yield varies from less than 500 gm.
Nellore Fig. 10.30	Based on the coat colour, three varieties of the breed are : "Palla" – completely white or white with light brown spots on head, neck, back and legs; "Jodipi/ Jodimpu" – white with black spots, particularly around the lips, eyes and lower jaw, but also on belly and legs; and "Dora" – completely brown. Animals are relatively tall with little hair except at brisket, withers and breech; Rams are horned but ewes are almost always polled; ears are long and drooping; tail is short and thin; 86% of the animals carry wattles.	Fleece is extremely coarse, hairy and open. This breed is famous for mutton. Ewes are not prolific and late maturing.
Mandya Fig. 10.31	Relatively small in size; white in colour, but in some cases, the face is light brown, which may extend up to the neck; possess a compact body with a typical reversed U-shaped wedge from the rear. Ears are long, leaf-like and drooping, tail is short and thin; a large percentage of the animals carry wattles; slightly Roman nose; both sexes are polled; coat is extremely coarse and hairy. There is a high incidence of cryptorchidism in the male.	It is the best meat breed of the country.

Table 10.7 : Popular Indian breeds of sheep and their uses (contd.)

Sheep breed	Description	Uses
Hassan	Small-sized; white body with light brown or black spots; ears are medium long and drooping; 39% of males are horned; females are usually polled. Belly and legs are generally devoid of wool.	Fleece is white, extremely coarse and open. Mostly used for meat.
Mecheri Fig. 10.32	Medium-sized; light brown in colour; ears are medium sized; both sexes are polled; tail is short and thin. Body is covered with very short hair which is not shorn.	Its skin is of very high quality. This breed also has good dressing percentage (55%).
Vembur	Tall; colour is white with irregular fawn patches all over the body; ears are medium-sized and drooping; tail is short and thin; males are horned but ewes are polled; body is covered with very short hair which is not shorn.	Used mainly for meat purpose.
Coimbatore Fig. 10.33	Medium-sized; white colour with black or brown spots; ears are medium in size and directed outward and backward; tail is short and thin; 38% males are horned but females are polled.	Fleece is white, coarse, hairy and open. Used for meat purpose.
Nilgiri	Medium-sized; body colour is generally white with exceptions having brown patches on face and body; face line is convex, giving a typical Roman nose. Ears are broad, flat and drooping; males have horn buds and scurs but the females are polled. Tail is medium and thin.	Fleece is fine and dense. The breed produces fine fleece but there is little organised shearing and marketing of its wool. The breed is mostly maintained for manure by tea planters and other flock owners.
Ramand White Fig. 10.34	Medium-sized; predominantly white, though some animals have fawn or black markings over the body. Ears are medium-sized and directed outward and downward. Males have twisted horns but females are polled. Tail is short and thin.	Mainly used for meat.
Madras Red Fig. 10.35	Medium-sized; body colour is predominantly brown and its intensity varies from light tan to dark brown. Some animals have white markings on the forehead, inside the thighs and on the lower abdomen. Ears are medium-long. Males have twisted horns but females are polled. Body is covered with very short hair which is not shorn.	This breed is used for meat purpose. Dressing percentage is about 40%.
Tiruchi Black	Small-sized. Completely black body. Males are horned but ewes are polled. Ears are short and directed downward and forward. Tail is short and thin.	Fleece is extremely coarse, hairy and open. Good for meat purpose also.
Kenguri (Tenguri)	Medium-sized. Body colour is mostly dark brown, but colours ranging from white to black with spots of different shades are also not uncommon. Males are horned but the females are generally polled.	It is a meat breed.

Table 10.7 : Popular Indian breeds of sheep and their uses (contd.)		
Sheep breed	**Description**	**Uses**
Chottanagpuri	Small, light-weight animals. Light grey and brown in colour. Ears are small and parallel to the head. Tail is short and thin. Both sexes are polled.	Fleece is coarse, hairy and open and is generally shorn thrice in a year.
Shahabadi Fig. 10.36	Medium-sized, leggy animals. Fleece colour is mostly grey, sometimes with black spots. Ears are medium sized and drooping. Tail is extremely long and thin. Both sexes are polled. Belly and legs are devoid of wool.	Fleece is extremely coarse, hairy and open. Annual yield varies between 500 gm to 1 kg.
Balangir	Medium-sized. White or light brown or of mixed colours; few animals are black. Ears are small and stumpy. Males are horned and females are polled. Tail is medium long and thin. Belly and legs are devoid of wool.	Fleece is extremely coarse, hairy and open and used for carpet industry. The breed is used for meat also.
Ganjam Fig. 10.37	Medium-sized. Coat-colour ranging from brown to dark tan; some have white spots on the face and body. Ears are medium-sized and drooping. Nose line is slightly convex. Tail is medium long and thin. Males are horned but females are polled.	Fleece is short and hairy and is not shorn. Used for meat purpose also.
Bonpala	Tall, leggy and well-built. Fleece colour ranges from complete white to complete black with a number of intermediary tones. Ears are small and tubular. Both sexes are horned. Tail is short and thin. Belly and legs are devoid of wool.	Fleece is coarse, hairy and open. This breed is used for meat and carpet wool.
Gaddi (Bhadarwah) Fig. 10.38	Medium-sized. Usually white, although tan, brown and black and mixtures of these are also seen. All males and 10-15% females are horned. Tail is short and thin.	Fleece is relatively fine, lustrous and dense. It is clipped thrice in a year. The undercoat is used for manufacturing Kulu Shawls and blankets.
Rampur Bushair	Medium-sized. Fleece colour is predominantly white, but brown, black and tan colour are also seen in varying proportions. Ears are long and drooping. Roman nose. Males are horned but most of the females are polled. Legs, belly and face are devoid of wool.	Fleece is of medium quality and dense. This breed is also good for meat.
Bhakarwal Fig. 10.39	Medium-sized. Roman nose. Generally white, although coloured fleeces are occasionally observed; all animals are spotted, fawn or grey. Rams are horned and ewes polled. Ears are long and drooping. Tail is short and thin.	Fleece is coarse and open, used for blanket and carpet industries.
Poonchi	Animals are similar in appearance to Gaddi except being lighter in weight. Predominantly white in colour but spotted sheep varying from brown to light black are also seen. Ears are medium long. Tail is short and thin. Legs are short giving a low-set conformation.	Wool is of medium to fine quality, mostly white in colour. It is clipped twice or thrice in a year with annual yield of 1.6 kg/sheep.

Table 10.7 : Popular Indian breeds of sheep and their uses (contd.)

Sheep breed	Description	Uses
Gurez Fig. 10.40	Largest sheep breed of J & K. Generally white in colour, although some animals are brown or black or have brown or black spots. A small proportion of animals have small, pointed horns. Tail is short and thin. Ears are long, thin and pointed.	Fleece is generally coarse and hairy and suitable for carpet making.
Changthangi Fig. 10.41	Strongly built, large framed. Good fleece cover of an extraordinarily long staple.	Wool is of a good carpet/medium apparel quality. Animals are usually shorn twice a year, generally in May/June and September/October.

Fig. 10.20. Nali sheep

Fig. 10.22. Marwari sheep

Fig. 10.21. Chokla sheep

Fig. 10.23. Magra sheep

Fig. 10.24. Jaisalmeri sheep

Fig. 10.25. Malpura sheep

Fig. 10.26. Patanwadi sheep

Fig. 10.27. Muzaffarnagari sheep

Fig. 10.28. Deccani sheep

Fig. 10.29. Bellary sheep

Fig. 10.30. Nellore sheep

Fig. 10.31. Mandya sheep

Fig. 10.32. Mecheri sheep

Fig. 10.33. Coimbatore sheep

Fig. 10.34. Ramand White sheep

Fig. 10.35. Madras Red sheep

Fig. 10.36. Shahabadi sheep

Fig. 10.37. Ganjam sheep

Fig. 10.38. Gaddi (Bhadarwah) sheep

Fig. 10.39. Bhakarwal sheep

Fig. 10.40. Gurez sheep

Fig. 10.41. Changthangi sheep

II. EXOTIC BREEDS OF SHEEP

In India few exotic breeds of sheep have been introduced with a view to improve the quality as well as quantity of wool and mutton from sheep farms. The exotic breeds are classified into three categories, viz., fine wool breeds, mutton breeds and dual purpose breeds. A brief description about these breeds is as follows :

A. Fine wool breeds

Their fleece is heavy, dense and of good quality, so these breeds produce a fine and crimpy wool. They have strong banding instinct and can graze on poor quality range. Following fine wool breeds are popular in India :

(1) **Merino** : This most popular fine wool breed traces to the Spanish Merino type. Selection within the Merino group has resulted in a large variety of breeds and strains distributed throughout the world. The Merino is a thin-tailed, fine-wooled breed that favours aridity of environment. They are good grazers and are able to forage over large areas of poor grasslands. Merinos in general have strong constitutions and are noted for hardiness and longevity. Their face and legs are white and the skin is pink. Rams mostly have heavy spiral horns, whereas the ewes are polled. The head and legs are generally covered by wool. Mature rams weigh about some 75 kg and the ewes 65 kg. Ram's height is about 70 cm and that of ewes 60 cm. Fleece yield varies widely depending on environmental conditions and time of breeding but averages 4 to 5 kg for rams and 3 to 4 kg for ewes. The staple length is 5 to 10 cm. The Merino, tends to be a seasonal breeder, and hence, has low prolificacy.

(2) **Rambouillet** : The Rambouillet is a descendant of the old Spanish Merino. It was developed as a breed in France. Modern Rombouillets are large, rugged, fast-growing sheep. The rams may have large spiral horns or are polled. The ewes are polled. They have large heads with white hair around the nose and ears. The face and legs are white, skin is pink. Mature rams in good condition with full fleece weigh from 100 to 125 kg and ewes from 60 to 90 kg.

(3) **Polworth** : This breed was evolved by mating Lincoln Merino ewes with Merino rams. They have fairly level frame, clear eyes, soft face, pink rose body colour which is sometimes mottled. They may be horned or polled. Although the animals are bulky in appearance, yet they have symmetrical lines. Their fleece is of even quality and average staple length is not less than 10 cm. The value of the fleece runs very close to Merino. They produce mutton of almost desirable quality.

B. Mutton Breeds

Specialised mutton breeds that mature fast, have high prolificacy, higher body weight gains, high feed conversions, and high carcass yield and produce good quality mutton. India has imported these mutton sheep breeds mostly from U.K. and other countries to improve mutton production in the indigenous breeds. Following mutton breeds are popular in India.

(1) **Suffolk** : Suffolk was developed in south-eastern England. They have straight legs and are of large size. Their face, ears and legs are very black in colour. The head and ears are entirely free from wool and black hair extends to a line on back of the base of the ears. The breed is also noted for its alertness and activity. Both rams and ewes are polled. The fleece is moderately short, dense and fine. The greasy wool yield is 2.75 to 3.25 kg annually. This breed is the first and the foremost mutton breed. The mature rams in good condition weigh from 100 to 135 kg and ewes from 70 to 100 kg. The ewes are very prolific and excellent milkers.

(2) **Dorset** : The breed is native to southern England. There are horned and polled strains of Dorset. The lace, ears and legs are white in colour and practically free from wool. The ears are of medium size, thin, silky and carried well forward. The nostrils, lips and skin are pink. The hooves are white. They produce a carcass of medium weight, and superior quality meat. The weight of fleece is 2.75 to 3.25 kg, and the wool is short, close, and fine in texture. A mature ram in good condition weighs 80 to 110 kg and ewe from 50 to 80 kg. The lambs weigh about 18 to 22 kg at 9 months of age. The

breed is prolific, hard and is capable of doing well under most conditions.

(3) **South down** : These breeds have excellent mutton conformation. The animals are low set, compact, wide and deep with legs set wide apart, and have broad head with a wool cap that comes just below the eyes. The face, ears and legs are light brown and the skin is bright pink in colour. The breed is polled although scars are sometimes found on rams. The eyes are large, bright and prominent, and the ears are of medium size and covered with short wool. The ewes are not too prolific and are only average milkers. The fleece is short, close, fairly dense and of fine quality. The annual greasy fleece weight is around 2.25 to 3.25 kg. The breed is early maturing. Mature rams in good condition weigh about 15 kg at 3 months, 22 kg at 6 months and 27 kg at 9 months age.

C. Dual-purpose breeds

The dual purpose breeds yield meat and wool of good quality.

(1) **Corriedale** : The Corriedale breed had originated in New Zealand, where both mutton and wool production are sought. Lincolon and Leicaster rams were crossed with Merino ewes and by interbreeding and careful selection. Corriedale was established that produced a good balance of mutton and wool. The face, ears and legs of these animals are covered with white hair, although black spots are sometimes present. Both sexes are polled, although rams sometimes have horns. Mature rams in good condition weigh 80 to 110 kg and ewes from 55 to 85 kg. On an average, they produce 4.5 to 5.5 kg of wool annually. The wool is generally noted for exceptional length, brightness, softness and for a very distinct crimp. The ewes are fair in prolificacy and milking ability. In India Corriedale have mostly been imported from Australia.

D. Pelt breeds

Pelt sheep are maintained for their lamb pelts which are used for garments. They are generally of poor mutton quality and the fleece from the mature animals are of coarse carpet type and thus of relatively lower value. Karakul is the chief pelt type breed in India.

(1) **Karakul** : The Karakuls were first imported in 1975 by India. Their performance was studied under the hot-arid conditions at Bikaner (Rajasthan) and cold-arid conditions at other parts of Jammu and Kashmir. It is very well adapted to extreme climate conditions and sparse vegetation resources. The mature Karakul animals are angular, upstanding and of very poor mutton type conformation. The rams have horns, but the ewes are polled. Rams weigh about 90 kg and ewes about 65 kg on an average. The face, ears and legs are black or brown. The ears are drooping.

The lamb pelts are collected through slaughtering of lambs within 24 to 48 hours of birth. In India, lamb pelts are in good demand in Jammu and Kashmir for production of various products like caps, collars, ladies and children coats, etc.

REARING OF SHEEP IN FARM

(1) **Selection and purchase of animal** : It is desirable to purchase healthy animals of 12-18 months of age.Crossbred sheep are available for purchase from State Government/Government of India sheep breeding farms. Ewes (female sheep) can be purchased in regular sheep markets or from breeders in villages, while rams (male sheep) of exotic/crossbred from Government farms. The animals purchased have to be identified by fixing ear tags.

(2) **Housing and shelter management** : Normally sheep do not require elaborate housing facilities but minimum provisions will definitely increase productivity, especially protection against inclement weather conditions (sun, rain and winds) and predation.

(i) Shed could be provided with gunny bags or temporary or removable protections made of thatching materials and bamboos. The roof of the shed should be made of asbestos sheet supported by tubular or angular steel. Wooden rafters and thatching materials could also be used (Fig. 10.42).

Fig. 10.42. Sheep farm

(ii) Exotic breeds should be provided with 0.9-1.1 m^2, and native and crossbred sheep with 0.8-0.9 m^2 space per head. Sheds measuring 18m × 6m can accommodate about 120 sheep.

(iii) A chain link fencing or thorny bush enclosure of 12 m × 6 m can be provided for night paddocking of sheep on each side of the shed.

(3) **Feeding management** : Sheep can survive well on natural grasses, shrubs and farm wastes. Maintaining sheep on concentrate is not the natural method. But increasing unavailability of adequate feeding and foraging areas poses a serious threat to sheep production. So, now it is obligatory to develop arid and semi-waste lands into pastures to provide the sheep top feed through plantation of fodder trees and grain shrubs.

Since sheeps are reared in flocks, so it is very difficult to feed either ewe or ram individually. So bulk of energy and other nutrient requirements are provided mostly from daily grazing. However, feed requirement of ewe flock varies during different stages of breeding cycle as follows :

(i) **Feeding during resting period** : Ewes, not engaged in breeding purposes, are usually kept on a low level of feeding. If these ewes are maintaining their body weight to the extent of about 0.02 to 0.03 kg/day, they are on maintenance ratio. If they loose this ratio, their feed will be supplemented with additional concentrate mixture.

(ii) **Feeding during gestation period** : As the breeding season draws near, the ewes will again be put on better pastures. The amount of nutrients as shown in table 10.8 should be increased by 50 per cent to meet the requirement of fast growing foetus during last six weeks of gestation (total gestation period varies from 144 to 152 days). During late stage of pregnancy, the ewes feed will be supplemented with 250 to 300 gm of any concentrate feed having about 25% DCP (Dry Crude Protein) and 75% TDN (Total Digestable Nutrients) to meet the demands of unborn lamb. This supplementation is known as "Steaming up".

(iii) **Feeding during suckling period** : After giving birth of lambs (baby sheep), the mother ewes must be properly fed so that they can produce sufficient quantities of milk to nurse the new born at least for 3 months. The supplemented feed at this

Table 10.8 : Nutrients requirement of non lactating early pregnant ewe

Body weight (kg)	Dry matter intake (kg)	DCP (gm)	TDN (gm)	ME (M. cal)	Ca (gm)	P (gm)
20	0.56	25	280	1.03	1.7	1.0
25	0.70	31	350	1.27	2.1	1.6
30	0.83	37	415	1.52	2.4	1.9
35	0.95	42	475	1.72	2.6	2.1
40	1.06	47	530	1.93	2.9	2.3
45	1.17	51	585	2.09	3.2	2.5

(DCP : Dry crude protein; TDN : Total digestable nutrients; ME : Metabolisable energy)

stage must be increased up to 500 gm either as gram chuni or a mixture of cereal grains and wheat bran in the ratio of 2 : 1 along with legume hay from 1-2 kg. Mineral like cobalt salt other than common salt must be added with feed of lactating ewes.

For rams (male sheep), feeding system is same as that of ewes. During breeding season, they need supplementary feed from 250 to 500 gm of grain mixture consisting of crushed gram 2 parts, wheat bran 1 part and salt 1 part daily. Overfatty ram should have reduced ration and plenty exercise.

(4) **Breeding management** : Sheep generally attain reproduction maturity at about 2 years of age. According to breed and locality, this age, however, can vary. Female sheep or ewes are polyestrous. The cycle repeates every 16 days and heat period lasts for about 30 hours. In India, three main mating seasons are observed for sheep; viz., Summer (March-April), Autumn (June–July) and post-monsoon (September-October). In plains, mating is high during autumn, while summer is the successful mating season in hilly areas. In farms, ewes need some special cares prior to breeding as follows :

(i) **Extra feeding or flushing** : Two to three weeks prior to breeding season, ewes are given extra grain or lush pasture in order to make their ovary fully matured. Feeding about 250 gm grain per day has shown to increase in lamb crop by twin birth.

(ii) **Removal of wool** : For successful mating, locks of wool or tag should be removed from all over body, particularly at the region of reproductive organs of both ewe and ram. This process is referred to as *ringing* or *tagging*.

(iii) **Eyeing** : In some breeds, wool around the eyes may hide the vision resulting in wool blindness. During mating season, this wool should be clipped away.

(iv) **Marking** : Each mating pair of ram and ewe should be marked separately in the flock. It will help the farmers to identify the sterile sheep if present in the flock.

(v) **Synchronisation of heat** : It is now possible to make most of the adult ewe of the flock in heat or estrus stage within a desirable period. Such synchronisation of estrus or heat can be done by various methods. The most common practice is to insert a pessary containing a synthetic hormone, called *Cronolone* into the vagina of ewe by using a special instrument. The hormone is gradually absorbed through vaginal wall into the blood and prevent ovulation. The pessary is then removed after certain period (15-17 days) by pulling the string attached to it. Within few days, most of the ewes come on heat when mating can be done by selected ram.

Synchronisation process is preferred as it helps the farm owner to plan the lambing time in suitable season and for easier management of flock. It also helps in breeding process by AI or hand mating. By adopting this method, farmers can catch the best/particular marketing time of sheep.

Mating schedule : Mating in sheep is usually done by following systems :

(i) **Flock system** : Here, during mating season, rams are allowed to mate the ewes at will in the flock. However, fighting is the main chaos arised in this system. To avoid this, smaller mating flock (30-40 ewes per ram) is made, and young rams are selected for aged ewes while experienced rams for maiden ewes.

(ii) **Pen system** : In this system, selected ewes and rams are kept in pen during night for mating. This system is mainly followed in stall fed sheep farm.

(iii) **Hand mating** : Ewes in estrus are first selected by the vasectomised rams who cannot fertilise but attempt to mate them. These ewes are then kept in separate pen with the selected ram for mating. However, this system is time consuming, so not useful in commercial farms. But it is useful for any experimental farm.

(iv) **Artificial insemination (AI)** : In India AI is not so popular in commercial farms, except Rajasthan, because of poor viability of sheep-sperm and preservation problems. However, in view of cross breeding profit, AI is now being tried for sheep breeding in many parts of the country.

Fig. 10.43. Feeding of pregnant ewes

(5) **Caring of pregnant, parturient and lactating ewes** : Careful management of the pregnant, parturient and lactating ewes will have a marked influence on the percentage of lambs dropped and reared successfully. So, the following steps may be taken to afford proper attention to these animals.

(i) Pregnant ewes should not be handled too frequently.

(ii) Advanced pregnant ewes should be given effective care in their feeding (Fig. 10.43) and be separated from the main flock.

(iii) Extra feed during the later part of pregnancy (3-4 weeks before parturition) will be beneficial for the pre-parturient ewes in improving milk production of ewes, birth weight and growth of lambs.

(iv) Early maturing breeds have slightly shorter gestation period, so gestation length which ranges from 142 to over 150 days should be counted properly.

(v) Parturient ewes should be protected from cold and chilly weather.

(6) **Care at lambing/parturition** : An ewe about to parturate prefers to leave the flock. She is restless, the udder is often distended and external genitals are in a flushed and flaccid condition. Generally in a healthy ewe parturition is normal. Still the following precautions may be taken during and after parturition.

(i) The ewe should be watched for dystokia or difficult birth. In case of dystokia, veterinary aid and advice from an experienced shepherd or stockman should be taken.

(ii) Maiden ewes in poor condition or small-framed ewes mated to big rams will generally have difficulty in parturition and will have to be assisted.

(iii) Newborn lambs, after being licked by the mother, generally stand on their legs and start seeking for teats and suckle milk. If they are not able to do so after sufficient time, provide help to them in suckling colostrum (the first milk) which is very essential for health and survivability of lambs.

(iv) Newborn lambs should be protected from cold, rain and winds.

(v) Artificial milk feeding or foster mother can be arranged for disowned or orphan lambs. Goats can serve as excellent foster mother, but ewes which have lost their lamb early after birth may also be utilised.

(vi) The naval cord of the lamb should be ligated, and antiseptically dressed.

(vii) Lambs should not be handled too frequently immediately after birth and let the dams (the female parent) lick and recognise them properly.

(viii) Newborn lambs should be allowed to be with their mothers all 24 hours for the first week or so.

(ix) Sufficient quantity of good quality hay and concentrates should be given (if possible) to the lactating ewes (Fig. 10.44) for meeting nutritional requirements for early lactation.

(x) Plenty of clean fresh drinking water should be provided as the lactating ewes drink surprisingly higher amount of water during lactation.

(7) **Care of lambs** : The lamb should be taken care of at maximum extent during early period of life, which will ensure better survival. The following steps may be taken for ensuring better growth and survival.

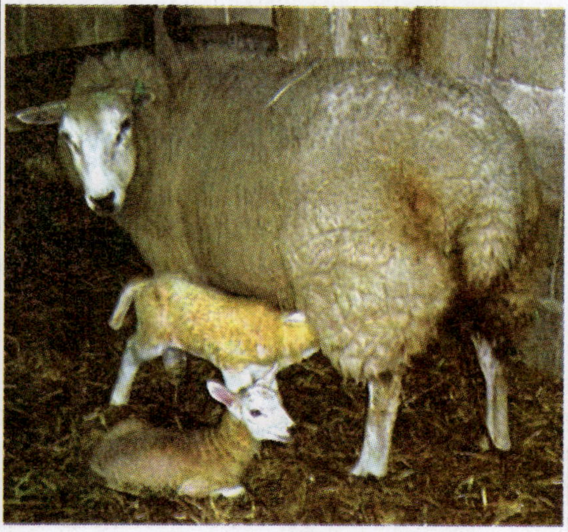

Fig. 10.44. Lactating ewe

(i) Proper suckling of lambs should be ensured. Examine udders for blindness of teats or mastitis*.

(ii) Indifferent ewes should be kept away from lambs and suckling of such lambs should be arranged by foster mother.

(iii) Creep feed (good quality hay with or without concentrate mixture) will be provided to suckling lambs in addition to suckling of milk from the tenth day to weaning age.

(iv) If possible green leguminous fodder or fresh tree leaves should be provided to lambs to nibble during suckling period.

(v) Lambs may be ear-tagged or tattooed on the ear for identification (tattooing forceps and ear-tagging forceps should be cleaned and sterilised at the time of use). Tail docking and castration may also be done in the first week or so by placing elastrator (strong rubber band) at the intervertebral space and not on the vertebra.

(vi) During castration the lambs should be kept on perfectly dry, clean and hygienic site so as to minimize the risk of attack from tetanus.

(8) **Weaning and care of weaners** : The management of weaners plays an important part in good sheep husbandry. The following steps are important in proper care and management of weaners.

(i) Weaning should preferably be done at 90 days after birth, although in breeds with low milk production or where re-breeding is desired, it can be done around 60 days after birth.

(ii) During weaning, lambs should be taken away from ewes for grazing separately for few days and then mixed with main flock. Supplementary feeding and good clean pastures for growing weaners should be provided.

(iii) Weaned lambs should be treated against gastro-intestinal parasites by first month, and vaccinated against enterotoxaemia and sheep-pox.

* Mastitis : Inflammaton of breast tissues.

(iv) Weaners should not be grazed on poor burry and thorny types of pasture, since it could cause skin irritation, injury to the eyes and damage to wool.

(v) They should be protected against vagaries of climates and predation.

(9) **Culling** : Culling of sheep is very important for the development of a good flock. It helps to remove undesirable animals and breeding only from those which are most approximating the ideal sheep. About 10-20 per cent culling should be practised annually to develop a good flock. The flock size should be maintained by replacing culled ewes by ewe lambs born in the flock.

(10) **Identification** : The shepherds commonly practise notching or punching holes in the ears for identification of lambs. Tattooing is also satisfactory but is more expensive. Metal or plastic ear-tags with stamped letters and numbers are most suitable although they are relatively expensive and heavy for the ears of the smaller native sheep. These ear-tags are applied with the help of a clincher.

(11) **Dipping** : To control the ectoparasites the sheep should be dipped a few weeks after shearing when they have grown sufficient new wool to hold the chemical substance. There are standard designs for sheep dips and there are many products effective against ectoparasites (see Table 10.9). A foot-bath may also be provided at the entrance of the farm to prevent the spread of contagious diseases like foot-and-mouth and foot rot.

Table 10.9 : List of dips required for the treatment of sheep from ectoparasites

Ectoparasite	Suggested component of dip
Maggots	Dry sulphur dip
Lice	DDT, Dieldrin
Ticks	BHC, arsenic compound
Scab	Lime sulphur solution

(12) **Maintenance of records** : It is very essential to maintain the necessary records at an organised (experimental or commercial) sheep farm to know about the inputs and outputs. This helps in working out the economy of sheep production per unit of area per animal. The following records should be maintained : livestock strength, breeding, lambing, shearing, wool production, mortality, purchase of animal feeds, medicines, sale of animals and wool.

(13) **Shearing** : Shearing is done mechanically either with clippers, a pair of scissors or by power-operated machines depending upon the size of operations. Most flocks are usually shorn twice a year, i.e. March-April after the winter and September-October after the rains. In some states like Jammu & Kashmir and Rajasthan, sheep are shorn thrice a year.

(14) **Marketing of wool and animals** :

The wool and meat sectors in the country are small and widely scattered. These are the sectors, where organised sector and decentralised sector run complimentary to each other towards meeting the requirements of all sections of the domestic market as well as export. The sector is very loosely knit and heterogenous. Taking the due cognizance of these complexities in the wool sector, the Government of India has constituted the Central Wool Development Board (CWDB) under the Ministry of Textiles for the overall growth and development of this sector. The State Governments have also set up separate corporation/federation to encourage the sector.

Shepherds generally market wool and animals, through rural agents on the basis of rough estimates of weights. The animals are also sold in village weekly markets, where shepherds are exploited to a great extent.

State Governments through sheep corporation, board and federation, have also set up sheep co-operative societies which purchase wool and animals on weight basis and shepherds are paid better prices.

(15) **Seasonal management** : In India, sheeps are available in 4 district zones with different topographical and climatic conditions. Moreover, sheep rearing in most places is exclusively a pastoral enterprise. So, seasonal changes greatly affect such pasture land. Therefore, some special practices should be followed round the year at each sheep-

zones for efficient sheep management. A brief discussion follows :

(i) **Winter season (December-February)** : In temperate and north-western regions weather is extremely cold, whereas in eastern and southern regions it is moderately cold. Grazing condition deteriorates due to poor growth of grass. So all sheep, especially rams and ewes should be provided with supplementary feedings and lambs should be protected against cold.

(ii) **Spring season (March-April)** : Spring is mild, heading for the hot season. After harvesting Rabi crops, stubble grazing is available in pasture. In northern, southern and eastern regions, mating is started. So farmers have to put the ram with ewes which are on heat for successful mating. Before breeding, other operations include : (1) Shearing of flock – it is done immediately after winter; prior to clipping, dung and other falty materials should be removed from fleece. (2) Dipping – it is done by careful lifting of individual sheep into dipping bath of proper size, filled with antiparasitic solution in water for about 2 minutes (see Table 10.9). Dipping in spring prevents the sheep from sheep-blowfly attack, eradicate common external parasites like tick, lice, etc. and also clean the skin from waste materials. Dipping may be repeated at suitable intervals and prior to dipping sheep should be offered water drink, otherwise they can drink the dip. Dipping and shearing the flock prior to breeding make the sheep more active and improve their fertility also. In addition to above operations, vaccination against many infections especially entero-toxaemia is done during spring.

(iii) **Hot season (May-June)** : Due to increased temperature, scarcity of green grasses are found in northern, southern and eastern regions. However, stubble of harvested Rabi crops may provide some feed to sheep. The main operational schedules in these areas include : (1) caring of pregnant ewes – in large flock, ewes close to lambing should be protected from inclement weather. Their feet and udder should be kept free from any infection. (2) steaming up – as the ewes' weight increases, they need to be fed with concentrate in trough with at least 0.5 m of trough space per ewe to prevent knocking each other while feeding. This steaming up can be achieved by feeding the concentrate in the following schedule :

(1) 6-4 weeks prior to lambing : 200 gm concentrate + normal grazing/day/ewe.

(2) 4-2 weeks prior to lambing : 300 gm concentrate + normal grazing/day/ewe.

(3) 2-0 weeks prior to lambing : 450 gm concentrate + normal grazing/day/ewe.

Such concentrate will contain :

(a) Cereal grains – 40 parts + Wheat bran – 30 parts or linseed cake – 20 parts or

(b) Barley, Dal or Maize – 45 parts + sorghum grain – 35 parts + oil cakes – 20 parts. To either mixture, 10 parts of any crude dairy protein should be mixed.

After lambing, supplementary feeding with concentrate should be gradually diminished by 8-10 weeks.

In temperate regions, however, grasses continue to grow. So sheeps of this region do not need such supplementary feeding. However, shepherd should be careful in rearing the lambs like marking and castration of them, as lambs have already born in this region.

(iv) **Rainy season (July to September)** : The pastures show abundant growth during rainy season. In northern, southern and eastern regions, reseeding of pastures can be done during early half. Later half is time of lambing. So newly born lambs should be given special care. Deworming of all sheep can be done by antihelminthics. In temperate region, rainfal ranges from moderate to heavy, so the flocks reach the high alpine for grazing. Deworming of sheeps, pairing of overgrown hoofs and preparation of ewes for breeding are the main managerial practices in temperate sheep farms.

(v) **Autumn season (October to November)** : Autum is good for grazing of sheep in all regions. Sheep can also perform stubble grazing after harvesting of Kharif crops. Sometimes, they are migrated to remote pastures. Common practices of sheep rearing include shearing and dipping of sheep for temperate regions and weaning and castration for other regions.

(16) **Health management** : In farms, the sheep may suffer from various diseases that can contaminate others living together. So health care should be of primary measure in any sheep farm management.

SHEEP DISEASES AND CURE

(A) Sheep-pox

Epidemiology : Sheep-pox is a highly contagious disease. It causes a mortality of 20 to 50 per cent in animals below the age of 6 months and causes damage to the wool and skin in adults. The disease is transmissible to other goats in contact, but not to other species of animals. It, however, spreads slowly.

Symptoms : The disease is characterised by high fever and symptoms of pneumonia and acute enteritis. Skin lesions appear particularly in parts free from wool, notably around the eyes, inner side of the thigh, udder and under surface of the tail. The internal organs such as trachea, lungs, kidneys and intestine are also affected.The disease results in emaciation and, as already mentioned, frequent deaths of affected animals.

Treatment, prevention and control : The diseased animal should be treated with palliatives. The infected litter should be burnt and the bedding changed every day. Affected animals should be kept on soft diet. The ulcers on the skin should be washed with potassium permanganate lotion and dusted with boric acid. Strict hygienic measures, like vaccination should be adopted. In about 15 to 20 days, the animals become resistant to the disease.

(B) Brucellosis of sheep

Transmission : The mode of entry is by ingestion or via conjunctiva. The aborted foetus, vaginal discharge and milk from infected sheep contain a large number of organisms.

Symptoms : In infected sheep, state of abortion may occur followed by a quiescent period during which a few abortions occur. The aborted animals do not breed. After 2 years or more another abortion storm is likely to occur.

Diagnosis, treatment and control : It is not possible to diagnose brucellosis on the basis of symptoms alone. The suspicion is aroused when humans in contact suffer from undulant fever and there is poor breeding record in sheep herd and evidence of mastitis. The diagnosis can be done by the isolation of organisms and by serological tests.

Treatment is based on proper hygiene, vaccination, testing and disposal. Good management practice is essential.

(C) Tetanus

This is an infectious, non-febrile disease of animals and men, and is characterised by spasmodic tetany and hyperaesthesia. This disease is prevalent all over the world.

Transmission : Infection takes place by contamination of wounds. Deep punctured wounds provide favourable conditions for the spores to germinate, multiply and produce toxin which is subsequently absorbed in the animal body. The microorganism is present in soil and in animal faeces, and is carried into the wound by a penetrating object.

Symptoms : Tetanus affects many species of domesticated animals but occurs particularly in horses and lambs; less frequently in adult sheep, goats, cattle, pigs, dogs and cats. The initial symptoms are mild stiffness and an unwillingness to move. More severe symptoms develop after 12-24 hours of infection which are stiffness of limbs, neck, head, tail, and twitching of muscles. Mastication becomes very difficult because mouth cannot be opened, hence the name **lockjaw**.

Treatment : The treatment is carried out by first injecting antitoxin, then treating the wound. Penicillin parenterally is beneficial. Muscular relaxation is achieved by injection of relaxants. The animal should be kept in a dark room and fed with the help of stomach tube.

(D) Listeriosis

Transmission : The infectious organisms are excreted in the faeces, urine, aborted foetus, uterine discharge and milk of infected animals. The blood sucking arthropods may spread infection since organisms have been isolated from cattle ticks and tabanid flies.

Symptoms : In farm animals, the disease occurs towards the end of winter or early spring. The first signs of meningoencephalitis are stiffness of neck, incoordinated movement of limbs and tendency to move in circles or to lean against a fence or wall. Abortions in cattle usually occur after 4-8 months of pregnancy and at a comparatively later stage in sheep.

Treatment : Tetracyclines are very effective in meningoencephalities of cattle, less so in sheep. The recovery rate depends on the speed with which the treatment is commenced.

(E) Campylobactor Abortion (Vibriosis)

Transmission : Transmission occurs by coitus. The organism survives in low temperature used during semen storage.

Symptoms : Among sheep the disease is characterised by abortion, occurring towards the end of gestation. Usually abortion is preceded by vaginal discharge for several days. The aborted foetus is edematous with petechial haemorrhages on serous surfaces and necrotic foci in the liver.

Treatment : Abortion rate can be reduced by antibiotic therapy, and particularly by using chlorotetracycline and concurrently with the development of specific immunity.

(F) Johne's Disease

Johne's disease is a specific chronic contagious enteritis of cattle, sheep, goat, buffaloes and occasionally of pigs.

Transmission : Under natural conditions, the disease spreads by ingestion of feed and water contaminated by the faeces of infected animals. The infection occurs mostly in the early month of life. The incubation period extends from 12 months to several years.

Treatment : The organisms are more resistant to chemotherapeutic agents *in vitro* than Mycotuberculosis. Because of this, the practical utility of treatment in clinical cases is poor.

CONSTRAINTS IN SHEEP PRODUCTION IN INDIA

Various factors like price level, trade, capital and labour always come in the way of increasing efficiency of sheep production. These constraints are being discussed one by one.

(1) **Wool price** : Though India is famous for quality wool fabric production, the prices of indigenous raw wool paid to the shepherds are ridiculously low. There are many reasons for this, of which predominance of middlemen in wool trade seems to be the real cause of poor income and constraints in sheep production.

(2) **Poor management** : The biological potential of sheep is higher than cattle. Still sheep producers are not able to make the best use of sheep's potential. The low production rate of sheep is because of poor flock management which makes low return of investment.

(3) **Grazing problems** : Regions with large sheep population have differences in grazing land and available feed during rainy and dry season. During the drought, farmers are often forced to reduce the number of sheep by forced sale even at a low price. This can be prevented by introducing some alternative feed supplements during the drought at moderate price.

(4) **Trade** : India exports sheep meat mainly to Arabian countries. The international mutton price is subjected to erratic changes which drastically affect the sheep trade.

(5) **International restrictions** : Import and export policies on sheep product also limit the meat supply by several factors and thereby influence the sheep production in India. Few of such factors are as follows : (a) health and sanitation restriction. (b) anticipated increases in domestic demand. (c) influence of international market price. (d) seasonal fluctuations in supply create problems in shipping and coordination with trading countries, and (e) lack of international grading and packing system.

(6) **Capital** : Capital investment is a must for making desired improvement in sheep production. In

India, mostly landless and poor people are involved in sheep rearing who hardly can get the required financial support.

(7) **Nomadic production system** : The nomadic nature of sheep flocks in India also limits sheep production. Sheep owners being poor, ignorant and wandering, cannot afford to get sufficient training and income for improving productivity of the sheep flock.

(8) **Risk as constraint** : Sheep are living in harsh environment conditions. Adverse weather conditions, diseases, etc. can cause death or forced selling of sheep at cheaper rates.

(9) **Skilled labour** : Recent advances in sheep rearing like synchronisation of estrus with breeding season, increasing ovulation for twins and triplets production, production of lambing from off-season breeding, etc. require highly skilled labour for feeding, breeding, housing, fattening of lambs and disease control along with more sophisticated management skills. The ordinary shepherd cannot access such advanced management.

(10) **Poor administration** : Poor administration in forest range pastures, common panchayat land or canal and roadside pasture, harassments and illegal hindrances, etc. discourage sheep owners. They are compelled indirectly to abandon their sheep flocks and often forced to pay much more than official requirements for grazing in common pasture lands.

SHEEP DEVELOPMENT IN INDIA

Till date, Indian sheep population rank 4th in world sheep population and 40 country breeds are available in different agro-ecological regions of the country. In India, sheep are mostly reared for wool and meat. Sheep skins and manure constitute important sources of earning, the latter particularly in southern India. Milk from sheep is of limited importance and that too in very limited areas of Jammu and Kashmir, Rajasthan and Gujarat. Indian sheep are not regarded as dairy sheep. The production characteristics of Indian sheep with respect to quality and quantity of meat and wool are still very poor in comparison to sheep of other developed countries. Realising the importance of sheep in agrarian economy, different development programmes have been formulated under successive Five Year Plans for **Sheep and wool development** in the country during the post-independent period. Some of such programmes are discussed below :

(1) **Establishment of different centres** : The central government established CSWRI (Central Sheep and Wool Research Institue) at Avikanagar, Rajasthan and its regional stations, under UNDP/GOI project, to undertake fundamental and applied research in sheep production and wool utilisation. During the fourth plan, a large sheep-breeding farm was established in collaboration with the Australian Government, at Hissar, for pure-breeding Corriedale sheep. Seven more such farms have been established in Jammu & Kashmir, Uttar Pradesh, Madhya Pradesh, Bihar, Andhra Pradesh and Karnataka, to produce exotic pure-bred or cross-bred rams. During the fifth plan, a large number of breeding farms were established in the central and state sectors to produce genetically superior breeding stocks. It was also planned to reorganise and strengthen the existing sheep-breeding farms in the states as well as to expand and reorganise sheep and wool extension centres.

(2) **Sheep development programmes** : The National Commission on Agriculture (NCA) reviewed the previous sheep and goat development activities and made recommendations on the approach as well as on organisation with a view to implementation of various development programmes. A number of sheep development programmes were undertaken under specialised programmes, such as DPAP, small-farmer (SF), marginal-farmer (MF) and agricultural labourer schemes.

(3) **Breeding strategies** : Different breeding strategies are followed for improvement of sheep products in different regions of the country. In the north temperate and northwestern regions, it involves breeding for apparel wool sheep through cross-breeding of indigenous breeds with exotic fine-wool breeds. For the northwestern and central peninsular

regions and Bihar, selection among better carpet-wool breeds and crossing of extremely coarse and hairy indigenous breeds with exotic fine-wool and dual-purpose breeds to improve carpet-wool production and quality, and mutton production has been recommended. For improving mutton production in the southern peninsular region, the strategy contemplates selection within better indigenous breeds such as Nellore and Mandya, and upgrading of inferior breeds with these two breeds.

(4) **Health coverage of sheep** : Sheep are frequently exposed to internal parasitic infestations since they graze closer to the ground. It causes economic losses to sheep farmers due to increase mortality in sheep. In view of this, different sheep development corporations are supplying antihelminthics free of cost to the farmers through the veterinary institutions of Department of Animal Husbandry and Veterinary Services.

(5) **Establishment of sheep and wool producers co-operative societies** : Karnataka sheep and wool development corporation is engaged in organising sheep farmers on the other lines of co-operative principles and is encouraging establishment of sheep and wool producers co-operative societies. It is providing share capital to each of the sheep producers co-operative society, organising scientific sheep husbandry training camps, providing mechanical wool shearing facilities, etc. The societies shall refund the share capital provided by the corporation after a period of five years in 5 equal instalments.

(6) **Mechanical wool shearing programme** : Mechanical wool shearing not only increases the staple length of the wool but also fetches better rate for the wool produced. So Wool Development Corporation is providing mechanical wool shearing services to sheep farmers at a very low cost.

(7) **Training programmes** : Sheep husbandry has been a traditional occupation in the country and sheep farmers are highly superstitious regarding their own sheep management systems. This has been the root cause for the reduced return among the sheep farmers. Hence, free training camps in scientific sheep husbandry and fodder production are carried out in different parts of the country.

(8) **Subsidy schemes for sheep farming** : Karnataka sheep and wool development corporation is providing subsidy to women self-help groups of scheduled caste and scheduled caste women/men members of the sheep and wool producers co-operative societies for sheep farming.

(9) **Conservation programme** : Indiscriminate cross-breeding during the last few years has significantly reduced the number of few important indigenous breeds, particularly those from Jammu & Kashmir, where almost all the native breeds of sheep have been involved in large-scale cross-breeding with exotic fine-wool breeds. Other such breeds are the Magra, Pugal and Chokla breeds of Rajasthan. All these breeds were declared as endangered and their population is trying to be increased and maintained.

(10) **Development of three-breed-cross** : A three-breed-cross is developed from Malpura, Garole and Patanwadi. This breed is capable of achieving higher market weight, better milk production ability and giving multiple births leading to enhanced economic return.

(11) **Implementation of advanced reproductive technology** : Advanced reproductive technologies like estrous synchronisation, artificial insemination, embryo transfer techniques, in vitro fertilisation, cryopreservation of gametes, etc. are now being implemented in commercial sheep farms from research laboratories.

(12) **Genetic improvement** : An integrated approach of selection and breeding based on molecular and phenotypic information of the traits are thought to be used for desired and rapid genetic gain in meat and carpet wool production. This will be achieved through : Identification of major/candidate genes and SNPs (Singe Nucleotide Polymorphisms) followed by validation of their association with mutton/wool production traits in sheep.

Substantial genetic variation exists between and within breeds as regard to resistance to diseases of parasitic and microbial origins which will be exploited to develop disease resistant sheep. Molecular mechanism of resistance/susceptibility and delineation of markers to identify them through the single nucleotide polymorphisms (SNPs) will be incorporated in the selection programme.

Table 10.10 : Some comparative aspects between goat and sheep

	Goat	Sheep
Taxonomy	**Goats (*Capra hircus*) has 60 chromosomes**	**Sheep (*Ovis aries*) have 54 chromosomes**
Morphology	A goat's tail goes up (unless it is sick, frightened, or in distress). Most goats have hair coats that do not require shearing or combing. The goat does not have cut lip. Most goats naturally have horns, which are more narrow, upright, and less curved than sheep horns. Some goats have beards. Male goats have glands beneath their tail. They develop a distinct odour as they grow in sexual maturity. The odour is very strong during the mating season.	Sheep tails hang down and are often docked (shortened) for health and sanitary reasons. Most sheep grow wool coats that need to be sheared at least annually. Sheep have an upper lip that is divided by a distinct philtrum (groove). Sheep have face or tear glands beneath their eyes and foot or scent glands between the toes. Many breeds of sheep are naturally hornless (polled). Some sheep have manes. Sheep tend to curl their horns in loops on the sides of their heads. Sexually mature rams have much less of an odour, if they have an odour at all.
Reproduction	The estrous (heat) cycle of the doe (female goat) averages 21 days. Goats are much easier to artificially inseminate (AI) than sheep because the does show distinct signs of estrous which make it easy to determine the time of insemination. Male goats have an offensive odour during the mating season. Though it varies by breed, goats tend to be less seasonal and more prolific than sheep.	The estrous (heat) cycle of the ewe averages 17 days. Sheep have a more complicated cervix which makes passage of an insemination device very difficult. Sheep show few visible signs of estrous as compared to goats. Male sheep or rams do not have an offensive odor during the mating season. Sheep are comparatively less prolific than goat.
Foraging behaviour	Goats are natural browsers, preferring to eat leaves, twigs, vines, and shrubs. They are very agile and will stand on their hind legs to reach vegetation. Goats like to eat the tops of plants. Goats require a more nutritious diet.	Sheep are grazers, preferring to eat short, tender grasses and clover. Their dietary preference is forbs and they like to graze close to the soil surface.
Behaviour	Goats are naturally curious and independent. Goats will seek shelter more readily than sheep. In a fight, a goat (buck) will rear up on his hind legs and come down forcibly to butt heads.	Sheep tend to be more distant and aloof. Sheep have a stronger flocking instinct and become very agitated if they are separated from the rest of the flock. It is easier to keep sheep inside a fence than goats. In a fight, a ram will back up and charge to butt heads.This fighting behaviour favours the ram.

Continued to next page

Table 10.10 : Some comparative aspects between goat and sheep (contd.)

	Goat	Sheep
Nutrition	Goats have higher maintenance requirements because they are not able to digest the cellulose of plant cell walls. Grain-feeding is less likely to be profitable in goat production. Goats deposit fat around their internal organs before depositing external fat over their back, ribs, and loin. Goats require more copper in their diet than sheep and are not very sensitive to copper toxicity.	Sheep tend to grow much faster than goats, no matter what the diet is. They convert feed more efficiently. Sheep deposit external fat before depositing internal fat. Some of the wool breeds deposit fat around their organs similarly to goats. Sheep have a narrow tolerance for excess copper in their diet, though toxic levels depend upon the level of other minerals in their diets. For this reason, it is generally recommended that sheep not to be fed grain and mineral mixer that have been formulated for other livestock, as these feeds likely have copper added to them. It is not advisable to use copper as a deworming agent for sheep.
Social dominance	Due to their more aggressive behaviour, adult goats will usually dominate sheep, especially if the goats have horns.	However, when young bucks and rams are maintained together, rams will dominate because the ram will pre-emptively strike the buck in the abdomen while the buck is still in the act of rearing up.
Commercial aspects	Goat farming is 10 times more profitable than sheep farming. Goat gives birth 4 kids in a year. Goat matures in 300 days. They are highly resistant to diseases as compared to sheep. Goat meat has very high demand. Marketing of goats is very easy as compared to sheep. Goats require less fodder as compared to sheep. In the end, the profit obtained by selling goats is much higher than that obtained by selling wool and sheep.	Sheep gives birth to 1 kid in a year. Maturity period of sheep is 390 days. In India, sheep meat has comparatively less demand. Sheep requires much fodder than goat for feeding. The only plus point with sheeps is that, they produce 2-3 times more wool in a year than the goats.

Chapter Eleven : Pig Farming

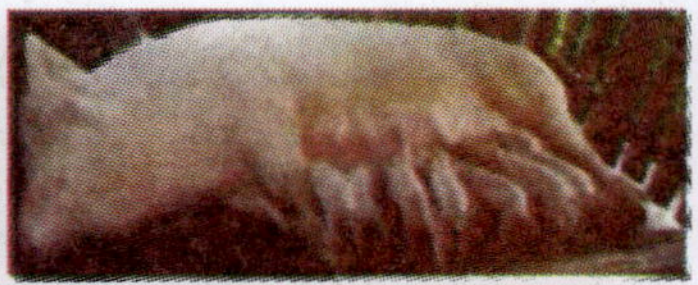

PIG FARMING – THE PIGGERY

INTRODUCTION

Our country is facing the challenges of securing food. So nutritional security to fast growing population needs an integrated approach for livestock farming. Piggery is most potential source of meat production among the various livestock species, and more efficient feed converters after the broiler. Apart from providing meat, it is also a source of bristles and manure. Pig farming can provide employment opportunities to seasonally employed rural farmers and supplementary income to improve their living standards.

Pigs are regarded as an animal which are roaming on the dirty places and eating faeces and other waste materials. Occasionally chasing human and hurting them badly. General people of India for a long time were avoiding it and kept it away from home and locality. These are not popular as source of protein supplement. Even today this is not taken as food by majority of the Indians, because of their dirty appearance and behaviour. However, gradually the scenario is changing. Nowadays these are cultured in a good hygienic farms, and good varieties of pig-breeds are imported. The well-dressed hams are gradually accepted by the general people. It can be predicted that it will be a means of livelihood for some unemployed youths very soon. Pig farming is very important component in North East India. Out of total pig population in India, 28% are grown in this region. Good quality breeds are available at ICAR Research Complex for North Eastern Hill Region, Barapani–793 103, Meghalaya.

Taxonomic features

The systematic position of the swine (after J. Z. Young, 1981) is as follows :

Phylum — *Chordata,*
Sub-phylum — *Verebrata/Crainata*
Super-class — *Gnathostomata*
Class — *Mammalia*
Sub-class — *Theria*
Infra-class — *Eutheria*
Cohort — *Ferungulata*
Super order — *Paraxonia*
Order — *Artiodactyla*
Family — *Suidae*
Genus — *Sus* Linn

The European breeds of domestic swine were derived from the local wild pig, *Sus scrofa,* whereas the breeds in the Far Eastern parts of the globe were derived from another wild pig, *Sus vittatus.* The two types inbred readily. The modern breeds of pig evolved from different crossings between the two original types and the present day domestic pig, *Sus domesticus* is the result of thousands of years of evolution through gradual domestication.

Advantages of pig rearing

(1) Pigs convert inedible feeds, forages, certain grain by-products obtained from mills, meat by-products, damaged feeds and garbage into valuable nutritious meat. Most of these feeds are either not

edible or not very palatable to human beings. The pig has got highest feed conversion efficiency, i.e., they produce more live weight gain from a given weight of feed than any other class of meat producing animals except broilers.

(2) Pig grows fast and is a prolific breeder, farrowing 10 to 12 piglets at a time. It is capable of producing two litters per year under optimal management conditions. A sow can be bred as early as 8-9 months of age and can farrow twice in a year.

(3) With a small investment on building and equipment, proper feeding and sound disease control programme, the farmer can profitably utilise his time and labour in this subsidiary occupation.

(4) The carcass return is quite high, i.e., 65-80 per cent of live body weight. Pigs are well known for their meat yield. In comparison to other livestock species whose dressing yields may not exceed 65%, greater can be obtained from pig.

(5) Pork is most nutritious with high fat and low water content and has got better energy value than that of other meats. It is rich in vitamins like thiamin, Niacin and riboflavin.

(6) The faeces of pigs are used as a manure to maintain soil fertility. Pigs manure is widely used as fertilizer for agriculture farms and fish ponds.

(7) Pigs store fat rapidly for which there is an increasing demand of pig meat/fat from poultry feed, soap, paints and other chemical industries.

(8) Pig farming provides quick returns since the marketable weight of fatteners can be achieved within a period of 6-8 months.

(9) There is good demand from domestic as well as export market for pig products such as pork, bacon, ham, sausages, lard, etc.

Target group of pig farming

Small and landless farmers are suitable for pig farming, because it requires minimum land space and financial investment. It provides part-time earning for educated youth having agriculture as occupation. Uneducated youth can also be involved in pig culture, because it is easy to handle pigs with minimum involvement. Pig farming provides employment of urban unemployed women, because pigs can be maintained by women.

BREEDS OF PIGS USED IN INDIA

The indigenous pig has been used for pig production for a long period of time. It is small in size. Improved breeds are now being used for grading up the pig production in the rural areas. The imported pig breeds reared in India are as follows :

(1) Large White Yorkshire

It is the most extensively used exotic breed in India (Fig.11.1). The Large White Yorkshire is a popular English Bacon breed which had its origin nearly a century ago in Yorkshire and neighbouring countries in northern England. Yorkshire sows are

Some common terms in relation to pig

Details	Expression	Details	Expression
Species commonly called as	Swine or Sus	Castrated male	Hog or stag or barrow
Group of animals	Drove or stock or herd	Castrated female	Spayed
Adult male	Boar	Female with its offspring	Suckling
Adult female	Sow	Act of parturition	Farrowing
Young male	Boarling	Act of mating	Coupling
Young female	Gilt	Sound produced	Grunting
Newborn one	Piglet or pigling	Pregnancy	Gestation

Fig. 11.1. Large white Yorkshire pig

noted as good mothers. They not only farrow and raise large litters, but they are great milkers.

These are entirely white in colour. Black pigment spots, called "freckles", do not constitute a defect, though they are frowned upon by breeders.

Head is moderately long, face slightly dished, snout broad, wide between the ears. Neck is long, fine and proportionately full to shoulders with wide and deep chest. Back is long, level and wide from neck to rump. Legs are straight and well set, level with the outside of the body, with flat bone. Mature boars weight varies from about 300 to 450 kilograms, while average sows weight from 250 to 350 kg.

Carcass quality possess necessary length, gives first grade bacon, good ham and comparatively light forequarters.

This variety gets popularity because it is excellently used for the purpose of cross-breeding. It is also a prolific breeder. It provides a good amount of meat for consumption. Mature boar weigh about 300-400 kg and mature sow is 230-320 kg in weight.

(2) Landrace

The origin of this breed is in Denmark, where it has been bred and fed to produce the highest quality bacon in the world. The animal is with long body, large drooping ears and long snout (Fig. 11.2). The breed is white in colour, although black skin spots 'freckles' are rather common. The breed is characterised by its long, deep side; square ham; relatively short legs; trim jowl; heavy lop ears. The carcass is more lean than that of the meat. There is less back fat and lard. The breed is noted for prolificacy and for efficiency of feed utilisation. This is an excellent breed for cross-breeding experiments. In U.S.A. the breed has been extensively used for developing cross-breed foundation. In India the Landrace stock has been imported with the same objective. Ham quality is equal to white Yorkshire breed. Mature boars weigh 270-360 kg and mature sow weighs 200-320 kg.

(3) Middle White Yorkshire

This is extensively used in some areas of India (Fig. 11.3). This variety grows rapidly and gives

Fig. 11.2. Landrace

Fig. 11.3. Middle white Yorkshire

good dressing percentage. It is moderately prolific breeder as compared with Large White Yorkshire. This breed can provide good quantity of ham. This breed was evolved in Yorkshire of Northern England by crossing the Large White with smaller breed of Yorkshire extraction. The breed is accepted as an excellent pork pig, reaching slaughter weight early and with a high percentage of lean meat to bone. The breed is hardy, grows rapidly, but is not as prolific as the Large White. In India the breed has been extensively used for improving the indigenous stock in rural areas.

Their skin is white in colour, free from coarseness, wrinkles or any spots. Head is moderately short, straight jaw, short snout. Neck is fairly light, medium length, proportionately and on standing shoulders evenly set. Back is long and level to root of the tail with well sprung ribs. Legs are straight and fairly short, well set apart, fine and flat bone upon toes and a good walker. The adult males weigh about 250–340 kg and females about 180–270 kg.

(4) Berkshire

The breed originated in and takes its name from Berkshire, a country in south-central England, which was the centre of its development. The old English hog, a descendant of the boar, served as foundation stock and these early animals were further improved by introducing Chinese, Siamese and Neopolitan blood. The breed is now highly valued as producer of good quality pork (Fig. 11.4). In India especially in the South the breed is popular for upgrading programmes.

Colour is black with "six white points", that is white on feet, nose and tail, but more than 10 per cent white of an animal disqualifies it. Head has a short head with dished face and its ears are erect. Long, level arched narrow back. Legs are straight and well set long legs. Full grown boars weigh between 275 – 375 kg, Sows weigh from 200–290 kg. Its reputation throughout the world has been for meat, well-balanced carcass with a high cut-out value.

(5) Hampshire

The Hampshire pig is a domestic swine breed characterised by erect ears and a black body with a whitish band around the middle, covering the front legs. The American National Swine Registry notes this as the fourth "most recorded breed" of pigs in the United States, and probably the oldest American breed of hogs. Importations of this hog breed were thought to have been made from Hampshire in England between 1827 and 1839. Residents of Hampshire are often colloquially referred to as 'Hampshire Hogs' (Fig. 11.5), a name which goes back at least to the 1790s. Hampshire hogs are noted for being well-muscled and rapid growers, and for exhibiting good carcass quality when used as meat animals. When used as breeding stock, the sows of

Fig. 11.4. Berkshire pigs

Fig. 11.5. Hampshire Hog

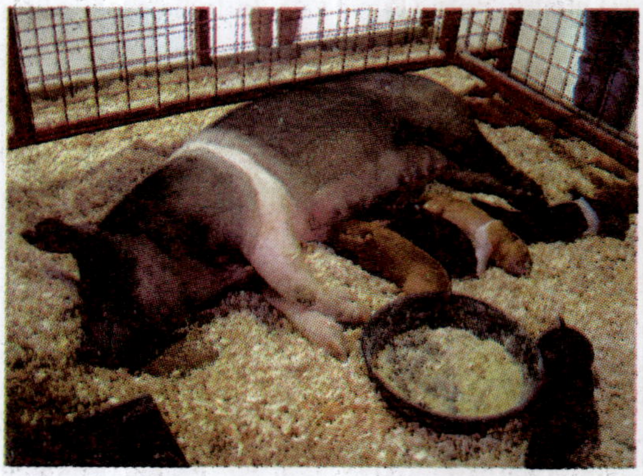

Fig. 11.6. Hampshire Hog with piglets

this breed have been praised for their capacity as mothers, having "extra longevity in the sow" (Fig. 11.6). Hampshire are good tempered; they do not grow as fast as many cross-breeds, but they do grow faster than Yorkshires.

(6) Tamworth

The breed originates from central England. The breed derives its name from the town of Tamworth, which is located on the river Thames. Golden red colour varies from light to dark. This is an extremely good bacon type pig. The individuals are long-legged with long, smooth sides and strong backs. The head is strikingly long and narrow with a long snout and fairly large ears that are carried somewhat erect. The carcass produces bacon on the finest quality. The sows are prolific and careful mothers, and the pigs are excellent foragers. Mature boars weigh up to 350 kg and the mature sows from 250 to 300 kg. The breed has been widely used for cross-breeding purposes in the tropics, particularly in South-east Asia. They thrive under close confinement feeding, but they must be well managed and fed. Landrace is susceptible to sunburn.

(7) Duroc

North-eastern United States gave rise to one of the most popular U.S. breeds—the Duroc (Fig. 11.7), which is the outcome of blending of two breeds, Jersey Red and Duroc of New York. The breed is moderately red coloured with shades varying from a golden to cherry red colour. The Duroc is noted for excellent rate of gain and feed efficiency. Maturing early, the Duroc sow has large litter and is a good mother. The carcass is considered as a good meat type. The weight of mature boar is about 400 kg and of sow it is normally 350 kg. The breed is now popular both in South East Asia and in the American tropics mostly due to its colour, hardiness and fast growth.

Fig. 11.7. Duroc

(8) Chester White

The breed had its origin in the south-eastern Pennsylvania, principally in Chester and Della counties of U.S.A. The foundation stock included important pigs of English Yorkshire, Lincolnshire arid Cheshire breeds. As the name indicates, the breed is white in colour with some bluish spots sometimes found on the skin. Chester White sows are very prolific. The pigs are good feeders; they mature early and make good gains.

(9) Hereford

It is one of the newer breeds of swine. The breed was originated in Missouri in the U.S.A. The most distinctive characteristic of the Hereford breed is their colour marking, which is similar to that of Hereford cattle. Two-third of it is red in colour, either light or dark with white face. The white colour must appear on at least two feet and extend up to an inch or more above the hoof. In size, the Hereford is smaller than the other breeds of swine and in general it is of compact type.

BREEDS OF INDIA

(1) Indigenous breed

In the North-east states of India this type of breed is found. The local pigs in Nagaland, Mizoram, Manipur, and Arunachal Pradesh are small in size, hairy and black in colour with elongated face and short ears. They are potbellied (Fig. 11.8) and reported to have early age maturity, fast farrowing, and shorter inter-farrowing intervals. Some of them are white striped on the neck or shoulders and spotted on the fore-head. These pigs survive without much care especially in feeding and disease management. The Khasi, Garo, Mizo, Naga, Arunachal pigs are similar in this regard and this pig breeds are scavenging type.

Fig. 11.8. Indigenous

(2) Ghungroo

It is an indigenous strain of pig first reported from North Bengal. It is popular among the local people because of high prolificacy and ability to sustain in low input system. This breed/strain produces high quality pork utilising agricultural by-products and kitchen wastes. Ghungroo are mostly black coloured with typical Bulldog face appearance, with a litter size of 6-12 piglets, individually weigh about 1.0 kg at birth and 7.0-10.0 kg at weaning (Fig. 11.9). Both sexes are very much docile and easy to handle. At National Research Centre on Pig, Rani, Guwahati, Ghungroo pigs are being maintained under intensive system of rearing with standard breeding, feeding and management system. Their evaluation for genetic potential for use in future breeding programmes is in progress and this indigenous strain is performing very well in terms of productive and reproductive efficiency. Some of the selected sows have delivered litter size as high as 17 piglets at birth as compared to the other indigenous strains of pigs.

SELECTION OF PIGS FOR FARMING FROM THE AVAILABLE BREED

There are important characteristics which need to be considered in developing a good sow herd. The farmer should choose a good size of litters, their strength and vigour of litters will also be noticed. Information on milking ability of the mother should

Fig. 11.9. Ghungroo with piglets

be considered before purchasing the litter. Grain and feed efficiency, fertility, selection of individual animals from a herd is more important than the selection of a particular breed. Each producer at the time of setting up his herd should purchase his animals from a reliable disease-free herd and should obtain as much information of the animals as possible. Once the herd is established the selection of the gilts and boars for replacement in the breeding herd should be based on the types and performance.

In selection of a gilt or sow the primary aim is to secure a female that will produce large litters of fast growing pigs capable of being fattened to marketable weights at an age of six months or less. In the practical swine enterprise, 'growing-fatting' generally refers to that period from weaning to market weight of almost 100 kg. In selecting a gilt or sow **on the basis of appearance**, the following main points are considered : (1) general form or types; (2) size or weight for age; (3) development in the regions of high-priced cuts of pork; (4) quality.

(1) **General form or types** : As for form or type, a desirable gilt or sow presents a well-balanced appearance (Fig. 11.10). From the side view, the top line appears as a strong uniform arch; the underline is straight; the legs are medium in length; the sides are deep and long. As viewed from the front or rear, the shoulders, back, loin and rump are fairly wide, and the width is carried uniformly from front to rear.

(2) **Size or weight for age** : Gilts do not show as much depth of body as sows and consequently appear more "leggy" at the younger age. A desirable animal has a large heart girth, as indicated by the distance around the body just behind the shoulders. The head is trim in appearance with good width at the snout and between the eyes. The face is medium in length, varying somewhat with the breed.

Fig. 11.10. Yorkshire gilt

(3) **Development in the regions of high-priced cuts of pork** : A good gilt or sow shows a reasonable amount of development in the regions of the high priced cuts of pork chops, which come from the back and loin; ham, which comes from the rear quarters; and bacon, which comes from the sides. Therefore, selection of animals having wide, full hams is essential. Preference should be given to a body that is wide over the back and loin and long and deep.

Pedigree : A pedigree is the statement of an animal's ancestry. In addition to the names and herd-book numbers of the animals in each of several generations, usually it includes also the date of birth of each animal and the name of the breeder. It may extend back to the beginning of a breed's record history, or only for two or three generations.

In a real sense, a pedigree is good or bad, according to as the individuals in it are good or bad. The better the individuals, the more uniform they are in the type of performance desired, the more prepotent and uniform the breeding qualities of the individual may be expected to be.

(4) **Quality – Performance or productive ability** : It is desirable to consider actual performance, or productive ability of an animal. In the case of a gilt or a sow, one measures efficiency of rate of gain as shown by its weight for its age. In the case of a sow, information on the number of pigs in each litter that she has farrowed, number of piglets raised per litter and the weight of the litter should be considered.

Prepotency or transmitting ability : Prepotency is the ability to transmit characteristics to the offspring to a marked degree. It refers to either male or female. From a genetic standpoint, there are two requisites that an animal must possess in order to be prepotent : (a) dominance and (b) homozygosity. If a given boar or sow possesses a great number of genes that are completely dominant for desirable type and performance and if the animal is relatively homozygous, in that case it can be said that the animal is a prepotent.

A desirable boar is larger for his age and heavier of bone than gilt, and he shows masculinity and ruggedness, rather than feminity. As with gilt, a boar should be 4 to 6 months of age at the time of selection since serious defects in type are not so likely to develop after this age is reached.

Selection of gilts (female)

Selection of gilts for the breeding herd should be made at market weight, i.e., when the animals weigh about 90 kg. Gilts should be selected from sows which have consistently farrowed and weaned large litters (Figure 11.10). They will reach market in minimum time and have desirable market type. It would be desirable to choose gilts whose litter and other full sibs have given good performance in daily weight gain and feed conversion efficiency.

Selection of boars (male)

Selection of boar is extremely important, particularly for a small breeding farm or unit. The boar should be purchased from a breeder or a farm maintaining adequate information on its performance. The boar should be selected from a dam which has consistently farrowed and weaned high litters. A good boar which attains weight of 90 kg in about 5-6 months, will be of good type and will be strong on feet and legs (Fig. 11.11). The feed conversion from weaning to 90 kg weight would be the most desirable.

Fig. 11.11. Yorkshire boar

Fig. 11.12. Large White Yorkshire gilt with piglets

Points to be considered while replacing boars and gilts

The mother of the pig to be selected should have had large litters of 8 piglets or more (Fig. 11.12). The weaning weight (at 56th day) of a litter in case of a gilt selection should have been 120 kg and in the case of sow it should be not less than 150 kg. The gilt or the boar should have reached a body weight of about 90 kg in about 6 months. The pig should have adequate length and depth of the body, thick well-muscled hams should be firm and trim. The pig should have sound feet and legs. For gilt back fat thickness of 4 cm or less and for boars 3.2 cm or less is ideal. Gilts should have a minimum of 12 evenly spaced, functional teats. An animal with blind teats should be avoided as there will be little or no milk from these teats and the defect is heritable. Negative blood test for both brucellosis and leptospirosis should be made during selection and the pigs should be vaccinated against swine fever. Pigs should be free from other diseases and physical defects

FEEDING MANAGEMENT

During formulation of supplement food most economical ingredients should be selected. Grains like maize, sorghum, oat, other millets, wheat and rice should form the basic ingredients. Protein supplements like oil cakes, fish meal and meat meal should be provided. No vitamin supplements are necessary if the pigs are allowed to pasture or are fed fresh green legumes. Vitamin B_{12} supplement would be necessary if little or no animal protein is fed.

Antibiotic supplements at the rate of 11 mg of antibiotic per kilogram of ration and mineral supplements should be provided.

In Tables 11.1 and 11.2, details of specific formulation of the various requirements in the formation of creep, grower and finisher rations for pigs are given.

The farmer should prepare the complete ration recommended (Table 11.2) for different classes and provide the pigs the amount they will eat without waste two or three times daily. The following is the approximate amount of dry feed the pigs will consume per day. All grains in mixed feeds should be ground. Generally feeding in the form of wet mash is not superior to (slop feeding) dry feeding. Slop feeding requires more time and excessive labour. If a ration is fairly high in fibre, pelleting the feed may increase the rate and efficiency of gain in weight. Pelleting may also decrease the amount of feed that is wasted. It is important not to overfeed sows which have been bred. Overfat sows are apt to produce weak pigs and crush more piglets at farrowing. Sows

Table 11.1 : Various requirements in the formation of creep, grower and finisher rations for pigs

Nutrients		Creep feed (up to weaning)	Grower ration (20-40 kg)	Finisher ration (40-90 kg)
Protein supplement (%)	Oil cakes	16-18	14-16	13-14
	Animal protein	8-10	4	2
Grains (Maize, sorghum, millets or combination of grains) (%)		60-65	50-55	40-50
Wheat bran or rice bran (%)		5	10	20
Lucerne meal (%) if available		—	5-8	—
Mineral mixture (%)		0.5	0.5	0.5
Antibiotic supplement (mg)		40	20	10

Table 11.2 : The composition of the concentrate feed for various age groups of pigs

Ingredients	Creep feed (14th to 56th day)	Grower ration (up to 40 kg)	Finisher ration (40-90 kg)	Pregnant and nursing sows
Maize or sorghum or broken wheat, broken rice and barley in convenient combinations (%)	65	50	50	50
Oil cakes (groundnut oil cake, soybean oil-cake, sesame oil cake, linseed oilcake (%)	14	18	20	20
Molasses (%)	5	5	5	5
Wheat bran or rice bran (%)	10	15	20	18
Fish meal or meat meal or cooked offal, skim milk powder, dairy wastes (%)	5	5	3	5
Mineral mixture (%)	1	4	1.5	1.5
Salt (%)	—	3	0.5	0.5

should gain about 35 kg and gilts about 55 kg from breeding to farrowing.

HOUSING OF PIGS

Good housing with adequate accommodation incorporating all essential requirements of pigs must be provided so that the animals can grow quickly and efficiently. This is only possible through scientific housing which includes the provision for fresh air, exercise, sunlight protection from inclement weather conditions.

Pigs are very much susceptible to extremes of heat and cold. Pigs are poorly provided with heat regulating mechanism and sweating only takes place from the snout. So it is difficult to keep them cool in hot weather. Temperature in a pig house is, therefore, one of the most important factors to be taken into

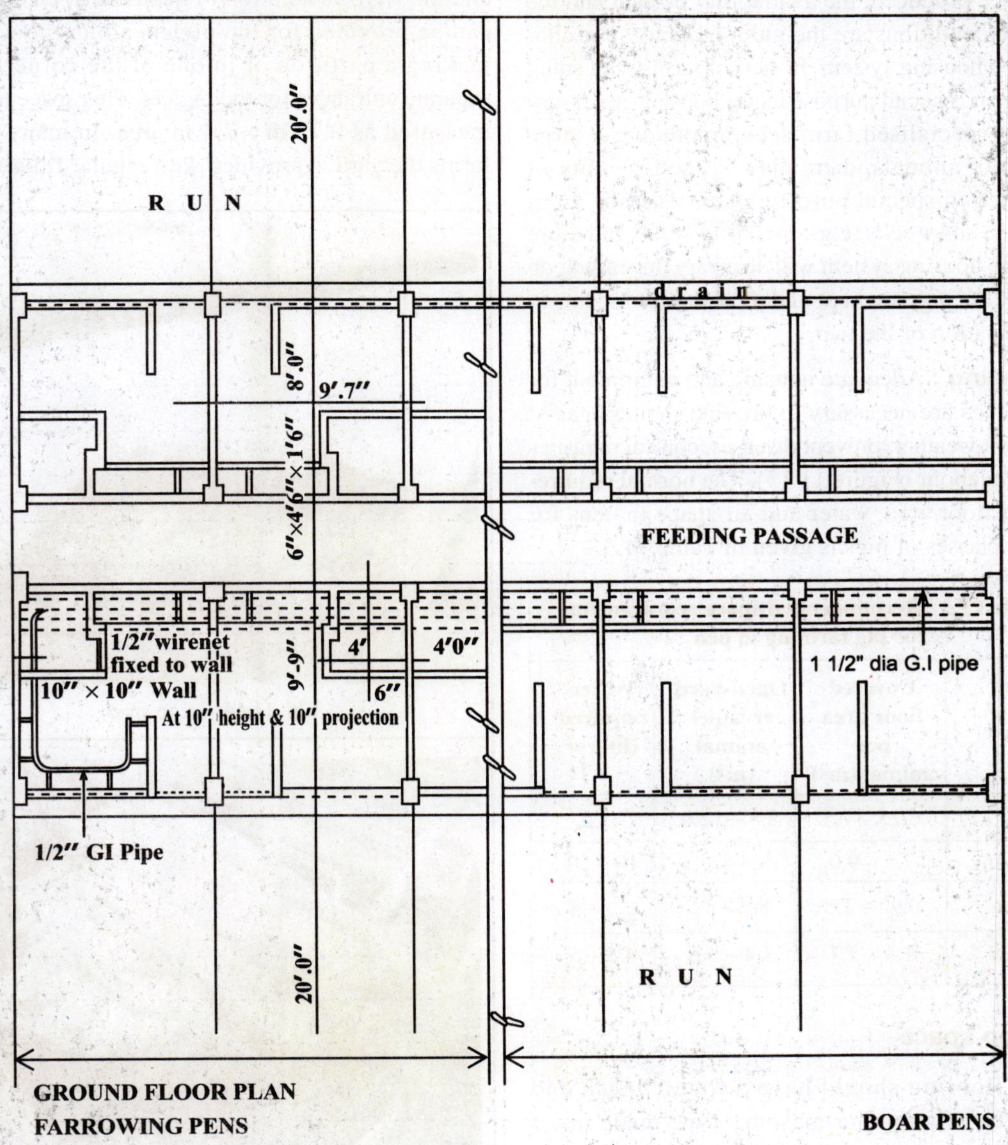

Fig. 11.13. Sketch of housing plan of a modern pig farm

consideration while erecting a pig house particularly in tropical countries.

SYSTEMS OF HOUSING PIGS

Pigs are kept under two systems — (a) **open air system** and (b) **indoor system**. A combination of the two is, however, followed in most places in this country. Both these systems have advantages and disadvantages. The size of the enterprise, the type of pigs to be produced, the availability of land and the climatic conditions are the guiding factors for choosing the housing system. In case of farms with small holdings no special purpose in pig house is necessary but in a specialised farm accommodating a large number of animals, there may be the necessity of constructing special purpose houses for different categories of stocks, e.g., breeding stock and store pigs. The housing system will also vary depending on whether the pigs are kept under pasture or indoor, or a combination of the two.

Housing : Adequate housing and equipment for raising pigs are necessary to provide shelter against inclement weather, prevent diseases, control parasites and save labour (Figure 11.13). The normal requirement of floor area, water and air space in pens for various classes of pigs is given in Table 11.3.

Table 11.3 : Recommended space requirement for pig farming in pen

Class of animals	Covered floor area per animal (m²)	Open-yard area per animal (m²)	Water required (liters)
Boar	6.25 – 7.5	8.8 – 12.0	45.5
Farrowing	7.5 – 9.0	8.8 – 12	18 – 22
Weaner	0.96 – 1.8	8.8 – 12	3.5 – 4
Dry sow	1.8 – 2.7	1.4 – 1.8	4.5 – 5

Creep space

The flooring should have a rough finish and should be of a regular masonry type made up of waterproof cement mortar (Figure 11.14). Proper drains should be provided so that the effluents are disposed of. Generally under village conditions the housing can be made up of pens measuring 3 m × 2.4 m or 3 m × 3 m with an open yard of nearly the same dimension or in some cases slightly longer. Walls should be 1.2–1.5 m high from the floor. For the purposes of farrowing some of the pens could be converted into farrowing pens (Figure 11.15) by providing guard rails made up of G.I. pipes of 5 cm diameter along the walls, 20–25 cm from the ground and the wall. In addition to guard rails, creep space can be provided for the piglets along the wall by making a partition or in one of the corners with separate entrances for the piglets. This space usually measured as 0.75 m × 2.4 m area. In many of the farms the yard is provided with regular flooring.

Fig. 11.14. Creep space

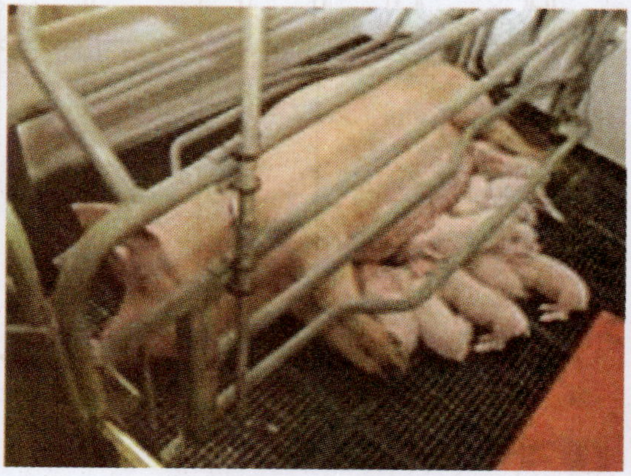

Fig. 11.15. Farrowing pen

Prolonged exposure of exotic breeds of pigs to bright sunshine may cause pigs to become overheated even during moderate weather. Shade helps in preventing deaths and increasing production efficiency during hot weather. Planting of trees in the neighborhood of pens for reducing the intensity of heat is recommended. But it is not desirable to plant large trees for giving regular shade because they permit rapid build-up of parasite levels.

LOCATION OF THE FARM

The farm should be located near the city where there is a heavy demand of pork products. This is particularly necessary to avoid the cost of transport of feed and other stuff as also for disposal of animals. Where pedigree pigs are sold in large numbers it is advantageous to locate the house near a railway station. Proximity to public utilities like electricity, gas and water supply will be an added help to the farm. At the same time the farm should not be situated in the municipal areas where there might be objection in future from health authorities as also for future possibilities of expansion.

Layout of building has a marked influence on efficient working. The most important item to be considered is the facility for proper sanitation with a good drainage system. The building should be situated in such a way that doors and windows will receive maximum sunlight and the pigs will get shelter from prevailing direction of the wind.

Wallows

Pigs have very few sweat glands. In areas having warm weather mature breeding animals and fattening animals need a wallow during summer months. Instead of permitting unsanitary wallows a masonry wallow with proper drainage would be desirable. The size of the wallow will depend upon the number and size of the animals.

Age of breeding stock

Well-developed gilts may as a general rule breed to farrow when 12-14 months old. This depends more on development than on age. Gilts should weigh at least 120 kg before breeding. Ovulation rate

Table 11.4 : Breeding Management

Age to breed gilts	8 months
Weight of breed gilts	100–120 kg
Length of heat period	2–3 days
Best time to breed in heat period	Gilts – first day, Sows — Second day
Number of services per sow	2 services at an interval of 12–14 hours
Period of oestrous cycle	18–24 days (Average 21 days)
Occurrence of heat after weaning	2–10 days
Gestation period	114 days

increases during successive oestrous periods (up to fifth) following puberty. Thus it is advantageous to delay the breeding of gilts until the second or third oestrous. Litter size increases on an average in succeeding pregnancies up to 5th or 6th litter. It is therefore, advantageous to cull the sow from a breeding herd or a commercial herd after her fifth or sixth litter as the litter size goes down thereafter. The detail of breeding management is given in Table 16.4.

Detection of heat

The average length of oestrous cycle in pigs is 21 days. The oestrous symptoms last for five to seven days beginning with vulvar swelling and vaginal discharge. In true oestrous there is frequent urination, reduced appetite, mounting and standing for service detected by the erection of ears and immobility when normal pressure is applied to the back. The application of pressure on the back is used to determine the correct breeding time. Animals with a predisposition for weak oestrous should be brought near the boar to exhibit heat symptoms a little more clearly.

Best time for breeding is during the latter half of the first day or early on the second day of oestrous. In many cases the gilts and sows continue to exhibit the standing heat on the next day. In these cases the animals should be rebred and the interval in the case of rebreeding should be 12–14 hours. This procedure will ensure a high conception rate in the herd.

Sows come into oestrous one to four days after farrowing but they should not be bred at this time. Sows may also come into heat two to ten days after weaning and may be bred at this time. But better results can be obtained by breeding them in the second post lactation oestrous. The animals which have been bred should be observed for the appearance of subsequent oestrous. If sows not conceived even after successful mating with a boar in two continuous oestrous cycles it is desirable to cull them from the herd.

Flushing

It is the method of feeding sows and gilts before breeding. A good grower ration fed to sows and gilts seven to ten days before breeding helps in increased ovulation rates in them. After breeding sows and gilts should be fed a limited but well balanced ration until the last six weeks of pregnancy and then full feeding should be resumed.

Care and management of pregnant animals

The gestation period of sow varies from 109–120 days with an average of 114 days. Pregnant animals should be housed in groups in separate enclosures to avoid fighting which may result in abortion. It would also be advisable to house pregnant gilts and sows in separate groups during gestation. About 3 m^2 of dry housing should be available for each sow. The pregnant animals should be allowed to move about every day in the morning on a free range or a pasture, if available. A pasture area is presumed to be clean, if a cultivated crop was raised.

Management at farrowing

Farrowing time is the critical time in pig production. Death rate is high during farrowing and the first week after farrowing. Sows may be farrowed in pens equipped with guard rails and a creep space in farrowing crates or in farrowing stalls. A pen equipped with guard rails and a creep space is adequate. The pen should be maintained at 24ºC to 28ºC until the piglets is three or four days old and at 18ºC to 22ºC until the piglets are approximately six weeks old. The heat lamps should be hung 45 cm from the floor and suitably protected. The farrowing pens should be thoroughly cleaned before the sow is brought in. This will prevent a large number of diseases of piglets. The sow should bring to the farrowing pen at least one week prior to farrowing so that it becomes familiar with the surroundings. She should be washed thoroughly before being brought to the farrowing pen. The feed ration should be made bulky by substituting one-third of the regular ration with wheat bran. The amount of ration fed should also be reduced by one third till the sow farrows. The sow should be watched closely for determining the approximate time of farrowing and feed should not be given 12 hours before farrowing.

Care during farrowing

An attendant should be on hand when the sow farrows. Otherwise manypiglets will die. It takes generally 2 to 4 hours for complete farrowing to take place. The piglets should be removed as they are farrowed and kept warm in the creep space until farrowing is complete. Each piglet should be cleaned of all mucus to ensure that the breathing passages are clear. The navel cord should be tied 2-5 cm away from the navel, cut with a disinfected pair of scissors and the stumps painted with iodine. Piglets should be allowed to be nursed after birth. In about 2 days they settle down to their individual teats. They nurse 8-10 times in 24 hours in the initial period. Trampling by the sow should be prevented during the first two weeks.

Removal of needle teeth

Piglets are born with four pairs of sharp teeth (needle teeth), with two pairs on each jaw. They are of no practical use to the piglets and they may irritate the sow's udder during nursing or cause injury to other piglets. Clipping of these teeth shortly after birth will prevent the injury of the udder caused by the needle teeth.

Anaemia in piglets

Anaemia is a common nutritional disease in piglets. This condition can be prevented and cured by supplying iron either orally or by injection. Oral administration consists of spraying or swabbing the sow's udder with a saturated solution of ferrous sulphate (0.5 kg of ferrous sulphate in 10 liters of hot water). This solution must be applied daily from birth until the piglets start eating creep feed. Intra-muscular infection of iron-dextran compounds is the more effective method of preventing anaemia.

Raising orphan piglets

The death of a sow after farrowing, mastitis, lactation failure of litters larger than the sow is able to raise, resulting in orphan pigs. If another sow has farrowed within a short time of the first farrow, the orphan piglets may be transferred to her. This transfer must be made within a few days after farrowing because those sections of sow's udder, not used, soon cease producing milk. To ensure acceptance of new pigs the sow should be separated from her own litter for short time and then the new piglets are brought to her and a disinfectant or other material should be sprinkled on all the piglets to mask the odour.

Orphan piglets can also be raised with milk replacer. Milk replacer consists of one egg yolk thoroughly mixed with one litre of cow milk. This mixture supplies a well-balanced diet except for iron. To compensate for the lack of iron one-eighth tea spoonfull of ferrous sulphate may be added to one litre of milk. An injection with iron compound may also be used.

Castration

The male piglets not selected for breeding may be castrated when they are three to four weeks old.

Management during lactation

Sows and gilts should be fed with good judgment during lactation. When fully fed a sow needs about 0.5 kg of feed per piglet she is nursing in addition to her own requirements.

Separation of piglets from mother (Weaning)

Normal weaning age of piglets is 8 weeks. The sow should be separated from the piglets for a few hours each day to prevent stress of weaning and feed is reduced gradually. The piglets should be dewormed after 2 weeks of weaning. The piglets should be gradually shifted from 18 per cent protein creep feed to 16 per cent grower ration over a period of two weeks. Group of 20 piglets of more or less the same age should be housed in each pen.

Prevention and control of swine diseases

- All pigs should be vaccinated against swine fever at the age of 2-4 weeks. Breeding pigs should be tested for brucellosis and leptospirosis. As a routine measure all young pigs at the time of weaning should be inoculated against swine fever.
- Animals for the farm should be purchased from disease-free herds. Newly purchased animals should be isolated from the other animals in the farm for a period of three to four weeks. No visitor will be allowed visiting the farm. Those stys or pig houses cleared of the animals are kept empty for three to four weeks for destruction of micro-organisms causing the disease.

ECONOMY

The pig population of the country is 13.291 million as per 1997 livestock census. Provisional result of census from states reveals that it constitutes around 1.30% of the total world's population. During 2011-12 the production of pork and pork products were estimated to be 672550 metric tons, with 3.03% growth rate ir last decade. It comprised over 38% of the total world meat product. Indian share in piggery meat production moderately increased from 0.53% in 1981 to 0.63% in 2002 to 0.75% in 2012. The contribution of pork products in terms of value works out to 0.80% of total livestock products and 4.32% of the meat and meat products. The contribution of pigs to Indian exports is very poor. About 934 tons of pork and pork products were exported during 1995-96. The value of pork and pork products exported is Rs. 262.00 lakh against the total value of Rs. 61604.00 lakh on account of meat and meat products export.

The pig farming constitutes the livelihood of rural poor belonging to the lowest socio-economic strata and they have no means to undertake scientific pig farming with improved foundation stock, proper housing, feeding and management. Therefore, suitable schemes to popularise the scientific pig breeding cum rearing of meat producing animals with adequate financial provisions are necessary to modernise the Indian pig industry and to improve the productivity of small sized rural pig farms.

In view of the importance of pig farming in terms of its contribution to rural poor and possible potentials for pig rearing in our country, Government of India has initiated measures to promote the pig farming on scientific lines under its five year plans. The first step towards this direction is establishment of eight bacon factories and organisation of pig production in rural areas attached to bacon factories. In order to make available good foundation stock, regional pig breeding stations were established for each bacon factory. Further expansion of pig breeding programmes paved the way for establishment of 115 pig breeding farms (1992-93) throughout the country.

SPACE FOR NOTES

Chapter Twelve : Rabbit Farming

INTRODUCTION

Rabbit farming has grown from raising a few rabbits for family consumption to large commercial operations with hundreds of rabbits. Expansion is much simpler than other livestock alternatives because a large range of existing facilities can be applied for rabbits. In developing countries such as India, where enormous meat shortages exist, the potential for rabbit production is greatest. It is an important livestock activity with great scope as it is relatively easy, rewarding and takes little space compared to other animal farms. Rabbit farming can also provide a very valuable source of income in the rural parts of India. Additionally, a large number of rabbits are also required for medical and cosmetic research centres every year.

Rabbit farming or Cuniculture is the agricultural practice of breeding and raising domestic rabbits, usually for their meat, fur, or wool. The distribution of rabbit farming varies across the globe and while it is on the decline in some nations, in others it is expanding.

HISTORY

Rabbit keeping in commercial base started in the last century. The scarcity of food following first and second World Wars pave the way for extensive rabbit rearing. During those times, rabbit's flesh was denoted as 'Underground Mutton' and served as a source of food for rural families. Later on, F.A.O. through different package programmes encouraged the farmers for rabbit farming. Thus, rabbit farming gradually spread to different countries. In India, from early sixties rabbit farming gained importance in places like Himachal Pradesh, Jammu and Kashmir through introduction of German Angora breed. In 1977, the government of India took much interest in rabbit raising in different states and established a Central Research Station to encourage rabbit farming and to envisage research programme to mitigate the problems in rabbit farming. Now rabbit farming has been established for wool production in Sikkim, Arunachal Pradesh and as a broiler industry in areas of temperate and subtropical climate such as West Bengal, Assam, Manipur, Andhra Pradesh, Tamil Nadu, Kerala and Karnataka.

Taxonomic position of rabbits

Phylum : Chordata

Class : Mammalia

Order : Lagomorpha

Family : Leporidae

Rabbits are small mammals in the family Leporidae; there are eight different genera in this family, classified as rabbits which are distributed in different parts of the world, viz., *Pentalagus, Bunolagu, Nesolagus, Romerolagu, Brachylagus, Sylvilagus, Orytolagus* and *Poelagus.*

Rabbit breeds : Rabbits come in an array of breeds with different colours, sizes, shapes and coat types. Throughout the world, different varieties of domestic rabbit were created through selective breeding or natural selection. Most of the breeds were originally developed for one of the following purposes : meat, pelt, wools, show or biological research. Here few popular rabbit breeds reared in Indian farms are being described (Table 12.1).

Table 12.1 : Name and description of some rabbit breeds popular in India

Breeds	Description
Soviet Chinchilla	This breed was evolved in erstwhile USSR. Adults weigh 4.5 to 5 kg. Though this breed is reared for meat, its fur has demands in fur crafts (Fig. 12.1).
Grey Giant	This breed is also a native of erstwhile USSR. Adults weigh 4.5 to 5 kg. Due to the resemblance of its fur with that of hare, it is often mistaken as hare. It is also reared for meat and fur skin (Fig. 12.2).
Newzeland White	This breed was evolved in England. Fur is white and skin is albino. The eye colour is red due to the absence of melanin pigment. Adult weight is 4.5 to 5 kg. Meat and fur skin are the main products (Fig. 12.3).
White Giant	White Giant also originated in the erstwhile Soviet Union. It is almost similar in appearance to Newzeland White. Colour of the fur is white, while that of the eyes and skin is red. The length of hind limbs as well as body size is larger than that of Newzeland White (Fig. 12.4).
Angora	Angora is a very ancient breed of small rabbit weighing around 3 kilograms. It is a wool type rabbit with white fur. Annual wool yield recorded is between 300-1000 gm in 3 to 4 clippings (Fig. 12.5).
Cross-breed	The crosses between the above mentioned breeds and local types have been produced a number of cross-bred rabbit. They are found to be highly adaptive to Indian conditions. The adult weighs 4 to 4.5 kg and breeds all through the year. Colour of fur is not uniform (Fig. 12.6).

Fig. 12.1. Soviet Chinchilla

Fig. 12.2. Grey Giant

ADVANTAGES OF RABBIT FARMING

(1) In general, rabbit breeds eight times in a year.

(2) Female may produce five to eight youngs which attain sexual maturity in four to six months. Life span of rabbit is about seven to eight years.

(3) Slaughter weight of rabbit is about 2 kg, which can be achieved in 12 to 15 weeks.

(4) Rabbits have got fast reproductive process and are able to remate within 24 hours of giving birth (kindling). Their gestation period is about 31 days.

Fig. 12.3. Newzeland white

Fig. 12.4. White Giant

Fig. 12.5. Angora rabbit

(5) They consume a large amount of forages from diverse origins and so can be reared on small amounts of costly concentrates.

Fig. 12.6. A cross-bred rabbit

(6) They can be reared in the kitchen garden/ backyard of farmer's house.

(7) Apart from providing meat and wool, rabbits also provide income from manure, etc.

(8) Residual feed, together with rabbit manure is highly suitable for vermicompost, which in turn provides excellent manure for fertilising the field.

(9) Initial investment cost for rabbit farm or rabbitry is low.

(10) Quick returns of invested money, i.e., six months after the establishment of farm.

LIFE HISTORY OF RABBITS

Wild rabbits are found in many places around the world. They are present in many habitats, including woods, forests, grasslands, deserts and wetlands. In the wild, rabbits live in groups of up to 70 in number. They make homes under the ground in a series of tunnels called burrows. A group of burrows is called a **warren**. Rabbits are herbivores. They pass two different kinds of droppings. **Faecal pellets** are hard, round droppings that are normally produced from wastes of indigestable plant fibres. **Cecotropes**, on the other hand, are soft, dark-coloured droppings produced as a result of incomplete digestion of nutrient-rich food by cecum, which rabbits re-ingest to obtain extra nutrients. Consuming cecotropes is known as **coprophagy**.

Fig. 12.7. A lop rabbit with long drooping ears

Body and Appearance (in general) : A rabbit has long ears, which can be more than 10 cm long. Some kinds of rabbit have ears called lops which flop down (e.g., Holand flop rabbit; Fig. 12.7). Rabbits have large, strong hind legs. Each front foot has five toes, and each hind foot has four. They move on the tips of their toes. Their body can be 20 cm to 50 cm long and weighs up to 5 kg. Their fur is generally long and soft, but varies according to the breed and can be brown, grey, creamy or white in colour. They have small tail, sometimes white on top. In their natural habitats, the rabbits show excellent hearing and speed. Rabbits have two sets of incisor teeth, one behind the other. Their teeth grow constantly.

Life Cycle : Male rabbits are also referred to as **bucks**, females are **does**, and newborns are **kits** or **bunnies**. They attain sexual maturity in five to six months.

In females ovulation occurs only after mating. A female rabbit is, therefore, considered to be in heat when she accepts service (an act of mating by male) and in diestrus when she refuses the buck. The sexual behaviour of a female rabbit is thus very special. She has no reproductive cycle but can stay in heat for several continuous days. After mating with a buck, a doe is pregnant for about 30 days and gives birth to a litter* of about 4 to 12 kits. Rabbits can breed eight times a year. They have got fast reproductive process and are able to remate within 24 hours of giving birth (kindling). Usually baby rabbits are kindled during the night and are born small, blind and hairless. Bunnies open their eyes when 10 days old. A day later they begin to hear. They feed on very nutritious milk from their mother's body for about 4-5 weeks. When they are about 3 weeks old, they leave the nest in search of feed.

Rabbits live for 8-10 years, although it has been reported to live longer.

Cannibalism : Cannibalism or eating organisms of same species is also shown by rabbits. Does sometimes eat their young, which may result from a ration inadequate in either quantity or quality or from the nervousness of a doe disturbed after kindling. Does usually do not kill and eat healthy bunnies, but limit their cannibalism to those born dead or those that are injured and have died. Proper feeding and handling during pregnancy can prevent this tendency.

RABBIT FARMING

Rabbit house : A good housing facility with adequate arrangement of ventilation is the most important factor in rabbit house. Good house should have proper cages or hutches and arrangement for feeding, watering and cleaning. It is important to keep the broiler rabbit in one roof and to prevent them from running away. There are several ways of housing rabbits. This depends on financial involvement and the climate of the place. In backyard rabbit farming, locally available materials are used. The commonly used materials include tin cans, bamboos, old boxes, wood, bricks, asbestos sheets, etc.

Rearing management : Rabbits can be reared in two systems.

(i) **Deep litter system** : In this system, small number (not more than 30) of rabbits are reared in an enclosure within the farm house (Fig. 12.8). To prevent rabbits to dig burrows, the floor is made up of concrete. Litter materials such as paddy husk, paddy straw or wood shavings are put on the floor up to 4-6 inches. Male and female rabbits are housed separately. Deep litter system is not suitable for intensive system of rearing. Rabbits reared in deep litter system are more susceptible to diseases.

* A litter is the offspring at one birth of mother animal.

Fig. 12.8. Rabbit farming in deep litter system

Management of bannies are also very difficult in deep litter system.

(ii) **Cage system** : In farm house, rabbits can be reared by keeping them in diffrent cages. (a) **Adult cage** : Cages for doe are usually of 90 cm in length and 70 cm in breadth and 50 cm in height and those of buck are usually of 60 cm in length, 60 cm in breadth and 45 cm in height. Cages are made up of wood, bamboo, or weld mesh. The size of wire mesh should be 1 cm × 1 cm for the bottom and 2.5 cm × 2.5 cm for the sides. From the ground to the bottom of cage, there must be a space of 75 cm to 90 cm. The leg of the cages should be protected by metal guards to prevent nuisance caused by rats, snakes, etc. The cages should be placed in cooler surroundings.

(b) **Cage of young** : Young rabbits after weaning are kept in groups in cages or pens (Fig. 12.9). Each such cage/pen measures about 1.5 m in length, 1.2 m

Fig. 12.9. Typical rabbit cages in rabbit house

Fig. 12.10. Nest box

in breadth and 0.5 m in height. Several such cages may be placed closed together on floor or stand in the rabbit house. About 20 weaning rabbits can be reared in each cage/pen. Once the young attains maturity, they may be housed singly, especially adult males. If they are kept together, they will fight and get injured.

(c) **Cage for kindling** : The cage of pregnant doe is almost like that of above cage. But the bottom and sides of such cage should be made of weld mesh of 1.5 cm × 1.5 cm. It will prevent the young bunnies to come out of the cage.

Nest box : To provide safe and calm environment during kindling, a mother doe should always have access to a nesting box, kept inside the kindling cage. Usual size of such box is 22 inches in length, 12 inches in breadth and 12 inches in height. Sizes can vary but should always provide enough space for each doe to keep her litter close together. Nest boxes are designed to open in the upper portion (Fig. 12.10). The bottom of the box should be made up of weld mesh of 1.5 × 1.5 cm. A circular hole of about 15 cm diameter is made (10 cm up from the bottom line) on any one side of the box. This hole allows the doe to move from the box to kindling cage. Designing the hole 10 cm up from the bottom line prevents the kits coming out from the box. Nest boxes can be made of galvanised iron, wood or metal sheets but never of plastic materials. **Hutches** : Rabbits can also be reared outdoor by keeping them in hutches that are placed in sheltered location with slopping waterproof roofing. Hutch is made up of wood/ asbestos and weld wire meshes (Fig. 12.11). It should be well protected from predators. Litter materials like those of deep litter system, may sometimes be used on the floor of hutch.Such hutch can be kept in backyard also.

Fig. 12.11. An outdoor rabbit hutch

FEEDING MANAGEMENT

Food and nutrients : Rabbits relish all types of grains (sorghum, bajra and other grains) and legumes. Green fodders, kitchen wastes such as carrots and cabbage leaves and other vegetable wastes are also relished by rabbits. Different types of rabbit feed with components and nutrient requirements during different stages of life are listed in Table 12.2 and 12.3 respectively.

Feeder and water troughs : Rabbit feeders or food pots are generally made out of aluminum or galvanised iron sheets. Feeders may be designed in such a way (usually J-shaped), that they can be attached to the front panel of the cage and can be filled from outside without opening the doors of the hutch or cage (Fig.12.12). For watering automatic pipeline or bottle watering system are used widely in commercial rabbit farms. Glazed earthenware pots can also be used for watering in cages. To reduce the investment cost, feed and water can also be served in cups.

Feeding schedule : The teeth of rabbits are continuously growing. Therefore, rearing rabbits with concentrate feed alone is impossible. The time

Table 12.2 : Different types of foodstuffs suitable for feeding rabbits

Types of feed	Foodstuffs
(1) **Roughages Dry Bulky feeds**	Hay made from grass or legume (Alfalfa)
(2) **Fresh Bulky feeds** :	
(a) Grasses :	Guinea grass, Napier grass
(b) Green cereal plants :	Maize, Sorghum
(c) Legumes :	Alfalfa, Berseem
(d) Roots :	Turnips, radish, carrots, sweet potatoes, yams
(e) Green vegetables and other plants :	Green leaves of cabbage, spinach, lettuce, etc.
(3) **Concentrates** :	
(a) Energy supplements	Banana leaves, various weeds, local grass.
(b) Protein supplements	Maize, Sorghum, millets, barley, soya flour, groundnut cake, cotton seed cake, meat, fish meal.

Table 12.3 : Nutrient requirement of rabbits during different stages of life

Details of nutrients	For growth	For maintenance	For pregnancy	For lactation
Digestible energy (kcal)	2500	2300	2500	2500
Protein (%)	18	16	17	19
Fibre (%)	10-13	13-14	10-13	10-13
Fat (%)	2	2	2	2

Fig. 12.12. Food and water devices for rabbit in cage

schedule for feeding should be maintained strictly. If there is late in feeding rabbits, they become restless, which in turn reduces their body weight. Rabbits generally do not take feed during daytime due to high temperature. They are active during night and are fed with green fodder which they devour without wastage. On the other hand, concentrate feed in the form of pellets should be given in the morning. Clean freshwater should be provided to rabbits during all times of the day. Amount of different food required by rabbits during different stages of life are listed in Table 12.4.

Table 12.4
Feed requirements of rabbits during different stages of life (gm/day)

Particulars	Concen-trates	Hay	Greens + Veg/ Fruit, garden waste
Breeding stock (male/female)	280	80	Free feeding
Weaners (7-12 weeks)	60	30	Free feeding
Growers (13-24 weeks)	90	30-40	Free feeding
Adult rabbits (above 24 weeks)	140	50-60	Free feeding

BREEDING MANAGEMENT

Breeding practices : For producing small sized breeds, females or does are selected at the age of 5.5 months and for medium sized breed, 6 months old females are selected. For all cases, male or buck should be at least 6 months old.

Mating : The doe is always introduced into the buck's cage for mating, because if a buck is put into the doe's cage, she might fight with him to protect her cage. If the doe is on heat, she will let the buck mate her. After mating the doe is returned to her hutch/cage. They should not be left together, even if mating has not occured. It is better to return the doe to her hutch and to bring her back again to the buck after few hours or on the next day. Breeding is normally done in early morning or late afternoon.

Breeding hazards and management

Breeding failures : A doe may occasionally fail to become pregnant. The causes of such failure are listed below.

(1) **Age** : The buck or the doe may be too young or too old. Such rabbits should not be selected for breeding.

(2) **On heat** : The doe may not be on heat. So, after four days she might be tried again.

(3) **False pregnancy** : If the doe starts to pull her fur out, 16 to 20 days after mating, she has probably developed a false pregnancy.

(4) **Physical condition** : A rabbit which is too fat or too thin is less likely to breed than a normal rabbit. In such cases, feeding schedule should be changed.

(5) **Disease or injury** : A rabbit showing signs of disease or injury should not be used for breeding. It should be treated or culled depending on the condition and may be used for breeding after complete recovery.

(6) **Moulting** : Rabbits which are moulting (renewal of hair) are less likely to breed successfully.

(7) **Sterility** : Sometimes a buck or doe may be completely sterile. Records can be used to find out which rabbit is sterile.

(8) **Hot weather** : During very hot weather when temperatures reach above 30°C, especially when relative humidity is also high, breeding can be impaired. These conditions cause stress and can cause temporary sterility in bucks.

(9) **Disturbance to doe** : If the doe is disturbed, she may kindle on the hutch floor (without a nest) and the litter may die from exposure. If she is disturbed after the litter is born and jumps into the nest box, she may stamp with her back feet and injure or kill the new born.

Selection of reproduction method : For successful rabbit production, the choice of reproduction method must be preceded by careful study and planning. There are three basic methods of reproduction – (1) extensive, (2) semi-intensive, and (3) intensive.

(1) **Extensive reproduction method** : In this system, the breeder fully utilizes the does' maternal instincts by allowing them to nurse their young for 5-6 weeks, and rebreeding them soon after weaning. Does are, therefore, serviced once every two and a half months.

(2) **Semi-intensive method** : Here the does are serviced 10-20 days after kindling and the young are weaned for 4-5 weeks. So the doe is pregnant while still nursing. The most important phase of embryo development takes place during the slump in milk production, which may have even ceased. So there is no real competition between the demands of gestation and lactation. As these does never have a resting period they need sufficient and balanced concentrate feed.

(3) **Intensive method** : Here the does are reserviced just after kindling, taking advantage of the fact that they are then in the heat. Weaning should take place 4 weeks at the latest (usually 26–28 days).

Considering the greater nutritional needs of the pregnant doe, especially one which is lactating, semi-intensive and intensive reproduction systems are only suitable where does get the right quantity and quality of feed. If these conditions are not met, the does will usually accept the male, but will abort. Abortion extends the interval between litters to make it as long as the extensive breeding interval. In wet tropical climate, with balanced feeding, it is possible for a doe to produce 40 -50 young a year by a combination of intensive and semi-intensive methods of reproduction.

Management of pregnant doe : If the pregnant doe has no previous litter, she can be kept in a smaller, single size hutch/cage for the first three weeks. At least a week before kindling, the doe should be placed into a breeding hutch/cage where she will give birth. At the same time a nest box and suitable soft materials for making a nest, such as grass, hay or straw should be kept ready. Twenty-eight days after breeding, the nest box is placed in the doe's hutch. Few days before kindling she starts to make the nest. The day before kindling, the doe will pull out some of her fur from her belly around her teats. She will use this to give a fine lining to the nest.

Feeding the pregnant doe : During pregnancy, feeding of the doe should be gradually increased during the last 10 -14 days of pregnancy to fulfill the demand of growing embryos. Good feeding at this time is likely to increase the birth weight of the young rabbits and the milk yield of the mother.

Management during kindling : The birth of the babies or bunnies is known as 'kindling' or littering. Kindling often takes place in the evening or at night. It lasts for 15 to 30 minutes depending on the size of the litter. As each bunny is born, the mother licks it and allows it to suckle. A doe cannot successfully rear more than eight young, because she has only eight teats. So if more than eight bunnies are born, strongest eight young rabbits from the litter are kept with the doe. Others are taken away and either fostered or culled.

Sometimes due to some disturbances, a doe may not make a nest at all and simply gives birth on the floor of the hutch. Such bunnies are likely to die from cold or due to negligence. If they are placed in the nest box with some soft material, there is a chance that the doe will accept them. If not, the only alternative is fostering.

Sometimes a doe may kill her young (cannibalism) or abandon them for no apparent reason. If it happens twice with the same doe, she can be classed as 'bad mother' and should be separated from the herd and culled, because this characteristic can pass on to her offsprings. The nest box and the bunnies should be protected from predators, otherwise, the doe may get upset or frightened and may kill or abandon the bunnies. Again, during the kindling time, except the caretaker, other persons should not be allowed to enter the rabbitry

Management of young litter : A few hours after kindling, the nest box is removed from the cage and the litters are checked carefully. Few bunnies from a large litter can be transferred to a foster mother to ensure uniform growth and development at weaning time. To facilitate acceptance by the foster mother, Vicks Vaporab can be rubbed on the nose of mother as well as the bunnies to deaden their sense of smell. This will enable a successful acceptance by the doe.

Caring during lactation : Doe's milk is the only feed for the bunnies during the first two or three weeks of their lives. Their development during this important period depends on the mother's milk supply. So doe should be well fed during the last half of pregnancy and lactation. On the other hand, a doe may occasionally fail to produce milk. In such cases, the young will starve to death within 2 or 3 days unless the condition is noted and the young are transferred to foster mothers.

Management during weaning : Weaning is the time when the young are separated from the doe. Normal weaning time is at 4 to 6 weeks of age when the young rabbit's live weight is over 500 gm. During the weaning period the young gradually give up milk for solid feed. During weaning young rabbits in the litter are withdrawn at the same time and placed 6 to 8 per cage in the area set aside for fattening. Alternatively, the doe (mother) may be removed from the young rabbits. The young rabbits begin to eat solid feed supplied to them at 18th to 20th days and at 30th day the doe's milk provides no more than 20% of the daily dry-matter intake. In this time they are more susceptible to diseases. Cooled boiled drinking water and hydrogen peroxide at the rate of 1 ml per litre of water can be used for disinfection. From weaning time, males and females are kept as well as reared separately.

DISEASE MANAGEMENT

Some of the common diseases incurred by rabbits and their control measures are given in Table 12.5.

Common symptoms in diseased rabbits :

- Dull and depressed.
- Reduction in the body weight and emaciation.
- Heavy loss of hair.
- No active movement of rabbits. Usually they stay in one particular place in the cage.
- Reduction of feed intake.
- Watery or mucus discharge from the eye, nose, anus and mouth.
- Increased body temperature and respiratory rate.

General hygienic measures to be followed to control disease in rabbitry

- ♦ Rabbit farm should be located in elevated place with good ventilation.
- ♦ The cages should be kept very clean.
- ♦ There should be trees around the rabbit shed.
- ♦ White washing should be done twice a year.
- ♦ Lime solution should be applied below the cages twice a week.
- ♦ In summer seasons water may be sprinkled.

RABBITRY RECORDS

Keeping record is a great help in day-to-day management of rabbitry. A quick glance at the record will enable sound management. One should design its own record cards to meet the needs. The rabbitry record should contain the following informations :

(1) Doe number or name,

(2) Date of mating or remating,

(3) Date of expected litter,

(4) Date of litter production,

Table 12.5 : Common diseases of rabbits, their symptoms and control

Sl. No.	Name of the disease	Important symptoms	Control measures
1.	Coccidiosis	Diarrhoea, loss of appetite and dehydration. May cause death in 24-36 hours.	(i) Coccidiostats like Sulphaquinoxaline + Sulphamerazine @ 0.02 to 0.10% in drinking water (ii) Strict hygienic measures
2.	Body mange (Ear canker)	Intense purity, scratching, scaling of the skin, loss of fur, and animal becomes weak.	(i) Application of ascabiol lotion (ii) Ivermectin injection 0.02 ml/kg body wt. (S/C) (iii) Strict hygienic measures.
3.	Enteritis complex	Subnormal temperature, crouched posture, roughened hair coat, gelatinous mucus with dry feaces and bloat.	(i) No successful treatment (ii) Tetracycline in feed @100 gm/tonne feed (iii) Adequate amount of fibre should be given.
4.	Pasteurellosis (Snuffles)	Discharges from nostrils, wiping of nose with paws, loss of appetite, high rise of temperature.	(i) 400000 IU of penicillin and 0.5 g streptomycin given (IM).
5.	Shorehock	Inflamed areas or sores on the under surface of hind paws, weakness and dehydration.	(i) Sores should be opened and antiseptic dressing should be applied. Soft bedding should be provided during treatment period.
6.	Hind quarter paralysis	Sudden jerk, excitement, paralysis, loss of control over urination and defication.	(i) Affected animals should be slaughtered.
7.	Hairball occlusion	Wool get accumulated in stomach and blocks normal passage of food.	(i) Treat the animal with mineral oil or neopeptin or fresh pineapple juice.
8.	Wryneck	Permanent turning of neck to one side, unable to eat, weakness.	Affected animals should be slaughtered.

(5) Number of bunnies in the litter : alive and dead,

(6) Date of recrossing,

(7) Date of weaning,

(8) Number weaned,

(9) Doe breeding record,

(10) Litter record,

(11) Buck breeding record,

(12) Financial records (including utilities expenditures and sales).

USES OF RABBITS

Meat : Fresh or frozen rabbit meat is sold all year round. It can be used in similar ways as chicken is used.

Wool : Angora rabbits are mainly raised for their wool, which is known for its quality. Angora rabbits are of many types. The differences are mainly in the amount of wool produced and the percentage of guard hair in the wool. Among the various types, German Angora is the best and annually yields 1000 to 1200 gm of wool under ideal management practices.

Laboratory : Rabbits are widely used for biological and medicine laboratories. Such rabbit farming is governed by specific guidelines

Manure as a fertilizer : As the cost of fertiliser product increases, so does the value of rabbit manure. Fifteen does, two bucks, and their litters will produce approximately one tonne of manure a year. Rabbit manure is drier than poultry manure. Rabbit manure contains approximately 1.3% N, 0.9% P, 1.0% K and fetches a good price in the market.

SPACE FOR NOTES

Chapter Thirteen : Vermiculture and Vermi-composting

INTRODUCTION

Vermiculture is a newly emerging subject of biological technology. This deals with the earthworms of the world. The tragedy is that almost 50% of the earthworm species has remained undescribed till today. Whatever described, all is not suitable for vermiculture. It has been envisaged that the life history pattern and behaviour study are necessary for defining an earthworm suitable for culture. Among 3627 identified species, hardly 50 to 60 species have been studied thoroughly. Therefore, there remains ample scope for further research, which may contribute a lot for vermicomposting in the days to come. Moreover, vermicomposting can also be utilised in large scale waste management.

What is vermiculture?

Collection, breeding and successful rearing of earthworm for generations in an artificial media or laboratory is known as vermiculture.

What is vermicomposting?

Composting is the process of bioconversion of organic matter by heterotrophic micro-organisms (bacteria, fungi, actinomycetes and protozoa) into humus-like material.

The process of composting, using various worms, usually red wigglers, white worms, and other earthworms, to create a heterogeneous mixture of decomposing vegetable or food waste, bedding materials, and **vermicast** is called **vermicomposting**.

Vermicast is also known as **worm castings**, **worm humus** or **worm manure**. It is the end-product of the breakdown of organic matter by an earthworm. These castings have been shown to contain reduced levels of contaminants and a higher saturation of nutrients than what organic materials do before vermicomposting. It contains water-soluble nutrients, nutrient-rich organic fertiliser and soil conditioner. This vermicompost is an excellent bio-organic manure.

Systematic Position :

(After – E.E. Ruppert and R.D. Barnes, 1994)

Phylum : Annelida

Class : Oligochaeta

Vermibed : Vermibed (*vermes* = earthworms; *bed* = bedding) is the actual layer of good moist loamy soil placed at the bottom, about 150 to 200 mm thick above and a thin layer (50 mm) of broken bricks and coarse sand below. Earthworms are introduced into the loamy soil, which the worms will inhabit as their home. About 100 earthworms (a combination of epigeics and anecics) are introduced into a compost pit with a vermibed. The vermibed should always be kept moist, but should never be flooded.

SUITABLE SPECIES OR BREEDS

All earthworms are not suitable for vermicomposting. Only those that live on the soil surface and thrive on organic wastes are suitable. About 3627 species of earthworms have been identified in the world. Out of it only 3 to 5 per cent species are capable of vermicomposting.

Suitable worms for vermicomposting possess following characteristics.

(1) The worm should exhibit high biomass consumption together with a high efficiency of conver-

sion of ingested biomass to body proteins, a physiological trait required for achieving high growth rate.

2. The worm should have wider range of tolerance to environmental factors including adaptation to feed on a variety of organic residues.

3. Worm should produce large number of cocoons at a time and should have rapid population growth.

4. Life cycle should consist a longer period for adults.

5. The worm should be disease resistant.

The earthworm species most often used for composting is the Red Wiggler (*Eisenia foetida* or *Eisenia andrei*), European night crawlers (*Eisenia hortensis*), Red earthworm or dilong (China) (*Lumbricus rubellus*), etc. Among these breeds red earthworm is less well adapted to the shallow compost bin than *Eisenia*. Blue worms (*Perionyx excavatus*) may be used in the tropics.

In India usually following earthworms are cultured for vermicomposting : *Perionyx excavatus, Lampito mauritii, Drawida willsi, Polypheretima elongata, etc*. These species are commonly found in organic-rich soils and live in rotting vegetation, compost, and manure piles. They may be an invasive species in some areas. As they are shallow-dwellers and feed on decomposing plant matter in the soil, they adapt easily to living on food or plant waste, in the confines of a worm bin.

Composting worms can be obtained through online order or from nursery mail-order suppliers or angling (fishing) shops where they are sold as bait. They can also be collected from compost and manure piles. These species are not the same worms that are found in ordinary soil or on pavement when the soil is flooded with water.

Classification of earthworms on the basis of their habitat

(a) Epigeic earthworms

The earthworms which live on the surface of the soil in leaf litter are called **epigeic** earthworms. These species tend not to make burrows but live in and feed on the leaf litter. Epigeic earthworms are bright red or reddish-brown in colour, but they are not stripy.

Fig. 13.1. *Lumbricus castaneus,* an epigeic earthworm

Epigeic earthworm species include *Dendrobaena octaedra, D. attemsi, Dendrodrilus rubidus, Eiseniella tetraedra, Heliodrilus oculatus, Lumbricus rubellus, L. castaneus (Fig. 13.1), L. festivus, L. friendi, Satchellius mammalis*

(b) Endogeic earthworms

The earthworms those make horizontal burrows through the soil to move around and to feed and reuse these burrows to a certain extent are called **endogeic** earthworms. Some can burrow very deeply in the soil. Endogeic earthworms live in and feed on the soil. These earthworms are pale grey, pale pink, green or blue coloured.

Endogeic earthworm species include *Allolobophora chlorotica* (Fig. 13.2), *Apporectodea caliginosa, A. icterica, A. rosea, Murchieona muldali, Octolasion cyaneum* and *O. tyrtaeum*

Fig. 13.2. *Allolobophora chlorotica,* an endogeic earthworm

Fig. 13.3. *Apporectodea longa,* an anecic earthworm

(c) **Anecic earthworms**

The earthworms which make permanent vertical burrows in soil are called **anecic** earthworms. They feed on leaves on the soil surface that they drag into their burrows. They also cast on the surface, and these casts can quite often be seen in grasslands. They also make middens (piles of casts) around the entrance to their burrows. Anecic species are the largest species of earthworms. They are darkly coloured at the head end (red or brown) and have paler tails.

Anecic earthworm species include *Lumbricus terrestris* and *Apporectodea longa* (Fig. 13.3).

NON-SUITABLE BREEDS

Lumbricus terrestris [Night crawlers (US) or common earthworm (UK)] are not recommended as they burrow deeper than most compost bins can accommodate. *Perionyx excavatus* worms are also not suitable for worm compost in bins, but scientists opine that this species is amongst the best suited for large scale vermicomposting in tropical climate, i.e., East &West Himalayan states, Pilibhit, Bengal and little Andaman islands.

LARGE SCALE CULTURE

Large scale vermicomposting is practised in developed countries like Canada, Italy, Japan, Malaysia, the Philippines, and the United States. The vermicompost may be used for farming, landscaping, to create compost tea, or for sale. Some of these operations produce worms for bait and/or home vermicomposting.

There are three main methods of large-scale vermiculture — (1) **Windrow system,** (2) **Raised bed,** (3) **Flow-through system.**

(1) **Windrow system of earthworm farming**

In this type of farming, windrows are usually constructed on a concrete surface. It may be in a pit or may be on soil surface. Windrows are linear piles on the ground containing feedstocks up to 3 feet high. If the floor is made up of cement, then earthworm bed will be prepared with soil, straw, leaves, fertilizer, etc. They are being used extensively both in the open and under cover, but require either a lot of land or large buildings. During composting process the whole pile is covered by a PVC sheet to prevent predators and environmental hazards (Fig. 13.4).

Windrow system is used for the production of compost by piling organic matter or biodegradable waste, such as animal manure and crop residues, in long rows (**windrows**). This method is suited to producing large volumes of compost. After composting, these rows are generally turned to an organic material with improved porosity and oxygen content. Mixing in or removal of moisture, and redistribution of cooler and hotter portions of the pile makes the compost suitable for agricultural use. Windrow composting is a commonly used in farm scale

Fig. 13.4. Window system of vermicomposting

composting method. Composting process control parameters include the initial ratios of carbon and nitrogen rich materials, the amount of bulking agent added to assure air porosity, the pile size, moisture content, and turning frequency. The temperatures of the windrows must be measured and logged constantly to determine the optimum time to turn the windrows for quicker compost production.

(2) Raised bed earthworm farming

In this type of farming, soil of the floor will be raised with the help of axes to prepare a bed. Bedding is any material that provides a relatively stable habitat to worms. A good bedding for vermicomposting should possess :

(a) **High absorbancy** : As worms breath through skin, the bedding should be able to absorb and retain adequate water.

(b) **Good bulking potential** : It is the property which ensures proper oxygen supply to the worms.

(c) **Low nitrogen content (High Carbon : Nitrogen ratio)** : Although worms consume their bedding as it breaks down, it is very important that this be a slow process. High protein/nitrogen levels can result in rapid degradation and associated heating may be fatal to worms.

Some common bedding materials are given in the table 13.1.

On the top of the bed, **worm chow** of about one inch thickness will be spread. Worm chows are artificial feed for the earthworm. The **ingredients of worm chow** are – ground corn, ground soybean hulls, wheat middling, dehydrated alfalfa, cane molasses, calcium carbonate, porcine meat meal, dehulled soybean meal, ground oats, ground wheat, di-calcium phosphate, mono-calcium phosphate, fish meal, dried beet pulp, wheat germ, corn gluten meal, salt, soybean oil, porcine animal fat preserved with BHA, folic acid, choline chloride, DL-alpha tocopheryl acetate, riboflavin, pyridoxine hydrochloride, nicotinic acid, menadione dimethyl-pyrimidinol bi-sulphate, calcium pantothenate, vitamin B-12 supplement, vitamin A acetate, manganous oxide, zinc oxide, ferrous carbonate, copper, sulfate, zinc sulfate, calcium iodate, cobalt carbonate. The worm chow is commercially available in the market.

A location of raised bed farming is preferably a shady spot so that one can keep the bed somewhat moist. If it is a full sun area there should be provision to cover with a few inches of mulching material to protect earthworms from the sun and to retain moisture. Leaves, newspaper, straw will work fine for this

Table 13.1 : Some common bedding materials

Bedding material	Absorbancy	Bulking potential	Carbon : Nitrogen ratio
1. Dry loose leaves	Poor-medium	Poor-medium	40 – 80
2. Shrub Trimmings	Poor	Good	53
3. Sawdust	Poor-medium	Poor-medium	142 – 750
4. Cow manure	Medium–Good	Good	22 – 56
5. Straw-general	Poor	Medium-good	48 – 150
6. Wheat straw	Poor	Medium-good	100 – 150
7. News paper	Good	Medium	170
8. Corrugated card board	Good	Medium	563
9. Softwood, chips,	Poor	Good	212 – 1313
10. Paper from municipal waste stream	Medium–Good	Medium	127 – 178

purpose. The most convenient width for the short sides of an earthworm bed is 3 feet. For length, worm growers generally construct beds of at least 8 feet. If beds are longer than 8 feet, some growers like to install dividers every 8 to 10 feet for ease in dividing, harvesting, cleaning, or feeding.

The raised bed depth should be 12 to 24 inches. If continuous freezing or excessively hot temperatures occur in the area, consider building beds 12 to 24 inches below ground where the constant ground temperature will keep the worms from freezing or overheating. The underground bed will be made by digging soil of suitable length (Fig 13.23). The wall of the pit should be covered with PVC lining. Then from below the layers of various thickness of coconut fibres (2 inches), wood dust (2 inches), agro waste and dung (15 inches), and leaf or straw (6 inches) are placed. The ideal distance between beds is 3 feet. This allows room for manually or mechanically operated equipment to be used between the beds for feeding, harvesting, or cleaning.

(3) **Flow through system of earthworm farming**

This system is well suited to indoor facilities. In indoor space the concrete surface is constructed in the pit to prevent predators gaining access to the worm population. The dimension of the pit should be 3 feet × 8 feet. The feed is initially dumped in one corner of the pit (Fig. 13.5). The worms are introduced into the pile of the feed and kept for some days. Morning or evening is the best time to add worms. The lengthwise direction of the earthworm beds and their shelters should parallel the prevailing winds. For example, if the wind generally blows from west to east, the beds should be laid out in a west-east direction. This will prevent intense winds from hitting the largest part of the shelter and will help prevent covers, if they are used, from blowing off. When the bed is rich with casts then another stock of feed is placed in front of the previous dump, so that the worms can migrate to the new feed (Fig. 13.5). The castings can be removed from the previous dump. The flow through system eliminates the need to separate worms from the casting before packaging, because the worms are already moved forward to the new feed pile.

Construction of vermicompost pits in rural Bengal

In rural Bengal, pits are usually constructed in a shaded area, either under a tree or under a shed. There are two types of pits, which are constructed depending upon the requirement and space availability of the composters — (i) Four-tank system and (ii) Two-tank system. Each compost pit of any convenient dimension can be dug.

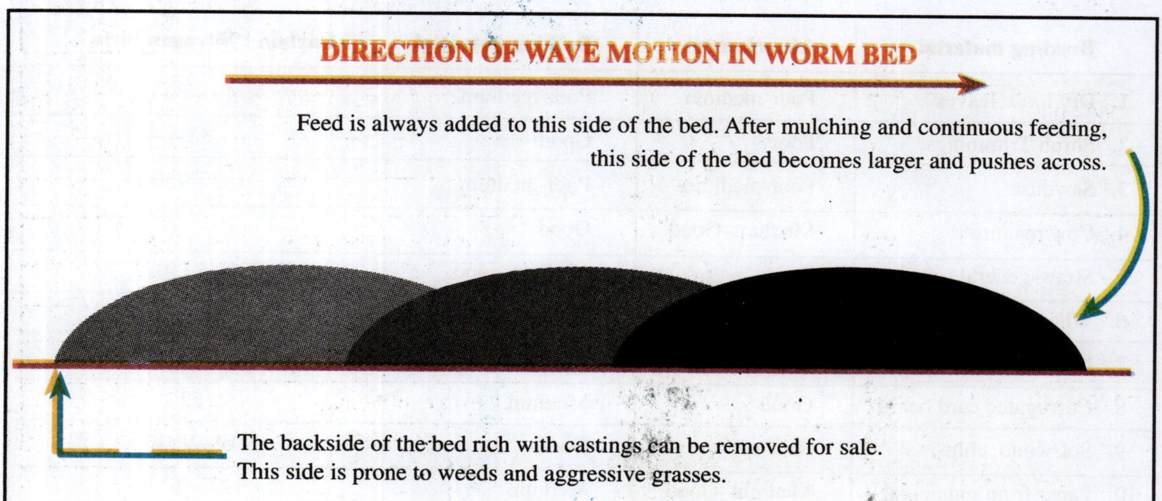

Fig. 13.5. Movement of castings through a worm bed

Fig. 13.6. Brick construction of four-tank system for vermicomposting

(i) Four-tank system

A tank of 4 m × 4 m × 1 m (length × breadth × height) dimension is preferably made under the shade of a tree (Fig. 13.6). This is then divided into four equal parts with brick walls that have vents to facilitate aeration as well as migration of earthworms from one tank to another. This unit is designed especially for the small farmer who approximately collects 20 to 30 kg of cattle or farm waste per day. After construction, the tanks should be marked as 1, 2, 3 and 4. The loading of bio-dung and vermibed should be carefully maintained. The schedule of loading the unit is described in the Table 13.2.

To simplify the loading procedure for composting in rural sectors where the availability of organic material is insufficient, a four-tank system can be set up based on a combination of bio-dung composting method and vermitech technique. This process enables continuous compost production using cattle dung produced daily at cattle sheds, weeds, leaf litter and other farm waste.

(ii) Two-tank system

In this system a smaller tank of 1 m × 1 m × 1 m (length × breadth × height) dimension is made with the brick above ground under shade (Fig. 13.7). The tank is divided into two equal halved units vertically by a wall containing vents. The two-tank system is recommended for little household garbage or less quantity of farm waste.

Table 13.2 : Schedule of maintenance of four-tank composting system

Period in days	Tank	Process
000 - 030	01	Collection of biomass and cattle dung
030 - 060	01	Soaking of biomass with water, cattle dung slurry, and covering it with black polythene sheet. This is called bio-dung preparation.
	02	Collection of biomass
060 - 090	01	Inoculation of earthworms
	02	Biodung preparation
	03	Biomass collection
090 - 120	01	Vermicompost ready and migration of earthworms from pit 1 to pit 2
	02	Vermicomposting
	03	Biodung preparation
	04	Biomass collection
120 - 140	01	Harvesting of compost and collection of biomass
	02	Vermicompost ready and migration of earthworms from pit 2 to pit 3.
	03	Vermicomposting
	04	Bio-dung preparation

Fig. 13.7. Two-tank brick composting system constructed under the shed of a tree

Schedule of maintenance of Two tank composting system

An average family produces 250 to 500 gm of garbage that is added daily into one of the tanks. After a few days when a layer is formed, 15 to 20 cm dry/green leaves and a thin layer of soil are placed to cover it. Another layering of waste, over a period of time, follows this and it takes about two months for the tank to be filled. This is then covered with a black polythene sheet. The waste is now added to the second tank. The polythene sheet is removed after 15 to 20 days, allowed to cool for a day, and about 150 to 200 locally collected earthworms are released into the biomass. It takes approximately 45 to 60 days for the biomass to be converted into vermicompost. In the meantime the second tank gets filled and starts decomposing. The earthworms from the first tank by this time start migrating into the second tank through the vents. The vermicompost is harvested from the first tank, which is again made ready for getting filled.

Requirements for the outdoor vermicomposting

Outdoor plan requires about 10-20% more worms than the indoor method, as there are more predators and other climatic variables in open tank system. The outdoor system requires sufficient shaded land space. The most convenient pit of easily manageable size is 2m × 1m × 0.75m. A tank may be constructed with brick and mortar with proper water outlets, or a plastic crate (600 mm × 300 mm × 300 mm) with holes drilled at the bottom. The pit is made ready by setting vermibed (Fig. 13.22). The vermibed is sprinkled with water at regular interval. Then composter earthworm is introduced (Fig. 13.20). After few days when the compost is ready for harvesting, the vermicompost is collected and composting material is replaced (Fig. 13.21).

Collection of earthworms

Earthworms can be collected from neighbourhood or can be obtained from nurseries. One can also contact the local agricultural department, municipalities, and nurseries of the forest departments for earthworms. Since earthworms have acquired economic importance, one can also check in the firms/horticultures/nurseries, which deal in selling of earthworms. But it is always desirable to collect it from composter themselves from the locality. The composter should locate the earth-worm castings over the soil of his/her locality. The stem of the "bish katali tree" (*Polygonum hydropiper*), is crushed and an aqueous extract made. The spray of this aqueous extract will force the earthworm to come out of their hideout. Then the earthworms could be collected and washed in water for 10 to 15 minutes for vermicomposting.

Efficient management tips

Handful lumps of fresh cattle dung should be placed at random over the vermibed. The compost pit should be layered to about 50 mm with dry leaves or preferably chopped hay/straw. For the next 30 days the pit is kept moist by watering it whenever necessary. The bed should neither be dry nor soggy. The pit is then covered with coconut or Palmyra leaves or an old jute (gunny) bag to discourage birds. Plastic sheets on the bed should be avoided as they trap heat. After the first 30 days, wet organic waste of animal and/or plant origin from the kitchen or hotel or hostel or farm, that has been pre-digested, is spread over it to a thickness of about 50 mm. This can be repeated twice a week. All these organic wastes can be turned over or mixed periodically with a pickaxe or a spade. Care should be taken not to disturb the vermibed in which the worms live. Garbage should be added till the compost pit is nearly full. Pit should be kept moist for another 30 to 45 days. Turning over the material in the pit should be done with care avoiding injury to the worms. Turning over can be done on every fifth or seventh day with the help of a forked spade.

Regular watering should be done to keep the right amount of moisture in the pits. In 60 to 90 days time the compost should be ready as indicated by the presence of earthworm castings (vermicompost) on the top of the bed. The compost should be turned occasionally since this allows for aeration. If the weather is very dry it should be dampened periodically. The pile should be moist, not wet and soggy. Vermicompost can now be harvested from the bin/pit.

SMALL SCALE CULTURE

Construction of vermicompost pits in Backyard or indoor spaces

Bins can be made from different unused household materials. Among various materials three types of bins built from different materials are shown in Fig. 13.8. These are Cinder blocks, Refrigerator shell, PVC drum. The idea of using a refrigerator shell is proved good. Worms in bins need three basic things: food, ventilation and drainage. The back of the fridge is perforated (Fig. 13.9), which will be the bottom of the worm bin to allow for drainage. Ventilation will come through the top. The bin is placed on broken cinder blocks to elevate it and improve drainage (Fig. 13.10). The leachate that seeps out is full of nutrients for plants. It may be drained into the soil to feed plants in the garden. The making of vermibed requires some steps to follow. First, a layer of shredded newspaper or other cellulose-rich material like coconut husk is placed at the bottom (Fig. 13.11). Then on it is sprinkled the "worm poop" soil through which worms come in. Then the worms are released in the bed. The next layer is food scraps. On top of it all, is placed another layer of shredded newspaper and covered with a damp towel.

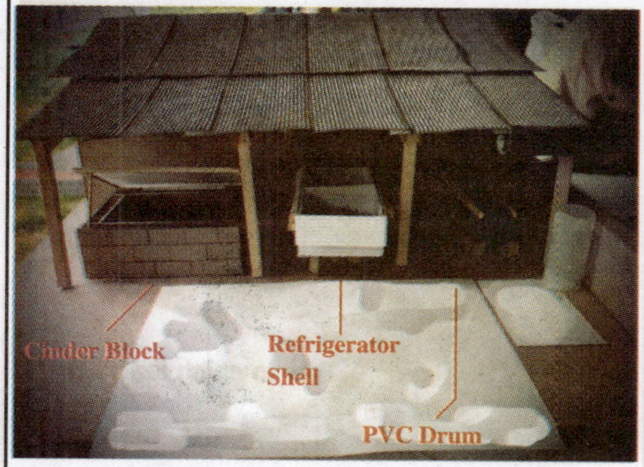

Fig. 13.8. Three types of worm bins prepared from different unused materials

Fig. 13.9. Perforated under surface of the bins for allowing drainage and ventilation

Vermiculture with Red Wigglers (*Eisenia fetida*) and Indian Blue Worms (*Perionyx excavatus*) reduces household waste and converts it into fertiliser. The figure 13.12 shows a handful of two worms. They are very difficult to differentiate but both are wriggling around.

One can also start vermiculture in a small area on a diet of watermelon rinds, lime and mango peels in our country near small garden (Fig. 13.13). One should avoid meats and fatty foods because they can smell bad. Papaya seeds can make the worms sterile, so it also should be avoided. The worms should be fed on alternate days. They need food at least once per week, and more the feeding, the better will be the

Fig. 13.10. Blocks of broken bricks and other garbage are used for elevating the bin and to catch leachate.

Fig. 13.12. Showing two varieties : Red Wigg' .is (*Eisenia fetida*) and Indian Blue Worms (*Perionyx excavatus*) used for vermi-compost.

Fig. 13.11. A layer of shredded newspaper or other cellulose-rich material is being placed and on top of it "worm poop" soil.

Fig. 13.13. Here the worms are kept on a diet of water-melon rinds, lime and mango peels, a few yard clippings and coffee grounds. Meats and fatty foods are avoided because they can smell bad and papaya seeds can make the worms sterile.

result. They can eat food up to their own weight every day. The shredded newspaper should be made damp with water at regular interval.

FEED STOCK FOR SMALL SCALE CULTURE

Small-scale vermicomposting is well suited to turn kitchen waste into high-quality soil amendments, where space is limited. Worms can decompose organic matter without the additional human physical effort. Such systems usually use kitchen and garden waste, utilising earthworms and other micro-organisms to digest organic wastes, such as kitchen scraps. This includes all fruits and vegetables (including citrus and other "high acid" foods); vegetable and fruit peels and ends; coffee grounds and filters; tea bags (even those with high tannin levels); grains, such as bread, cracker and cereal (including mouldy and stale); eggshells (rinsed off); leaves and grass clippings (not sprayed with pesticides).

Composting worms, which are detritivorous (eaters of trash), such as the red wiggler *Eisenia fetida* are **epigeic** — (1) Indian *Perionyx excavatus* (Fig. 13.14) is epigeic type earthworm. Its ectodermal layer is provided with dense pigmented granules, which makes its colour deep dark. These animals are also called as litter species. Common earthworms, e.g.,

Eutyphoeus gammiei, Lampito mauritii, Metaphire houlleti, and *Lumbricus terrestris* are **anecic** (deep burrowing) species and hence unsuitable for use in a closed system. Other soil species that contribute include insects, other worms and moulds. (2) There is another variety of earthworm, known as **endogeic** variety. These are living in the deep and horizontal burrow of the soil. They prefer soil with inorganic materials. Their body is lightly pigmented. *Metaphire posthuma* (Fig. 13.17), *Pheretima elongata, Pontoscolex corithrurus,* are species of endogeic variety. These are not suitable for vermi-composting.

FEEDSTOCK FOR LARGE-SCALE CULTURE

There are few food wastes that earthworms cannot compost, although meat waste and dairy products are likely to putrefy, and in outdoor bins can attract vermin. Green waste should be added in moderation to avoid heating the bin. Such vermi-composting systems need reliable sources of large quantities of food. This system presently uses dairy cow or pig manure; sewage sludge; agricultural waste; food processing and grocery waste; cafeteria waste; grass clippings and woodchips as feedstock.

CLIMATE AND TEMPERATURE

The most common worms used in composting systems is red worms [*Eisenia foetida* (Fig. 13.15 and 13.16), *Eisenia andrei*, and *Lumbricus rubellus*] who feed most rapidly at a temperature of 15°C–25°C. They can survive at 10°C. Temperature above 30°C may harm them. This temperature range means that indoor vermi-composting with red worms is suitable in tropical climates. Other worms like *Perionyx excavatus* are suitable for warmer climates. If a worm bin is kept outside, it should be placed in a sheltered position away from direct sunlight and insulated against frost in winter. It is necessary to monitor the temperature of large-scale bin systems. In warmer climate *Eiseina foetida* and *Perionyx excavatus* are used together (Fig. 13.18) for successful vermi-composting.

HARVESTING

Vermi-compost is ready for harvest when it contains few-to-no scraps of uneaten food or bedding. There are several methods of harvesting from small-scale systems: *"dump and hand sort"*, *"let the worms do the sorting"*, *"alternate containers"* and *"divide and dump"*. These differ on the amount of time and labour involved and whether the vermi-composter wants to save as many worms as possible from being

Fig. 13.14. Indian Blue Worms *(Perionyx excavatus)*

Fig. 13.15. Indian Red Wigglers *(Eisenia fetida)*

Fig. 13.16. *Eisenia fetida* in copulating state

Fig. 13.17. A common earthworm, *Metaphire posthuma* in clump

Fig. 13.18. Two varieties of earthworms : Red Wigglers *(Eisenia foetida)* and Indian Blue Worms *(Perionyx excavatus)* are clumped together.

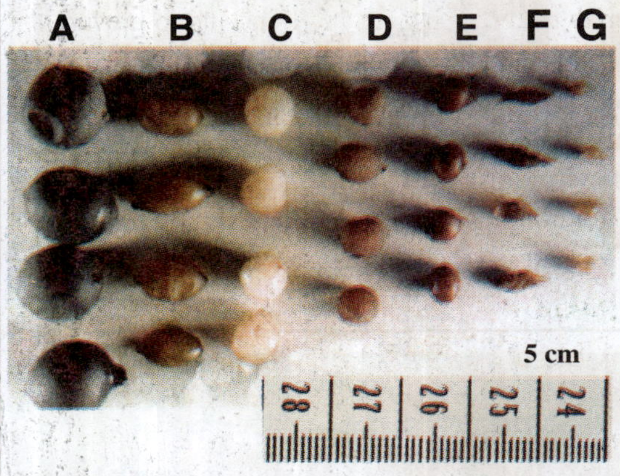

Fig. 13.19. Cocoons of different Indian earthworms —
A. *Eutyphoesus gammiei* B. *Lampito mauriti*
C. *Poutoscolex corithrurus* D. *Polypheritima elongata*
E. *Drawida nepalensis* F. *Perionyx excavatus*
G. *Dichogaster naodigliani*

trapped in the harvested compost. It will be wise to try to pick out as many eggs/cocoons as possible and return them to the bin during harvesting. Cocoons are small, lemon-shaped yellowish objects that can usually be seen pretty easily with the naked eye and picked out.

The top of the bed is harvested from pit in case of large-scale culture. The material should be placed in a heap in the sun so that most of the worms move down to the cool base of the heap. The compost is then sieved before being packed. The earthworms and the thicker material, which remains on top of the sieve, will be returned back in the bin and the process starts again.

PROPERTIES

Vermi-compost has been shown to be richer in many nutrients than compost produced by other composting methods. It also has outperformed a commercial plant medium with nutrients added, but needed adjustment for magnesium and pH. However, other studies have shown that the effects of homemade, backyard vermi-compost compared to municipal compost were lower in terms of soil microbial biomass, soil microbial activity, and yields of a species of ryegrass.

Further, one study concluded that the differences between methods of composting were in large part due to the feedstock, and therefore, no generalisations can be made between composts made from varying materials. It is rich in microbial life which converts nutrients already present in the soil into plant-available forms. Unlike other compost, worm castings also contain worm mucus which prevent nutrients from washing away with the first watering and holds moisture better than plain soil

BENEFITS OF VERMI-COMPOST

Soil quality

- Improves soils physical structure
- Enriches soil with micro-organisms (adding enzymes such as phosphatase and cellulase)

- Microbial activity in worm castings is 10 to 20 times higher than that the worm ingests in the soil and organic matter.
- Attracts deep-burrowing earthworms already present in the soil.
- Improves water holding capacity due to the presence of the wound mucus.

Plant growth

- Enhances germination, plant growth, and crop yield.
- Improves root growth and structure.
- Enriches soil with microorganisms (adding plant hormones such as auxins and gibberellic acid).

Economic

- Bio-wastes conversion reduces waste flow to landfills.
- Elimination of bio-wastes from the waste stream reduces contamination of other recyclables collected in a single bin (a common problem in communities practising single-stream recycling).
- Creates low-skill jobs at local level.
- Low capital investment and relatively simple technologies make vermi-composting practical for less-developed agricultural regions.

Environmental

- Helps to close the "metabolic gap" through recycling waste on-site.
- Production reduces greenhouse gas emissions such as methane and nitric oxide (produced in landfills or incinerators when not composted or through methane harvest).

As fertilizer

- Vermi-compost can be mixed directly into the soil, or steeped in water and made into a worm tea by mixing some vermi-compost in water, bubbling in oxygen with a small air pump, and steeping for a number of hours or days.
- The microbial activity of the compost is greater if it is aerated during this period. The resulting liquid is used as a fertilizer or sprayed on the plants.
- The waste liquid or leachate, that drains into the bottom of some vermi-composting systems as water-rich breakdown foods. It is best applied back to the bin when added moisture is needed.
- The pH, nutrient, and microbial content of these fertilizers vary upon the inputs fed to worms. Pulverised lime stone or calcium carbonate can be added to the system to raise the pH for making a sustainable environment.

PROBLEMS

Smell : When closed, a well-maintained bin is odourless; when opened, it should have little smell; if any, the smell is earthy. Worms require gaseous oxygen. Oxygen can be provided by air holes in the bin, occasional stirring of bin contents, and removal of some bin contents if they become too deep or too wet. If decomposition becomes anaerobic from excess feedstock added to the bin in wet conditions, or layers of food waste have become too deep, the bin will begin to smell like ammonia.

Moisture : If decomposition becomes anaerobic, then to restore healthy conditions and to prevent the worms from dying, the smelly, excess waste water must be removed and the bin is returned to normal moisture level. This is done by reducing the addition of food scraps which has high moisture content, and then adding fresh, dry bedding (such as shredded newspaper) into the bin and mixing it well.

Pests : Pests such as rodents and flies are attracted by certain materials and odours, usually from large amounts of kitchen waste, particularly meat. Eliminating the use of meat or dairy product in a worm bin decreases the possibility of pests. In warm weather, fruit and vinegar flies breed in the bins if fruit and vegetable waste is not thoroughly covered with bedding. This problem can be avoided by thoroughly covering the waste by at least 2 inches of bedding. Maintaining the correct pH (close to neutral)

Fig. 13.20. Introduction of composter earthworm in decomposing vermibed

Fig. 13.21. Replacement of organic composting material

Fig. 13.22. Preparing vermibed with waste organic materials

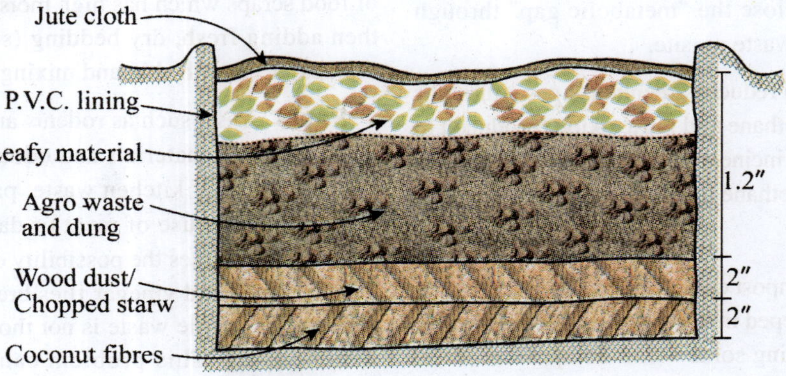

Fig. 13.23. Vermi-composting process

and water content of the bin can help avoid these pests as well.

Escaping of worms : Worms generally stay in the bin, but may try to leave the bin when first introduced, or often after a rainstorm when outside humidity is high. Maintaining adequate conditions in the worm bin and putting a light over the bin when the worms are first introduced, eliminates this problem.

Nutrient levels : Commercial vermi-composters test, and may amend their products to produce consistent quality and results. Because the small-scale and home systems use a varied mix of feedstocks, the nitrogen, potassium and phosphorus content of the resulting vermi-compost will also be inconsistent. NPK testing may be helpful before the vermi-compost or tea is applied to the garden. In order to avoid over-fertilisation issues, such as nitrogen burn, vermi-compost can be diluted as a tea 50:50 with water, or as a solid can be mixed in 50:50 with potting soil. The mucus produced creates a natural time release fertilizer which cannot burn plants.

SPACE FOR NOTES

SPACE FOR NOTES

Bibliography

Agarwal, H. P. and Seth, M. K. 2009. Sericulture in India. Daya Publishing House, New Delhi.

Alagarswami, K. 1991. Production of cultured pearls. Publication and information division, ICAR, Krishi Anusandhan Bhavan, New Delhi.

Alagarswami, K. and Dharmaraj, S. 1984. Manual on pearl culture techniques. Central Marine Fisheries Research Institute. ICAR, Cochin, India.

Atwal, A. S. 1986. Agricultural pests of India and South-east Asia. Kalyani Publishers, New Delhi.

Banerjee, G. C. 2004. A Textbook of Animal Husbandary. Oxford and IBH Publishing Co. Pvt. Ltd., New Delhi, 8th ed.

Beach, C. B. 1914. Emu, 'The New Student's Reference Work'. F. E. Compton and Company, Chicago, USA.

Bhatnagar, R. K. and Palta, R. K. 1996. Earthworm: vermiculture and vermicomposting. Kalyani Publishers, New Delhi.

Bhattacharya, S. 2006. Cattle ownership makes it a man's world. New scientist, USA.

Bhattyacharya, K. N. 2009. Pranividya (Bengali). Allied Book Publisher, Calcutta.

Biswas, S. 2011. Earthworm species and gutflora associates. Lambert Academic Publication, ebook.

Buchanan, M. A., Russell, E. and Block, S. D. 1988. Earthworms in environmental and waste management. SPB Academic Publishers, Netherlands.

Chattopadhyay, D. 2011. Snatak Pranividya (Bengali). Book Syndicate Pvt Ltd., Calcutta.

Chattapadhyay, S. 2011. Introduction to lac and lac culture. Department of Forest Biology and Tree Improvement Faculty of Forestry, Birsa Agricultural University, Kanke, Ranchi.

Chattopadhyay, S. B. 2000. Principles and procedures of plant protection. Oxford & IBH Publishing Co., Calcutta.

Chaudhuri, H. and Alikunhi, K. H. 1957. Observation on the spawning in Indian Carps by hormone injection. Curr. Sci. 26(12) : 381-382.

Chaudhuri, P. S. 2006. Kechor jiban baichitra o kecho prajukti (Bengali). Gyan Bichitra Prakashani, Kolkata.

Chondar, S. L. 1999. Biology of Finfish and Shellfish. SCSC Publishers, 58/B, Nilmoni Mullick Lane, Howrah, West Bengal, India.

Clay, J. 2004. World Agriculture and the Environment: A Commodity-by-Commodity Guide to Impacts and Practices. Island Press, Washington, USA.

Clutton-Brock, J. 1999. A Natural History of Domesticated Mammals. Cambridge University Press, Cambridge, UK.

Edwards, C. A. and Bohlen, P. J. 1996. Biology and ecology of earthworms. Chapman and Hall, London.

Edwards, C. A. 1998. Earthworm ecology. CRC Press, LLC, Florida.

Gajalakshmi, S. and Abbasi, S. A. 2004. Earthworms and vermicomposting. Indian J. Biotech. 3 : 486-494.

Ganga, G. and Chetty, J. S. 1997. An Introduction to Sericulture. Oxford and IBH Publishing Co. Pvt. Ltd., Calcutta.

Garg, V. K., Gupta, R. and Yadav, A. 2010. Vermicomposting technology for solid waste management. J. Env. Sc. and Eng.; Guru Jainbeswar University, Haryana.

Garside, E.T. 1959. Some effects of oxygen in relation to temperature on the development of lake trout embryos. Can. J. Zool., 37 : 689-698.

Ghorai, N. 1995. Lac culture in India. International Books and Periodicals, New Delhi.

Ghosh, G. K. 1994. Bee-keeping in India. APH Publishing Co., New Delhi.

Heaf, P. and Heaf, D. 2009. Bee-keeping for all. Hafan, Cae Llwyd, Gwynedd, UK.

Hill, D. S. 1983. Agricultural insect pests of the tropics and their control. Cambridge University Press, Cambridge, UK.

Hill, D. S. and Waller, J. M. 1988. Pests and diseases of tropical crops. Longman Scientific & Technical, UK.

Huffman, B. 2006. The ultimate ungulate. Ultimate Ungulate.com.

ICAR. 1979. Characteristics of cattle and buffalo breeds in India. New Delhi, India.

ICAR. 1990. Handbook of animal husbandry. New Delhi.

Indian Poultry Industry Yearbook. 1994. 10th edition, New Delhi, India.

INORA. 1994. Earthworm Biotechnology. The Institute of Natural Organic Agriculture Report, Pune, India.

Integrated Sustainable Energy and Ecological Development Association (INSEEDA). 2010. Duck Rearing — an Introduction. Booklet No. 222, India.

Invasive Species Specialist Group (ISSG). 2005. Bos taurus. Global Invasive Species Database — e report.

Ismail, S. A. 1997. Vermicology — the biology of earthworm. Orient Longman, Hyderabad. India.

Jana, B. B., Sarkar, G. and Kundu, G. 1985. Pattern of Physico-chemical changes in the water during hatching of Indian carp eggs in a Chinese hatchery system. Aquaculture, 47 : 89-96.

Jhingran, V. G. 1983. Fish and Fisheries of India. Hindustan Publishing Corporation (India). Jawahar Nagar, New Delhi.

Jhingran, V. G. and Sehgal, K. L. 1978. Coldwater Fisheries of India. Inland Fisheries Society of India, Barrackpore, West Bengal, India.

Jhingran, V. G. and Sharma, B. K. 1980. Integrated Live-stock-Fish Farming in India. Proceedings of Conference of Integrated Agriculture-Aquaculture Farming Systems, Manila, Philippines.

Kale, R. D. 1998. Earthworms : Cinderella of organic farming. Prism Books Pvt. Ltd., Bangalore.

Kar, N. 2002. Animal Husbandry and rural development. Deep & Deep Publications, Lucknow, India.

Kinne, O. and Kinne, E. M. 1962. Rates of development in embryos of a cyprinodont fish exposed to different temperature — Salinity — Oxygen combinations. Can. J. Zool., 40 : 231-253.

Kranz, J., Schmutterer, H. and Koch, W. 1978. Diseases, pests and weeds in Tropical crops. John Willey, USA.

Kumar, C. A. 1994. State-of-the-art report on vermiculture in India. Council for Advancement of People's Action and Rural Technology (CAPART) : 1-60.

Kurian, C. V. and Sebastian, V. O. 1976. Prawns and Prawn Fisheries of India. Hindustan Publishing Corporation, Jawahar Nagar, Delhi.

Lovell, T. 2009. Nutrition Feeding of Fish. Springer Pvt. Limited, Akash Deep Building, Barakhamba Road, New Delhi.

Mason, M. 1996. A World Dictionary of Live-stock Breeds, Types and Varieties, 4th Edition. Wallington, UK.

Mathur, V. B. and Qudri, S. M. H. 2010. Mountages, Mounting and Harvesting technology for quality cocoon producer. CSRTI, Mysore, India.

NABARD 2007. Report on Piggery. National Bank for Agriculture and Rural Development, India.

NABARD 2009. Poultry farming in U.P., National Bank for Agriculture and Rural Development, India.

NABARD 2010. Report on Duck rearing. National Bank for Agriculture and Rural Development, India.

Nageswara, A. R. 1997. Comparison of growth performances of Khaki Campbell, Indian local ducks and their reciprocal crosses. Proceedings of the 11th European symposium on waterfowl, France.

Narasaiah, M. L. 2003. Problems and Prospects of Sericulture. Discovery Publishing House, New Delhi.

Nowak, R. M. and Paradiso, J. L. 1983. Walker's Mannuals of the World. The Johns Hopkins University Press, Baltimore, Maryland, USA.

Oguchi, S., Kamihira, M. and Iigima, S. 1998. Quail embryo culture using artificial vessels. Animal cell technology : Basic and applied aspects. Springer—Science & Business Media, Dordrecht.

O'Malley, P. 1997. Emu Farming, in 'The New Rural Industries'. Rural Industries Research & Development Corporation, India.

Pal, G., Bhagat, M. L. and Bhattacharya, A. 2009. Economics and Resource use efficiency of lac cultivation in Jharkhand. Indian Journal of Forestry, 32 : 95-98.

Philip, C. 2007. The Barefoot Bee-keeper. ebook. publisher, Lulu.com.

Phillips, F. E. 2008. Bee-keeping. MacMillan & Co., New York.

Pillay, T. V. R. 2005. Aquaculture Principles and Practices. Fishing News Books, Blackwell Science Ltd., Garington Road, Oxford.

Pradhan, S. 1998. Insect pests of crops. NBT, India.

Purdy, H. R., Dawes, R. J. and Hough, R. 2008. Breeds of Cattle. TRS Pub. Corpn.

Quadri, S. M. M. 2012. A new dimension in the promotion of bivoltine sericulture. CSTRI, Mysore.

Rajendran, S. 1994. Exploiting earthworms for fertilizers. Down to Earth. 2 (21) : 22-23.

Rath, S. 1998. The Complete Cow. Voyageur Press, Minnesota, USA.

Raudiansky, S. 1992. The Covenant of the Wild. William Morrow and Company, Inc., New York.

Roger, S., Pullin, V. and Shehadeh, Z. H. 1980. International center for living aquatic resources management. Manila, Philippines.

Roy, G. A. 1949. Diseases of adult bees. Dadant and Sons, New Delhi.

Saravanam, L. 2012. Dont't invest in Emu farms, say Salem authorities. The Times of India, New Delhi.

Sathe, S. 1995. Small holder duck rearing in India. Proceedings of the 10th European Symposium on Water fowl, Halle (Saale), Germany.

Scholten, B. A. 2011. Operation Flood, Food Aid and Development. Pub. I. B. Tauris, New York.

Sharma, K. K., Jaiswal, A. K. and Kumar, K. K. 2006. Role of Lac culture in Biodiversity conservation : issues at stake and conservation strategy. Current Science, 91 : 894-898.

Singh, D. and Singh, D. P. 2006. A handbook of bee-keeping. Riddlin Int., New Delhi.

Singh, S. 1962. Bee-keeping in India. ICAR Publications, India.

Sulaiman, E. and Pillai, B. V. 2005. Working of Poultry Industry in India. J. V. Pub. Home, India.

Tekriwal, K. L. and Rao, A. A. 1999. Ornamental Aquarium Fish of India. Kingdom Books, Water-looville, England.

Thear, K. 2005. Keeping Quail : A guide to domestic and commercial management. Broad Leys Publishing Ltd., UK.

Timbrell, J. 2002. Introduction to toxicology. Talyor and Francis, New York.

Van Embden, H. F. 1992. Pest Control. Cambridge University Press, UK.

Van Embden, H. F. and Peakall, D. B. 1996. Beyond silent spring—integrated pest management and chemical safety. Chapman & Hall, London.

Voelker, W. 1986. The Natural History of Living Mammals. Plexus Publishing, New Jersey, USA.

Walker, C. H., Hopkin, S. P., Sibly, R. M. and Peakall, D. B. 2001. Principles of ecotoxicology. Talyor and Francis, New York.

Wedmore, W. 2001. Manual of Bee-keeping. Taylor, London.

Wen, H., Lan, X., Zhang, Y., Zhao, T., Wang, Y., Kajiura, Z. and Nakagaki, M. 2010. Transgenic Silkworms (*Bombyx mori*) produce recombinant spider dragline silk in cocoons. Mol. Biol. Rep., 37 : 1815-1821.

Young, J. Z. 1981. The Life of Vertebrates. ELBS.

Website:

Bhumika Animal Breeder Farm, M.D. Central Silk Board (SB), 2014.

Central Silk Technological Research Institute, Bangalore (CSTRI), 2014.

Central Sericulture Germplasm Resource Centre, 2014.

Commercial Goat Farming in India, 2014.

Department of Sericulture, Government of Karnataka, 2014.

Goat India., 2014.

Jharkhand State Co-operative Lac Marketing and Procurement Federation Ltd., 2014.

Organic Farming in India, 2014.

Indian Livestock Farms, 2014.

Indian Institute of Natural Resin and Gum, 2014.

International Livestock Research Institute, 2011.

Transgenic Silkworm Research Centre, Tokyo, 2014.
University of Agriculture and Technology, 2014.

www.catletoday.com

www.indiansilk.kar.nic.in

www.rashmesiri.kar.nic.in

www.silk.csrtimys.res.in

www.ilri.ernet.in

www.news.jharkhand.org.in

www.goatindia.com

www.nabard.org

www.goatwala.com

www.agricultureinformation.com

www.indianlivestockfarms.com/rabbit/goat

www.agritech.tnau.ac.in

www.tao.org/docrep

www.indg.in/quailfarming/farm/agrionnoffrabfarmnest.com

www.roysfarm.com/quailfarm.in

www.bhumikaanimalbreeder.com/rabbit farming

www.indiamart.com/rabbit

www. csb.gov.in

Index

B